Dynamic Stability of Hydraulic Gates and Engineering for Flood Prevention

Noriaki Ishii
Osaka Electro-Communication University, Japan

Keiko Anami
Osaka Electro-Communication University, Japan

Charles W. Knisely
Bucknell University, USA

A volume in the Advances in Civil and Industrial Engineering (ACIE) Book Series

IGI Global
DISSEMINATOR OF KNOWLEDGE

Published in the United States of America by
IGI Global
Engineering Science Reference (an imprint of IGI Global)
701 E. Chocolate Avenue
Hershey PA, USA 17033
Tel: 717-533-8845
Fax: 717-533-8661
E-mail: cust@igi-global.com
Web site: http://www.igi-global.com

Library of Congress Cataloging-in-Publication Data

Names: Ishii, Noriaki, 1946- author. | Anami, Keiko, 1974- author. | Knisely, Charles W., author.
Title: Dynamic stability of hydraulic gates and engineering for flood prevention / by Noriaki Ishii, Keiko Anami, and Charles W. Knisely.
Description: Hershey, PA : Engineering Science Reference, [2018] | Includes bibliographical references.
Identifiers: LCCN 2017013574| ISBN 9781522530794 (hardcover) | ISBN 9781522530800 (ebook)
Subjects: LCSH: Sluice gates--Stability.
Classification: LCC TC553 .I84 2018 | DDC 627/.42--dc23 LC record available at https://lccn.loc.gov/2017013574

This book is published in the IGI Global book series Advances in Civil and Industrial Engineering (ACIE) (ISSN: 2326-6139; eISSN: 2326-6155)

British Cataloguing in Publication Data
A Cataloguing in Publication record for this book is available from the British Library.

For electronic access to this publication, please contact: eresources@igi-global.com.

Advances in Civil and Industrial Engineering (ACIE) Book Series

Ioan Constantin Dima
University Valahia of Târgovişte, Romania

ISSN:2326-6139
EISSN:2326-6155

Mission

Private and public sector infrastructures begin to age, or require change in the face of developing technologies, the fields of civil and industrial engineering have become increasingly important as a method to mitigate and manage these changes. As governments and the public at large begin to grapple with climate change and growing populations, civil engineering has become more interdisciplinary and the need for publications that discuss the rapid changes and advancements in the field have become more in-demand. Additionally, private corporations and companies are facing similar changes and challenges, with the pressure for new and innovative methods being placed on those involved in industrial engineering.

The **Advances in Civil and Industrial Engineering (ACIE) Book Series** aims to present research and methodology that will provide solutions and discussions to meet such needs. The latest methodologies, applications, tools, and analysis will be published through the books included in **ACIE** in order to keep the available research in civil and industrial engineering as current and timely as possible.

Coverage

- Structural Engineering
- Construction Engineering
- Hydraulic Engineering
- Quality Engineering
- Urban Engineering
- Materials Management
- Operations research
- Production Planning and Control
- Earthquake engineering
- Coastal Engineering

IGI Global is currently accepting manuscripts for publication within this series. To submit a proposal for a volume in this series, please contact our Acquisition Editors at Acquisitions@igi-global.com or visit: http://www.igi-global.com/publish/.

The Advances in Civil and Industrial Engineering (ACIE) Book Series (ISSN 2326-6139) is published by IGI Global, 701 E. Chocolate Avenue, Hershey, PA 17033-1240, USA, www.igi-global.com. This series is composed of titles available for purchase individually; each title is edited to be contextually exclusive from any other title within the series. For pricing and ordering information please visit http://www.igi-global.com/book-series/advances-civil-industrial-engineering/73673. Postmaster: Send all address changes to above address.

Titles in this Series

For a list of additional titles in this series, please visit: www.igi-global.com/book-series

Recent Advances in Applied Thermal Imaging for Industrial Applications
V. Santhi (VIT University, India)
Engineering Science Reference • copyright 2017 • 306pp • H/C (ISBN: 9781522524236) • US $205.00 (our price)

Performance-Based Seismic Design of Concrete Structures and Infrastructures
Vagelis Plevris (Oslo and Akershus University College of Applied Sciences, Norway) Georgia Kremmyda (University of Warwick, UK) and Yasin Fahjan (Gebze Technical University, Turkey)
Engineering Science Reference • copyright 2017 • 320pp • H/C (ISBN: 9781522520894) • US $205.00 (our price)

Engineering Tools and Solutions for Sustainable Transportation Planning
Hermann Knoflacher (Vienna University of Technology, Austria) and Ebru V. Ocalir-Akunal (Gazi University, Turkey)
Engineering Science Reference • copyright 2017 • 374pp • H/C (ISBN: 9781522521167) • US $210.00 (our price)

Design Solutions and Innovations in Temporary Structures
Robert Beale (Independent Researcher, UK) and João André (Portuguese National Laboratory for Civil Engineering, Portugal)
Engineering Science Reference • copyright 2017 • 503pp • H/C (ISBN: 9781522521990) • US $200.00 (our price)

Modeling and Simulation Techniques in Structural Engineering
Pijush Samui (National Institute of Technology Patna, India) Subrata Chakraborty (Indian Institute of Engineering Science and Technology (IIEST), Shibpur, India) and Dookie Kim (Kunsan National University, South Korea)
Engineering Science Reference • copyright 2017 • 524pp • H/C (ISBN: 9781522505884) • US $220.00 (our price)

Computational Modeling of Masonry Structures Using the Discrete Element Method
Vasilis Sarhosis (Newcastle University, UK) Katalin Bagi (Budapest University of Technology and Economics, Hungary) José V. Lemos (National Laboratory for Civil Engineering, Portugal) and Gabriele Milani (Technical University in Milan, Italy)
Engineering Science Reference • copyright 2016 • 505pp • H/C (ISBN: 9781522502319) • US $210.00 (our price)

Handbook of Research on Emerging Innovations in Rail Transportation Engineering
B. Umesh Rai (Chennai Metro Rail Limited, India)
Engineering Science Reference • copyright 2016 • 664pp • H/C (ISBN: 9781522500841) • US $235.00 (our price)

IGI Global
DISSEMINATOR OF KNOWLEDGE

701 East Chocolate Avenue, Hershey, PA 17033, USA
Tel: 717-533-8845 x100 • Fax: 717-533-8661
E-Mail: cust@igi-global.com • www.igi-global.com

Table of Contents

Foreword v

Preface vii

Acknowledgment x

Section 1
Introduction and Fundamentals

Chapter 1
Introduction: History of Gate Failures and Overview of Vibration Mechanisms 1
OVERVIEW OF HISTORICAL RECORD OF DAM AND GATE FAILURES 1
PREREQUISITE KNOWLEDGE NEEDED FOR GATE INSTABILITY MECHANISMS 18
CONCEPTUAL OVERVIEW OF MECHANISMS OF GATE INSTABILITY 20
CONCLUSION 32

Chapter 2
Vibrations: With Resonator and Stability Concepts 44
INTRODUCTION 44
SINGLE DEGREE OF FREEDOM (SDOF) DAMPED SYSTEMS 45
TWO DEGREES-OF-FREEDOM SYSTEM 50
PARAMETRIC COUPLING 60
MULTIPLE DEGREES-OF-FREEDOM (MDOF) SYSTEM AND NORMAL MODES 64
CONTINUOUS SYSTEM AND MODE SHAPE 64
ENERGY FLOW IN VIBRATIONS 69
DAMPING 72
SDOF SYSTEM IN FLOWING FLUID 74
FREQUENCY POWER SPECTRA 75
EXPERIMENTAL MODAL ANALYSIS 76
RESONATORS 77
U-TUBE RESONATORS 78
HELMHOLTZ RESONATORS 80
STANDING WAVE RESONATORS 81
STABILITY CONCEPTS 86
CONCLUSION 89

Chapter 3
Fluid Dynamics: Basic Concepts, Gate Discharge, and Flow Stability 94
INTRODUCTION 94
FLUID PROPERTIES 95
STATIC GATE LOADS 98
CONSERVED PROPERTIES 102
GATE DISCHARGE COEFFICIENTS 113
POTENTIAL FLOW 120
HYDRODNAMIC INSTABILITIES AND TRANSITION TO TURBULENCE 122
CONCLUSION 133

Chapter 4
Rayleigh Wave Theory: Gate Excitation Mechanism 140
INTRODUCTION 140
RAYLEIGH WAVE THEORY 143
INITIAL AND BOUNDARY VALUE PROBLEM FOR DISPERSIVE WATER WAVES INDUCED BY GATE VIBRATIONS 151
WATER EXCITATION EFFECTS 152
THEORETICAL DEVELOPMENT OF RAYLEIGH WAVE THEORY TO PREDICT FLOW-RATE VARIATION PRESSURE 154
WATER EXCITATION EFFECTS ON A GATE ACTING AS PRESS-SHUT DEVICE 189
EXPERIMENTAL CONFIRMATION OF THEORETICAL ANALYSIS 195
CONCLUSION 200

Chapter 5
Streamwise Gate Vibration: Added Mass and Damping 204
INTRODUCTION 204
THEORETICAL ANALYSIS OF PUSH-AND-DRAW PRESSURE 205
PUSH-AND-DRAW PRESSURE 212
ADDED MASS EFFECTS AND SMALL WAVE-RADIATION DAMPING 218
EXPERIMENTAL CONFIRMATION OF THEORETICAL ANALYSIS 220
DIMENSIONLESS ADDED MASS AND FLUID-DAMPING COEFFICIENTS 222
SUMMARY OF FLOW-INDUCED PRESSURE LOADING ON GATES DUE TO DISCHARGE FLUCTUATIONS AND STREAMWISE VIBRATIONS 225
POTENTIAL FOR SELF-EXCITED VIBRATION 226
CONCLUSION 227

Section 2
Gate Excitation Mechanisms

Chapter 6
Overflow Gates: Nappe Oscillations and Vibrations 231
INTRODUCTION 231
OVERFLOW GATES 232
NAPPE OSCILLATIONS 236

VIBRATION OF INFLATABLE WEIRS 268
FLAP GATE VIBRATION 269
CONCLUSION 287

Chapter 7
Flow-Induced Vibration of Long-Span Gates: One and Two Degrees-of-Freedom 294
INTRODUCTION 294
RESULTANT HYDRODYNAMIC PRESSURE INDUCED BY STREAMWISE GATE VIBRATION 298
EQUATION OF MOTION 301
IDENTIFICATION OF ADDED MASS, FLUID EXCITATION AND FLUID DAMPING 302
VIBRATION FREQUENCY AND NET FLUID-EXCITATION RATIO 307
EXPERIMENTAL VERIFICATION OF VIBRATION FREQUENCY AND FLUID-EXCITATION RATIOS 310
DYNAMIC INSTABILITY OF SCHINZNACH AND ARA RIVER LONG-SPAN GATES 314
VISCOUS FLOW EFFECTS ON GATES WITH UNDERFLOW 316
GATES WITH TWO DEGREES-OF FREEDOM IN UNDERFLOW 336
LONG-SPAN GATES WITH SIMULTANEOUS OVER- AND UNDERFLOW 358
CONCLUSION 378

Chapter 8
Eccentricity Instability: A Mechanism for Tainter (Radial) Gate Vibration 387
INTRODUCTION 387
HYDRODYNAMIC PRESSURE INDUCED BY RIGID-BODY ROTARY OSCILLATION 390
EQUATION OF MOTION 392
EXCITATION RATIO AND VIBRATION FREQUENCY 395
A DESIGN CRITERION FOR DYNAMIC STABILITY OF ECCENTRIC GATES 399
COMPARISON OF THEORY WITH EXPERIMENTAL RESULTS 401
CONCLUSION 403

Chapter 9
Pressure on a Submerged Plate Undergoing Streamwise Rotation: Theory and Validating Experiments 407
INTRODUCTION 407
VERTICAL, FLAT, AND RIGID WEIR PLATE 408
INCLINED CIRCULAR-ARC SKINPLATE 436
CONCLUSION 459

Chapter 10
Theory of Self-Excited Coupled-Mode Vibration of Tainter Gates: A Concern for Gate Designers 463
INTRODUCTION 463
EQUATIONS OF MOTION 467
APPROXIMATE SOLUTION 476
CHARACTERISTICS OF COUPLED-MODE SELF-EXCITED VIBRATION 482
ENERGY TRANSFER FROM VIBRATION TRAJECTORIES 487
INSTABILITY OF THE FOLSOM GATE FOR CONDITIONS AT FAILURE 490
CONCLUSION 494

Section 3
Validation and Applications

Chapter 11
Measurement of Coupled-Mode Instability on Model Scale: Details of Validation........................... 499
INTRODUCTION 499
TWO-DIMENSIONAL MODEL STUDIES 500
THREE-DIMENSIONAL MODEL STUDIES 510
CONCLUSION 518

Chapter 12
Wachi Dam and Folsom Dam: Coupled-Mode Instabilities – Mechanism for Failures 521
INTRODUCTION 521
FOLSOM DAM TAINTER GATE 522
WACHI DAM TAINTER GATE 553
CONCLUSION 558
CONCLUSION 559

Chapter 13
Analogies Between Coupled-Mode Gate Vibration and Coupled-Mode Flutter: The Need for Dynamic Design.. 565
INTRODUCTION 565
ANALYSIS OF THE TACOMA NARROWS BRIDGE FAILURE 567
FLUTTER THEORY FOR AIRFOILS 573
BLUFF BODY FLUTTER ANALYSIS 581
COMPARISON OF SELF-EXCITED COUPLED-MODE GATE INSTABILITY AND TWO-DIMENSIONAL BRIDGE FLUTTER 586
CONCLUSION 589

Chapter 14
Full-Scale Gate Measurements: Extended Validation of Theory 593
INTRODUCTION 593
TAINTER GATE "A" 594
TAINTER GATE "B" 602
TAINTER GATE "C" 614
CONCLUSION 623

Chapter 15

Reflections and Conclusions 626
PREAMBLE 626
INTRODUCTION 626
BEGINNINGS 627
REFLECTIONS ON THE PROCESS 627
IN-WATER FREQUENCY AND DYNAMIC PRESSURES 629
SUMMER 1995: FAILURE DRIVES SUCCESS 630
DEVELOPMENT OF DESIGN STAGE DYNAMIC STABILITY ANALYSIS 631
REFLECTIONS BY PROFESSOR EMERITUS ISHII 646
CLOSING 649
APPENDIX: CLOSING ADDRESS FOR 2001 INTERNATIONAL SYMPOSIUM 649

About the Authors 654

Index 656

Foreword

Dams are used worldwide to create reservoirs for flood protection, irrigation and recreation, as well as to provide the hydrostatic head required for hydroelectric power generation. In most cases, these dams must have flow control structures in the form of hydraulic gates and valves so that the dam structures can reliably serve their intended purpose. Often, there are enormous energies involved in the controlled flow and failures of the flow control devices can be catastrophic, possibly leading to incalculable loss of life and property. Clearly, reliable design of the flow control structures is essential and that is the subject of this book.

Hydraulic structures subjected to high energy flows will invariably vibrate in response to the random turbulence in the flow, but this excitation mechanism is usually managed with relative ease. The real problems arise when structural motion induces fluid forces which enhance the structural response. In this way, the fluid-structural coupling transfers energy from the flow to the structure and a flow-induced structural instability develops. As with airplane wing flutter or suspension bridge deck galloping, these phenomena can be very destructive and must be understood and eliminated at the design stage.

The theory developed in the book is rigorous but should be readily accessible to graduate engineers, especially with the background fundamentals provided in Chapters 1 to 5 of the book. Indeed, some graduate students and practitioners may find the chapters on the fundamentals unnecessary to understand the development (Chapters 6 to 10) and application (Chapters 11 to 15) of the theory. The theory is nicely complemented with numerous illustrative figures as well as photographs of operating gates in service. The emphasis on rigorous mathematical development, physical insight, and practical applications is refreshing. The authors are well placed to prepare this monograph, having spent their careers working on the subject problems and being able to draw heavily on their own research and experience in its preparation.

The mechanisms causing flow-induced vibrations can be very elusive and even counterintuitive. Coupled mode flutter was discovered to cause catastrophic wing failures of early monoplanes during World War I and it took some 20 years to understand the mechanism and to develop a predictive theory. The recent failures of the San Onofre Nuclear Power Station steam generators demonstrates that 40 years of research has been insufficient to enable completely reliable design of these components against flow-induced vibrations. In this book the authors have shown convincingly that, after nearly 75 years of research, there is little consensus on the mechanism which caused the infamous Tacoma Bridge failure. The lessons are clear. Flow-induced vibration problems can be extremely difficult to understand and, thus, to eliminate at the design stage. That said, failure to consider them at the design stage within the context of existing knowledge and experience can have catastrophic consequences.

This book is a very useful addition to the flow-induced vibration literature, filling a current void in relation to hydraulic structures and providing an excellent guide to the related literature. It will be a useful reference for graduate students and engineers working in the field of fluid-structure interaction, and an essential read for engineers carrying out research on dynamic instabilities of hydraulic gate structures or charged with the responsibility for designing them against these potentially dangerous excitation mechanisms.

D. S. Weaver
McMaster University, Canada

Preface

Why are we writing a book about fluid dynamic mechanisms of hydraulic gate instabilities? Are there not enough of such books already available? The answer to the second question is a rather uncertain maybe. Maybe there are enough books about the classical gate vibrations such as the vertical vibration of vertical lift gates. What is missing is a comprehensive overview of the developments in gate instability mechanisms in the past approximately 35 years. Interestingly, the date on the reprint of the book *Flow-Induced Vibration: An Engineering Guide* by Naudascher and Rockwell is 1994 and the following year saw the failure of the 87-ton Folsom Dam Tainter gate during routine operation.

The failure at the Folsom Dam was classified in the forensics report as a static failure due to excessive trunnion pin friction, despite eyewitness testimony by the operator who expressly notes "vibration and rumbling." There is an assertion in the literature (Lewin, 1998, 2001) that among the gates he was considering "only the gate at the Washi [sic] Dam in Japan failed due to hydrodynamic causes. . ." and "The collapse of the gate of the Washi [sic] Dam is the only recorded case of complete failure of a gate due to vibration. . ." Interestingly and despite eyewitness testimony of vibrations, the official report on the 1967 Wachi gate failure does not attribute the failure to hydrodynamically-induced vibration. No mechanism for vibration of such a large gate was known at the time. The mechanism that Lewin cites for this failure, the eccentricity instability mechanism, was proposed almost ten years after the failure. The gate design companies in Japan endorsed eccentricity instability as the failure mechanism since it absolved them of any liability. There was no eccentricity in their design specifications for the gate. The eccentricity instability mechanism, however, was shown to be inoperative for this gate due to insufficient eccentricity between the trunnion pin center and the skinplate center. In fairness to Prof. Lewin, we need to note that much of what is now known about the self-excited coupled-mode dynamic instability mechanism for Tainter gates was not available in the open literature in 1998, and perhaps not widely available in 2001. The authors suspect that Prof. Lewin would write his statement differently today given the availability of details of the coupled-mode dynamic instability mechanism.

Gate designers need to be better informed about dynamic instability mechanisms. Once a statement such as that by Lewin is printed, it becomes increasingly more challenging to convince designers to look rationally at the new developments that contradict such a printed statement. Our concern is that it will take another, perhaps more disastrous, failure of a Tainter gate before the hydraulic gate community develops a collective curiosity about the more complex issues behind a dynamic instability mechanism that may require a triggering event such as a corroded trunnion pin or large earthquake- or landslide-induced surface wave to initiate its occurrence.

Preface

Initially, the focus of this book was to be on just the mechanisms uncovered in the past 35 years by the senior author, Noriaki Ishii in conjunction with the other two authors, Keiko Anami and Charles Knisely, along with many masters' and doctoral students at the Osaka Electro-Communication University in Osaka, Japan. Noriaki Ishii was a member of the forensics team that examined the Folsom Dam gate failure. Despite many evenings spent on the dam crest trying to resolve the conflict between the eyewitness testimony and the mechanisms of vibration known at the time, he had to admit short-term defeat and accept the team assessment of the cause of the failure. At the time of the forensic analysis, no mechanism of self-excited vibration for a Tainter gate, open 0.76 m, was known. All known mechanisms at the time were applicable to small gate openings, submerged discharge, or trunnion eccentricity, none of which pertained to the Folsom case. In analyzing the Wachi Dam gate failure, Noriaki Ishii came to understand that no single degree-of-freedom model, other than one with eccentricity, could correctly describe the observed vibrations at the Wachi Dam. Experience in mode-coupling for long-span gates several decades after the Wachi failure guided the search for the mechanism that might have driven the observed vibration of the Folsom Dam gate, and retrospectively that of the Wachi Dam gate. In the 10 years after the Folsom gate failure, through his experience and insight gained in trying to understand the Wachi gate failure and other types of gate vibration, Noriaki Ishii lead the effort to identify a vibration mechanism that does pertain to Tainter gates with large gate openings. That mechanism is the self-excited coupled-mode dynamic instability.

We were encouraged by reviewers and colleagues to broaden the topics considered in the book, and so we arrived at the current configuration:

- An introductory chapter.
- Two chapters on broad overviews of fundamentals intended for entry-level engineers.
- Two chapters on more advanced concepts of dispersive wave, fluid excitation due to flow rate variation, added mass and damping in fluids.
- One chapter on established instability mechanisms (nappe oscillation and vibration of flap gates).
- Two chapters on other gate instability mechanisms (coupled-mode vibration of long-span gates and eccentricity instability).
- Five chapters on details of the mechanism of the self-excited coupled-mode dynamic instability including test results from five full-scale gates that were tested for stability.
- One chapter examining similarities and difference between coupled-mode Tainter gate instability and coupled-mode bridge flutter.
- One concluding chapter, summarizing the efforts to find the mechanism that sustains self-excited coupled-mode vibrations of Tainter gates, that well might have been involved in the Folsom Dam gate, possibly causing or contributing to the failure of the gate. Fittingly, this chapter contains the nascent draft of guidelines for dynamic designed Tainter gates.

The topics in this book are the accumulation of knowledge from many years of collaborative efforts. The content of the last six chapters is essential to any aspiring gate design engineer. Just as bridge engineers learned to accommodate dynamic wind excitation in the design phase after the Tacoma Bridge failure, so too must gate design engineers learn to accommodate both static and dynamic stability in the wake of the Folsom Dam gate failure.

Currently, large hydro projects continue to come on line around the world. We fear that without considering the dynamic instability mechanisms in this text, gate designers will bear witness to another gate failure due to some form of triggering excitation. Our hope would be that by explaining the theory in as much detail as possible, gate design engineers would adopt a dynamic design criteria that could avert such a future disaster.

REFERENCES

Lewin, J. (1998). Hazard and reliability of hydraulic equipment for dams. In *The Prospect for Reservoirs in the 21st Century: Proceedings of the Tenth Conference of the BDS*. Thomas Telford. doi:10.1680/tpfrit21c.27046.0004

Lewin, J. (2001). Failures and vulnerabilities of reservoir control structures. In *Learning from construction failures: Applied forensic engineering*. New York, NY: Wiley.

Acknowledgment

Lewisburg, PA, USA and Osaka, Japan

March 2017

The authors are indebted to those who taught them.

Noriaki Ishii gratefully acknowledges the instruction and collegial support of Prof. K. Imaichi (Osaka University), Prof. T. Uematsu (Osaka University), Prof. E. Naudascher (University of Karlsruhe) and Senior Executive Manager M. Adachi (Panasonic Company).

Keiko Anami extends her thanks to Prof. N. Ishii. and to Prof. I. Ushiyama, Chairman of the Board of Directors and Institute Head, Ashikaga Institute of Technology, who provided her with both collegial and financial support while she was in Ashikaga.

Charles Knisely extends his thanks for the formal teaching and collegial support of Prof. D. Rockwell (Lehigh University), and the informal collegial teaching and assistance of Prof. S. Ziada (McMaster University), Dr. A. Schachenmann (Sulzer AG, Switzerland), Prof. D. Weaver (McMaster University), Prof. E. Naudascher (University of Karlsruhe), Prof. M. Morkovin (Illinois Institute of Technology), Dr. W. Blake (David Taylor Research Center), Dr. J. Gerschfeld (David Taylor Research Center), Prof. H. Nakagawa (Kyoto University), Prof. M. Matsumoto (Kyoto University), Prof. W. Frank (University of Siegen, Germany), Prof. K. Alexander (University of Canterbury, Christchurch, NZ), Prof. A. da Silveira (Universidade Federal de Uberlândia,Brazil) and Prof. F. Mariano (Universidade Federal de Goiás, Brazil).

We all would also like to acknowledge the support of our university administrators, including Chairman Kuniya Fukuda at Osaka Electro-Communication University, President Izumi Ushiyama at Ashikaga Institute of Technology, and President John Bravman at Bucknell University.

The authors would like to express their sincere gratitude to our collaborators who made substantial contributions to the material in this book, including the collection, analysis and reporting of significant data from model scale experiments and field tests for the flow-induced vibration of full-scale Tainter gates.

One particular collaborator we wish to thank is Prof. Akinori Nakata of Osaka Electro-Communication University, whose PhD dissertation on "A Study on Flow-Induced Vibration of Shell-Type Long-Span Gates" was completed in March 1996, under Prof. Ishii.

We would also be remiss in not thanking our families for the time we begged, borrowed, and stole from them to complete this manuscript and to undertake the many projects that contributed to it. Thank you for your patience, your understanding, your support and above all your love.

Lewisburg, PA, USA and Osaka, Japan

March 2017

Section 1
Introduction and Fundamentals

Chapter 1
Introduction:
History of Gate Failures and Overview of Vibration Mechanisms

ABSTRACT

In this chapter a summary of the historical record of dam and gate failures is compiled. Once the reader is familiar with the failure statistics, a concise summary of the prerequisite topics such as press-shut and press-open devices needed to understand the analysis of gate failures is presented. Subsequently, a non-mathematical conceptual descriptions of various excitation mechanisms, such as discharge fluctuations, shear layers instability, nappe instability, coupled-mode long-span gate instability, eccentricity instability and coupled-mode Tainter gate instability that excite gate vibration are presented to set the stage for more analytical treatment in subsequent chapters.

OVERVIEW OF HISTORICAL RECORD OF DAM AND GATE FAILURES

Hydraulic gates are used to contain municipal water reservoirs, regulate water levels for hydroelectric power plants, and control discharge to minimize large scale flooding. The failure of these hydraulic gates can be disastrous, in terms of potential loss of life, potential destruction of downstream property, potential loss of generating capacity, as well as lost reservoir capacity in regions with limited seasonal rainfall. Ultimately, gate failure has the potential to lead to destruction of the entire dam. Gate failure is included in the statistics of dam failures or dam incidents, even when the failure of the gate does not result in the destruction of the dam. Gate failure is grouped with other spillway issues, and, thus, it is not obvious precisely how many dam failures result from or are initiated by gate failures.

Dam Failures

Dams are generally classified as embankment (earthfill, rockfill, lay-core, roller compacted concrete), concrete gravity, or arch dams (single arch or compound arch). Figure 1 shows simplified schematics of these types of dams. Failure modes, design criteria and design procedures for dams are provided in

DOI: 10.4018/978-1-5225-3079-4.ch001

Novak et al. (2001). Despite the best intentions of dam designers, some dams fail nonetheless, on very limited occasions usually under very exceptional circumstances.

According to the National Performance of Dams Program (NPDP, 2010) at Stanford University, there were over 1400 dam failures in the U.S.A. from 1848 to 2005. The yearly average number of failures increased to around 20 per year after about 1980 and resulted in over 4000 cumulative fatalities. Since 1985, there have been very few fatalities (graphical presentation of the data can be found at NPDP Historic Record, 2007).

Worldwide 2.20% of dams built before 1950 failed (ICOLD, 1995). For dams built after 1950, the failure rate declines to less than 0.5%. Regan (2010) notes that failure may occur any time from first filling to much later in the dam's life. The Teton Dam failed on first filling, while the Kantale Dam in Sri Lanka did not fail until after 117 years of successful operation. Regan (2010) tabulated statistics for 4000 dams in 87 countries with 1158 incidents or failures for which the age of the dam was known. The data Regan (2010) presents are consistent with the ICOLD (1995) data that showed failures are most

Figure 1. (a) Embankment dam (earthfill, rockfill, clay-core, concrete-core); (b) gravity dam; and (c) arch dam (single and compound arch)
After Novak et al. (2001).

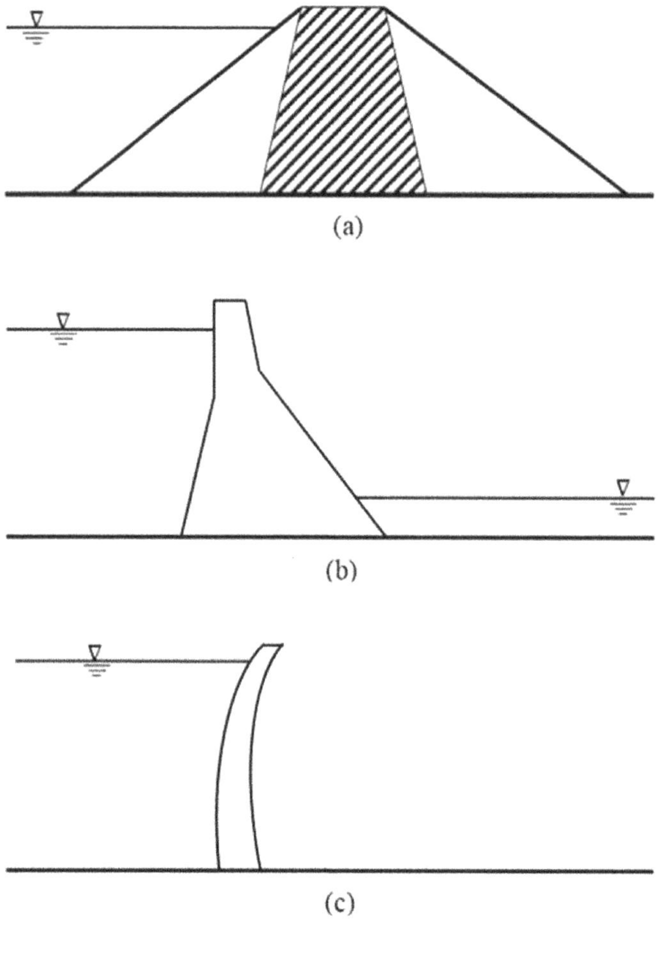

common during the first year and often during the initial filling. Regan (2010) also found that about 31% of dam safety incidents occurred within the first five years after construction was completed. He found that earthen and rockfill dams are more susceptible to failure in the first five years (42% suffered incidents) than either arch dams (29% incident rate) or gravity dams (18% incident rate).

For those dams that survived their first five years, Regan (2010) found that earthfill dams are statistically more likely than average to experience failure, while gravity dams are less likely than average to fail. Rockfill and arch dams are closer to average than either the gravity or earthfill dams. Regan also found that 47 of 115 structural safety incidents with gravity dams, after the first five years, involved spillway gates (approximately 41%). Failures of dams which passed the century mark before their failure (as cited in Vogel, 2009b) include the Emery Dam in 1966, which failed after 116 years of operation; the Spaulding Pond (Mohegan Park) Dam, which failed in 1963, and the Warmwithens Dam, which failed in 1970, both after 110 years; and the Center Creek Dam, which failed in 1973 after 104 years. At the extreme end of the longevity curve is the Kantalai Dam which failed in 1986 after 1400 years of operation. Thandaveswara (2010) provides case studies of dam failures in India. Lukman et al. (2011) provide an overview of dam failures in Nigeria. An exhaustive compilation of failures is available in Vogel (2009a). Other compilations of dam failures can be found in NSWDSC (2010b), Christian & Baecher (2002), Graham (1999), Table of Dam Failures (2014), and ASDSO (2014). Donnelly & Morgenroth (2005) relate historical failure rates to risk and proactive safety measures. Foster et al. (2000) provide statistics for failures of embankment dams. Chanson (2000) summarizes failure data for weirs and stepped spillways.

Embankment dams are frequently used as tailings dams containing waste from mining operations. These embankment dams are often formed from the coarser fractions of the tailings or other residual materials from the mining operations (Azam & Li, 2010). These embankment dams have failed at a rate of about 2 per year over the past 25 years, a reduction from the approximate 5 failures per year over the preceding 3 decades, from 1960 to 1990 (Azam & Li, 2010; WISE-Uranium, 2016). Azam & Li (2010) conclude the rate of failure of tailings dams over the past 100 years is about 1.2%, a figure they claim is more than two orders of magnitude greater than the failure rate of water retention dams.

The reasons for the relatively higher failure rate of tailings dams can be traced to several factors: the low quality fill material; sequential raisings of dam height; lack of applicable design requirements and/or guidelines, especially in developing countries; and the expense of maintaining the dams after mine closure (Azam & Li, 2010)

Tailings dam failures are often accompanied by fatalities and long-term environmental consequences due to heavy metals and other contaminating products in the residual mining wastes. One of the two most recent failures was the largest such structure to fail. The 2015 collapse of the tailings dams at Bento Rodrigues in the state of Minas Gerais in Brazil resulted in at least 17 deaths, a loss 158 homes and environmental degradation from the dam site to well into the Atlantic Ocean, more than 660 km away (WISE-Uranium, 2016). The plume of red-brown fluid was visible from space as it dispersed into the Atlantic Ocean. The two companies involved will spend about $2.6 billion on the recovery efforts and clean up (Kiernan, 2016).

Further information on tailings dam failures can be found in ICOLD (2001), Bowker & Chambers (2015), as well as in the sources cited in the previous three paragraphs along with additional citations in all of these sources. Tailings dams do not typically have spillways and gates, and will not be considered further in this book.

The most common cause of dam failure is overtopping due to excessive rainfall rates, due to operator error, or due to an ill-advised choice of design capacity. According to Christian & Baecher (2002), and sources cited therein, overtopping is the reason 40% of all dams failed. Excessive rainfall, even when operational procedures are correctly followed, may result in flow rates that exceed the capacity of the dam, leading to static failure resulting from overtopping the dam. McCully (2005) discusses perhaps the most costly overtopping failure in terms of human lives: the 1975 failure of as many as 62 dams in the Henan Province of China during Typhoon Nina. McCully writes, "As many as 230,000 people died in this domino-effect collapse of dams on the Huai River, some 85,000 in the flood waves and the rest from resulting epidemics and famine."

The second most common cause of dam failures is foundation failures due to static loading of the dam base, seepage, or erosion. For example, the failure of the Malpasset Dam in France in 1959 resulted in 450 casualties. Several sources (DBSA, 2010; Christian & Baecher, 2002; McCully, 2005; Lempérière, 2003; and many others, all citing ICOLD, 1995) suggest that 30% of all dam failures are foundation failures. Forensic investigations of these failures conducted immediately after the failures provided some sense of the root cause of the failure, but in retrospect, other causes have been identified, some even decades after the failure (see, for example, Hendron & Patton, 1987; Seed & Duncan, 1987; Habib, 1987; and Wittke & Leonards, 1987). Retrospective examination of these failures has provided a more complete understanding of their causes than was available in the initial forensic investigations, thereby providing input to the design process to avoid the same problems in the future.

Guidelines for dam design (for example, USACE, 2010; NZSOLD, 2000; NSWDSC, 2010a; NYSDEC, 1989; and, TCEQ, 2009; among many others) have evolved in response to these static dam failures in an attempt to assure the static stability of dams. Since every dam site has unique topography, unique geology, and unique hydrology, designers are still forced to make decisions concerning the probable maximum precipitation and to select a corresponding design capacity for the dam. While the design guidelines are still evolving, the decline in post-1950 failure rates is evidence of their collective success (ICOLD, 1995). In addition, Lempérière (2003) cites a failure rate of about 3% for pre-1980 earthfill dams in developing countries, with the rate falling after 1980.

Despite the apparent success of these guidelines, building completely safe dams is not possible. Any attempt to do so would certainly prove to be economically unviable. Lempérière (2003) lists some of the challenges facing a dam designer:

1. Each dam is unique with unique consequences of failure
2. All dams must resist enormous water forces
3. Leakage must be minimized in all dams
4. Each dam has unique geology that must be understood by the designer and incorporated into the foundation design
5. Economic considerations must be made regarding the use of available local materials as well as the logistics of transporting these materials
6. Schemes must be devised to control flooding during construction
7. Local seismology must be assessed and incorporated into the design
8. Unique environmental impacts of each reservoir must be assessed

Christian & Baecher (2002) discuss the design view of the engineer versus the economist. Their view is that the engineer wants a 100% certainty of dam viability, while the economist is the one making the risk versus cost decision. Ultimately, every project will need to work within a budget regardless of whether the construction is by a for-profit private entity or by a non-profit governmental entity. Dam construction always requires an acceptance of some level of risk; construction cost must be balanced with reliability and safety.

At times, the risk may be substantially greater than the designer might suspect when contractors surreptitiously substitute sub-standard materials to enhance their company's profitability. For this reason, inspection and oversight during construction are essential. Unwittingly, designers might also be designing dams with a greater probability of future failure due to climate changes associated with global warming. Christian & Baecher (2002) discuss the concept of designing for a 1000-year flood based on a maximum of 250 years of data, and usually even less. The U.S. has existed as a country for less than 250 years, and most stream records extend only for somewhere between 50 and 100 years into the past. Thus predicting floods with return periods longer than about 200 years is a form of guess work and is extremely risky. To exacerbate the matter further, the current rate of climate change with its associated statistical increase in the frequency of severe, intense storms makes it difficult to assess the performance of dams in another 20 to 50 years. As the collective understanding of global climate change increases, it may become necessary to further revise the design rainfall estimates and to design dams for greater discharge capacities in some regions while perhaps decreasing expected discharge in others. A further complication is the post-construction upstream changes in land use that can significantly affect the rate of run-off. See Lempérière and Vigny (2005) for a discussion of the options for establishing the design discharge capacity.

In areas prone to earthquakes, dam designers take into account dynamic seismic loading in addition to static loading. Earthquake-resistant design guidelines have been formulated (FEMA, 2005; ANCOLD, 1998; and others). The unpredictability of earthquakes makes it very challenging to determine if these guidelines are sufficient, but there has been no known death resulting from an earthquake-induced failure of a large dam (Dam Safety, 2010). More than one thousand embankment dams and reservoirs suffered damage due to the Wenchuan earthquake (Richter scale 8.0 on 12 May 2008) in China and the Maule earthquake (Richter scale 8.8 on 27 Feb 2010) in Chile, but no large dam failed (Dam Safety, 2010). Hall (1984) provides additional data on how concrete dams faired during earthquakes:

1. In the 1906 San Francisco earthquake, the 145-foot high, curved gravity Crystal Springs Dam was located adjacent to the fault line and survived undamaged
2. The Pacoima Dam, a 370-foot high arched dam with 10-foot wall thickness at the top and 99-foot thickness at the base, survived peak accelerations of 1.25 g during the magnitude 6.4 San Fernando earthquake in 1971
3. The Hsinfengkiang Dam, a 344-foot high concrete dam near Canton, China, developed longitudinal cracking in the upper portion of the dam during the 6.1 magnitude earthquake (suspected reservoir-induced seismicity, see the following paragraph) in 1961, but the dam did not fail
4. A magnitude 6.5 earthquake near Poona, India in 1965 (also suspected of being a reservoir-induced seismic event) caused longitudinal cracking in the upper portions of the 338-foot high, concrete gravity Koyna Dam. Although the dam remained standing, the earthquake killed 180 people and 2200 were injured (Hinks & Gosschalk, 1993).

Dam engineers are aware of the risk of reservoir-induced seismicity (RIS) or reservoir-triggered seismicity (RTS), in which the filling of the reservoir triggers seismic activity, as discussed in a variety of sources such as Jauhari (1999), Hinks & Gosschalk (1993), Gosschalk et al. (1994), and Dam Safety (2010). At least 46 suspected cases of reservoir-induced seismicity have been reported; 11 of those cases were confirmed by consensus (Hinks & Gosschalk, 1993). Reservoir-induced seismicity usually appears within the first four years after the initial filling of a reservoir (Gosschalk et al., 1993). RIS is usually restricted to dam heights greater than 100 m or reservoirs greater than 500 x 10^6 m^3 in tectonically sensitive regions. However, there are locations where reservoir-induced seismicity has occurred at smaller dams and reservoirs (ICOLD, 1989).

In general, however, Hinks & Gosschalk (1993) conclude that most dams fared well in earthquakes. Failures were due overwhelmingly to liquefaction of material in the dam or in the foundation of the dam. Those that did experience failure did not do so immediately after the earthquake, but rather after a delay of up to 24 hours. In his review of reservoir vulnerabilities, Lewin (2001), citing Williams (1996), argues the need to discharge water quickly from the reservoir to avoid overtopping as discharge through the turbines is reduced in the event of a massive post-earthquake power load reduction at hydroelectric stations.

In their overview of why dams fail, Christian & Baecher (2002) argue that engineers by nature want to be sure that their designs function properly. The authors explain the concepts of Probable Maximum Flood (PMF) and Probable Maximum Precipitation (PMP) used in the U.S. to design discharge capacities of reservoir, gates, and spillways. Although they note that no gate in the U.S. has ever experienced a PMF, the most common cause of dam failure is overtopping. Christian & Baecher note this inconsistency and suggest that some overtopping failures occur before completion of the dam project, others occur in older dams with inadequate design capacities for their spillways, and others are at tailings dams and inadequately maintained older dams. Vogel (2009b) notes that loss of storage capacity due to silting in the absence of a sediment management program will also increase the susceptibility of a dam to overtopping. In addition, some dams fail due to overtopping resulting from operator error (Sagar & Hutton, 2000; Graham & Hilldale, 2002).

Christian & Baecher (2002) suggest that forensics teams must thoroughly investigate all possible causes, not just the first plausible one, when a dam fails. The St. Francis Dam failure in 1928 illustrates this point. The failure of this dam was initially attributed to rock slaking (falling apart of rocks when submerged in water) on the right side of the channel. At the time, the proposed failure mechanism seemed so obvious that no one saw any need for deeper investigation. It was not until more than half a century later that Rogers (1995) argued (convincingly, according to Christian & Baecher) that the failure was actually initiated on the left side of the dam by uplift pressure from the rising water, which reactivated a pre-historic landslide that formed the left dam abutment. When the dam failed in 1928, no one was aware of the preexisting geologic conditions. Christian & Baecher speculate that if the St. Francis Dam failure had been correctly assessed in 1928, perhaps the Malpasset Dam failure in 1959 could have been prevented. It took five years of investigation for the forensics team to conclude that the Malpasset Dam failed due to geologic conditions similar to those at St. Francis. Knowledge of those conditions would have been invaluable when the Malpasset Dam was designed.

The generalized take away message from the discussion of the St. Francis Dam failure is to beware of complacently accepting the first identifiable cause of a failure. A more careful investigation of all possible mechanisms, more than just the simplest or most obvious one, may save lives and considerable expense in the future. We will return to this point later when discussing the Folsom Dam gate failure.

A similar retrospective look at dam failures was taken at the 1985 International Workshop on Dam Failures at Purdue University. The failures at Baldwin Hills in 1963, at Vaiont in 1963, at Malpasset in 1959, and at Teton in 1976 were discussed individually and then collectively at the workshop. Papers by Seed & Duncan (1987), Penman (1987), and Sherard (1987) re-examined the Teton failure. Habib (1987) and Wittke & Leonards (1987) both re-examined the Malpasset failure. Hendron & Patton (1987), Nonveiller (1987), and Belloni & Stefani (1987) re-examined the Vaiont catastrophe, and Scott (1987), Leps (1987), and Wilson (1987) re-examined the Baldwin Hills reservoir failure. Other workshop presentations and written commentary along with the records of the final panel discussion were printed in Volume 24 of *Engineering Geology* (1987). A shorter summary of the workshop is given in Four Dam Failures (1985). Many of the comments found in these discussions of dam failures are applicable to a discussion of gate failure, the focus of this book.

Participants at the 1985 workshop were unable to reach consensus on all but a few points regarding the failure of these four dams. However, many general principles were established, which could help to prevent such incidents in the future. Four comments from the symposium are relevant to the subsequent discussions of gate failures and gate vibration mechanisms:

The state of the art [now] is far advanced over what it was, say, [35 years ago], but the profession can ill afford to be complacent about the state of current knowledge. A thorough grasp of past experiences is good insurance against future catastrophic failures. (Symposium on dam failures -- panel discussion, 1987: Leonards, p. 611)

[N]ature always gives us a warning of failures but we may have to listen very closely to hear the warning. When a failure occurs we'd have to be deaf not to hear the lesson. And that's the value of failures - that each contains some lesson, possibly not [a] new lesson, but if it's an old lesson it is one that we haven't learned well enough. (Symposium on dam failures -- panel discussion, 1987: Duncan, p. 541)

The failure of a dam is the result of a complex concordance of causes. We very often find several causes, even if not all of them are brought to light with certainty. Still more difficult is to establish the "primary" cause, or the triggering mechanism, because the concept of a primary cause is in itself debatable when subsidiary causes leading to the failure are required. (Symposium on dam failures -- panel discussion, 1987: Londe, p. 542)

Lastly, there is the importance of charity. I don't think we should be overly critical of those who design or are charged with the failures we've talked about. We should rather think that advances in engineering are made by pushing forward the boundary between what we can calculate and what we cannot calculate. (Symposium on dam failures -- panel discussion, 1987: Milligan, p. 574)

Gate Failures

The focus of this introduction will now shift from the dam as a whole to the gates that control the efflux from the reservoir. An overview of several gate types used to regulate the outflow from dams is shown in Figure 2.

Failure of a gate may be classified as a dam failure if there are no means for emergency closure of the opening or if there is overtopping of a closed gate. If there are back-up procedures that limit the discharge, then the gate failure is considered an incident. In general, gate failures are included as spillway failures in the ICOLD (1995) data, making it difficult to distinguish gate failures from other spillway incidents. Lewin (2001) writes that gate hazards and gate reliability have received more attention recently than

Figure 2. Simplified schematics of various gate types frequently used as outlet control structures for spillways and weirs
After Naudascher, 1991.

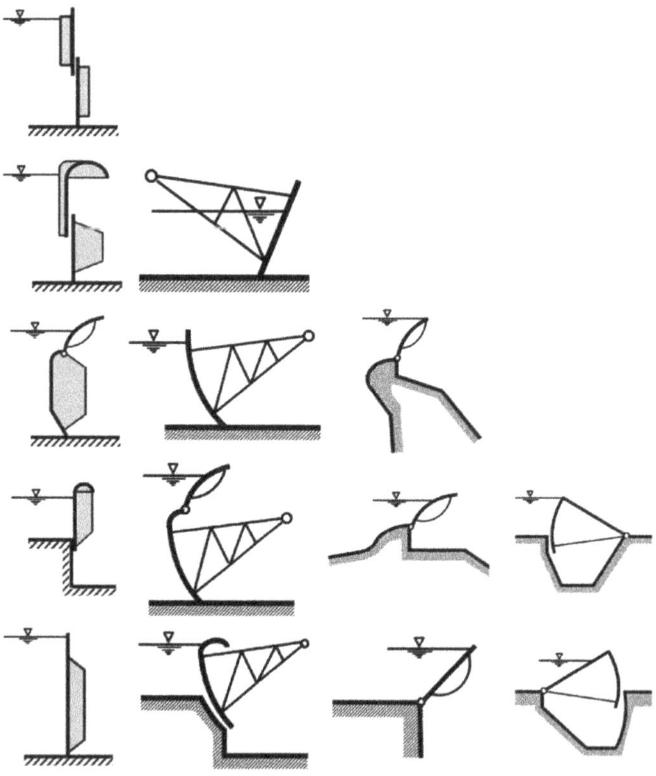

concerns for the overall dam. One estimate of gate incident rates has already been mentioned: 47 out of 115 structural incidents at gravity dams are gate-related (41% of structural incidents, see Regan, 2010). An additional estimate of gate failure rates can be found in Lewin (2001), who reported that 2% of the 240 dams had malfunctioning gates. A study by the U.S. National Research Council (NRC, 1983, p. 11) shows a gate incident rate of 2.4% of 285 dams. Hobbs & Azavedo (2000) list six gate failures. Graham & Hilldale (2001, 2002) summarize gate failures and incidents. Lewin et al. (2003) tabulate eleven gate failures and categorize gate failures and incidents by probable causes. They note that power interruption (power failure or power supply failure) was the most common cause of gate failure, occurring in four of the eleven documented failures. Gate vibration is cited as the cause of three of the eleven gate failures. If the Folsom Dam gate failure were counted as a vibration-induced failure (as we believe), then the number of failures due to vibration would be the same as the number due to power interruption documented by Lewin et al. (2003).

Frizell (2005) presents the results of a survey of U.S. Bureau of Reclamation field operating personnel on concerns or problems they might be having with the hydraulic structure infrastructure in their charge. The top three most problematic spillway features were

1. The chute or tunnel invert (22%),
2. The stilling basin/plunge pool (21%), and
3. The gates (19%).

The outlet works feature found to be most problematic by 27% of the respondents was the category "gates and valves." Frizell lists as the top structural issues related to spillway and outlet work needing further investigation:

1. Deterioration of gates and surfaces surrounding gates.
2. Operation of gates by wire ropes or chains.
3. Erosion in the channels downstream from the energy dissipation structures or from the release points of the structures.

Clearly, the field operating personnel would welcome additional research on gates and gate support structures.

Gate problems have been investigated by designers, manufacturers, and operating agencies using models, prototypes, and even full-scale gates when these problems have arisen at specific gate installations. Several of these cases have made their way into the open literature. Pickering (1971), Schmidgall (1972), and Pugh (1977) report on investigations (undertaken by the U.S. Army Corps of Engineers Waterways Experiment Station in Vicksburg, MS) of vibration of spillway gates on the Arkansas River in Arkansas and Oklahoma. On the Arkansas River there are multiple dams and locks, several of which had problems with the vibration of their Tainter gates. Using a 1:12 scale to investigate the gate vibrations and examine alternative designs, Schmidgall (1972) concluded that the major causes of the vibration were the gate lip and bottom seal designs. A retrospective analysis of the mode shapes of the vibrating skinplates shown in Schmidgall (1972) suggests that while seal vibration may have served as the trigger, the self-excited coupled-mode instability may have played a role in these vibrations. This mechanism would not be identified for more than another 20 years, however.

Vibrations of the gates at the Barkley Dam in Kentucky were investigated, again by the U.S. Army Corps of Engineers Waterways Experiment Station, and reported by Hart & Hite (1979a, 1979b). The U.S. Army Corps investigated the vibrations of the 50- by 55-ft spillway Tainter gates with submerged discharge when the gate lip was submerged by high tailwater. Field tests were conducted on two gates using pressure transducers and accelerometers. Many gate components were found to vibrate at the same frequency as the gate lip pressure fluctuations. This frequency corresponded to the natural frequency of at least one gate component. Vibration amplitudes were found to decrease with increasing tailwater. Removal of the lip led to significant attenuation of the vibrations. Remedial measures in design and operation of the gates were recommended.

Other practical gate-related flow-induced vibration examples are found in the *Proceedings of the 1979 IAHR Practical Experiences With Flow-Induced Vibrations Symposium* (Naudascher & Rockwell, 1980), as well as in the Proceedings of the ICOLD Congresses held over the past thirty years (see ICOLD, 1982 to 2012).

Several gate incidents and failures are summarized in Cassidy (2000) based on reports presented at the 20th ICOLD Congress in 2000. Based on the report by Lima Da Silva et al. (2000), Cassidy reports that in July 1994 at the Itaipu Dam on the border between Brazil and Paraguay, a rod failure occurred in a hydraulic cylinder while one of the radial gates was being closed. Subsequently bulkheads were installed upstream of the gate to allow for maintenance of the gate. Detailed investigations suggested gate racking may have been the cause, resulting in gate vibrations, which then overloaded the hydraulic cylinders, leading to the failure of one of the rods. Cassidy also reports that a similar failure apparently occurred at the Tucurui Dam (also in Brazil) in 1995.

Another failure summary provided by Cassidy (based on the report by Jalal Zadeh & Fouladi Nashta, 2000) concerned severe damage to the intermediate spillway radial gate at the 106 m high Sefidrud Dam in Iran in 1990 when it was subjected to a magnitude 7.3 earthquake. The accelerations from the earthquake, estimated at 6.0 g, produced a dynamic load that caused the trunnion arms to buckle even though the reservoir was 6 m below its maximum normal level.

An additional gate failure reported by Cassidy and also cited in Graham & Hilldale (2001, 2002), is the October 1990 failure of one of the seventeen large 15.0 x 13.5 m radial gates at the Singur Dam in Andhra Pradesh, India due to errors in fabrication and construction. Cassidy, citing Mande et al. (2000), reports that a suspected lapse in inspection was the cause of this failure. Graham & Hilldale (2002) report that, upon initial filling with the water level still 3 m below the design level, welds holding the left side trunnion arms to the tie flats failed. The gate was dislodged, but held by the right side trunnion arm for 22 hours before collapsing and being washed away. Cassidy emphasizes the importance of using only top quality materials and workmanship in the construction of large dams. The growing tendency to make inspection a responsibility of the contractor who invariably has a financial conflict of interest is not conducive to good quality control, according to Cassidy.

Cassidy (2000) and Graham & Hilldale (2001, 2002) discussed, as well, the failure at the Picote Dam on the Douro River in Portugal. On February 16, 1966, during a flood and while the gate raising was being operated by remote control, debris accumulated on the left chain resulting in a loss of articulation of the chain links and the subsequent failure of the left chain hoist motor. The right motor kept turning. The skewing and subsequent warping of the gate led to the failure of the right chain hoist motor as well. The gate dropped to the sill due to its own weight, tearing the trunnion girders from the supporting structures, and was washed downstream. The two adjacent gates were in the open position and were rendered inoperative by the loss of support of one end frame. Dam failure did not occur.

In 1996, rainfall in the Saguenay River basin in the province of Quebec in Canada produced conditions that exceeded the province's 1000-year flood. As Cassidy (2000) writes, gates failed for a variety of reasons that included overtopping, due in part to jammed stop logs upstream of the gates, broken or defective hoist mechanisms compounded by the added loading of the water on top of the gates, and the vast debris flow. Gates also failed when power outages prevented the hoist motors from operating. Some dam sites were inaccessible and key personnel could not reach the dams due to the flooding. Because of this flood, major dam safety legislation was recommended and adopted in Canada and elsewhere.

Yet another incident occurred at the San Teresa Dam, located on the Tormes River in Spain. In 1963, just three years after completion of the dam, the malfunction of the automatic controls on five of the 52.5-foot (16-meter) long Tainter gates resulted in the gates failing to open and being overtopped. Foundation erosion followed the overtopping (Graham & Hilldale, 2001; 2002).

Sagar & Hutton (2000) discuss their experiences with Tainter gate vibration problems in India, the Caribbean, and Indonesia. In India, a Tainter gate with simultaneous over- and underflow vibrated to failure and was washed away. The authors did not identify the site, but the incident is reminiscent of the failure at Khadakwasla Dam in 1961. In the Caribbean, an unnamed Tainter gate failed when the primary and back-up power supplies failed during a hurricane, resulting in overtopping of the gates. Some gate members and hoists were damaged. In Indonesia, gate vibrations were so severe that they frightened personnel working on the spillway bridge and a nearby footbridge. Sagar & Hutton attribute the vibration to an inappropriate design of the bottom geometry and the gate seal. Their suggested remedy was to prevent jet impingement on gate members by using a relatively thin rectangular gate bottom seal instead of a J-type seal. Information on designing gate lips to avoid vibration problems can be found in Sagar (1999).

A general observation contained in Cassidy's report reinforces the observations concerning the gates on the Arkansas River and at the Barkley Dam. Radial gates have a tendency to vibrate at small gate openings due to seal instability, further exacerbated by slight submergence of the gates. Cassidy (2000) cites the report by Berridge (2000) that recommends maximizing the angle between the upstream gate face and the spillway.

Another observation by Cassidy (2000), again citing other congress attendees (Dias & Gusmao, 2000; Sagar, 2000; and Jalal Zadeh & Fouladi Nashta, 2000), is that overtopping of large radial gates should be avoided. With large overflows serious vibrations can ensue and frequently cause damage. Where there is a large amount of debris accumulation upstream of the gate, a movable flap might be provided to pass the trash. Sagar (2000) argues for a fixed chute, because hinges on a movable flap can become corroded or jammed if not properly maintained.

A listing of the gate failures cited in the preceding paragraphs, as well others, has been assembled in Table 1. The failures are listed in chronological order for all gates for which we were able to identify the dates of failure. In the 1990s there was an average of about one gate failure per year.

Based on the reports of Sagar, Hutton, and other consulting engineers throughout the ICOLD Congress proceedings, there is evidence that more gate failures and vibration problems have occurred than have been reported. Lewin et al. (2003) state that Lewin has investigated more gate failures due to vibrations than any other cause.

Many cases of gate vibration have involved Tainter gates, which is not surprising given the number of Tainter gates installed as spillway gates. The reason Tainter gates are preferred as spillway gates is that they can be easily opened and closed with a relatively small expenditure of power. However, the same logic should then clearly suggest that they will be more susceptible to vibration than heavier gates that require larger energy expenditure to open. In fact, as discussed in Sagar & Hutton (2000), Pickering (1971), Hart & Hite (1979), and many other sources, the vibration propensity of Tainter gates at small opening is quite well known. In the case of the gate at the Schinznach Dam in Switzerland, no cure could be found to eliminate the excitation of the gate vibration, so the gate was fitted with a damper to control the amplitude of the vibration (Petrikat, 1980).

As shown in Table1, gate failure and gate vibrations are world-wide problems of seemingly increasing frequency. Table 1 and sources cited therein show that failures and vibration problems have occurred in Canada, Switzerland, Sweden, Hungary, Slovenia, South Africa, Brazil, Australia, India, Indonesia, England (see Noble & Lewin, 2000), and in the Caribbean, as well as in several other countries. However, the two largest gate failures, one already attributed to vibration and the other currently considered a static failure (contrary to substantial evidence of vibration to be presented later in this book), occurred at the Wachi Dam in Japan in 1967 and at the Folsom Dam in the USA in 1995. In both failures, the gates were radial (Tainter) gates and vibrations were noted in eyewitness reports.

A 35-ton Tainter gate at the Wachi Dam in Japan failed on July 2, 1967. The dam is located on the upper reaches of the Yura River in Kyoto Prefecture. The report by the committee investigating the failure (Yano, 1968) provided a summary of the failure. While the No. 3 spillway gate was being closed from a 30-cm gate opening, the No. 4 gate was being opened. The No. 3 gate suddenly failed with a loud

Table 1. Gate failures in chronological order

Dam	Date	River	Short Description	Source(s)
Khadakwasla	1961	Khadakwasla Reservoir (India)	Breach of upstream Panshet Dam released flood wave into downstream reservoir when Khadkawasla reservoir was already full, and gates were nearly full open; simultaneous over- and underflow of gates; vibration of the structure; failure within 4 hours.	Than (2010) Vog (2009b)
San Teresa	1963	Tormes (Spain)	Malfunction of automatic controls of five 52.5-foot (16-meter) long Tainter gates led to overtopping and foundation erosion.	G&H (2002)
Picote	1966	Douro (Portugal)	Debris accumulation in hoist chains, loss of articulation of links, left side motor failure, warping of gate, right side motor failure.	G&H (2002) Cass (2000)
Wachi	1967	Yura (Japan)	While No. 3 gate was being shut down from a gate opening of about 30 cm and No. 4 gate was being opened, No. 3 spillway gate suddenly underwent vibration and failed; gate along with the overhead bridge was carried about 136 m downstream, leaving a giant hole at the gate bay.	Yano (1968)
Chikkahole	1972	Chikkahole (India)	Power failure leading to gate malfunction resulted in overtopping and failure.	G&H (2002)
Banqiao and Shimantan	1975	Huai (China)	During a typhoon, reservoir filled to near maximum capacity; sluice gates were opened, but were partly blocked with sediment, leading to overtopping of gates and failure of dam. Water surged over the downstream valleys and plains at nearly 50 km/hr. Smaller Shimantan Dam collapsed shortly afterwards; as many as 62 dams collapsed.	McC (2005)
Goulburn	1978	Goulburn (Australia)	Gate failure caused by corrosion leading to dam foundation stability problem.	Than (2010) Chan (2000)
Machhu II	1979	Machhu (India)	Abnormal floods, electrical power failure, delayed opening of gates, inadequate spillway capacity resulted in overtopping of the embankment; maximum depth of 6.1 m of water over the crest and within two hours, the dam failed.	H&A (2000) Than (2010)
Tous	1982	Jucar (Spain)	500-year flood, electrical power outage, gates closed and inoperable, gates overtopped and dam failure.	G&H (2002)
Upriver	1986	Spokane (USA)	Lightning strike caused generator fault; turbine closure produced water surge; attempts to open radial gates failed; river breached closure embankments and the dam abutments of Powerhouse #1.	G&H (2002)
Guernsey	1986	North Platte (USA)	Debris left inside the gate resulted in plugged drain lines, permitting accumulation of water, resulting in loss of buoyancy. Gate could be opened about half way for 7 hours before debris was cleared. No further damage reported.	G&H (2002)
Seton	1989	Seton (Canada)	Opened automatically when hoist motor turned on due to frozen water in electrical conduit pushing the leads upward, forcing the contacts to energize motor, raising gate past full open position, causing structural damage to gate.	G&H (2002)
Sefidrud	1990	Sefidrud (Iran)	Gate arms buckled after a magnitude 7.3 earthquake.	Cass (2010)
Singur	1990	Godavari (India)	Welds failed on left trunnion arm causing left hoist motor to fail, warping gate, causing right hoist motor to fail. Gate crashed to sill and swept downstream.	G&H (2002) Cass (2000)
Belci	1991	Tazldu (Romania)	Exceptional magnitude torrential rainfall; radial and flap gates on spillway were deployed; only one radial gate could be opened 40 cm; electrical power outage; gates could not be opened; dam failed.	G&H (2002) H&A (2000)
Tarbela	1992	Indus (Pakistan)	Hoist ropes on radial gates failed after gate jammed during lowering operation; the far right gate became stuck in the open position, approx. 11.2 cm above sill; malfunction caused by insufficient clearance for thermal expansion.	G&H (2002) H&A (2000)

continued on following page

Table 1. Continued

Dam	Date	River	Short Description	Source(s)
Mavčiče	1993	Sava (Slovenia)	Accidental gate opening: uncontrolled automatic start of the oil pump opened hydraulic valve for raising radial gate on the automatic gate system.	G&H (2002) H&A (2000)
Tirlyan	1994	Tributary of Belaja (Russia)	3 days of heavy rain; malfunction of radial gates during a flood resulted in overtopping and failure.	G&H (2002)
Folsom	1995	American (USA)	During routine gate opening, operator noted vibration, stopped gate; diagonal brace failed; gate open around left side; vibration suspected to have played a role.	G&H (2002) H&A (2000) Cass (2000)
Saguenay	1996	Saguenay (Canada)	280 mm of rain in 72-hour period; high river flows overtopped several dams in the region; spillway gates partially blocked with floating and sunken debris brought to the surface by high flows; blockage reduced the capacity of the spillway gates, which contributed to overtopping and failure of dams.	G&H (2002)
Cresta	1997	North Fork Feather (USA)	37.8 meter long and 8.5 meter high drum gate failed; a combination of plumbing failures led to flooding of the floatation chamber and subsequent failure.	G&H (2002)
Spanish Fork	2000	Spanish Fork (USA)	Undersized bearing on trunnion failed on first loading. Gate operated for 14 years without trunnion constraint before failing.	Hoff (2000)
Gouin	2002	St. Maurice (Canada)	Unstable separation point with flow control jumping between the upstream beam flange and the gate lip; resulting pressure fluctuations along lower face of the web of the beam were responsible for failure; cross bottom beam was torn apart by the vibrations.	Carb (2005)
Hebgen	2008	Madison (Montana)	Stoplog failure, releasing 3,400 cu ft (normal discharge: 900 cu ft) water, causing 1' rise in Madison River.	Lord (2010); DamFail (2016))
Lake Delhi	2010	Maquoketa (Iowa)	Heavy rains, gate malfunction contributed to failure by overtopping	Fied (2011); DamFail (2016)
Ord	Not found	Ord (Australia)	Gate transmitter failure; control contact failure; gates fail to open and were overtopped.	H&A (2000)

Guide to references: Carb (2005) = Carballada et al. (2005); Chan (2000) = Chanson (2000); Cass (2000) = Cassidy (2000); DamFail (2016) = Dam Failure (2016); Fied (2011) = Fiedler et al. (2011); H&A (2000) = Hobbs & Azavedo (2000); G&H (2002) = Graham & Hilldale (2001, 2002); Hoff (2000) = Hoffmann (2000); Lord (2010) = Lord et al. (20101); McC (2005) = McCully (2005); Than (2010) = Thandaveswara (2010); Vog (2009b) = Vogel (2009b); Yano (1968) = Yano (1968)

noise and was swept far downstream. The operational records of the No. 3 gate showed that it had been opened 56 times with gate openings in the range of 3 cm to 10 cm for 63% of the 110 hours of operation.

The 1967 failure led to the formulation of a mechanism of instability known as "eccentricity instability," which depends on an eccentricity between the skinplate center and the trunnion center (see Ishii & Imaichi, 1977; Ishii et al., 1977; Ishii et al., 1980; Ishii & Naudascher, 1984). After the mechanism had been postulated, however, a very careful examination of the detailed gate drawings revealed only a very small eccentricity (about 72 mm), which was insufficient to drive the eccentricity instability mechanism on its own. After the eccentricity instability mechanism was eliminated, there was no other mechanism of vibration known at the time. Meanwhile, guidelines to avoid eccentricity instability were developed (Ishii & Naudascher, 1992) for gates designed with an eccentricity. A number of gates were intentionally designed with an eccentricity either to assure the closure and sealing of the gate or to reduce the effort required to raise the gate. It would take almost 40 years before a vibration mechanism explaining the failure of the Wachi gate was formulated, coming about only because another gate, the Folsom Dam gate, failed.

In July of 1995, a massive 87-ton Tainter gate failed at the Folsom Dam on the American River in California. The operator's eyewitness report (as cited in Ishii, 1995a) was very definitive.

When spill gate Number 3 reached around 2½ feet I felt an unusual vibration. I immediately pushed the stop button and started to check the gate. The vibration intensified and as I turned to check the gate, I could see the right side, looking down stream, of the gate separating from the structure and water coming out around the gate...

At the time of the failure, no known vibration mechanism for a gate with such a large gate opening had been identified. However, a finite element method (FEM) calculation at that time showed that the stress in a diagonal brace exceeded the carrying capacity of the brace. Based in part on this evidence, the USBR forensics team (USBR, 1996) concluded that the gate failure was due to excessive static loading on the brace. The loading was traced to increased friction of the trunnion pin due to inadequate maintenance. Later we will critically examine the evidence leading to the Report's conclusions. At this juncture, however, it seems pertinent to mention Milligan's comment at the 1985 Workshop on dam failures:

I don't think we should be overly critical of those who design or are charged with the failures we've talked about. We should rather think that advances in engineering are made by pushing forward the boundary between what we can calculate and what we cannot calculate.

In other words, had the calculation methods available now been known then, the Report may have addressed the actual causes, not just the symptoms, of the gate failure. Both before and even after the Report was published, there was some speculation that vibrations played a role in the Folsom failure (Cassidy, 2000). According to Cassidy, the gate was open 0.73 m with a head of 12.2 m at the time of the failure. The chain links were stiff and corroded and resulted in a jerking motion when the gate was raised and lowered. The jerking motion of the chains might have triggered gate vibrations, but the preliminary suppositions about the possibility of the jerking of the chain inducing vibrations were dismissed since there was, at the time, no known mechanism to sustain the vibrations. Cassidy cited Todd (1999) as the source of his information, yet Todd himself concluded that the flowing water increased the damping by an order of magnitude, and rejected the possibility that vibrations played a role in the failure. Todd essentially supported the forensics team's conclusion that the failure was due to the moment induced by trunnion friction. Todd mentioned that there had been some initial concern that the lightweight gate design may have played a role in the vibrations, but that the gate was dynamically stable. Despite Todd's assurances, discussion about vibrations at Folsom continued.

In a USBR report issued a year later, Graham & Hilldale (2001, 2002) discussed the Folsom failure in the section of their report entitled "Spillway Gates Failed, Resulting in the Release of Water." They noted that there are often multiple contributing factors to such gate failures. In the case of the Folsom Dam, they specifically mentioned that the failure was precipitated by more than corrosion, more than a lack of maintenance, and more than a lack of lubrication. They postulated that the gate was likely overstressed because of the high reservoir level and that flow-induced vibrations, as well as trunnion pin corrosion, may have contributed to the failure.

In the same year, Lewin (2001) basically repeated Todd's (1999) summary of the reported sequence of events leading to the failure, noting that failure occurred when the connection of a brace in the radial arm sheared, but leaving unaddressed the cause of the loading.

The forensics report's conclusions led gate owners to immediately inspect the trunnion pins in their installations (see reports by Hutton (1999) and Edwards et al. (1998), among others). While this focus on trunnion pin lubrication was most certainly a wise precaution, we must be wary of treating symptoms instead of causes. Unknowingly, owners and operators may actually increase the risk of vibration by reducing trunnion pin friction. If the side seal friction becomes inoperative due to a large, sudden gate displacement (called hard excitation), trunnion friction may be the only friction maintaining gate stability. A reduction in trunnion friction can potentially lead to a gate being more susceptible to certain vibration mechanisms.

Design retrofits were made to the remaining Folsom gates after the failure (Cassidy, 2000 and Lewin, 2001). These gates were significantly stiffened and the replacement gate was much stiffer than the failed gate. Field tests carried out in 1998 confirmed that the stiffened gates did not vibrate (Cassidy citing Todd, 1999). However, these studies did not include the self-excited dynamic instability mechanism with hard excitation.

According to Lewin (2001), the failure at Folsom could have been predicted and it could have been avoided. He asks rhetorically if there were no engineers who might have made different choices in the design and material selection, had they not had false confidence in the number of apparently successful implementations of the same scheme. Lewin tells us that even after the Folsom failure, some engineers continued to use the same bad design, with the justification that they had extensive experience with the design and the gates performed satisfactorily for many years. Lewin's comment is one worth thinking about: "Sometimes it requires a second disaster before a risk is consistently recognized." Our hope is that the research presented in this book might avert a second disaster.

In the course of analyzing the Folsom Dam gate failure over the past decade, we uncovered a previously unknown mechanism of gate failure, which we call self-excited coupled-mode dynamic instability (also called, in shortened form, coupled-mode instability) which may, at times, require hard excitation. Hard excitation means the gate must be displaced beyond a certain threshold before the mechanism becomes activated. We suggest that both the Wachi and Folsom gates failed due to this coupled-mode dynamic instability mechanism. Dynamic stability of gates is more difficult to assess than static stability because of the interaction of structural dynamics with the fluid dynamics of the flow under and/or over the gate.

Lewin (2001) notes that hydraulic gates are designed to control the flow of water. Therefore, hydrodynamics must be the starting point in their design. Multiple sources of hydrodynamic excitation can be found in flow past these control gates, including eddy formation and vortex shedding from protruding members, fluctuating discharge, leakage flows, unsteady flow separation, and reattachment due to poor design. Damping forces due to side seal friction, trunnion friction, and, to a lesser extent, inherent structural damping are also present. In some cases, the water flow may also serve to provide damping. In other cases, however, the fluid flow can provide excitation that may become dominant. Lewin notes the challenges of quantifying all possible hydrodynamic influences, despite the extensive technical literature available. Design guidelines are intended to provide a best practice approach to accommodating these hydrodynamic forces.

The mechanisms of hydraulic gate vibration (other mechanisms of vibration are known besides coupled-mode instability) can involve multiple vibration modes and/or resonators interacting through the coupling provided by the hydrodynamic pressure fluctuations that are amplified by gate motion and fluctuating fluid flow. An understanding of the mechanisms underlying these potential dynamic instabilities is essential for the long-term safe design of hydraulic gates. The path to understanding these mechanisms has many twists and turns. It is far from straightforward.

Attributing the Folsom gate failure to static loading was so plausible and the loading of the strut to trunnion friction so demonstrable that any further investigation of the failure seems unwarranted. Yet would it not have been prudent for the forensics team to have considered other plausible explanations, including a coupled-mode vibration mechanism, to avert a future failure?

We agree that the poorly maintained and corroded trunnion pin indeed played a substantial role in the Folsom failure. We concur that there were elevated stresses in the cross-members, but we find that the levels were insufficient for static failure. Our models demonstrate that failure should have occurred when the gate was first opened, when the loading was 12% higher than at an opening of 0.76 m where the failure actually occurred. In fact, Edwards et al. (1998) came to the same conclusion: "The most severe loads on the gate usually occur during initial opening when the hydrostatic load and the trunnion friction are maximized. Spillway gate hoists must have the capacity to lift the gate weight plus overcome seal friction and trunnion friction." Todd (1997) concurs as well: "The greatest loading applied to the gate, discounting vibration, occurs when the gate is just being raised from the sill. Under these conditions there are the following loads: water static load; dead weight of the gate; friction from the side seals; chain loads; and trunnion friction moment." If the greatest static load occurs upon incipient opening, why did the gate not fail immediately? Why could it be opened more than 0.7 m before it failed?

In our opinion, the corroded trunnion indeed hindered the opening of the gate, but when the torque exerted on the trunnion by winding the hoist chain overcame the static friction of the corroded trunnion pin, the gate naturally "jumped" due to the release of the counteracting trunnion torque. The sudden change in mean gate position meant that the damping of the side seals (when considered as amplitude-dependent Coulomb damping) fell below the level needed to maintain the quasi-static equilibrium of the gate. With the low level of side seal friction, the coupled-mode vibration mechanism became active and the gate quickly reached vibration amplitude levels sufficient to break the joint holding the diagonal member in place. The estimated required vibrational displacement of the skinplate along its centerline was about 8 mm.

Guidelines have been developed for the design of hydraulic gates (for example, USACE, 2010), as well as for the dams themselves. Such guidelines have certainly been effective in reducing the number of gate failures. The focus of these guidelines has been and continues to be the static loading and earthquake response of the gate. Earthquake loading is assumed to be a horizontal acceleration; however, the actual seismic loads on the gates will likely be much larger, as indicated by measurements on the 111 m high Pacoima Dam during the magnitude 6.8 earthquake in 1994 in Northridge. Sarkaria (1997) found that the vertical and horizontal accelerations were of comparable magnitude and both can be equal to the horizontal peak ground acceleration (PGA). He finds that magnification factors at the gate crests can range from 1.5 to 3.5, depending on the height of the dam. Thus the seismic loads on the gates and trunnion anchorages will likely be much larger than those indicated by the horizontal peak ground acceleration. Furthermore, the transverse component could contribute to buckling of structural members or cocking of the gate in the gate bay.

None of the guidelines appear to recommend dynamic analysis of the gate design, other than the response of the gate to horizontal seismic excitation. The effects of hydrodynamic excitation on gate dynamics are currently not included in the design guidelines. Dynamic analysis is missing in large part due to the lack of knowledge of the hydrodynamic loading and its interaction with the modal vibrations of the gate.

The failure at the Folsom Dam sparked strong interest in understanding how gate reliability affects the overall reliability of dams (see for example Levin et al., 2003; Baecher & Christian, 2000; and many others). In several presentations of seismic analysis of Tainter gates to satisfy FERC regulations (Edwards et al., 1998; Edwards & Planck, 1998; Benson, 2000; among others), the gate was assumed to be closed and/or the possibility of rigid-body rotation around the trunnion pin was not considered. The finite element models appeared to have no spring element representing the supporting chains or cables. Our studies of gate models and tests on full-scale gates show that the rigid-body rotation around the trunnion pin appears to play a substantial role in the instability of the gate. Exposing the gate to a horizontal acceleration response spectrum (ARS) while the gate is closed will undoubtedly provide very different conclusions than when the gate is open and able to undergo rigid-body rotation. Without a predictive capability for the potential coupling between multiple vibration modes due to the combination of rigid body rotation and added mass effects, the analysis program will not correctly determine the coupled response. Even though the current seismic analysis method might predict stability, there could be a dynamic instability of the gate due to modal coupling.

Coupling can be explained using a simple example. Assume the steel gate is initially in an open equilibrium position, as shown in Figure 3a. The gate has a curved surface facing the water, called the skinplate, and arms that support the skinplate, one on each side of the channel, from a downstream position. It can be opened by rotating the arms and the skinplate about the downstream support points. Now assume the gate experiences a horizontal step input or a harmonic forcing function that drives a dynamic response in the form of a bending deformation of the skinplate due to the dynamic loading, as shown in Figure 3b. An examination of the sketches before deformation and after deformation reveals a change in the mass moment of inertia about the trunnion pin. If the frequency of the dynamic

Figure 3. A simplified example with exaggerated deformations explaining coupling due to a horizontal seismic input force: (a) equilibrium position with gate open when a dynamic horizontal force loads the skinplate; (b) downstream dynamic skinplate deformation due to dynamic loading with gate initial position as reference; (c) [actually occurs simultaneously with (b)] rigid-body rotational motion due to the unbalanced mass moment of inertia, reducing the discharge area; (d) gate returning to its initial position, but momentum carries it past equilibrium to a location with greater gate opening (indicated by long dashed lines). Note that the gate moves downward and downstream from (a) to (c) and then upward and upstream from (c) to (d), following a press-shut trajectory indicating energy input to the structural motion.

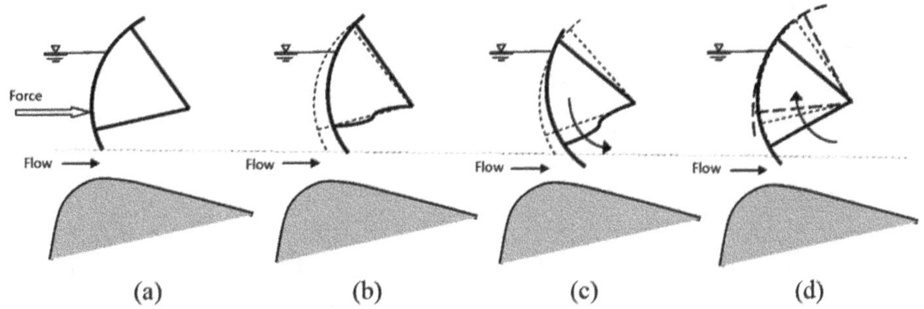

deformation of the skinplate approaches that of the rigid-body rotation around the skinplate, the initial horizontal loading has produced a torsional displacement as in Figure 3c. The torsional displacement, in turn, drives a fluctuating discharge, creating a pressure loading that sustains and amplifies the initial skinplate deformation, leading to increased upward torsional displacement as in Figure 3d, as the skinplate displacement returns and passes its original position. The repeated cyclic increase in skinplate amplitude in both modes quickly leads to failure.

Accepting that we might be only examining symptoms instead of understanding the problem, an example comes to mind. The failure of the gate at Sefidrud in Iran would be consistent with the analysis in this example. However, this is only an assessment of symptoms and a careful investigation would be needed to decide on the true nature of the failure at Sefidrud Dam.

Proper maintenance of the trunnion may be sufficient to guarantee the stability of the gates during routine gate operation, but other sources of step load inputs to the gate could also be potential triggers for the coupled-mode mechanism, as explained in detail later in this book. With a different triggering mechanism, reduced trunnion friction could also lead to a greater instability.

Earthquakes may generate a seismic seich due to the ground motion. The impact of such a seich upon an open gate would provide the hard excitation needed for the onset of coupled-mode self-excited gate vibration. Hinks & Gosschalk (1993) described the effects of a 7.5 to 7.8 magnitude earthquake with a maximum vertical displacement of the ground surface of about 4.5 to 5.5 m near the Hebgen Dam in Montana. The main fault line was within 215 m of the dam and the reservoir behind the 35 m high embankment dam was full when the earthquake struck. A wave was generated by the earthquake, producing an overtopping flow at a depth of 1 m for about 10 minutes. The wave receded and travelled to the other end of the reservoir, returning after about ten minutes to again overtop the dam for another extended period. The back and forth of the seismically generated wave recurred to produce at least four overtoppings of the dam crest. The authors concluded that fault ruptures near a reservoir can induce significant seiches. These seismically-induced seiches could clearly serve as the hard excitation needed for the coupled-mode mechanism if the gate were open at the time of the earthquake.

Gate design today is reminiscent of bridge design before the failure of the Tacoma Narrows Bridge in 1942. The focus is on static design, with the assumption that when a structure is sufficiently massive and rigid, dynamic analysis is unnecessary. Design engineers need a method for evaluating the susceptibility of their design to dynamic fluid-structure interaction. We present details of such a method in later chapters in this book, see Chapters 10 and 14 for the method and its validation on full-scale gates.

PREREQUISITE KNOWLEDGE NEEDED FOR GATE INSTABILITY MECHANISMS

Fundamental engineering principles needed to understand the mechanisms of gate vibration are covered in this section and presented in more detail in Chapters 2 and 3. Readers who are familiar with these topics can proceed directly to Chapter 4. Those with no previous experience may wish to consult the references provided at the end of the Chapters 2 and 3.

Mechanical Vibrations

In the realm of mechanical vibration, knowledge of the free and forced vibrations of one degree-of-freedom damped systems (often abbreviated as single degree-of-freedom systems, or sdof systems) is needed. Based on the sdof analysis, vibrations in systems with two degree-of-freedom and normal modes can be analyzed. Subsequently, the analysis can be generalized to include the vibration of distributed systems. It is necessary to understand modal coupling, where one mode serves as the forcing for the other and vice-versa, to comprehend the mode-coupled vibration of long-span gates and Tainter gates. Knowledge of mode shapes, modal damping, and system response to forced vibration will be needed as well. Familiarity with the concept of hard excitation will be needed to understand how the Folsom Dam could be stable and operate for so long without a major vibration problem.

Resonators

Some of the mechanisms of vibration require an understanding of resonators. An understanding of the Helmholtz resonator (excitation of a closed volume with a narrow opening, like blowing over the open end of a necked bottle) is essential to understanding nappe oscillations in the presence of some aeration. Standing acoustic waves can always serve as excitation for both structural and fluid systems. A hydrodynamic application of a standing acoustic wave is noise generated by a leaking valve in a water piping system. Propagating acoustic waves can serve in certain circumstances as energy dissipaters. Surface waves can interact with vortex shedding and exacerbate vibration problems.

Fluid Dynamics/Hydrodynamics

Control of the discharge from a reservoir using the principles of fluid dynamics (called hydrodynamics when discussing water flow) is the *raison d'etre* for hydraulic structures, and at the same time the fluid flow is the source of the forces that can cause these structures to vibrate. The energy needed for vibration comes from the kinetic, potential, and pressure energies of the flowing water. Lewin (2001) notes that the design of a gate must start with the hydrodynamics. Concepts needed in hydrodynamics are the conservation of mass and momentum. Conservation of energy concepts are needed in understanding sources of excitation for flow-induced vibration.

An undergraduate course in fluid dynamics should prepare most readers for the hydrodynamics covered in this book. Some of the topics we will discuss are

- Non-dimensional coefficients and the use of Reynolds number and Froude numbers in formulating dynamic similarity of flows
- Use of the extended or modified Bernoulli equation to account for frictional losses in terms of loss coefficients
- The role of pressure in unsteady systems such as the propagating pressure wave in the water hammer problem
- The transient response of a fluid system to a sudden change in its boundary conditions. This response can be sufficient to trigger an instability if there is a resonator present that will reinforce and amplify the pressure disturbance.
- Potential flow theory applied to dissipative wave theory
- Stability of systems and their response to small perturbations. In particular, the suggestion that the Folsom gate could be a friction-maintained meta-stable system will require a fundamental understanding of stability concepts.

We refer readers interested in an overview of flow-induced vibration and the vibration of hydraulic gates to compilations such as Kolkman (1976, 1984a, 1984b), Blake (1986a, 1986b), Naudascher & Rockwell (1994), Lewin (1995, 2008), ICOLD (1996), Blevins (2001), Kolkman & Jongeling (2007a), and Kaneko et al. (2014). Table 2 provides readers with references to other flow-induced vibrations sources. In this book, we focus on the application of these concepts to understanding the instability mechanisms for hydraulic gates, particularly on developments over the past 35 years.

CONCEPTUAL OVERVIEW OF MECHANISMS OF GATE INSTABILITY

Gate failures can result from both static and dynamic forces. The static failure mechanisms are relatively well known; existing design guidelines for hydraulic gates were formulated in response to past static failures. Dynamic failures can result from a variety of mechanisms. Some types of instability are only applicable to specific gate geometries.

In this section we will review some of the known mechanisms of dynamic gate instability. In all cases, a flow-induced force component acts in phase with the velocity of the gate as it undergoes vibration. A force in phase with the gate velocity will supply energy to the gate structure until an amplitude is reached where, on a per cycle basis, the dissipation matches the added energy input or until the gate ultimately fails.

Extraneously-induced excitation (EIE) results from an externally driven source of fluctuating fluid force such as freestream turbulence, surface waves, rotating mechanical machinery, acoustic waves, earthquakes, and oscillating flow. The driving force must be, by definition, independent of the motion of the gate. The response of the gate depends on how susceptible the gate is to vibration at a given frequency (called the gate's admittance function) and the distribution of energy at each wavelength in the excitation (called the spectral forcing function). Kolkman & Jongeling (2007) note that turbulence is seldom a problem with sustained flow-induced vibration of hydraulic gates. Other than the concern for earthquake excitation, we will not discuss EIE in further detail. Additional information on EIE can be found in the sources cited in Table 2.

Press-Shut and Press-Open Devices

If the base of an inclined plate is directed toward the reservoir as in Figure 4a, a sudden increase in the reservoir depth would tend to add a force to the top side of the plate forcing it to close if it were elastic or elastically mounted. Any device in which the flow area and the discharge tend to decrease with an increase in the reservoir pressure is called a press-shut device. Conversely, if the plate is inclined with its base pointed away from the reservoir, as in Figure 4b, a sudden increase in reservoir depth would tend to lift the plate upward (again assuming it is elastic or elastically mounted). Any device in which the flow area and discharge tend to increase with an increase in reservoir pressure is called a press-open device. Clearly, a press-shut device is statically stable; it will tend to close with a slow gradual increase in pressure. A press-open device, on the other hand, is clearly statically unstable. It will tend to open with a slow gradual increase in the reservoir pressure. Surprisingly, there are installed gates that are designed as press-open devices and the operators complain constantly about difficulties in sealing the gate when it is closed.

Table 2. Concordance of flow-induced vibration topics in various sources

Topic	Naudascher and Rockwell (1994)	Kolkman and Jongeling (2007a)	Lewin (1995)	Kaneko et al. (2014)	ICOLD (1996)	Ishii, Anami, and Knisely	Blevins (2001)	Blake (1986a, b)	Other
Turbulence Excitation (EIE)	91-105	4-5; 75-78	167	58-60; 64-69; 118-127	85	None	239-304	497-654	
Press-open & Press-shut Instabilities with Fluctuating Discharge	228-237	84-94; 99-102	183-184	130-154; 249-255	79-80; 105-106; 109-113	Ch. 1; Ch. 7	408-411	None	
Seal Vibration	357-360	173-174; 210-221; 249-250	177-179	346	53	Ch. 1; Ch. 7	None	None	Sagar (1990)
Shear Layer Separation and Reattachment/ Shear Layer Instabilities	110-176	5-7; 78-83	185-186	57-70; 74-78; 257-264	89-93	Ch. 1; Ch. 7	377-380	130-211	
Bi-stable Flow Instability	146-150	None	None	None	129-133	Ch.1	None	None	
Vertical Vibration of Gates	342-351	78-83; 94-99; 102-104	172	None	106-108	Ch. 1; Ch. 7	None	None	
Horizontal Vibration of Gates	352-356	99-102	171	None	109-113	Ch.1; Ch. 7	None	None	Jongeling (1988, 1989)
Coupled Mode of 2 DOF Flat Bottomed Gates		32	None	None	141	Ch.1; Ch. 7	None	None	Billeter & Staubli (2000)
Nappe Instability/ Overflow Gates	142-144; 173-174	104-107; 221-232	None	None	123-127; 233	Ch. 6	None	None	Chanson
Simultaneous Over- and Under-flow	361-364	None	184	None	117	Ch. 7	None	None	
Long Span Coupled-Mode Instability	238-240	174-177	None	None	233; 247	Ch. 7	None	None	
Eccentricity instability of Tainter Gates	236-237	171-173	None	None	None	Ch 8	409	None	
Coupled-Mode Instability Tainter Gates (flutter)	None	146-147 (flutter)	None	132-136; (flutter)	None	Ch 10, 11, 12, 13, 14	130-139 (flutter)	None	

A very simple example of a press-open device can be made by holding two pieces of paper, aligned with each other and with each of the top corners clamped between the index finger and thumb in each hand. Bring the clamped sheets to your lips and blow on the edge of the paper. You will hear a high pitched screech. You can vary the frequency of the screech by changing the tension in the paper by exerting a greater outward force on the paper. A more melodious application of the same concept can be achieved with the double reed of the oboe.

Figure 4. (a) Press-shut plate gate, and (b) a press-open plate gate

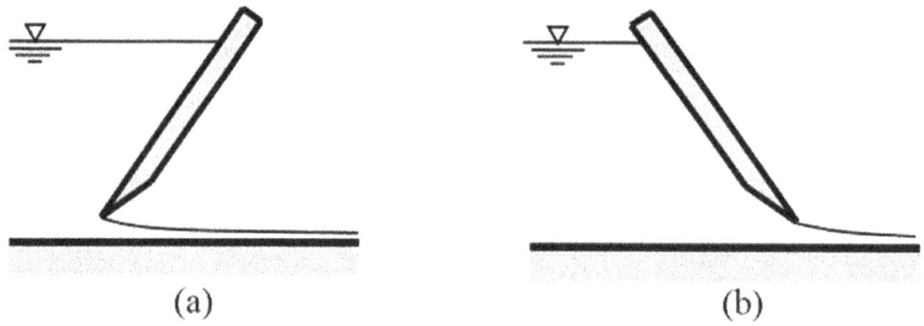

To understand the generation of the vibrations induced by an initially steady flow through a stationary but flexible opening, it is necessary to consider the fluid dynamic aspect of these press-open and press-shut devices.

Discharge (Flow Rate) Fluctuations

Flowing water possesses specific (i.e., per unit mass) pressure energy (p/ρ), specific kinetic energy ($V^2/2$), and specific potential energy (gz), but for elevation changes of only a few meters, the changes in potential energy are invariably negligible. A sudden change in the flow rate will require a small but finite amount of time to propagate through the fluid system. Locally, energy and mass must be conserved, even as the system is adjusting to the new conditions. In the time interval required to communicate this condition change via pressure waves throughout the system, the change in flow rate will be accompanied by a corresponding change in the local pressure. If the pressure change is such that it produces a net force in the same direction as the velocity of the closing structure, there will be energy transfer to the closing structure, forcing it past its equilibrium position. The structure will continue its travel until the restoring spring force (assumed to be proportional to structural displacement) matches the displacing force and the structure then begins to return towards equilibrium. Moving towards equilibrium means the device is opening and the discharge is now increasing, producing a further variation in the local pressure. If the local pressure again produces a net force in phase with the structure's velocity, the opening will again continue past the equilibrium point, until the elastic forces grow to a sufficient magnitude to stop the structural displacement. If the energy added to the structure in each cycle of motion exceeds the dissipation due to damping, the structural amplitude will be amplified with each cycle. With sufficient energy input, the forces on the structure will exceed the loading capacity of the structure and failure results.

Several examples of press-open and press-shut devices with fluctuating discharge will be presented in Chapters 5 and 7. For gates with small openings, the fluctuating discharge can become as large as or larger than the steady discharge, resulting in potentially destructive vibrations. The above conceptual explanation of the mechanism of discharge fluctuations in gates with small openings should be sufficient to understand Naudascher and Rockwell's guideline relative to fluctuating discharge:

All flow systems with structural movements that are accompanied by appreciable variations in flow rate and/or head should be assessed for possible MIE [movement-induced excitation].

A related guideline from Naudascher & Rockwell specifically concerns press-open devices with a substantial upstream volume and a very short exit passageway for the fluid (what they call "negligible fluid inertia"):

Valves controlling the outflow from a Helmholtz resonator of negligible fluid inertia are susceptible to movement-induced excitation if they act as press-open devices.

Naudascher & Rockwell also offer a guideline for press-shut devices, which have relatively long exit passageways ("large fluid-inertia effects from the adjacent flow passage"):

Devices controlling flow through small openings are susceptible to movement-induced excitation if they are press-shut devices and exposed to large fluid-inertia effects from the adjacent flow passage.

Further details and examples of gates and valves acting as press-open and press-shut devices will be presented in Chapter 7. Lewin (1983) finds seal problems to be the most common cause of gate vibrations. Using the guidelines for press-open and press-shut devices from Naudascher & Rockwell, it is possible to re-configure the seal arrangement to avoid vibration. Aspects of flat bottomed gates, which act as neither press-shut nor press-open devices, and yet still have vibration problems can be found in Chapter 7.

Shear Layer Dynamics: Separation, Reattachment, Instability, and Bi-Stable Flows

Viscosity is a fluid property that serves to complicate fluid flow phenomena, and yet is essential for many of the positive applications of fluid dynamics (pumping, sailing, and flying, among others) in the modern world. Viscosity is the property responsible for the formation of hydrodynamic boundary layers, the regions near solid surfaces where the fluid velocity adjusts from that of the freestream to that of the bounding solid surface. When there is a sudden discontinuity in the solid surface or even just when the solid surface curves away from the flow direction too quickly, the fluid flow departs from the surface and follows a path that is not bounded by a solid surface. The departure from the surface is called separation, and the term *free shear layer* is used to describe the region where the fluid velocity adjusts from that of the freestream to that of the stagnant or recirculating bounding fluid region. A further change in the bounding flow passageways can cause the fluid to approach and again flow along the solid surface. The location where this happens is called the reattachment point.

Both hydrodynamic boundary layers and free shear layers can undergo a transition process from laminar to turbulent flow. For free shear layers the transition from laminar to turbulent is characterized by swirling regions of fluid called eddies or vortices. These vortices are generated in a somewhat organized manner, and they can become substantially more focused through the vibration of the bounding surface or the impingement of the shear layer vortices on a downstream structure. Figure 5a shows a free shear layer separating from an upstream corner with no downstream impingement corner present. The swirling flow pattern is identifiable, but it is clearly less concentrated that the swirling pattern seen in Figure 5b where the flow impinges on a downstream corner and, through a feedback process, intensifies the inherent instability of the shear layer, as evidenced by the accompanying velocity spectra. The

Figure 5. Enhancement of shear layer instability through impingement on a downstream structure, including visualization of a separating shear layer (upper left) and an impinging shear layer (upper right) as well as corresponding spectra of velocity fluctuations at the edge of the shear layer (u/U =0.95) for eight locations downstream of separation; $Re_{\theta_o} = 243$, $L/\theta_o = 144$
Spectra and photographs from Rockwell & Knisely, 1979.

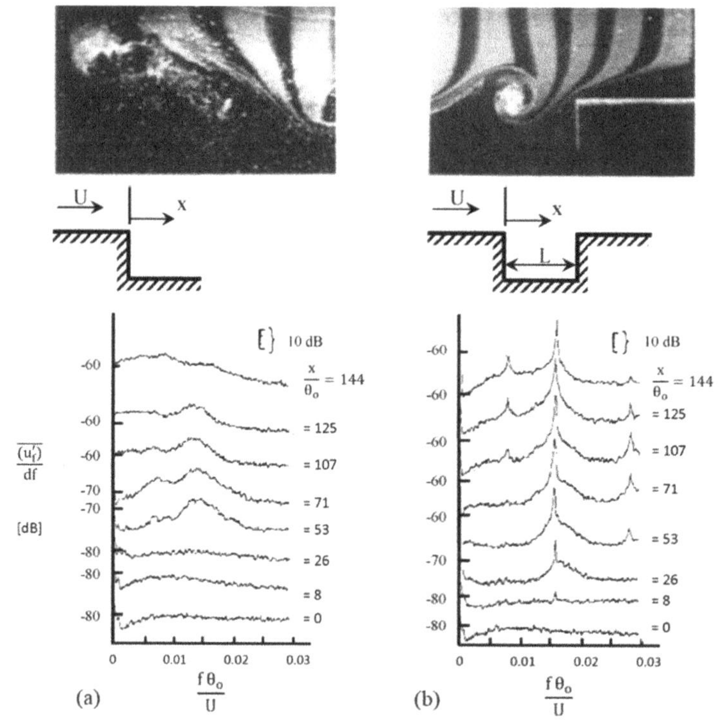

mechanism of excitation through the impingement of these vortices will be denoted as impinging leading edge vortex excitation (ILEV excitation).

A gate designed such that it has a clearly defined separation point and an impingement edge for shear layer vortices can be susceptible to ILEV excitation. The feedback process from impingement, shown schematically in Figure 6, focuses the shear layer velocity fluctuations and the corresponding cavity pressure fluctuations to a spectral peak as in Figure 5b. If the gate has a component with a compatible natural frequency, vibrations can result. Ishii & Knisely (1989, 1990) provide evidence of gate excitation by vortices shed from the gate lip when the gate discharge is submerged.

Shear layers that intermittently re-attach and then detach again, such as flow over a moderate thickness flat-bottomed gate, as shown in Figure 7b, can also produce a fluctuating pressure beneath the gate. The flow is said to oscillate between two bi-stable states, one as a reattaching shear layer, similar to that for a wider gate, as shown in Figure 7a, and the other as a separated shear layer similar to that for a narrow gate, as shown in Figure 7c. Naudascher & Rockwell (1994) describe several additional examples of flows with bi-stable states in their Section 6.6.

If the frequency of the pressure fluctuation matches a structural frequency of the gate and is of sufficient magnitude, the periodic switching between two bi-stable states can produce gate vibrations. The fluctuating shear layer can also act like a control valve, modulating the area available for flow of

Figure 6. Gate with cavity type impinging shear layer generating gate vibrations (not meant to represent a real gate, just to illustrate the concept of instability excitation)

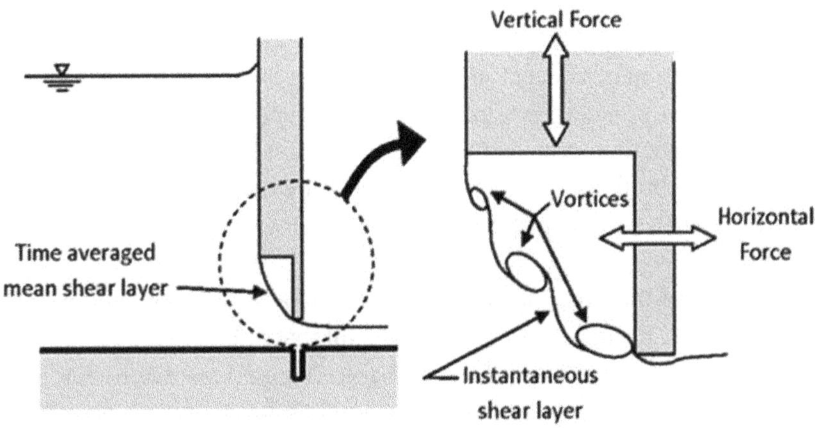

Figure 7. Periodic reattachment of the separated shear layer beneath a gate can create a strong fluctuating pressure and is a possible source of gate excitation. For large values of gate width, L, as shown in (a), and for small gate widths, as shown in (c), the mean shear layer position is stable. At intermediate lengths, movement of the shear layer can drive fluctuating discharge.
After Naudascher & Locher, 1974.

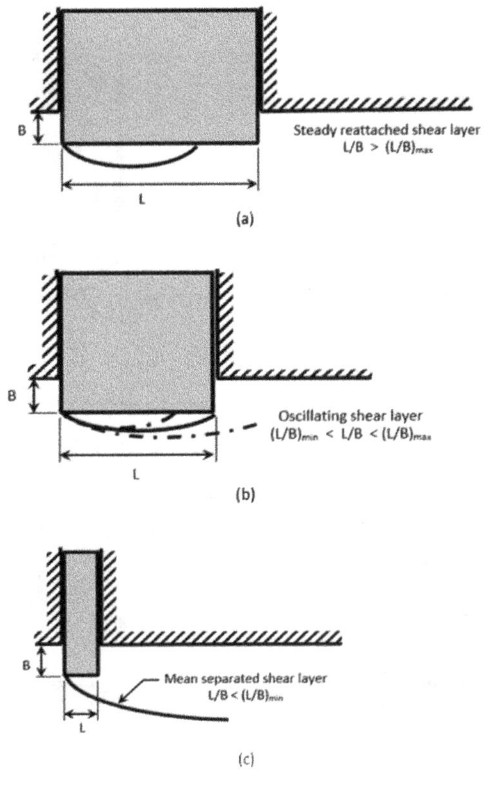

the freestream past the gate especially when the channel beneath the gate is narrow. For gates at small openings, the oscillating shear layer can become a controlling influence on discharge (Ishii & Knisely, 1989; 1990).

The process of bi-stable oscillation is illustrated in Figure 8, showing the growth of the closed recirculation region, the formation of concentrated vorticity that is then shed into the wake region as the recirculation region opens to the wake. Pressure fluctuations accompany the opening and closing of the recirculation, creating a fluctuating force on the bounding surface.

Studies of vertical gate vibrations have been undertaken by many investigators, among them Kolkman (1963), Hardwick (1974), Kolkman & Vrijer (1977), Thang & Naudascher (1986), Thang (1990), and Kanne et al. (1991).

Kanne et al. (1991) found that the vortex interaction was also accompanied by discharge fluctuations and fluid inertia effects. Kanne et al. (1991) maintain that at large gate openings, the mechanism of vibration is galloping, but still augmented by discharge fluctuations due to gate motion and inertia effects. Galloping is a term taken from the low frequency wind-induced vibration of iced-over power lines. The shape of the structure is such that small movements of the structure, or small changes in the angle of the approaching flow, result in very different flow patterns around the structure producing a pressure force in phase with the structure's velocity.

Figure 8. Closed and open recirculation region of bi-stable flow regime, showing progression from (a) minimum recirculation length, (b) growth of the recirculation region, (c) vortex growth in recirculation bubble, and (d) opening of the recirculation region to the wake
After Partenscky, 1964.

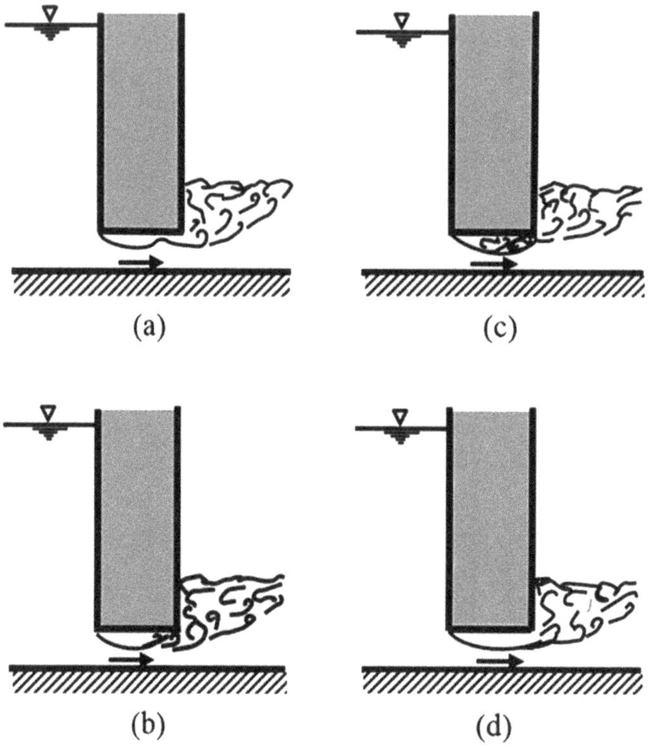

Jongeling (1988; 1989) reported on the streamwise (horizontal vibrations) response of a gate due to shear layer vortices. He found that streamwise (or in-flow) vibration of flat-bottomed gates with submerged discharge was due to vortices shed from the leading edge, which did not impinge on the bottom of the gate. This process, called leading-edge vortex shedding (LEVS), is shown schematically in Figure 9a for leading edge shedding with horizontal gate vibration, and in Figure 9b for LEVS driving vertical gate vibration. As shown in Figure 9a the excitation is termed Body-Resonant-LEVS (BR-LEVS) meaning the vortex shedding is at the same frequency as the gate vibration. Jongeling determined that the pressure loading very near the bottom of the downstream vertical surface of the gate is in phase with the gate's velocity, leading to energy transfer to the structure. Jongeling found three distinct modes of streamwise vibration, depending on the approximate number of wavelengths of the oscillating shear layer over the width of the gate. Husain et al. (1998) offer criteria for classifying the flow beneath oscillating sluice gates.

Thang & Naudascher (1986) reported that for flat-bottomed vertical gates with submerged discharge, vertical vibrations of the gate were driven by the interaction of leading edge vortices with the underside of the gate, similar to the shear layer in Figure 9c. Two parameters govern the vibrations. One is a dimensionless parameter called the reduced velocity ($V_r = V/f_nL$ where V is the flow velocity, f_n is the natural frequency of the body and L is a characteristic length of the body). The other parameter is called the mass-damping parameter, a parameter that combines mass and damping effects into a parameter that tells how much effort is needed to get the body to vibrate (both parameters will be discussed further in Chapter 2). The amplitude of vibration was found to depend on the downstream submergence, decreasing with increasing submergence.

Billeter & Staubli (2000) studied excitation due to leading edge shear layer vortices acting on a flat-bottomed gate with two degrees-of-freedom and submerged discharge. They found a rich palette of possible vibration mechanisms, depending on the ratio of the natural frequencies in the horizontal and

Figure 9. Mechanisms of (a) body resonant, leading edge vortex shedding (BR-LEVS) producing horizontal gate vibration with the wavelength of the shed vortices determined in part by the frequency of the horizontal gate vibration, (b) leading edge vortex shedding (LEVS) producing a vertical vibration (occurs for $d < \lambda_{LEVS}$), and (c) impinging leading edge vortex (ILEV) excitation producing vertical vibration (occurs for $d > \lambda_{ILEV}$)
After Billeter, 2004.

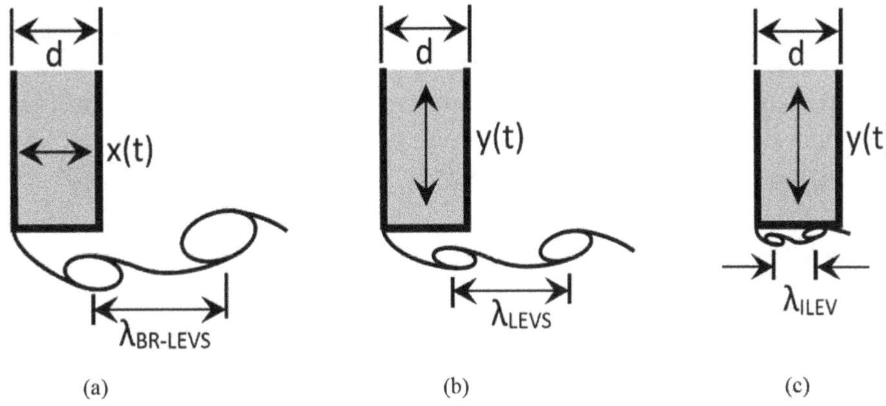

vertical direction, as well as on the reduced velocity. They found that gates with two degrees-of-freedom can vibrate in (a) only the horizontal direction, or (b) only in the vertical direction, or (c) in both directions concurrently, either as a coupled system or as two uncoupled motions. In some cases involving coupling, the gate behaved as a press-open device and in others as a press-shut device. Further details of the work of Billeter & Staubli (2000) will be reviewed in Chapter 7 where their graphical mapping of these vibration mechanisms can be found. Billeter (2004) further clarified the details of the ILEV mechanism and the BR-LEVS mechanism.

Nappe Instability

When water flows over a weir, a gate or another surface (such as a fountain), the water may form a thin sheet that traps a volume of air against the structure. The thin sheet of water, called the nappe, can potentially oscillate. These nappe oscillations can be noisy, resulting in complaints from nearby residents; these oscillations may ultimately threaten the reliability of the structure. Gates with nappe vibrations have been reported for overflow leaf gates (Nakagawa et al., 1989) and flap gates (Partenscky, 1964; Partenscky & Khloeung, 1967; Partenscky & Swain, 1971; Anami et al., 2000; 2014). Chanson (2000) studied nappe formation in weir structures in stepped spillways. Schematic diagrams of oscillating nappes for a leaf gate and a flap gate are shown in Figure 10. The nappe is unstable in a way that is reminiscent of shear layer instability in that the nappe also forms an undulating pattern, transverse to the flow direction. The undulation of the nappe alternately compresses and expands the volume of the trapped air, thereby producing a fluctuating pressure inside the trapped air cavity. The trapped air volume has a fundamental frequency that relates to its volume. If the nappe frequency and the air volume frequency are approximately the same, the motion of one drives the motion of the other. The fluctuating air pressure drives the nappe at the point where it separates from the structural surface, setting in motion the next cycle of nappe undulations. If the frequency and phasing of the air pressure fluctuation driving the nappe undulation at the separation point is compatible with the natural nappe frequency, the amplitude of the nappe undulations continues to grow. The radiated sound can exceed 100 dBA. If the supporting structure is a flap gate with a frequency close to that of the nappe instability, the structural vibration can also couple with either the nappe instability or the air volume resonator, or both.

The mechanics of nappe oscillations have been investigated by Petrikat (1958), Cremer & Ising (1972), Treiber (1974), Kolkman (1974), Schwartz (1964a, 1964b, 1966), Nishikawa (1978); Sumi & Okubo (1988); Nakagawa et al. (1988); Casperson (1993a, 1993b) and Schmid & Henningson (2002).

The classical countermeasure for nappe oscillations is to aerate the nappe by either supplying an external air vent of sufficient capacity or by placing splitters along the weir crests to provide openings of sufficient area in the nappe to permit the air to flow in and out without resonant compression.

Knisely (1989) analyzed the sound radiated from a nappe. He then presented a novel countermeasure for nappe oscillations on the laboratory scale (Knisely, 1994). These and other related works will be the topic of Chapter 6.

Simultaneous Overflow and Underflow

Gates exposed to simultaneous over- and underflow are susceptible to many of the same mechanisms as gates with just underflow or gates with just overflow. In addition, they can also be subject to other mechanisms, such as vortex shedding or coupled-mode vibration. Vortex shedding is the organized release

Figure 10. Nappe oscillations in flow over (a) a leaf gate, and (b) a flap gate

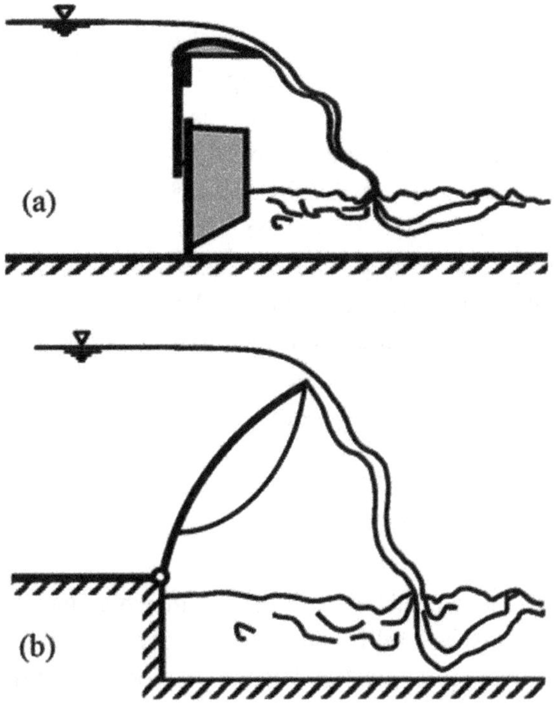

of the vorticity ("swirling fluid") generated in the flow around a body as the shear layers separate on opposite sides of the body, usually forming a pattern of swirling patches of fluid with alternating swirl directions. Petrikat (1958) noted that when a segment weir was exposed to simultaneous overflow and underflow, "the vortices caused by the overflow and underflow induced vibration in the same manner as on the cylinder. . . The magnitude of the forces thus generated was such that the test rig was destroyed."

Coupled-mode vibration was mentioned in the context of flat-bottomed gate in our discussion of Billeter & Staubli (2000). Coupled-mode vibration is of particular importance as it relates to long-span gates and to Tainter gates. When two or more modes couple, motion in one direction produces forces that drive the other mode. Each cycle of motion feeds the next, and the amplitude of the two modes quickly grows. Husain (1999) found that the fluctuating pressure and force coefficients along both the top and bottom surfaces can increase by about 10% to 50% due to the interaction of the overflow with the underflow. In cases where the dominant flow frequencies become resonant with a natural frequency of the gate, these fluctuating forces can be magnified by a factor ranging from 1.5 to 2.5. Simultaneous over- and underflow was the cause of the gate failure in Khadakwasla, India.

Ishii et al. (1995) reported on the vibration characteristics of an operational full-scale shell-type long-span gate with simultaneous over- and underflow. They found the vibrations occurred at small gate openings when the streamwise and vertical bending mode frequencies coincided. "[T]he gate underwent a coupled-mode vibration with a vibration trajectory that could be characterized as an inclined ellipse, traversed in the counterclockwise direction. Such motion, it is argued, is typical of gates behaving as coupled-mode press-shut devices. This type of vibration involves the coupling of the horizontal and vertical bending vibrations through the hydrodynamic forces acting on the gate crest and on the inclined weir plate in the vertical direction and on the vertical and inclined weir plates in the horizontal direction."

Coupled-Mode Long-Span Gate Instability

Ishii & Knisely (1992) investigated the dynamics of a shell-type long-span gate, shown as a schematic cross-section with underflow in Figure 11. With submerged discharge, the shell-type gate can potentially oscillate in either the streamwise or the vertical directions, as was also found by Billeter & Staubli (2000). Furthermore, and still in agreement with the findings of Billeter & Staubli, the gate is also susceptible to a coupled-mode mechanism when the in-water horizontal and vertical frequencies coalesce. In coupled-mode vibration, the gate behaves as a press-shut device. Later, Ishii et al. (1995) provided full-scale data on the mechanism of vibration of a shell-type long-span gate with simultaneous over- and underflow as was discussed in the preceding section. Any criteria for vibration-free operation of gates in underflow must be adjusted for the case of simultaneous over- and underflow because with water above the gate, the vertical vibration frequency will be strongly affected by the added mass effects of the water. Details of coupled-mode vibration of long-span gates, both with underflow and with simultaneous over- and underflow, will be presented in Chapter 7.

Eccentricity Instability

After the 1967 failure of the Tainter gate at the Wachi Dam, Imaichi & Ishii (1977) and Ishii, et al. (1977, 1980) formulated the "eccentricity instability" mechanism. This instability mechanism depends upon the eccentricity between the loading on the skinplate and the location of the trunnion pin to provide a sufficient moment (defined as the force on the skinplate times the eccentricity) that de-stabilizes the gate after a small perturbation. There are still Tainter gates in operation with a designed eccentricity between the trunnion center and the skinplate center. Ishii & Naudascher (1992) provided guidelines for the allowable limits on eccentricity for the gate to remain stable. Chapter 8 will provide a detailed explanation of this eccentricity instability mechanism.

Figure 11. Shell-type long-span gate schematic cross section, with underflow only and submerged discharge

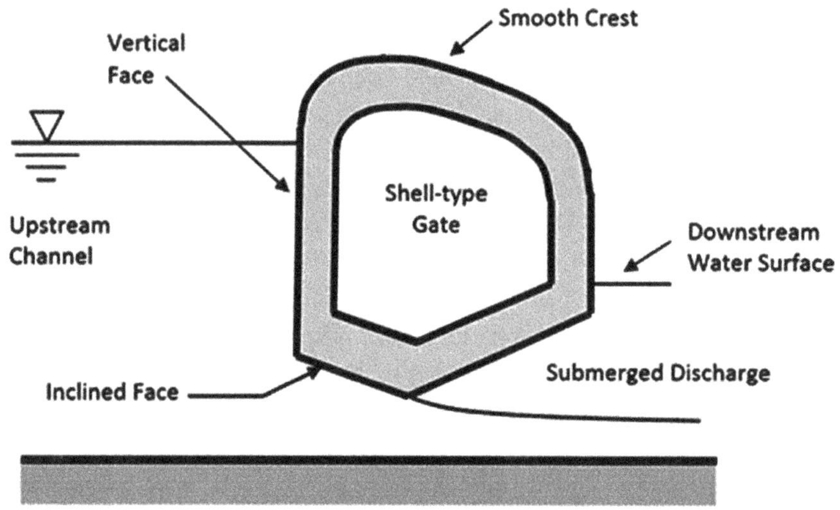

Coupled-Mode Self-Excited Tainter Gate Instability

The process of determining the cause of the Folsom Dam gate failure was a like a fascinating detective story. Initially, finding the solution was like trying to assemble a jigsaw puzzle with some of the pieces missing, and without an image of what the finished picture should look like. Ishii (1995a) suspected the eccentricity instability mechanism may have been involved, but the eccentricity of the Folsom gate was far smaller than the 0.42 m that would have been needed to actuate this mechanism. With almost a meter gate opening, the usual small-gate-opening mechanisms were not effective either. And yet, there was unequivocal testimony from the operator of rumbling and vibrations, building in intensity, just before the gate failed.

From our previous work on the coupling of vertical and horizontal bending mode vibrations of the long-span gates, we understood the coupled-mode mechanism. What seemed to be missing was the corresponding second mode to effectively couple with the rigid body rotation of the gate around the trunnion pin. Through experimental modal analysis on one of the remaining original Folsom gates, Ishii (1995b) uncovered the predominant bending mode of the skinplate. In air, the frequency of the skinplate mode was about a factor of four greater than the frequency of the rigid body rotation of the gate around the trunnion pin. Due to the massive size of the gate and its relatively lightweight design, the mass of the water immediately adjacent to the gate that had to be set into motion if the skinplate moved was substantially greater than the mass of the gate itself. In the end, the added mass due to the water environment shifted the bending frequency of the gate to a value just slightly less than the frequency of the rigid body rotation. Ishii (1997) presented these findings to the USBR in 1997 in the hope that the official forensic report might be appended with the new data. The USBR did not append the report.

The theory and laboratory validation of the coupled-mode instability mechanism in Anami's doctoral thesis (Anami, 2002) and in peer-reviewed papers that came out of her thesis work (Anami & Ishii, 1999; 2000; 2003) were the basis of the continued work in uncovering the mechanism of the coupled-mode dynamic instability.

The coupled-mode mechanism that we propose relies on the coupling of the modes through the pressure on the skinplate, producing skinplate deformation that yields a net change in the moment of inertia about the trunnion pin (see Figure 3). The unbalanced moment of inertia produces a rotation about the trunnion pin, which affects the instantaneous discharge through the gate. This discharge fluctuation is accompanied by a pressure fluctuation that is in phase with the velocity of the skinplate. Energy flows from the fluid to the skinplate motion, driving the skinplate elastic deformation to a greater amplitude. A greater skinplate amplitude produces a greater unbalanced moment of inertia until the elastic forces become large enough to bring the motion of the skinplate to rest. When the elastic deformation of the skinplate is at its maximum, the elastic restoring forces are unbalanced. The direction of the skinplate motion and, quickly thereafter, the direction of the rigid body motion, reverse themselves and the process continues in reverse, again with the pressure forces driving the skinplate and the skinplate driving the rotation mode. The rotation mode results in pressure fluctuations that continue to force the elastic deformation of the skinplate.

The truly challenging part of the formulation of this mechanism was assembling the mathematical description of the process in a way that yielded a method to predict the stability of a given installation. The self-excited coupled-mode Tainter gate instability mechanism is the focus of Chapters 10 through 14.

CONCLUSION

Over the last fifty years, there have been numerous investigations of gate vibrations, many of which have been previously compiled in Kolkman (1976, 1984a, 1984b), Naudascher & Rockwell (1994), ICOLD (1996), Lewin (2001b), and Kolkman & Jongeling (2007a). Naudascher & Rockwell (1994) organized their study of flow-induced vibration by excitation mechanism, which is an excellent way to gain an overall understanding of the field of flow-induced vibrations. They classified the excitation mechanisms as (1) Extraneously-Induced Excitation (EIE), (2) Instability-Induced Excitation (IIE), or (3) Movement-Induced Excitation (MIE). Subsequent publications (ICOLD, 1996; Lewin, 1995, 2001, 2008; and Kolkman & Jongeling, 2007) have adopted the Naudascher & Rockwell classification scheme of vibration problems (or some variant thereof).

Our book is intended to complement these important works, building on existing knowledge to prevent gate failures in the future. Echoing the sentiment expressed at the 1987 Symposium on Dam Failures, as engineers, we cannot afford to be complacent. Every past failure provides an opportunity to learn more about static and dynamic forces, especially those that affect the interaction between the water and the gates that control its movement.

Through persistent research efforts over a period of more than ten years, a mechanism for the vibration that may have caused the Folsom Dam gate failure has been identified as a viable basis for gate vibration at large gate openings. It is the intent of this book to provide sufficient guidance that a recently graduated engineering student can, through careful study of the book, gain an understanding of the vibration mechanism and the process leading to its formulation.

REFERENCES

ASDSO. (2014). *Dam Failures and Incidents*. Association of State Dam Safety Officials. Retrieved from http://www.damsafety.org/news/?p=412f29c8-3fd8-4529-b5c9-8d47364c1f3e

Anami, K. (2005, December). *Flow-Induced Coupled-Mode Self Excited Vibration of Large-Scaled Tainter gates*. Unpublished doctoral dissertation, Osaka Electro-Communication University, Osaka, Japan.

Anami, K., & Ishii, N. (1999). Flow-induced coupled-mode-vibration of Folsom Dam Tainter gates in California. In M. J. Pettigrew & M. P. Paidoussis (Eds.), *Flow-Induced Vibration: The 1999 ASME Pressure Vessels and Piping* (Vol. 396, pp. 343–350). New York, NY: American Society of Mechanical Engineers.

Anami, K., & Ishii, N. (2000). Flow-induced dynamic instability closely related to Folsom Dam Tainter gate failure in California. In S. Ziada & T. Staubli (Eds.), *Flow-Induced Vibration* (pp. 205–212). Rotterdam, NL: Balkema.

Anami, K., & Ishii, N. (2003). Model tests for non-eccentricity dynamic instability closely related to Folsom dam Tainter gate failure. *Proceedings of the ASME 2003 Pressure Vessels and Piping Conference* (pp. 213-222). New York, NY: American Society of Mechanical Engineers. doi:10.1115/PVP2003-2092

Anami, K., Ishii, N., & Knisely, C. W. (2014a). Retrospective consideration of a plausible vibration mechanism for the failure of the Folsom Dam Tainter gate. *International Journal of Mechanical Engineering and Robotics Research, 3*(4), 314–345.

Anami, K., Ishii, N., & Knisely, C. W. (2014b). Theory of self-excited coupled-mode dynamic instability of Tainter gates. *International Journal Mechanical. Engineering and Robotics Research, 3*(4), 678-707. Retrieved from http://ijmerr.com/currentissue.php

Anami, K., Ishii, N., Knisely, C. W., & Matsumoto, Y. (2000). Measured and calculated vibration characteristics of a full-scale flap gate undergoing self-excited vibration. *Proceedings of ASME Pressure Vessels and Piping Conference* (Vol. 414, pp. 23-30). New York, NY: American Society of Mechanical Engineers.

Anami, K., Ishii, N., Knisely, C. W., & Matsumoto, Y. (2014). Field test validation of analytical model for vibration characteristics of a flap gate undergoing self-excited vibration. *Intern. J. Mech. Engineering and Robotics Research, 3*(4), 1-15. Retrieved from http://ijmerr.com/currentissue.php

ANCOLD. (1998). *Guidelines for Design of Dams for Earthquakes.* Hobart, Australia: Australian National Committee on Large Dams.

Azam, S., & Li, Q. (2010). Tailings dam failures: A review of the last one hundred years. *Geotechnical News, (*December 2010), 50-53.

Baecher, G. B., & Christian, J. T. (2000). The practice of risk analysis and the safety of dams. *Proceedings of the Conference on Geotechnical Engineering.* Cairo, Egypt: Cairo University.

Benson, S. (2000). Tainter gate inspection for the FERC part 12 inspection of Morony Dam. In *Annual USCOLD Lecture Series: The Challenge of the 21st Century* (pp. 343–356). Seattle, MA: United States Committee on Large Dams.

Berridge, R. (2000). Gated spillway on dams and other release facilities. In *Transactions of the International Congress on Large Dams, Proceedings of the 20th ICOLD Congress* (*Vol. 4*, pp. 661-676)., Paris: International Commission on Large Dams.

Belloni, L. G., & Stefani, R. (1987). The Vajont slide: Instrumentation – past experience and the modern approach. *Engineering Geology, 24*(1-4), 445–474. doi:10.1016/0013-7952(87)90079-2

Billeter, P. (2004). Properties of single shear layer instabilities and vortex-excitation mechanisms of thick plates. *Journal of Fluids and Structures, 19*(3), 335–348. doi:10.1016/j.jfluidstructs.2003.12.008

Billeter, P., & Staubli, T. (2000). Flow-induced multiple-mode vibrations of gates with submerged discharge. *Journal of Fluids and Structures, 14*(3), 323–338. doi:10.1006/jfls.1999.0274

Blake, W. K. (1986a). Mechanics of Flow-induced Sound and Vibration: Vol. 1. *General Concepts and Elementary Sources.* New York, NY: Academic.

Blake, W. K. (1986b). Mechanics of Flow-induced Sound and Vibration: Vol. 2. *Complex Flow-Structure Interactions.* New York, NY: Academic.

Blevins, R. D. (2001). *Flow-Induced Vibrations (2nd Reprint Edition)*. Malabar, FL: Krieger Press.

Bowker, L. N., & Chambers, D. M. (2015). The risk, public liability, & economics of tailings storage facility failures. Retrieved from https://www.earthworksaction.org/files/pubs-others/BowkerChambers-RiskPublicLiability_EconomicsOfTailingsStorageFacility%20Failures-23Jul15.pdf

Carballada, L., Hang, S., Lemay, M., Stirbu, M., Holder, G., & Collin, Y. (2005). Gate failure, numerical and physical modeling at Gouin Dam. In S.K. Chakrabarti, S, Hernández, & C.A. Brebbia (Eds.), Fluid Structure Interaction and Moving Boundary Problems (pp. 131-140). Southampton: WIT.

Casperson, L. W. (1993a). Fluttering fountains. *Journal of Sound and Vibration*, *162*(2), 251–262. doi:10.1006/jsvi.1993.1117

Casperson, L. W. (1993b). Fluttering fountains: Simplified models. *Journal of Applied Physics*, *74*(8), 4894–4898. doi:10.1063/1.354321

Cassidy, J. (2000). Summary of Sessions on Question 79. *Proceedings of the 20th ICOLD Congress*, Paris: International Commission Large Dams.

Chanson, H. (2000). A review of accidents and failures of stepped spillways and weirs. In *Proceedings of the Institution Civil Engineers, Water and Maritime Engineering* (Vol. 142, pp. 177-188). London, UK: Thomas Telford Services Ltd. doi:10.1680/wame.2000.142.4.177

Christian, J. T., & Baecher, G. B. (2002). Dam Failure. Retrieved from http://ceae.colorado.edu/~amadei/CVEN5768/PDF/Damfailure.pdf

Cremer, L., & Ising, H. (1972). Selbsterregte Schwingung von Überfallstrahlen. *Acustica*, *27*(2), 94–107.

Failures, D. (2016). Dam failures, dam incidents (near failures). Association of State Dam Safety Officials, Retrieved from http://www.damsafety.org/media/Documents/PDF/US_FailuresIncidents.pdf

Safety, D. (2010). Dam safety and earthquakes. *International Water Power and Dam Construction*, September. Retrieved from http://www.waterpowermagazine.com/features/featuredam-safety-and-earthquakes/

DBSA. (2010). Guide to best practice in the operation, maintenance and safety of dams. Development Bank of Southern Africa. Retrieved from http://www.dbsa.org/EN/About-Us/Publications/Documents/Best%20Practise%20Dams.pdf

Dias, C. C., & Gusmao, L. (2000). L'expérience portugaise dans le domaine des équipements hydromécaniques d'évacuateurs de crue. *Proceedings of the 20th ICOLD Congress*. Paris: International Commission Large Dams.

Donnelly, C. R., & Morgenroth, M. (2005). Risky business for dams. *International Water Power and Dam Construction*. Retrieved from http://www.waterpowermagazine.com/features/featurerisky-business-for-dams/

Duncan, J. M. (1987). Symposium on dam failures– panel discussion. *Engineering Geology*, *24*(1-4), 541–574. doi:10.1016/0013-7952(87)90086-X

Edwards, W. D., & Planck, S. A. 1998. Recreating the past: inspection and analysis of spillway radial gates. In *Annual Conference Proceedings Association of State Dam Safety Officials, Inc.* (pp. 467-477) Lexington, KY: Association of State Dam Safety Officials.

Edwards, W. D., Planck, S. A., & Powers, A. C. (1998). *Analysis and inspection of spillway gates: what is your risk? Proceedings of the 18th Annual USCOLD Lecture Series: Managing the Risks of Dam Project Development* (pp. 495–510). Buffalo, NY: United States Committee on Large Dams.

Geology, E. (1987). [complete volume]. *Engineering Geology, 24.*

FEMA. (2005). *Federal Guidelines: Earthquake Analysis and Design of Dams.* Washington, DC: Federal Emergency Management Agency.

Fiedler, W., King, W., Schwanz, N., Garton, J., & McDaniel, L. (2011). Dam safety: What happened to Lake Delhi Dam? *Hydro Review, 30*(6). Retrieved from http://www.hydroworld.com/articles/hr/print/volume-30/issue-6.html

Foster, M., Fell, R., & Spannagle, M. (2000). The statistics of embankment dams failures and accidents. *Canadian Geotechnical Journal, 37*(5), 1000–1024. doi:10.1139/t00-030

Four Dam Failures. (1985). Four major dam failures re-examined. *Waterpower and Dam Construction, November*, 33-46.

Frizell, K. H. (2005). *Spillway, Outlet Works, and Water Conveyance Structure Needs – Survey Results.* Hydraulic Laboratory Report HL-2005-03. Denver CO: U.S. Bureau of Reclamation.

Gosschalk, E. M., Severn, R. T., Charles, J. A., & Hinks, J. L. (1994). An engineering guide to seismic risk to dams in the United Kingdom, and its international relevance. *Soil Dynamics and Earthquake Engineering, 13*(3), 163–179. doi:10.1016/0267-7261(94)90015-9

Graham, W. P. (1999). *A Procedure for Estimating Loss of Life Caused by Dam Failure.* U.S. Bureau of Reclamation, Dam Safety Office, Report DSO-99-06. Denver, CO: U.S. Bureau of Reclamation.

Graham, W. P., & Hilldale, R. C. (2001). *Spillway Gate Failure or Misoperation: Representative Case Histories.* Dam Safety Office, Report DSO-01-01. Denver, CO: U.S. Bureau of Reclamation. Retrieved from http://www.usbr.gov/ssle/damsafety/TechDev/DSOTechDev/DSO-01-01.pdf

Graham, W. P., & Hilldale, R. C. (2002). Spillway gate failure or misoperation: representative case histories. *Water Operation and Maintenance Bulletin,* No. 201, U.S. Bureau of Reclamation, 1-12. Retrieved from http://www.usbr.gov/pps/WaterOandMBulletins/eng2000.htm

Habib, P. (1987). The Malpasset Dam failure. *Engineering Geology, 24*(1-4), 331–338. doi:10.1016/0013-7952(87)90070-6

Hall, J. F. (1984). Dams and earthquake safety. *Engineering and Science, 47*(5), 17–20.

Hardwick, J. D. (1974). Flow-induced vibration of vertical-lift gate. *Journal of the Hydraulics Division, 100*(5), 631–644.

Hart, E. D., & Hite, J. E. (1979a). Barkley Dam gate vibrations. In E. Naudascher & D. Rockwell (Eds.), *Practical Experiences with Flow-Induced Vibrations* (pp. 498–501). Berlin: Springer.

Hart, E. D., & Hite, J. E. (1979b). *Prototype Gate Vibration Tests, Barkley Dam, Cumberland River, Kentucky* (Technical Report HL-79-8). Hydraulics Laboratory. Vicksburg, MS: U. S. Army Engineer Waterways Experiment Station.

Hendron, A. J., & Patton, F. D. (1987). The Vaiont slide—a geotechnical analysis based on new geologic observations of the failure surface. *Engineering Geology*, *24*, 475-491.

Hinks, J. L., & Gosschalk, E. M. (1993). Dams and earthquakes—a review. *Dam Engineering*, *IV*(1), 9–26.

Hobbs, G., & Azavedo, D. (2000). Spillway gate reliability. *Proceedings of ANCOLD 2000 Conference* (pp. 47-62). Cairns, Australia: Australian National Committee on Large Dams.

Hoffmann, P. (2000). Radial gate, held in place without bolts for 14 years, finally fails. *Water Operation and Maintenance Bulletin, 194*, 21-27. Denver, CO: U.S. Bureau of Reclamation. Retrieved July 1, 2014 from http://www.usbr.gov.infrastr/waterbull

Husain, D. (1999). Dynamic study of simultaneous underflow-overflow past low head sluice gates. *Journal of King Abdulaziz University: Engineering Science*, *11*(1), 51–67. doi:10.4197/Eng.11-1.4

Husain, D., Negm, A.-A. M., & Alhamid, A. A. (1998). Criteria for dynamic flow behavior below vibrating sluice gates. *Journal of King Abdulaziz University: Engineering Science*, *10*(2), 77–91. doi:10.4197/Eng.10-2.5

Hutton, C. C. (1999). Tainter gate testing and evaluation. In Waterpower'99: Hydro's Future: Technology, Markets, and Policy (pp. 1-10). American Society of Civil Engineers. doi:10.1061/40440(1999)61

ICOLD. (1982 to 2012) Proceedings from ICOLD Congresses 14 to 24. Paris: International Commission on Large Dams. Retrieved from http://www.icold-cigb.org/GB/Publications/reports.asp)

ICOLD. (1989). Selecting seismic parameters for large dam – Guidelines. ICOLD Bulletin 72. Paris: International Commission on Large Dams.

ICOLD. (1995). *Dam Failures–Statistical Analyses, ICOLD Bulletin 99*. Paris: International Commission on Large Dams.

ICOLD. (1996). *Vibration for Hydraulic Equipment for Dams; Review and Recommendations, ICOLD Bulletin 102*. Paris: International Commission on Large Dams.

ICOLD. (2001). *Tailings Dams - Risk of Dangerous Occurrences, Lessons Learnt from Practical Experiences. ICOLD Bulletin 121*. Paris: International Commission on Large Dams.

Ishii, N. (1995a, August 24). *Folsom Dam Gate Failure Evaluation and Suggestions*. Report submitted to U.S. Bureau of Reclamation.

Ishii, N. (1995b, November 8). *Folsom Dam Gate Failure Evaluation Based on Modal Analysis and Suggestion.* Report submitted to U.S. Bureau of Reclamation.

Ishii, N. (1997, July 17). *Dynamic Stability Evaluation for Folsom Dam Radial Gate and Its Inherent Non-Eccentricity Instability.* Report submitted to U.S. Bureau of Reclamation.

Ishii, N., Imaichi, K., & Takemoto, S. (1977). Water waves and structural loads caused by a periodic change of discharge from a sluice gate. *Bulletin of the Japan Society of Mechanical Engineers, 20*(6), 998–1007. doi:10.1299/jsme1958.20.998

Ishii, N., Imaichi, K., & Hirose, A. (1977). Instability of elastically suspended Tainter gate system caused by surface waves on the reservoir of a dam. *Transactions of the ASME, Journal of Fluid Engineering, Ser. I, 99*(4), 699-709. doi:10.1115/1.3448888

Ishii, N., Imaichi, K., & Hirose, A. (1980). Dynamic instability of Tainter gates. In E. Naudascher & D. Rockwell (Eds.), *Practical Experiences with Flow-Induced Vibrations* (pp. 452–460). Berlin: Springer. doi:10.1007/978-3-642-81528-7_49

Ishii, N., & Knisely, C. W. (1989). Shear layer dynamics and unsteady discharge beneath a long span gate undergoing streamwise vibrations. In Č. Maksimović & M. Radojković (Eds.), *Proceedings of Hydrocomp 89* (pp. 500–509). Rotterdam, NL: Elsevier.

Ishii, N., & Knisely, C. W. (1990). Shear layer behaviour under a streamwise vibrating gate and its significance to gate vibration. [Series III]. *JSME International Journal, 33*, 131–138.

Ishii, N., & Knisely, C. W. (1992). Flow-induced vibration of shell-type long-span gates. *Journal of Fluids and Structures, 6*(6), 681–703. doi:10.1016/0889-9746(92)90003-L

Ishii, N., Knisely, C. W., & Nakata, A. (1995). Field study of a long-span shell-type gate undergoing flow-induced vibration. *Journal of Fluids and Structures, 9*(1), 19–41. doi:10.1006/jfls.1995.1002

Ishii, N., & Naudascher, E. (1984). Field tests on natural vibration modes of a Tainter gate. In K. V. H. Smith (Ed.), *Channels and Channel Control Structures* (pp. 209–222). Berlin: Springer. doi:10.1007/978-3-662-11300-4_17

Ishii, N., & Naudascher, E. (1992). A design criterion for dynamic stability of Tainter gates. *Journal of Fluids and Structures, 6*(1), 67–84. doi:10.1016/0889-9746(92)90056-9

Jalal Zadeh, A. A., & Fouladi Nashta, C. (2000). Measures on gated spillway of Sefidrud and Marun Dams – Case studies. *Proceedings of the 20th ICOLD Congress.* Paris: International Commission on Large Dams.

Jauhari, V. P. (1999). Operation, monitoring and decommissioning of large dams in India. *World Commission on Dams.* Cape Town, South Africa. Retrieved from http://probeinternational.org/library/wp-content/uploads/2011/02/Dams-in-India.pdf

Jongeling, T. H. G. (1988). Flow-induced self-excited in-flow vibrations of gate plates. *Journal of Fluids and Structures, 2*(6), 541–566. doi:10.1016/S0889-9746(88)80022-7

Jongeling, T. H. G. (1989). Self-excited in-flow gate vibrations. *Proceedings of the International Symposium on Hydraulics for High Dams*, Beijing, China (pp. 562-569).

Kaneko, S., Nakamura, T., Inada, F., Kato, M., Ishihara, K., Nishihara, T., & Mureithi, N. W. (2014). *Flow-Induced Vibrations, Classifications and Lessons from Practical Experiences* (2nd ed.). New York, NY: Academic.

Kanne, S., Naudascher, E., & Wang, Y. (1991). On the mechanism of self-excited vertical vibration of underflow gates. *Proceedings International Conference on Flow-Induced Vibration*, Brighton (Vol. 1991, pp. 405-410). London: Institute of Mechanical Engineers.

Kiernan, P. (2016). Mining dams grow to colossal heights, and so do the risks. Wall Street Journal. Retrieved from http://www.wsj.com/articles/brazils-samarco-disaster-mining-dams-grow-to-colossal-heights-and-so-do-the-risks-1459782411

Knisely, C. W. (1989). On the acoustics of nappe oscillation. In Č. Maksimović & M. Radojković (Eds.), *Proceedings of Hydrocomp 89* (pp. 510–519). Rotterdam, NL: Elsevier.

Knisely, C. W. (1994). Nappe oscillation countermeasure. In *Contributed Papers in Fluids Engineering 1994: Proceedings of 1994 ASME Fluids Engineering Division Summer Meeting* (FED Vol. 182, pp. 17-24). New York, NY: American Society of Mechanical Engineers.

Kolkman, P. A. (1963). *Analysis of vibration measurements on an underflow type of gate. Publication, No. 33.* Delft, NL: Delft Hydraulics Laboratory.

Kolkman, P. A. (1974). Instability of a vertical water-curtain closing an air chamber. In E. Naudascher, (Ed.), *Flow-Induced Structural Vibrations, IUTAM/IAHR Symposium, Karlsruhe, Germany*, (pp 17-33). Heidelberg: Springer. doi:10.1007/978-3-642-48279-3_2

Kolkman, P. A. (1976). *Flow-induced gate vibrations. Publication, No. 164.* Delft, NL: Delft Hydraulics Laboratory.

Kolkman, P. A. (1984a). Vibrations of hydraulic structures. In P. Novak (Ed.), *Developments in Hydraulic Engineering 2* (pp. 1–53). London, UK: Elsevier Applied Science.

Kolkman, P. A. (1984b). Gate Vibrations. In P. Novak (Ed.), *Developments in Hydraulic Engineering 2* (pp. 55–112). London, UK: Elsevier Applied Science.

Kolkman, P. A., & Jongeling, T. H. G. (2007). *Dynamic Behaviour of Hydraulic Structures: Part A Structures in Flowing Fluid.* Delft Hydraulics. Retrieved from http://www.deltares.nl/en/publications?search-expression=kolkman&page=1

Kolkman, P. A., & Vrijer, A. (1977). Gate edge suction as a cause of self-exciting vertical vibrations. *Proceedings of the 17th Congress of the International Assoc. for Hydraulic Research*, Baden-Baden (*Vol. 1*).

Lempérière, F. (2013). Design and construction of dams, reservoirs, and balancing lakes. Retrieved from http://www.hydrocoop.org/dam-design-construction-reservoirs-balancing-lakes/print/

Lempérière, F., & Vigny, J. P. (2005). Shall we forget the traditional 'Design flood'? *Hydropower and Dams*, 2. Retrieved from http://www.hydrocoop.org/shall-we-forget-the-traditional-design-flood/print/

Leonards, G. A. (1987). Overview and personal commentary. *Engineering Geology*, *24*(1-4), 577–612. doi:10.1016/0013-7952(87)90088-3

Leps, T. M. (1987). Failure of Baldwin Hills Reservoir, 1963 – Interpretation of step-by-step failure sequence. *Engineering Geology*, *24*(1-4), 83–88. doi:10.1016/0013-7952(87)90050-0

Lewin, J. (1983). Vibration of hydraulic gates. *Journal of the Institution of Water Engineers and Scientists*, *37*, 165–179.

Lewin, J. (1995). *Hydraulic gates and valves in free surface flow and submerged outlets*. London, England: Thomas Telford.

Lewin, J. (2001). Failures and vulnerabilities of reservoir control structures. In P. Campbell (Ed.), *Learning from Construction Failures: Applied Forensic Engineering* (pp. 168–193). New York, NY: Wiley.

Lewin, J. (2008). Spillway gate design features which can cause vibration. *Proceedings of the Ensuring Reservoir Safety into the Future, 15th British Dams Society Conference*, Warwick. Thomas Telford. doi:10.1680/ersitf.35225.0012

Lewin, J., Ballard, G., Bowles, D.S. (2003, April). Spillway gate reliability in the context of overall dam failure risk. *Proc. 2003 USSD Annual Lecture*, Charleston, South Carolina. doi:10.1680/ersitf.35225.0012

Lima Da Silva, C. A., Garcete, R. A., Da Rosa, E., Fancello, E. A., & Bernardini, P. A. N. (2000). Failure and repair of hoist rod of a very large radial spillway gate – Itaipu hydroelectric power plant. *Proceedings of the 20th ICOLD Congress* (*Vol. 4*, pp.709-732). Paris: International Commission on Large Dams.

Londe, P. (1987). Symposium on dam failures -- panel discussion. *Engineering Geology*, *24*, 542.

Lord, D. W., Clarkson, K., Harris, C., & Stilwell, J. (2010). Hebgen intake stoplog failure - lessons learned. *Proc. U.S. Society on Dams 2010* (pp. 897-912). Retrieved from http://ussdams.com/proceedings/2010Proc/897-912.pdf

Lukman, S., Otun, J. A., Adie, D. B., Ismail, A., & Oke, I. A. (2011). A brief assessment of a dam and its failure prevention. *Journal of Failure Analysis and Prevention*, *11*(2), 97–109. doi:10.1007/s11668-010-9420-1

Mande, O. D., Mehta, D. K., & Kumar, H. (2000). Failure of a spillway radial gate and its restoration for Singur Dam, India. *Proceedings of the 20th ICOLD Congress* (pp. 547-562). Paris: International Commission on Large Dams.

McCully, P. (2005, June). And the walls came tumbling down: Dam safety concerns grow in wake of failures, changing climate. *World Rivers Review*. Retrieved from http://www.internationalrivers.org/and-the-walls-came-tumbling-down

Milligan, V. (1987). Symposium on dam failures -- panel discussion. *Engineering Geology*, *24*, 574.

Nakagawa, H., Takebayashi, S., & Mukuno, S. (1988). A study on low frequency noise of water fall. *Proceedings of the International Symposium on Hydraulics for High Dams* (pp. 1206-1217).

Naudascher, E. (1961).Vibration of gates due to underflow-overflow. *J. Hydraulics Div., Proc. ASCE*, 87 (HY5), 63-86.

Naudascher, E. (1991). Hydrodynamic Forces (IAHR/AIRH Hydraulics Structures Manual, Vol. 3), Rotterdam: A.A. Balkema.

Naudascher, E., & Locher, F. A. (1974). Flow-induced forces on protruding walls. *J. Hydraulics, Div, Proc. ASCE*, 100(2), 295-313.

Naudascher, E., & Rockwell, D. (1994). *Flow-Induced Vibrations: An Engineering Guide; IAHR Hydraulic Structures Design Manual*. Rotterdam: A.A. Balkema.

NSWDSC. (2010a). *General Dam Safety Guide*. New South Wales Dam Safety Committee, Report DSC3G. Retrieved from http://www.damsafety.nsw.gov.au/DSC/Download/Info_Sheets_PDF/Dam/DSC3G.pdf

NSWDSC. (2010b). *Reliability of Spillway Control Systems*. New South Wales Dam Safety Committee Report DSC3D. Retrieved from http://www.damsafety.nsw.gov.au/DSC/Download/Info_Sheets_PDF/Dam/DSC3D.pdf

NYSDEC. (1989). Guidelines for Design of Dams. New York State Department of Environmental Conservation, Division of Water, Bureau of Flood Protection, Dam Safety Section.

NZSOLD. (2000). *New Zealand Dam Safety Guidelines*. Wellington, New Zealand: NZSOLD, New Zealand Society on Large Dams.

Nishikawa, H. (1978). A study on nappe oscillation and associated noise. *Proceedings of the 26th Japan National Congress for Applied Mechanics*, Tokyo (pp. 537-548).

Noble, M., & Lewin, J. (2000). Three cases of gate vibration. In P. Tedd (Ed.), *British Dam Society, Proc. 11th Conference, Bath*. London, UK: Thomas Telford.

Nonveiller, E. (1987). The Vajont Reservoir slope failure. *Engineering Geology*, *24*(1-4), 493–512. doi:10.1016/0013-7952(87)90081-0

Novak, P., Moffat, A. I. B., Nalluri, C., & Narayanan, R. (2001). *Hydraulic Structures* (3rd ed.). CRC.

NPDP (National Performance of Dams Project). (2010). Retrieved from http://npdp.stanford.edu/index.html

Stanford Historic Record NPDP. (2007). Historic record of dam performance. Retrieved from http://npdp.stanford.edu/npdphome/Historic%20Performance%20of%20Dams.pdf

NRC. (1983). *Safety of Dams – Flood and Earthquake Criteria*. Committee on Safety Criteria for Dams, Water Science and Technology Board, National Research Council. Retrieved from http://www.nap.edu/catalog/288.html

Partenscky, H. W. (1964). *Forces dynamiques aggissant sur les portes des écluses lors de la derivation partielle des crues* [Ph.D. Dissertation]. Université de Toulouse.

Partenscky, H. W., & Khloeung, I. S. (1967). Oscillations de lames déversantes non aérées. *Proceedings 12th IAHR Congress*, Fort Collins, CO.

Partenscky, H. W., & Swain, A. (1971). Theoretical analysis of flap gate vibration. *Proceedings of the 14th IAHR Congress*, Paris.

Penman, A. D. M. (1987). Teton investigation – a review of existing findings. *Engineering Geology*, *24*(1-4), 221–237. doi:10.1016/0013-7952(87)90063-9

Petrikat, K. (1958). Vibration tests on weirs and bottom gates. *Journal of Water Power*, *10*, (February) 53-57; (March) 99-104; (April) 147-149; (May) 190-197.

Petrikat, K. (1980). Seal vibration. In E. Naudascher & D. Rockwell (Eds.), *Practical Experiences with Flow-Induced Vibrations* (pp. 476-497). Heidelberg: Springer.

Pickering, G. A. (1971). Spillway Gate Vibrations on Arkansas River Dams, Arkansas and Oklahoma (TR H-71-5). U.S. Army Engineer, Waterways Experiment Station.

Pugh, C. A. (1977). Spillway Vibration, Pressure, and Velocity Measurements, Ozark Lock and Dam, Arkansas River, Arkansas (TR-H-77-6). U.S. Army Engineer, Waterways Experiment Station.

Regan, P. J. (2010). Dams and civil structures: an examination of dam failures vs. age of dams. Hydro Review, *29*(4), 62. Retrieved from http://www.hydroworld.com/articles/hr/print/volume-29/issue-4/articles/dams--civil-structures.html

Rockwell, D. O., & Knisely, C. W. (1979). The organized nature of flow impingement upon a corner. *Journal of Fluid Mechanics*, *93*(03), 413–432. doi:10.1017/S0022112079002573

Rogers, J. D. (1995). A man, a dam and a disaster: Mulholland and the St. Francis Dam. In D.B. Nunis, Jr. (Ed.), The St. Francis Dam Disaster Revisited. Historical Society of Southern California, Los Angeles Ventura County Museum of History & Art, Ventura.

Sagar, B. T. A. (1999). Gate lip hydraulics. *Proceedings of the 28th IAHR Congress*, Graz, Austria.

Sagar, B. T. A. (2000). Problems with Tainter gates on spillways. *Proceedings of the. 20th ICOLD Congress*, International Commission on Large Dams, Paris.

Sagar, B.T.A., & Hutton, C.C. (2000). Tainter gate operation and management. In *Annual USCOLD Lecture Series: The Challenge of the 21st Century* (pp. 93-106).

Sarkaria, G. S. (1997). Ensuring the safety of spillway radial gates. Water Operation and Maintenance Bulletin, 181, 9-16.

Schmid, P. J., & Henningson, D. S. (2002). On the stability of a falling liquid curtain. *Journal of Fluid Mechanics*, *463*, 163–171. doi:10.1017/S002211200200873X

Schmidgall, T. (1972). Spillway gate vibrations on Arkansas River dams. *J. Hydraulics Div., Proc. of ASCE* (pp. 219-238).

Schwartz, H. I. (1964a). Nappe oscillation. *J. Hydraulics Div., Proc. ASCE* (pp. 129-143).

Schwartz, H. I. (1964b). Projected nappes subject to harmonic pressures. *Proceedings - Institution of Civil Engineers*, *28*(3), 313–326. doi:10.1680/iicep.1964.10086

Schwartz, H. I. (1966). Edgetones and nappe oscillation. *The Journal of the Acoustical Society of America*, *39*(3), 579–582. doi:10.1121/1.1909930

Scott, R. F. (1987). The Baldwin Hills reservoir failure. *Engineering Geology*, *24*(1-4), 155–159. doi:10.1016/0013-7952(87)90057-3

Seed, H. B., & Duncan, J. M. (1981), The Teton Dam failure – a retrospective review. *Proceedings of 10th International Conference on Soil Mechanics and Foundation Engineering*, Stockholm (Vol. 4, pp. 219-238).

Seed, H. B., & Duncan, J. M. (1987). The failure of Teton Dam. *Engineering Geology*, *24*(1-4), 173–205. doi:10.1016/0013-7952(87)90060-3

Sherard, J. L. (1987). Lessons learned from the Teton Dam failure. *Engineering Geology*, *24*(1-4), 239–256. doi:10.1016/0013-7952(87)90064-0

Sumi, T., & Okubo, S. (1988). Vibration characteristics of nappe over flap gate (in Japanese). *Proc. 43rd Annual Meeting of JSCE*, Hiroshima.

Symposium on Dam Failures -- Panel Discussion. (1987). *Engineering Geology*, *24*, 541-574.

Table of Dam Failures. (2014). Table of Dam Failures. Retrieved from http://retrocee.engr.ucdavis.edu/faculty/lund/dams/Dam_History_Page/Failures.htm

TCEQ. (2009). *Design and Construction Guidelines for Dams in Texas. Texas Commission on Environmental Quality, Dam Safety Program*. Field Operations Support Division.

Thandaveswara, B. S. (2010). History of dam failures. Retrieved from http://nptel.iitm.ac.in/courses/IIT-MADRAS/Hydraulics/pdfs/Unit41/41_2.pdf

Thang, N. D. (1990). Gate vibrations due to unstable flow separation. *J. Hydraulic Engineering, Proc. of ASCE (Vol.* 116, pp. 342-361).

Thang, N. D., & Naudascher, E. (1986). Self-excited vibrations of vertical lift gates. *IAHR Journal of Hydraulic Research*, *24*(5), 391–404. doi:10.1080/00221688609499316

Todd, R. (1997). Folsom spillway gate failure and lessons learned. Energy and Water; Sustainable Development, Proceedings of Theme D, Water for a Changing Global Community (pp. 66-71). ASCE.

Todd, R. (1999, May 17-21). Spillway radial gate failure at Folsom Dam, California. *Proceedings of the 19th Annual USCOLD Lecture Series*, Atlanta, Georgia (pp. 15-29).

Treiber, B. (1974), Theoretical study of nappe oscillation. In E. Naudascher (Ed.), *Flow-Induced Structural Vibrations, IUTAM/IAHR Symposium, Karlsruhe, Germany*, (pp. 34-46). Heidelberg: Springer.

USACE. (2010). *Engineering Manuals EM 1110-2-XXXX*. Retrieved 12 October 2010 from http://140.194.76.129/publications/eng-manuals/. Includes EM 1110-2-2701 056 CECW-ED Vertical Lift Gates 30 Nov 97

EM 1110-2-2702 112 CECW-ET Design of Spillway Tainter Gates 01 Jan 00;

EM 1110-2-2703 272 CECW-ED Lock Gates and Operating Equipment 30 Jun 94;

EM 1110-2-2105 006 CECW-ED Design of Hydraulic Steel Structures Change 1 31 May 94; EM 1110-2-2200 088 CECW-ED Gravity Dam Design 30 Jun 95;

EM 1110-2-2201 238 CECW-ED Arch Dam Design 31 May 94; and others.

USBR. (1996). *Report on Spillway Gate 3 Failure, Folsom Dam.* Sacramento, CA: U.S. Bureau of Reclamation, Mid-Pacific Region.

Vogel, A. (Ed.). (2009a). *Bibliography of the History of Dam Failures (BHDF).* Vienna, Austria: Risk Assessment International.

Vogel, A. (2009b). Dam failures in the sight of long term behaviour – lessons learnt. *Proceeding of the. 2nd International Conference on the Long Term Behavior of Dams (LTBD09)*, Graz, Austria.

Williams, I. S. (1996). After the earthquake – ensuring that the spillway can be operated when it really counts. Proceedings of the 7th Hydro Power Engineering Exchange, Hamilton, New Zealand.

Wilson, S. D. (1987). The Baldwin Hills reservoir failure. *Engineering Geology*, *24*(1-4), 161–166. doi:10.1016/0013-7952(87)90058-5

WISE-Uranium. (2016). Chronology of major tailings dam failures. WISE Uranium Project. Retrieved from http://www.wise-uranium.org/mdaf.html

Wittke, W., & Leonards, G. A. (1987). Modified hypothesis for the failure at Maupasset Dam. *Engineering Geology*, *24*(1-4), 367–394. doi:10.1016/0013-7952(87)90073-1

Yano, K. (1968).Wachi Dam Tainter gate failure (in Japanese). In Annual report of Disaster Prevention Research Institute, Kyoto University (No. 11, Issue B, pp. 203-219). Kyoto, Japan: Kyoto University.

Chapter 2 Vibrations: With Resonator and Stability Concepts

ABSTRACT

Vibration concepts are reviewed. Single degree-of-freedom vibration (SDOF) are analyzed. Subsequently, the analysis is extended to two degrees-of-freedom (2DOF) systems and coupling in a 2DOF system. The analysis of parametric coupling is introduced. Two sections on energy flow and the modeling of damping follow. Normal modes and mode shapes for systems with multiple degrees-of-freedom (MDOF) will then be considered. By generalizing MDOF systems to continuous systems, we can analyze bending modes in plates. Experimental modal analysis is introduced to prepare the reader for later application of this technique to full-scale operational gates. The second section of this chapter reviews fundamental concepts of fluid-structure systems with resonance. The chapter concludes with a short discussion of stability concepts.

INTRODUCTION

Knowledge of vibration concepts is needed to describe the motion resulting from gate excitation. In this chapter a short review of these vibration concepts will be presented. A more thorough treatment can be found in textbooks on mechanical vibration, including Timoshenko et al. (1974), Den Hartog (1985), Steidel (1989), Thomson & Dahleh (1998), Palm (2007), Inman (2008), Meirovitch (1986), Timoshenko (2008), Thorby (2008), and Rao (2010). More advanced topics are covered in Meirovitch (1967) and Ewins (2000). Further information concerning structural vibrations in flowing fluids can be found in Blevins (1990), Blake (1986), Naudascher & Rockwell (1994), and Kolkman & Jongeling (2007).

In this chapter, we will first analyze single degree-of-freedom vibration (SDOF) and then build on these concepts to examine two degrees-of-freedom (2DOF) systems and coupling in a 2DOF system. In particular, we will introduce the analysis of parametric coupling, as it is essential to the development of the theory of self-excited coupled mode vibration of Tainter gates. Two additional sections on energy flow and the modeling of damping are important relative to gate vibrations. Normal modes and mode shapes for systems with multiple degrees-of-freedom (MDOF) will then be considered. By generalizing

DOI: 10.4018/978-1-5225-3079-4.ch002

MDOF systems to continuous systems, we can analyze bending modes in plates, such as the skinplate of a Tainter gate. We also introduce experimental modal analysis to prepare the reader for later application of this technique to full-scale operational gates.

When the enclosed or surrounding fluid participates in the vibration, the system vibration may experience resonance with the fluid pressure fluctuations, meaning that the fluid pressure fluctuations maintain or amplify the vibration. The second section of this chapter provides fundamental concepts of fluid-structure systems with resonance.

Vibrations in a system with energy input from the enclosed or surrounding fluid may be dissipated (meaning the systems is stable), may be sustained at a certain amplitude (indicating neutral stability), or may be amplified (causing the system to become unstable and possibly fail), depending upon the system's damping which serves to dissipate the system energy. A short discussion of stability concepts is included as the final section of this chapter.

SINGLE DEGREE OF FREEDOM (SDOF) DAMPED SYSTEMS

A sufficiently compact body can be modeled as a concentrated point mass, m. When the concentrated mass is suspended on a linear spring with a spring constant, k, and exposed to viscous damping (proportional to and opposing the body velocity), c, the equation of motion for the natural vibration of the body with a single degree-of-freedom can be derived from $\sum \vec{F} = m\vec{a}$ yielding

$$m\ddot{x} + c\dot{x} + kx = 0 \tag{1}$$

After dividing by m and introducing the symbols $\omega_n^2 = k / m$ to represent the square of the undamped natural frequency, and $\zeta = \dfrac{c}{2\sqrt{mk}}$ to represent the damping ratio, Equation 1 can be re-written as

$$\ddot{x} + 2\zeta\omega_n\dot{x} + \omega_n^2 x = 0 \tag{2a}$$

If the system is driven by a sinusoidal force given by $F(t) = F_o \sin(\omega_f t)$, the equation of motion becomes

$$\ddot{x} + 2\zeta\omega_n\dot{x} + \omega_n^2 x = \frac{F_o}{m}\sin(\omega_f t) \tag{2b}$$

For a system consisting a mass and its associated mass moment of inertia I_θ, a torsional spring and a torsional damper with a single torsional degree-of-freedom, the equation of motion derived from $\sum M = I_\theta\ddot{\theta}$ takes an analogous form to that in Equation 2b:

$$\ddot{\theta} + 2\zeta_\theta\omega_n\dot{\theta} + \omega_n^2\theta = \frac{M_o}{I_\theta}\sin(\omega_f t) \tag{2c}$$

where θ is the torsional displacement; $\zeta_\theta = c_\theta / (2 I_\theta \omega_n)$ is the torsional damping ratio; $\omega_n^2 = k_\theta / I_\theta$ is the undamped natural torsional frequency; c_θ is the torsional damping factor; k_θ is the torsional spring constant; I_θ is the mass moment of inertia of the body; and $M(t) = M_o \sin(\omega_f t)$ is the harmonic forcing moment.

Free Vibration After an Initial Disturbance

A lightly damped system with $\zeta < 1.0$ is said to be under-damped. In such a case, the system described by Equation 2a will exhibit an oscillatory response if the mass is given *either* an initial displacement or an initial velocity or *both* an initial displacement and an initial velocity. Systems in which $\zeta = 1.0$ are called critically damped. When $\zeta > 1.0$, the system is said to be over-damped. Neither critically damped nor over-damped systems exhibit oscillatory responses and will therefore not be considered further in this book.

The natural frequency of vibration of the system is affected by the amount of damping. The damped natural frequency is given by

$$\omega_d = \omega_n \sqrt{1 - \zeta^2} \tag{3}$$

Note in the limit as the damping goes to zero, the damped natural frequency approaches the natural frequency. For non-zero damping, the damped natural frequency will be less than the natural frequency.

The response of the system described in Equation 2b to a disturbance can be written in terms of the damping ratio, the natural frequency, and the damped natural frequency as

$$x(t) = e^{-\zeta\omega_n t}\left(A e^{i\omega_d t} + B e^{-i\omega_d t}\right) \tag{4}$$

where *A* and *B* are constants dependent upon the initial conditions for displacement and velocity.

The rate of decay from one peak to the next, called the logarithmic decrement, can be determined experimentally and then used to determine the system damping ratio. By measuring the peak amplitude at some point in the process, denoted A_i, and then again after *n* cycles, denoted A_n, the log decrement, δ, can be calculated, and is related to the damping factor as follows:

$$\delta = \frac{1}{n}\ln\left(\frac{A_i}{A_n}\right) = \frac{2\pi\zeta}{\sqrt{1-\zeta^2}} \tag{5}$$

Forced Vibration of an SDOF System

A single degree-of-freedom system exposed to a sinusoidal fluctuating force, described by Equation 2b, will follow the imposed force after an initial transient, vibrating at the forcing frequency with an amplitude and with a phase shift relative to the forcing function. The steady-state system response to harmonic forcing $F(t) = F_o \sin(\omega_f t)$, in which F_o is the magnitude of the force and ω_f is the frequency of the force, can be written as

$$x(t) = X\left(\frac{\omega_f}{\omega_n}\right)\sin\left(\omega_f - \phi\left(\frac{\omega_f}{\omega_n}\right)\right) \tag{6}$$

in which the amplitude X and phase ϕ are given in terms of the frequency ratio, ω_f / ω_n, by

$$X\left(\frac{\omega_f}{\omega_n}\right) = \frac{F_o/k}{\sqrt{\left[1-\left(\frac{\omega_f}{\omega_n}\right)^2\right]^2 + \left[2\zeta\left(\frac{\omega_f}{\omega_n}\right)\right]^2}} \tag{7}$$

$$\phi\left(\frac{\omega_f}{\omega_n}\right) = \tan^{-1}\left[\frac{2\zeta\left(\frac{\omega_f}{\omega_n}\right)}{1-\left(\frac{\omega_f}{\omega_n}\right)^2}\right] \tag{8}$$

The graphical representations of Equations 7 and 8 are given in Figure 1. Please note that for frequency ratios, $\omega_f / \omega_n < 1.0$ the system's amplitude is larger than it would have been if the force had been applied statically, that is, the quantity $X / (F_o / k)$ takes on values greater than 1.0.

A similar type of response would be obtained for any SDOF system, governed by the SDOF differential equation of motion, Equation 2b.

Mass on a Massless Rod

As an example of an SDOF system, consider a mass that is permitted to slide along a fixed, frictionless rod. The mass is assumed to be constrained by two springs, as shown in Figure 2. (The dampers, also called dashpots, are not shown in Figure 2 to avoid cluttering the figure.) Assuming the individual spring constants are each *k*/2 and the damping of each dashpot is *c*/2, the equation of motion for this mass-spring-damper system in the radial direction for a *fixed rod* would be that given in Equations 1 and 2a. The system response to sinusoidal forcing in the radial direction would be the same as that given in Equation 6.

Figure 1. (a) Amplitude and (b) phase response of a SDOF system for sinusoidal forcing for multiple values of the damping ratio, ranging from 1% to 40%

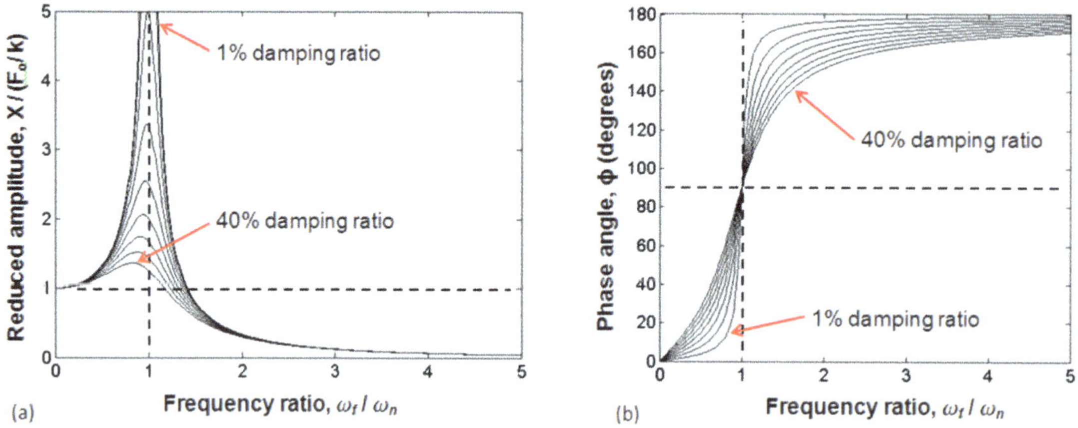

Figure 2. A point mass on a fixed, massless rod with spring constants k/2 and dashpots with damping coefficients c/2 (dashpots not shown to avoid clutter)

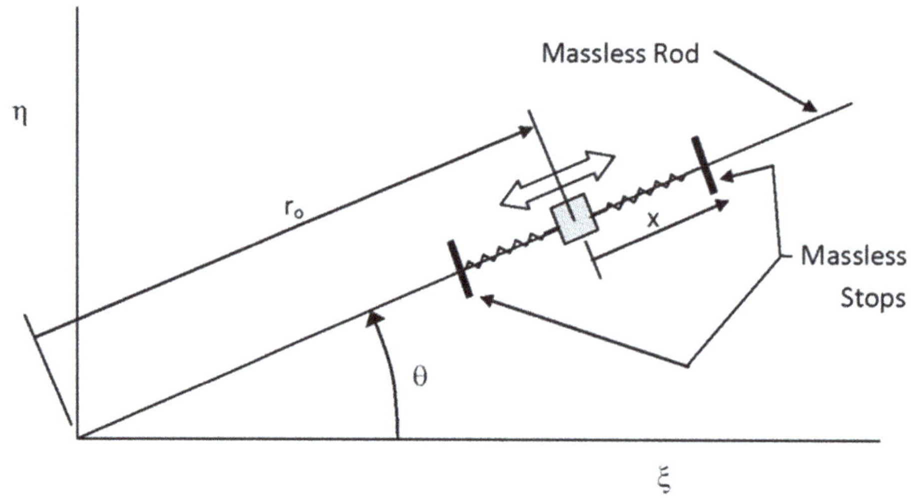

Now assume the mass is fixed to the massless rod (the limit for infinite spring constants), but the rod is free to rotate about the origin subject to torsional spring and damping forces. Assume the torsional spring constant is k_θ and the torsional damping constant is c_θ. The equation of motion can be obtained from $\sum M = \frac{d}{dt}\left(I_\theta \dot{\theta}\right) = I_\theta \ddot{\theta}$ provided I_θ, the mass moment of inertia, remains constant. In the present case, $I_\theta = m r_o^2$. The resulting equation of motion can then be written as

$$I_\theta \ddot{\theta} + c_\theta \dot{\theta} + k_\theta \theta = 0 \tag{9}$$

Assume further that a sinusoidal force, $F(t) = F_o \sin(w_f t)$, is applied to the mass at a distance r_o from the origin in the *tangential* direction (that is, normal to the rod), creating a sinusoidal moment $M(t) = F(t) r_o$. Dynamic equilibrium requires that

$$I_\theta \ddot{\theta} + c_\theta \dot{\theta} + k_\theta \theta = r_o F_o \sin(\omega_f t) \tag{10}$$

After dividing by I_θ and introducing the symbols $\omega_n^2 = k_\theta / I_\theta$ to represent the square of the natural torsional frequency and $\zeta_\theta = \frac{c_\theta}{2\sqrt{I_\theta k_\theta}}$ to represent the torsional damping ratio, Equation 10 can be re-written as

$$\ddot{\theta} + 2\zeta_\theta \omega_n \dot{\theta} + \omega_n^2 \theta = \frac{F_o r_o}{I_\theta} \sin(\omega_f t) \tag{11}$$

The solution to Equation 11 will look quite similar to that for the previously considered rectilinear vibration of the mass along the rod, given in Equation 6.

$$\theta(t) = \Theta\left(\frac{\omega_f}{\omega_n}\right) \sin\left(\omega_f - \phi_\theta\left(\frac{\omega_f}{\omega_n}\right)\right) \tag{12}$$

The amplitude Θ and the phase ϕ_θ will also be analogous to those found for the rectilinear motion:

$$\Theta\left(\frac{\omega_f}{\omega_n}\right) = \frac{F_o r_o}{k_\theta} \frac{1}{\sqrt{\left[1 - \left(\frac{\omega_f}{\omega_n}\right)^2\right]^2 + \left[2\zeta_\theta\left(\frac{\omega_f}{\omega_n}\right)\right]^2}} \tag{13}$$

and

$$\phi_\theta\left(\frac{\omega_f}{\omega_n}\right) = \tan^{-1}\left[\frac{2\zeta_\theta\left(\frac{\omega_f}{\omega_n}\right)}{1 - \left(\frac{\omega_f}{\omega_n}\right)^2}\right] \tag{14}$$

This geometry will be considered again later in this chapter in the section on parametric coupling, in which both the mass and the rod are allowed to move simultaneously.

TWO DEGREES-OF-FREEDOM SYSTEM

Two degrees-of-freedom (2DOF) systems generally consist of two masses that can move relative to one another or a single mass that can undergo motion in two directions. Several examples of 2DOF systems are given in Figure 3. The coupling of a vertical (heaving) mode with a torsional mode, as shown in Figure 3c, is the usual simplified formulation of classic coupled-mode flutter of an airfoil.

As an example of a 2DOF coupled system, consider a system consisting of two masses free to slide rectilinearly along a frictionless surface, but constrained by springs and dampers as shown in Figure 4. In Figure 4, the displacement of m_1 is x_1 and the displacement of m_2 is x_2. Each of the two masses experiences a net force that is dependent upon both its own displacement and velocity and the displacement and velocity of the other mass. By drawing free-body diagrams and summing the forces with a consistent assumption about displacement magnitudes and directions, the equations of motion for the two masses can be shown to be

$$m_1\ddot{x}_1 + \left(c_1 + c_2\right)\dot{x}_1 + \left(k_1 + k_2\right)x_1 - c_2\dot{x}_2 - k_2x_2 = 0 \tag{15a}$$

$$m_2\ddot{x}_2 + \left(c_2 + c_3\right)\dot{x}_2 + \left(k_2 + k_3\right)x_2 - c_2\dot{x}_1 - k_2x_1 = 0 \tag{15b}$$

These two equations can also be derived from LaGrange's equation; for more complicated systems, LaGrange's equation is recommended. Equations 15a and 15b can also be written as a single matrix equation in the following form:

$$\underbrace{\begin{bmatrix} m_1 & 0 \\ 0 & m_2 \end{bmatrix}}_{\text{mass matrix}}\begin{Bmatrix} \ddot{x}_1 \\ \ddot{x}_2 \end{Bmatrix} + \underbrace{\begin{bmatrix} c_1 + c_2 & -c_2 \\ -c_2 & c_2 + c_3 \end{bmatrix}}_{\text{damping matrix}}\begin{Bmatrix} \dot{x}_1 \\ \dot{x}_2 \end{Bmatrix} + \underbrace{\begin{bmatrix} k_1 + k_2 & -k_2 \\ -k_2 & k_2 + k_3 \end{bmatrix}}_{\text{stiffness matrix}}\begin{Bmatrix} x_1 \\ x_2 \end{Bmatrix} = 0 \tag{16}$$

If the *mass matrix* has non-zero off-diagonal terms, the system is *dynamically coupled*. Stiffness (or spring) or *static coupling* is present when the off-diagonal terms in the *stiffness matrix* take on non-zero

Figure 3. Typical two degrees-of-freedom systems: (a) Two mass-spring rectilinear system with spring coupling; (b) One mass with vertical and horizontal spring coupling; (c) One mass with vertical and torsional spring coupling; (d) Double pendulum

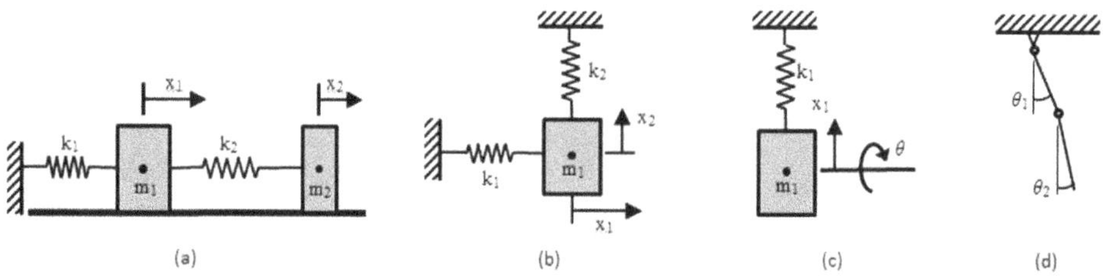

Figure 4. Two degrees-of-freedom system with two masses, three springs, and three dampers

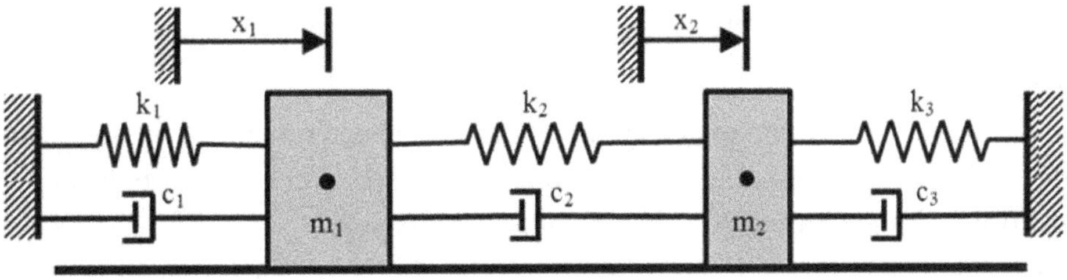

values. In Equation 16, there is also coupling through the damping terms since the *damping matrix* has non-zero off-diagonal terms.

The inclusion of damping in the formulation of these equations of motion results in complex characteristic roots. For small values of damping, the effect on frequency is small and the modal frequencies are approximately correct even with damping neglected. For this reason, the undamped free vibration of the two-mass system will be considered to illustrate the concept of normal modes and mode-shape.

Free Vibration of Spring-Coupled 2DOF System

As an introduction to the response of a 2DOF system, we will consider the free vibration of the undamped spring-coupled system, and subsequently an approximate solution for the free vibration of the system with small damping. For the undamped system, the damping coefficients $c_1 = c_2 = c_3 = 0$.

With the damping matrix set to zero, Equation 16 becomes

$$\begin{bmatrix} m_1 & 0 \\ 0 & m_2 \end{bmatrix} \begin{Bmatrix} \ddot{x}_1 \\ \ddot{x}_2 \end{Bmatrix} + \begin{bmatrix} k_1 + k_2 & -k_2 \\ -k_2 & k_2 + k_3 \end{bmatrix} \begin{Bmatrix} x_1 \\ x_2 \end{Bmatrix} = 0 \tag{17}$$

There are many approaches to finding a solution to Equation 17. The most straight-forward approach is to assume the two masses must oscillate at the same frequency and pass through their respective equilibrium positions simultaneously. This set of assumptions is characteristic of normal modes. Let the assumed solution be

$$x_1 = A_1 \cos \omega t \text{ or } x_1 = A_1 e^{i\omega t} \tag{18a}$$

$$x_2 = A_2 \cos \omega t \text{ or } x_2 = A_2 e^{i\omega t} \tag{18b}$$

Substituting the assumed solution, Equations 18a and b, into Equation 17 and simplifying yields the following matrix equation:

$$\begin{bmatrix} k_1 + k_2 - \omega^2 m_1 & -k_2 \\ -k_2 & k_2 + k_3 - \omega^2 m_2 \end{bmatrix} \begin{Bmatrix} A_1 \\ A_2 \end{Bmatrix} = 0 \tag{19}$$

For either functional representation (cos or exp), the second derivative will be the original function multiplied by $-\omega^2$, and upon substitution of the functions and the second derivatives into the equation of motion, the original function can be eliminated through division.

One trivial solution to Equation 19 is $A_1 = A_2 = 0$, but this is just the formulation of static equilibrium. If, however, the determinant is zero, then any combination of A_1 and A_2 will satisfy the equation of motion. Setting the determinant to zero yields

$$m_1 m_2 \omega^4 - \left(m_1 k_2 + m_1 k_3 + m_2 k_1 + m_2 k_2\right)\omega^2 + k_1 k_2 + k_1 k_3 + k_2 k_3 = 0 \tag{20}$$

Solving Equation 20 as a quadratic equation in ω^2 provides the frequencies of vibration, given as

$$\omega_1^2, \omega_2^2 = \frac{1}{2}\left\{\frac{\left(k_1 + k_2\right)m_2 + \left(k_2 + k_3\right)m_1}{m_1 m_2}\right\} \mp \frac{1}{2}\left[\left\{\frac{\left(k_1 + k_2\right)m_2 + \left(k_2 + k_3\right)m_1}{m_1 m_2}\right\}^2 - 4\left\{\frac{\left(k_1 + k_2\right)\left(k_2 + k_3\right) - k_2^2}{m_1 m_2}\right\}\right]^{1/2} \tag{21}$$

To simplify the solution for frequencies of vibration, a few specific cases will be considered. The simplest case is when $m_1 = m_2 = m$ and $k_1 = k_2 = k_3 = k$. With these simplifications, Equation 17 becomes

$$\begin{bmatrix} m & 0 \\ 0 & m \end{bmatrix} \begin{Bmatrix} \ddot{x}_1 \\ \ddot{x}_2 \end{Bmatrix} + \begin{bmatrix} 2k & -k \\ -k & 2k \end{bmatrix} \begin{Bmatrix} x_1 \\ x_2 \end{Bmatrix} = \begin{bmatrix} 0 \\ 0 \end{bmatrix} \tag{22}$$

Equation 19 becomes

$$\begin{bmatrix} 2k - \omega^2 m & -k \\ -k & 2k - \omega^2 m \end{bmatrix} \begin{Bmatrix} A_1 \\ A_2 \end{Bmatrix} = \begin{bmatrix} 0 \\ 0 \end{bmatrix} \tag{23}$$

and Equation 20 becomes

$$\omega^4 - \frac{4k\omega^2}{m} + \frac{3k^2}{m^2} = 0 \tag{24}$$

which has the following four roots:

$$\omega_1 = \pm\sqrt{\frac{k}{m}} \text{ and } \omega_2 = \pm\sqrt{\frac{3k}{m}} \tag{25}$$

Only the positive roots are physically realistic. The vibration is a real function of time at the positive frequency in each mode. See Meirovich (2010) for a more formal discussion of this point.

To find the ratio of the amplitudes A_2 / A_1, recall that the assumed solutions for x_1 and x_2 must satisfy both of the equations represented by matrix Equation 19. Solving the two equations for the ratio of A_2 to A_1 for this case yields

$$r_1 = \left(\frac{A_2}{A_1}\right)_1 = \frac{2k - \omega^2 m}{k} = \frac{k}{2k - \omega^2 m} = 1 \text{ for } \omega_1 = \sqrt{\frac{k}{m}} \tag{26a}$$

and

$$r_2 = \left(\frac{A_2}{A_1}\right)_2 = \frac{2k - \omega^2 m}{k} = \frac{k}{2k - \omega^2 m} = -1 \text{ for } \omega_2 = \sqrt{\frac{3k}{m}} \tag{26b}$$

Equations 26a and 26b tell us that for the first mode, when $\omega_1 = \sqrt{\frac{k}{m}}$, the two masses move in phase and at the same amplitude, and for the second mode, at the higher frequency $\omega_2 = \sqrt{\frac{3k}{m}}$, the two masses move in opposite directions, coming closer together and then farther apart in each cycle of vibration. This relative motion of the two masses is portrayed in Figure 5, where the first and second mode maximum displacements are shown relative to their respective equilibrium positions.

Figure 5. Maximum modal displacements of the two masses in the undamped 2DOF system for $m_1 = m_2 = m$ and $k_1 = k_2 = k_3 = k$

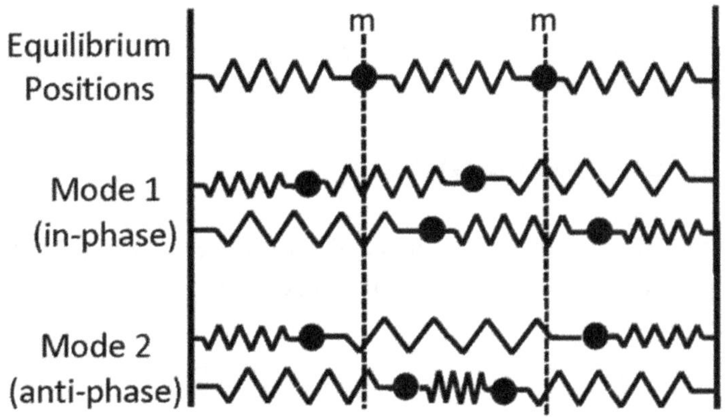

The mode shape is a non-dimensional presentation of the amplitude of each system mass, usually normalized so that the first mass m_1, or the mass with the greatest amplitude, has an amplitude of 1.0. The actual displacements of each mass after an arbitrary initial disturbance depend on the initial conditions. Displacing both masses an equal distance in the same direction and releasing them with no initial velocity will excite first mode vibration. Similarly, equidistant displacement of the masses in opposite directions and releasing them from rest will excite second mode vibrations. Any other initial conditions will result in a combination of the two modal vibrations.

The analysis can be repeated for other mass-spring combinations. If $m_1 = m$ and $m_2 = 2m$, and the three spring constants are still equal, the two modal frequencies would be

$$\omega_1 = 0.796\sqrt{\frac{k}{m}} \text{ and } \omega_2 = 1.538\sqrt{\frac{k}{m}} \tag{27}$$

and the amplitude ratios for the two modes would be

$$r_1 = \left(\frac{A_2}{A_1}\right)_1 = 1.366 \text{ and } r_2 = \left(\frac{A_2}{A_1}\right)_2 = -0.366 \tag{28}$$

For a third example, let $m_1 = m, m_2 = 2m, k_1 = k_2 = k,$ and $k_3 = 2k$. The two modal frequencies in this case are

$$\omega_1 = \sqrt{\frac{k}{m}} \text{ and } \omega_2 = 1.581\sqrt{\frac{k}{m}} \tag{29}$$

and the amplitude ratios for the two modes become

$$r_1 = \left(\frac{A_2}{A_1}\right)_1 = 1.0 \text{ and } r_2 = \left(\frac{A_2}{A_1}\right)_2 = -0.50 \tag{30}$$

Figure 6 shows modal diagrams for the first two cases considered. The modal diagram for the third case is left as an exercise for the reader.

Note in the second modes, where the masses move in opposition, there will be a node on the second spring, that is, a point on the spring between the two masses that does not move.

2DOF System With Torsional Vibration

With no damping included in the analysis, the choice of coordinate system, as shown in Figure 7, affects whether a 2DOF system is a coupled system, or if it can be treated as two SDOF systems.

Figure 6. Modal diagrams normalized by the first mode amplitude A_1 for the first two systems considered in the text: (a) equal masses, all spring constants equal, no damping, and (b) mass 2 is double mass 1, all spring constants equal, no damping

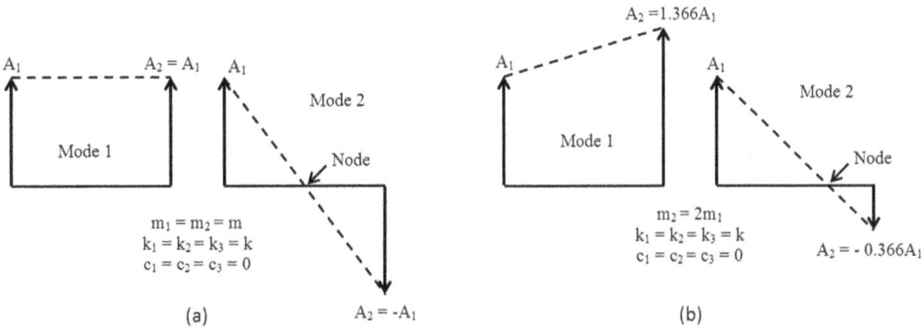

In Figure 7a, the coordinate system is placed at the center of mass for the beam. The translational displacement of the center of mass is uncoupled from the rotation of the beam. From a free-body diagram, the following equations of motion can be derived:

$$\begin{bmatrix} m & 0 \\ 0 & J_{CG} \end{bmatrix} \begin{Bmatrix} \ddot{x} \\ \ddot{\theta} \end{Bmatrix} + \begin{bmatrix} (k_1 + k_2) & (k_2 L_2 - k_1 L_1) \\ (k_2 L_2 - k_1 L_1) & (k_1 L_1^2 + k_2 L_2^2) \end{bmatrix} \begin{Bmatrix} x \\ \theta \end{Bmatrix} = \begin{Bmatrix} 0 \\ 0 \end{Bmatrix} \tag{31}$$

where J_{CG} is the mass moment of inertia about the center of gravity. The mass matrix in Equation 31 is a diagonal matrix, indicating no dynamic coupling.

Next, consider the coordinate system located at a unique point P, as in Figure 7b, where $k_1 L_3 = k_2 L_4$ and a force acting normal to the beam axis will produce only pure translation since the moments remain balanced. In this case, the system is dynamically coupled as seen from the diagonal stiffness matrix in the equations of motion as given by

$$\begin{bmatrix} m & me \\ me & J_P \end{bmatrix} \begin{Bmatrix} \ddot{x}_P \\ \ddot{\theta} \end{Bmatrix} + \begin{bmatrix} (k_1 + k_2) & 0 \\ 0 & (k_1 L_3^2 + k_2 L_4^2) \end{bmatrix} \begin{Bmatrix} x_P \\ \theta \end{Bmatrix} = \begin{Bmatrix} 0 \\ 0 \end{Bmatrix} \tag{32}$$

J_P is the mass moment of inertia about point *P*, and *e* is the distance from *P* to the center of gravity.

When the coordinate system is placed at the left end of the beam, at $x = x_1$, as in Figure 7c, the system has both static and dynamic coupling, as can be seen from the non-zero off-diagonal terms in both the mass and stiffness matrices in the equations of motion

$$\begin{bmatrix} m & mL_1 \\ mL_1 & J_1 \end{bmatrix} \begin{Bmatrix} \ddot{x}_1 \\ \ddot{\theta} \end{Bmatrix} + \begin{bmatrix} (k_1 + k_2) & k_2 L \\ k_2 L & (k_2 L^2) \end{bmatrix} \begin{Bmatrix} x_1 \\ \theta \end{Bmatrix} = \begin{Bmatrix} 0 \\ 0 \end{Bmatrix} \tag{33}$$

Figure 7. An undamped 2DOF system may be coupled or uncoupled depending on the choice of coordinate system: (a) Coordinates $x - \theta$ placed at the center of mass, (b) Coordinates $x - \theta$ placed at point P where an applied force results only in translation, (c) Coordinates $x - \theta$ placed at the left end of the beam

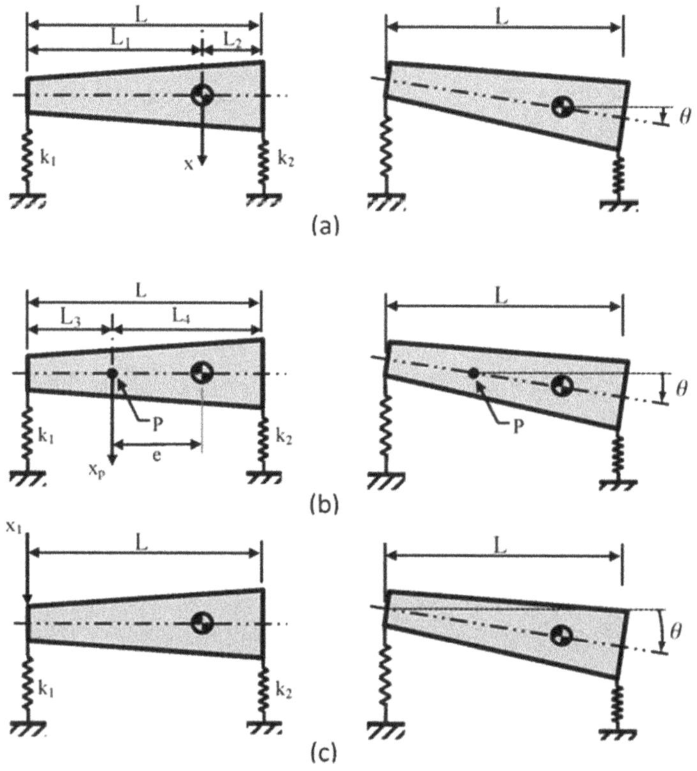

where J_1 is the mass moment of inertia about point 1 at the left end of the beam.

For undamped 2DOF systems, it is possible to find a coordinate system that decouples the two modes of vibration. Such coordinates are called the principal coordinates. See Thorby (2008) for a detailed explanation of the process of finding principal coordinates. For damped systems, it may not be possible to find principal coordinates if there is coupling through the damping terms.

Finding the principal coordinates permits the analyst to use the SDOF solutions for each of the modes of the system and to compute the response to any input as the summation of the modal responses.

Damped 2DOF System

Timoshenko (2008), Thorby (2008), and Vierck (1967), among others, provide insight into the added difficulties when damping is included in the analysis of 2DOF and multi-DOF systems. Even when the equations of motion are transformed into principal coordinates, the damping matrix can have non-zero off-diagonal terms, coupling the two modes. In cases where the damping ratio, $\zeta = \dfrac{c}{2\sqrt{mk}}$, for each mode is small, the damped natural frequencies will be very close to the undamped natural frequencies.

Damping ratios for many gates are of the order 1% $(\zeta \approx 0.01)$ or smaller, so often the damped solution is not required and the undamped frequencies suffice.

In this analysis of damping, please refer to the system shown in Figure 4, with the equations of motion repeated here for convenience:

$$\begin{bmatrix} m_1 & 0 \\ 0 & m_2 \end{bmatrix} \begin{Bmatrix} \ddot{x}_1 \\ \ddot{x}_2 \end{Bmatrix} + \begin{bmatrix} c_1 + c_2 & -c_2 \\ -c_2 & c_2 + c_3 \end{bmatrix} \begin{Bmatrix} \dot{x}_1 \\ \dot{x}_2 \end{Bmatrix} + \begin{bmatrix} k_1 + k_2 & -k_2 \\ -k_2 & k_2 + k_3 \end{bmatrix} \begin{Bmatrix} x_1 \\ x_2 \end{Bmatrix} = 0 \tag{16}$$

Assumed solutions in the form

$$x_1 = A_1 e^{st} \tag{34a}$$

$$x_2 = A_2 e^{st} \tag{34b}$$

when substituted into Equation 16, yield the following equations:

$$\begin{bmatrix} s^2 m_1 & 0 \\ 0 & s^2 m_2 \end{bmatrix} \begin{Bmatrix} A_1 \\ A_2 \end{Bmatrix} + \begin{bmatrix} s\left(c_1 + c_2\right) & -sc_2 \\ -sc_2 & s\left(c_2 + c_3\right) \end{bmatrix} \begin{Bmatrix} A_1 \\ A_2 \end{Bmatrix} + \begin{bmatrix} k_1 + k_2 & -k_2 \\ -k_2 & k_2 + k_3 \end{bmatrix} \begin{Bmatrix} A_1 \\ A_2 \end{Bmatrix} = 0 \tag{35a}$$

Or, by combining the coefficients

$$\begin{bmatrix} s^2 m_1 + s\left(c_1 + c_2\right) + k_1 + k_2 & -\left(sc_2 + k_2\right) \\ -\left(sc_2 + k_2\right) & s^2 m_2 + s\left(c_2 + c_3\right) + k_2 + k_3 \end{bmatrix} \begin{Bmatrix} A_1 \\ A_2 \end{Bmatrix} = 0 \tag{35b}$$

Setting the determinant of the coefficient matrix equal to zero and then expanding yields the following fourth order polynomial for *s*:

$$\begin{aligned} &m_1 m_2 s^4 + \left[m_1\left(c_2 + c_3\right) + m_2\left(c_1 + c_2\right)\right]s^3 + \begin{bmatrix} m_1\left(k_2 + k_3\right) + m_2\left(k_1 + k_2\right) \\ +\left(c_1 c_2 + c_1 c_3 + c_2 c_3\right) \end{bmatrix} s^2 \\ &+\left[\left(k_2 c_1 + k_3 c_1 + k_3 c_2\right) + \left(k_1 c_2 + k_1 c_3 + k_2 c_3\right)\right]s + \left(k_1 k_2 + k_1 k_3 + k_2 k_3\right) = 0 \end{aligned} \tag{36}$$

With only positive coefficients, this polynomial cannot have any positive real roots. Three other options remain:

Option 1: There may be two pairs of complex conjugate roots, which will give exponentially decaying sinusoids at two distinct frequencies. This option is the usual result with small values of damping.

Option 2: There may be two negative real roots and one pair of complex conjugate roots, which yields a single decaying sinusoid superposed with a non-oscillatory decay. This option is the usual result for higher levels of damping, but still less than critical damping.

Option 3: There may be four negative real roots, which lead only to non-periodic behavior. The system returns to equilibrium without vibration. This option is not considered further in this book.

Since the gates to be discussed in this book typically have small damping, the focus of the following discussion will be on Option 1. Vierck (1967) provides details of the solution for the roots of Equation 36. For lightly damped systems, there are two pairs of complex conjugate roots given as

$$s_1 = -p_1 + iq_1$$

$$s_2 = -p_1 - iq_1$$

$$s_3 = -p_2 + iq_2$$

$$s_4 = -p_2 - iq_2 \quad (37)$$

The damped solutions will be of the form

$$x_1 = A_{11}e^{-p_1 t}\sin\left(q_1 t + \phi_{11}\right) + A_{12}e^{-p_2 t}\sin\left(q_2 t + \phi_{12}\right) \quad (38a)$$

$$x_2 = A_{12}e^{-p_1 t}\sin\left(q_1 t + \phi_{21}\right) + A_{22}e^{-p_2 t}\sin\left(q_2 t + \phi_{22}\right) \quad (38b)$$

The function x_1 is the displacement of m_1 and x_2 is the displacement of m_2. The coefficients A_{11} and A_{21}, A_{12} and A_{22} and the phase angles ϕ_{11} and ϕ_{21}, and ϕ_{12} and ϕ_{22} are related pairs of values. The four initial conditions are thus sufficient to evaluate the constants.

For lightly damped systems, p_1 and p_2 will be small and the frequencies q_1 and q_2 will approach the undamped natural frequencies for the same system with no damping. The solution, however, is not truly a normal mode since the phases of the two sinusoidal components of the motion of the two masses may vary. The two masses may pass through their equilibrium points at different times in the cycle of motion.

The introduction of damping results in a phase difference between the sinusoidal components that may differ from 0° or 180°. For multi-DOF systems this phase can be measured experimentally and then used in the construction of the experimental mode shape. One method of validating a finite element model of a system is to compare the computed mode shapes and modal frequencies with their measured counterparts.

Forced Vibration of 2DOF System

Assume now that the masses in Figure 4 are acted on by sinusoidal forces, with $F_1(t)$ acting on m_1 and $F_2(t)$ acting on m_2. These forcing functions can be incorporated into the equations of motion to yield

$$\begin{bmatrix} m_1 & 0 \\ 0 & m_2 \end{bmatrix} \begin{Bmatrix} \ddot{x}_1 \\ \ddot{x}_2 \end{Bmatrix} + \begin{bmatrix} c_1 + c_2 & -c_2 \\ -c_2 & c_2 + c_3 \end{bmatrix} \begin{Bmatrix} \dot{x}_1 \\ \dot{x}_2 \end{Bmatrix} + \begin{bmatrix} k_1 + k_2 & -k_2 \\ -k_2 & k_2 + k_3 \end{bmatrix} \begin{Bmatrix} x_1 \\ x_2 \end{Bmatrix} = \begin{Bmatrix} F_1(t) \\ F_2(t) \end{Bmatrix} \tag{39}$$

Thomson and Dahleh (1998) show that for an undamped system $\left(c_1 = c_2 = c_3 = 0\right)$, consisting of two masses and three springs being forced by a single sinusoidal force $F_1(\sin \omega t)$ acting on m_1 with $F_2 = 0$, the steady state amplitudes of the two masses can be given as

$$\begin{Bmatrix} X_1 \\ X_2 \end{Bmatrix} = \frac{1}{\left|Z(\omega)\right|} \begin{bmatrix} \left(k_{22} - m_2\omega^2\right) & -k_{12} \\ -k_{21} & \left(k_{11} - m_1\omega^2\right) \end{bmatrix} \begin{Bmatrix} F_1 \\ 0 \end{Bmatrix} \tag{40}$$

in which $\left|Z(\omega)\right| = m_1 m_2 \left(\omega_1^2 - \omega^2\right)\left(\omega_2^2 - \omega^2\right)$ with ω_1 and ω_2 representing the two modal frequencies.

The individual amplitudes can be written as

$$X_1 = \frac{\left(k_{12} - m_2\omega^2\right) F_1}{m_1 m_2 \left(\omega_1^2 - \omega^2\right)\left(\omega_2^2 - \omega^2\right)} \tag{41a}$$

$$X_2 = \frac{-k_{21} F_1}{m_1 m_2 \left(\omega_1^2 - \omega^2\right)\left(\omega_2^2 - \omega^2\right)} \tag{41b}$$

When $m_1 = m_2 = m$ and $k_1 = k_2 = k_3 = k$, the frequency of the first mode is $\omega_1^2 = k / m$ and that of the second mode is $\omega_2^2 = 3k / m$. In this case, Equations 41a and b can be reduced to the following solutions:

$$\frac{X_1}{\left(F_1 / k\right)} = \frac{\left(2 - \frac{\omega^2}{\omega_1^2}\right)}{\left(1 - \frac{\omega^2}{\omega_1^2}\right)\left(3 - \frac{\omega^2}{\omega_1^2}\right)} \tag{42a}$$

$$\frac{X_2}{\left(F_1\big/k\right)} = \frac{1}{\left(1-\frac{\omega^2}{\omega_1^2}\right)\left(3-\frac{\omega^2}{\omega_1^2}\right)} \tag{42b}$$

The solutions given in Equations 42a and b are plotted in Figure 8.

Note that unlike the SDOF system, not all responses for $r < 1$ are greater than the equivalent static deflection. Also note that m_1 remains stationary when this system is driven at the frequency $\omega = \sqrt{2}\omega_1$.

PARAMETRIC COUPLING

Stable Parametric Vibrations

If the mass in Figure 2 is permitted to vibrate rectilinearly along the rod and the rod is permitted to rotate simultaneously, the situation becomes more complicated because the motion of the mass in the radial direction, denoted as the x-direction, changes the moment of inertia and the rotation of the rod in the θ-direction induces a centrifugal force that acts in the radial x-direction. It is now a two degrees-of-freedom (2DOF) system with parametric coupling.

The equation of motion in the x-direction has an added force term, $mr\dot{\theta}^2$, the centrifugal force due to the motion in the θ-direction. After summing the forces and dividing by the mass, the equation of motion becomes

Figure 8. Forced vibration response of an undamped 2DOF system with $m_1 = m_2 = m$ and three springs with $k_1 = k_2 = k_3 = k$. Solid line is the response of m_1; dashed line is that of m_2.

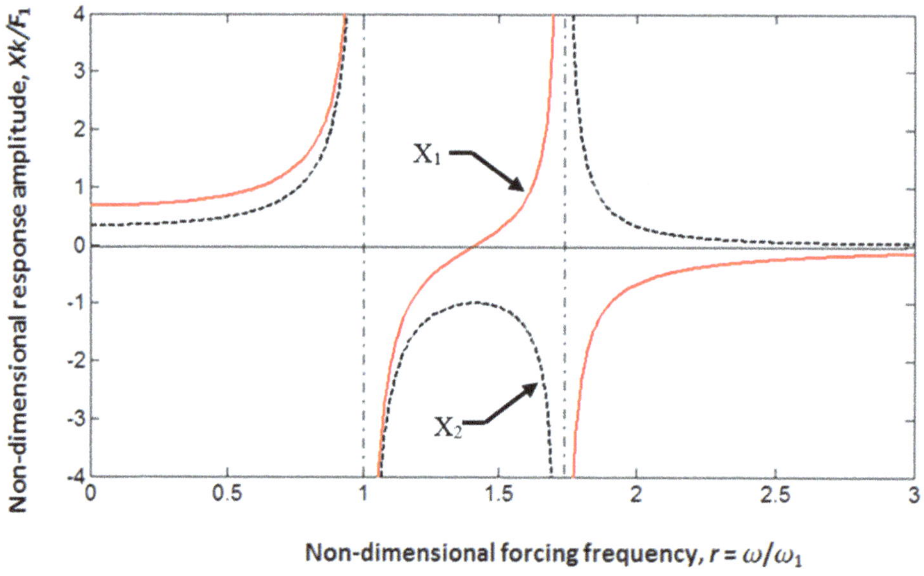

$$\ddot{x} + 2\zeta\omega_n \dot{x} + \omega_n^2 x - \left(r_o + x\right)\dot{\theta}^2 = 0 \tag{43}$$

In addition, there is also an added term in the tangential direction (*i.e.*, θ -direction) due to the changing mass moment of inertia:

$$\ddot{\theta} + 2\zeta\omega_{n\theta}\dot{\theta} + \omega_{n\theta}^2 + gx\cos\left(\theta\right) / \left(r_o + x\right)^2 + 2\dot{\theta}\dot{x} / \left(r_o + x\right) = 0 \tag{44}$$

To demonstrate the coupling, Equations 43 and 44 were simulated in SIMULINK for a set of conditions. In the simulation the rectilinear frequency was chosen as 5 Hz ($\omega_n = 31.42$ rad/s) and the location of the mass along the rod, r_o, as 10 m. The rectilinear damping ratio ζ was chosen as 0.01. For the tangential direction, $\omega_{n\theta} = \left(1 + \varepsilon\right)\omega_n$ where $-0.05 \leq \varepsilon \leq 0.05$ and tangential damping was chosen as $\zeta_t = 0.025$.

The initial conditions were a rectilinear displacement $\left(x = x_o\right)$, with no initial rectilinear velocity and no tangential displacement or velocity. However, physically displacing the mass along the rod axis will produce a concurrent static displacement in the rotational direction, which depends upon the displacement of the mass, the length of the rod, and the torsional spring constant. Figure 9 shows the rectilinear and rotational response of the system after the initial rectilinear displacement. Note the small, initially growing, transient tangential response to the rectilinear perturbation. In the present simulation, this is due to the unbalanced moment about the origin due to the position of the mass. We will return to this type of coupling later in the book in Chapters 10 to 14 when we consider the bending mode of the skinplate coupling with the rotational mode of a Tainter gate. In that case, the gate motion produces a fluctuating discharge that provides a force in phase with the skinplate velocity, leading to self-sustained vibration, and the skinplate displacement provides the driving force for the rotational gate vibration.

Unstable Coupling Through the Moment of Inertia

To simulate a bit more complicated coupling, two additional forces will be included in the *x*-direction equation of motion; one that is in phase with the rectilinear velocity, or leading the rectilinear displacement by 90°, and the other in phase with the rotational velocity. The first force can be represented as $-A_1\dot{x}$ and the second as $-A_2\dot{\theta}$ in the *x*-direction equation of motion. The equation of motion in the *x*-direction then becomes

$$\ddot{x} + \left(2\zeta\omega_n - A_1\right)\dot{x} + \omega_n^2 x - \left(r_o + x\right)\dot{\theta}^2 - A_2\dot{\theta} = 0 \tag{45}$$

In this chapter, no physical significance is ascribed to these additional forces. The present model is applicable to a gate impacted by a horizontal force (previously portrayed in Figure 3 in Chapter 1). The mathematics demonstrate that a torsional deflection can result from such an initial rectilinear deflection. In Chapter 10, we will show that a torsional gate deflection results in a discharge fluctuation, which may drive the initial rectilinear deflection, producing a potential coupled-mode instability.

Figure 10 shows the system trajectory for three different values of A_1:

Figure 9. Response of the coupled rectilinear-rotational system to an initial rectilinear displacement: (a) time-dependent rectilinear displacement, and (b) time-dependent rotational displacement (r_o *= 10m,* z *= 0.01,* z_θ *= 0.025,* w_n *= 10p = 5Hz,* x_o *= 4m, integration time 0.001 s)*

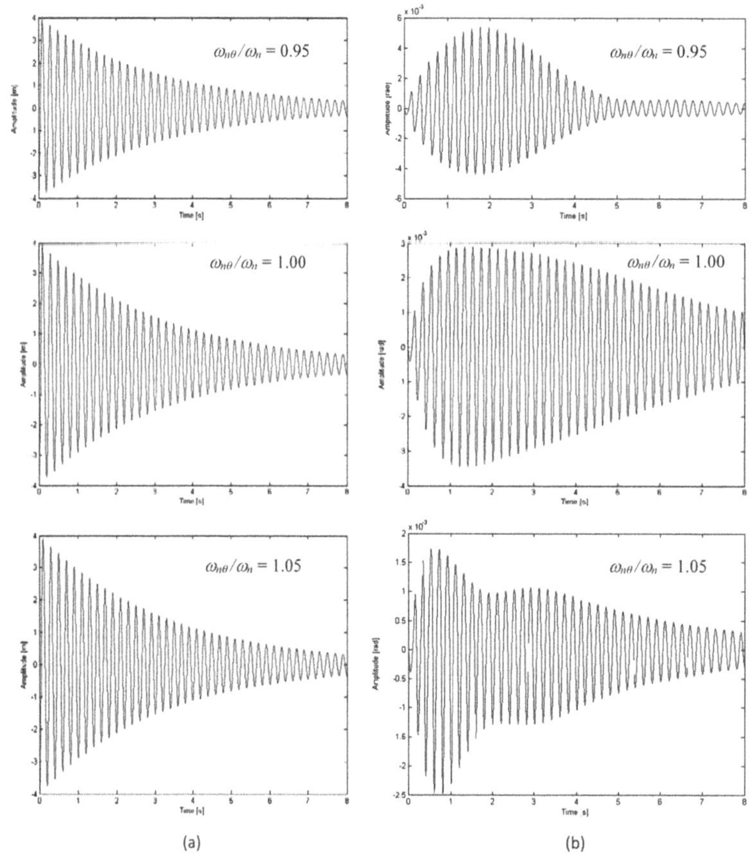

$$A_1 < 2\zeta\omega_n$$

$$A_1 = 2\zeta\omega_n$$

$$A_1 > 2\zeta\omega_n$$

all with $A_2 = 0$. In Chapter 10, a similar response will be found for a Tainter gate under certain conditions involving gate submergence, gate opening, and the natural frequency of the skinplate bending mode relative to the solid body rotation frequency of the gate.

In Figure 10a, with $A_1 < 2\zeta\omega_n$, the trajectory is an ellipse that diminishes in amplitude with time. In Figure 10b, with $A_1 = 2\zeta\omega_n$, the trajectory becomes a stable ellipse indicating that, on a per cycle basis, the system is dissipating the same amount of energy as is added to the vibration from the flow.

In Figure 10c, with $A_1 > 2\zeta\omega_n$, the loop increases in amplitude indicating an instability and an impending failure. Interestingly the growth in the rectilinear direction is not accompanied by a growth in the rotational direction until the last 5 or 6 cycles shown. The direction of rotation of the trajectory changes from counter-clockwise to clockwise as the rotational displacement begins to grow.

Rotating Rod With Oscillating Masses

In a thematically related paper, Butikov (2004) considers a torsional spring oscillator consisting of a rod with two masses symmetrically displaced from the center of the rod. The masses are forced to oscillate symmetrically about the midpoint of the rod, providing a periodic variation in the moment of inertia for the torsion spring oscillator.

Figure 10. (a) Stable system trajectory for $A_1 < 2\zeta\omega_n$; (b) metastable system trajectory for $A_1 = 2\zeta\omega_n$; (c) unstable system trajectory for $A_1 > 2\zeta\omega_n$. Note the scales are not uniform.

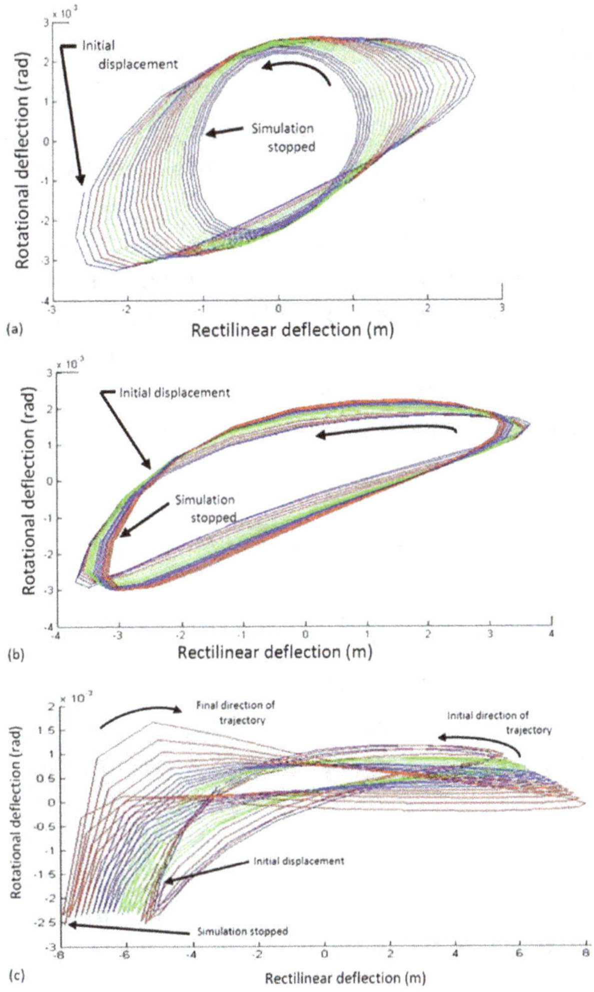

Butikov examines the stability of the torsional oscillations as a function of the modulation frequency of the symmetrically disposed masses on the rod. Although his system does not precisely correspond to Tainter gate vibration, it does illustrate the concept of parametric resonance.

A distinct difference exists between the parametric resonance considered by Butikov and the resonance previously considered in the forced vibration of the coupled 2DOF system. The modulation frequency in parametric resonance does not have to be at the natural frequency of the oscillator. If the system has even just weakly oscillatory behavior, parametric resonance is possible if the modulation amplitude exceeds the threshold value needed to overcome the system damping and the modulation frequency is at twice the system frequency, equal to the system frequency, or is at any even sub-multiple of the system frequency. Once the threshold amplitude is exceeded, the torsional response grows exponentially. The most unstable response occurs when the modulation frequency is twice the system natural frequency. Butikov's paper can provide the reader with a better understanding of parametric resonance; the mechanism of coupled-mode self-excited vibration of Tainter gates involves the interaction of a parametric resonance with flow-induced loading of the gate's skinplate.

MULTIPLE DEGREES-OF-FREEDOM (MDOF) SYSTEM AND NORMAL MODES

The extension from two degrees-of-freedom (2DOF) to multiple degrees-of-freedom (MDOF) is quite easily accomplished using the same matrix notation as used for 2DOF systems. For MDOF the mass, damping, and stiffness matrices are $n \times n$ matrices where n is the number of degrees-of-freedom needed to describe the system. The acceleration, velocity, displacement, and forcing vectors are $1 \times n$ column matrices.

The normal mode formulation is also employed in MDOF systems. For interested readers, we recommend Thorby's (2008) well-written explanation of the transformation from global to modal coordinates. With the prevalence of the use of the finite element method for structural dynamic analysis, mode shapes and frequencies can be readily extracted from a well-constructed finite element model of the MDOF system. As previously noted, matching the calculated mode shapes and frequencies with the corresponding experimentally-obtained data is a way to validate the finite element model.

CONTINUOUS SYSTEM AND MODE SHAPE

For a system of lumped, discrete masses, there are natural frequencies and mode shapes that correspond in number to the number of degrees-of-freedom. As the system becomes a continuous distribution of mass over a spatial extent, the number of mode shapes and frequencies becomes infinite. In the following section, three specific systems will be considered: a cable (or string) in tension, a cantilever beam, and a two-dimensional rectangular plate. For other geometries, we refer the reader to Blevins (1979).

Cable in Tension

A cable or a string in tension is a readily understood vibrating system in the context of stringed musical instruments such as violins, banjos, and guitars. The string is tuned by adjusting the tension in the string, so *a priori* most people are aware that tension affects the frequency of vibration. The mathematics validates this understanding.

Assume the string is of uniform diameter, with a length L, under a tension T, and having a mass of ρ_l per unit length. If a small section of the string, dx, is assumed to have some deflection y, a force balance on the small section may be written as

$$\frac{\partial^2 y}{\partial x^2} = \frac{\rho_l}{T}\frac{\partial^2 y}{\partial t^2} \tag{46}$$

It is customary to define $\frac{T}{\rho_l} = c^2$ where c is the wave velocity (or celerity). (Note that this use of c is unrelated to the previous use of c as a damping coefficient.) Equation 46 can then

$$\frac{\partial^2 y}{\partial x^2} = \frac{1}{c^2}\frac{\partial^2 y}{\partial t^2} \tag{47}$$

To solve this equation, called the one-dimensional wave equation, two initial conditions and two boundary conditions are needed. The usual initial conditions are that the string is plucked (given an initial displacement) with zero initial velocity. More generally, $y_o(x)$ is an assumed initial displacement and $\dot{y}_o(x)$ is an assumed initial velocity. The usual boundary conditions for a stringed musical instrument are that the string is rigidly fixed at both ends, a fixed-fixed boundary condition. This boundary condition would also be appropriate for a hoist cable on a gate that is rigidly held on the take-up spool and assumed fixed to a point on the gate.

The solution for the free vibration of a fixed-fixed string or cable is

$$y(x,t) = \sum_{n=1}^{\infty} \sin\frac{n\pi x}{L}\left[A_n \cos\left(\frac{nc\pi t}{L}\right) + B_n \sin\left(\frac{nc\pi t}{L}\right)\right] \tag{48}$$

where the coefficients A_n and B_n are determined by the initial conditions as follows:

$$A_n = \frac{2}{L}\int_0^L y_o(x)\sin\frac{n\pi x}{L}dx \text{ and } B_n = \frac{2}{nc\pi}\int_0^L \dot{y}_o(x)\sin\frac{n\pi x}{L}dx \tag{49}$$

The spatial distribution $\sin\left(\frac{n\pi x}{L}\right)$ defines the mode shape. The first five mode shapes are shown in Figure 11. The first mode $(n = 1)$ is a half sine wave; the second mode is a full sine wave with one node at the midpoint $(L/2)$, and each successively higher mode adds another half sine wave with another node. The frequency of vibration of the nth mode is

Figure 11. First five mode shapes for a cable

$\omega_n = \left(n c \pi / L\right)$ or $f_n = \omega_n / 2\pi = \left(nc / 2L\right)$.

Recalling the definition of $c^2 = \dfrac{T}{\rho_l}$, the frequency of the nth mode becomes $f_n = \dfrac{n}{2L}\sqrt{\dfrac{T}{\rho_l}}$, confirming our original *a priori* expectation.

Cantilever Beam (Fixed-Free)

The next geometry can be thought of as a clamped ruler, hereafter called a cantilever beam. The beam is assumed uniform, slender, and homogeneous with no axial loading. As with the cable, the equilibrium equation for a differential length of the beam, *dx*, will be written, yielding a partial differential equation (second order in time and fourth order in the spatial coordinate) that must be solved subject to two initial conditions and four boundary conditions. Two boundary conditions are at one end, here called $x = 0$, where the displacement and the slope of the beam are required to be zero. The remaining two boundary conditions are at the other end of the beam, where the bending moment must be zero and the rate of change of bending moment with x must also be zero. The equation for the transverse free vibration of such a beam, in terms of the transverse deflection $u\left(x,t\right)$, is

$$\frac{\partial^2 u}{\partial t^2} + c^2 \frac{\partial^4 u}{\partial x^4} = 0 \tag{50}$$

where the celerity of the wave, c, depends upon the Young's modulus, E, the assumed constant moment of inertia I, the material density ρ, and the assumed uniform cross-sectional area of the beam, A:

$$c^2 = \frac{EI}{\rho A} \tag{51}$$

Equation 50 can be solved by the separation of variable technique; for details of the solution, see Palm (2007) or another vibration textbook. The solution results in a characteristic equation that can be solved for the eigenvalues for the vibration of the cantilever beam. The mode shapes and frequencies are in terms of the eigenvalues (denoted as β_n) that, for the cantilever beam, must satisfy the following equation:

$$\cosh \beta L \cos \beta L = -1 \tag{52}$$

The first five roots of Equation 52 and the corresponding modal frequency constants, γ_n, as given in Palm (2007), are shown in Table 1.

The frequency of vibration of the cantilever beam in the nth mode is related to the value of γ_n given in Table 1, and can be written as

$$f_n = \frac{\gamma_n}{2\pi L^2}\sqrt{\frac{EI}{\rho A}} = \frac{\gamma_n}{2\pi}\sqrt{\frac{EI}{mL^3}} \tag{53}$$

where $m = \rho AL$ is the mass of the beam.

The mode shapes of the beam are given by

$$F_n\left(x\right) = \cosh \beta_n x - \cos \beta_n x - \left(\frac{\sinh \beta_n L - \sin \beta_n L}{\cosh \beta_n L + \cos \beta_n L}\right)\left(\sinh \beta_n x - \sin \beta_n x\right) \tag{54}$$

Figure 12 shows the first five mode shapes for a cantilever beam. Again, the number of nodes is one less than the mode number. An example of cantilever beam modal vibration is the movement of a vertical automobile radio antenna mounted on the hood of older-style cars. Often these antennae vibrate in the second or third mode transverse to the oncoming wind, depending on vehicle velocity and the size of the antenna.

Understanding the vibration of a cantilever beam will aid in the understanding of the vibration of a flap gate discussed in Chapter 6.

Table 1. Values of spatial and temporal frequency constants for a cantilever beam

Mode	1	2	3	4	5
$\beta_n L$	1.875	4.694	7.855	10.996	14.137
$\gamma_n = \beta_n^{\,2} L^2$	3.516	22.034	61.701	120.91	199.85

Figure 12. First five modes of free vibration of a cantilever beam

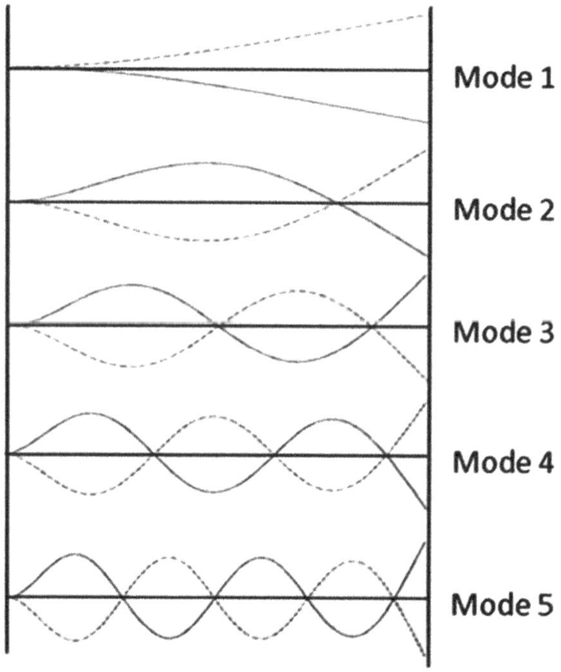

Rectangular Plate

Another structure that will be considered in detail is the two-dimensional rectangular plate. Knowledge of plate vibration will be useful later in this book when we examine the dynamics of weir plates and skinplates subjected to fluid flow. According to Blevins (1979), a general closed form solution for the vibration frequencies and mode shapes of a flat plate does not exist. The usual assumption is that the mode shape for plate vibration can be written as the product of the solutions for a beam in the *x*-direction and a beam in the *y*-direction. For a plate of dimension *a* by *b* with thickness *h*, with the sides $\left(x = 0 \text{ and } x = a\right)$ simply supported and with the top and bottom $\left(y = 0 \text{ and } y{=}b\right)$ free, Blevins (1979) (citing Leissa, 1973) gives the following expression for the mode shapes of the plate:

$$w_{ij}\left(x, y\right) = \sin\frac{i\pi x}{a} F_j\left(y\right) \tag{55}$$

where $F_j\left(y\right)$ is the mode shape of a free-free beam (also provided by Blevins, 1979 in his Table 8.1), given as

$$F_j\left(y\right) = \cosh\beta_j y + \cos\beta_j y - \left(\frac{\cosh\beta_j b - \cos\beta_j b}{\sinh\beta_j b - \sin\beta_j b}\right)\left(\sinh\beta_j y + \sin\beta_j y\right) \tag{56}$$

in which β_j are the roots of the following transcendental equation

$$\cosh \beta b \cos \beta b = 1 \tag{57}$$

In later chapters, the form of the y-direction mode will be approximated as a rigid body rotation around a horizontal line at a specified location $y_{rot} = r_s$. In that case, the $F_j(y)$ is described by the rotational angle of the plate, ψ, and the distance from the point where $y_{rot} = r_s$

$$F_{rot}(y) = |y - r_s| \sin \psi \tag{58}$$

Blevins (1979) provides the following formula for estimating the frequency of vibration of the plate (simply supported at $x = 0$ and $x = a$, free at $y = 0$ and $y = b$)

$$f_{i,j} = \frac{\lambda_{i,j}^2}{2\pi a^2}\left[\frac{Eh^3}{12\gamma\left(1-\nu^2\right)}\right]^{1/2} \tag{59}$$

where $\lambda_{i,j}^2$ depends on the aspect ratio a / b of the rectangular plate, E is the modulus of elasticity, γ is the mass per unit area of the plate, and ν is Poisson's ratio. For the (1, 1) mode, the value of $\lambda_{i,j}^2$ ranges from 9.760 for $a / b = 0.4$ to 9.484 for $a / b = 2.5$.

ENERGY FLOW IN VIBRATIONS

To understand the interaction of forces and displacements, we introduce the concept of a rotating vector. As shown in Figure 13, the oscillatory displacement of a point mass, $x = X \sin(\omega t)$, can be represented by a vector rotating around the origin. The rotational speed of the vector represents the frequency of vibration and the projection of the rotating displacement vector onto the x-axis, $X \sin\left(\omega t\right)$, represents the instantaneous magnitude of the vibration. Since the velocity vector is found by differentiating the oscillating displacement, it can be written as

$$\dot{x} = \omega X \cos(\omega t) = \omega X \sin(\omega t + \pi / 4).$$

In Figure 13, the velocity of the point mass is represented by a constant length vector ωX rotating at the prescribed frequency and leading the displacement by 90°. The projection of the rotating velocity vector along the x-axis represents the instantaneous amplitude of the velocity, $\omega X \cos(\omega t)$. The acceleration would be a third vector, 180° out of phase with the displacement and of magnitude $\omega^2 X$. The velocity and acceleration vectors are included in Figure 14.

Figure 13. Rotating vector representing a sinusoidal oscillation

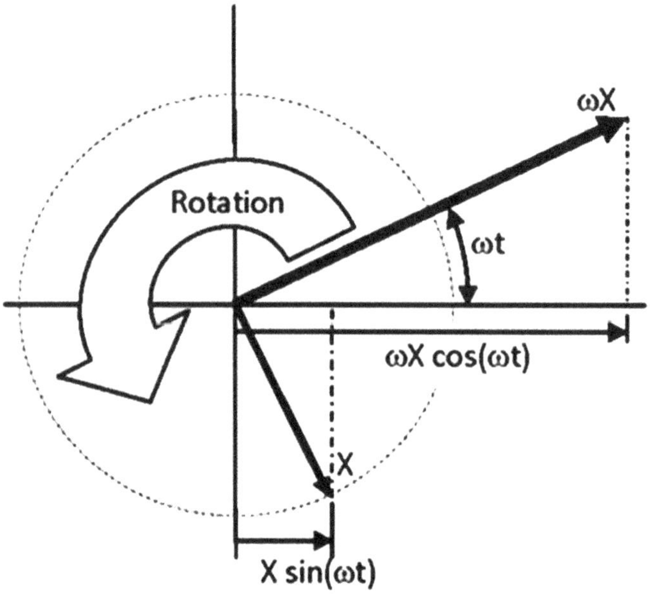

Figure 14. (a) Rotating vector diagram showing exciting force, spring force, damping force and inertia force along with displacement, velocity and acceleration for an SDOF system. (b) Force-displacement diagram showing energy added by the exciting force and energy dissipated by the damping force
After Naudascher and Rockwell, 1994.

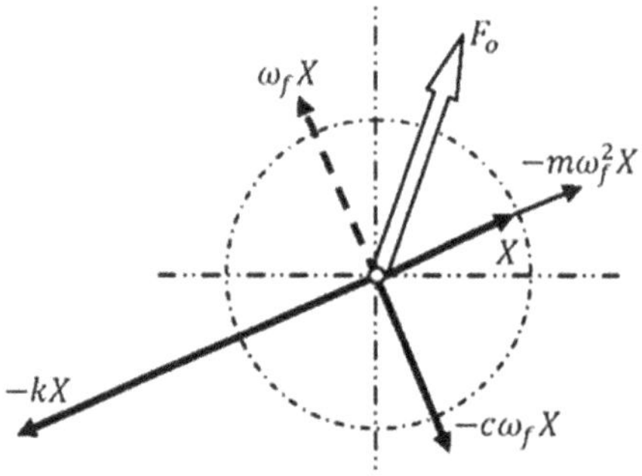

(a)

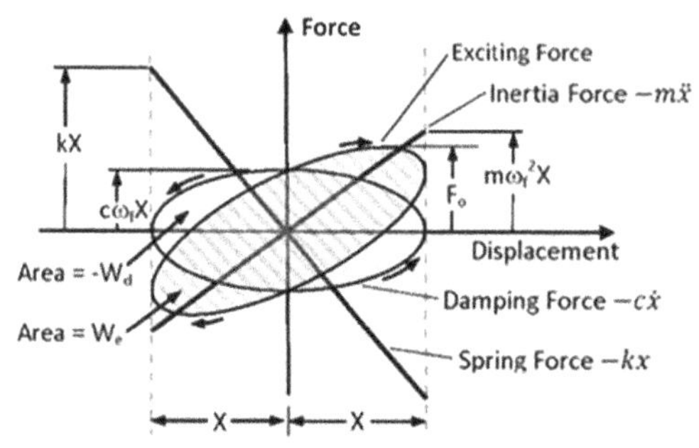

(b)

Figure 14a shows the rotating vector diagram for an SDOF system excited by a sinusoidal force F_o. The equation of motion includes the inertia force vector $-m\ddot{x}$, the spring force $-kx$ and the damping force $-c\dot{x}$, as well as the exciting force. As required by dynamic equilibrium, adding all of the force vectors (excitation, damping, restoring, and inertia forces) in a tip-to-tail fashion results in zero net force on the structure.

The magnitude of each of these forces when plotted as a function of the displacement provides a picture of the energy flow for the cycle. In Figure 14b the spring force and the inertia force are lines with no enclosed area indicating that they neither supply nor consume energy. The exciting force traces an inclined elliptical pattern when plotted versus displacement (the inclination angle is the phase lag between the imposed force and the system response). The loop is traversed in the clockwise direction which is interpreted as energy added to the vibration. The damping force, on the other hand, traces a horizontal elliptical pattern (because of the 90° phasing between velocity and displacement) which is traversed in the counter-clockwise direction, and is indicative of energy dissipation.

The areas enclosed by the exciting force ellipse and the damping force ellipse will be identical when the system is in dynamic equilibrium. If the exciting force ellipse has a greater area, the system amplitude will grow. Conversely, if the damping ellipse area is greater, the amplitude will decay. The system equilibrium position can be determined by plotting both the excitation energy and the dissipation energy versus system amplitude, as shown schematically in Figure 15a. The system steady state amplitude is the amplitude at which these two curves cross. In theory, there could be multiple crossings if there were different dissipation mechanisms that were amplitude dependent (see the next section on damping). Instability of the system will occur if the excitation energy supplied exceeds the work of the dissipative processes and there is a perturbation of sufficient amplitude. The system in Figure 15b shows a possible equilibrium amplitude, but the system will most likely become unstable if an impact or another event produces a displacement larger than the indicated equilibrium amplitude.

Figure 15. Energy-displacement diagrams showing (a) the equilibrium amplitude for a stable system and, (b) a system that is stable for small amplitude perturbations, but which becomes unstable with a larger perturbation (hard excitation)

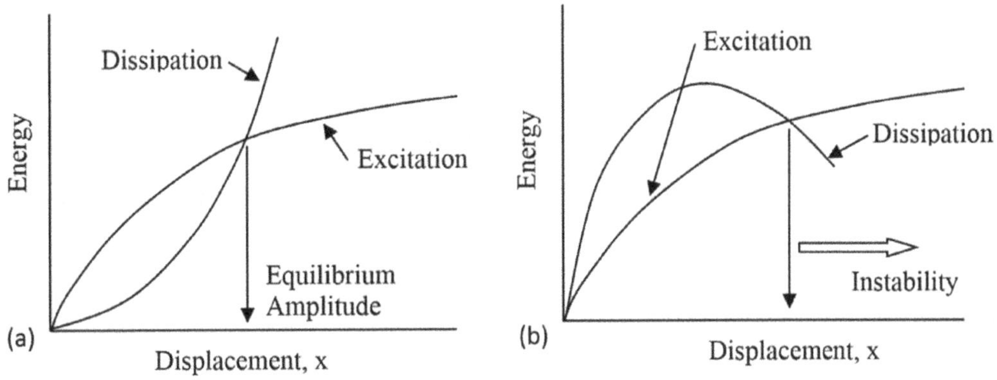

DAMPING

Viscous Damping

In this section we will present the mathematics leading to Figures 14 and 15 and then contrast this analysis with that for Coulomb damping and hysteretic damping. Assume that the viscously damped SDOF system responds to forcing with $x(t) = X\sin(\omega_f t)$. The damping force is $F_d = c\dot{x} = c\omega_f X\cos(\omega_f t)$. The rate of dissipation of energy is the product of this force times the velocity, or $\dot{E} = c\dot{x}^2 = c\omega_f^2 X^2\cos^2(\omega_f t)$ Over one cycle of vibration, the energy dissipated by the viscous damping is

$$E = \int_0^{2\pi/\omega_f} \dot{E}dt = \int_0^{2\pi/\omega_f} c\omega_f^2 X^2\cos^2(\omega_f t)\,dt = \pi c\omega_f X^2 \tag{60}$$

Subsequently, expressions for equivalent viscous damping will be defined for other forms of damping by equating their energy dissipation per cycle to that calculated using Equation 60. An alternative calculation of the damping coefficient can be undertaken if the energy dissipated per cycle is known:

$$c_e = E / \pi\omega_f X^2 \tag{61}$$

The equation of the elliptic damping path shown in Figure 14b can be obtained by squaring the damping force

$$F_d^2 = c^2\dot{x}^2 = c^2\omega_f^2 X^2\cos^2(\omega_f t) = c^2\omega_f^2 X^2\left(1 - \sin^2(\omega_f t)\right)$$

and then solving for a unity coefficient on the right, as follows:

$$\left(\frac{F_d}{c\omega_f X}\right)^2 + \sin^2(\omega_f t) = \left(\frac{F_d}{c\omega_f X}\right)^2 + \left(\frac{x}{X}\right)^2 = 1 \tag{62}$$

which is the equation of an ellipse with semi-major and semi-minor axes of $c\omega_f X$ and X, respectively, as shown in Figure 14b. The area of this ellipse is $\pi(c\omega_f X)(X) = \pi c\omega_f X^2$, which agrees with Equation 60.

Coulomb Damping

When one solid surface slides past another, the friction-induced damping is frequently modeled as Coulomb damping (see Palm, 2007; Timoshenko, 2008; and Rao, 2010 for further details). The friction force, which always opposes the relative motion of the two surfaces, can be written as μN where μ is the coefficient of friction and N is the normal force exerted on the surface. When there is sinusoidal relative motion of one surface given by $x(t) = X \sin(\omega_f t + \phi)$ the work done to overcome friction for each quarter cycle is $\mu N X$ (*i.e.*, force times distance) and the energy dissipated per cycle is four times the work per quarter cycle. Setting this energy per cycle equal to the energy dissipated by an equivalent viscous dissipation makes it possible to define an equivalent viscous damping coefficient, $c_{e,C}$.

$$c_{e,C} = \frac{4\mu N X}{\pi \omega_f X^2} = \frac{4\mu N}{\pi \omega_f X} \tag{63}$$

It is important to note that the equivalent viscous damping coefficient for Coulomb damping exhibits a decreasing magnitude as the amplitude of vibration increases. Ultimately, this characteristic leads to systems that become unstable when subjected to an impulsive displacement that exceeds a given threshold because the Coulomb damping cannot dissipate the input energy.

An equivalent viscous damping ratio can be obtained by placing this equivalent damping coefficient into the defining equation for damping ratio:

$$\zeta_{e,C} = \frac{c_{e,C}}{2\sqrt{mk}} = \frac{c_{e,C}}{2m\omega_n} = \frac{2\mu N}{\pi m X \omega_f \omega_n} \tag{64}$$

Substituting the equivalent damping ratio into the previously developed equation of motion permits an approximate calculation of the system's response to a disturbance.

Hysteretic Damping

When materials are cyclically loaded and unloaded, energy is absorbed due to internal deformations such as grain boundary slipping or other internal processes. The loading and unloading process produces an inclined closed curve in the force-displacement plane that is traversed in a clockwise direction. This difference in path caused by the application or removal of the load is called hysteresis. It is usually modeled as a spring and damper acting in parallel. Rao (2010) and Palm (2007) both develop expressions for the equivalent damping for systems exhibiting hysteresis. Rao (2010) expresses the hysteresis damping in terms of hysteretic log decrement, where

$$\delta = \ln\left(\frac{A_i}{A_{i+1}}\right) \approx \ln(1 + \pi\beta) \approx \pi\beta = \pi\frac{h}{k} \tag{65}$$

The parameters β and h are assumed to be constants that provide a measure of the hysteretic damping. The equivalent viscous damping coefficient and equivalent damping ratio can be defined in terms of either β or h

$$c_{e,h} = \frac{\beta}{k\omega_f} = \frac{h}{\omega_f} \tag{66}$$

and

$$\zeta_{e,h} = \frac{c_{e,h}}{2\sqrt{mk}} = \frac{c_{e,h}}{2m\omega_n} = \frac{\beta}{2} = \frac{h}{2k} \tag{67}$$

Hysteretic Damping Maintained Stability

Hysteretic damping plays a significant role in certain hydraulic gate installations. In installations with music note bottom seals, the deformation of the seal as the gate impinges upon the sill during low amplitude vibration extracts sufficient energy to maintain a constant small amplitude vibration. In laboratory studies of such gates, the gate has been shown to undergo the same steady small amplitude vibration. While undergoing these small amplitude vibrations with hysteretic damping maintained stability, the gate, however, quickly becomes unstable if given an additional perturbation of sufficient amplitude. These studies will be presented in more detail in Chapter 11.

SDOF SYSTEM IN FLOWING FLUID

In a flowing fluid, the effects of the fluid mass on the vibrating system must be included in the analysis. This is done formally by incorporating added mass, added damping and added rigidity into the equation of motion. For a SDOF system this then means that the equation of motion can be written as

$$\left(m + \Delta m\right)\ddot{x} + \left(c + \Delta c\right)\dot{x} + \left(k + \Delta k\right)x = F\left(t, x, \dot{x}\right) \tag{68}$$

In practical terms, in a flowing fluid, it is impossible to separate the effects of added mass, Δm, from those due to added stiffness, Δk, without a complete theoretical model of the fluid-structure interaction. The added mass and added stiffness effects are lumped into the added mass coefficient. See Naudascher and Rockwell (1994) and Kolkman and Jongeling (2007) for further details on the effects of added mass, damping, and stiffness. Equation 68 can be re-written with the added stiffness and added mass effects incorporated into the added mass coefficient, as follows:

$$\left(m + \Delta m\right)\ddot{x} + \left(c + \Delta c\right)\dot{x} + \left(k\right)x = F\left(t, x, \dot{x}\right) \tag{69}$$

The main effect of the added mass is a reduction (often a substantial reduction) in the natural frequency of the system. In a fluid, the natural frequency becomes

$$\omega_n = \sqrt{k / (m + \Delta m)} \text{ rad/s or } f_n = (1 / 2\pi)\sqrt{k / (m + \Delta m)} \text{ Hz} \tag{70}$$

Added mass effects are often neglected for structures vibrating in air since the density of air is several orders of magnitude less than that of most solids. In water, the effects of the fluid mass cannot be neglected because the density of water is of the same order of magnitude as many solids.

Added mass can also serve as a coupling term for certain structural geometries. Motion in one direction can, through the added mass effect, create pressures that can drive a vibration in the other direction. Details of added mass coupling can be found in Naudascher & Rockwell (1994) and in Kolkman & Jongeling (2007). Chapter 5 will provide a detailed analysis of added mass effects for gates in flowing water.

If a body vibrates in a compressible fluid or near the free surface of a liquid, the progressive waves produced by the body's movement can actually act as added damping in the structural vibration. Details of the added damping due to free surface wave generation will be considered in detail in Chapter 4.

FREQUENCY POWER SPECTRA

In a fluid flow system the excitation is seldom a clean sine wave. Often there may be multiple frequency components present from several sources along with broadband noise from fluid turbulence. The fluctuating signal to be analyzed might be the acceleration, velocity, or displacement of the body, or the fluctuating fluid velocity or the pressure fluctuations on a solid surface. The most intuitive way to determine the frequency components in a signal is to use the analog process shown in the block diagram in Figure 16. This analog process is the old-fashioned method of measuring the power spectrum. With this method, the signal, $g(t)$, is fed to a narrow-band bandpass filter with a selectable center frequency f_c and a narrow constant bandwidth Δf. The filtered signal containing fluctuating components at frequency $f_c \pm \Delta f / 2$, is then fed to a squaring circuit and averaged. After the values of the averaged squared signal and the center frequency are recorded, the filter center frequency is incremented to the next center frequency and the process is repeated. The value of $\overline{\{g'(f_c)\}^2}$, the mean-square fluctuation at frequency f_c, when plotted versus f_c produces the power spectrum of the time signal.

For at least the past thirty years, essentially all power spectral signal processing has been done using fast Fourier transforms. The fast Fourier transform (FFT) is an efficient algorithm for implementing the discrete Fourier transform (the digital version of the Fourier transform). The Fourier transform is written as

Figure 16. Analog method of generating the power spectrum of a time signal

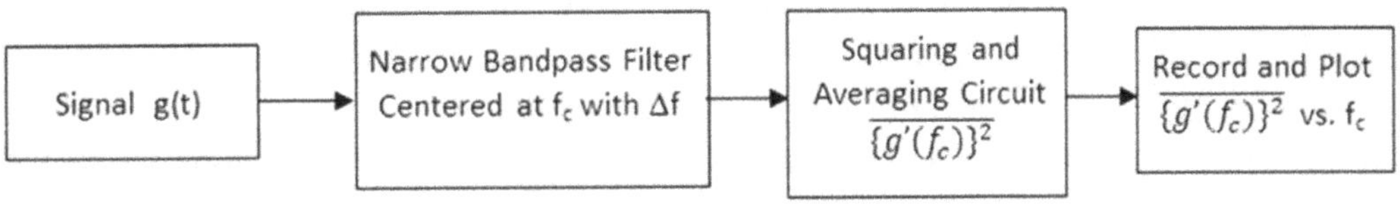

$$G(\omega) = \int_{-\infty}^{\infty} g(t) e^{-i\omega t} dt = C(\omega) + iQ(\omega) \tag{71}$$

where C(ω) is called the co-spectrum and Q(ω) is called the quad-spectrum. The power spectrum of the signal is formed by summing the squares of the real and imaginary components.

$$S_{gg}(\omega) = C^2(\omega) + Q^2(\omega) \tag{72}$$

The sampling frequency in the discrete implementation of the Fourier transform must be at least twice the highest frequency present in the time signal to avoid aliasing. The input signal must be filtered to remove all frequency components above one half of the sampling frequency. In addition, an inherent assumption in using the Fourier transform is that the signal is repetitive in time. To avoid spurious components it is necessary to select a windowing function to pre-multiply the signal so it has zero initial and final values with zero initial and final slopes. It is then necessary to correct the spectral value for the effects of the diminished amplitudes in the beginning and ending portions of the signal. Further details of signal processing can be found in Bendat & Piersol (2010).

One application of spectral analysis is determining the frequency content of force loading on a structure. Knowing the input force spectrum, the system response can be calculated by a point-by-point multiplication of the input force spectrum by the frequency response transfer function (that resembles Figure 1 for a SDOF system). For a lightly-damped system with a strong peak at its natural frequency, the response spectrum can exhibit a peak even when the input force spectrum shows broadband noise.

In this book, spectral analysis will be used primarily to determine experimentally the structural mode shapes for gate components. This application of spectral analysis to determine mode shapes, called experimental modal analysis, has been essential in uncovering the mechanisms of gate vibration that are presented in this book.

EXPERIMENTAL MODAL ANALYSIS

Determining the mode shape and frequencies of vibration of a large complicated structure such as the skinplate of a gate is necessarily a challenging assignment. The engineer has only a few limited options. First, a finite element analysis model (FEM) of the structure can be constructed to analyze the structural dynamics. Building the model is time consuming, but once the model is constructed, minor changes in the structure and its support can be incorporated to determine how a proposed change might affect the structural dynamics. One of the most challenging aspects of a finite element analysis is validating the dynamic predictions. Ultimately, one needs to have experimental data concerning the structural dynamics to validate the numerical calculations. Selection of different types of elements available in an FEM may yield differing calculated system dynamics. Without experimental validation, it is exceedingly difficult to select one prediction over another.

The most usual method of providing validation of a finite element model is to compare the predicted dynamic structural response to an experimentally measured dynamic structural response. The dynamic structural response is usually determined using experimental modal analysis. Modal analysis requires a substantial investments, both in equipment and in personnel time, but achieves reasonable results. A brief overview of experimental modal testing follows; details can be found in Ewins (2000).

To apply experimental modal analysis to a structure, calibrated accelerometers are placed over the surface of the structure, being careful that the added mass due to the instruments does not substantially alter the system mass. The frequency response of the accelerometers needs to be substantially higher than the highest expected structural frequency. Then the structure is subjected to excitation, often with a mechanical shaker providing a controlled frequency and amplitude of excitation. In this process, the test is repeated as the shaker is driven through the frequency range of interest. The accelerometer spectra vary substantially depending on the frequency of excitation. For larger structures, an impact hammer works better than a mechanical shaker for providing excitation. An impact hammer is a hammer of known mass with a load cell in-line with the striking surface of the hammer so that the impulsive loading imparted to the structure can be recorded. Regardless of excitation source, the structure must be tested repeatedly to obtain statistically averaged data at each frequency in the range of expected modal frequencies. After the spectral energy input and the spectral response of the structure are recorded, the transfer function can be calculated by dividing the response spectrum by the input spectrum. Each frequency that shows coherent amplitude and phase responses is then considered as a potential modal frequency.

The challenge in modal analysis is to correctly interpret the resulting spectral functions. Many versions of commercially available software incorporate an inverse modal indicator function (IMIF) that assists the user in determining if the peak in a spectral response has physical significance or if it is a spurious peak lacking the expected phase and amplitude coherence with neighboring measurement locations. Once the peaks have been correctly identified, the amplitude and phase are plotted relative to the locations of measuring points, permitting the mode shape to be determined at the selected frequency. The tests should be repeated for better statistical reliability. This requires many accelerometers to be positioned at different locations on a large structure, such as a gate, exciting the structure using a mechanical shaker or an impact hammer, and then measuring the response. Then the accelerometers must be repositioned at different locations and the process repeated, because it would be prohibitively expensive to purchase all of the accelerometers and processing channels needed to test an entire gate simultaneously.

For massive gates, the challenge is to input sufficient energy to excite the relevant modes of vibration. When we tested massive, full-scale gates, we developed a bolt-breaking technique that involves installing a machined bolt (resembling a tensile test specimen) between the gate sill and the bottom horizontal girder of the gate. The gate operator opens the gate in increments to tension the bolt to its maximum value. Opening the gate to the next increment causes a sudden failure of the bolt, which imparts a step input into the gate, which has sufficient energy to excite the vibration modes of interest. This method allowed us to validate the theoretical models that have been developed for the coupled-mode self-excited instability mechanism.

RESONATORS

The vibrations to this point involve the vibration of structures. Under certain circumstances, the fluid (air and water) contained within or bounded by the solid surfaces, may be capable of sustaining oscillating pressures. Some cases may involve a mass of fluid acting as the system mass in a mass-spring damper system, such as the U-tube resonator and the Helmholtz resonator. In other cases, standing acoustic waves or surface waves on a free surface may be excited through interactions with an adjacent resonator.

Naudascher and Rockwell (1994) provide a comprehensive overview of fluid resonators. These fluid resonators can drive or be driven by structural vibration and fluid instabilities.

A fluid resonator can be considered a discrete mass resonator, if the fluid mass moves in unison at all locations. If the direction and amplitude of fluid motion varies spatially, the fluid resonator is treated as a distributed mass system (Naudascher and Rockwell, 1994). We will consider four specific fluid resonators that under specified circumstances can interact with the structural vibration of the gate to enhance the potential for instability. The first two resonators, the U-tube resonator and the Helmholtz resonator, are discrete mass resonators, while the second two, standing water waves and standing acoustic waves, are distributed mass resonators.

U-TUBE RESONATORS

A U-tube resonator is simple to construct, requiring only a tube bent or formed into a U-shape and filled with a liquid to a certain level, as shown in Figure 17. Under equilibrium conditions, the liquid in the two legs of the tube will be at the same level. By applying a differential pressure across the two legs, one can draw the liquid to a higher level on one side while decreasing the fluid level in the other leg by an equal amount (assuming uniform diameter). A sudden release of the pressure difference, analogous to releasing the displaced mass in a spring-mass system, will allow the fluid to return to equilibrium in an oscillatory manner, much like the mass on a spring.

By calculating the maximum potential energy and the maximum kinetic energy and setting the two maximum values equal, the equation for the frequency of the resulting oscillation can be found:

Figure 17. U-tube resonator

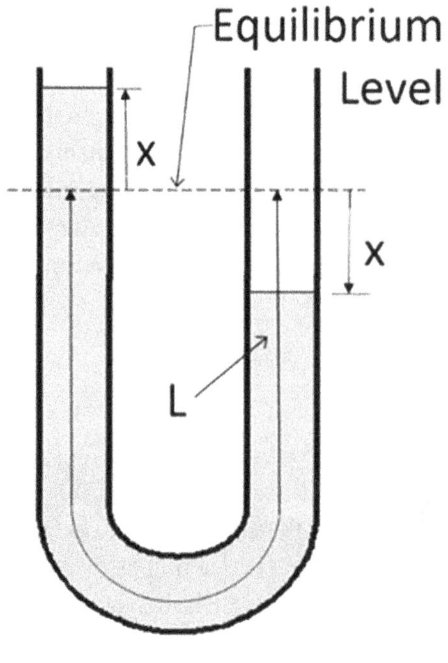

$$\omega = 2\pi f = \sqrt{\frac{2g}{L}} \tag{73}$$

where g is the acceleration of gravity and L is the total length of the fluid in the tube (see Figure 17). Blevins (1979) provides a compilation of frequencies and mode shapes for various U-tube geometries, ranging from a basin or tank with an attached tube to two interconnected basins of differing size with several different connecting tube geometries. In all cases, the fluid is assumed to move with the same velocity at all radial locations in the tube and the damping effects of viscosity have been neglected. Naudascher and Rockwell (1994) provide an expression for the frequency of a generalized U-tube resonator of variable area, as shown in Figure 18.

The oscillation frequency is

$$\omega = 2\pi f = \sqrt{\frac{g\left(1 + {A_3}/{A_1}\right)}{L_1\left({A_2}/{A_1}\right) + L_2 + L_3\left({A_2}/{A_3}\right)}} \tag{74}$$

Zahariea (2010), among others, took into account viscous damping for a constant diameter U-tube. The damping ratio can be calculated as

Figure 18. Generalized U-tube resonator
After Naudascher and Rockwell, 1994.

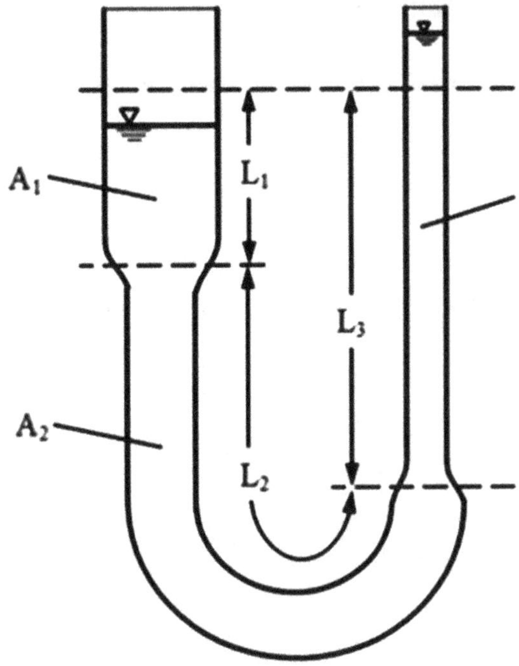

$$\zeta = \frac{8\mu L\sqrt{2g / L}}{\rho g d^2} = \frac{8\mu}{\rho d^2}\sqrt{\frac{2L}{g}} \tag{75}$$

where d is the diameter of the U-tube, μ and ρ are the dynamic viscosity and the density of the manometer fluid, respectively, and L is the length of the fluid column. As is usual in discrete mass systems, the damped natural frequency is given as

$$\omega_d = \omega_n\sqrt{1-\zeta^2} \tag{76}$$

HELMHOLTZ RESONATORS

A Helmholtz resonator relies on the fluid compressibility in an enclosed volume to provide a restoring spring force. It is usually represented as an enclosed volume, V, connected to the environment by a neck of area $A=\pi d_n^{\,2}/4$, of diameter d_n and of length L_n as shown schematically in Figure 19. A Helmholtz resonator can be excited by cross-flow over the open end of the neck, as demonstrated by blowing across the end of a bottle to generate a low frequency tone. The low frequency tone, well below that of the lowest frequency standing acoustic wave, is characteristic of Helmholtz resonators.

When the wavelength, λ, of the tonal sound is such that $\lambda \gg L_n$, $\lambda \gg A^{1/2}$, and $\lambda \gg V_n^{\,1/3}$, the frequency of Helmholtz resonators can be approximated as that of a mass-spring system where the mass is the mass of fluid in the neck (the shaded areas in Figure 19) and the spring is the compressibility of the gas volume as the mass in the neck moves in and out. The radiation of sound from the open end of the tube requires an effective length. Kinsler et al. (1999) provide equations for the effective length, depending on whether the open end of the neck is unflanged, as in Figure 19a, or flanged, as in Figure 19b.

For an unflanged neck, such as that shown in Figure 19a, the effective length is given by

$$L' = L_n + 0.725 d_n\left(unflanged\right) \tag{77}$$

Figure 19. Schematic of (a) an unflanged Helmholtz resonator, and (b) a flanged Helmholtz resonator

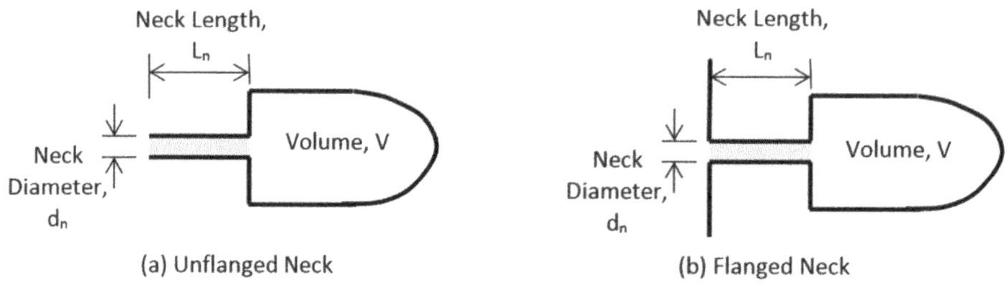

For the flanged case, as shown in Figure 19b, which can also be used in the case of no neck ($L_n = 0$), the effective length L' is given by

$$L' = L_n + 0.85d_n \left(flanged\right) \tag{78}$$

With the appropriate effective length the frequency of a Helmholtz resonator can be calculated by

$$\omega_{HR} = 2\pi f_{HR} = c\sqrt{\frac{A}{L'V}} \tag{79}$$

where c is the speed of sound for air (or for the surrounding medium if it is not air).

Blevins (1979) supplies expressions for the frequencies of more complicated Helmholtz resonator geometries. One that will be of interest in nappe instability is the Helmholtz resonator having a single volume with multiple openings (n in number), each with neck area A_i and physical neck length L_i with effective length L_i'. In this case, the natural frequency of the Helmholtz resonator can be given as

$$\omega_{HRn} = 2\pi f_{HRn} = c\sqrt{\sum_{i=1}^{n}\left(\frac{A_i}{L_i'V}\right)} \tag{80}$$

Naudascher and Rockwell (1994) review the application of a concept similar to the acoustic Helmholtz resonator to free surface conditions to describe surge tanks or open basins. These systems, while of possible interest, will not be reviewed here. We refer interested readers to Naudascher and Rockwell (1994) and the references cited therein.

STANDING WAVE RESONATORS

The next type of resonator is the standing wave resonator. The wave may be a free surface wave on a body of water or an acoustic wave in an enclosed air space (or other gas). The two types of standing waves are quite similar in some respects but also substantially different in others.

Standing Water Waves

Water with a free surface can oscillate if exposed to a forcing function. If a tank containing a liquid with the free surface is accelerated in the horizontal direction, Newton's Second Law tells us the water surface will be inclined at an angle θ such that $\tan\theta = a_x / g$, where a_x is the horizontal acceleration. If there is a periodic reversal of the acceleration, a sloshing mode can be created where the liquid sloshes back and forth in the tank at the imposed frequency. Once the water surface has been perturbed and the forcing function is removed, the free surface will eventually settle back to a horizontal surface after a period of oscillation. The frequency and wavelength of the oscillations depend primarily upon the geometry of the bounding surface with a secondary dependence on the depth.

In a long lake with an atmospheric disturbance at one end, a depression or upwelling of the water surface can result in the reverse effect at the opposite end of the lake. A standing wave is created that can oscillate back and forth for many periods before it is damped. The standing wave created in this manner is called a seiche and can be dangerous for unsuspecting visitors to the lake. According Hughes (1965) and ISGS (2009), among others, a 10-ft seiche struck the shore of Lake Michigan in 1954, killing seven or eight people (depending on the source).

The standing wave problem is described in terms of the velocity potential that must satisfy the LaPlace equation subject to the boundary conditions at the two sides of the channel, the constant pressure condition at the free surface and the boundary condition requiring the flow along the bottom of the channel to be parallel to the wall. Irrotational flow is assumed. The linear solution to the potential flow equation is valid only when the wave height is substantially smaller than the wavelength and the depth. See Stoker (1957), Méhauté (1976), and Sorensen (2010) for the details of the analysis.

Knowing that a wave incident on the wall must be reflected, we can *a priori* recognize that the relationship between the wavelength, λ, for the mode *n* wave and the channel width, *L*, must be given by

$$L = n\frac{\lambda}{2} \tag{81}$$

The first three modes on a transverse standing wave in a channel are shown in Figure 20.

As with all wave phenomena, the frequency and wavelength are related. Their product is the convective speed (celerity) of the wave

$$f\lambda = \frac{\omega}{2\pi}\lambda = c \tag{82}$$

Figure 20. First three modes of a standing wave across a channel

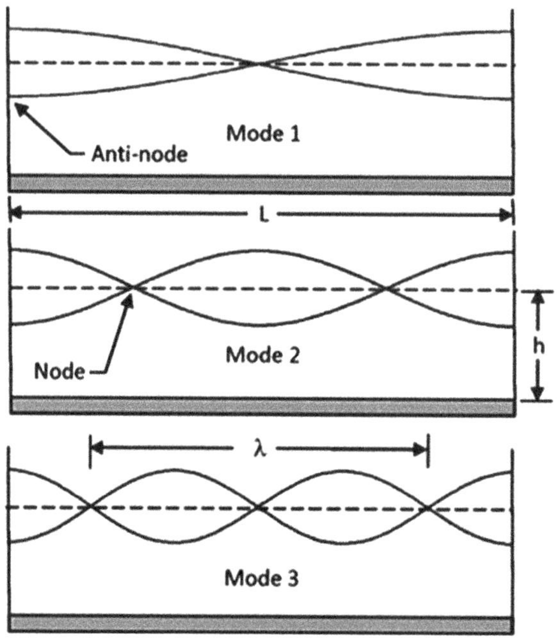

The wave speed (celerity) is given as

$$c = \frac{g\tau}{2\pi}\tanh\left(\frac{2\pi h}{\lambda}\right) = \frac{g}{\omega}\tanh\left(kh\right) \tag{83}$$

where τ is the period, h is the depth of the channel, $\omega = 2\pi f = 2\pi / \tau$ is the radian frequency and $k = 2\pi / \lambda$ is the wave number. The wavelength can be written in terms of the period, recognizing that $1 / f = \tau$:

$$\lambda = \frac{c}{f} = \frac{g\tau^2}{2\pi}\tanh\left(\frac{2\pi h}{\lambda}\right) \tag{84}$$

Alternatively, Equation 84 can be solved for the frequency of the wave

$$f = \frac{\omega}{2\pi} = \sqrt{\frac{g}{2\pi\lambda}\tanh\left(\frac{2\pi h}{\lambda}\right)} \tag{85}$$

Thus, by selecting the mode *n* for a channel of width *L*, the wavelength can be calculated as $\lambda = 2L / n$ from Equation 81. Substituting the value of λ into Equation 85 provides the frequency of the wave oscillation. If needed, the celerity can be then calculated from either Equation 82 or Equation 83.

For shallow waves ($h / \lambda < 1 / 20$) the calculation is simpler; since $\tanh(2\pi h / \lambda) \approx 2\pi h / \lambda$, Equation 83 reduces to

$$c = \sqrt{gh} \tag{86}$$

The wave frequency is given as $f = c / \lambda$.

The interaction of vortex shedding from bridge piers can serve as a driving force for transverse standing surface waves in channels, changing a smooth flowing channel into one with side to side standing waves with amplitudes as much as 35% of mean depth (Zima and Ackermann, 2002). The interaction can occur when the ratio of the vortex shedding frequency to the wave frequency is in the range 0.7 to 1.3. Zima and Ackermann (2002) proposed a correlation of the wave amplitude generated by vortex shedding from arrays of cylinders in the channel as

$$\frac{A}{h} = K\frac{N}{St^2}\frac{D}{P}\left(\frac{D}{L}\right)^2 \tag{87}$$

in which *A* is the maximum wave amplitude, *h* is the mean flow depth, $K = 2.255$ is a constant, *N* is the number of cylinders in each transverse row, *D* is the cylinder diameter, *P* is the longitudinal pitch between rows, and *St* is the Strouhal number based on cylinder diameter (for the given longitudinal and transverse pitch). The correlation of Zima and Ackermann was shown to be valid only for first mode waves with transverse pitches T/D < 5 by Ghomeshi et al. (2007), who proposed two more-general correlations from the statistical analysis of their data which they claim is general for any mode n.

For transverse pitch ratios (T/D) less than 5.0:

$$\frac{A}{h} = 0.078N^{-1/3}\left(\frac{D}{L}\right)^{-1/3}\left(\frac{D}{P}\right)^{1/2} St^{-1/2}n^{2/3} \text{ for T/D} < 5 \tag{88}$$

and for transverse pitch ratios of 5.0 or greater:

$$\frac{A}{h} = 1.41N^{2/3}\left(\frac{D}{L}\right)\left(\frac{D}{P}\right)^{1/2} St^{-1/2}n^{2/3} \text{ for T/D} \geq 5 \tag{89}$$

The induced standing wave has the potential to drive other vibration-susceptible systems, such as gates and valves, under certain conditions. This type of resonator would be of special concern for a side discharge weir or gate with piers or other obstructions in the channel. Hardwick, et al. (1979) traced a gate vibration problem at the El Chocon hydroelectric power station in Argentina to vortex shedding from a central pier in the channel before the intake to the gate. The vortex shedding frequency was very close to the resonant frequency of the penstock, exciting the first waterhammer mode of the penstock. The combined vortex shedding-waterhammer system then supplied the forcing for the gate vibrations. Although this study did not describe the presence of a transverse standing wave in the upstream channel, the circumstances lead one to speculate that standing waves *may* have been involved in the mechanism of vibration.

The standing wave in the penstock was modeled by Hardwick, et al. (1979) as an organ-pipe resonance, which leads us to the next topic, standing acoustic waves.

Standing Acoustic Waves

In any enclosed volume an acoustic wave, once generated, will reflect off of the enclosing walls, and even off of openings due to the impedance change between the enclosed air space and the opening. It is beyond the scope of this book to develop fully the acoustic theory. Instead we will draw upon the results to be able to apply them to problems related to gate instability. We refer the reader to the following texts for more details: Dowling & Ffowcs-Williams (1983); Morse & Ingard (1987); Kinsler et al. (1999); Blackstock (2000), as well as handbooks by Crocker (1997; 1998), Vér & Beranek (2005); and Schoeder (2007), among many others.

The simplest standing wave is a one-dimensional wave in a tube of small diameter, D, relative to its length. Figure 21 shows the mode shapes for the first four acoustic modes in closed-closed tubes, in open-open tubes and in open-closed tubes. For closed-closed tubes, the acoustic standing wave mode shapes look the same as the standing water wave mode shapes, as previously shown in Figure 20. The acoustic wave, however, is an axial compression and expansion (rarefaction) that travels back and forth in the tube, while the water wave is a periodic transverse change in water elevation. For tubes with an open end, the boundary condition requires that the fluctuating pressure amplitude approach zero at the open end of the tube. Due to radiation and mass loading effects, just as was the case with the Helmholtz

Figure 21. Acoustic standing wave pressure amplitudes in tubes: (a) Closed-closed tube, (b) Open-closed tube, and (c) Open-open tube
After Naudascher and Rockwell (1994).

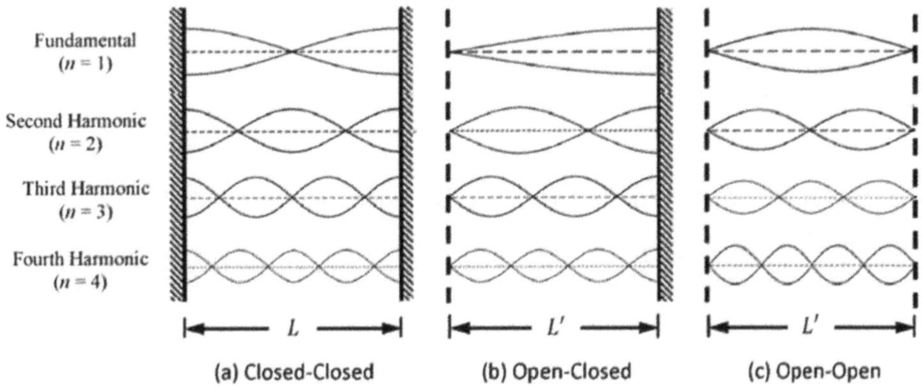

resonator, a corrected tube length is used when calculating standing acoustic waves in tubes with open ends. For a small cylindrical tube, the end correction for a single open end is $\Delta L = 0.3D$. For a tube with two open ends (an open-open tube) the end corrections are additive so the effective length is the physical length plus 0.6D.

For the closed-closed tube in mode *n*, the wavelength is

$$\lambda = 2L / n \tag{90a}$$

For the open-closed tube, with an effective length $L' = L + 0.3D$, the wavelength is

$$\lambda = 4L' / \left(2n - 1\right) \tag{90b}$$

For the open-open tube, with an effective length $L' = L + 0.6D$, the wavelength is

$$\lambda = 2L' / n \tag{90c}$$

From basic thermodynamics of ideal gases, the acoustic wave speed, or speed of sound, in a gas depends on the temperature (see, for example, Moran & Shapiro, 2014; Çengel & Boles, 2014):

$$c_a = \sqrt{kRT} \tag{91}$$

A standard accepted value for c_a for air at 20°C is 343 m/s. Once wavelength and speed of sound have been determined, the frequency can easily be calculated as $f = c / \lambda$.

In two-dimensional or three-dimensional rectangular regions, the normal modes of the volume can be calculated assuming a superposition of the normal modes in each of the coordinate directions. The wavelength in each direction can be determined using the appropriate form of Equation 90, and using the speed of sound calculated from Equation 91, the resonant frequency of the (n_x, n_y, n_z), mode is

$$f = c_a\left(\left(\frac{1}{\lambda_x}\right)^2 + \left(\frac{1}{\lambda_y}\right)^2 + \left(\frac{1}{\lambda_z}\right)^2\right)^{1/2} \quad (92)$$

If all the boundaries are composed of rigid walls, the frequency of the (n_x, n_y, n_z) mode can be found by substituting Equation 90a into Equation 92:

$$f = \frac{c_a}{2}\left(\left(\frac{n_x}{L_x}\right)^2 + \left(\frac{n_y}{L_y}\right)^2 + \left(\frac{n_z}{L_z}\right)^2\right)^{1/2} \quad (93)$$

where $L_x, L_y,$ and L_z are the dimensions of the bounding walls.

In addition, non-orthogonal modes may co-exist as discussed in the previously cited texts on acoustics, but will not be considered in this book.

When gate structures vibrate in enclosed air spaces or bound an enclosed air space, one should check for possible interactions with an acoustic mode of the air volume. The lowest frequency acoustic mode for the air space can be determined by assigning a value of 1 for the mode numbers into Equation 93. Usually the lowest frequency acoustic mode is substantially higher than the frequency of structural vibration, but a check is always prudent if there is any doubt.

For regular non-rectangular geometries, there are additional expressions for mode shapes and frequencies available in acoustic texts as well as in Naudascher and Rockwell (1994). For irregular shaped volumes, numerical calculation of the normal modes and frequencies is necessary if there is a need for the frequencies and mode shapes of standing waves. The minimum frequency of a standing acoustic wave can be calculated from the maximum physical dimension of the space as

$$f_{\min} = \frac{c_a}{2L_{\max}} \quad (94)$$

If this minimum frequency is not close to any structural frequencies, further calculation of the acoustic modes is unwarranted.

STABILITY CONCEPTS

In most cases, stability analysis addresses the response of a system subjected to a small disturbance. The analysis can consider a static disturbance to detect *static instability*; or the analysis can be assume a small dynamic disturbance at a specified frequency to detect any *dynamic instabilities* of the system or structure. In dynamic stability analysis, the small disturbance, called the perturbation, is formulated as a function of frequency, and the system response if plotted as a function of frequency. In fluid-flow systems, wavenumber or convective speed of the perturbation is also specified and the disturbance is assumed to grow either spatially or temporally. More details on fluid flow instability are presented in Chapter 3.

Typically, in dynamic stability analysis, the perturbation is added to the solution for the natural (unforced) modes of vibration of the structure. The composite solution, consisting of the perturbation plus unforced solution substituted into the differential equation of motion. After carrying out the differential operations on the composite solution, the original differential equation is subtracted, leaving the differential perturbation equation that describes the system response. The perturbation equation is linearized, assuming small amplitude disturbances.

The solution to the stability equation will show one of three results:

1. The amplitude of the perturbation at a specified frequency may decrease in time indicating stability at that specific frequency.
2. The amplitude of the perturbation remains constant in time, indicating neutral stability at that specific frequency.
3. The amplitude of the perturbation at a specified frequency may increase in time; growth of the disturbance indicates instability at that specific frequency.

The resulting linearized instability equation is limited to small amplitude disturbances. Once instability commences, nonlinear effects must be included in any subsequent analysis. Small amplitude stability does not guarantee system stability for larger amplitude disturbances. Figure 22d shows an example of a system stable to small amplitude static disturbances, but unstable for disturbances with larger amplitude.

Stability analysis is also undertaken for the solutions to differential equations (that may or may not represent the dynamics of real systems). For linear systems, *Nyquist stability* or *Roth-Hurwitz stability* methods are frequently used to assess the stability of the solution. For nonlinear equations *Lyupanov stability* analysis is frequently used. Control systems engineering is another area in which stability analysis is critical. Phillips & Parr (2010), Nise (2015), and Dorf & Bishop (2010), among others, incorporate stability analysis in design of feedback control systems and provide further details on linear and nonlinear system stability. Strogatz (2014) provides useful analysis methods for nonlinear system dynamics.

Static Stability

Static stability can be understood by a simple diagram shown below in Figure 22. For a structure with continuous concave walls, Figure 22a, moving the ball to a new location and releasing it cannot cause the ball to do anything but return to its original position. The system is stable. The system in Figure 22b is unstable; the ball will leave the convex surface with even the slightest displacement. In Figure 22c, a *neutrally stable* system is portrayed; the ball will not move regardless of where it is placed on the surface. The situation in Figure 22d shows a system which is stable for a small displacement of the ball and is unstable for larger displacements. The system is Figure 22d is said to be metastable or conditionally stable.

Traditionally, civil engineering design considered, and in many sub-fields still considers, only the static stability of structures. Bažant (2000) and Bažant & Cedolin (2010) provide an overview and details, respectively, of static stability analysis for structures, as well as dynamic stability considerations.

Figure 22. Conceptual model of stability showing (a) a stable system, (b) an unstable system, (c) a neutrally stable system, and (d) a metastable or conditionally stable system, stable for very small disturbance, but unstable for larger disturbances
After White, 2006.

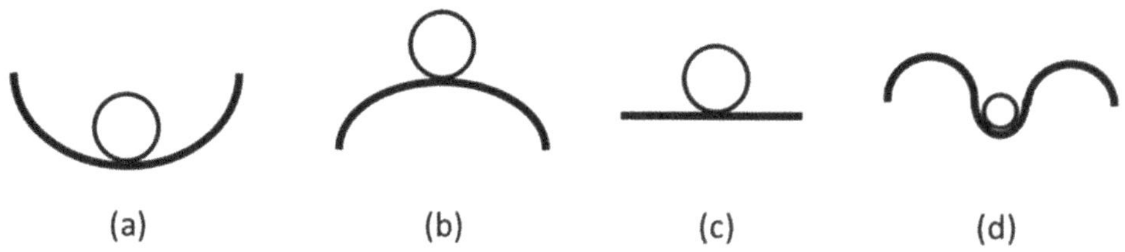

Static instabilities include buckling of columns, as well as the deformation and failure of frames, arches and trusses. Course notes by Felipa (2015) also provide an overview of both static and dynamic stability analysis. Kanvinde (2003) provides guidance on shaker table testing and dynamic analysis of structures. The application of stability analysis to an automatic control system for a hydraulic gate and channel was presented in the thesis of de Graaff (1998).

Static Stability of Hydraulic Gates

Similar to the static stability portrayed in Figure 22, the static stability of a hydraulic gate can be represented in a simplified diagram shown as Figure 23. Here the gate is represented by a flat, rigid plate. If the plate is inclined in the downstream direction, as in Figure 23a, any increase in the upstream water level will only increase the hydrostatic pressure, with a component acting downward to maintain the gate seal. The gate in Figure 23a is an example of a press-shut structure.

The same gate, when inclined in the upstream direction as in Figure 23b will tend to open when the upstream water level exceeds the depth d_o at which the vertical component of the force due to hydrostatic pressure loading exceeds the weight of the gate structure, or the moment due to the hydrostatic pressure exceeds the restraining moment. Such a gate is a press-open device. Press-open devices are statically unstable.

Figure 23. Static stability of hydraulic gates (a) statically stable (press-shut gate), (b) statically unstable gate (press open gate), (c) statically neutral (vertical flat-bottomed gate with level sill)

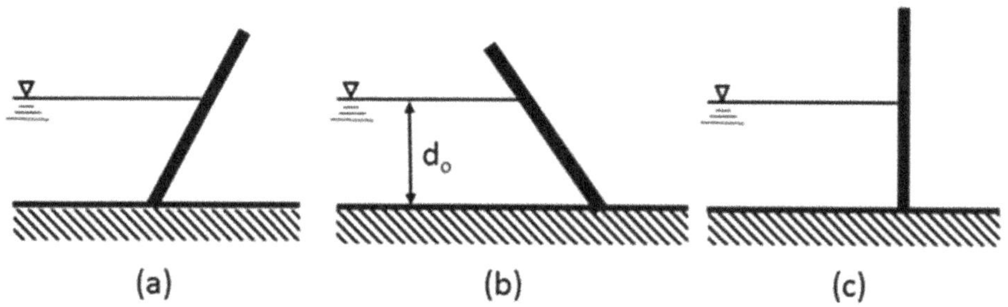

The gate in Figure 23c is neutrally stable provided the sill is horizontal as shown. A slight inclination of the sill can make the gate statically stable or statically unstable depending on whether the angle between the sill and the gate is greater than or less than 90°, respectively.

Dynamic Stability of Hydraulic Gates

The dynamic stability of hydraulic gates is the topic of this current book. It is considerable more involved than static stability. Previously cited guidelines (see Chapter 1) concerning the possibility of dynamic instability, given in Naudascher & Rockwell (1994), are worth repeating:

1. All systems with structural movements accompanied by discharge fluctuations should be carefully examined for movement-induced instability.
2. Press-shut flow control devices (gates, valves, etc.) with flow through small openings and large fluid inertia are susceptible to movement-induced instability.

Certain types of gates are susceptible to other specific mechanisms. Flat-bottomed gates may be susceptible to instability excitation due to shed vortices, as discussed in Chapter 1. Figures 8 and 9 in Chapter 1 show examples of both vertical and streamwise gate excitation resulting from vortex shedding beneath flat-bottomed gates. Tainter gates (radial gates) when operated at small gate openings are susceptible to small-amplitude vibration due to the fluctuating discharge. Tainter gates with trunnion eccentricity are susceptible to eccentricity instability to be presented in Chapter 8. Flap gates with overflow are susceptible to vibrations coupled with nappe instabilities (see Chapter 6).

Another type of dynamic gate instability occurs when the system's conditional stability relies on side-seal friction to maintain small amplitude stability, called *friction-maintained* stability. Such gates may be susceptible to dynamic instability when exposed to a sudden large-amplitude disturbance (Anami et al., 2011). Such a disturbance is termed *hard excitation.* Examples of hard excitation for galloping of prismatic bodies can be found in Novak (1969, 1970, and 1972) and summarized in Naudascher & Rockwell (1994).

The details for assessing the dynamic stability of hydraulic gates are quite complicated, drawing on many aspects of fluid dynamics, wave dynamics, structural dynamics, and mathematics. The subsequent chapters will contribute additional tools for use in the analysis of gate instabilities.

CONCLUSION

In this chapter, we have introduced the reader to aspects of mechanical vibration theory, a brief overview of resonator concepts, and a descriptive discussion of stability concepts. All of these topics are prerequisite to understanding the interaction between fluid flow and vibrations. We refer the reader to the cited references for further theoretical details that will be used as needed in this text, but their derivations are beyond the scope of this book.

Coupling between the rectilinear motion of the mass on a rod and the rotational motion of the rod around the origin is a fundamental concept needed to understand the coupled-mode self-excitation mechanism for Tainter gates. This simple model will be expanded to encompass the modal deformation of the skinplate in the radial direction and the rotation of the gate around the trunnion pin. In particular, the results shown in this chapter for Coulomb damping and hysteretic damping will be used to explain the concept of friction-maintained stability. Gate designers need to understand these dissipative mechanisms in order to design gates that are stable even to a large perturbations.

Resonator concepts will be the basis for much of the discussion of nappe oscillations of overflow weirs, flap gates with simultaneous over- and underflow in Chapter 6.

More details of stability analysis of laminar flows will be presented in Chapter 3 on fluid dynamic concepts. Stability considerations will also be employed in arriving at the theory of self-excited coupled-mode gate vibrations and will also be featured in the comparison of these gate vibrations with coupled-mode flutter of bluff structures exposed to air flow in Chapter 13.

REFERENCES

Anami, K., Ishii, N., Knisely, C. W., Tsuji, T., Oku, T., & Sato, S. (2011). Friction-maintained dynamic stability. In *Vibration Problems ICOVP 2011* (pp. 779–785). Dordrecht, NL: Springer Netherlands. doi:10.1007/978-94-007-2069-5_105

Bažant, Z. P. (2000). Structural stability. *International Journal of Solids and Structures*, *37*(1), 55–67. doi:10.1016/S0020-7683(99)00078-5

Bažant, Z. P., & Cedolin, L. (2010). *Stability of Structures: Elastic, Inelastic, Fracture and Damage Theories*. London, UK: World Scientific.

Bendat, J. S., & Piersol, A.G. (2010). Random Data: Analysis and Measurement Procedure (4th ed.). Hoboken, NJ: John Wiley. doi:10.1142/7828

Blackstock, D. T. (2000). *Fundamentals of Physical Acoustics*. Hoboken, NJ: Wiley-Interscience.

Blake, W. (1986). *Mechanics of Flow-induced Sound and Vibration* (Vol. I and II). Cambridge, MA: Academic Press.

Blevins, R. (1979). *Formulas for Natural Frequency and Mode Shape*. Malabar, FL: Krieger Publishing Company.

Blevins, R. (1990). *Flow-induced Vibration* (2nd ed.). New York, NY: Van Nostrand Reinhold Co.

Butikov, E. I. (2004). Parametric excitation of a linear oscillator. *European Journal of Physics*, *25*(4), 535–554. doi:10.1088/0143-0807/25/4/009

Çengel, Y. A., & Boles, M. A. (2014). *Thermodynamics: An Engineering Approach* (8th ed.). New York, NY: McGraw-Hill.

Crocker, M. J. (1997). *Encyclopedia of Acoustics (4-volume set)*. Hoboken, NJ: Wiley-Interscience.

Crocker, M. J. (1998). *Handbook of Acoustics*. Hoboken, NJ: Wiley-Interscience.

de Graaff, B. J. A. (1998). *Stability Analysis of Vflugter Gate*. Unpublished Master's Thesis, TU Delft, Delft, NL. Retrieved from http://repository.tudelft.nl/islandora/object/uuid:7a5ecb16-defb-44a7-9b39-6e3cdf4131c7?collection=research

Den Hartog, J. P. (1985). *Mechanical Vibrations*. Mineola, NY: Dover.

Dorf, R. C., & Bishop, R. H. (2010). *Modern Control Systems* (12th ed.). Upper Saddle River, NJ: Pearson.

Dowling, A. P., & Ffowcs-Williams, J. E. (1983). *Sound and Sources of Sound*. Hemstead, UK: Ellis Horwood Ltd.

Ewins, D. J. (2000). *Modal Testing: Theory, Practice, and Application* (2nd ed.). Baldock, UK: Research Studies Press.

Fahy, F. J. (2000). *Foundations of Engineering Acoustics*. Cambridge, MA: Academic Press.

Felippa, C. (2015). *Introduction to aerospace structures (ASEN 3112) - Fall 2015, Department of Aerospace Engineering Sciences, University of Colorado at Boulder*. Retrieved from http://www.colorado.edu/engineering/CAS/courses.d/Structures.d/Home.html

Ghomeshi, M., Mortazavi-Dorcheh, S. A., & Falconer, R. (2007). Amplitude of wave formation by vortex shedding in open channels. *Journal of Applied Sciences*, *7*(24), 3927–3934. doi:10.3923/jas.2007.3927.3934

Hardwick, J. D., Kenn, M. J., & Mee, W. T. (1980). Gate vibration at El Chocon hydro-power scheme, Argentina. In E. Naudascher (Ed.), *1979 IUTAM/IAHR Symposium, Karlsruhe, Germany* (pp. 476-497). Heidelberg: Springer. doi:10.1007/978-3-642-81528-7_47

Hughes, L. (1965). Prediction of surges in the southern basin of Lake Michigan. *Monthly Weather Review*, *93*(5), 292–296. doi:10.1175/1520-0493(1965)093<0292:TPOSIT>2.3.CO;2

Inman, D. J. (2008). *Engineering Vibration* (3rd ed.). Englewood Cliffs, NJ: Prentice Hall.

ISGS. (2009). *Seiches: Sudden, large waves a Lake Michigan danger*. Retrieved November 30, 2010, from http://www.isgs.illinois.edu/sections/engin-coast/lakemich-coastal-seiches.shtml

Kanvinde, A. M. (2003). Methods to evaluate the dynamic stability of structures-shake table tests and nonlinear dynamic analyses. *Proceedings of the EERI Meeting, Portland*. Retrieved from http://citeseerx.ist.psu.edu/viewdoc/download?doi=10.1.1.125.417&rep=rep1&type=pdf

Kinsler, L. E., Frey, A. R., Coppens, A. B., & Sanders, J. V. (1999). *Fundamentals of Acoustics* (4th ed.). Hoboken, NJ: Wiley.

Kolkman, P. A., & Jongeling, T. H. G. (2007). *Dynamic Behaviour of Hydraulic Structures: Part A Structures in Flowing Fluid*. Retrieved from http://www.deltares.nl/en/publications?search-expression=kolkman&page=1

Le Méhauté, B. (1976). *An Introduction to Hydrodynamics and Water Waves*. New York, NY: Springer. doi:10.1007/978-3-642-85567-2

Leissa, A. W. (1973). The free vibration of rectangular plates. *Journal of Sound and Vibration*, *31*(3), 257–293. doi:10.1016/S0022-460X(73)80371-2

Meirovitch, L. (1967). *Analytical Methods in Vibrations (Fifteenth printing, 1993)*. New York, NY: Macmillan.

Meirovitch, L. (1986). *Elements of Vibration Analysis* (2nd ed.). New York, NY: McGraw-Hill.

Moran, M. J., & Shapiro, H. N. (2014). *Fundamentals of Engineering Thermodynamics* (8th ed.). Hoboken, NJ: Wiley.

Morse, P. M., & Ingard, K. U. (1987). *Theoretical Acoustics*. Princeton, NJ: Princeton University Press.

Naudascher, E., & Rockwell, D. (1994). *Flow-Induced Vibrations: An Engineering Guide*. Mineola, NY: Dover.

Nise, N. S. (2015). *Control Systems Engineering* (7th ed.). Hoboken, NJ: Wiley.

Novak, M. (1969). Aeroelastic galloping of prismatic bodies. *Journal of the Engineering Mechanics Division*, *95*(1), 115–142.

Novak, M. (1972). Galloping oscillations of prismatic structures. *Journal of the Engineering Mechanics Division*, *98*(1), 27–46.

Palm, W. J. III. (2007). *Mechanical Vibration*. Hoboken, NJ: John Wiley.

Phillips, C. L., & Parr, J. (2010). *Feedback Control Systems* (5th ed.). Upper Saddle River, NJ: Pearson.

Rao, S. S. (2010). *Mechanical Vibrations* (5th ed.). Englewood Cliffs, NJ: Prentice Hall.

Schroeder, M. (2007). *Springer Handbook of Acoustics*. Heidelberg: Springer.

Sorensen, R. M. (2010). *Basic Coastal Engineering* (3rd ed.). Heidelberg: Springer.

Steidel, R. F. (1989). *An Introduction to Mechanical Vibrations* (3rd ed.). Hoboken, NJ: John Wiley.

Stoker, J. J. (1957). *Water Waves: The Mathematical Theory with Applications*. Hoboken, NJ: Interscience.

Strogatz, S. H. (2014). *Nonlinear Dynamics and Chaos: With Applications to Physics, Biology, Chemistry, and Engineering (Studies in Nonlinearity)* (2nd ed.). Boulder, CO: Westview Press.

Thomson, W. T., & Dahleh, M. D. (1998). *Theory of Vibrations with Applications* (5th ed.). Englewood Cliffs, NJ: Prentice Hall.

Thorby, D. (2008). *Structural Dynamics and Vibrations in Practice*. Oxford, UK: Butterworth-Heinemann/Elsevier.

Timoshenko, S. (2008). *Vibration Problems in Engineering* (2nd ed.). Garberville, CA: Wolfenden Press.

Timoshenko, S., Young, D. H., & Weaver, W. (1974). *Vibration Problems in Engineering* (4th ed.). Hoboken, NJ: John Wiley.

Vér, I. L., & Beranek, L. L. (Eds.). (2005). *Noise and Vibration Control Engineering: Principles and Applications* (2nd ed.). Hoboken, NJ: Wiley. doi:10.1002/9780470172568

Vierck, R. K. (1967). *Vibration Analysis*. Scranton, PA: International Textbook.

White, F. M. (2005). *Viscous Fluid Flow* (3rd ed.). New York, NY: McGraw Hill.

Zahariea, D. (2010). Dynamic response of the U-tube liquid manometer with equal diameter columns. *Proceedings of the 25th IAHR Symposium on Hydraulic Machinery and Systems*, Timişoara, Romania. Retrieved from http://iopscience.iop.org/

Zima, L., & Ackerman, N. L. (2002). Wave generation in open channels by vortex shedding from channel obstructions. *Journal of Hydraulic Engineering*, *128*(6), 596–603. doi:10.1061/(ASCE)0733-9429(2002)128:6(596)

Chapter 3
Fluid Dynamics:
Basic Concepts, Gate Discharge, and Flow Stability

ASTRACT

A review of basic fluid dynamics is presented in this chapter. Fluid static loading of hydraulic gates is examined. The focus in the present context will be on one-dimensional, incompressible flow of Newtonian fluids (air and water). Viscous effects will be included as loss coefficients in pressure drop calculations through ducts and channels. Discharge coefficients of hydraulics gates are presented to account for viscous effects in the flow past these gates. More advanced concepts related to the instabilities of boundary layers and free shear layers, and transition to turbulence will be introduced briefly and references provided for further investigation by the interested reader. Readers are encouraged to review additional fluid dynamic concepts using the text with which they are most comfortable.

INTRODUCTION

A review of basic fluid dynamics is presented in this chapter. Fluid *mechanics* deals with the forces exerted by a fluid and the forces needed to set a fluid in motion. Fluid *dynamics*, on the other hand, implies that the fluid is moving; however, fluid statics is taken into account by including a zero value for the fluid velocity. While there are indeed nuance differences, the terms fluid mechanics and fluid dynamics are often used interchangeably. The focus in the present context will be on one-dimensional, incompressible flow of Newtonian fluids (air and water). Viscous effects will be included as loss coefficients in pressure drop calculations through ducts and channels. A short review of potential flow will be included. More advanced concepts related to the instabilities of boundary layers and free shear layers, and transition to turbulence will be introduced briefly and references provided for further investigation by the interested reader.

There is an array of undergraduate textbooks on fluid mechanics, including those by Pritchard (2010), White (2010), Çengel & Cimbala (2009), Munson et al. (2009), and Crowe et al. (2008). This list is not inclusive, so students are encouraged to use the text with which they are most comfortable.

DOI: 10.4018/978-1-5225-3079-4.ch003

FLUID PROPERTIES

The properties of fluids that are relevant to gate vibrations include:

- Fluid density, ρ (mass per unit volume)
- Specific weight, $\gamma = \rho g$ (weight per unit volume)
- Specific gravity, *SG* (ratio of the density of the given fluid to that of water under standard conditions)
- Dynamic viscosity, μ (resistance to flow due to an imposed shear stress)
- Kinematic viscosity, $\mu = \mu / \rho$ (a measure of the fluid's ability to diffuse momentum)

Density, Specific Weight, and SG: For all liquids and for gases flowing at relatively low velocities, the density is assumed constant and the fluid is said to be *incompressible*. For gases, the ideal gas law, $\rho = p / RT$ (with p = absolute pressure, R = the gas constant for each specific gas, and T = absolute temperature), is used to determine the gas density. For gases, the changes in pressure and temperature due to gas compressibility effects can usually be neglected for Mach numbers (see Table 1 for definition) less than about 0.3. Unless otherwise specifically noted, this book assumes fluids are incompressible. Once density is known, the specific weight and specific gravity can be found by multiplying by the acceleration or gravity, or by dividing by the standard density of water (1000 kg/m^3 or 62.4 lbm/ft^3), respectively.

Viscosity: For gases, the dynamic viscosity, μ, increases with increasing temperature; for liquids, dynamic viscosity decreases with increasing temperature. Dynamic viscosity does not vary significantly with pressure. Since the density of gases varies with pressure, however, kinematic viscosity ($\mu = \mu/\rho$) of gases depends on pressure as well as on temperature. The kinematic viscosity of liquids is predominantly a function of temperature alone for a wide range of moderate pressures, up to hundreds to thousands of atmospheres for most liquids (Avallone & Baumeister, 1987).

In the following discussion, and for the most part throughout the text, the working coordinate system is assumed to be a Cartesian system with coordinate directions (x, y, z) and corresponding velocities (u, v, w).

For Newtonian fluids, the dynamic viscosity is the proportionality coefficient between shear stress and strain rate $\frac{du}{dy} + \frac{dv}{dx}$ or, more generally, between the stress tensor, $\hat{\boldsymbol{\tau}}$, and the strain rate tensor, $\nabla \vec{\boldsymbol{V}}$:

$$\tau = \mu \left(\frac{du}{dy} + \frac{dv}{dx} \right) \text{or } \hat{\boldsymbol{\tau}} = \mu \nabla \vec{\boldsymbol{V}} \tag{1}$$

An additional fluid property that is significant in small-scale model studies and at gas-liquid interfaces, is the surface tension σ (the force needed per unit width to maintain an interfacial film of fluid). The surface tension depends on the material of the wetted surface and/or the fluid on the other side of the surface interface. Values can be found in Avallone & Baumeister (1987) and sources cited therein.

Thermodynamic Properties of Fluids

The fluid dynamics properties described above might be considered a subset of a broader group of properties. Thermodynamic properties of interest in this book include

- Pressure
- Temperature
- Specific volume
- Internal energy
- Enthalpy
- Entropy
- Bulk modulus
- Specific heats
- Ratio of specific heats.

In addition, the kinetic energy ($\frac{1}{2}mV^2$, where *m* is the mass and *V* is the fluid velocity) and the potential energy ($mg\Delta z$, where Δz is the elevation above a defined reference level) are also of interest from the energy perspective.

Pressure can be given as gage pressure (when measured relative to atmospheric pressure), or as absolute pressure when atmospheric pressure is included. In English units these pressures are usually denoted as *psig* (pounds force per square inch gage) and *psia* (pounds force per square inch absolute), respectively. In SI the standard pressure unit is *kPa* which might have a comma followed by a *g* or an *a,* if there is any doubt about whether the pressure is gage or absolute. Gage pressure plus atmospheric pressure is the absolute pressure:

$$p_{absolute} = p_{gage} + p_{atm} \tag{2}$$

Temperature also has relative and absolute scales; °C and or °F are both relative to the ice point and the boiling point of water at standard atmospheric pressure. Absolute scales in Kelvin K and Rankine R are related to the relative scales by additive constants as follows:

$$T(K) = T(°C) + 273.15 \tag{3a}$$

$$T(R) = T(°F) + 459.67 \tag{3b}$$

Specific volume *v* is defined as volume per mass, an intensive property, and is the reciprocal of density. The specific internal energy, given by the symbol *e* (on a per unit mass basis), is the sum of microscopic translational, rotational and vibrational kinetic energy of the atoms comprising the material and varies linearly with temperature for incompressible materials of constant phase. In many thermodynamics textbooks, the specific internal energy is given as *u* and the specific enthalpy by *h*. We will use

u as a component of fluid velocity and h as an elevation. Specific enthalpy, given by the symbol i, is the sum of the internal energy, e, and the flow work term (the product of pressure and specific volume, pv).

$$i = e + pv \tag{4a}$$

and

$$\Delta i = \Delta e + p\Delta v + v\Delta p \tag{4b}$$

Entropy, given by the symbol s, is a more abstract property defined in terms of the heat transfer during a quasi-static reversible process. Since entropy cannot be measured directly, it is often determined through application of the *T-ds* equations, which are given as

$$Tds = de + pdv \tag{5a}$$

and

$$Tds = di - vdp \tag{5b}$$

Other properties that will be needed on occasion include the bulk modulus E, a measure of the compressibility of the fluid) and the ratio of specific heats (especially for gas processes). The bulk modulus is defined as either the isothermal bulk modulus E_T or the isentropic bulk modulus E_s as follows:

$$E_T = -v\left(\frac{dp}{dv}\right)_{T=const} \tag{6a}$$

and

$$E_s = -v\left(\frac{dp}{dv}\right)_{s=const} \tag{6b}$$

For liquids, the derivative dp / dv is approximated as a difference ratio $\Delta p / \Delta v$ since a very large pressure is needed to produce a small volume change.

Specific heat is the amount of energy needed to increase a unit mass of a given substance by one degree. The heat addition can be done at either constant volume, with the corresponding value is denoted as c_v, or, at constant pressure, with the value denoted as c_p. The ratio of specific heats is given by $k = c_p / c_v$. The parameter k is sometimes also called the isentropic exponent since for an isentropic process of an ideal gas $pv^k = \text{constant}$. This ratio, k, is a number a bit larger than 1 for gases. For liquids (incompressible substances), k =1.0.

For ideal gases, the ideal gas law provides a relationship between thermodynamics properties, p, v, and T:

$$pv = RT \tag{7}$$

in which R is the gas constant for the gas of interest and is given by $R = R_u / M$ where R_u is the universal gas constant and M is the equivalent molar mass of the gas. The pressure and temperature must be in absolute units.

Dimensional Analysis

With multiple parameters influencing a given process, it is very advantageous to consider the similitude of the process in terms of *dimensionless ratios* of the dimensional parameters. Using methods of dimensional analysis it is possible to reduce the number of parameters needed to describe the process, thereby reducing the number of experiments needed to confirm the behavior of the system during the process.

There are multiple ways of undertaking dimensional analysis from the intuitive trial and error approach to the very structured Buckingham Pi theorem. In the end, all methods can be adapted to produce the usual menu of dimensionless parameters used in fluid dynamic analyses (see introductory fluid mechanics textbooks for details). Table 1 provides a list of the commonly used dimensionless groups and their physical interpretation. As with all dimensionless quantities, these groups can be raised to powers and multiplied or divided by one another to produce an almost endless variety of groups.

STATIC GATE LOADS

When the gate is closed, it must be able to withstand the hydrostatic load. In usual gate design guidelines, static stability is of primary concern.

Hydrostatic Pressure

Taking the positive z-direction as the upwards pointing vertical, the differential pressure in a fluid control volume of $dxdydz$ undergoing accelerations in the negative directions relative to the coordinate axes (i.e., $-a_x$ in the x-direction, and so on), can be represented as

$$dp = -\rho\left[a_x dx + a_y dy + \left(a_z + g\right)dz\right] \tag{8}$$

When the fluid is stationary, the non-gravitational accelerations are all zero, leaving

$$dp = -\rho g dz \tag{9a}$$

or for an assumed constant value of density, after integration

$$\Delta p = -\rho g \Delta z \tag{9b}$$

Table 1. Dimensionless groups used frequently in fluid dynamics and flow-induced vibrations

Dimensionless Parameter	Usual Definition	Physical Interpretation	Applicability
Reynolds Number, Re	$\rho VD / \mu$ or VD / ν	Inertia Force/ Viscous Force	All types of viscous flow phenomena
Froude Number, Fr	$V / \sqrt{gl}$ or V^2 / gl	Inertia Force/ Gravity Force	Free surface flows
Euler Number, Eu Pressure coefficient, c_p	$P / \rho V^2$ or $\Delta P / \rho V^2$	Pressure force/ Inertia force	Flow driven by pressure
Cauchy Number, Ca	$\rho V^2 / E_s$ or $\rho V^2 / E_T$	Inertia Force/ Compression Force	Compressible flows
		E_s is isentropic bulk modulus, and E_T is isothermal bulk modulus)	
Weber Number, We	$\rho V^2 / \sigma$ (σ is surface tension)	Inertia Force/ Surface Tension Force	Free surface flows influenced by surface tension
Mach Number, Ma	V / c (c is speed of sound)	Fluid Velocity/ Speed of Sound	Compressible flows
Strouhal Number, St* or Reduced Frequency, ω_R	$St = fL / V$ (f is frequency) $\omega_R = 2\pi fL / V$	Frequency/ (Convective Frequency)	Periodic flows, fluid structure interaction
		Characteristic Body Length/ Wavelength of Disturbance	
Reduced Velocity, V_R	$V_R = V/Lf = \tau / (L / V)$	Period of Vibration/ Convective Time Scale	Flow-induced vibrations
Scruton Number, *Sc*	$Sc = 4\zeta / m_r$ or $Sc = 4\pi\zeta / m_r$	Damping Ratio/ Mass Ratio	Flow-induced vibrations
Force Coefficient, C_F	$F / \left(\frac{1}{2} \rho V^2 A\right)$	Force/Dynamic Force	Hydrodynamics and aerodynamics
Lift Coefficient, C_L	$F_L / \left(\frac{1}{2} \rho V^2 A\right)$		
Drag Coefficient, C_D	$F_D / \left(\frac{1}{2} \rho V^2 A\right)$		
Discharge Coefficient, C_d	$Q_{actual} / Q_{theoretical}$	Correction for non-ideal behavior	Permits use of ideal fluid calculations with non-ideal flows

*Some prefer to use *St* only for vortex shedding and use reduced velocity as a dimensionless frequency parameter. Readers should carefully read the context of usage.

Equations 9a and 9b are frequently called the hydrostatic law. These equations are used in calculating the hydrostatic load on structures, such as gates submerged in water.

As an example consider a plane gate surface that is *L* wide and *H* high and is submerged in water at an angle θ relative to the vertical direction, as shown in Figure 1. The right side of the gate is exposed to atmospheric pressure, which acts to balance the atmospheric pressure component on the left side, so we only need to calculate the loading due to the hydrostatic pressure (or use the gage pressure). The gate height *H* is not the controlling dimension. The area loaded by the water is determined by the water depth *h* and the inclination angle, provided that the height $H > h / \cos\theta$. For an increment *-dz* in the vertical direction, the length along the gate surface is $dz / \cos\theta$. The net force is then the integral of the gage pressure over the differential area $dA = Ldz / \cos\theta$:

$$F = \int_0^h p_{gage}\left(z\right) L\left(\frac{-dz}{\cos\theta}\right) = \frac{L}{\cos\theta}\int_0^h \rho gz dz = \frac{\rho g h^2 L}{2\cos\theta} \tag{10}$$

The point of application of the equivalent concentrated force can be found by equating the moment of the distributed force about the point where $z = 0$ on the gate and the moment of the equivalent concentrated force at some distance along the surface of the gate s_F:

$$s_F = \frac{\int_0^h \left(\rho g z^2 / \cos^2\theta\right) L dz}{F} = \frac{2h}{3\cos\theta} \tag{11}$$

Alternatively, the point of application of the equivalent concentrated force could be found by the intersection of a horizontal line through the centroid of the loading diagram with the gate surface.

The analysis method would be similar for curved gate surfaces, but the integration would become more complicated. Gates with water above the surface of the gate must include the vertical component of the water loading. Most gates with water above the gate will not be loaded hydrostatically since the water would usually be flowing over the gate. However, a gate with a flap mounted on the top could certainly have a hydrostatic load due to the water on the top surface of the main gate.

Figure 1. Hydrostatic pressure acting on an inclined gate surface

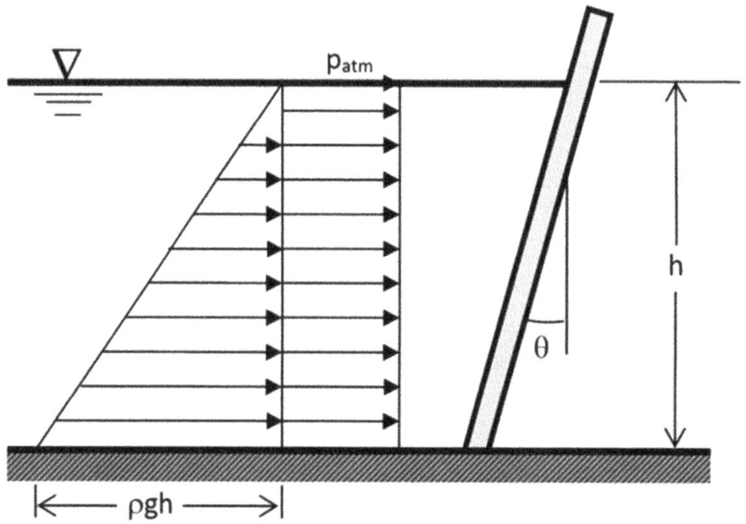

Buoyancy

Another hydrostatic force component is the force due to buoyancy when the structure displaces water. For example, long-span shell-type gates usually require consideration of buoyancy because of the hollow air space inside these types of gates, as shown in Figure 2. The vertical component of the force obtained by the rigorous integration of pressure over the wetted surface as in Equation 9 would provide the buoyant force for an object floating on the surface. Alternatively, we could use Archimedes' principle that tells us the buoyant force is equal to the weight of the displaced liquid. Figure 2a shows a simplified long-span shell-type gate cross-section where it is assumed that the gate is not a self-flooding type of gate and that it has been lowered into a static reservoir. The gate shell is assumed to be water-tight. Archimedes' law tells us the buoyant force is equal to the weight of the displaced water. If the weight of the gate is not greater than the buoyant force, the gate will not close. In the case of the simplified gate shown in Figure 2, the volume per unit length of the gate can be approximated as the sum of the rectangular area plus the triangular area. The buoyant force is then

$$F_B = \rho g L \left(W H_1 + W H_2 / 2 \right) \tag{12}$$

More realistically, in Figure 2b the same gate is now installed in a river with a reservoir only on the upstream (left) side of the gate. The hydrostatic forces now supply a net moment that will try to tip the gate about the gate lip. The vertical face has a horizontal force of magnitude $F_H = \rho g H_1^2 L / 2$ applied $2H_1 / 3$ below the reservoir surface. The inclined lower surface experiences a force normal to the surface which has a vertical component due to buoyancy as well as the horizontal hydrostatic pressure. All force components impart a clockwise overturning moment about the gate lip. A force and moment balance with the inclusion of the support structure forces and resisting moments would be necessary to determine if the forces were sufficient to overturn the gate.

Figure 2. (a) Simplified closed long-span shell-type gate in a static reservoir; (b) a closed long-span shell-type gate with a reservoir only on the left side

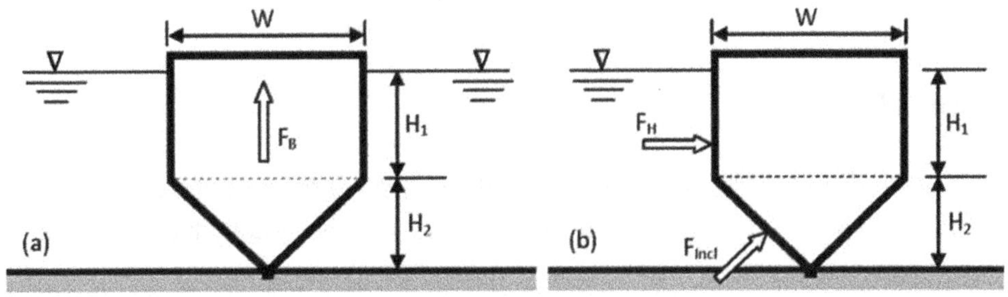

CONSERVED PROPERTIES

In contrast to static systems, systems that involve fluid flow require calculations that are based on three conservation laws: conservation of mass, conservation of energy, and conservation of momentum. These calculations might be used in an engineering design analysis to provide the following information:

- Optimum conduit sizing
- Flow rate for specified heads
- Pressure drops for specified flow rates
- Restraining loads on structural supports
- Fluid heat loss or heat gain through the conduit
- Energy input rates to pumps, fans, or compressors
- Energy output rates from turbines

Conservation of Mass

In classical mechanics, mass must be conserved. This principle can be stated in words as "the mass flow rate into the system minus the mass flow rate out of the system is the rate of accumulation of mass in the system." In a steady flow system (no accumulation of mass), conservation of mass (also called continuity) requires that the mass flow rate entering the system be the same as the mass flow rate leaving the system.

For a simple single-inlet (subscript 1), single-outlet (subscript 2) system with incompressible fluid in steady flow, mass flow rate is given by

$$\dot{m}_1 = \rho A_1 V_1 = \rho A_2 V_2 = \dot{m}_2 \tag{13a}$$

For a differential volume $dxdydz$ in an incompressible fluid, conservation of mass requires that

$$\frac{\partial u}{\partial x} + \frac{\partial v}{\partial y} + \frac{\partial w}{\partial z} = 0 \tag{13b}$$

where *u, v,* and *w* are the velocity components in the *x, y,* and *z* directions, respectively.

For an incompressible fluid with constant density, Equation 13a yields the following relationship between velocities and areas:

$$V_2 / V_1 = A_1 / A_2 \tag{14}$$

Conservation of Energy

In any system, the First Law of Thermodynamics tells us that energy must be conserved. The energy present at the start of a process plus the net inflow of energy into the system during the process must be the final energy of the system at the end of the process. For steady flow in a simple system with a single inlet and a single outlet the First Law can be written as

$$q - w = \Delta i + \Delta ke + \Delta pe \tag{15}$$

where q is the heat transfer per unit mass (assumed positive into the system), w is the work done by the system (work output is assumed to be positive), Δi is the change in enthalpy per unit mass, Δke is the change in kinetic energy per unit mass, and Δpe is the change in potential energy per unit mass.

Application of Equation 15 to the isothermal (T = constant) flow of an incompressible ($v = 1 / \rho = \text{constant}$) fluid through a pipe yields the Bernoulli equation, in several forms:

$$v\Delta p + \Delta ke + \Delta pe = 0 \tag{16a}$$

which can be re-written as

$$p_1 / \rho + ½V_1^2 + gz_1 = p_2 / \rho + ½V_2^2 + gz_2 \tag{16b}$$

or, after dividing each term by the acceleration of gravity,

$$p_1 / \rho g + ½V_1^2 / g + z_1 = p_2 / \rho g + ½V_2^2 / g + z_2 \tag{16c}$$

Note that all terms in Equation 16c are head terms and have units of length. The first term on each side is the pressure head, the second term is the velocity head and the last term is the elevation head. Implicit in the assumption of isothermal flow of an incompressible fluid is that the flow must be isentropic. As can be seen from Equation 5a, if $de = 0$ (constant temperature) and $dv = 0$ (incompressible), then ds must also be zero, and therefore the flow is isentropic. For this reason q must also be zero.

In a real flow, the isothermal assumption is only approximately correct as the frictional dissipation of energy will result in a very small temperature increase. The method of dealing with these frictional effects is to represent the dissipated energy as "lost" energy. In a flow system, each pipe fitting and each valve as well as the wall friction loss can be represented by loss coefficients that when summed and multiplied by the velocity head yield the total head loss. These losses are then, in effect, included in the energy balance that is represented by a modified form of the Bernoulli equation to be developed in the next section of text.

Internal Flow

To calculate the friction loss in a pipe or a duct, the friction factor must be found. To determine the friction factor, the Reynolds number and the surface roughness (assuming the flow is turbulent) must be known. The procedure for finding the friction factor can be found in all undergraduate fluids textbooks. For the purpose of the following discussion, we assume that the reader can find the friction factor for a given flow using the Moody diagram, the iterative Colebrook equation, or the less precise, explicit Haaland equation. The Colebrook equation requires an initial guess for the friction factor and then one or more subsequent iterations to check the convergence.

$$\frac{1}{\sqrt{f}} = -2\log_{10}\left(\frac{\varepsilon / D_h}{3.7} + \frac{2.51}{\mathrm{Re}\sqrt{f}}\right)\text{(Colebrook)} \tag{17a}$$

$$\frac{1}{\sqrt{f}} = -1.8\log_{10}\left(\frac{\varepsilon / D_h}{3.7} + \frac{6.9}{\mathrm{Re}}\right)\text{(Haaland)} \tag{17b}$$

In Equations 17a and 17b, ε is the pipe wall roughness, D_h is the hydraulic diameter defined as 4 times duct area/duct perimeter so that for a circular pipe $D_h = D$. *Re* is the Reynolds number, a dimensionless grouping formed from velocity, *V*, kinematic viscosity, ν, and hydraulic diameter, such that $Re = VD_h / \nu$.

When the explicit Haaland equation, Equation 17b, is used to obtain a first estimate of *f* which is subsequently used in the right-hand side of the Colebrook equation, Equation 17a, the need for further iteration is usually eliminated.

If the friction factor for a pipe segment of length L_j and diameter D_j is f_j, then the friction head loss due to fluid flow through the pipe segment with a mean velocity of V_j is given as

$$h_{L,j} = f_j\left(L_j / D_j\right)\left(V_j^2 / 2g\right) = f_j\left(L_j / D_j\right)\left(A_1 / A_j\right)^2\left(V_1^2 / 2g\right) \tag{18}$$

where V_1 is the mean inlet velocity to the piping system where the pipe area is A_1. The loss coefficients for fittings and valves in the pipe are tabulated in fluid dynamics textbooks and handbooks, usually in terms of a loss coefficient *K*. The additional head losses due to the fittings and valves, frequently called the minor losses, are given as

$$\sum\left(K_i V_i^2 / 2g\right) = \sum\left\{K_i\left(A_1 / A_i\right)^2\right\}\left(V_1^2 / 2g\right)$$

where the summation is the sum of the losses for all the fittings and valves. Accounting for the head losses, the modified Bernoulli equation for a pipe with multiple segments of differing diameters with various fittings is obtained:

$$\frac{p_1}{g} + \frac{\alpha_1 V_1^2}{2g} + z_1 = \frac{p_2}{g} + \frac{\alpha_2 V_2^2}{2g} + z_2 + \left\{\Sigma\left[f_j\left(L_j / D_j\right)\left(\frac{A_1}{A_j}\right)^2\right] + \Sigma\left[K_i\left(\frac{A_1}{A_i}\right)^2\right]\right\}\left(\frac{V_1^2}{2g}\right) \tag{19}$$

The last two terms in Equation 19 account for the head loss due to wall friction from section 1 to section 2 and the head loss due to all fittings, respectively. The sum of these terms is called the friction head, $h_{friction}$.

The two coefficients α_1 and α_2 in Equation 19 are kinetic energy correction factors accounting for the difference in kinetic energy flux in the actual velocity profile compared to that of the assumed uniform profile at inlet (state 1) and exit (state 2). For an incompressible fluid

$$\alpha = \frac{1}{A}\int\left(\frac{V(r)}{V_m}\right)^3 dA = \frac{1}{R}\int_0^R\left(\frac{V(r)}{V_m}\right)^3 rdr \tag{20}$$

where V_m is the mean velocity obtained from $V_m = \dot{m}/\rho A$. For parabolic laminar flow profiles α = 2.0 and for turbulent flow profiles α ranges from 1.03 to 1.3. This correction factor is often neglected in pipe flow calculations with turbulent flow. Later will we introduce a momentum correction factor β. The calculation of α and β for open channel flow through an irregular stream bed is discussed in several texts (Chow, 1959; French, 1986; Chanson, 2004; and Thandaveswara, 2010; among others).

If there is energy input by a pump or energy extraction by a turbine, Equation 18 can be modified to include the pump head input and the turbine head output as follows:

$$\frac{p_1}{g} + \frac{\alpha_1 V_1^2}{2g} + z_1 + h_{pump} = \frac{p_2}{g} + \frac{\alpha_2 V_2^2}{2g} + z_2 + h_{friction} + h_{turbine} \tag{21}$$

For the simplified case of a constant diameter horizontal pipe with turbulent flow at both inlet and outlet (so α can be ignored), no pump or turbine and no fittings or valves, Equation 21 reduces to the Darcy-Weisbach equation:

$$\frac{\Delta p}{\rho g} = \frac{p_1 - p_2}{\rho g} = f\frac{L}{D}\frac{V_m^2}{2g} \tag{22}$$

Open Channel Flow

For open channel flow, the hydraulic radius is used as the characteristic length scale in calculating the Reynolds number. The hydraulic radius, R_h, is defined as the flow area divided by the wetted perimeter. Using the hydraulic radius as the length dimension limits laminar flow to Re < 500. In a uniform channel flow, the water surface is inclined with the same slope as the channel bed. The inclined flow requires that the component of gravitational acceleration in the flow direction to be balanced by the frictional resistance in the channel, so that the average water velocity remains constant. The bed slope is given by $S_b = \sin(\theta_b)$ where θ_b is the angle between the bed and the horizontal direction, as shown in Figure 3.

Chezy (1775) was the first to propose a formulation for the velocity of a uniform open channel flow with a given hydraulic radius (R_h) and bed slope (S_b):

$$V = C\sqrt{R_h S_b} \tag{23}$$

Figure 3. Uniform open channel flow

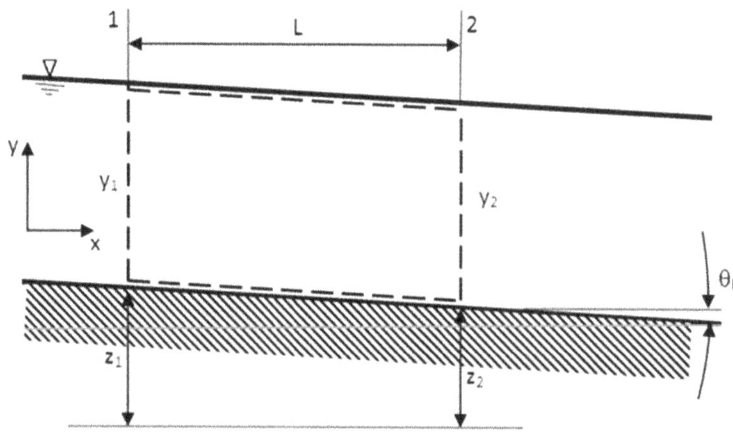

where C is called the Chezy factor, with tabulated empirical values in textbooks and handbooks. The value of C varies with the hydraulic radius (R_h) and the bed surface roughness, and ranges from about 22 to 220.

More commonly, the later formulation by Manning (1890) is used to calculate the flow velocity given the depth, the channel shape and the bed slope:

$$V = \frac{1.49}{n} R_h^{2/3} S_b^{1/2} \; (R_h \text{ in feet}) \tag{24a}$$

$$V = \frac{1}{n} R_h^{2/3} S_b^{1/2} \; (R_h \text{ in meters}) \tag{24b}$$

in which n is called Manning's n, with tabulated empirical values depending upon bed material and its surface finish, and ranging usually between 0.01 to 0.06.

Alternatively, Equation 22, the Darcy-Weisbach equation, can also be applied to open channel flow with the hydraulic radius used as the length scale, to yield:

$$V = \sqrt{\frac{8gR_h S_b}{f}} \tag{25}$$

where f is the Darcy-Weisbach friction factor. The Moody diagram, the Haaland equation or the Colebrook equation can be used to determine f by replacing the pipe diameter D with $4R_h$ and the relative roughness ε/D with as $\varepsilon/4R_h$. The relationship between Chezy's constant, Manning's n and the friction factor f can be found by dividing the left sides of Equations 23, 24 and 25 by $\sqrt{R_h S_b}$:

$$\frac{V}{\sqrt{R_h S_b}} = C = \frac{1.49}{n} R_h^{1/6} = \sqrt{\frac{8g}{f}} \tag{26}$$

While f is a non-dimensional quantity, the values of C and n must obviously be dimensional from the form of Equation 26. The tabulated values of C and n (dependent on viscosity which is a function of temperature) are not generally recognized as functions of the water temperature. This usually negligibly small temperature dependence over the temperature range of naturally flowing water is included in the viscosity values used in the determination of f. Caution is advised when using pipe flow roughness data for open channel flow.

Torricelli Equation

Another application of the Bernoulli equation, Equation 16b, is the calculation of the discharge velocity from a tank. Initially, the flow will be assumed ideal and subsequently a correction factor will be included. Figure 4 shows a long tank with a sharp edged circular orifice. The tank is assumed to be sufficiently large or have sufficient in-flow to maintain the free surface at a fixed level.

Consider points 1 and 2, shown in Figure 4, both of which are at atmospheric pressure. With the assumption that $V_1 = 0$, Equation 16b reduces to

$$V_2 = \sqrt{2g\Delta h} \tag{27a}$$

and is known as the Torricelli equation.

Real fluids will continue to contract after exiting the nozzle. The minimum cross-sectional area in the jet issuing from the tank is called the *vena contracta*. To account for this contraction as well as very small viscous effects, a discharge coefficient, written as C_d, is introduced into the equation for volumetric discharge Q from the tank as

Figure 4. Discharge from a tank

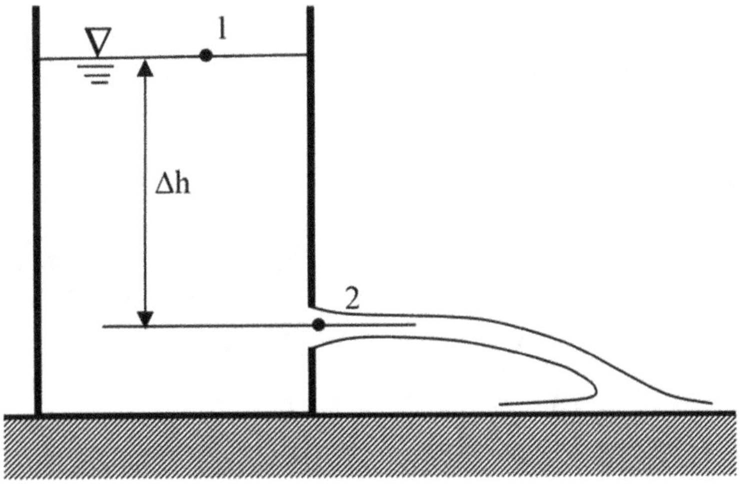

$$Q = A_2 V_2 = A_2 C_d \sqrt{2g\Delta h} \tag{27b}$$

The exact value of the discharge coefficient depends on the detailed geometry of the discharge orifice; however, in the absence of other information, a value of about 0.6 to 0.62 would be a reasonable approximation.

Overflow Weirs

Using an analysis similar to that used for the derivation of the Torricelli equation, the discharge from an overflow weir notch of arbitrary geometry can be analyzed. Figure 5 shows a side and front view of water flowing through a weir notch with a width, *w(h)*, that varies with the distance beneath the upstream free surface. The incremental area *wdh* can be treated as an orifice.

Equation 27b when applied to this area yields a differential volumetric discharge given by

$$dQ = V\left(h\right) dA = C_d w dh \sqrt{2gh} \tag{28a}$$

Integrating Equation 28a using the specified *w(h)* function from the free surface to the bottom of the weir notch provides an expression for the weir discharge:

$$Q = \int_0^H dQ = C_{d,avg} \sqrt{2g} \int_0^H w\sqrt{h} dh \tag{28b}$$

For a rectangular weir notch, w = constant = W, permitting the analytical integration of Equation 28b to yield the following expression for the volumetric discharge for a sharp-edged rectangular notched weir:

$$Q_{Rect} = \int_0^H dQ = C_{d,avg} W \sqrt{2g} \int_0^H \sqrt{h} dh = \frac{2}{3} C_d W \sqrt{2gH^3} \tag{28c}$$

Figure 5. Overflow discharge through a weir notch of arbitrary width, side and front view

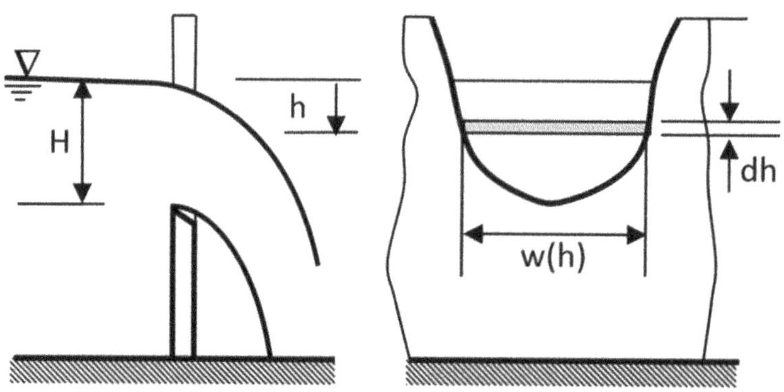

The discharge coefficient depends on the upstream channel depth beneath the bottom of the notch and the cross-stream weir width relative to the channel width. Bos (1989) provides a comprehensive overview of discharge measurements using weirs and orifices. For fully contracted, sharp-edged rectangular weirs, his recommended equation for discharge is

$$Q = C_{de}\frac{2}{3}\sqrt{2g}W_e H_e^{1.5} \tag{29}$$

where C_{de} is the effective discharge coefficient, W_e is the effective transverse weir dimension, and H_e is the effective head above the weir. The simplicity of this equation is somewhat deceiving since the values for C_{de} must be found either graphically or computationally as a function of the upstream channel geometry and the weir dimensions relative to the channel dimensions.

A fully contracted weir is one in which the walls and bed of the bounding upstream channel are sufficiently far from the weir that the nappe characteristics are not affected by the upstream channel boundaries. For a fully contracted weir, certain conditions must be met. Figure 6 shows the relevant geometry and notation, while Table 2 provides a list of conditions for a weir installation to be considered fully contracted.

The effective weir width is the physical weir width plus an empirical correction factor, that is, $W_e = W_w + K_w$. The correction factor K_w is given in Figure 7 as a function of the ratio of weir width to channel width. The effective head is $H_e = H_1 + 0.001$ m. The value of C_{de} is determined using the empirical coefficients listed in Table 3 as functions of the ratio W_w / W_1 in the following equation:

Figure 6. Fully-contracted sharp-edged rectangular notch weir
After Bos, 1989.

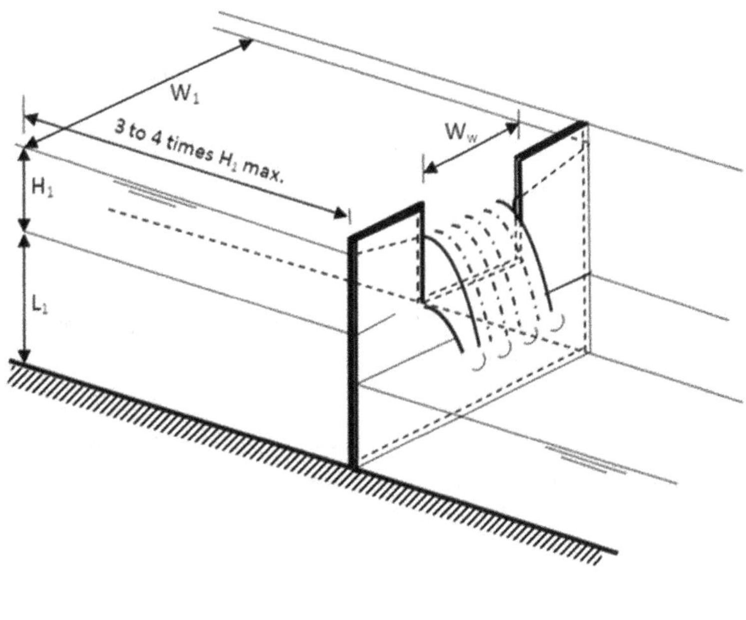

Table 2. Conditions for fully-contracted sharp-crested rectangular weir

Weir and channel width relationship	$W_1 - W_w \geq 4H_1$
Head to depth	$H_1 / L_1 \leq 0.5$
Head	$0.07\text{m} \leq H_1 < 0.60\text{m}$
Weir width	$W_w \geq 0.3\text{m}$
Depth beneath weir	$L_1 \geq 0.30\text{m}$

Figure 7. Additive correction factor (in m) for effective weir width for fully-contracted sharp-edged rectangular weir
After Bos, 1989.

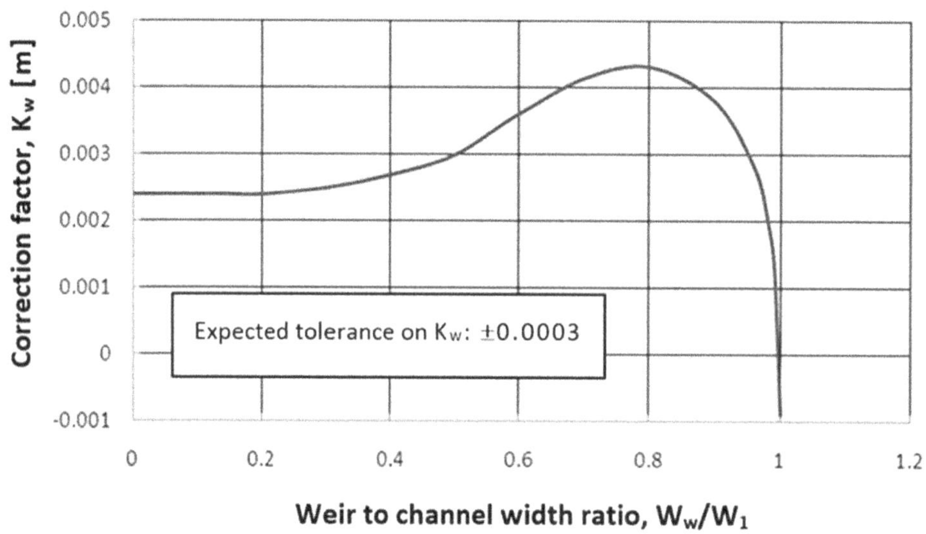

$$C_{de} = C_1 + C_2 \frac{H_1}{L_1} \tag{30}$$

The corrections to the weir width and the channel head are intended to account for viscous and surface tension effects, according to Bos (1989). For all rectangular weir types, Bos (1989) recommends straight upstream rectangular channel geometry, but also notes that when the upstream channel cross-sectional area is greater than ten times the weir notch area (H_1W_w), the upstream approach velocity will be negligibly small and, if the weir is fully contracted, the upstream channel geometry is unimportant.

Bos (1989) also presents a review of V-notched weirs. Only 90° V-notched weirs will be considered in the present chapter.

Table 3. Coefficents for use in discharge coefficient calculation with Equation 30

W_w/W_1	C_1	C_2
0	0.587	-0.0023
0.1	0.588	-0.0021
0.2	0.589	-0.0018
0.3	0.590	0.0020
0.4	0.591	0.0058
0.5	0.592	0.0110
0.6	0.593	0.0180
0.7	0.595	0.0300
0.8	0.597	0.0450
0.9	0.599	0.0640
1.0	0.602	0.0750

V-notched weirs are classified as partially contracted or fully contracted, according to the conditions presented in Table 4. Using a similar formulation to that for rectangular weirs, the discharge from a V-notch weir with a notch angle of θ is given by

$$Q = C_{de} \frac{8}{15} \sqrt{2g} \tan \frac{\theta}{2} H_e^{2.5} \tag{31}$$

Here H_e is the effective head and is given by $H_e = H_1 + K_H$. For fully contracted 90° V-notch sharp-edged weirs, when the ratios $H_1 / L_1 \leq 0.4$ and $H_1 / W_1 \leq 0.2$, the value for K_H is 0.8 mm and the value for C_{de} is 0.578. For other notch angles and flow conditions, see Bos (1989) for graphs to determine the appropriate values of K_H and C_{de}.

Table 4. Conditions for partially-contracted and fully-contracted V-notched weirs from Bos (1989)

Parameter	Partially Contracted	Fully Contracted
Head to depth ratio	$H_1 / L_1 \leq 1.2$	$H_1 / L_1 \leq 0.4$
Head to breadth ratio	$H_1 / W_1 \leq 0.4$	$H_1 / W_1 \leq 0.2$
Head	$0.05\text{m} \leq H_1 \leq 0.60\text{m}$	$0.05\text{m} \leq H_1 \leq 0.38\text{m}$
Upstream depth beneath weir	$L_1 \geq 0.1\text{m}$	$L_1 \geq 0.45\text{m}$
Upstream channel breadth	$B_1 \geq 0.60\text{m}$	$B_1 \geq 0.90\text{m}$

Conservation of Momentum

The momentum equation for an incompressible fluid flow can be derived in integral form or in differential form. Basic fluids textbooks show that the two formulations are equivalent. In integral form, the momentum equation is

$$\Sigma \vec{F} = \frac{\partial}{\partial t} \iiint \vec{V} \rho d\forall + \iint \vec{V} \rho \vec{V} \cdot d\vec{A} \tag{32a}$$

where the forces on the left hand side are the forces exerted on the boundaries of the fluid. The symbol $\vec{V}$ is the vector velocity and the symbol $\forall$ is used to represent the volume over which the integration is carried out.

The first integral is over the entire control volume and represents the time rate of change of vector momentum in the control volume. The second integral is over the surface of the control volume and represents the net efflux of momentum. In steady flow, the first integral is zero. The steady flow components of Equation 32a in the *x*- and *y*-directions in a system with a single inlet and a single outlet are

$$\Sigma F_x = \dot{m}\left(\beta_2 V_{2x} - \beta_1 V_{1x}\right) \tag{32b}$$

$$\Sigma F_y = \dot{m}\left(\beta_2 V_{2y} - \beta_1 V_{1y}\right) \tag{32c}$$

where the momentum correction factor (also called the Bossinesq coefficient), β, accounts for the difference in the integral of the square of velocity relative to the square of the integral of velocity, and is defined as

$$\beta = \frac{1}{A} \iint \left(v / V_{avg}\right)^2 dA \tag{32d}$$

For a uniform velocity profile, the value of β is 1.0.

The forces acting on the control volume include both surface forces and body forces. For a one-dimensional flow of an incompressible, inviscid fluid subject only to pressure and gravitational forces, Equation 32 reduces to Equation 16, the Bernoulli equation.

The momentum equation for a differential control volume $dxdydz$ in an incompressible Newtonian fluid can be written as

$$\frac{d\vec{V}}{dt} + \left(\vec{V} \cdot \nabla\right)\vec{V} = \frac{-1}{\rho} \nabla p + \frac{\mu}{\rho} \nabla^2 \vec{V} \tag{33a}$$

or as separate components in the x, y and z directions:

$$\frac{\partial u}{\partial t}+u\frac{\partial u}{\partial x}+v\frac{\partial u}{\partial y}+w\frac{\partial u}{\partial z}=\frac{-1}{\rho}\frac{\partial p}{\partial x}+\frac{\mu}{\rho}\left(\frac{\partial^2 u}{\partial x^2}+\frac{\partial^2 u}{\partial y^2}+\frac{\partial^2 u}{\partial z^2}\right) \tag{33b}$$

$$\frac{\partial v}{\partial t}+u\frac{\partial v}{\partial x}+v\frac{\partial v}{\partial y}+w\frac{\partial v}{\partial z}=\frac{-1}{\rho}\frac{\partial p}{\partial y}+\frac{\mu}{\rho}\left(\frac{\partial^2 v}{\partial x^2}+\frac{\partial^2 v}{\partial y^2}+\frac{\partial^2 v}{\partial z^2}\right) \tag{33c}$$

$$\frac{\partial w}{\partial t}+u\frac{\partial w}{\partial x}+v\frac{\partial w}{\partial y}+w\frac{\partial w}{\partial z}=\frac{-1}{\rho}\frac{\partial p}{\partial z}+\frac{\mu}{\rho}\left(\frac{\partial^2 w}{\partial x^2}+\frac{\partial^2 w}{\partial y^2}+\frac{\partial^2 w}{\partial z^2}\right) \tag{33d}$$

where again u, v, and w are the x, y and z velocity components. In hydraulics, Equations 33a-d (the Navier-Stokes equations) are seldom applied directly, but serve as a theoretical basis for boundary layer analysis and are used in computational analysis of hydraulic structures.

Prandtl (1904) showed how the Navier-Stokes equations can be simplified for two-dimensional flow near a solid surface, in which case the pressure is a function of only the x-coordinate and is impressed from the inviscid freestream. Prandtl's steady flow boundary layer momentum equation for an incompressible fluid is

$$u\frac{\partial u}{\partial x}+v\frac{\partial u}{\partial y}=\frac{-1}{\rho}\frac{dp}{dx}+\frac{\mu}{\rho}\left(\frac{\partial^2 u}{\partial y^2}\right) \tag{34}$$

The pressure term is a known function of the streamwise distance, so the two unknowns are u and v. The continuity equation provides a second equation for u and v.

When the boundary layer equation is solved simultaneously with the continuity equation, the streamwise velocity component u and the normal velocity component v can be determined as functions of the similarity parameter that includes the distance downstream and the normal distance from the wall. Further details of boundary layer theory can be found in Schlichting & Gersten (2000), White (2006), Panton (2005), and many other intermediate and advanced level fluid dynamics textbooks.

GATE DISCHARGE COEFFICIENTS

To compute forces acting on gates, it is necessary to know the discharge past the gate. The flow past hydraulic gates with underflow is in many respects similar to the flow in orifices and weirs just considered. However, gates designed for overflow and those with simultaneous under- and overflow must be treated a bit differently. In this portion of the chapter, we will consider these general types of gates separately.

Discharge Beneath a Gate

Careful examination of that portion of Figure 4 that is above the horizontal line passing through point 2 is reminiscent of the flow pattern beneath a gate. A modified form of the Torricelli equation can also be applied to flow beneath a gate by adjusting the discharge coefficient C_d. Gate discharge can be calculated for a given flow geometry provided the correct discharge coefficient can be calculated or found from empirical data. The general form of the equation of calculating discharge is

$$q = Q / b = C_d a\sqrt{2gy_1} \tag{35}$$

in which q is the discharge per unit width, Q is the discharge, b is the channel width, a is the gate opening height, g is the gravitational acceleration, and y_1 is the upstream water head.

Kolkman (1994) argues for the use of the contraction coefficient, C_c, instead of the discharge coefficient, C_d. Novak et al. (2001) and Naudascher (1991) note that the value of C_c depends upon the Froude number, so the discharge must be known before determining C_c. Ultimately, if precision is required, an iterative determination of either C_c or C_d can be undertaken. For free discharge, these two coefficients are related:

$$C_d = \frac{C_c}{\sqrt{1 + C_c y / y_0}} \tag{36}$$

Details of discharge coefficient variation for various gate geometries with underflow, as well as gates with overflow (to be discussed subsequently), can be found in many sources (Rajaratnam & Subramanya, 1967; Miller, 1994; Naudascher, 1991; Naudascher & Rockwell, 1994; Lewin, 1995; Corsi & Schuler,1995 ; Replogle et al.,2004, Wahl, 2005; Sepúlveda et al., 2009; Belaud et al., 2009; Lozano et al., 2009; Cassan & Belaud, 2011; Habibzadeh et al., 2011; Oshima et al. (2013); Shayana & Farhoudi, 2013, among others). Example correlations of discharge coefficients or contraction coefficients will be presented. More precise tabulation of the discharge coefficients for specific gate types, can be found in the reference cited in the preceding sentence.

Discharge Beneath a Vertical Lift Gate

The vertical sluice gate discharge coefficients were tabulated by Henry (1950) and modified and update by several people since then. Figure 8 permits a rough estimation of C_d for a given vertical sharp-edged gate in terms of upstream head, downstream water depth, and gate opening. For discharge from high head gates in tunnels see Naudascher (1991) and sources cited therein.

Discharge Beneath a Tainter Gate

For gates with underflow in open channels, the downstream outflow may be free discharge or submerged discharge. Gentilini (1941) and Toch (1953) provided discharge characteristics for Tainter gates, which have been subsequently substantiated by additional testing by others (see Cozzo, 1978, for example).

Figure 8. Discharge coefficient for a vertical sluice gate
After Novak et al. (2001), original data from Henry (1950).

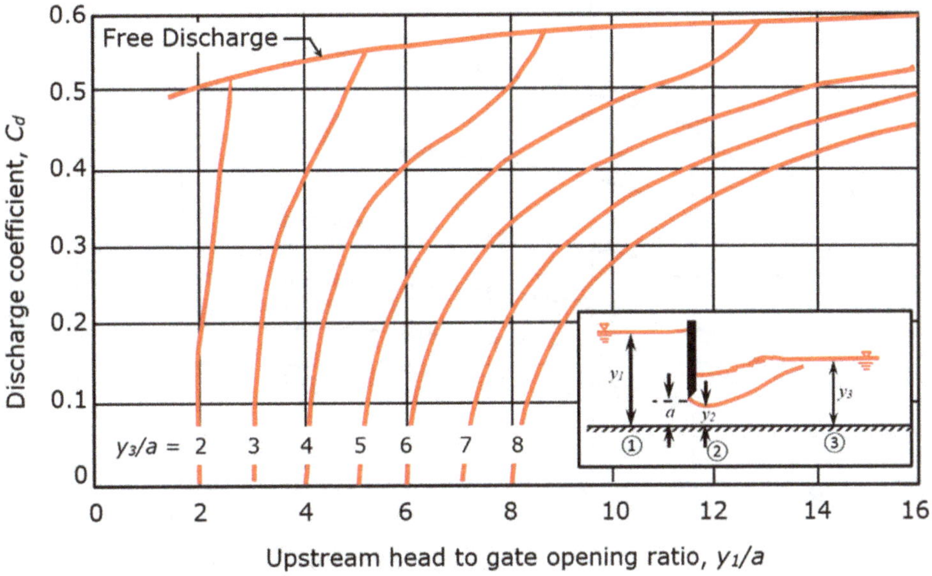

Tabulated empirical discharge coefficient values from Gentilini (1941) and Cozzo (1978) for inclined sluice gates and Tainter gates with free discharge are tabulated in Figure 9. The two sets of curve fits to discharge coefficient data, one set from Gentilini (1941) and the other from Cozzo (1978), were taken from Naudascher (1991) and Kolkman (1994), respectively. Typically, Tainter gates with free outflow usually have discharge coefficient between 0.5 and 0.7.

For Tainter gates or other gates with inclined surfaces (for C_c to within ±5%), Toch (1953) proposed the following equation for the contraction coefficient for free discharge beneath the gate with a flat apron.

$$C_c = 1 - 0.75\left(\frac{\theta^\circ}{90}\right) + 0.36\left(\frac{\theta^\circ}{90}\right)^2 \tag{37}$$

in which the angle q is defined as the angle between the tangent to the gate lip and the tangent to the gate apron.

For Tainter gates on a nappe-shaped spillway, a polynomial curve can be fit to the USACE (1977) data provided in Kolkman (1994) and Lewin (1995), for gates located between 0.1 and 0.3 design heads downstream of the gate crest. The discharge coefficient is a function of the angle, θ between the spillway crest tangent and the skinplate lip tangent:

$$C_d = 0.2025\left(\frac{\theta^\circ}{90}\right)^2 - 0.2376\left(\frac{\theta^\circ}{90}\right) + 0.7404 \tag{38}$$

For gates located at the spillway crest, the discharge coefficient correlation is

Figure 9. Discharge coefficient values for free discharge from Tainter gates
Solid lines represent data from Cozzo (1978) and dashed lines represent data from Gentilini (1941).

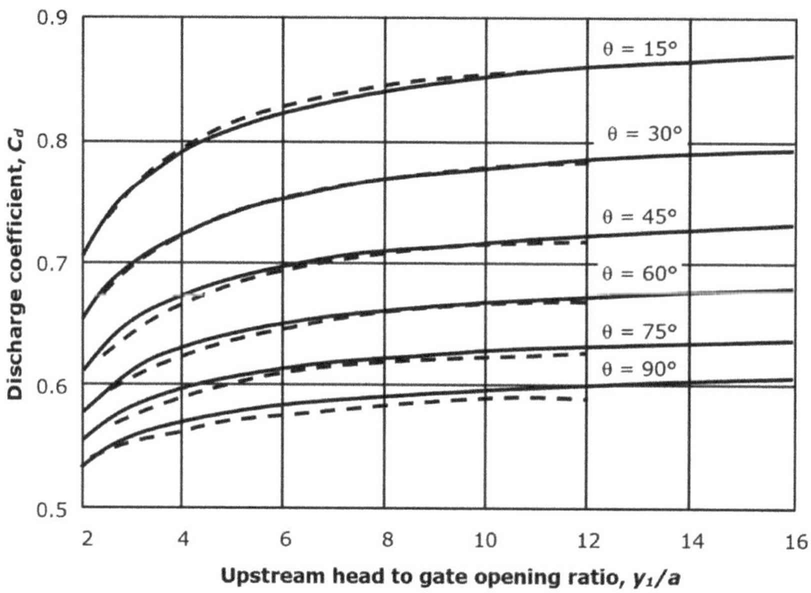

$$C_d = 0.13365\left(\frac{\theta^\circ}{90}\right)^2 - 0.1278\left(\frac{\theta^\circ}{90}\right) + 0.6789 \tag{39}$$

In both cases, the discharge can be calculated using Equation 35, where a is the gate opening normal to the spillway surface and y_1 is the upstream head on the gate.

Naudascher (1991), citing original work from Pulina & Voigt (1981), notes that for reverse Tainter gates with planar skinplates and curved sills beneath the gate followed by a depressed stilling basin, the discharge capacity increases relative to that for a planar discharge sluiceway. The contraction coefficient becomes essentially independent of the skinplate angle with the sill. The discharge coefficient data found in Naudascher (1991) for a reverse Tainter gate with a planar skinplate and a sill along with a depressed stilling basin can be represented with the following equation within about ±3.5%:

$$C_d = 0.538 + 0.125\left\{\frac{(y_0 - w)}{a}\right\}^{0.33} \tag{40}$$

where y_0 is the upstream energy head, w is crest height above the upstream bed, and a is the gate opening above the sill crest.

Discharge coefficients for submerged Tainter gates can be found in Chow (1973), Naudascher (1991), Kolkman (1994), Lewin (1995), Clemmens (2003), and sources cited in these references. Submerged discharge coefficients tend to range from 0.3 to 0.6, but also depend upon details of the seal, the radius of the Tainter gate skinplate, the height of the trunnion pin relative to downstream sluiceway, the gate pier geometry upstream of the gate, the geometry of the sill, and the position of the gate relative to the spillway crest.

Discharge Over a Gate or Weir

For overflow gates, the discharge is usually calculated in terms of the upstream elevation above the gate level. Two possible formulations for flow over a vertical leaf gate are

$$q = \frac{Q}{b} = C_d \frac{2}{3}\sqrt{2g h^{3/2}} = C_d^* \frac{2}{3}\sqrt{2g H^{3/2}} \tag{41}$$

in which h is the upstream head relative to overflow elevation and H is the total head relative to overflow elevation. The value of the discharge coefficient is typically between 0.6 and 1.1, but may be larger or smaller depending on geometric details. The aeration of the overflow nappe affects the values of C_d and C_d*.

For a fully-aerated vertical thin plate weir or gate, Hager (1994) lists seven commonly used correlations. One of the most commonly accepted is that from Rehbock (1929):

$$C_d = 0.4023\left[1 + \frac{0.0011}{h}\right]^{3/2}\left[1 + 0.125\left(\frac{h}{w}\right)\right] \tag{42}$$

in which h is the upstream head relative to overflow elevation and w is the weir or gate height. The correlation is valid when 0.03 m $\le h \le$ 0.75 m, $w >$ 0.30 m, $h/w < 1$, and the channel width $b >$ 0.30 m.

Another frequently used correlation for a fully-aerated vertical thin plate weir or gate is the correlation from Kindsvater & Carter (1959):

$$C_d = 0.401\left[1 + \frac{0.0009}{h}\right]^{3/2}\left[1 + 0.135\left(\frac{h}{w}\right)\right] \tag{43}$$

A third option, cited by Hager (1994) as being based on "new and reliable data," is from the Hydraulic Research Station:

$$C_d = 0.3988\left[1 + \frac{0.001}{h}\right]^{3/2}\left[1 + 0.150\left(\frac{h}{w}\right)\right] \tag{44}$$

which is valid for $h >$ 0.02 m, $w \ge$ 0.15 m, $h/w \le 2.2$, and the location upstream of the weir for measurement of head $L_h <$ 3.0 w.

For flap gates, Castro-Delgado (1984) undertook an extensive study, reviewed previous data, undertook both experimental and computational studies, and provided design guidelines. His presentation of previous work from Langer (1952) and Bazin (1892) is shown in Figure 10. Additional details from

Figure 10. Discharge coefficient for an inclined planar flap gate in a flat-bottomed channel with free overflow

Data points and dashed line from Langer (1952); solid line with scatter range represents data from Bazin (1892); after Castro-Delgado (1984).

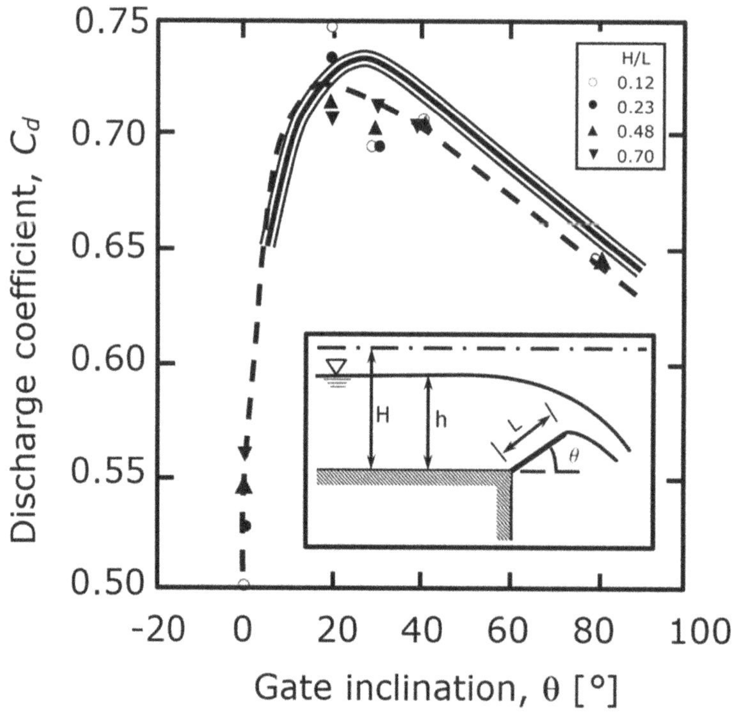

Castro-Delgado's work can be found in Castro-Delgado et al. (1983), Castro-Delgado & Celik (1984), and Naudascher (1991).

Combined Over- and Underflow Discharge

Negm et al. (2002) concluded that formulating simultaneous over- and underflow with a single discharge coefficient is not reasonable. Their attempted formulation gave unacceptably large errors. Instead, they formulated simultaneous discharge in terms of ratios of geometric parameters, *H/d*, *h/b* and *y/d* as shown in Figure 11.

The following equation proposed by Negm et al. (2002) is within ±4% of their measured data for simultaneous over- and underflow.

$$\frac{q_t}{\sqrt{2g}d^{1.5}} = -0.3863 + 0.8764\left(\frac{H}{d}\right) - 0.1494\left(\frac{h}{b}\right) - 0.7341\left(\frac{y}{d}\right) \tag{45}$$

Equation 45 is valid for $0.47 \leq y/d \leq 4.0$; $0.65 \leq b/d \leq 5.0$; $1.8 \leq H/d \leq 7.5$ for bed slopes from 0 to 1.61%.

Figure 11. Nomenclature used by Negm et al. (2002) for developing discharge formulations for simultaneous over- and underflow

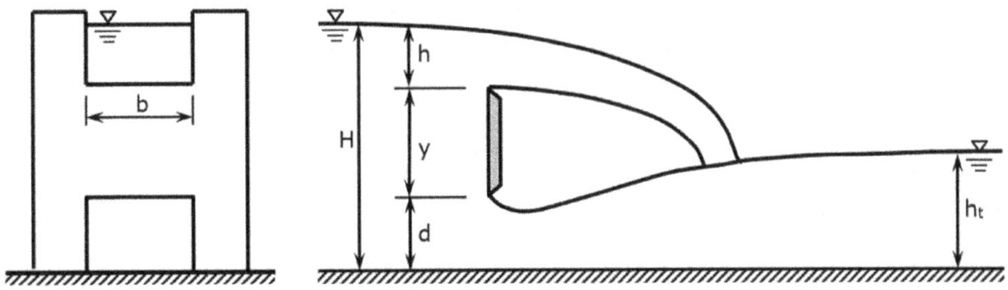

The same authors also cite their previous work (Negm et al., 2000) where they combined the overflow weir discharge coefficient with the gate discharge coefficient for different overflow and underflow geometrical shapes (triangular and rectangular), incorporating an interaction function that depends on whether the discharge is free or submerged:

$$Q_p = F\left(\frac{2}{3}C_{dw}b_e\sqrt{2g}h_e^{1.5} + C_{dg}db\sqrt{2g(H-h_d)}\right) \tag{46}$$

where the weir discharge coefficient, C_{dw}, is from the Kindsvater & Carter (1959) correlation, Equation 43. The gate discharge coefficient is from Rajaratnam & Subramanya (1967). The parameters b_e and h_e are the effective weir width and the effective head on the weir. The value of h_d is given as the product of the contraction coefficient C_c and the gate opening *d*. In the case of submerged discharge, the value of h_d is the depth of the supercritical flow just downstream of the gate. The interaction function *F* is given as

$$F = 0.833 + 0.0306\left(\frac{H}{d}\right)\text{(free discharge)} \tag{47}$$

or

$$F = 1.0486 + 0.00084\left(\frac{H}{d}\right)\text{(submerged discharge)} \tag{48}$$

Equations 47 and 48 are valid for $1.5 < H/d < 6.0$ and $2.4 < H/d < 7.5$, respectively.

The values for the interaction function *F* may be assumed to be average values of $F = 0.933$ for free discharge or $F = 1.052$ for submerged discharge with the predicted discharge still within 5% of the measured values.

Recently, the discharge calculation from Rajaratnam & Subramanya (1967) for flow under sluice gates has been updated to include losses (Habibzadeh et al., 2011).

POTENTIAL FLOW

For flows with high Reynolds numbers, it is often possible to obtain reasonable estimates for flow velocities and pressures by assuming an infinite Reynolds number, which would correspond to a vanishing value of viscosity. In the limit of an infinite Reynolds number, for a two-dimensional flow of an incompressible fluid, the Navier-Stokes equations (Equations 4.33a-d) reduce to the Euler equations

$$\frac{\partial u}{\partial t}+u\frac{\partial u}{\partial x}+v\frac{\partial u}{\partial y}=\frac{-1}{\rho}\frac{\partial p}{\partial x} \tag{49a}$$

$$\frac{\partial v}{\partial t}+u\frac{\partial v}{\partial x}+v\frac{\partial v}{\partial y}=\frac{-1}{\rho}\frac{\partial p}{\partial y} \tag{49b}$$

It is possible to formulate a stream function that will provide streamlines when it is set equal to a constant. The velocity components can be defined in terms of the stream function, ψ, as

$$u=\frac{\partial \psi}{\partial y} \text{ and } v=\frac{-\partial \psi}{\partial x} \tag{50}$$

which when substituted for *u* and *v* will satisfy the continuity equation.

For flow of an inviscid fluid, the flow must be irrotational, which requires $\frac{\partial u}{\partial y}-\frac{\partial v}{\partial x}=0$, or in terms of the stream function

$$\frac{\partial}{\partial y}\left(\frac{\partial \psi}{\partial y}\right)-\frac{\partial}{\partial x}\left(-\frac{\partial \psi}{\partial x}\right)=\nabla^2\psi=0 \tag{51}$$

The stream function for an inviscid fluid must thus satisfy the LaPlace equation, as shown in Equation 51.

Fluid flow is driven by a potential that often is either an elevation or a pressure head. A potential function can be defined in terms of its relationship to the velocity components:

$$u=\frac{\partial \phi}{\partial x} \text{ and } v=\frac{\partial \phi}{\partial y} \tag{52}$$

Substituting these velocity components into the continuity equation yields

$$\frac{\partial}{\partial x}\left(\frac{\partial \phi}{\partial x}\right)+\frac{\partial}{\partial y}\left(\frac{\partial \phi}{\partial y}\right)=\nabla^2\phi=0 \tag{53}$$

Thus, the potential function must also satisfy the LaPlace equation, as shown in Equation 53.

With these definitions, streamlines and the equipotential lines form a grid of orthogonal curvilinear lines, often called a flow net. As an example of a flow net, consider the flow beneath a sluice gate shown in Figure 12. From the spacing between streamlines, it is possible to calculate the local velocity. Knowing the velocity, the pressure can be calculated from the Bernoulli equation. With known pressures on the gate surface, the total hydrodynamic force on the gate can be calculated. If viscous forces are significant, the potential flow pressure distribution along the surface provides the external pressure gradient for calculating the boundary layer near the surface of the structure.

Boundary layer corrections are often necessary to extrapolate accurately results from small-scale model studies to full-scale prototypes. For further information on the scaling of C_d with Reynolds number, see Varschney (1977) and Kobus (1985).

Prior to the advent of digital computing, the curvilinear squares shown in Figure 12 were commonly drawn at a drawing board with pencil and eraser. The engineer would draw a first guess at the placement of the streamlines and the equipotential lines and then check by measuring the diagonals of the curvilinear squares. Any square for which the diagonals were not equal would be revised and a new iteration of the flow net would be generated. The process was repeated until each curvilinear square possessed two diagonals of equal length. Today streamlines and potential lines are most commonly generated using either finite difference or finite element computer programs.

Simple analytical potential flow solutions include uniform flow, source and sink flow, point vortex, and a doublet. Due to the linearity of the LaPlace equation these simple solutions can be combined to produce other solutions such as the Rankine half-body, the Rankine oval, flow past a cylinder, flow past a rotating cylinder and so on.

Further, these fundamental solutions can be transformed in the complex plane to produce additional solutions such as the Joukowski airfoil, flow over a plate and many additional flows. Numerical methods permit potential flow calculations for essentially all practical geometries of interest in hydraulics. For an extensive presentation of the calculation of steady hydrodynamic forces on hydraulic structures, we refer the reader to Naudascher (1991), Erbisti (2004), and Lewin (1995).

Figure 12. Potential flow net for flow beneath a gate
After Rouse, 1946.

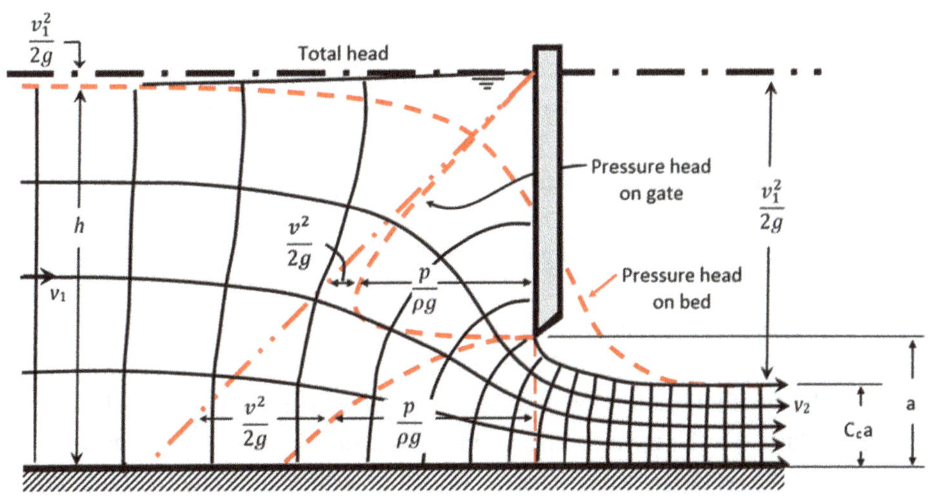

In aerodynamics, frequent use is made of potential line vortices, which can be thought of as a continuous distribution of a vortex over a spatial extent. An important vortex theorem is that a vortex line cannot begin or end in a fluid domain.

A further useful result from potential flow is known as the Biot-Savart Law of Induction, which has its origins in electro-magnetism theory. A vortex filament of length $d\vec{s}$ of strength Γ will induce a velocity at a distance $\vec{r}$ that is given by

$$d\vec{V} = \frac{\Gamma}{4\pi} \frac{d\vec{s} \times \vec{r}}{\left|\vec{r}^{\,3}\right|} \tag{54}$$

This induction plays an essential role in the selective amplification of relatively broadband shear layer instability to produce a sharp spectral peak when shear layer vortices impinge on downstream surfaces and undergo a sudden change in their circulation, as will be discussed in the next section on hydrodynamic instabilities and transition to turbulence.

Potential flow will be used subsequently in Chapters 4, 5, and 7-10 to develop calculation methods for added mass and added fluid damping due to wave radiation. The added mass and added damping parameters are essential to a fundamental understanding of hydraulic gate instabilities.

HYDRODNAMIC INSTABILITIES AND TRANSITION TO TURBULENCE

Understanding the transition of fluid flow from a laminar to a turbulent state is one of the last unsolved problems in classical physics. Even though we now have methods of calculating fluid instabilities, we have no complete theory capable of predicting the entire transition process. Fluid instability excitation can act on various parts of hydraulic gates, and different gates can respond differently to various types of fluid instabilities.

Shear Layer Instability and Flow Periodicity

First we will consider the fundamental case of viscous shear flow instability, such as that found in a boundary layer. Then we will look at the inviscid instability of a single free shear layer and the interactions between closely spaced shear layers such as those found in planar jets. After discussing shear layer instabilities, we will describe the generation of another type of instability, the formation of the Kármán vortex street by the shedding of vortices in the wake of a bluff body and . Finally, we will touch on additional fluid dynamic sources of periodicity, specifically, the interfacial instability of a liquid sheet in an air environment, the switching of the flow between two bi-stable states, and the separation and reattachment of a shear layer near the trailing edge of a body.

Analysis of Shear Layer Instability

To determine the stability of a viscous flow with a known two-dimensional mean flow solution, [U(y), W(y)], the flow is perturbed by an infinitesimally small disturbance at an assumed frequency and wavelength. The disturbance can grow, remain steady, or decay. Growth of the disturbance is interpreted as instability, decay as stability; when the disturbance amplitude remains constant the flow is said to be neutrally stable. The locus of points of neutral stability forms a boundary between regions of stability and regions of instability, called the neutral stability curve.

To perform an instability analysis, introduce harmonic disturbances into the Navier-Stokes equations, in terms of travelling waves with amplitudes that can vary transverse to the flow (in the *y*-direction) and that propagate at an angle ϕ relative to the *x*-direction, given as follows:

$$\left(\hat{u}, \hat{v}, \hat{w}, \hat{p}\right) = \left[u\left(y\right), v\left(y\right), w\left(y\right), p\left(y\right)\right] \exp\left[i\alpha(x\cos\phi + z\sin\phi - ct\right] \tag{55}$$

After introducing these disturbances, the two new functions so created are denoted as the *x*- and *z*-components. They can be combined to create a one-dimensional flow that is in the same direction as ϕ.

Let

$$U_0 = U\cos\phi + W\sin\phi \text{ and } u_0 = u\cos\phi + w\sin\phi \tag{56}$$

Here the mean flow is denoted by upper case symbols and the fluctuating flow by the corresponding lower case symbols. If the mean flow in the *z*-direction is zero ($W = 0$), the most unstable flow is when $\phi = 0$ since $U\cos\phi$ is never greater than *U*. This conclusion is consistent with Squire's theorem of instability that tells us the most unstable waves are those that are in the same direction as the mean flow.

Algebraic manipulation the continuity equation and the *x*- and *y*-momentum equations, and subsequent linearization (removal of product terms of the fluctuating quantities), permits derivation of the Orr-Sommerfeld equation:

$$\left(U - c\right)\left(v'' - \alpha^2 v\right) - U''v + \frac{iv}{\alpha}\left(v'''' - 2\alpha^2 v'' + \alpha^4 v\right) = 0 \tag{57}$$

where primes denote differentiation with respect to *y*.

The boundary conditions for Equation 57 must agree with the known mean profile for which the stability analysis is being done.

For free shear layers (jets, wakes, and mixing layers):

$$v\left(\pm\infty\right) = 0 \text{ and } v'\left(\pm\infty\right) = 0 \tag{58a}$$

For boundary layers:

$$v\left(0\right) = v'\left(0\right) = 0 \text{ and } v\left(\infty\right) = v'\left(\infty\right) = 0 \tag{58b}$$

Schematics of several common shear flows are shown in Figure 13: a mixing layer with a discontinuous change in velocity (Figure 13a); a mixing layer with linear shear (Figure 13b); a mixing layer with a hyperbolic tangent distribution (Figure 13c); a jet with a hyperbolic secant squared profile (Figure 13d); a wake with a Gaussian wake deficit (Figure 13e); and the laminar Blasius boundary layer profile (Figure 13f).

The solutions of Equation 57 are assumed to be of the form of a traveling wave growing exponentially in time with a complex wave speed ($c = c_r + ic_i$ and $\omega = \alpha c_r$)

$$\boldsymbol{u}(x,y,t) = \boldsymbol{u}(y)\exp\left[i\alpha\left(x-\left(c_r + ic_i\right)t\right)\right] \tag{58c}$$

or, growing exponentially in space with a complex wavenumber ($\alpha = \alpha_r + i\alpha_i$):

$$\boldsymbol{u}(x,y,t) = \boldsymbol{u}(y)\exp\left\{i\left[\left(\alpha_r + i\alpha_i\right)x - \omega t\right]\right\} \tag{58d}$$

The two assumed solutions are not equivalent except if $c_i \ll c_r$ or $\alpha_i \ll \alpha_r$, in which case

$$\left(\alpha_i\right)_{\text{spatial}} \cong \left(\frac{\alpha c_i}{c_r}\right)_{\text{temporal}} \tag{59a}$$

Figure 13. Schematics of shear layers profiles for which stability analyses have been undertaken: (a) discontinuous mixing layer; (b) linear mixing layer; (c) hyperbolic tangent mixing layer; (d) planar jet with secant-squared profile; (e) planar wake with a Gaussian wake deficit; and (f) laminar Blasius boundary layer profile
After Blake, 1986a.

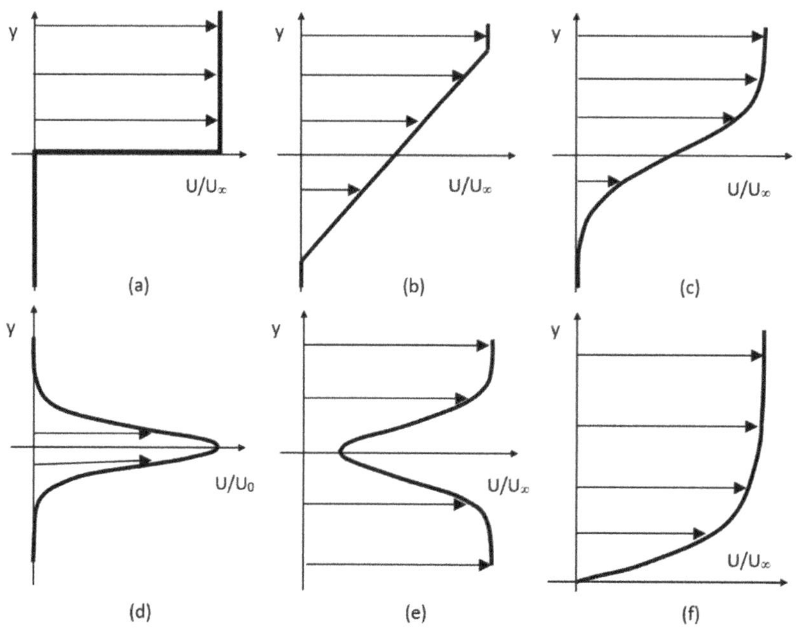

$$\left(\alpha_r\right)_{\text{spatial}} \cong \left(\alpha\right)_{\text{temporal}} \tag{59b}$$

By applying the appropriate boundary conditions, as given in Equations 58a and 58b, the resulting solution is given in terms of eigenfunctions that depend on the Reynolds number (*Re*), the wavenumber (α), and the wave speed (*c*).

Assuming either spatial $\left(\alpha = \alpha_r + i\alpha_i\right)$ or temporal $\left(c = c_r + ic_i\right)$ growth or decay of the perturbation, a neutral stability curve can be found where $\alpha_i = 0$ for spatial neutral stability or $c_i = 0$ for temporal neutral stability. These curves, when plotted on nondimensional wavenumber versus Reynolds number axes, divide the plane into regions of stability and regions of instability. Figure 14 shows temporal neutral stability curves for mean velocity profiles with inflection points (jets, wakes, and mixing layers) and those with no inflection point (boundary layers). The neutral stability lines separate the region of instability (inside the curve) from the region of stability (outside the curve). The rate at which a disturbance grows depends on the wavenumber of the disturbance. Figure 15 presents the growth rate for disturbances for each of the profiles shown in Figure 13. Note that the free shear layers with inflection points in their mean velocity profiles grow substantially faster than the wall-bounded flows with no inflection point. This difference in growth rates tells us that when laminar separation occurs, the resulting free shear layer will most likely become unstable at Reynolds numbers much lower than for the wall bounded flow from which it separates. In flow past bluff bodies, laminar separation may lead to a laminar separation bubble in which the separated laminar shear layer undergoes transition to turbulence, resulting in turbulent reattachment of the shear layer, possibly followed by a subsequent turbulent separation.

Also note that the planar jet profile (d, in Figure 13) has both symmetric and antisymmetric instability modes.

Figure 14. Temporal neutral stability curve schematic diagram for flows with and without inflection points in their mean velocity
After White, 2006.

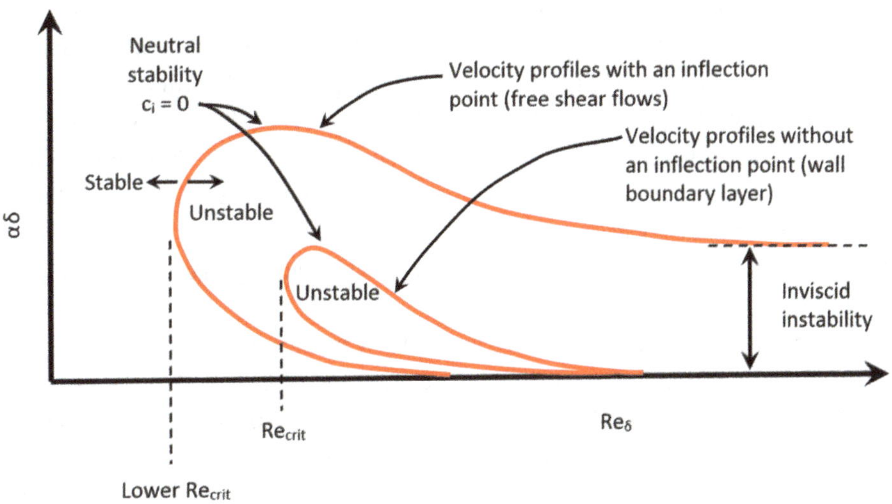

Figure 15. Spatial disturbance growth rates for the profiles shown in Figure 13
After Blake, 1986a.

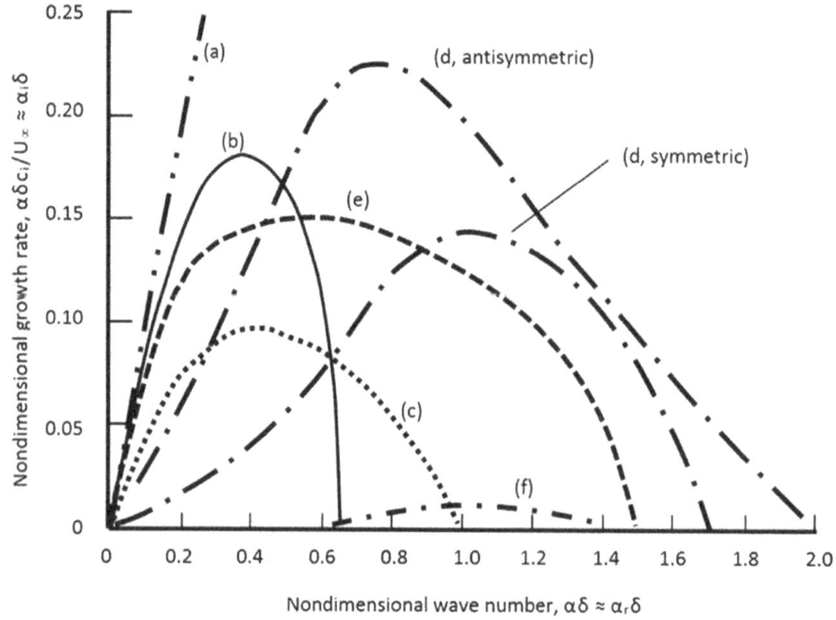

It is important to recall that the Orr-Sommerfeld equation is limited to small amplitude disturbances. Once instability commences, nonlinear effects must be included in any subsequent analysis of the process as the instability progresses towards turbulence.

For wall-bounded flow without inflection point in the velocity profile, once instability is initiated, Tollmien-Schlichting waves appear, followed by the onset and growth of turbulent spots, and ultimately the complete transition of the boundary layer to a fully developed turbulent flow. Further details on transition of boundary layers can be found in Schlichting & Gersten (2000), Blake (1986b), White (2006), Tennekes & Lumley (1972), Bradshaw (1971), Betchov & Criminale (1967), and others.

For free shear flows, the flow can become unstable for a range of frequencies with a relatively broad distribution of growth rates, as shown in Figure 15. This understanding of shear layer instability helps explain the impinging shear layer instability presented briefly in Figure 5 in Chapter 1, which is reproduced in this chapter as Figure 16 for convenience. Any periodic dynamic interaction of the shear layer vortices with a surrounding surface can provide Biot-Savart feedback to focus the instability at a particular frequency in this range. The two shear layers shown in Figure 16 have the same conditions at separation. The spectral velocity fluctuations along the edge of the shear layer with the impingement edge show a focusing of the energy at a single sharp-peaked frequency whereas the fluctuations at the same location in the shear layer without impingement show a broad haystack of frequencies at substantially lower amplitudes.

The planar mixing layer is a simple geometry, but similar instabilities occur in all separated laminar shear layers. The subsequent impingement of a separated shear layer can sharply focus the frequency content and amplify the disturbance. If the natural frequencies of solid structures bounding the flow are within the range of the fluid instability frequencies, the fluctuating pressure can produce structural vibration. The structural vibration can then further enhance the fluctuating fluid dynamic disturbance. This is an essential feature of fluid dynamic excitation of flow-induced vibration.

The two shear layers of a planar jet experience a similar instability. The growth rates of both the symmetric and antisymmetric modes, as shown in Figure 15, are higher and extend over a wider range frequencies relative to that for the planar mixing layer. A jet impinging on an edge can emit a strong tonal sound, known as the edge tone, as the vortices impinge on a downstream edge. If the edge vibrates, the additional feedback leads to potential failure if permitted to continue unchecked. Studies on the edge tone have been carried out by Nyborg (1954), Powell (1961), Karamcheti et al. (1969), Stegen & Karamcheti (1970), Rockwell (1972), Wooley & Karamcheti (1974), and others. Much of the available literature on edge tones, hole tones (a round jet impinging on an axisymmetric impingement surface), and that on jet-edge interactions is summarized in Blake (1986a) and in Naudascher & Rockwell (1994).

There are several noteworthy characteristics of impinging shear layer instability. The region of the shear layer just after separation is very susceptible to disturbances. When this susceptible region is perturbed by the upstream influence from the vortex impingement, as shown in the mechanism schematic in Figure 17(a), the vortex development of the entire shear layer becomes more organized. The shear layer vortices have a natural range of frequencies, but the tonal feedback signal focuses the oscillation at a particular frequency. The frequency of oscillation is known to depend on the upstream conditions of the flow prior to separation (boundary layer parameters, acceleration history, and angle of separation), as well as on the mean freestream velocity, and the impingement length. When the shear layer oscillation frequency data is plotted as a function of impingement length with a fixed freestream velocity, or as a function of

Figure 16. Feedback due to vortex impingement enhances instability and produces a sharp spectral peak
From Rockwell & Knisely, 1979.

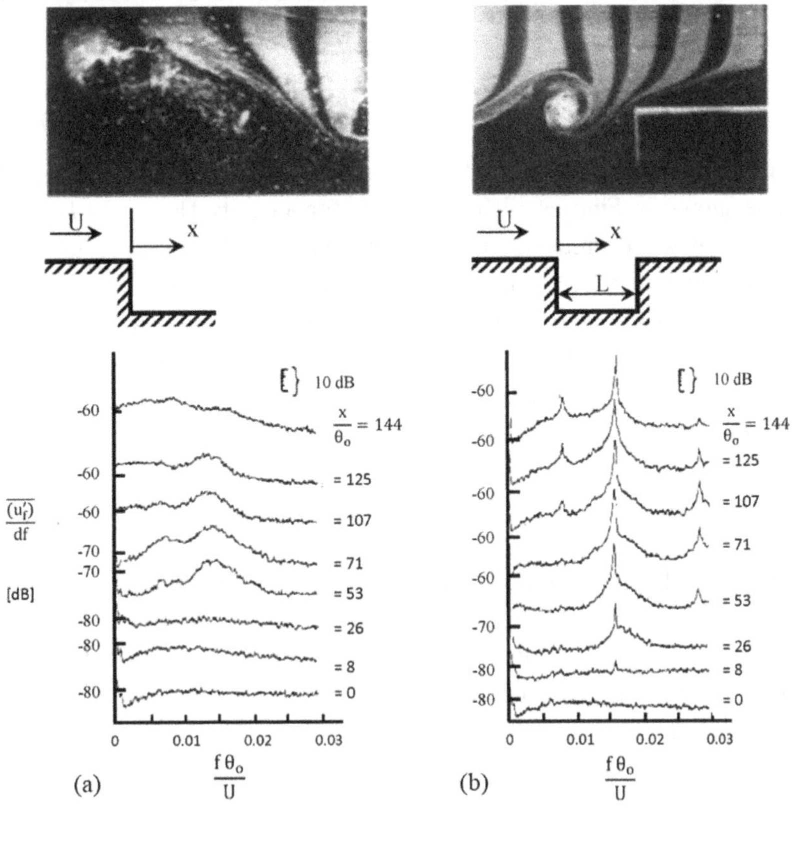

Figure 17. (a) Physical mechanism of impinging shear layer instability with frequency variation with (b) varying freestream velocity with fixed impingement length and (c) varying impingement length with fixed freestream velocity
From Knisely & Rockwell, 1982.

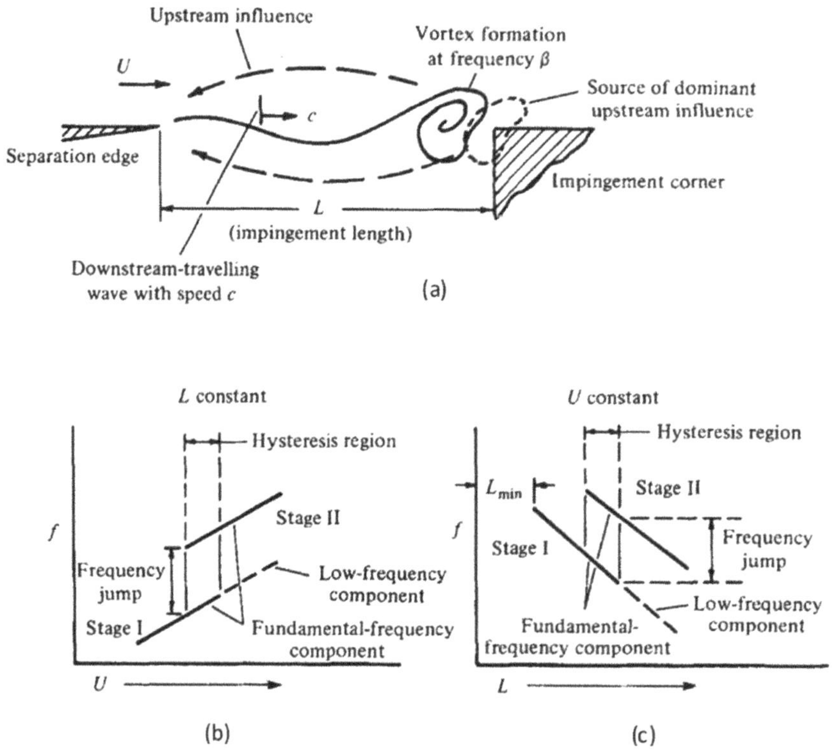

velocity for a fixed impingement length, the data tend to lie on straight lines with discontinuous jumps between the lines, as shown in Figures 17(b) and 17(c) for a cavity flow. The frequency data exhibit hysteresis with increasing and decreasing values of the length and velocity. The discontinuous jumps can be physically interpreted as the flow adjusting through the addition or deletion of a wavelength of the oscillation to the shear layer. The frequency jumps are accompanied by characteristic changes in the amplitude of pressure and velocity fluctuations, which reach their peak values and decline prior to the jump to the next higher mode. These frequency jumps and fluctuation amplitudes characteristics are commonly found features of impinging shear layer instability.

Vortex Shedding

Vortex shedding is another fluid instability usually involving the interaction of two shear layers in the wake of a body resulting in a periodic unsteady flow. The vortices separating from the body shear layers can induce vibration of the body. As was discussed in the section on fluid resonators in Chapter 2, vortex shedding can also couple with other fluid resonators such as free surface waves and standing acoustic waves to produce systemic excitation.

Vortex shedding occurs when the vorticity generated by the flow over the surface of the forebody collects in the separated regions on each side of the body, as shown in the top sketch in Figure 18 (defined as $\Delta t = 0$). After a small time increment (t_1), for reasons not yet completely understood, one of the two regions of oppositely signed circulation grows stronger than the other region. The stronger region dominates the near wake at time t_2 and draws the circulatory region from the opposite side toward the wake centerline, inducing a transverse force on the body. The stronger top-side vortex in Figure 18 is shed from the body at time t_3, permitting the near-wake domination by the bottom-side vortex, and the initiation of a new nascent top-side vortex which is subsequently drawn toward the wake centerline. At t_4 the bottom-side vortex is shed, establishing the ongoing periodic shedding of vortices as the process is repeated. Morkovin (1964) and Blevins (1990) provide an informative sketches of the development of vortex shedding past a circular cylinder as a function of the Reynolds number.

Many studies and reviews of vortex shedding have been presented in the literature. We refer the reader to the reviews by Marris (1964), King (1977), Sarpkaya (1979), Bearman (1984), Griffin & Hall (1991), Matsumoto (1999), and others. Kolkman & Jongeling (2007) compiled Strouhal numbers for a variety of rounded cross-sectional geometries (after the work by Delany & Sorensen, 1953).

Vortex shedding from rectangular cross-section structures is obviously of significance in bridge and building aerodynamics, as well as in certain hydraulic structures such as trash racks for hydraulic gates and for intake structures (Nguyen & Naudascher, 1991). Figure 19 shows the Strouhal number variation with the slenderness ratio (streamwise length to normal width) of rectangular cylinders with both square and rounded edges. The variation of the vortex shedding frequency with slenderness ratio has three dis-

Figure 18. Development of vortex shedding in a start-up flow with increasing time from top to bottom
After Kolkman & Jongeling, 2007.

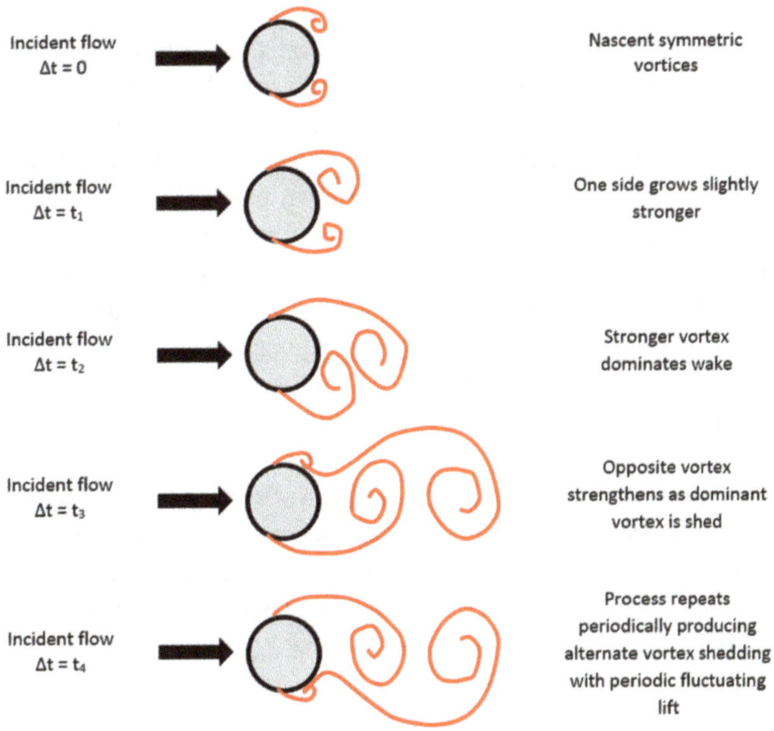

Figure 19. Strouhal numbers for vortex shedding from rectangular cylinders at zero angle of attack as a function of cylinder slenderness ratio
After Nguyen & Naudascher, 1991.

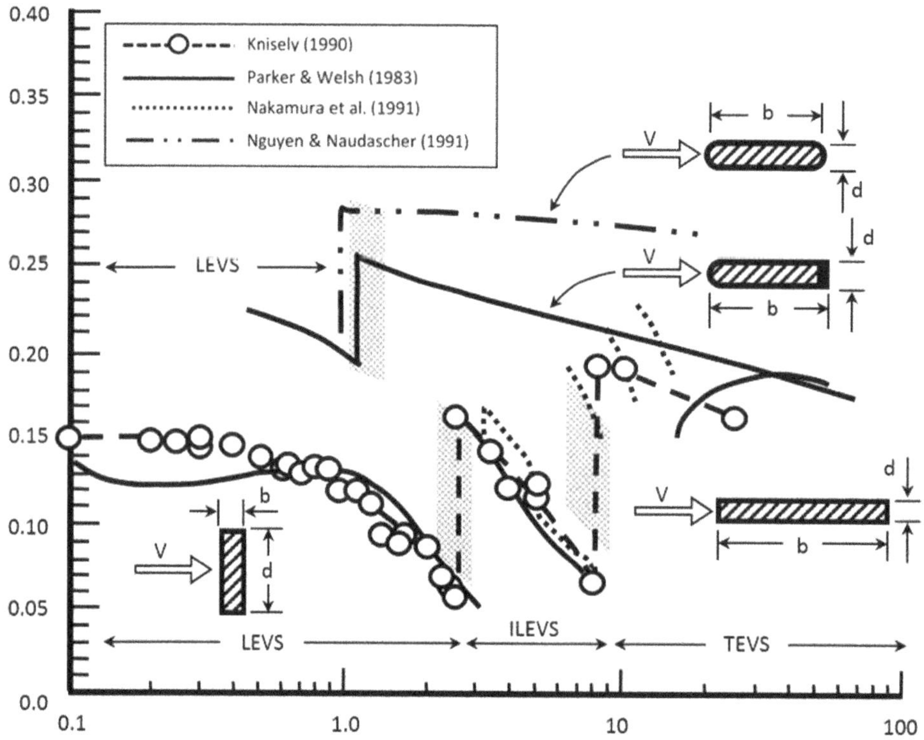

tinct regions. For a cylinder with a very small streamwise extent, as illustrated in the bottom left corner of Figure 19, the flow separates from the leading edges of the cylinder producing separated recirculation regions on the top and bottom surfaces. This region is denoted as leading-edge vortex shedding (LEVS),

With increasing slenderness ratio, the Strouhal frequency decreases, until the body is long enough for the separated shear layers to begin interacting with the downstream corners. For square-edged cylinder with a *b*/*d* ratio of about 2.5 to 3.0 (depending in Reynolds number, free stream turbulence level, and possibly aspect ratio), the Strouhal frequency experiences a discontinuous jump to a higher frequency as the interaction occurs. This jump, shown as a vertical dashed line a light gray shaded region. The shedding in this region is denoted as impinging leading edge vortex shedding. The separation bubbles along the top and bottom surfaces accumulate the front face vorticity, until a vortex is shed from the separation bubble. The shed vortex travels along the horizontal body surface, separates into the wake, triggering the release of the opposite signed vorticity, and forms a periodic vortex street in the wake of the body.

A second discontinuity occurs in the range of *b*/*d* values of about 8 to 9 (or larger at higher Reynolds numbers). The vortex shedding can be disrupted in this second gray-shaded region, perhaps extending from $b/d \approx 8$ to $b/d \approx 20$ at higher Reynolds numbers. This region of trailing edge vortex shedding (TEVS) has also been reported to have a series of smaller step discontinuities as indicated by the data of Nakamura et al. (1991).

The top two data traces in Figure 19 summarize data for cylinders with rounded leading edges (Parker & Welsh, 1983) and cylinders with both leading and trailing edges rounded (Nguyen & Naudascher, 1991), Rounding the leading edge reduces the extent of the LEVS region since the shear layer can more easily negotiate the less sudden change in surface direction. The discontinuity occurs at a b/d value of approximately 1.0. Note that in the TEVS region the Strouhal number trends for the square leading edge and rounded leading edge are consistent, suggesting the effect of the leading edge condition no longer plays a significant role. The rounded trailing edge data trend, on the other hand, appears to remain distinct from that of the other two geometries.

Data on Strouhal numbers for rectangular cylinders at incidence can be found in Knisely (1990) and Naudascher & Rockwell (1994).

While vibrations induced by vortex shedding can indeed be destructive, the mechanism has a self-limiting amplitude characteristic that typically keeps the amplitude of vibration of circular cylinders at about 1 to 1.2 cylinder diameters for lightly-damped, light-weight structures, although higher amplitudes have been reported. Further, vortex-shedding induced excitation occurs over a relatively narrow fluid velocity range for which the vortex-shedding frequency corresponds roughly to the structural vibration frequency. The region over which the structural frequency captures the vortex shedding frequency is called the lock-in or lock-on region. The extent of the lock-in region depends on the mass damping parameter, called the Scruton number. The Scruton number is defined ambiguously (differing by a factor of π) in the literature as either $Sc = 4\zeta/m_r$ or as $Sc = 4\pi\zeta/m_r$ where, in both cases, ζ is the structural damping ratio and m_r is the mass ratio. The mass ratio is defined as the structural mass per unit width of the flow divided by the reference mass of water, $m_r = m/\left(\rho d^2\right)$ with d as the characteristic length scale for the body. For details of the lock-in process, see Toebes (1969), Bearman (1984), Barbi et al. (1986), Williamson & Roshko (1988), Griffin & Hall (1991), Naudascher & Rockwell (1994), and Kolkman & Jongeling (2007), among many others.

The classical Karman vortex shedding resulting from the interaction of two bluff body shear layers may occur in the downstream wake of gates with simultaneous over-and underflow. Otherwise, Karman vortex shedding is seldom a direct factor in gates and weirs, but may well be a source of excitation of structural components such as cables, chains, and other structures appended to the main gate structure. Also, as previously discussed in Chapter 2, vortex excitation of standing waves on the free surface can produce systemic unsteadiness.

A similar type of vortex-shedding from a single shear has been reported for flat-bottomed gates with submerged discharge (Thang & Naudascher, 1986; Naudascher & Rockwell, 1994; Kolkman & Jongeling, 2007, Ishii & Knisely, 1989; 1990). These gate vibrations will be discussed more fully in Chapter 7 – Flow-Induced Vibration of Long-Span Gates.

In addition to the well-known cross-flow excitation, vortex shedding can also excite an inline response from a structure with small mass damping parameters and sufficient streamwise elasticity. Early studies by King et al. (1973), followed by reports from Ermshaus et al. (1985) and Obasaju et al. (1990; 1992), indicate that symmetric vortex shedding can excite an inline response from a variety of cross sections.

Interface Instability

The instability of a liquid layer issuing either from a slit or as a free overflow bears a pronounced similarity to that of a free shear layer. Figure 20 shows a schematic of an overflowing weir. If the air space is sufficiently aerated, the nappe will be stable. If the air space is not sufficiently aerated, the nappe will undergo a transverse oscillation (one instant of the deflection is shown schematically as the dashed line in Figure 20) that grows in amplitude towards the base of the nappe. Several authors have considered the stability of a liquid sheet, including Schwartz (1964a, 1964b, 1966), Binnie (1972), Cremer & Ising (1972), Kolkman (1974), Treiber (1974), and more recently Casperson (1993a. 1993b, 1994). All of them agree that without the spring effect of the trapped air space beneath the nappe, the interface instability is not possible.

Bos (1989) provides guidelines for the maximum amount of aeration needed to avoid a sub-atmospheric pressure in the air space beneath the nappe. French (1986) provides a worked example using Bos' correlation. Lewin (1995) provides a graph of air demand in terms of overflow depth and fall height. The volumetric air-flow demand to avoid nappe oscillation ranges from about 1/5 to 5 times the volume discharge of water. Insufficient aeration can result in the air space behaving as a Helmholtz resonator with nappe oscillations occurring despite the aeration.

The mechanism of nappe oscillations will be further explored in Chapter Six - Overflow Gates and Nappe Oscillations.

Bi-Stable Flow Instability

Bi-stable flow instability, the final flow instability to be considered in this chapter, was previously presented in Chapter 1. Bi-stable flow instability is the instability that results when a separated flow has an imbalance in the rate of fluid entrainment relative to the rate of fluid returned to the separated region. A commonly occurring bi-stable instability relative to hydraulic structures is the flow over an intermediate length forward step, as portrayed in Chapter 1, Figures 7 and 8 (see Naudascher & Locher,

Figure 20. Schematic of a nappe showing interface instability

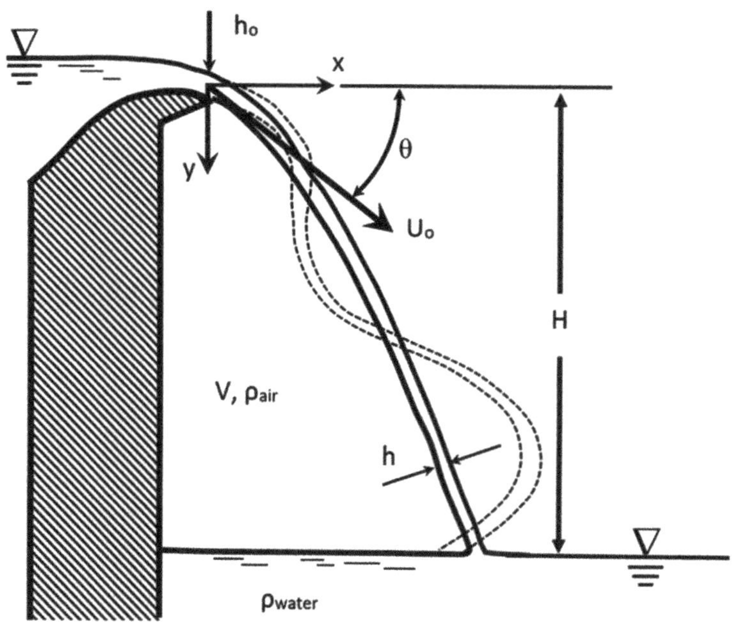

1974). Another occurrence of bi-stable instability is in the flow through an overly divergent diffuser-type passage (Fox & Kline, 1962; Smith & Kline, 1974).

The oscillations in bi-stable flow instability are at very low frequencies. The countermeasure for such instabilities is a system re-design that forces the flow to remain in one of the two stable patterns. Increased awareness of such bi-stable flow patterns on the part of hydraulic designers could avoid expensive retrofits.

Other bi-stable flow patterns have been reported for tandem cylinders and arrays of circular cylinders. See Zdravkovich (1977), Roberts (1966), and Naudascher & Rockwell (1994) for further discussion of such bi-stable flows of cylinder groups.

CONCLUSION

The fundamentals of fluid flow involving conservation of mass, momentum and energy have been reviewed. Applications of these principles to hydraulic structures have been highlighted and sources of added detailed information were provided. Subsequent chapters on added mass and radiation damping will build on these concepts, as will the chapters on specific types of hydraulic gate instabilities. The reader is encouraged to consult the cited references for details.

REFERENCES

Avallone, E. A., & Baumeister, T. (1996). *Marks' Standard Handbook for Mechanical Engineers* (10th ed.). New York, NY: McGraw-Hill Professional.

Barbi, C., Favier, D. P., Maresca, C. A., & Telionis, D. P. (1986). Vortex shedding and lock-on of a circular cylinder in oscillatory flow. *Journal of Fluid Mechanics*, *170*(1), 527–544. doi:10.1017/S0022112086001003

Bazin, H. (1892). Recent experiments on the flow of water over weirs, Annales du Ponts et Chaussees (translation by Arthur Marichal & John C. Trautwein). Proceedings of the Engineering Club of Philadelphia, PA, 9(4).

Bearman, P. W. (1984). Vortex shedding from oscillating bluff bodies. *Annual Review of Fluid Mechanics*, *16*(1), 195–222. doi:10.1146/annurev.fl.16.010184.001211

Belaud, G., Cassan, L., & Baume, J. P. (2009). Calculation of contraction coefficient under sluice gates and application to discharge measurement. *Journal of Hydraulic Engineering*, *135*(12), 1086–1091. doi:10.1061/(ASCE)HY.1943-7900.0000122

Betchov, W. O., & Criminale, R. (1967). *Stability of Parallel Flows*. Cambridge, MA: Academic.

Binnie, A. M. (1972). The stability of a falling sheet of water. *Proceedings of the Royal Society of London. Series A, Mathematical and Physical Sciences*, *326*(1565), 149–163. doi:10.1098/rspa.1972.0002

Blake, W. K. (1986a). Mechanics of Flow-induced Sound and Vibration: Vol. 1. *General Concepts and Elementary Sources*. Cambridge, MA: Academic.

Blake, W. K. (1986b). Mechanics of Flow-induced Sound and Vibration: Vol. 2. *Complex Flow-Structure Interactions*. Cambridge, MA: Academic.

Blevins, R. D. (2001). *Flow-Induced Vibrations* (2nd ed.). Malabar, FL: Krieger.

Bos, M. G. (1989). *Discharge Measurement Structures* (3rd ed.). Wageningen, NL: International Institute for Land Reclamation and Improvement/ILRI.

Bradshaw, P. (1971). *An Introduction to Turbulence and its Measurement*. Oxford, UK: Pergamon.

Casperson, L. W. (1993a). Fluttering fountains. *Journal of Sound and Vibration*, *162*(2), 251–262. doi:10.1006/jsvi.1993.1117

Casperson, L. W. (1993b). Fluttering fountains: Simplified models. *Journal of Applied Physics*, *74*(8), 4894–4898. doi:10.1063/1.354321

Casperson, L. W. (1994). Fluttering fountains: Stability criteria. *Journal of Applied Physics*, *75*(10), 4892–4894. doi:10.1063/1.355775

Cassan, L., & Belaud, G. (2011). Experimental and Numerical Investigation of Flow under Sluice Gates. *Journal of Hydraulic Engineering*, *138*(4), 367–373. doi:10.1061/(ASCE)HY.1943-7900.0000514

Castro-Delgado, M. (1983). *Abfluß- und Auflastbeiwerte für den Entwurf von Stauklappen* [*Discharge and load coefficient for flap gate design*]. Unpublished doctoral dissertation, University of Karlsruhe, Karlsruhe, Germany.

Castro-Delgado, M., & Celik, I. (1986). Analysis of free-surface flows past overflow gates using finite element method. *Computers & Fluids*, *14*(2), 159–169. doi:10.1016/0045-7930(86)90007-1

Castro-Delgado, M., Celik, I., & Herrling, B. (1984). Computation of the Potential Flow over Hinged-Leaf Gates. In *Computational Methods and Experimental Measurements* (pp. 163–175). Springer Berlin. doi:10.1007/978-3-662-06375-0_13

Çengel, Y., & Cimbala, J. (2009). *Fluid Mechanics with Student Resources DVD* (2nd ed.). New York, NY: McGraw-Hill.

Chanson, H. (2004). *The Hydraulics of Open Channel Flow: An Introduction* (2nd ed.). Rotterdam, NL: Elsevier.

Chow, V. T. (1959). *Open Channel Flow*. New York, NY: McGraw-Hill.

Clemmens, A. J., Strelkoff, T. S., & Replogle, J. A. (2003). Calibration of submerged radial gates. *Journal of Hydraulic Engineering*, *129*(9), 680–687. doi:10.1061/(ASCE)0733-9429(2003)129:9(680)

Corsi, S. R., & Schuler, J. G. (1995). *Discharge Ratings for Tainter Gates and Roller Gates at Lock And Dam No. 7 on the Mississippi River, La Crescent, Minnesota*. Water-Resources Investigations Report 95-4089: US Department of the Interior, US Geological Survey.

Cozzo, G. (1978). A formula to calculate the discharge coefficient of gates (in Italian). *L'Energia Elletrica*, *11-12*, 504–513.

Cremer, L., & Ising, H. (1972). Selbsterregte Schwingung von Überfallstrahlen (in German). *Acustica, 1*(27), 94–107.

Crowe, C. T., Elger, D. F., Roberson, J. A., & Williams, B. C. (2008). *Engineering Fluid Mechanics* (9th ed.). Hoboken, NJ: Wiley.

Delaney, N. K., & Sorenson, N. E. (1953). *Low Speed Drag of Cylinders of Various Shapes. NACA Technical Note 3038*. National Air & Space Administration.

Ermshaus, R., Knisely, C., & Naudascher, E. (1985). Flow visualization in the wake of three-dimensional bodies undergoing self-sustained oscillations. In *Optical Methods in Dynamics of Fluids and Solids* (pp. 217–223). Heidelberg: Springer. doi:10.1007/978-3-642-82459-3_28

Fox, R. W., & Kline, S. J. (1962). Flow regimes in curved subsonic diffusers. *Journal of Fluids Engineering, 84*(3), 303–312.

French, R. H. (1986). *Open-Channel Hydraulics*. New York: McGraw Hill.

Gentilini, B. (1941). Efflusso dalle luci soggiacenti alle paratoie piane inclinate e a settore. *L' Energia Elettrica, 18*(6), 361–380.

Griffin, O. M., & Hall, M. S. (1991). Review—vortex shedding lock-on and flow control in bluff body wakes. *Journal of Fluids Engineering, 113*(4), 526–537. doi:10.1115/1.2926511

Habibzadeh, A., Vatankhah, A. R., & Rajaratnam, N. (2011). Role of energy loss on discharge characteristics of sluice gates. *Journal of Hydraulic Engineering, 137*(9), 1079–1084. doi:10.1061/(ASCE)HY.1943-7900.0000406

Hager, W. H. (1994). Discharge measurement structures. In D. S. Miller (Ed.), *Discharge Characteristics, IAHR Hydraulic Structures Design Manual* (pp. 3–54). London, UK: Taylor and Francis.

Henry, H. R. (1950). Discussion of 'Diffusion of submerged jets'. *Transactions of the American Society of Civil Engineers, 115*(1), 687–693.

Ishii, N., & Knisely, C. W. (1989). Shear layer dynamics and unsteady discharge beneath a long span gate undergoing streamwise vibrations. In Č. Maksimović & M. Radojković (Eds.), *Proceedings of Hydrocomp 89* (pp. 500–509). Rotterdam, NL: Elsevier.

Ishii, N., & Knisely, C. W. (1990). Shear layer behaviour under a streamwise vibrating gate and its significance to gate vibration. *JSME International Journal, 33*, 131–138.

Karamcheti, K., Bauer, A. B., Shields, W. L., Stegen, G. R., & Woolley, J. P. (1969). Some features of an edge-tone flow field. *NASA Special Publication, SP-207*, 275–304.

Kindsvater, C. E., & Carter, R. W. (1959). Discharge characteristics of rectangular thin-plate weirs. *Transactions of the American Society of Civil Engineers, 124*(1), 772–801.

King, R. (1977). A review of vortex shedding research and its application. *Ocean Engineering, 4*(3), 141–171. doi:10.1016/0029-8018(77)90002-6

King, R., Prosser, M. J., & Johns, D. J. (1973). On vortex excitation of model piles in water. *Journal of Sound and Vibration, 29*(2), 169IN1-188IN2.

Knisely, C. W. (1990). Strouhal numbers of rectangular cylinders at incidence: A review and new data. *Journal of Fluids and Structures, 4*(4), 371–393. doi:10.1016/0889-9746(90)90137-T

Kobus, H. (Ed.). (1985, September 3-6). Symposium on Scale Effects in Modelling Hydraulic Structures (2nd ed.). Proceedings from the 1984 IAHR/DVWK Symposium on Scale Effects in Modelling Hydraulic Structures, Stuttgart, Germany. Institut für Wasserbau, University of Stuttgart.

Kolkman, P. A. (1974). Instability of a vertical water-curtain closing an air-chamber. In E. Naudascher (Ed.), *Flow-Induced Structural Vibration, IUTAM/IAHR Symposium* (pp. 17-33). Heidelberg: Springer. doi:10.1007/978-3-642-48279-3_2

Kolkman, P. A. (1994). Discharge relationships and component head losses in hydraulic structures. In D. S. Miller (Ed.), *Discharge Characteristics, IAHR Hydraulic Structures Design Manual* (pp. 55–151). London, UK: Taylor and Francis.

Kolkman, P. A., & Jongeling, T. H. G. (2007). *Dynamic Behaviour of Hydraulic Structures: Part A - Structures in Flowing Fluid*. Delft, NL: Delft Hydraulic.

Langer, W. (1952). Bestimmung der dynamischen Auflasten und Drehmomente an Stauklappen bei veränderlichen Klappenradien und Vorradien (in German). Unpublished doctoral dissertation, University of Karlsruhe, Karlsruhe, Germany.

Lewin, J. (1995). *Hydraulic Gates and Valves*. London, UK: Thomas Telford.

Lozano, D., Mateos, L., Merkley, G. P., & Clemmens, A. J. (2009). Field calibration of submerged sluice gates in irrigation canals. *J. Irrigation Drainage Eng., 135*(6), 763-772.

Lozano, D., Mateos, L., Merkley, G. P., & Clemmens, A. J. (2009). Field calibration of submerged sluice gates in irrigation canals. *Journal of Irrigation and Drainage Engineering, 135*(6), 763–772. doi:10.1061/(ASCE)IR.1943-4774.0000085

Marris, A. W. (1964). A review on vortex streets, periodic wakes, and induced vibration phenomena. *Journal of Fluids Engineering, 86*(2), 185–193.

Matsumoto, M. (1999). Vortex shedding of bluff bodies: A review. *Journal of Fluids and Structures, 13*(7), 791–811. doi:10.1006/jfls.1999.0249

Miller, D. S. (Ed.). (1994). *Discharge Characteristics, IAHR Hydraulics Design Manual Series*. London, UK: Taylor and Francis.

Morkovin, M. V. (1964). Flow around circular cylinder—a kaleidoscope of challenging fluid phenomena. *Proceedings of the ASME Symposium on Fully Separated Flows* (pp. 102-118).

Munson, B. R., Young, D. F., Okiishi, T. H., & Huebsch, W. W. (2009). *Fundamentals of Fluid Mechanics* (6th ed.). Hoboken, NJ: Wiley.

Nakamura, Y., Ohya, Y., & Tsuruta, H. (1991). Experiments on vortex shedding from flat plates with square leading and trailing edges. *Journal of Fluid Mechanics, 222*(-1), 437–447. doi:10.1017/S0022112091001167

Naudascher, E. (1991). *Hydrodynamic Forces, IAHR Hydraulics Design Manual Series*. London, UK: Taylor and Francis.

Naudascher, E., & Locher, F. A. (1974). Flow-induced forces on protruding walls. *Journal of the Hydraulics Division, 100*(2), 295–313.

Naudascher, E., & Rockwell, D. (1994). *Flow-Induced Vibrations: An Engineering Guide*. Mineola, NY: Dover.

Nguyen, T. D., & Naudascher, E. (1991). Vibration of beams and trashracks in parallel and inclined flows. *Journal of Hydraulic Engineering, 117*(8), 1056–1076. doi:10.1061/(ASCE)0733-9429(1991)117:8(1056)

Novak, P., Moffat, A. I. B., Nalluri, C., & Narayanan, R. (2001). *Hydraulic Structures* (3rd ed.). London, UK: Spon Press /Taylor and Francis.

Nyborg, W. L. (1954). Self-maintained oscillations of the jet in a jet-edge system. I. *The Journal of the Acoustical Society of America, 26*(2), 174–182. doi:10.1121/1.1907304

Obasaju, E. D., Ermshaus, R., & Naudascher, E. (1990). Vortex-induced streamwise oscillations of a square-section cylinder in a uniform stream. *Journal of Fluid Mechanics, 213*(-1), 171–189. doi:10.1017/S0022112090002270

Obasaju, E. D., Ermshaus, R., & Naudascher, E. (1992). Further results on the flow-induced streamwise vibration of cylinders. *Journal of Fluids and Structures, 6*(1), 51–66. doi:10.1016/0889-9746(92)90055-8

Oshima, M., Ishido, T. & Boiten, W. (2013). Discharge coefficient for full-width sharp-crested high weirs. *Journal of the. Japan Society of Civil Engineers, 1*(1), 360-365.

Panton, R. L. (2005). *Incompressible Flow* (3rd ed.). Hoboken, NJ: Wiley.

Powell, A. (1961). On the edgetone. *The Journal of the Acoustical Society of America, 33*(4), 395–409. doi:10.1121/1.1908677

Prandtl, L. (1928). Über Flüssigkeitsbewegung bei kleiner Reibung [Motion of fluids with very little viscosity]. NACA Technical Memorandum 452. (originally published 1904).

Pritchard, P. J., Leylegian, J. C., & Bhaskaran, R. (2016). *Fox and McDonald's Introduction to Fluid Mechanics* (9th ed.). Hoboken, NJ: Wiley.

Pulina, B., & Voigt, A. (1981). Hydrodynamische Belastung der Wehrverschlüsse an den Saarstufen Rehlingen, Mettlach und Schoden sowie Abflußleistung bei spezifischen Betriebsunfällen – Modelluntersuchungen [Hydrodynamic load of flood-gates and of flow on specific operations measured in the individual hydraulic models for the Saar barrages Rehlingen, Mettlach and Schoden]. Mitteilungsblatt Nr. 49. Karlsruhe, Germany: Bundesanstalt für Wasserbau. Retrieved from http://vzb.baw.de/publikationen/mitteilungsblaetter/0/mb49_Pulina_Voigt.pdf

Rajaratnam, N., & Subramanya, K. (1967). Flow equation for the sluice gate. *Journal of Irrigation and Drainage Engineering*, *93*(9), 167–186.

Rehbock, T. (1929). Wassermessungen mit scharfkantigen Überfallwehren (in German, translated title: Water measurement with sharp-edged overflow weirs). *Zeitschrift des Vereins Deutscher Ingeniure*, *73*(24), 817–823.

Replogle, J. A., Clemmens, A. J., & Pugh, C. A. (2004). Hydraulic design of flow measuring structures. In L. W. Mays (Ed.), *Hydraulic Design Handbook*. New York, NY: McGraw-Hill.

Roberts, B. W. (1966). Low Frequency, Aeroelastic Vibrations in a Cascade of Circular Cylinders. Mechanical Engineering Science Monograph No. 4. London, UK: Institution of Mechanical Engineers.

Rockwell, D. (1972). Transverse oscillations of a jet in a jet-splitter system. *Journal of Fluids Engineering*, *94*(3), 675–681.

Rockwell, D., & Knisely, C. W. (1979). The organized nature of flow impingement upon a corner. *Journal of Fluid Mechanics*, *93*(03), 413–432. doi:10.1017/S0022112079002573

Rouse, H. (1946). *Elementary Mechanics of Fluids*. Hoboken, NJ: Wiley.

Sarpkaya, T. (1979). Vortex-induced oscillations: A selective review. *Journal of Applied Mechanics*, *46*(2), 241–258. doi:10.1115/1.3424537

Schlichting, H., & Gersten, K. (2000). *Boundary Layer Theory* (8th ed.). Heidelberg: Springer. doi:10.1007/978-3-642-85829-1

Schwartz, H. I. (1964a). Nappe oscillation. *Journal of the Hydraulics Division*, *90*, 129–143.

Schwartz, H. I. (1964b). Projected nappes subject to harmonic pressures. *Proceedings - Institution of Civil Engineers*, *28*(3), 313–326. doi:10.1680/iicep.1964.10086

Schwartz, H. I. (1966). Edgetones and nappe oscillation. *The Journal of the Acoustical Society of America*, *39*(3), 579–582. doi:10.1121/1.1909930

Sepúlveda, C., Gómez, M., & Rodellar, J. (2009). Benchmark of discharge calibration methods for submerged sluice gates. *Journal of Irrigation and Drainage Engineering*, *135*(5), 676–682. doi:10.1061/(ASCE)IR.1943-4774.0000013

Shayana, H. K., & Farhoudi, J. (2013). Effective parameters for calculating discharge coefficient of sluice gates. *Flow Measurement and Instrumentation*, *33*, 96–105. doi:10.1016/j.flowmeasinst.2013.06.001

Smith, C. R., & Kline, S. J. (1974). An experimental investigation of the transitory stall regime in two-dimensional diffusers. *Journal of Fluids Engineering*, *96*(1), 11–15. doi:10.1115/1.3447086

Stegen, G. R., & Karamcheti, K. (1970). Multiple tone operation of edgetones. *Journal of Sound and Vibration*, *12*(3), 281–284. doi:10.1016/0022-460X(70)90072-6

Tennekes, H., & Lumley, J. L. (1972). *A First Course in Turbulence*. Boston: MIT Press.

Thandaveswara, B. S. (2010). *Hydraulics, Unit 33*. Retrieved from http://nptel.iitm.ac.in/courses/IIT-MADRAS/Hydraulics/pdfs/Unit33/33_1.pdf

Thang, N. D., & Naudascher, E. (1986). Vortex-excited vibrations of underflow gates. *Journal of Hydraulic Research*, *24*(2), 133–151. doi:10.1080/00221688609499327

Toch, A. (1953). Discharge characteristics of Tainter gates. *Proceedings of the American Society of Civil Engineers*, *79*(10), 1–20.

Toebes, G. (1969). The unsteady flow and wake near an oscillating cylinder. *Journal of Fluids Engineering*, *91*(3), 493–502.

Treiber, B. (1974). Theoretical study of nappe oscillation. In E. Naudascher (Ed.), *Flow-Induced Structural Vibration, IUTAM/IAHR Symposium* (pp. 34-46). Heidelberg: Springer.

USACE. (1977). *Hydraulic Design Criteria* (Vol. 2). Vicksburg, MS: U.S. Army Engineer Waterways Experiment Station.

Varschney, D. V. (1977, April). Model scale and discharge coefficient. *Water Power and Dam Construction*.

Wahl, T. L. (2005). Refined energy correction for calibration of submerged radial gates. *Journal of Hydraulic Engineering*, *131*(6), 457–466. doi:10.1061/(ASCE)0733-9429(2005)131:6(457)

White, F. M. (2006). *Viscous Fluid Flow* (3rd ed.). New York, NY: McGraw-Hill.

White, F. M. (2010). *Fluid Mechanics* (7th ed.). New York, NY: McGraw-Hill.

Williamson, C. H. K., & Roshko, A. (1988). Vortex formation in the wake of an oscillating cylinder. *Journal of Fluids and Structures*, *2*(4), 355–381. doi:10.1016/S0889-9746(88)90058-8

Woolley, J. P., & Karamcheti, K. (1974). Role of jet stability in edgetone generation. *AIAA Journal*, *12*(11), 1457–1458. doi:10.2514/3.49525

Zdravkovich, M. M. (1977). Review of flow interference between two circular cylinders in various arrangements. *Journal of Fluids Engineering*, *99*(4), 618–633. doi:10.1115/1.3448871

Chapter 4
Rayleigh Wave Theory:
Gate Excitation Mechanism

ABSTRACT

In this chapter, the mechanism for generating fluctuating reservoir pressures will be explored. The generation of surface waves and fluctuating pressures in the upstream reservoir due to gate motion will be examined. The energy input into surface waves is dissipated, or dispersed, as the waves travel away from the structure. These waves are called dispersive waves. In this chapter, the mathematical description of these dispersive waves will be developed. Streamwise vibration will always produce a periodic fluctuation of the reservoir pressure, thereby inducing dispersive water waves on the free surface. In addition, both the streamwise bending vibrations and the vertical vibrations of the gate can potentially produce flow rate variation beneath the gate resulting in fluctuating reservoir pressures, and inducing dispersive waves on the free surface. The theoretical framework for coupling pressure fluctuations and dispersive waves with gate vibration to supply energy to the gate motion will be developed.

INTRODUCTION

For hydraulic gates with free discharge and limited streamwise extent of the gate lip, viscous effects are usually negligible. In such cases, the dynamic forces acting on the gate are substantially due to pressure fluctuations in the upstream reservoir. In this chapter, the mechanism for generating fluctuating reservoir pressures will be explored.

Vibrating structures piercing the free surface of a fluid can induce surface waves that travel outward from the structure and carry energy away from the structure. The energy input into these waves is dissipated, or dispersed, as the waves travel away from the structure. These waves are called dispersive waves. In this chapter, the mathematical description of these dispersive waves will be developed.

Dispersive water waves emanating from a vibrating gate are shown in Figures 1 and 2. The Tainter gate, shown in Figure 1, is suspended on wire cables, which are elastic and play the role of a spring, permitting vibration of the whole gate as a rigid body around the trunnion pin center. The skinplate is sufficiently large and thin to undergo streamwise bending vibration with nodes on each spanwise side.

DOI: 10.4018/978-1-5225-3079-4.ch004

The long-span gate, shown in Figure 2, can also undergo streamwise bending vibration due to its limited streamwise thickness and its elasticity. It can also vibrate in the normal direction (up-and-down vibration), due both to the gate elasticity and to the spring effect of its wire suspension cables.

The whole gate vibration around the trunnion pin of the Tainter gate and, similarly, the up-and-down vibration of the long-span gate each cause a variation in the flow rate (also variously called 'fluctuating discharge,' 'discharge fluctuation,' or 'discharge variation') beneath the gate. When the long-span gate is inclined toward the downstream direction, or when the angle between the tangent to the Tainter gate skinplate and the gate crest tangent is less than 90°, the streamwise bending vibration of the gate also results in flow rate variation beneath the gate. Depending on the inclination of the skinplate or long-span gate, streamwise gate bending can result in a larger gate opening in the center of the channel, resulting in a discharge that varies with the streamwise motion of the gate.

The flow rate variation beneath the gate produces a fluctuation in the reservoir pressure, and induces upstream propagating dispersive waves on the free surface. A similar argument can be made for streamwise gate vibrations, which may produce discharge fluctuations beneath the gate depending on the gate orientation relative to the crest. However, in addition, streamwise vibration will always produce a periodic fluctuation of the reservoir pressure, due to the push-and-draw effect of the gate against the backwater, thereby inducing dispersive water waves on the free surface.

For thin long-span plates and thin Tainter gate skinplates, the reservoir flow field is independent of shed vortices; other viscous effects are sufficiently small to neglect. The fluctuating reservoir free surface due to small amplitude vibration of a gate can be considered, to a very good approximation, an inviscid flow to be analyzed with potential flow theory. However, in order to predict the dispersive nature of the water waves resulting from the damping effect of the fluid viscosity, the potential theory needs a certain modification to represent the damping effect.

Figure 1. Dispersive water waves can be generated by both normal (up-and-down) and streamwise vibrations of a Tainter gate on a dam crest with flow from right to left

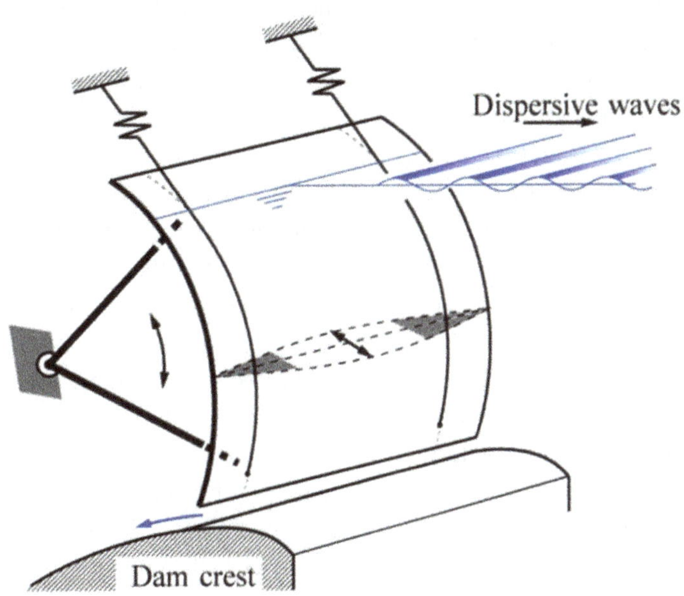

Figure 2. Dispersive water waves generated by normal (up-and-down) and streamwise vibrations of a long-span gate with flow from right to left

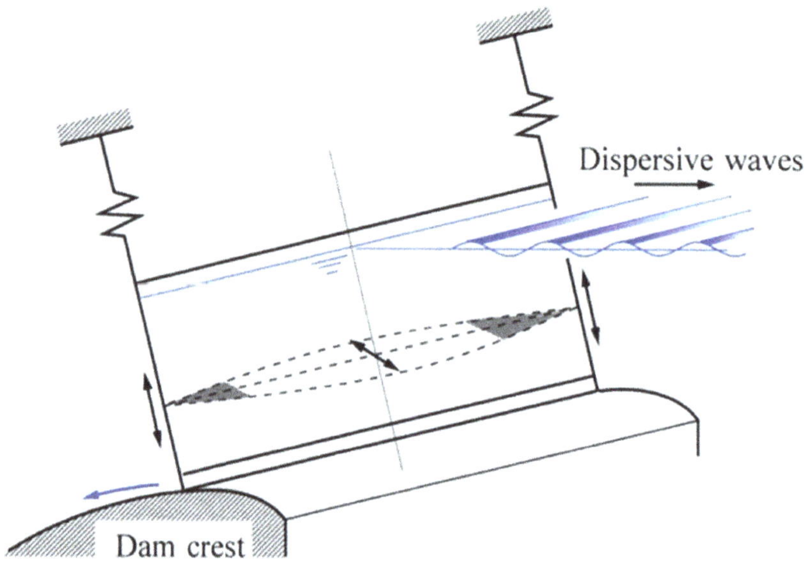

One way to include damping is to prescribe a Sommerfeld radiation condition (see Atkinson, 1947) as the distance to the upstream boundary approaches infinity. This boundary condition requires that the waves behave like outgoing progressive waves as they approach the far distant upstream boundary. However, the Sommerfeld boundary condition is limited to radiation in an unbounded fluid domain. If a rigid reflecting wall exists far upstream, it is clear that other conditions should be imposed at infinity in order to derive a solution which does not behave like an outgoing progressive wave.

Another clever approach, first presented by Lord Rayleigh (1883), is the "Rayleigh wave theory for dispersive water waves," in which a small dissipative body force proportional to the fluid relative velocity was introduced to represent crudely the effect of the small damping forces due to the fluid viscosity. Rayleigh's wave theory has the great mathematical advantage of being consistent with the existence of a velocity potential. The problem is solved with Rayleigh wave theory as an initial and boundary value problem by assuming the water over the whole fluid region to be initially at rest. Solutions are then derived in the limit of infinite time, without the necessity of imposing any radiation conditions. The upstream boundary must meet only the boundedness conditions. This method is relatively simple and realistic in predicting correct solutions for dispersive water wave problems such as the one we are trying to solve.

In the present chapter, we present the fundamental derivation of the equations for the Rayleigh wave theory in detail, in terms of the hydrodynamics of two-dimensional, irrotational, and incompressible fluid flow. The problem of dispersive water waves resulting from vibrations of Tainter and long-span gates can be reduced to an initial and boundary value problem that satisfies the Laplace equation for the velocity potential. The problem can be solved analytically and yields the hydrodynamic pressure fluctuation in the water, as well as a solution for the dispersive water waves.

Subsequently, the theoretical framework for the coupling of discharge fluctuations with gate vibration to supply energy to the gate motion will be developed. When a long-span gate is inclined in the downstream direction, or when the downstream angle between the tangent to the Tainter gate skinplate and the gate crest tangent is less than 90°, the streamwise bending vibration of the gate also results in flow-rate variation beneath the gate. All vibration modes for each of these gates can potentially induce a flow-rate variation pressure in the reservoir, which provides a significant energy source for the spontaneous gate vibration under certain geometrical and dynamic conditions. Such an excitation due to reservoir pressure fluctuation is referred to as the water excitation effect. The possibility of gate excitation due to these pressure fluctuations makes the development of a theoretical analysis of the flow-rate variation pressure essential for understanding the characteristics of this water excitation effect.

Finally, we will consider the added mass effect of the fluid on the structural vibrations and damping due to the water. The viscosity is expected to provide some damping, but the amplitude of the vibratory velocity of a massive gate will be relatively small, resulting in a very small viscous damping effect. Another factor in damping is the dissipation due to the radiation of surface waves. The wave radiation damping effect, as well as the added mass effect, will both be derived from the Rayleigh radiation theory developed earlier in the chapter. This analysis will show that the wave radiation damping effect approaches zero for most practical gates. Due to the small viscous and wave radiation damping effects, a massive gate may be susceptible to quickly growing, spontaneous vibration when an energy source due to flow-rate variation beneath the gate accompanies the gate vibration.

RAYLEIGH WAVE THEORY

Hydrodynamics of Two-Dimensional, Irrotational, and Incompressible Fluid Flow

Continuity Equation

The differential element of size d*X* by d*Y* in Figure 3 is assumed to be in a two-dimensional fluid flow. The symbols *u* and *v* represent the flow velocity in Cartesian coordinate directions *X* and *Y*, respectively.

Figure 3. Mass balance for a differential element in a two-dimensional fluid flow

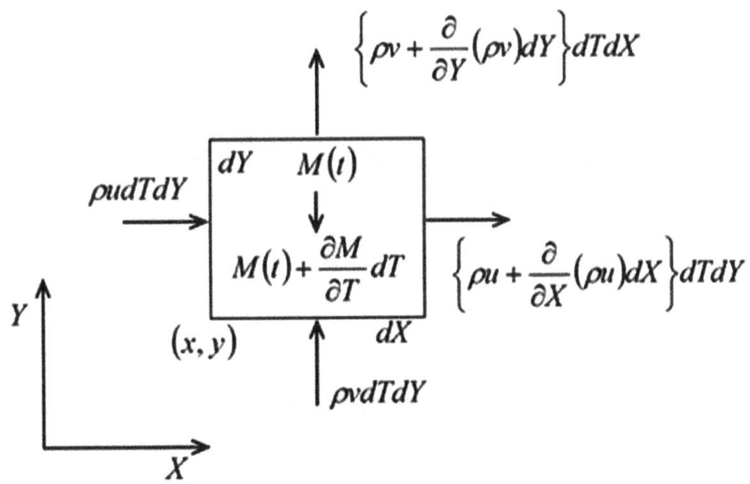

The specific mass (density) is represented by ρ [kg/m^3] and the mass in the differential element per unit length in the Z-direction is denoted by $M(t)$ which is given by $\rho dXdY$.

The continuity equation for the differential element can be derived from the following mass balance:

$$Mass\ flow\ in - Mass\ flow\ out = Rate\ of\ accumulation\ of\ mass\ in\ the\ element$$

The mass balance can be represented by the following mathematical expression:

$$-\frac{\partial(\rho u)}{\partial X}dXdYdT - \frac{\partial(\rho v)}{\partial Y}dXdYdT - \frac{\partial(\rho dXdY)}{\partial T}dT \quad (1)$$

which can be re-written as

$$\frac{\partial \rho}{\partial T} + \frac{\partial(\rho u)}{\partial X} + \frac{\partial(\rho v)}{\partial Y} = 0 \quad (2)$$

Equation 2 is the differential continuity equation for a compressible fluid. With the assumption that the fluid is incompressible, the continuity equation becomes

$$\frac{\partial u}{\partial X} + \frac{\partial v}{\partial Y} = 0 \quad (3)$$

Irrotational Flow

In Figure 4, the differential element $dXdY$ is shown after a translation and a deformation. The symbols Θ_1 and Θ_2 represent the rotation angles of dX and dY, respectively, and are given by the following expressions

$$\Theta_1 = \frac{\frac{\partial v}{\partial X}dXdT}{dX} \quad \text{and} \quad \Theta_2 = \frac{\frac{\partial u}{\partial Y}dYdT}{dY} \quad (4)$$

The resultant mean rotation, Ω, of the differential element is given by

$$\Omega = \frac{\frac{1}{2}(\Theta_1 - \Theta_2)}{dT} = \frac{1}{2}\left(\frac{\partial v}{\partial X} - \frac{\partial u}{\partial Y}\right) \quad (5)$$

The assumption of irrotational flow requires the rotation, Ω, to be zero, resulting in the following equation:

Figure 4. Translation and deformation of the differential element

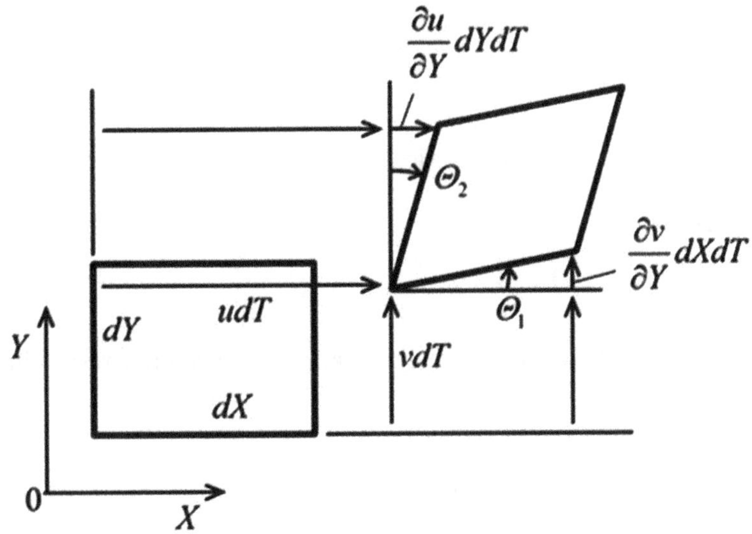

$$\frac{\partial v}{\partial X} - \frac{\partial u}{\partial Y} = 0 \tag{6}$$

Velocity Potential

The assumed irrotational flow must satisfy Equation 6. By defining a velocity potential Φ such that

$$u = \frac{\partial \Phi}{\partial X} \text{ and } v = \frac{\partial \Phi}{\partial Y} \tag{7}$$

The expressions for u and v, when substituted into Equation 6, show that

$$\frac{\partial v}{\partial X} - \frac{\partial u}{\partial Y} = \frac{\partial}{\partial X}\left(\frac{\partial \Phi}{\partial Y}\right) - \frac{\partial}{\partial Y}\left(\frac{\partial \Phi}{\partial X}\right) = 0$$

Thus, the irrotationality condition, Equation 6, is automatically satisfied.

Substitution of the expressions for velocity in terms of the velocity potential, Equations 7, into the continuity equation, Equation 3, shows that the velocity potential, Φ must satisfy the Laplace equation, variously written as

$$\frac{\partial^2 \Phi}{\partial X^2} + \frac{\partial^2 \Phi}{\partial Y^2} = 0 \text{ } or \text{ } \Phi_{XX} + \Phi_{YY} = 0 \text{ } or \text{ } \nabla^2 \Phi = 0 \tag{8}$$

The velocity potential, Φ, provides a solution to the Laplace equation, subject to the appropriate boundary conditions.

Stream Function

The continuity equation, Equation 3, can be satisfied by the introduction of a stream function, Ψ, defined as

$$u = \frac{\partial \Psi}{\partial Y} \text{ and } v = \frac{-\partial \Psi}{\partial X} \tag{9}$$

Substitution of Equation 9 into the irrotationality condition, Equation 6, results in the Laplace equation for stream function Ψ:

$$\frac{\partial^2 \Psi}{\partial X^2} + \frac{\partial^2 \Psi}{\partial Y^2} = 0 \text{ or } \Psi_{XX} + \Psi_{YY} = 0 \text{ or } \nabla^2 \Psi = 0 \tag{10}$$

Euler's Equations

Figure 5 again shows the differential element, dXdY, assumed to be in a two-dimensional, incompressible fluid flow. In Figure 5, $P(X, Y)$ represents the pressure and g is the gravitational acceleration. The force balances in the X and Y directions, respectively, are given as

$$-\rho dXdY\left(\frac{\partial u}{\partial T} + u\frac{\partial u}{\partial X} + v\frac{\partial u}{\partial Y}\right) + \left\{P\left(X,Y\right) - P\left(X + dX,Y\right)\right\}dY = 0 \tag{11a}$$

Figure 5. Forces on the differential element

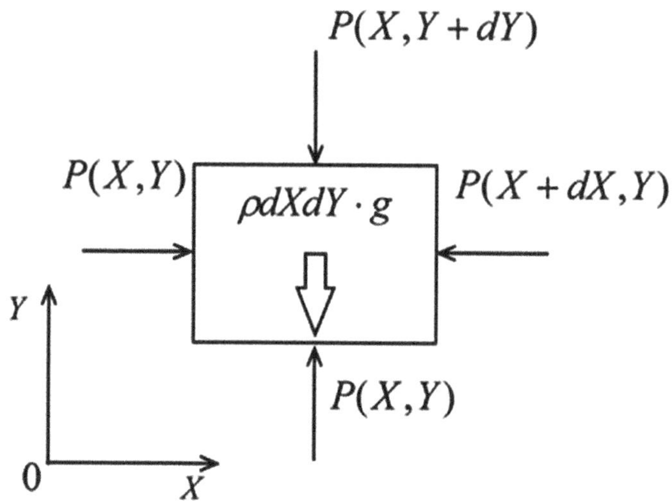

$$-\rho dXdY\left(\frac{\partial v}{\partial T}+u\frac{\partial v}{\partial X}+v\frac{\partial v}{\partial Y}\right)+\left\{P\left(X,Y\right)-P\left(X,Y+dY\right)\right\}dX-\rho gdXdY=0 \tag{11b}$$

which can be reduced to the following expressions:

$$\rho\left(\frac{\partial u}{\partial T}+u\frac{\partial u}{\partial X}+v\frac{\partial u}{\partial Y}\right)+\frac{\partial P}{\partial X}=0 \tag{12a}$$

$$\rho\left(\frac{\partial v}{\partial T}+u\frac{\partial v}{\partial X}+v\frac{\partial v}{\partial Y}\right)+\frac{\partial P}{\partial Y}+\rho g=0 \tag{12b}$$

Equations 12a and 12b are commonly called the Euler equations.

Pressure Equation

Substitution of the expressions for velocity in terms of the velocity potential, Equation 7, into the Euler equations, Equations 12a and b, results in the following expressions:

$$\rho\left(\Phi_{XT}+\Phi_{x}\Phi_{XX}+\Phi_{Y}\Phi_{XY}\right)+P_{X}=0 \tag{13a}$$

$$\rho\left(\Phi_{YT}+\Phi_{X}\Phi_{XY}+\Phi_{Y}\Phi_{YY}\right)+P_{Y}+\rho g=0 \tag{13b}$$

in which the subscripts imply differentiation with respect to the subscripted variable. Equations 13a and b can be re-written as

$$\frac{\partial}{\partial X}\left(\rho\Phi_{T}+\frac{\rho}{2}\Phi_{X}^{2}+\frac{\rho}{2}\Phi_{Y}^{2}+P+\rho gY\right)=0 \tag{14a}$$

$$\frac{\partial}{\partial Y}\left(\rho\Phi_{T}+\frac{\rho}{2}\Phi_{X}^{2}+\frac{\rho}{2}\Phi_{Y}^{2}+P+\rho gY\right)=0 \tag{14b}$$

Integrating Equations 14a and 14b over the *X* and *Y* plane results in the pressure equation in the following form:

$$\rho\Phi_{T}+\frac{\rho}{2}\left(\Phi_{X}^{2}+\Phi_{Y}^{2}\right)+P+\rho gY=0 \tag{15}$$

where the time-dependent integral constant was set to zero, to be consistent with the assumed zero value at the initial time.

Rayleigh's Dissipative Body Forces and Irrotational Fluid Motion

The concept of a dissipative body force proportional to the relative velocities in the X and Y directions was first introduced by Lord Rayleigh (1883). This body force in two dimensions is

$$\boldsymbol{f}_B = \tilde{\mu}\boldsymbol{V} = \tilde{\mu}u\vec{\boldsymbol{i}} + \tilde{\mu}v\vec{\boldsymbol{j}} \text{ [N/kg]} \tag{16}$$

The dissipative body force resists the deviation of any fluid particle from the state of uniform flow. The coefficient $\tilde{\mu}$ [1/s] is ultimately assumed to vanish. In the absence of dissipative forces, the problem for dispersive water waves to which we currently seek a solution is, to a certain extent, indeterminate. Lord Rayleigh explains in his paper that 'without them it did not seem easy to interpret the analytical expressions to which we are led." Lamb (1932) comments that this law of friction introduced by Rayleigh "does not profess to be altogether a natural one, but it serves to represent in a rough way the effect of small dissipative forces; and it has the great mathematical convenience in that it does not interfere with the irrotational character of the fluid motion."

Introducing the dissipative forces in Equation 16 into Equations 11a and 11b, the Euler equations can then be expressed as

$$\rho\left(\frac{\partial u}{\partial T} + u\frac{\partial u}{\partial X} + v\frac{\partial u}{\partial Y}\right) + \frac{\partial P}{\partial X} + \rho\tilde{\mu}u = 0 \tag{17a}$$

$$\rho\left(\frac{\partial v}{\partial T} + u\frac{\partial v}{\partial X} + v\frac{\partial v}{\partial Y}\right) + \frac{\partial P}{\partial Y} + \rho g + \rho\tilde{\mu}v = 0 \tag{17b}$$

Negation and differentiation of Equation 17a with respect to Y and differentiation of Equation 17b with respect to X results in the following expressions, respectively.

$$\rho\left(-u_{YT} - u_Y u_X - u u_{XY} - v_Y u_Y - v u_{YY}\right) - P_{XY} - \rho\tilde{\mu}u_Y = 0 \tag{18a}$$

$$\rho\left(v_{XT} + u_X v_X + u v_{XX} + v_X v_Y + v v_{XY}\right) + P_{XY} + \rho\tilde{\mu}v_X = 0 \tag{18b}$$

The summation of Equations 18a and 18b results in

$$\frac{\partial}{\partial T}\left(v_X - u_Y\right) + u\frac{\partial}{\partial X}\left(v_X - u_Y\right) + v\frac{\partial}{\partial Y}\left(v_X - u_Y\right) + \left(u_X + v_Y\right)\left(v_X - u_Y\right) + \tilde{\mu}\left(v_X - u_Y\right) = 0 \tag{19}$$

Denoting the fluid element's rotation as

$$\xi = v_X - u_Y \tag{20}$$

and using the substantial derivative, $\frac{D}{DT} = \frac{\partial}{\partial T} + u\frac{\partial}{\partial X} + v\frac{\partial}{\partial Y}$, Equation 19 can be rewritten as

$$\frac{D\xi}{DT} + \left(u_X + v_Y\right)\xi + \tilde{\mu}\xi = 0 \tag{21}$$

However, recalling Equation 3, the incompressible continuity equation, Equation 21 becomes

$$\frac{D\xi}{DT} + \tilde{\mu}\xi = 0 \tag{22}$$

The solution of Equation 22 is

$$\xi = ce^{-\tilde{\mu}T} \tag{23}$$

Note that as $T \to \infty$ the value of the rotation $\xi \to 0$, which means that long after the disturbance has been initiated ($T \to \infty$) the remaining flow field is irrotational ($\xi \to 0$). As a result, one may recognize that the introduction of dissipative forces, given by Equation 16, has not interfered with the irrotational character of the fluid motion. As long as the flow is irrotational, the velocity potential Φ as defined in Equation 7 exists, and satisfies the Laplace equation, Equation 8.

Pressure Equation with Dissipative Forces

If the dissipative body forces are included in the force balances and if the velocities in terms of the velocity potential as in Equation 7 are substituted into the Euler equations, Equations 17a and b, the modified forms of the pressure equations result:

$$\frac{\partial}{\partial X}\left(\rho\Phi_T + \frac{\rho}{2}\Phi_X^2 + \frac{\rho}{2}\Phi_Y^2 + P + \rho gY + \rho\tilde{\mu}\Phi\right) = 0 \tag{24a}$$

$$\frac{\partial}{\partial Y}\left(\rho\Phi_T + \frac{\rho}{2}\Phi_X^2 + \frac{\rho}{2}\Phi_Y^2 + P + \rho gY + \rho\tilde{\mu}\Phi\right) = 0 \tag{24b}$$

Integrating Equations 24a and 24b over the *X* and *Y* plane, the pressure equation can be derived in the following form:

$$\rho\Phi_T + \frac{\rho}{2}\left(\Phi_X^2 + \Phi_Y^2\right) + P + \rho gY + \rho\tilde{\mu}\Phi = 0 \tag{25}$$

where, again, the time-dependent integral constant is set to zero, to be consistent with the assumed zero value at the initial time.

Assuming a small fluctuation from the equilibrium state, the second order velocity potential terms in Equation 25 can be assumed to be negligible. With this assumption, the pressure fluctuation relative to its equilibrium state, $\Delta P_r = P + \rho gY,$, can be calculated by the following expression:

$$\Delta P_r - P + \rho gY = -\rho\Phi_T \tag{26}$$

where it has also been assumed that the dispersive forces ultimately vanish: $\tilde{\mu} \to 0$.

Boundary Condition on the Water Surface

The elevation of the water surface from its still water level is represented by $\tilde{\eta}(X;T)$, and thus the free surface is given by

$$Y = \tilde{\eta}\left(X;T\right) \tag{27}$$

Substitution of the surface elevation from Equation 27 into Equation 25, the pressure equation, on the free surface with pressure $P = 0$ results in the surface condition expressed as

$$\left[\rho\Phi_T + \frac{\rho}{2}\left(\Phi_X^2 + \Phi_Y^2\right) + \rho g\tilde{\eta} + \rho\tilde{\mu}\Phi\right]_{Y=\tilde{\eta}} = 0 \tag{28}$$

where the surface tension forces have been disregarded. Assuming a small fluctuation from the equilibrium state, the second order velocity potential terms in Equation 28 can be neglected, and therefore the surface condition in Equation 28 becomes

$$\left[\Phi_T + g\tilde{\eta} + \tilde{\mu}\Phi\right]_{Y=0} = 0 \tag{29}$$

which is called the dynamical condition on the free water surface.

Under the assumption that the dispersive forces ultimately vanish, $\tilde{\mu} \to 0$, the dynamical condition gives the surface elevation $\tilde{\eta}(X;T)$, as

$$\tilde{\eta} = -\frac{1}{g}\left[\Phi_T\right]_{Y=0} \tag{30}$$

The kinematic condition on the free surface is given by

$$\left[\frac{D\tilde{\eta}}{DT}\right]_{Y=\tilde{\eta}} = \left[\Phi_Y\right]_{Y=\tilde{\eta}} \tag{31a}$$

or, by expanding the substantial derivative,

$$\left[\tilde{\eta}_T + \Phi_X \tilde{\eta}_X + \Phi_Y \tilde{\eta}_Y\right]_{Y=\tilde{\eta}} = \left[\Phi_Y\right]_{Y=\tilde{\eta}} \tag{31b}$$

and requires that the vertical (up-and-down) velocity of the free surface, on the left-hand side of Equations 31a and b, is equal to the vertical velocity of a fluid particle on the free surface, on the right-hand side. The surface elevation $\tilde{\eta}(X;T)$ is independent of *Y*. The departure from the equilibrium state is assumed to be small. Under these restrictions, the kinematic condition becomes

$$\tilde{\eta}_T = \left[\Phi_Y\right]_{Y=0} \tag{32}$$

Differentiating the surface dynamical condition (Equation 29) with respect to time *T* and substituting the above kinematic condition (Equation 32) into the result yields the surface condition in the following form:

$$\left[\Phi_{TT} + \tilde{\mu}\Phi_T + g\Phi_Y\right]_{Y=0} = 0 \tag{33}$$

INITIAL AND BOUNDARY VALUE PROBLEM FOR DISPERSIVE WATER WAVES INDUCED BY GATE VIBRATIONS

At the onset of any gate vibration, the amplitude of vibration is very small. Because of this onset process, it is crucial to be able to determine the initial characteristics of vibration, such as the frequency, the damping ratio or the excitation ratio (negative damping ratio) in this linear region. These characteristics, in turn, are essential for assessing the dynamic stability of the gate. From this perspective, our focus is on small amplitude vibrations. The amplitudes are negligibly small relative to the span of Tainter gates and long-span gates. As a result, the flow field in the reservoirs behind the Tainter gate on the dam crest in Figure 1 and the long-span gate in Figure 2 can be treated as a two-dimensional field. Such a two-dimensional field is portrayed in Figure 6 for a Tainter gate. The figure shows the governing Laplace equation throughout the domain and the appropriate boundary conditions for the velocity potential.

The streamwise gate vibration and the flow rate variation under the gate can be represented by the velocity Φ_X as a boundary condition along or beneath the skinplate, which excites the reservoir flow to fluctuate, inducing the dispersive water waves on the water surface. Rayleigh wave theory assumes the water over the whole reservoir field to be originally at rest in its equilibrium state, with the following initial conditions for the velocity potential $\Phi\left(X, Y; T\right)$ prevailing:

$$\Phi\left(X,Y;0\right)=0 \text{ and } \Phi_T\left(X,Y;0\right)=0 \tag{34}$$

The solutions to the Laplace equation, Equation 8, can be determined in the limit as time goes to infinity.

In addition, Rayleigh wave theory imposes the boundedness conditions at upstream infinity and at infinite depth, as given in the following expressions:

$$\Phi\left(\infty,Y;T\right)=0 \text{ and } \Phi_X\left(\infty,Y;T\right)=0 \qquad \text{for upstream infinity}$$

$$\Phi\left(X,-\infty;T\right)=0 \text{ and } \Phi_Y\left(X,-\infty;T\right)=0 \qquad \text{for infinite depth}$$

$$\Phi_Y\left(X,-\infty;T\right)=0 \qquad \text{for finite depth} \tag{35}$$

On the water free surface, the surface condition in Equation 33 is naturally imposed.

WATER EXCITATION EFFECTS

The whole gate vibration of Tainter gates around the trunnion pin and the up-and-down vibration of long-span gates, shown in Figures 1 and 2, respectively, result in flow-rate variation in the beneath the gate. As discussed at the start of this chapter, all vibration modes for each of these gates can potentially induce a flow-rate variation pressure in the reservoir. This flow-rate variation pressure provides a significant energy source for the spontaneous gate vibration under certain geometrical and dynamic conditions. Such an excitation due to reservoir pressure fluctuation is referred to as the water excitation

Figure 6. Two-dimensional flow field of a reservoir dammed by a Tainter gate on the dam crest with direction of flow from right to left

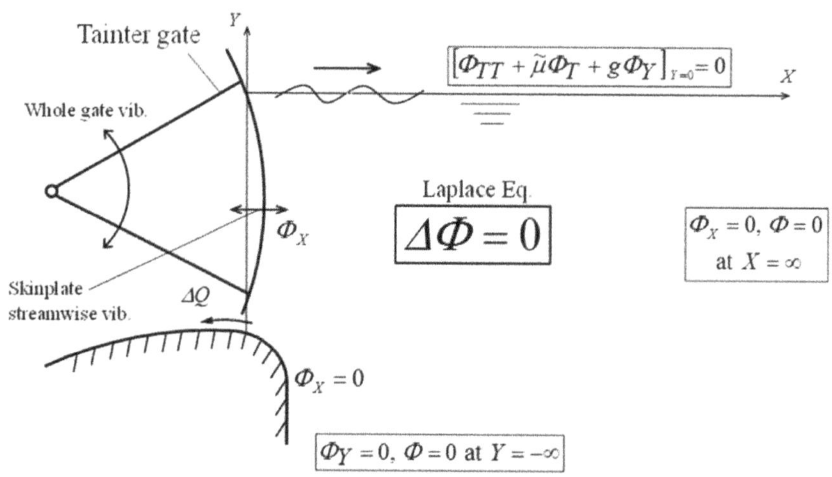

effect. In this section we will develop the analysis of gate excitation due to these pressure fluctuations. An understanding of the role of the flow-rate variation pressure is essential for understanding the characteristics of this water excitation effect.

Unfortunately, both the skinplate of a Tainter gate and the dam crest are usually curved, and the weir plate for a long-span gate is typically inclined in the downstream direction. These geometrical constraints make a purely theoretical analysis impossible. A reasonable simplification of the actual boundaries must be considered to permit a complete theoretical analysis of the pressure fluctuation in the reservoir. A bold assumption for the reservoir flow field is shown in Figure 7, where the gate plates and the reservoir wall are both assumed to be planar vertical surfaces. The X-axis is assumed to lie along the quiescent upstream water surface. Such a simplified analysis must be expected to result in discrepancies between measured pressure fluctuations in actual gate/reservoir geometries and the theoretical results obtained. The discrepancies can be accounted for by introducing correction factors determined from model experiments. Such a simplification of the reservoir flow field is an essential consideration in arriving at an analytical model in terms of the Rayleigh wave theory for the flow-rate variation pressure driving the water excitation effect.

To begin the analysis, a periodic forced discharge fluctuation beneath the vertical weir plate (sluice gate) will be introduced into the Rayleigh wave theory model to determine the characteristics of the water waves and structural loads on the sluice gate due to these discharge fluctuations. The analysis is restricted to small values of steady discharge beneath the sluice gate (*i.e.,* small gate openings) with small amplitude vertical oscillation of the gate. The analysis is undertaken for both an infinite and a finite depth reservoir.

The following assumptions are made in the analysis:

Figure 7. Simplified reservoir flow field and boundary conditions for flow-rate variation pressure with the direction of flow from right to left

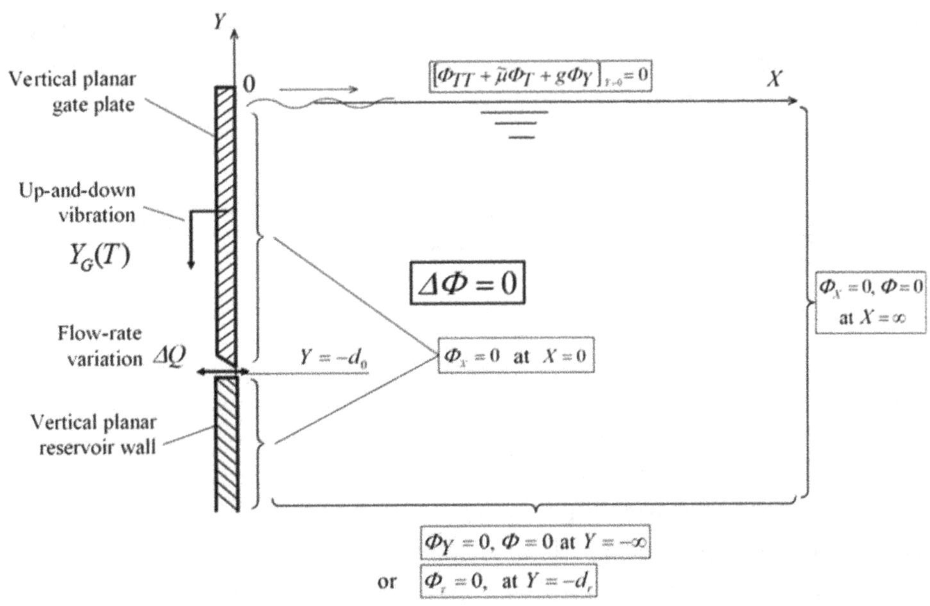

- Fluid flow in the reservoir is incompressible and irrotational.
- There are no obstacles obstructing the progress of water waves in the upper reaches of the reservoir.
- The oscillatory movement of the sluice gate (*i.e.,* the periodic change of the boundary near the gate opening) is replaced by a periodic change in the discharge flow through the gate opening.
- There is no phase delay between the periodic change of the sluice gate opening and the discharge fluctuation.

Subsequently, an analysis of the fluid excitation of gate vibration due to flow-rate variation pressure is undertaken. The analysis will provide the reader with the tools to understand the following observations:

- A basic parameter characterizing the water waves and structural loads is a Froude number. In the remainder of this book, this Froude number will be called *the basic Froude number*. It is defined as $F_0 = \sqrt{d_o / g}\Omega_0$, where d_0 is the depth of the gate opening, g is the acceleration of gravity, and Ω_0 is the circular frequency of gate vibration in air; and $F = \sqrt{d_o / g}\Omega$ is called the Froude number is based on the in-water circular frequency of gate vibration which is also the frequency of flow-rate variation. The Froude number represents the square root of the ratio of the inertia force to the gravity force of the fluid.
- The response characteristics of the water waves are analogous to those of a simple mass-spring-damper system undergoing forced vibration. Both systems display resonance behavior. When F_0 is equal to about 0.8, wave motions are in a resonant state, where the energy input by the flow-rate variation is consumed in producing larger amplitude water waves.
- With increasing Froude number, however, the water wave generation rapidly decreases and the energy input is confined to the reservoir flow field, producing a larger flow-rate variation pressure in the reservoir. The increased pressure may serve as excitation for the gate, and is known as the water excitation effect.

These theoretical results will be confirmed by experiments using a simple model of the reservoir and the sluice gate. The agreement between the model test results and the theoretical predictions demonstrate that Rayleigh's wave theory is applicable to the analysis of flow fields in reservoirs. The Rayleigh wave theory can be used to examine the dynamic loading on various gate types.

THEORETICAL DEVELOPMENT OF RAYLEIGH WAVE THEORY TO PREDICT FLOW-RATE VARIATION PRESSURE

Boundary and Initial Conditions

To analyze the incompressible flow resulting from periodic vertical oscillations of a sluice gate, an *X-Y* coordinate system is introduced with the positive *X*-axis extending upstream along the undisturbed free surface and the *Y*-axis extending vertically upward from the free surface at the location of the upstream face of the reservoir wall. Let the mean depth of the lower edge of the oscillatory sluice gate be d_0 and the mean height of the gate opening be b. The sluice gate undergoes periodic vertical oscillations, as shown in Figure 7, where the *downward* displacement $Y_G(T)$ is represented by

$$Y_G\left(T\right) = Y_0 e^{i\Omega T} \tag{36}$$

where Y_0 represents the amplitude of oscillation, Ω the circular frequency, and T the elapsed time. The time dependent discharge fluctuation through the gate opening, $\Delta Q(T)$, varies about the steady discharge Q_0. Hence, the total discharge quantity $Q(T)$ passing through the gate opening at any time is

$$Q\left(T\right) = Q_0 + \Delta Q\left(T\right) \tag{37}$$

where the discharge variation $\Delta Q(T)$ is given as

$$\Delta Q\left(T\right) = -qe^{i\Omega T} \tag{38}$$

When $T = 2\pi n / \Omega$ (n is an integer), the sluice gate is at its maximum downward displacement and the total discharge is at its minimum. Following from Equation 35 in Chapter 3, the amplitude q of the fluctuating discharge $\Delta Q(T)$ can be represented as

$$q = \delta\sqrt{gd_0} \text{ where } \delta = \sqrt{2}c_f Y_0 \tag{39}$$

where Y_0 is the amplitude of the sluice gate oscillation and c_f is the discharge coefficient for the sluice gate.

In Equation 38, the discharge fluctuation is assumed to be in-phase with the gate oscillation. This assumption was validated for vibration frequencies from 1.9 to 4.1 Hz by Ishii *et al.* (1976) (see the "Experimental Confirmation" section later in this chapter). The validity of Equation 38 must be determined for higher vibration frequencies through additional testing.

Ishii *et al.* also determined that c_f takes on a value very close to 1.0 for their sluice gate model. Based on this result for a sluice gate, Equation 38 is assumed to be valid for a long-span gate as well. Since the vibration of concern in this study is in the direction of gate support, more detailed model tests for c_f will be needed both for Tainter gates and for long-span gates, especially for any inclined long-span gate.

Let $\Phi^{(0)}(X,Y)$ denote the velocity potential for steady discharge Q_0 and $\Phi(X,Y;T)$ the unsteady velocity potential for the fluctuating discharge $\Delta Q(T)$. It follows then that for the total discharge $Q(T)$, the velocity potential, $\Phi^{(t)}(X,Y;T)$, is a linear combination of the steady and unsteady potentials:

$$\Phi^{(t)}\left(X,Y;T\right) = \Phi^{(0)}\left(X,Y\right) + \Phi\left(X,Y;T\right) \tag{40}$$

The velocity potential $\Phi^{(0)}(X,Y)$ must be a solution to the Laplace equation and can be obtained in the following forms for an infinite depth reservoir:

$$\Phi^{(0)}\left(X,Y\right)=(2/\pi)Q_0\int_0^{\infty}\left(e^{-SX}/S\right)\cos d_0 S\cos YSdS \tag{41a}$$

and for a finite depth reservoir of depth d_r:

$$\Phi^{(0)}\left(X,Y\right)=\left(Q_0/d_r\right)\left[X-\left(2/\pi\right)d_r\sum_{n=1}^{\infty}\left\{\frac{\exp\left(\frac{-n\pi X}{d_r}\right)}{n}\right\}\cos\frac{n\pi d_0}{d_r}\cos\frac{n\pi Y}{d_r}\right] \tag{41b}$$

The unsteady velocity potential $\Phi(X,Y;T)$ also must satisfy the Laplace equation over the domain $0\le X\le\infty$ and $\left(-\infty \text{ or } d_r\right)\le Y\le 0$ (depending on whether the reservoir is infinite or finite):

$$\Phi_{XX}+\Phi_{YY}=0 \tag{42}$$

Assuming the mean velocity $\Phi_Y^{(0)}$ on the free surface to be zero, the free surface condition for the unsteady potential takes the following form:

$$\left[\Phi_{TT}+\tilde{\mu}\Phi_T+g\Phi_Y\right]_{Y=0}=0 \tag{43}$$

which was previously presented as Equation 33. The coefficient $\tilde{\mu}$ in Equation 43, denotes the coefficient of the small fictitious dissipative body force that is proportional to velocity (Rayleigh, 1883).

To derive the boundary condition for the gate opening, the movement of the boundary in the neighborhood of the discharge opening will be neglected. The fluctuating discharge is assumed to be concentrated at the point $(X,Y)=(0,-d_0)$, as shown in Figure 8, one need only suppose that the height of gate opening $2a\to 0$ while the discharge velocity $\Phi_X(0,Y;T)\to\infty$ in such a way that the total change in discharge $\int_{-d_0-a}^{-d_0+a}\Phi_X\left(0,Y;T\right)dY$ tends to the finite limit $\Delta Q(T)$ given in Equation 38. The fluctuating discharge can then be written as

$$\Delta Q\left(T\right)=\int_{-d_0-a}^{-d_0+a}\Phi_X\left(0,Y;T\right)dY \tag{44}$$

The boundary conditions at $X=0$ and $X=\infty$ are

$$\Phi_X\left(0,Y;T\right)=0 \text{ for } \left(-\infty \text{ or } d_r\right)\le Y<\left(-d_0-a\right) \text{ and } \left(-d_0+a\right)<Y\le 0 \tag{45}$$

Figure 8. Modeling of (a) steady discharge and (b) discharge fluctuation beneath a sluice gate

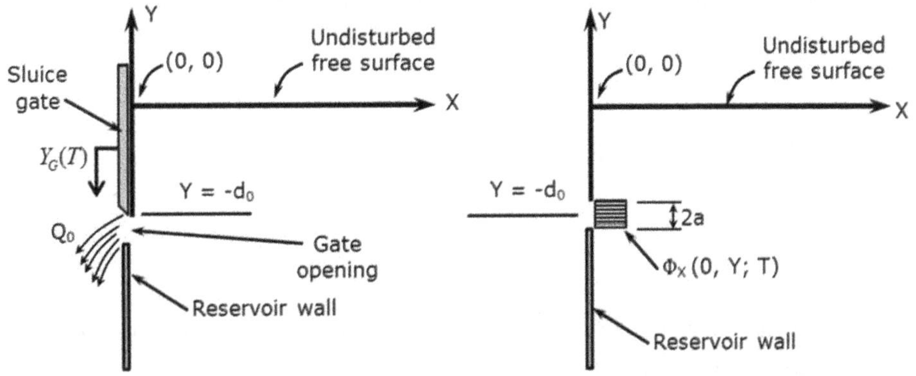

$$\Phi_X\left(\infty, Y; T\right) = 0 \text{ and } \Phi\left(\infty, Y; T\right) = 0 \tag{46}$$

The boundary conditions at the bottom of the reservoir for a reservoir with infinite depth are

$$\Phi_Y\left(X, -\infty; T\right) = 0 \text{ and } \Phi\left(X, -\infty; T\right) = 0 \tag{47}$$

and for a reservoir with finite depth, d_r

$$\Phi_Y\left(X, -d_r; T\right) = 0 \tag{48}$$

The boundedness condition, $\Phi\left(X, -\infty; T\right) = 0$ applies only in the case of an infinite reservoir.

Solution to the Laplace Equation

To obtain a solution to Equation 42 that satisfies the boundary conditions given in Equations 43 to 46 plus either Equation 47 or 48, the following Fourier cosine transform for the variable *X*,

$$\bar{\Phi}\left(S, Y; T\right) = \sqrt{2/\pi}\int_0^\infty \Phi\left(X, Y; T\right)\cos SXdX \tag{49}$$

is applied to Equation 42, yielding the following expression:

$$\bar{\Phi}_{YY}\left(S, Y; T\right) - S^2\bar{\Phi}\left(S, Y; T\right) = \sqrt{2/\pi}\Phi_X\left(0, Y; T\right) \tag{50}$$

where $\bar{\Phi}(S, Y; T)$ is the transform of $\Phi(X, Y; T)$ for the boundary conditions in Equation 46.

Solutions for the unsteady velocity potential, $\Phi(X,Y;T)$, will now be developed for the two reservoir depth conditions considered: the infinite depth reservoir and the reservoir of finite depth d_r.

For a Reservoir of Infinite Depth

The general solution of Equation 15 is given as

$$\bar{\Phi}\left(S,Y;T\right)=\bar{\Phi}^{(P)}\left(S,Y;T\right)+A\left(S;T\right)e^{SY} \tag{51}$$

where $\Phi^{(P)}(S,Y;T)$ is the particular solution and $A(S;T)$ is an arbitrary function. The particular solution must satisfy the following boundary conditions that correspond to Equation 12:

$$\bar{\Phi}_Y^{(P)}\left(S,-\infty;T\right)=0 \text{ and } \bar{\Phi}^{(P)}\left(S,-\infty;T\right)=0 \tag{52}$$

The following boundary condition at $Y=0$ can be imposed for the particular solution:

$$\bar{\Phi}_Y^{(P)}\left(S,0;T\right)=0 \tag{53}$$

By substituting Equation 51 into the free surface boundary condition, Equation 43, the following second-order ordinary differential equation is obtained:

$$A_{TT}\left(S;T\right)+\tilde{\mu}A_T\left(S;T\right)+gSA\left(S;T\right)=-\bar{\Phi}_{TT}^{(P)}\left(S,0;T\right)-\tilde{\mu}\bar{\Phi}_T^{(P)}\left(S,0;T\right) \tag{54}$$

The particular solution $\bar{\Phi}^{(P)}\left(S,Y;T\right)$ must satisfy Equation 50:

$$\bar{\Phi}_{yy}^{(P)}\left(S,Y;T\right)-S^2\bar{\Phi}^{(P)}\left(S,Y;T\right)=\sqrt{2/\pi}\Phi_X\left(0,Y;T\right) \tag{55}$$

To obtain a solution to Equation 55, the Fourier cosine transform for the variable Y:

$$\bar{\bar{\Phi}}^{(P)}\left(S,m;T\right)=\sqrt{2/\pi}\int_{-\infty}^{0}\bar{\Phi}^{(P)}\left(S,Y;T\right)\cos mYdY \tag{56}$$

is applied to Equation 55, yielding the following equation:

$$\left(m^2+S^2\right)\bar{\bar{\Phi}}^{(P)}\left(S,m;T\right)=-\left(2/\pi\right)\int_{-\infty}^{0}\Phi_X\left(0,Y;T\right)\cos mYdY \tag{57}$$

in which $\tilde{\bar{\Phi}}^{(P)}(S,m;T)$ is the transform of $\bar{\Phi}^{(P)}\left(S,Y;T\right)$ for the boundary conditions given in Equations 52 and 53. By making use of the boundary conditions in Equations 44, 45, and 38, it can be shown that

$$\begin{aligned}\int_{-\infty}^{0} \Phi_X\left(0,Y;T\right)\cos mYdY &= \int_{-d_0-a}^{-d_0+a} \Phi_X\left(0,Y;T\right)\cos mYdY \\ &= \int_{-d_0-a}^{-d_0+a} \Phi_X\left(0,Y;T\right)dY \times \cos md_0 \\ &= \Delta Q(T)\cos md_0\end{aligned}$$

where the first mean-value theorem was used. Using this above relationship with Equation 57 yields

$$\tilde{\bar{\Phi}}^{(P)}\left(S,m;T\right) = -\left(2/\pi\right)qe^{i\Omega T}\cos md_0/\left(m^2+S^2\right) \tag{58}$$

Applying the inverse transform for the variable *m*,

$$\bar{\Phi}^{(P)}\left(S,m;T\right) = \sqrt{2/\pi}\int_{-\infty}^{0}\tilde{\bar{\Phi}}^{(P)}\left(S,m;T\right)\cos mYdm$$

to Equation 58 results in the following integral representation of the particular solution $\bar{\Phi}^{(P)}\left(S,Y;T\right)$:

$$\bar{\Phi}^{(P)}(S,Y;T) = -(2/\pi)^{3/2}qe^{i\Omega T}\int_{-\infty}^{0}\frac{\cos md_0\cos mYdm}{m^2+S^2} \tag{59}$$

Substituting Equation 59 into Equation 54 yields the equation governing the arbitrary function $A\left(S;T\right)$.

$$A_{SS}(S;T)+\tilde{\mu}A_T(S;T)+gSA(S;T) = -\left(\frac{2}{\pi}\right)^{3/2}qe^{i\Omega T}\left(\Omega^2-i\tilde{\mu}\Omega\right)\int_{-\infty}^{0}\frac{\cos md_0dm}{m^2+S^2} \tag{60}$$

In order to derive the general solution of Equation 60, the initial conditions are needed, which can be prescribed at time $T=0$ as follows:

$$\Phi(X,0;0)=0,\Phi_T(X,0;0)=0 \tag{61a}$$

or

$$\bar{\Phi}(S,0;0)=0,\bar{\Phi}_T(S,0;0)=0 \tag{61b}$$

which states that the free surface is initially at rest in its horizontal equilibrium position. By substituting Equation 51 into Equation 61b, the following conditions are obtained:

$$\overline{\Phi}^{(P)}(S,0;0) + A(S;0)e^{SY} = 0, \quad \overline{\Phi}_T^{(P)}(S,0;0) + A_T(S;0) = 0 \tag{62}$$

By substituting the particular solution, Equation 59, into Equation 62, the initial conditions for the function $A(S;T)$ are obtained in the form:

$$A(S;0) = (2/\pi)^{3/2} q \int_{-\infty}^{0} \frac{\cos md_0 dm}{m^2+S^2} \text{ and } A_T(S;0) = (2/\pi)^{3/2} i\Omega q \int_{-\infty}^{0} \frac{\cos md_0 dm}{m^2+S^2} \tag{63}$$

Applying the Laplace transform to Equation 60, the transform $A(S;z)$ of $A(S;T)$ is found.

$$\bar{A}(S;\zeta) = \left(\frac{2}{\pi}\right)^{3/2} q \int_{-\infty}^{0} \frac{\cos md_0 dm}{m^2+S^2}\left[\frac{\zeta+\tilde{\mu}+i\Omega}{\zeta^2+\tilde{\mu}\zeta+gS} - \frac{\Omega^2 - i\tilde{\mu}\Omega}{(\zeta-i\Omega)(\zeta^2+\tilde{\mu}\zeta+gS)}\right] \tag{64}$$

where the initial conditions are given in Equation 63 were used. Applying the inverse transform based on the convolution theorem, the function $A(S;T)$ is obtained in the following integral representation:

$$A(S;T) = \left(\frac{2}{\pi}\right)^{\frac{3}{2}} q \int_{-\infty}^{0} \frac{\cos Pd_0 dP}{P^2+S^2} \cdot e^{-\frac{\tilde{\mu}}{2}T} \cdot \left(\cos\sqrt{gS - \frac{\tilde{\mu}^2}{4}} \cdot T + \frac{\tilde{\mu}/2 + i\Omega}{\Phi_0(S)} \sin \Phi_0(S)T\right)$$
$$- \left(\frac{2}{\pi}\right)^{\frac{3}{2}} q\left(\Omega^2 - i\tilde{\mu}\Omega\right) \int_{-\infty}^{0} \frac{\cos Pd_0 dP}{P^2+S^2} \cdot \frac{1}{\Phi_0(S)} \int_0^T e^{i\Omega\tilde{\tau}} \cdot e^{-\frac{\tilde{\mu}}{2}(T-\tilde{\tau})} \cdot \sin\Phi_0(S)\left(T-\tilde{\tau}\right) d\tilde{\tau} \tag{65a}$$

where the phase function $\Phi_0(S)$ is given by

$$\Phi_0(S) = \sqrt{gS - \tilde{\mu}^2/4} \tag{65b}$$

The integration with respect to $\tilde{\tau}$ from zero to T can be accomplished as follows:

$$\frac{1}{\Phi_0(S)} \int_0^T e^{i\Omega T} \cdot e^{-\frac{\tilde{\mu}}{2}(T-\tilde{\tau})} \cdot \sin\Phi_0(S)(T-\tilde{\tau}) d\tilde{\tau}$$
$$= \frac{1}{gS - \Omega^2 + i\tilde{\mu}\Omega} \cdot \left[e^{i\Omega T} - e^{-\frac{\tilde{\mu}}{2}T}\left\{\frac{\tilde{\mu}/2 + i\Omega}{\Phi_0(S)} \sin\Phi_0(S)T + \cos\Phi_0(S)T\right\}\right]$$

As a result, Equation 65a can be reduced to the following form:

$$A(S;T)=\left(2/\pi\right)^{3/2}q\left[\begin{array}{l}-\dfrac{\Omega^2-i\tilde{\mu}\Omega}{gS-\Omega^2+i\tilde{\mu}\Omega}\displaystyle\int_{-\infty}^{0}\frac{\cos md_0dm}{m^2+S^2}e^{i\Omega T}\\+\displaystyle\int_{-\infty}^{0}\frac{\cos md_0dm}{m^2+S^2}\frac{gSe^{-(\tilde{\mu}/2)T}}{gS-\Omega^2+i\tilde{\mu}\Omega}\left\{\cos\Phi_0\left(S\right)T+\frac{\tilde{\mu}/2+i\omega}{\Phi_0\left(S\right)}\sin\Phi_0\left(S\right)\right\}\end{array}\right] \tag{65c}$$

By substituting Equations 59 and 65c into Equation 51, the general solution $\bar{\Phi}(S,Y;T)$ is obtained in the form:

$$\begin{aligned}\bar{\Phi}\left(S,Y;T\right)=&-\left(2/\pi\right)^{3/2}q\left[\int_{-\infty}^{0}\frac{\cos md_0\cos mYdm}{m^2+S^2}+\frac{S_0e^{sy}}{S-S_0}\int_{-\infty}^{0}\frac{\cos md_0dm}{m^2+S^2}\right]e^{i\Omega T}\\&+\left(2/\pi\right)^{3/2}q\int_{-\infty}^{0}\frac{\cos md_0dm}{m^2+S^2}\frac{Se^{SY}e^{-(\tilde{\mu}/2)T}}{S-S_0}\left[\cos\Phi_0\left(S\right)T+\frac{\tilde{\mu}/2+i\Omega}{\Phi_0\left(S\right)}\sin\Phi_0\left(S\right)T\right]\end{aligned} \tag{66a}$$

where the constant S_0 is given by

$$S_0=\frac{\Omega^2-i\tilde{\mu}\Omega}{g} \tag{66b}$$

Now the inverse transform for S is applied to Equation 66a and the integrations on m are carried out, as follows:

$$\begin{aligned}\int_{-\infty}^{0}\frac{\cos md_0\cos mY}{m^2+S^2}dm&=\frac{1}{2}\int_{0}^{\infty}\frac{\cos\left(d_0+Y\right)m+\cos\left(d_0-Y\right)m}{m^2+S^2}dm\\&=\frac{\pi}{4S}\left(e^{-|d_0+Y|S}+e^{-|d_0-Y|S}\right)\end{aligned}$$

$$\int_{-\infty}^{0}\frac{\cos md_0dm}{m^2+S^2}=\frac{\pi}{2}\cdot\frac{1}{S}\cdot e^{-d_0S}$$

As a result, the velocity potential $\Phi\left(X,Y;T\right)$ is obtained in the form:

$$\begin{aligned}\Phi\left(X,Y,T\right)=&-\left(2/\pi\right)q\left[\frac{1}{2}\int_{0}^{\infty}\left(e^{-|d_0+Y|S}+e^{-|d_0-Y|S}\right)\frac{\cos SX}{S}dS+S_0\int_{0}^{\infty}\frac{e^{-(d_0-Y)S}\cos SX}{S\left(S-S_0\right)}dS\right]e^{i\Omega T}\\&+\left(2/\pi\right)q\int_{0}^{\infty}e^{-(d_0-Y)S}\cos SXdS\,\frac{e^{-(\tilde{\mu}/2)T}}{S-S_0}\left[\cos\Phi_0\left(S\right)T+\frac{\tilde{\mu}/2+i\Omega}{\Phi_0\left(S\right)}\sin\Phi_0\left(S\right)T\right]\end{aligned} \tag{67}$$

Moreover, the integration on S of the first and second terms can be carried out as follows:

$$\frac{1}{2}\int_0^\infty \frac{\cos SX}{S}\left(e^{-|d_0+Y|S}+e^{-|d_0-Y|S}\right)dS=\int_0^\infty \frac{e^{-SX}}{S}\cos d_0 S\cos YS$$

$$\int_0^\infty \frac{e^{-(d_0-Y)S}\cos SX}{S\left(S-S_0\right)}dS=-i\pi\frac{g}{\Omega^2-i\tilde{\mu}\Omega}\cdot e^{-(d_0-Y)S_0}\cdot e^{-\frac{\tilde{\mu}\Omega}{g}X}\cdot e^{-i\frac{\Omega^2}{g}X}$$

$$-\left[\int_0^\infty \frac{e^{-SX}\sin\left(d_0-Y\right)S}{S^2+S_0^{\,2}}dS+S_0\int_0^\infty \frac{e^{-SX}}{S}\cdot\frac{\cos\left(d_0-Y\right)S}{S^2+S_0^{\,2}}dS\right]$$

where the integrals on S from 0 to ∞ were expanded for integration in the complex plane, as shown in Figures 9a and 9b, where the singular points appear at $S = 0$ and $S = S_0$, given by Equation 66b. A residue at the singular point $S = S_0$ contributes to the second integral. Substituting the above integrations into Equation 67, $\Phi\left(X,Y;T\right)$ is obtained:

$$\begin{aligned}\Phi\left(X,Y;T\right)&=2iqe^{-(d_0-Y)S_0}e^{-iS_0X}e^{i\Omega T}\\&+\frac{2}{\pi}q\left[\int_0^\infty \frac{e^{-SX}}{S}\sin d_0S\sin YSdS-\int_0^\infty \frac{e^{-SX}}{S^2+S_0^2}\left\{S\cos\left(d_0-Y\right)S-S_0\sin\left(d_0-Y\right)S\right\}dS\right]e^{i\Omega T}\\&+\frac{2}{\pi}q\int_0^\infty \exp\left\{-\left(d_0-Y\right)S\right\}\cos SXdS\,\frac{e^{-(\tilde{\mu}/2)T}}{S-S_0}\left[\cos\Phi_0\left(S\right)T+\frac{\tilde{\mu}/2+i\Omega}{\Phi_0\left(S\right)}\sin\Phi_0\left(S\right)T\right]\end{aligned}\quad(68)$$

where the following integration also was used:

$$\begin{aligned}S_0^{\,2}\int_0^\infty \frac{e^{-SX}}{S}\cdot\frac{\cos\left(d_0-Y\right)S}{S^2+S_0^{\,2}}dS&=\int_0^\infty \frac{e^{-SX}}{S}\cos d_0S\cos YSdS\\&+\int_0^\infty \frac{e^{-SX}}{S}\sin d_0S\sin YSdS-\int_0^\infty \frac{S\cdot e^{-SX}\cos(d_0-Y)S}{S^2+S_0^{\,2}}dS\end{aligned}$$

By differentiating $\Phi\left(X,Y;T\right)$ with respect to the time T, $\Phi_T\left(X,Y;T\right)$ is obtained:

$$\begin{aligned}\Phi_T\left(X,Y;T\right)&=-2q\Omega e^{-(d_0-Y)S_0}e^{-iS_0X}e^{i\Omega T}\\&+\frac{2}{\pi}iq\Omega\left[\int_0^\infty \frac{e^{-SX}}{S}\sin d_0S\sin YSdS-\int_0^\infty \frac{e^{-SX}}{S^2+S_0^2}\left\{S\cos\left(d_0-Y\right)-S_0\sin\left(d_0-Y\right)S\right\}dS\right]e^{i\Omega T}\\&+\left(iqg/\pi\right)\left[\int_0^\infty \Psi_1\left(S\right)\exp\left\{i\Phi_1\left(S\right)T\right\}dT+\Psi_2\left(T\right)\exp\left\{i\Phi_2\left(S\right)T\right\}dS\right]\end{aligned}\quad(69)$$

where the phase functions $\Phi_1\left(S\right)$ and $\Phi_2\left(S\right)$, as well as the functions $\Psi_1\left(S\right)$ and $\Psi_2\left(S\right)$, are

Figure 9. Integration in the complex plane of (a) the first integral of Equation 67, and (b) the second integral of Equation 67

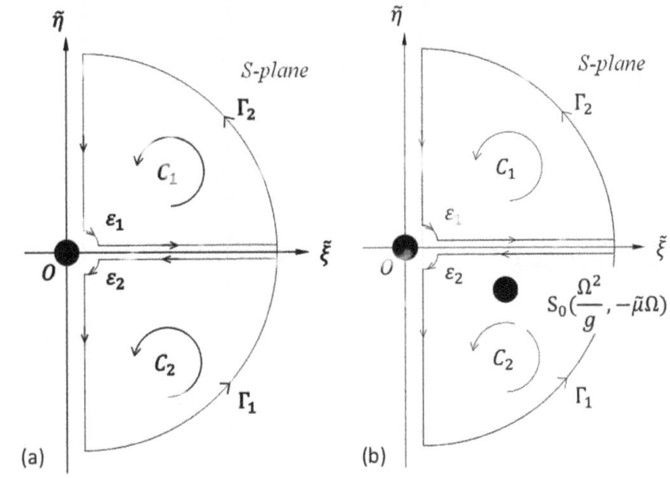

$$\left.\begin{aligned}&\Phi_1(S)=\Phi_0(S)+i\tilde{\mu}/2 \text{ and } -\Phi_2(S)=\Phi_0(S)-i\tilde{\mu}/2\\&\Psi_1(S)=e^{-(d_0-Y)S}\cos SX\Big/\left\{\Phi_0(S)-\Omega+\frac{\tilde{\mu}}{2}i\right\}\\&\Psi_2(S)=-e^{-(d_0-Y)S}\cos SX\Big/\left\{\Phi_0(S)+\Omega-\frac{\tilde{\mu}}{2}i\right\}\end{aligned}\right\} \tag{70}$$

All of the functions in Equation 70 are regular analytic functions of the variables in a domain containing the segment $[0,\infty)$. The imaginary parts of the phase functions $\Phi_1(S)$ and $\Phi_2(S)$ are positive. Moreover, the functions $\Psi_1(S)$ and $\Psi_2(S)$ have no stationary points on the integral domain. The third term in Equation 69 results from the initial conditions and, hence, could be expected to provide transients which die out like $1/T$ as $T\to\infty$, applying Kelvin's method of stationary phase (Stoker, 1957; Baker and Copson, 1939).

Thus, for large *T*, the following asymptotic representation for $\Phi_T(X,Y;T)$ is obtained:

$$\Phi_T(X,Y;T)=-2q\Omega e^{-(d_0-Y)S_0}e^{-iS_0X}e^{i\Omega T}+\frac{2}{\pi}iq\Omega\left[\begin{aligned}&\int_0^\infty\frac{e^{-SX}}{S}\sin d_0S\sin YSdS\\&-\int_0^\infty\frac{e^{-SX}}{S^2+S_0^{\,2}}\left\{S\cos(d_0-Y)S-S_0\sin(d_0-Y)Y\right\}dS\end{aligned}\right]e^{i\Omega T} \tag{71}$$

Since the positive constant $\tilde{\mu}$ representing the effect of dissipation is very small, *i.e.*, $\tilde{\mu}\ll\Omega$ the constant S_0 may be replaced by Ω^2/g. Then Equation 71 becomes:

$$\Phi_T\left(X,Y;T\right) = -2q\Omega e^{-\left(\Omega^2/g\right)d_0} e^{-\left(\tilde{\mu}\Omega/g\right)X} e^{\left(\Omega^2/g\right)Y} e^{-i\left[\left(\Omega^2/g\right)X-\Omega T\right]}$$
$$+\frac{2}{\pi}iq\Omega\left[\begin{array}{l}\displaystyle\int_0^\infty \frac{e^{-SX}}{S}\sin d_0 S \sin YSdS \\ \displaystyle -\int_0^\infty \frac{e^{-SX}}{S^2+\left(\Omega^2/g\right)^2}\left\{\begin{array}{l} S\cos\left(d_0-Y\right)S \\ -\left(\dfrac{\Omega}{g}\right)^2 \sin\left(d_0-Y\right)S\end{array}\right\}dS\end{array}\right]e^{i\Omega T} \tag{72}$$

After introducing the following dimensionless variables

$$x = X/d_0, \qquad y = Y/d_0, \qquad s = d_0 S, \qquad t = \Omega_0 T, \qquad \varphi = \Phi/q, \qquad \mu = \tilde{\mu}/\Omega_0 \tag{73}$$

where Ω_0 represents the circular frequency of the gate in air. With the introduction of the Froude number and the basic Froude number F_0, written respectively as

$$F = \sqrt{d_0/g}\,\Omega \text{ and } F_0 = \sqrt{d_0/g}\,\Omega_0 \tag{74}$$

Equation 72 can be reduced to the following dimensionless expression:

$$\varphi_t\left(x,y;t\right) = -2F_0 e^{-F_0^2} e^{-\mu F_0^2 x} e^{F_0^2 y} e^{-i\left(F_0^2 x - F_0 t\right)}$$
$$+\frac{2}{\pi}iF_0\left[\int_0^\infty \frac{e^{-sx}}{s}\sin s \sin ysds - \int_0^\infty \frac{e^{-sx}}{s^2+F_0^4}\left\{s\cos\left(1-y\right)s - F_0^2\sin\left(1-y\right)s\right\}ds\right]e^{iF_0 t} \tag{75}$$

For a Reservoir of Finite Depth

For a reservoir with a finite depth, d_r, the general solution of Equation 50 is given in the following form:

$$\bar{\Phi}\left(S,Y;T\right) = \bar{\Phi}^{(P)}\left(S,Y;T\right) + A\left(S;T\right)\cosh S(Y+d_r) \tag{76}$$

where $\bar{\Phi}^{(P)}\left(S,Y;T\right)$ is the particular solution and $A(S;T)$ is an arbitrary function. The homogeneous solution requires the hyperbolic cosine function to satisfy the boundary condition along the upstream channel floor. The particular solution must satisfy the following boundary conditions corresponding to Equation 48:

$$\bar{\Phi}_Y^{(P)}\left(S,-d_r;T\right) = 0 \tag{77}$$

The following boundary condition at $Y = 0$ can be imposed for the particular solution:

$$\bar{\Phi}_Y^{(P)}\left(S, 0; T\right) = 0 \tag{78}$$

The particular solution $\bar{\Phi}_{yy}^{(P)}\left(S, Y; T\right)$ must satisfy Equation 50:

$$\bar{\Phi}_{yy}^{(P)}\left(S, Y; T\right) - S^2 \bar{\Phi}^{(P)}\left(S, Y; T\right) = \sqrt{2/\pi}\,\Phi_X\left(0, Y; T\right) \tag{79}$$

To obtain a solution to Equation 79, the Fourier series transform for the variable Y:

$$\tilde{\bar{\Phi}}^{(P)}\left(S, m; T\right) = \sqrt{2/\pi}\int_{-\infty}^{0} \bar{\Phi}^{(P)}\left(S, Y; T\right)\cos mY dY \tag{80}$$

is applied to Equation 79, yielding the following equation:

$$\left(S^2 + \left(\frac{n\pi}{d_r}\right)^2\right)\tilde{\bar{\Phi}}^{(P)}\left(S, n; T\right) = -\sqrt{2/\pi}\int_{-d_r}^{0} \Phi_X\left(0, Y; T\right)\cos\frac{n\pi Y}{d_r}dY \tag{81}$$

in which $\tilde{\bar{\Phi}}^{(P)}\left(S, n; T\right)$ is the transform of $\bar{\Phi}^{(P)}\left(S, Y; T\right)$ and use has been made of the boundary conditions given in Equations 77 and 78. By making use of the boundary conditions in Equations 44 and 45, as well as Equation 38, it can be shown that

$$\begin{aligned}\int_{-d_r}^{0} \Phi_X\left(0, Y; T\right)\cos\frac{n\pi Y}{d_r}dY &= \int_{-d_0-a}^{-d_0+a} \Phi_X\left(0, Y; T\right)\cos mY dY \\ &= \int_{-d_0-a}^{-d_0+a} \Phi_X\left(0, Y; T\right)dY \times \cos\frac{n\pi d_0}{d_r} \\ &= \Delta Q(T)\cos\frac{n\pi d_0}{d_r}\end{aligned}$$

where, the first mean-value theorem was used. Using the above relationship with Equation 81 yields

$$\tilde{\bar{\Phi}}^{(P)}\left(S, n; T\right) = -\sqrt{\frac{2}{\pi}}\cdot\frac{1}{S^2 + \left(n\pi/d_r\right)^2}q\cdot e^{i\Omega T}\cos\frac{n\pi d_0}{d_r} \tag{82}$$

After that, the inverse Fourier series transform with respect to n,

$$\bar{\Phi}^{(p)}\left(S,Y;T\right)=\frac{1}{d_r}\tilde{\bar{\Phi}}^{(p)}\left(S,0;T\right)+\frac{2}{d_r}\sum_{n=1}^{\infty}\tilde{\bar{\Phi}}^{(p)}\left(S,n;T\right)\cos\frac{n\pi Y}{d_r} \tag{83}$$

is applied to Equation 82, with the following result:

$$\bar{\Phi}^{(P)}\left(S,Y;T\right)=-\sqrt{\frac{2}{\pi}}\cdot\frac{q\cdot e^{i\Omega T}}{d_r}\left\{\frac{1}{S^2}+2\sum_{n=1}^{\infty}\frac{\cos\left(n\pi d_0/d_r\right)}{S^2+\left(n\pi/d_r\right)^2}\cdot\cos\frac{n\pi Y}{d_r}\right\} \tag{84}$$

If the Fourier integral transform, Equation 49, is applied to the free-surface condition, Equation 43, and then the general solution given by Equation 76, is substituted into the resulting equation, the following equation determining the function $A\left(S;T\right)$ can be found:

$$A_{TT}\left(S;T\right)+\tilde{\mu}A_T\left(S;T\right)+gS\tanh Sd_r\cdot A\left(S;T\right)=-\frac{1}{\cosh Sd_r}\left\{\bar{\Phi}_{TT}^{(P)}\left(S,0;T\right)+\tilde{\mu}\bar{\Phi}_T^{(P)}\left(S,0;T\right)\right\} \tag{85}$$

Substitution of the particular solution, $\bar{\Phi}^{(p)}$ given by Equation 84, into Equation 85 yields:

$$A_{TT}(S;T)+\tilde{\mu}A_T(S;T)+gS\tanh Sd_r\cdot A(S;T)$$
$$=-\sqrt{\frac{2}{\pi}}\cdot\frac{\Omega^{2-}i\tilde{\mu}\Omega}{d_r\cosh Sd_r}\cdot qe^{i\Omega T}\cdot\left[\frac{1}{S^2}+2\sum_{n=1}^{\infty}\frac{\cos\left(n\pi d_0/d_r\right)}{S^2+\left(n\pi/d_r\right)^2}\right] \tag{86}$$

Substitution of the general solution, Equation 76, into the initial condition, Equation 61b, gives us:

$$\bar{\Phi}^{(P)}\left(S,Y;0\right)+A\left(S;0\right)\cosh S\left(Y+d_r\right)=0\text{ and }\bar{\Phi}_T^{(P)}\left(S,Y;0\right)+A_T\left(S;0\right)\cosh S\left(Y+d_r\right)=0 \tag{87}$$

from which the initial conditions for the function $A^*\left(S;T\right)$ can be determined:

$$A(S;0)=\sqrt{\frac{2}{\pi}}\cdot\frac{q}{d_r}\cdot\frac{1}{\cosh Sd_r}\left[\frac{1}{S^2}+2\sum_{n=1}^{\infty}\frac{\cos\left(n\pi d_0/d_r\right)}{S^2+(n\pi/d_r)^2}\right] \tag{88a}$$

$$A_T(S;0)=\sqrt{\frac{2}{\pi}}\cdot\frac{q}{d_r}\cdot\frac{i\Omega}{\cosh Sd_r}\left[\frac{1}{S^2}+2\sum_{n=1}^{\infty}\frac{\cos\left(n\pi d_0/d_r\right)}{S^2+(n\pi/d_r)^2}\right] \tag{88b}$$

Here the Laplace transform:

$$\bar{A}(S;\zeta) = \int_0^\infty e^{-\zeta T} A(S;T)\,dT \tag{89}$$

is applied to Equation 86 with the initial conditions, Equations 88 to yield the following result:

$$\begin{aligned}\bar{A}(S;\zeta) &= \sqrt{\frac{2}{\pi}}\cdot\frac{q}{d_r}\cdot\frac{\zeta+\tilde{\mu}+i\Omega}{\cosh Sd_r\cdot(\zeta^2+\tilde{\mu}\zeta+gS\tanh Sd_r)}\left[\frac{1}{S^2}+2\sum_{n=1}^{\infty}\frac{\cos(n\pi d_0/d_r)}{S^2+(n\pi/d_r)^2}\right]\\ &-\sqrt{\frac{2}{\pi}}\cdot\frac{q}{d_r}\cdot\frac{(\Omega 2-i\tilde{\mu}\Omega)/(\zeta-i\Omega)}{\cosh Sd_r\cdot(\zeta^2+\tilde{\mu}\zeta+gS\tanh Sd_r)}\cdot\left[\frac{1}{S^2}+2\sum_{n=1}^{\infty}\frac{\cos(n\pi d_0/d_r)}{S^2+(n\pi/d_r)^2}\right]\end{aligned} \tag{90}$$

The inverse Laplace transform of Equation 90 can be reduced to the form:

$$\begin{aligned}A(S;T) &= \sqrt{\frac{2}{\pi}}\cdot\frac{q}{d_r}\cdot\frac{1}{\cosh Sd_r}\left[\frac{1}{S^2}+2\sum_{n=1}^{\infty}\frac{\cos\left(n\pi d_0/d_r\right)}{S^2+(n\pi/d_r)^2}\right]\\ &\times\left[\begin{aligned}&e^{-\frac{\tilde{\mu}}{2}T}\left\{\cos\Phi_0(S)\cdot T+\frac{\tilde{\mu}/2+i\boldsymbol{\Omega}}{\Phi_0(S)}\cdot\sin\Phi_0(S)\cdot T\right\}\\ &-\frac{\Omega^2-i\tilde{\mu}\Omega}{\Phi_0(S)}\cdot\int_0^T e^{i\Omega\tau}\cdot e^{-\frac{\tilde{\mu}}{2}(T-\tau)}\sin\Phi_0(S)(T-\tau)d\tau\end{aligned}\right]\end{aligned} \tag{91}$$

The phase function, $\Phi_0\left(S\right)$, is given by

$$\Phi_0\left(S\right) = \sqrt{gS\tanh Sd_r - \tilde{\mu}^2/4} \tag{92}$$

The integration with respect to τ from zero to T can be accomplished as follows:

$$\begin{aligned}&\frac{1}{\Phi_0\left(S\right)}\int_0^T e^{i\Omega\tau}\cdot e^{-\frac{\tilde{\mu}}{2}(T-\tau)}\sin\Phi_0\left(S\right)\left(T-\tau\right)d\tau\\ &=\frac{1/g}{S\tanh Sd_r - S_0}\cdot\left[e^{i\Omega T}-e^{-\frac{\tilde{\mu}}{2}T}\left\{\frac{\left(i\Omega+\tilde{\mu}/2\right)\sin\Phi_0\left(S\right)T}{\Phi_0\left(S\right)}+\cos\Phi_0\left(S\right)T\right\}\right]\end{aligned}$$

As a result, Equation 91 can be reduced to the following form:

$$A\left(S;T\right)=\sqrt{\frac{2}{\pi}}\cdot\frac{q}{d_r}\cdot\frac{1}{\cosh Sd_r}\cdot\left[\frac{1}{S^2}+2\sum_{n=1}^{\infty}\frac{\cos n\pi d_0/d_r}{S^2+\left(n\pi/d_r\right)^2}\right]$$
$$\times\left[\begin{array}{l}-\dfrac{S_0}{S\tanh Sd_r-S_0}\cdot e^{i\Omega T}+\dfrac{e^{-\frac{\tilde{\mu}}{2}T}\cdot S\tanh Sd_r}{S\tanh Sd_r-S_0}\\ \times\left\{\cos\Phi_0\left(S\right)T+\dfrac{\tilde{\mu}/2+i\Omega}{\Phi_0\left(S\right)}\cdot\sin\Phi_0\left(S\right)T\right\}\end{array}\right] \tag{93}$$

in which, with the introduction of the small frictional coefficient $\tilde{\mu}\left(>0\right)$, the effect of the initial conditions dies out as time increases. The definition of S_0 is the same as previously given in Equation 66b.

Substitution of the particular solution, $\bar{\Phi}^{(p)}$, given by Equation 84, and the arbitrary function $A\left(S;T\right)$, given by Equation 93, into the general solution, Equation 76, yields the general solution of Equation 50, with the following form:

$$\bar{\Phi}\left(S,Y;T\right)=-\sqrt{\frac{2}{\pi}}\cdot\frac{q}{d_r}e^{i\Omega T}\left[\frac{1}{S^2}+2\sum_{n=1}^{\infty}\frac{\cos n\pi d_0/d_r}{S^2+\left(n\pi/d_r\right)^2}\cdot\cos\frac{n\pi Y}{d_r}\right]$$
$$+\sqrt{\frac{2}{\pi}}\cdot\frac{q}{d_r}\cdot\frac{\cosh S\left(Y+d_r\right)}{\cosh Sd_r}\cdot\left[\frac{1}{S^2}+2\sum_{n=1}^{\infty}\frac{\cos n\pi d_0/d_r}{S^2+\left(n\pi/d_r\right)^2}\right] \tag{94}$$
$$\times\left[-\frac{S_0}{S\tanh Sd_r-S_0}\cdot e^{i\Omega T}+\frac{e^{-\frac{\tilde{\mu}}{2}T}\cdot S\tanh Sd_r}{S\tanh Sd_r-S_0}\left\{\cos\Phi_0\left(S\right)\cdot T+\frac{\tilde{\mu}/2+i\Omega}{\Phi_0\left(S\right)}\cdot\sin\Phi_0\left(S\right)\cdot T\right\}\right]$$

Application of the inverse Fourier transform with respect to S, given as

$$\Phi\left(X,Y;T\right)=\sqrt{2/\pi}\int_0^{\infty}\bar{\Phi}\left(S,Y;T\right)\cos SXdS \tag{95}$$

to Equation 94 yields the velocity potential $\Phi\left(X,Y;T\right)$:

$$\Phi(X,Y;T) = -\frac{2}{\pi}\cdot\frac{q}{d_r}\cdot e^{i\Omega T}\int_0^\infty\left[\frac{1}{S^2}+2\sum_{n=1}^{\infty}\frac{\cos n\pi d_0/d_r\cdot\cos n\pi Y/d_r}{S^2+\left(n\pi/d_r\right)^2}\right]\cos SXdS$$
$$-\frac{2}{\pi}\cdot q\cdot\frac{S_0}{d_r}\cdot e^{i\Omega T}\int_0^\infty\left[\frac{1}{S^2}+2\sum_{n=1}^{\infty}\frac{\cos n\pi d_0/d_r}{S^2+\left(n\pi/d_r\right)^2}\right]\cdot\frac{\cosh S\left(Y+d_r\right)\cdot\cos SX}{W\left(S\right)\cdot\cosh Sd_r}\cdot dS$$
$$+\frac{2}{\pi}\cdot\frac{q}{d_r}\cdot e^{-\frac{\tilde{\mu}}{2}T}\int_0^\infty\left[\frac{1}{S^2}+2\sum_{n=1}^{\infty}\frac{\cos n\pi d_0/d_r}{S^2+\left(n\pi/d_r\right)^2}\right]\cdot\frac{\sinh Sdr\cosh S\left(Y+d_r\right)}{W\left(S\right)\cosh^2 Sd_r}\cdot S\cos SXdS \tag{96}$$
$$\times\left\{\cos\Phi_d\left(S\right)\cdot T+\frac{\tilde{\mu}/2+i\Omega}{\Phi_d\left(S\right)}\cdot\sin\Phi_d\left(S\right)\cdot T\right\}$$

where the function $W(S)$ is defined as

$$W\left(S\right) = S\tanh Sd_r - S_0 \tag{97}$$

The first and second integrals on S in Equation 96 can be integrated from zero to infinity on the real axis by expanding the integrals in the complex plane, as shown in Figure 10. The singular points appear at $S = 0$, S_0' and $S_j\left(j = 1,2,\ldots\right)$ given as follows:

$$S_0' = \tilde{\xi}_0 + i\tilde{\eta}_0,\text{ where }\tilde{\eta}_0 \ll 1.0 \tag{98a}$$

Figure 10. Integration in the complex plane of (a) the first integral in Equation 96, and (b) the second integral in Equation 96

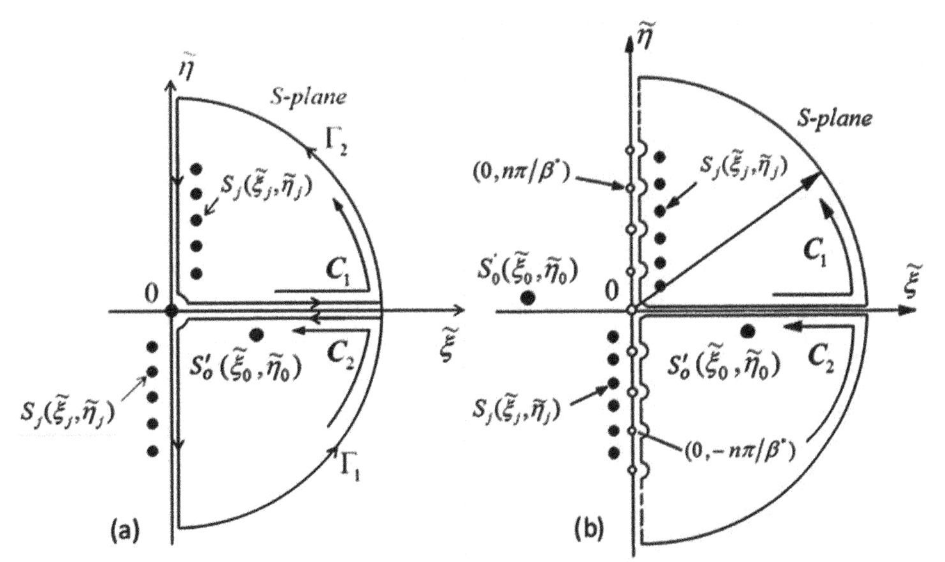

$$S_j = \tilde{\xi}_j + i\tilde{\eta}_j \left(j = 1, 2, \cdots\right), \text{where } \left|\tilde{\xi}_j\right| \ll 1.0 \tag{98b}$$

The equations determining the singular points S_0' and $S_j\left(j = 1, 2, \ldots\right)$ can be derived from Equation 97, resulting in the following expressions:

$$\tilde{\xi}_0 \tanh \tilde{\xi}_0 d_r - \frac{\Omega^2}{g} = 0 \text{ and } \tilde{\eta}_0 \tanh \tilde{\xi}_0 d_r - \frac{\tilde{\mu}\Omega}{g} = 0 \tag{99a}$$

$$\tilde{\xi}_j \tan \tilde{\eta}_j d_r + \frac{\mu\Omega}{g} = 0 \text{ and } \tilde{\eta}_j \tan \tilde{\eta}_j d_r + \frac{\Omega^2}{g} = 0 \tag{99b}$$

where $\eta_0, \tilde{\eta}_0$ and $\tilde{\xi}_j \; \left(j = 1, 2, \ldots\right)$ are small values representing the dispersive effect, based on the small fictitious dissipative body force coefficient $\tilde{\mu}$. In addition, they satisfy the following characteristics:

$$\tilde{\eta}_0 < 0 \text{ for } \tilde{\xi}_0 > 0 \text{ and } \tilde{\eta}_0 > 0 \text{ for } \tilde{\xi}_0 < 0 \tag{100a}$$

$$\tilde{\xi}_j > 0 \text{ for } \tilde{\eta}_j > 0 \text{ and } \tilde{\xi}_j < 0 \text{ for } \tilde{\eta}_j < 0 \tag{100b}$$

Solutions for $\tilde{\xi}_0$ and $\tilde{\eta}_j$ (j=1, 2, …) are schematically shown in Figure 11. Residues at the singular points S_0 and S_j (j=1, 2, …) contribute to the integrations. From the integrations in the complex planes, the integrals are reduced to the following expressions:

First integral term:

$$\int_0^{\infty}\left[\frac{1}{S^2} + 2\sum_{n=1}^{\infty}\frac{\cos n\pi d_0 / d_r \cdot \cos n\pi Y / d_r}{S^2 + \left(n\pi / d_r\right)^2}\right]\cos SXdS$$
$$= -\frac{\pi}{2}X + d_r \sum_{n=1}^{\infty}\frac{e^{-\frac{n\pi}{d_r}X}}{n}\cos\frac{n\pi d_0}{d_r}\cdot\cos\frac{n\pi Y}{d_r} + \frac{1}{\varepsilon} + O(\varepsilon) \tag{101}$$

Second integral term:

$$S_0\int_0^\infty\left[\frac{1}{S^2}+2\sum_{n=1}^{\infty}\frac{\cos n\pi d_0/d_r}{S^2+\left(n\pi/d_r\right)^2}\right]\cdot\frac{\cosh S\left(Y+d_r\right)\cdot\cos SX}{W\left(S\right)\cdot\cosh Sd_r}\cdot dS$$

$$=\frac{\pi}{2}X-d_r\sum_{n=1}^{\infty}\frac{e^{-\frac{n\pi}{d_r}X}}{n}\cos\frac{n\pi d_0}{d_r}\cdot\cos\frac{n\pi Y}{d_r}-\frac{1}{\varepsilon}-O(\varepsilon) \tag{102}$$

$$-\pi S_0\left[\begin{array}{l}\dfrac{e^{\tilde{\eta}_0 X}\cosh\tilde{\xi}_0\left(Y+d_r\right)}{W'\left(\tilde{\xi}_0\right)\cosh\tilde{\xi}_0 d_r}\left\{\dfrac{1}{\tilde{\xi}_0^2}+2\sum_{n=1}^{\infty}\dfrac{\cos n\pi d_0/d_r}{\tilde{\xi}_0^2+\left(n\pi/d_r\right)^2}\right\}ie^{-i\tilde{\xi}_0 X}\\ +\sum_{j=1}^{\infty}\dfrac{e^{(-\tilde{\eta}_j+i\tilde{\xi}_0 X)}\cos\tilde{\eta}_j\left(Y+d_r\right)}{W'\left(i\tilde{\eta}_j\right)/i\cdot\cos\tilde{\eta}_j d_r}\left\{\dfrac{1}{\tilde{\eta}_j^{\,2}}+2\sum_{n=1}^{\infty}\dfrac{\cos n\pi d_0/d_r}{\tilde{\eta}_j^{\,2}-\left(n\pi/d_r\right)^2}\right\}\end{array}\right]$$

where ε is a small positive value. The functions $W'(\tilde{\xi}_0)$ and $\eta_j, \tilde{\eta}_j$ are given by

$$W'\left(\tilde{\xi}_0\right)=\tanh\tilde{\xi}d_r+\tilde{\xi}_0 d_r/\cosh^2\tilde{\xi}_0 d_r \tag{103a}$$

$$W'\left(i\tilde{\eta}_j\right)/i=\tan\tilde{\eta}_j d_r+\tilde{\eta}_j d_r/\cos^2\tilde{\eta}_j d_r \tag{103b}$$

Substitution of Equations 98 and 99 into Equation 96 results in the solution to Equation 42 (the Laplace equation) satisfying the boundary conditions given in Equations 43 to 46 plus Equation 49 for a finite depth upstream channel, in the following series representation for large T:

Figure 11. Solutions of the transcendental equations for (a) $\tilde{\xi}_0\tanh\tilde{\xi}_0 d_r-\frac{\Omega^2}{g}=0$, *and (b)* $\tilde{\eta}_j\tan\tilde{\eta}_j d_r+\frac{\Omega^2}{g}=0$

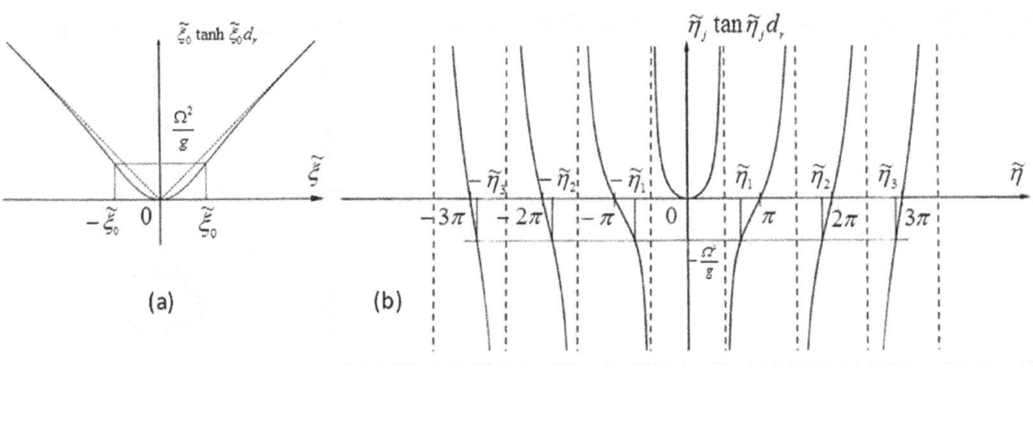

$$\Phi\left(X,Y;T\right)=\frac{2S_0}{d_r}q\left[\frac{e^{\tilde{\eta}_0 X}\cosh\tilde{\xi}_0\left(Y+d_r\right)}{W'\left(\tilde{\xi}_0\right)\cosh\tilde{\xi}_o d_r}\cdot\left\{\frac{1}{\tilde{\xi}_0^2}+2\sum_{n=1}^{\infty}\frac{\cos\frac{n\pi d_0}{d_r}}{\tilde{\xi}_0^2+\left(n\pi/d_r\right)^2}\right\}ie^{-i\left(\tilde{\xi}_0 X-\Omega T\right)}\right.$$
$$\left.+\sum_{j=1}^{\infty}\frac{e^{-\left(\tilde{\eta}_j+i\xi_j\right)X}\cos\tilde{\eta}_j\left(Y+d_r\right)}{W'\left(i\tilde{\eta}_j\right)/i\cos\tilde{\eta}_j\cdot d_r}\cdot\left\{\frac{1}{\tilde{\eta}_j^2}+2\sum_{n=1}^{\infty}\frac{\cos\frac{n\pi d_0}{d_r}}{\tilde{\eta}_j^2-\left(n\pi/d_r\right)^2}\right\}\cdot e^{i\Omega T}\right] \tag{104}$$

where the third term in Equation 96 is disregarded as a transient which dies out like $1/T$ as $T\to\infty$, applying Kelvin's method of stationary phase (Stoker, 1957; Baker and Copson, 1939).

By differentiating $\Phi\left(X,Y;T\right)$ given by Equation 104 with respect to time T, $\Phi_T\left(X,Y;T\right)$ is obtained:

$$\Phi_T\left(X,Y;T\right)=-\frac{2\Omega^3}{gd_r}q\left[\frac{e^{\tilde{\eta}_0 X}\cosh\tilde{\xi}_0\left(Y+d_r\right)}{\cosh\tilde{\xi}_0 d_r\cdot W'\left(\xi_0\right)}\left\{\frac{1}{\tilde{\xi}_0^2}+2\sum_{n=1}^{\infty}\frac{\cos\frac{n\pi d_0}{d_r}}{\tilde{\xi}_0^2+\left(n\pi/d_r\right)^2}\right\}e^{-i\left(\tilde{\xi}_0 X-\Omega T\right)}\right.$$
$$\left.-\sum_{j=1}^{\infty}\frac{e^{-\tilde{\eta}_j X}\cos\tilde{\eta}_j\left(Y+d\right)}{\cos\tilde{\eta}_j d_r\cdot W'\left(i\tilde{\eta}_j\right)/i}\left\{\frac{1}{\tilde{\eta}_j^2}+2\sum_{n=1}^{\infty}\frac{\cos\frac{n\pi d_0}{d_r}}{\tilde{\eta}_j^2-\left(n\pi/d_r\right)^2}\right\}ie^{i\Omega T}\right] \tag{105}$$

where the following approximations have been made, since $\tilde{\mu}$ and $\tilde{\xi}_j$ take on small values, corresponding to a small dispersive water effect:

$S_0\to\Omega^2/g$ and $-\tilde{\eta}_j+i\tilde{\xi}_j\to-\tilde{\eta}_j$

In order to rewrite Equation 105 in dimensionless form, along with the dimensionless variables in Equation 73, the following additional dimensionless variables are introduced:

$$\xi_0=d_0\tilde{\xi}_0,\qquad \eta_0=d_0\tilde{\eta}_0,\qquad \eta_j=d_0\tilde{\eta}_j\left(j=1,2,\cdots\right) \tag{106a}$$

$$\beta^*=\frac{d_r}{d_0} \tag{106b}$$

With the above dimensionless variables along with those in Equation 73, Equation 105 can be reduced to the following non-dimensional form:

$$\begin{aligned}\varphi_t\left(x,y;t\right) = -2\frac{F^3}{F_0\beta^*}\Bigg[&\frac{e^{\eta_0 x}\cosh\xi_0\left(y+\beta^*\right)}{\cosh\xi_0\beta^*\cdot W'\left(\xi_0\right)}\left\{\frac{1}{\xi_0^2}+2\sum_{n=1}^{\infty}\frac{\cos n\pi/\beta^*}{\xi_0^2+\left(n\pi/\beta^*\right)^2}\right\}e^{-i\left(\xi_0 x-\frac{F}{F_0}t\right)}\\ &-\sum_{j=1}^{\infty}\frac{e^{-\eta_j x}\cos\eta_j\left(y+\beta^*\right)}{\cos\eta_j\beta^*\cdot W'\left(i\eta_j\right)/i}\left\{\frac{1}{\eta_j^2}+2\sum_{n=1}^{\infty}\frac{\cos n\pi/\beta^*}{\eta_j^2-\left(n\pi/\beta^*\right)^2}\right\}ie^{i\frac{F}{F_0}t}\Bigg]\end{aligned} \tag{107}$$

where the functions $W'(\tilde{\xi}_0)$ and $W'(i\tilde{\eta}_j)/i$ given in Equations 103a and 103b can be re-written as

$$W'\left(\xi_0\right) = \tanh\xi\beta^* + \xi_0\beta^*/\cosh^2\xi_0\beta^* \tag{108a}$$

$$W'\left(i\eta_j\right)/i = \tan\eta_j\beta^* + \eta_j\beta^*/\cos^2\eta_j\beta^* \tag{108b}$$

The transcendental equations, Equations 99a and 99b, for determining the singular points S_0' and S_j $\left(j=1,2,\ldots\right)$ can be re-written as

$$\mu, \tilde{\mu} \text{ and } \eta_0 \tanh\xi_0\beta^* + \mu F_0 F = 0 \tag{109a}$$

$$\xi_j \tan\eta_j\beta^* + \mu F_0 F = 0 \text{ and } \eta_j \tan\eta_j\beta^* + F^2 = 0 \tag{109b}$$

Free Surface Elevation

The free surface elevation $\tilde{\eta}\left(X;T\right)$, given by Equation 30, can also be written in the following dimensionless form:

$$\eta\left(x;t\right) = \tilde{\eta}(X;T)/\delta \tag{110a}$$

and can also be represented as

$$\eta\left(x;t\right) = -\varphi_t\left(x,y;t\right)\Big|_{y=0} \tag{110b}$$

For a Reservoir of Infinite Depth

Substituting Equation 75 into Equation 110b yields $\eta\ (x;\ t)$ in the following form:

$$\eta_\infty\left(x;t\right) = 2Fe^{-F^2}e^{-\mu F_0 Fx}e^{-i\left(F^2x-\frac{F}{F_0}t\right)} + \frac{2}{\pi}iF\int_0^\infty \frac{e^{-sx}}{s^2+F^4}\left(s\cos s - F^2\sin s\right)dse^{-i\frac{F}{F_0}t} \tag{111}$$

The real part of Equation 111 provides an expression for the free surface elevation, given by

$$\eta_\infty\left(x;t\right) = 2Fe^{-F^2}e^{-\mu F_0 Fx}\cos\left(F^2x-\frac{F}{F_0}t\right) - \frac{2}{\pi}F\int_0^\infty \frac{e^{-sx}}{s^2+F^4}\left(s\cos s - F^2\sin s\right)ds\sin\frac{F}{F_0}t \tag{112}$$

which is generated by the real part of the gate vibration amplitude from Equation 36. The real part of the dimensionless form of the gate vibration amplitude, denoted as y_G, can be written as

$$y_G(t) \equiv Y_G(T)\,/\,Y_0 = \cos\frac{F}{F_0}t \tag{113}$$

Equation 112 can be interpreted to show that the free surface elevation consists of a progressive wave in the positive x-direction (the first term) and a standing wave (the second term). To distinguish clearly the properties of the progressive and the standing waves expressed by Equation 112, we will consider the two waves separately. The amplitude $A_\infty\left(F,x\right)$ of the progressive wave is

$$A_\infty\left(F,x\right) = 2Fe^{-F^2}e^{-\mu F_0 Fx} \tag{114}$$

and the amplitude, $B_\infty\left(F,x\right)$, of the standing wave is

$$B_\infty\left(F,x\right) = \frac{2}{\pi}F\int_0^\infty \frac{e^{-sx}}{s^2+F^4}\left(s\cos s - F^2\sin s\right)ds \tag{115}$$

These two amplitude expressions are plotted in Figures 12 and 13, respectively, with the Froude number, F, on the abscissa and the distance x from the sluice gate as the parameter in the graph. The amplitude $A_\infty\left(F,x\right)$ is practically independent of the distance x, since $\mu \ll 1$. The free surface elevation, $\eta_\infty\left(x;t\right)$, may be expressed in the following resultant form:

$$\eta_\infty\left(x;t\right) = \eta_{0\infty}(F,x)\cos\left(\frac{F}{F_0}t-\beta\right) \tag{116a}$$

Figure 12. Progressive water waves for infinite depth

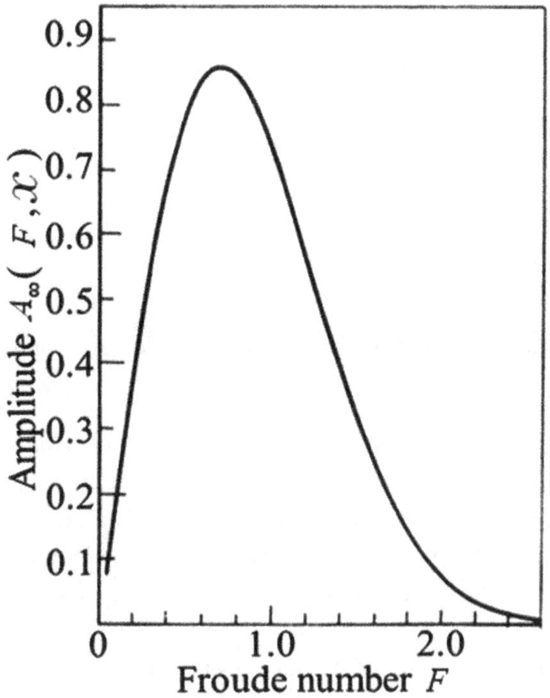

Figure 13. Standing water waves for infinite depth

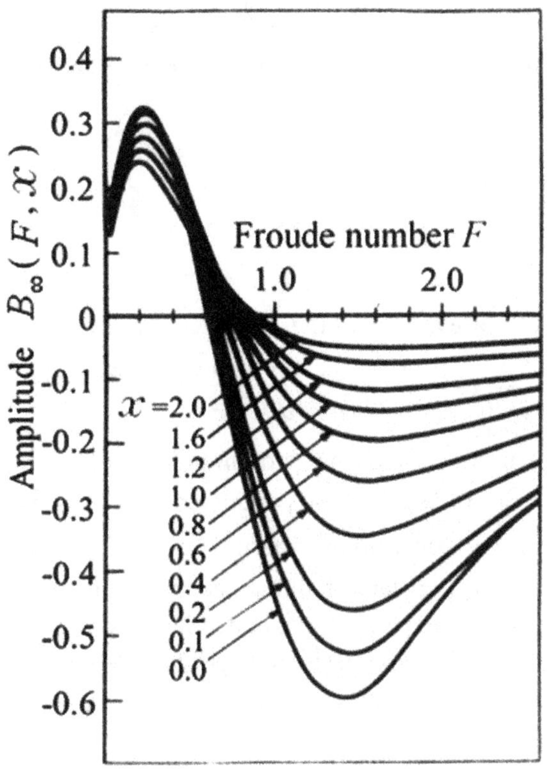

where

$$\eta_{0\infty}(F,x) = \left(A_\infty^2 - 2A_\infty B_\infty \sin F^2 x + B_\infty^2\right)^{1/2} \tag{116b}$$

$$\tan \beta_\infty(F,x) = \frac{A_\infty \sin F^2 x - B_\infty}{A_\infty \cos F^2 x} \tag{116c}$$

where $\eta_{0\infty}\left(F,x\right)$ denotes the amplitude of the free surface elevation and $\beta_\infty\left(F,x\right)$ is the phase lag relative to the motion of the sluice gate. The functions $\eta_{0\infty}\left(F,x\right)$ and $\beta_\infty\left(F,x\right)$ are shown in Figures 14 and 15 as functions of *F* with *x* as the parameter, as in the two previous figures.

Figures 14 and 15 portray the water waves induced on surface of the reservoir's backwater by the periodic vertical oscillations of the sluice gate. A fundamental parameter governing the characteristics of the water waves is the Froude number *F* defined by Equation 74. The response characteristics of the water waves are analogous to those of a simple mass-spring-damper system undergoing forced vibration. The analogy can be made with the submergence d_0 of the discharge opening acting as the mass, the gravitational acceleration *g* playing the role of the spring, and the wave speed $c=g/\Omega$ corresponding to the damping. When *F* is equal to about 0.8, the water waves fall into a resonant state, as is shown in Figure 14. The characteristics of this resonance chiefly depend on the progressive waves, shown in Figure 12.

The influence of the progressive wave on the composite wave characteristic diminishes rapidly as *F* increases. As the contribution of the progressive wave diminishes with increasing *F*, that of the standing wave increases, as seen in Figure 13. For large *F*, the phase lag $\beta_\infty\left(F,x\right)$ converges to 90°, as shown in Figure 15. It follows then that for large *F*, the asymptotic representation for the free surface elevation $\eta_\infty\left(x;t\right)$ is given by

$$\eta_\infty(x;t) \to B_\infty(F,x)\cos\left(\frac{F}{F_0}t - \pi/2\right) \tag{117}$$

The amplitude $\eta_{0\infty}$ and the phase lag β_∞ of the water waves are shown in Figures 16 and 17 in terms of the dimensionless distance $x' = x/\lambda$ on the abscissa with the parameter is *F;* the symbol λ is the dimensionless wavelength $\lambda = 2\pi / F^2 = 2\pi g / \left(d_0 \Omega^2\right)$.

Surface waveforms and their upstream propagation in time are shown in Figure 18 for $F = 1.4$, $F = 1.9$, and $F = 2.4$. Again the abscissa is *x*′, but the parameter is the dimensionless time $t_i = (i/10)T_\tau$, where *i* is an integer and T_τ is the dimensionless period $2\pi F_0 / F$. Note that with increasing Froude number (proportional to frequency) the wave amplitude decreases increasingly faster in both space and time. In Figure 18a the wave amplitude persists at t_5 and at larger *x*′ locations, while in Figure 18b and, to a greater extent, in Figure 18c, with progressively larger Froude numbers, the amplitude is more attenuated by time t_5 and disappears for the larger values of *x*′.

Figure 14. Amplitude of water waves for infinite depth
After Ishii et al. (1977b) © 1977 JSME. Used with permission.

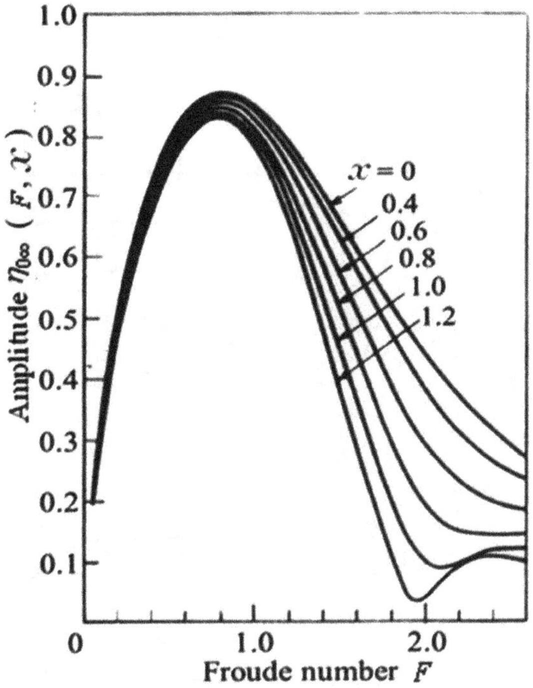

Figure 15. Phase of water waves for infinite depth
After Ishii et al. (1977b) © 1977 JSME. Used with permission.

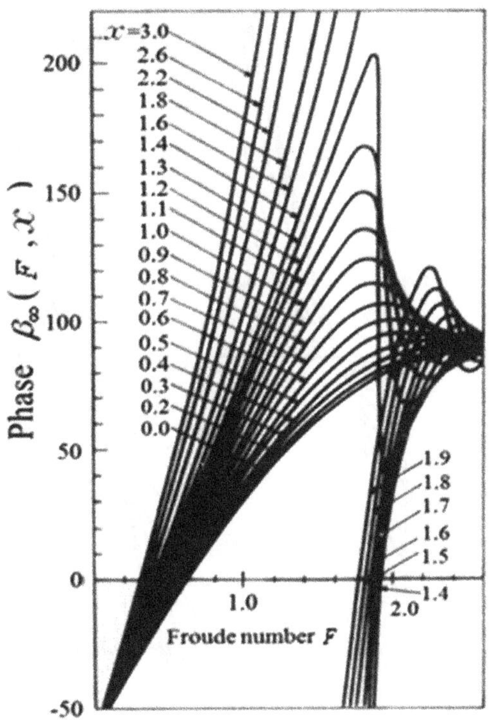

Figure 16. Amplitude of water waves for infinite depth
After Ishii et al. (1977b) © 1977 JSME. Used with permission.

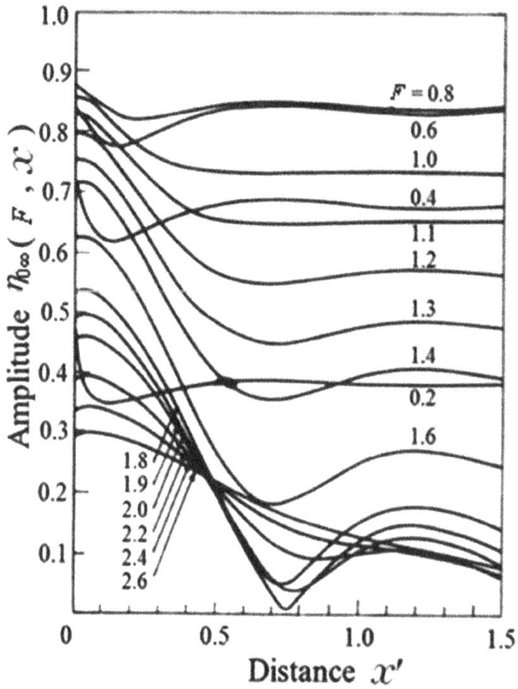

Figure 17. Phase of water waves for infinite depth
After Ishii et al. (1977b) © 1977 JSME. Used with permission.

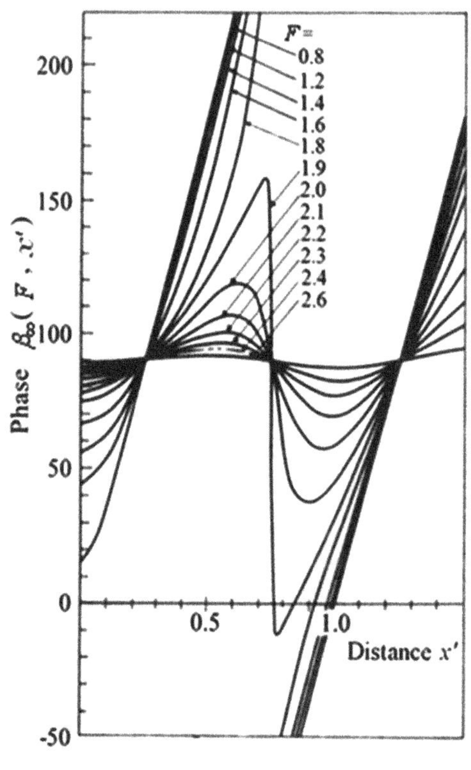

Figure 18. Waveform propagation over time; $t_i = (i/10)T_T$ *where i is an integer*
After Ishii et al. (1977b) © 1977 JSME. Used with permission.

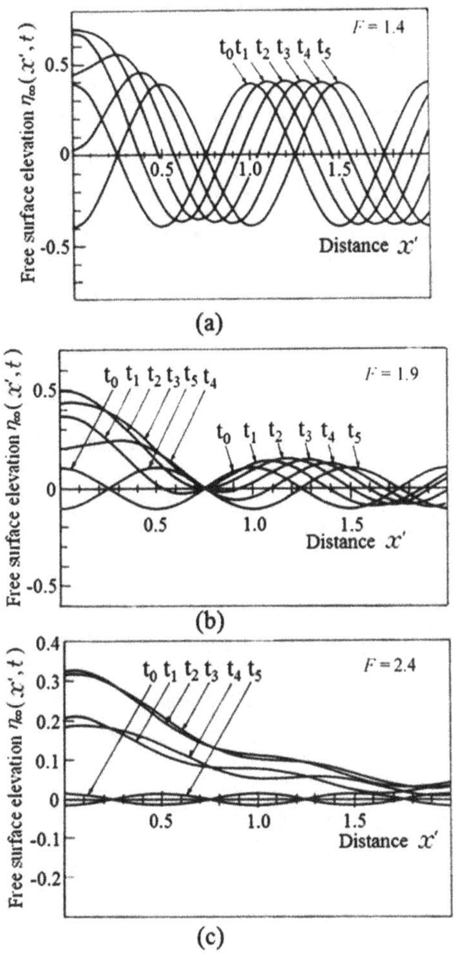

Several observations can be made concerning Figures 11 to 18:

- When F is approximately equal to 0.8, water waves with large amplitudes, which remain nearly constant with x', are generated (see Figures14 and 16). The waves progress in the positive x' -direction with nearly constant wave speed, as indicated by the almost constant rate of rate of phase change with distance (see Figures 15 and 17).
- When F is smaller than about 0.8, rapid changes of the amplitude $\eta_{0\infty}$ occur only for the very small x' values (Figures 14 and 16). The amplitude $\eta_{0\infty}$ for other values of x' remains nearly uniform. The wave speed is uniform and increases with increasing F (Figures 15 and 17).
- When F is larger than about 0.8, the effect of the standing waves becomes increasingly more dominant relative to the progressive waves with increasing F values. The maximum amplitude $\eta_{0\infty}$ is at $x' = 0$, with initially decreasing amplitudes with increasing x'. When x' is larger than a certain

value, however, the amplitude $\eta_{0\infty}$ has a tendency to increase for an intermediate range of F values. Such behavior is most obvious in Figure 16 for $F_0 = 1.9$ where the amplitude $\eta_{0\infty}$ approaches zero at $x' = 3/4$ as the contributions of the progressive and the standing waves destructively interfere. At the same time, the rapid change of the phase lag β is seen at $x' = 3/4$ (Figure 17). Though the point $x' = 3/4$ may seem to be a node of the standing wave, the waves are passing through the point and progressing in the positive x'-direction, as seen in Figure 18b.

- When F exceeds 1.9, such properties of the water waves weaken. For an adequately large value of F, the amplitude $\eta_{0\infty}$ uniformly decreases, as x' increases. Hence, the water waves behave like a standing wave which has a node at $x' = \infty$, as seen in Figure 18c.

For a Finite Depth Reservoir

A procedure similar to that for the infinite reservoir can be followed. Substituting $\varphi_t\left(x, y; t\right)$ given by Equation 107 into Equation 110b, the dimensionless free surface elevation $\eta\left(x; t\right)$ for a finite depth reservoir can be derived as

$$\begin{aligned}\eta(x;t) = 2\frac{F^3}{\beta^*}\Bigg[&\frac{e^{\eta_0 x}}{W'\left(\xi_0\right)}\left\{\frac{1}{\xi_0^2} + 2\sum_{n=1}^{\infty}\frac{\cos n\pi/\beta^*}{\xi_0^2 + \left(n\pi/\beta^*\right)^2}\right\}e^{-i\left(\xi_0 x - \frac{F}{F_0}t\right)} \\ &-\sum_{j=1}^{\infty}\frac{e^{-\eta_j x}}{W'\left(i\eta_j\right)/i}\left\{\frac{1}{\eta_j^2} + 2\sum_{n=1}^{\infty}\frac{\cos n\pi/\beta^*}{\eta_j^2 - \left(n\pi/\beta^*\right)^2}\right\}ie^{i\frac{F}{F_0}t}\Bigg]\end{aligned} \tag{118}$$

Taking the real part of Equation 118, the free surface elevation for the gate vibration given by Equation 113 can be reduced to

$$\begin{aligned}\eta(x;t) = 2\frac{F^3}{\beta^*}\Bigg[&\frac{e^{\eta_0 x}}{W'\left(\xi_0\right)}\left\{\frac{1}{\xi_0^2} + 2\sum_{n=1}^{\infty}\frac{\cos n\pi/\beta^*}{\xi_0^2 + \left(n\pi/\beta^*\right)^2}\right\}\cos\left(\xi_0 x - \frac{F}{F_0}t\right) \\ &+\sum_{j=1}^{\infty}\frac{e^{-\eta_j x}}{W'\left(i\eta_j\right)/i}\left\{\frac{1}{\eta_j^2} + 2\sum_{n=1}^{\infty}\frac{\cos n\pi/\beta^*}{\eta_j^2 - \left(n\pi/\beta^*\right)^2}\right\}\sin\frac{F}{F_0}t\Bigg]\end{aligned} \tag{119}$$

Recall that the depth ratio β^* was defined in Equation 106b and $1/\beta^*$, to be used later in this section, is thus the ratio of the gate submergence depth d_0 to the upstream reservoir depth d_r.

Equation 119 indicates that the free surface elevation consists of a progressive wave in the positive x-direction and a standing wave. To distinguish clearly the properties of the progressive and the standing waves expressed by Equation 119, we will consider the two waves separately. The amplitude $A(F, x, b^*)$ of the progressive wave is

$$A\left(F,x,\beta^*\right)=2F^3\frac{F^3}{\beta^*}\frac{e^{\eta_0' x}}{W'\left(\xi_0\right)}\left\{1/\xi_0'^2+2\sum_{n=1}^{\infty}\frac{\cos n\pi\alpha}{\xi_0'^2+\left(n\pi\alpha\right)^2}\right\} \tag{120}$$

and the amplitude, $B\left(F,x\right)$, of the standing wave is

$$B\left(F,x,\beta^*\right)=2\frac{F^3}{\beta^*}\sum_{j=1}^{\infty}\frac{e^{-\eta_j x}}{W'\left(i\eta_j\right)/i}\left\{\frac{1}{\eta_j^2}+2\sum_{n=1}^{\infty}\frac{\cos n\pi/\beta^*}{\eta_j^2-\left(n\pi/\beta^*\right)^2}\right\} \tag{121}$$

The free surface elevation $\eta\left(x;t\right)$ may be expressed in the following resultant form:

$$\eta\left(x;t\right)=\eta_0\left(F,x,\beta^*\right)\cos\left(\frac{F}{F_0}t-\beta\right) \tag{122a}$$

where the amplitude $\eta_0(F,x,b^*)$ and the phase lag $\beta(F_0,x,b^*)$ of the surface wave are given by

$$\eta_0\left(F,x,\beta^*\right)=\left(A^2-2AB\sin\xi_0 x+B^2\right)^{1/2} \tag{122b}$$

$$\tan\beta\left(F,x,\beta^*\right)=\frac{A\sin\xi_0 x-B}{A\cos\xi_0 x} \tag{122c}$$

Graphs of the amplitude $\eta_0\left(F,x,b^*\right)$ and the phase lag $\beta\left(F_0,x,b^*\right)$ of the surface waves on a finite depth reservoir at the upstream location x = 0.1 are presented in Figures 19 and 20, respectively, as functions of the Froude number with the depth ratio, β^*, as the parameter.

From Figures 19 and 20, the depth of the reservoir has almost no influence on either the wave amplitude or the phase lag for Froude numbers greater than 1.0. For Froude numbers less than 1.0, the reservoir depth affects both the wave amplitude and the phase lag, with both becoming increasingly larger as the depth ratio increases from 0 (the case of infinite reservoir) to 0.5.

Most large scale gates will have Froude numbers substantially greater than 1.0, suggesting the finite depth of the reservoir has a very limited impact on the surface waves generated by the vertical oscillation of a sluice gate. The finite depth does, however, have an effect of the reservoir pressure fluctuation due to flow-rate variation, as presented in the following section.

Flow-Rate Variation Pressure

The flow-rate variation pressure $\Delta P\left(X,Y;T\right)$, given by Equation 26, when made dimensionless, can be written as follows:

Figure 19. Amplitude of surface waves on a finite depth reservoir
After Ishii et al. (1977b) © 1977 JSME. Used with permission.

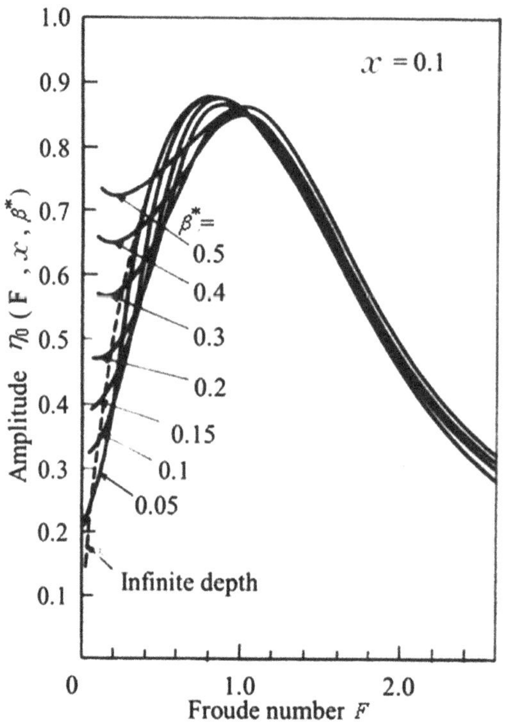

Figure 20. Phase lag of surface waves on a finite depth reservoir
After Ishii et al. (1977b) © 1977 JSME. Used with permission.

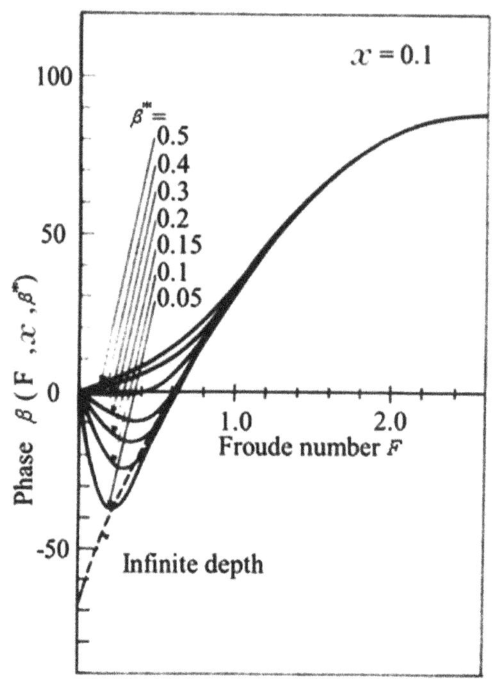

$$p_r(x,y;t) \equiv \frac{P_r/\rho g}{\delta} = \frac{P_r}{\sqrt{2c_f}\rho g Y_0} \text{ or } \frac{1}{\sqrt{2c_f}} \cdot \frac{P_r/\rho g}{Y_0} \tag{123a}$$

which can then be reduced to

$$p_r(x,y;t) = -F_0 \varphi_t(x,y;t) \tag{123b}$$

For a Reservoir of Infinite Depth

Substituting the time-derivative of fluctuating velocity potential for a reservoir of infinite water depth, given by Equation 75, into Equation 123b, the flow-rate variation pressure $p_{r\infty}(x,y;t)$ in a reservoir of infinite depth can be reduced to the following expression:

$$p_{r\infty}(x,y;t) = 2Fe^{-F^2} e^{-\mu F_0 Fx} e^{F^2 y} e^{-i\left(F^2 x - \frac{F}{F_0}t\right)}$$
$$+\frac{2}{\pi}F\left[-\int_0^\infty \frac{e^{-sx}}{s}\sin s \sin ys\, ds + \int_0^\infty \frac{e^{-sx}}{s^2+F^4}\left\{s\cos\left(1-y\right)s - F^2\sin\left(1-y\right)s\right\}ds\right] ie^{i\frac{F}{F_0}t} \tag{124}$$

The real part of the Equation 124 can be resolved to yield the following expression for the flow-rate variation pressure $p_{r\infty}(x,y;t)$:

$$p_{r\infty}(x,y;t) = p_{rp\infty}(F,x,y)\cos\left(F^2 x - \frac{F}{F_0}t\right) + p_{rs\infty}(F,x,y)\cos\left(\frac{F}{F_0}t + \frac{\pi}{2}\right) \tag{125a}$$

This pressure is induced by the gate vibration, as given by the real part of Equation 36 or in dimensionless form in Equation 11. The functions $p_{rp\infty}(F,x,y)$ and $p_{rs\infty}(F,x,y)$ are the amplitudes of the progressive and standing wave pressure waves. After some manipulation, they can be expressed as

$$p_{rp\infty}(F,x,y) = 2Fe^{-F^2} e^{-\mu F_0 Fx} e^{F^2 y} \tag{125b}$$

$$p_{rs\infty}(F,x,y) = \frac{2}{\pi}F\left[-\int_0^\infty \frac{e^{-sx}}{s}\sin s \sin ys ds + \int_0^\infty \frac{e^{-sx}}{s^2+F^4}\left\{s\cos(1-y)s - F^2\sin(1-y)s\right\}ds\right] \tag{125c}$$

The amplitude $p_{rp\infty}(F,x,y)$ decays in *x*, due to the presence of μ (>0) representing the dissipative effect of the water motion, approaching zero far upstream ($x = \infty$).

The resultant oscillatory pressure wave $p_{r\infty}\left(x,y;t\right)$ may be expressed in terms of amplitude and phase lag as follows:

$$p_{r\infty}(x,y;t)=p_{r0\infty}(F,x,y)\cos\left\{\frac{F}{F_0}t-\beta_{rp\infty}(F,x,y)\right\} \tag{126a}$$

where the resultant amplitude $p_{r0\infty}\left(F,x,y\right)$ and the phase lag $\beta_{r\infty}\left(F,x,y\right)$ relative to the gate vibration in Equation 113, are given by the following expressions:

$$p_{r0\infty}(F,x,y)=\sqrt{{p_{rp\infty}}^2-2p_{rp\infty}p_{rs\infty}\sin F^2x+{p_{rs\infty}}^2} \tag{126b}$$

$$\beta_{r\infty}(F,x,y)=\tan^{-1}\left(\frac{p_{rp\infty}\sin F^2x-p_{rs\infty}}{p_{rp\infty}\cos F^2x}\right) \tag{126c}$$

The most enlightening aspect of the flow-rate variation pressure is the pressure load acting on the sluice gate, which can be calculated at $x = 0$, that is, as $p_{r\infty}\left(0,y;t\right)$ in which the progressive and standing wave pressure amplitudes $p_{rp\infty}\left(F,0,y\right)$ and $p_{rs\infty}\left(F,0,y\right)$, are given by the following expressions, from Equations 125b and 125c, respectively:

$$p_{rp\infty}(F,0,y)=2Fe^{-F^2}e^{F^2y}\,' \tag{125b}$$

$$\begin{aligned}p_{rs\infty}(F,0,y)&=\frac{2}{\pi}F\left[-\int_0^\infty\frac{1}{s}\sin s\sin ysds+\int_0^\infty\frac{1}{s^2+F^4}\left\{s\cos(1-y)s-F^2\sin(1-y)s\right\}ds\right]\text{or}\\&=\frac{2}{\pi}F\left[-\frac{1}{2}\ln\frac{1+y}{1-y}+\int_0^\infty\frac{1}{s^2+F^4}\left\{s\cos(1-y)s-F^2\sin(1-y)s\right\}ds\right]\end{aligned}\,' \tag{125c}$$

The component pressure amplitudes calculated from Equations 125b′ and 125c′ of the progressive and standing waves are plotted in Figures 21a and 21b, respectively, as functions of the Froude number F with the dimensionless depth y as the parameter. The depth $y = 0$ corresponds to the free surface and the depth $y = -1$ to the gate opening.

The resultant pressure $p_{r\infty}\left(0,y;t\right)$ amplitude and phase lag $\beta_{r\infty}\left(F,0,y\right)$ are plotted in Figures 22a and 22b, respectively, as functions of the Froude number F with the dimensionless depth y as the parameter.

Alternatively, Figures 23a and 23b show the same data as functions of the dimensionless depth y with the Froude number F as the parameter.

Figure 21. Dependence of flow-rate-variation pressure component amplitudes on Froude number showing (a) the progressive wave pressure amplitude, $p_{rp\infty}\left(F,0,y\right)$, and (b) the standing wave pressure amplitude, $p_{rs\infty}\left(F,0,y\right)$, both at x = 0 for a reservoir of infinite depth

After Ishii et al. (1977b) © 1977 JSME. Used with permission.

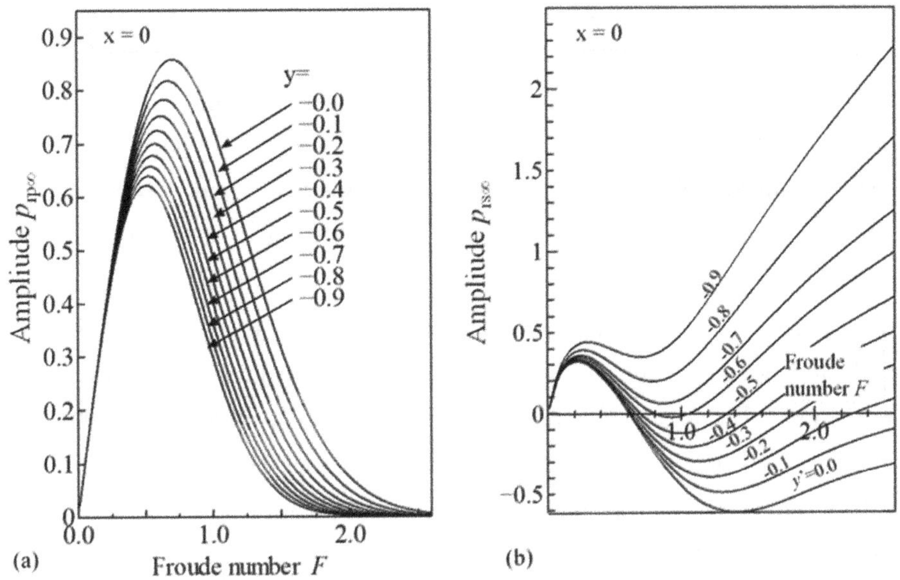

The flow-rate variation pressure at $y = 0$ (on the free surface) in Figure 21a, exhibits characteristics similar to those for the surface elevation of water waves at x=0 in Figures 12 to 15. This similarity is expected, since Equations 125b′ and 125c′ with $y = 0$ for the progressive and standing wave pressure amplitudes result in the same pressures as in Equations 114 and 115 with $x = 0$ for the surface elevations, respectively. With increasing Froude number F, the progressive wave pressure component amplitude also increases rapidly corresponding to the water surface elevations. The progressive wave pressure amplitude reaches a peak value between $F = 0.5$ to 0.8, due to water surface waves, and then rapidly decreases to zero for F values greater than about 2.5. With increasing submergence, that is, as y ranges from 0 to -1, the progressive wave pressure amplitude $p_{rp\infty}\left(F,0,y\right)$ uniformly decreases with the local maximum occurring at lower values of F. It should be noted that the pressure amplitude remains positive for all F and y values, as shown in Figure 21a. Thus, from the first term in Equation 125a, the progressive wave pressure component is in phase with the gate vibration as given in Equation 113.

The standing-wave pressure component $p_{rs\infty}\left(F,0,y\right)$, shown in Figure 21b, is positive for all water depths y, when F is smaller than about 0.6. Thus, for these small values of F, the standing-wave pressure component lags the gate vibration by 90° according to the second term in Equation 125a. When F is greater than about 0.6, however, the standing wave pressure component is positive on the lower half of the gate (for $-1{,}0 < y < -0.5$), but negative on the upper half of the gate ($-0.5 < y < 0$). Thus, the pressure on the lower half of the gate lags the vibration by 90°, while the pressure on the upper half leads the vibration by 90°. As F increases, the positive pressure values increase and the region of positive pressure region on the lower portion of the gate extends further upward toward the free surface. The negative pressure moves closer to zero (decreasing magnitude of negative pressure) and the region of negative pressure shrinks to be just a small region near the free surface.

The pressure we measure on the gate surface in the laboratory is the flow-rate variation pressure, the composite of the progressive and standing wave pressure components, which are plotted in Figures 23a and 24a. Several observations can be made about the resultant flow-rate variation pressure. With increasing Froude number *F*, the standing wave pressure wave $p_{rs\infty}(F,0,y)$ is very large in close proximity to the gate-opening at *y* = -1. Further, the flow-rate variation pressure amplitude $p_{r0\infty}(F,0,y)$ increases rapidly on the lower portion of the gate with depth beyond a certain value which moves closer to the free surface with increasing Froude number. The phase $\beta_{r\infty}(F,0,y)$ converges to -90° except at small *y* values close to the free surface. Near the free surface, *y* = 0, on the other hand, the amplitude $p_{r0\infty}(F,0,y)$ increases due to water surface waves, and the phase $\beta_{r\infty}(F,0,y)$ is positive when *F* is larger than about 0.6 and approaches +90° for larger values of *F*. Hence, the amplitude $p_{r0\infty}(F,0,y)$ has a minimum between the free surface and the gate-opening, as shown in Figure 23a. For *F* > 0.6, the phase $\beta_{r\infty}(F,0,y)$ changes from -90° to a positive value near the depth at which the structural load is a minimum, due to the water surface waves. The region of negative $\beta_{rp\infty}(F,0,y)$ values expands and that of the positive $\beta_{r\infty}(F,0,y)$ contracts as *F* increases. When *F* is smaller than about 0.6, the phase $\beta_{r\infty}(F,0,y)$ is negative for all *y* values and the amplitude, $p_{r0\infty}(F,0,y)$, is close to uniform for all *y* values except for the rapid amplitude change near the gate-opening.

Figure 22. Froude number dependence of (a) the composite flow-rate-variation pressure amplitude, and (b) the phase lag relative to gate motion at x = 0 for a reservoir of infinite depth
After Ishii et al. (1977b) © 1977 JSME. Used with permission.

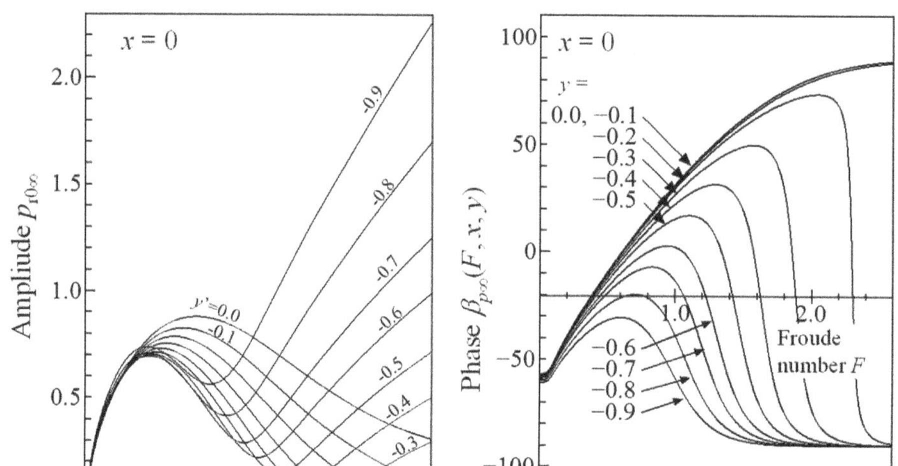

Figure 23. (a) Vertical distribution of pressure amplitude, $p_{r0\infty}(F,0,y)$ and (b) vertical distribution of phase, $\beta_{r\infty}(F,0,y)$ both at x=0 for Froude numbers less than 2.0 for a reservoir of infinite depth
After Ishii et al. (1977b) © 1977 JSME. Used with permission.

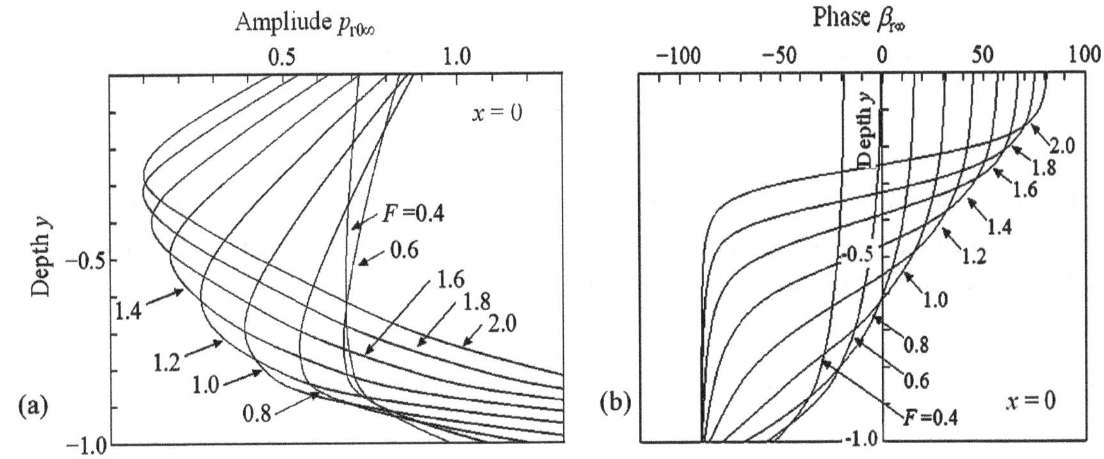

Figure 24. (a) Vertical distribution of pressure amplitude, $p_{r0\infty}(F,0,y)$ and (b) vertical distribution of phase, $\beta_{r\infty}(F,0,y)$ both at x=0 for Froude numbers from 2.0 to 20 for a reservoir of infinite depth
After Ishii et al. (1977b) © 1977 JSME. Used with permission.

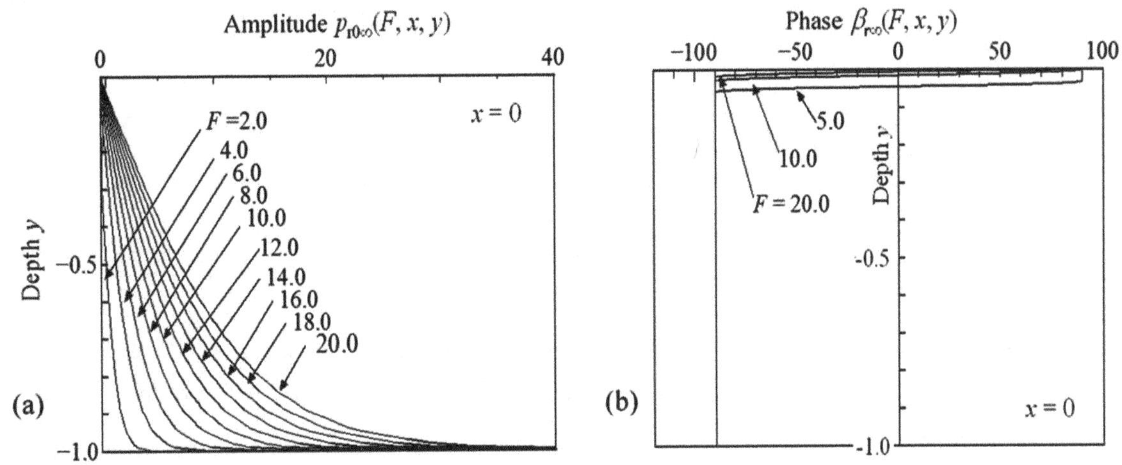

For a Reservoir of Finite Depth

Substitution of Equation 107 into Equation 123b yields the flow-rate variation pressure $p_r(x,y;t)$ for a finite depth reservoir, which can be reduced to the following expression:

$$p_r(x,y;t) = 2\frac{F^3}{\beta^*}\left[\frac{e^{\eta_0 x}\cosh\xi_0\left(y+\beta^*\right)}{\cosh\xi_0\beta^*\cdot W'\left(\xi_0\right)}\left\{\frac{1}{\xi_0^2}+2\sum_{n=1}^{\infty}\frac{\cos n\pi/\beta^*}{\xi_0^2+\left(n\pi/\beta^*\right)^2}\right\}e^{-i\left(\xi_0 x-\frac{F}{F_0}t\right)}\right.$$
$$\left.-\sum_{j=1}^{\infty}\frac{e^{-\eta_j x}\cos\eta_j\left(y+\beta^*\right)}{\cos\eta_j\beta^*\cdot W'\left(i\eta_j\right)/i}\left\{\frac{1}{\eta_j^2}+2\sum_{n=1}^{\infty}\frac{\cos n\pi/\beta^*}{\eta_j^2-\left(n\pi/\beta^*\right)^2}\right\}ie^{i\frac{F}{F_0}t}\right] \tag{127}$$

The real part of Equation 127 is the flow-rate variation pressure $p_r\left(x,y;t\right)$ induced by the real part of the gate vibration, given in Equation 36, or in dimensionless form in Equation 113. This flow-rate variation pressure can be reduced to the following from:

$$p_r(x,y;t) = p_{rp}(F,x,y,\beta^*)\cos\left(F^2x-\frac{F}{F_0}t\right)+p_{rs}(F,x,y,\beta^*)\cos\left(\frac{F}{F_0}t+\frac{\pi}{2}\right) \tag{128a}$$

The functions $p_{rp}(F,x,y,\beta^*)$ and $p_{rs}(F,x,y,\beta^*)$ are the amplitudes of the progressive and standing wave pressure waves. After some manipulation, they can be expressed as

$$p_{rp}(F,x,y,\beta^*) = 2\frac{F^3}{\beta^*}\frac{e^{\eta_0 x}\cosh\xi_0\left(y+\beta^*\right)}{\cosh\xi_0\beta^*\cdot W'\left(\xi_0\right)}\left\{\frac{1}{\xi_0^2}+2\sum_{n=1}^{\infty}\frac{\cos n\pi/\beta^*}{\xi_0^2+\left(n\pi/\beta^*\right)^2}\right\} \tag{128b}$$

$$p_{rs}(F,x,y,\beta^*) = -2\frac{F^3}{\beta^*}\sum_{j=1}^{\infty}\frac{e^{-\eta_j x}\cos\eta_j\left(y+\beta^*\right)}{\cos\eta_j\beta^*\cdot W'\left(i\eta_j\right)/i}\left\{\frac{1}{\eta_j^2}+2\sum_{n=1}^{\infty}\frac{\cos n\pi/\beta^*}{\eta_j^2-\left(n\pi/\beta^*\right)^2}\right\} \tag{128c}$$

The amplitude $p_{rp}(F,x,y,\beta^*)$ in Equation 128b decays exponentially in x since the coefficient in the exponential term is negative. This exponential decay reflects the dispersive effect of the progressive wave. The amplitude $p_{rp}(F,x,y,\beta^*)$ approaches zero as $x\to\infty$.

The resultant pressure wave $p_r(x, y; t)$ can be expressed in terms of the amplitude $p_{r0}(F,x,y,\beta^*)$ and the phase $\beta_r(F,x,y,\beta^*)$ relative to the movement of the gate as follows:

$$p_r(x,y;t) = p_{r0}(F,x,y,\beta^*)\cos\left(\frac{F}{F_0}t-\beta_r\right) \tag{129a}$$

$$p_{r0}(F,x,y,\beta^*) = \sqrt{p_{rp}^{\ 2} - 2p_{rp}p_{rs}\sin F^2x + p_{rs}^{\ 2}} \tag{129b}$$

$$\tan\beta_r(F,x,y,\beta^*) = \frac{p_{rp}\sin F^2x - p_{rs}}{p_{rp}\cos F^2x} \tag{129c}$$

The resultant pressure amplitude $p_{r0}(F,x,y,\beta^*)$ acting on the gate at $x = 0$ is plotted as a function of the water depth y with the depth ratio $1/\beta^*$ as the parameter in Figure 25 for three different Froude numbers. The Froude number takes the value 5 in Figure 25a, the value of 10 in Figure 25b, and the value 20 in Figure 25c. From the definition of the depth ratio β^* in Equation 110b, the inverse of depth ratio, $1/\beta^*$, is seen to represent the ratio of the gate submergence depth d_0 to the upstream reservoir depth d_r. Therefore, the case of $1/\beta^* = 0$ represents an infinite depth of reservoir; on the other hand, when the gate is directly installed on a horizontal dam sill $d_r = d_0$, and $1/\beta^* = 1.0$.

Figure 25 shows that the flow-rate variation pressure significantly increases with decreasing upstream reservoir depth. The progressive wave pressure component decreases rapidly with increasing Froude number, and hence this increment in pressure amplitude is caused by the standing wave pressure component. As the upstream reservoir depth decreases, the flow-rate variation can more effectively induce a standing wave pressure fluctuation. The reservoir depth thus plays a significant role in the production of the flow-rate variation pressure. This standing wave pressure component can effectively become an energy source to excite spontaneous gate vibration when the gate acts as a press-shut devices, as explained in the following section.

WATER EXCITATION EFFECTS ON A GATE ACTING AS PRESS-SHUT DEVICE

The term "press-shut device" was introduced in Chapter 1. A hydraulic gate acts as a press-shut device when the gate discharge opening decreases with an increase in the hydraulic static pressure load. Gates that act as press-shut devices will always be statically stable. For example, a Tainter gate may be designed such that the circular arc weir plate center is below the trunnion pin center. The hydrostatic load acts normal to the weir plate surface with all loads passing through the weir plate center, producing a static moment about the trunnion pin that causes the gate to close, as shown in Figure 26.

One of the observations in Chapter 1 was that press-shut devices regulating flow through small openings are potentially susceptible to movement-induced excitation. This excitation can (and will in Chapters 5, 7, 9 and 10) be related, in part, to the flow-rate variation pressure. It follows from the observation in Chapter 1, that any statically stable gate may be susceptible to dynamic instability at small gate openings. To the contrary, any gate that acts as a press-open device will be statically unstable. An example of a statically unstable gate would be a Tainter gate in which the circular arc weir plate center is above the trunnion pin center, such that the hydrostatic pressure would induce an opening moment about the trunnion pin. Usually statically unstable press-open devices are not susceptible to dynamic instability.

Another press-shut gate, shown in Figure 27, is the reverse Tainter gate installed on the Aare River near Schinznach, Switzerland. This reverse Tainter gate is a long-span gate with a span 22.5 m and a height of 4.2 m. The gate is inclined in the downstream direction so the hydrostatic pressure acts to

Figure 25. Vertical distribution of composite flow-rate-variation pressure amplitude, p_{r0} (F, x =0, y, β), acting on the gate at x = 0 for (a) a Froude number F = 5, (b) a Froude number F = 10 and (c) a Froude number F = 20*

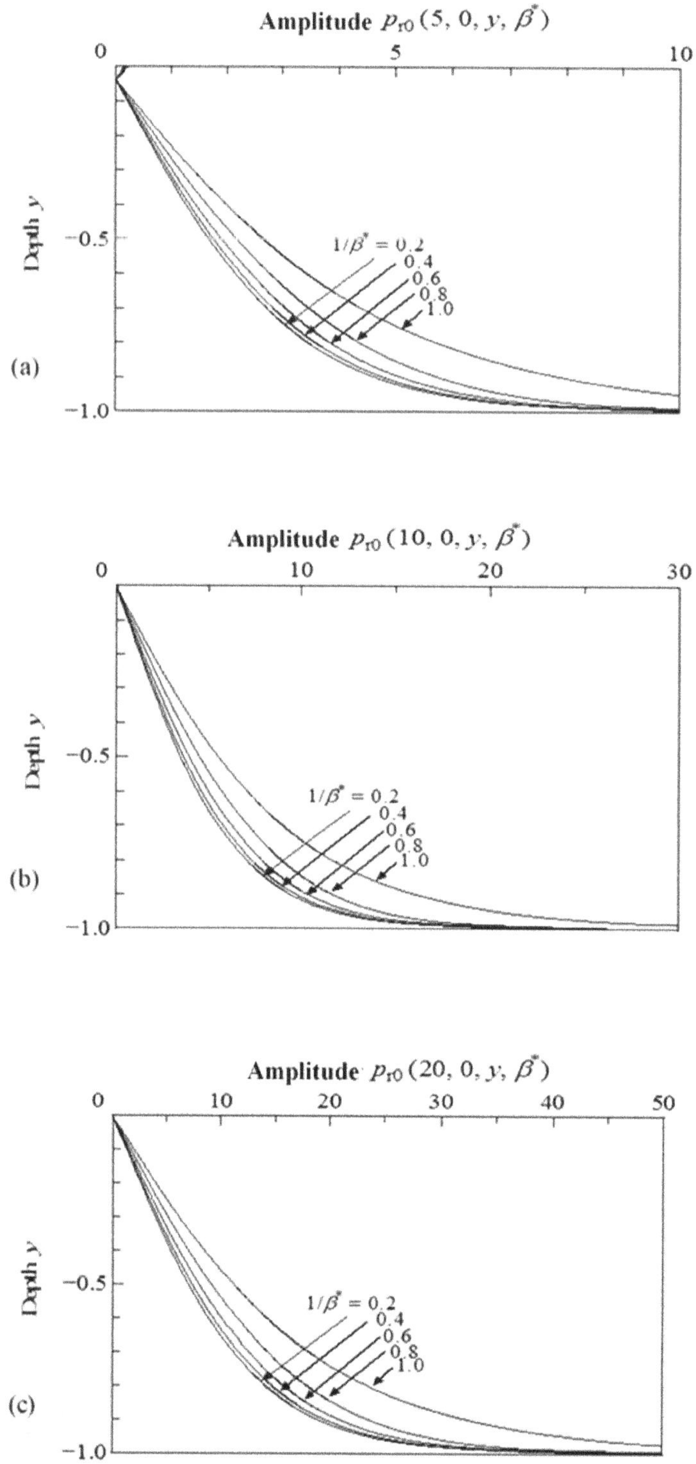

Figure 26. Tainter gate acting as a press-shut device with the skinplate center beneath the trunnion pin center

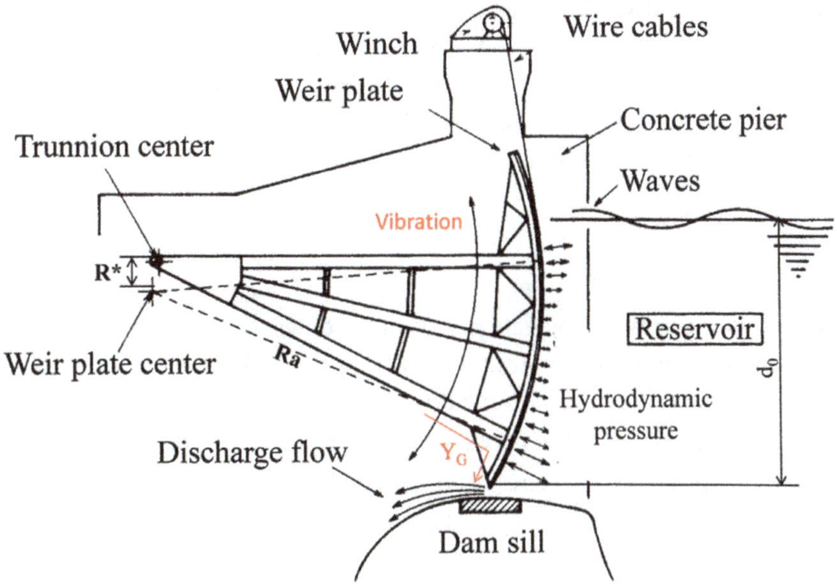

Figure 27. Schematic of a long-span gate (reverse Tainter gate) at the Schinznach Weir with the skinplate inclined in the downstream direction acting as a press-shut device
After Petrikat (1980).

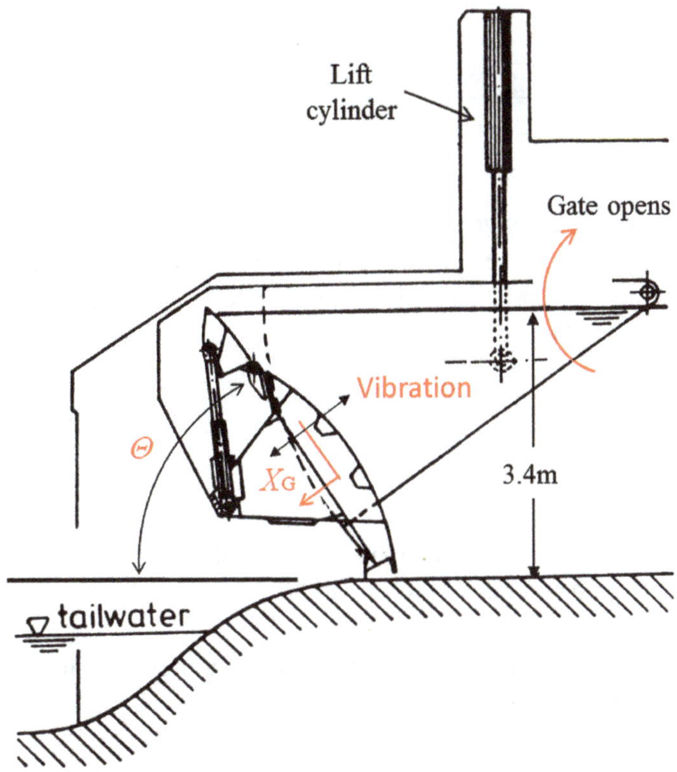

displace the gate downward and downstream due to its bending flexibility, resulting in a decrease in the opening, and thereby making it a press-shut device. As might be expected, the Schinznach gate experienced intense dynamic instability. The 100-ton structure underwent very intense streamwise vibration at a frequency of 4.1 Hz. In 1971, a similar long span gate on the Ara River in Japan, weighing 42 tons, also experienced very intense flow–induced vibrations (Petrikat, 1980).

The displacement X_G for the long-span Schinznach gate, shown in Figure 27, is geometrically related to the change in gate opening, Y_G, as follows:

$$Y_G(T) = X_G(T)\cos\Theta \tag{130}$$

where X_G is defined as positive when the gate moves in the downstream direction and Θ represents the angle of the weir plate relative to the horizontal dam sill. If $\Theta = 0$, the weir-plate displacement of the Schinznach gate, X_G, becomes equivalent to the downward skinplate-displacement of the Tainter gate, Y_G, shown in Figure 26. Using the amplitude of the Schinznach gate displacement, $\bar{X}_G$, as the representative length, the above expression can be reduced to the following dimensionless expression:

$$\frac{Y_G(T)}{\bar{X}_G\cos\Theta} = \frac{X_G(T)}{\bar{X}_G} \equiv x_G(t) \tag{130'}$$

Comparison of Equation 130′ with Equation 113 shows that the amplitude $\bar{X}_G \cos\Theta$ in Equation 130′ corresponds to Y_0 in Equation113. It then follows from Equation 123a that the reduced flow-rate variation pressure for the Schinznach gate can be written as

$$p_r(x,y;t) \equiv \frac{P_r}{\sqrt{2c_f}\rho g\bar{X}_G\cos\Theta} = \frac{1}{\sqrt{2c_f}\cos\Theta}\cdot\frac{P_r/\rho g}{\bar{X}_G} \tag{131}$$

When the flow-rate variation pressure P_r for the Schinznach gate is non-dimensionalized as in Equation 131, the non-dimensional pressure p_r will be given by the same expressions already found in either Equation 125 for the infinite depth or Equation 128 for the finite depth channel. In either case it is assumed that the weir plate is perpendicular to the horizontal sill and the gate undergoes harmonic motion given as

$$x_G(t) = \cos\left(\frac{F}{F_0}t\right) \tag{132}$$

The effect of inclination of the weir plate upon the flow-rate variation pressure can be corrected by the introduction of a correction coefficient which will be determined by comparing the theoretical and experimental results for model gates.

To consider the physical meaning of the flow-rate variation pressure on gate vibration we will re-write the expression for flow-rate variation pressure acting on the gate in a finite depth channel using Equation 128a with $x = 0$:

$$\begin{aligned} p_r(0,y;t) &= p_{rp}(F,0,y,\beta^*)\cos\frac{F}{F_0}t + p_{rs}(F,0,y,\beta^*)\cos\left(\frac{F}{F_0}t+\frac{\pi}{2}\right) \text{ or} \\ &= p_{rp}(F,0,y,\beta^*)\cos\frac{F}{F_0}t + p_{rs}(F,0,y,\beta^*)\left(-\sin\frac{F}{F_0}t\right) \end{aligned} \tag{133}$$

where the progressive wave pressure amplitude p_{rp} is given by Equation 128b with x=0 and the standing wave pressure amplitude p_{rs} is by Equation 128c with x=0. The vertical distributions of these two pressure amplitudes along the gate surface are shown in Figure 28, with $1/\beta^* = d_0/d_r$ set to 1.0, corresponding to the geometry shown in Figure 27. The progressive wave pressure amplitude p_{rp}, shown in the Figure 28a, takes on a small value which rapidly decreases with increasing Froude number F. In contrast, the standing wave pressure amplitude p_{rs}, shown in the Figure 28b, takes on a large value which significantly increases with increasing Froude number F.

A better understanding of the effects of the progressive and standing wave pressure components can be obtained by comparing the time-dependent pressure component amplitudes with the concurrent gate displacement and velocity. Such a comparison is undertaken in Figure 29. The gate displacement x_G is shown by the solid black line, the pressure load on the gate is by the dashed line with arrows, and the gate velocity by the dashed line. This schematic correlation provides insight on very important effects of the flow-rate variation pressure.

Phase relationships between the dynamics of the vibrating system and the fluctuating force acting on the system were considered in Chapter 2. Figure 14 in Chapter 2 can be consulted for a quick refresher of these phase relationships. The phase relationship between the pressure fluctuation and the gate displacement or velocity tells us what role each fluctuating pressure component plays.

In Figure 29a, the progressive wave pressure fluctuates in phase with the gate displacement, indicating that it cannot serve as an energy source to sustain the structural vibration. In Figure 29b, however, the standing wave pressure fluctuates in phase with the gate velocity. The standing wave pressure fluctua-

Figure 28. Vertical distributions of amplitudes of flow-rate-variation pressure components for x = 0 and 1/β =1.0 showing (a) the amplitude $p_{rp}(F, 0, y, 1.0)$ of the progressive (travelling) wave pressure component, and (b) the amplitude $p_{rs}(F, 0, y, 1.0)$ of the standing wave pressure component*

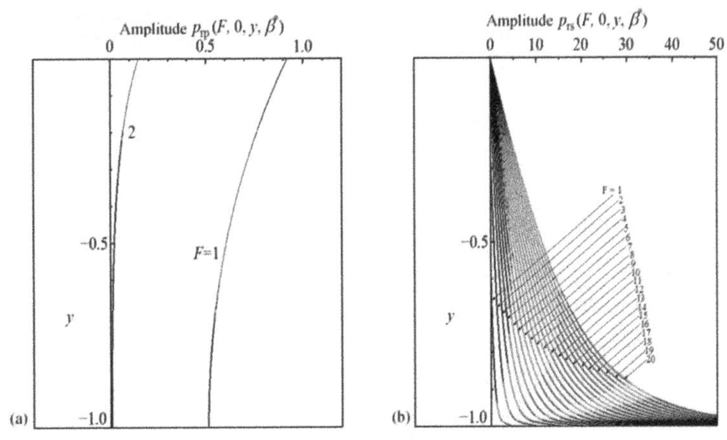

Figure 29. Phase relationships between fluctuating pressure amplitudes and gate displacement and velocity for (a) the flow-rate variation progressive wave pressure component, and (b) the flow-rate variation standing wave pressure component

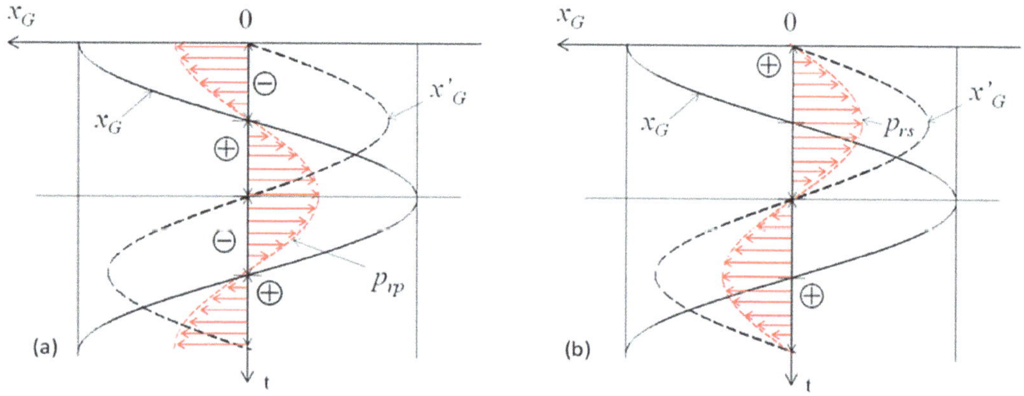

tion reaches its peak value just as the gate velocity is at its maximum and it reaches its lowest value just as the gate velocity is at its minimum. This phasing makes sense from basic fluid mechanics. When the gate velocity is at its maximum, the rate of decrease in the gate opening area is at a maximum and, correspondingly, the rate of decrease of the discharge will also be at its highest. When the rate of decrease in discharge is maximum, the fluctuating flow-rate leaving the reservoir reaches its minimum, resulting in the highest flow-rate variation pressure. Similar results pertain in the opposite direction of gate motion with the maximum rate of opening, the maximum rate of increase in discharge, the flow-rate reaching a maximum, and standing wave flow-rate variation pressure reaching its minimum value.

The water excitation of the gate is produced by the fluctuating standing wave pressure component that lags the gate vibration $x_G(t)$ by 90°, as shown in Equation 133 and in Figure 29b. For the first half-cycle as the gate moves upstream, the pressure load is negative, and pulls the gate in the direction it is moving and supplies energy to the gate vibration. For the second half-cycle as the gate moves downstream, the positive pressure acts on the gate, again, in the direction it is moving and again supplies energy to the gate motion. In this way, a potentially large amount of energy is supplied to the gate over the entire cycle of vibration, increasing the gate vibration amplitude in each cycle. This water excitation effect on the gate can be substantial since the standing wave pressure amplitude exhibits a very large value, as shown in Figure 28b, especially at the larger Froude numbers that are typical of full-scale gate operations. The energy supply per unit area of the gate can also be understood as the product of the fluctuating pressure and the gate vibration velocity, which exhibits a positive value over the full cycle. The standing wave pressure amplitude does, however, have a very small negative value very close to the free surface, as y approaches zero. This small region of negative pressure has a damping effect on the gate vibration, but its amplitude and domain rapidly decrease with increasing Froude number F, resulting in a negligible damping effect at Froude numbers typical of full-scale gates.

On the other hand, the first term in Equation 133 represents the progressive wave pressure component which is in phase with the gate displacement $x_G(t)$ given by Equation 132. The phase of the progressive wave pressure component relative to gate displacement and gate velocity is shown in Figure 29a. The product of the progressive wave pressure component and the vibration velocity changes sign every quarter cycle. As a result, the energy supplied and the energy consumed are equal, resulting in zero energy supply for the full cycle. This progressive wave pressure component produces a small added mass effect on the gate vibration, which is negligible, however, since it appears over a limited range of small Froude numbers and rapidly decreases with increasing Froude number.

For a better understanding, Equation 133 can be re-written as follows:

$$p_r(0,y;t) = -\left(\frac{F_0}{F}\right)^2 p_{rp}\left(F,0,y,\beta^*\right)\cdot x''_G + \left(\frac{F_0}{F}\right) p_{rs}\left(F,0,y,\beta^*\right)\cdot x'_G \tag{134}$$

The first term on the right side is the progressive wave pressure component times the gate acceleration, producing a small added mass effect on the gate vibration. The second term on the right is the standing wave pressure component times the gate velocity, producing the water excitation effect on the gate vibration.

A similar analysis of the Tainter gate acting as a press-shut device, shown in Figure 26, yields the same results.

EXPERIMENTAL CONFIRMATION OF THEORETICAL ANALYSIS

Experiments to confirm the theoretical results were undertaken using a simple model of a dam and sluice gate, shown in Figure 30. The sluice gate, at the left of the figure, undergoes vertical oscillation driven by a DC motor. The vibration amplitude was set at 1 mm and the mean gate opening was 2mm to satisfy the assumption in the theory that the steady discharge from the gate opening is sufficiently small.

The maximum amplitude of the water waves generated on the free surface of the reservoir was 3 mm. The distance of the gate opening from the bottom of the reservoir was 340 mm. The weir plate ② was 10 mm thick. The reservoir was 1500 mm long, 500 mm high and 40 mm wide.

The water circulation system consisted of a pump ⑩, a large inlet tank ④, the reservoir ③ and a discharge storage tank ⑫. The water level of the reservoir was kept constant. Water waves were measured with a handmade resistance wire gauge. The fluctuating hydrodynamic pressure was measured with a pressure transducer with 8 pressure taps (TOYODA PMS 0.5H) mounted vertically on the upstream weir surface.

In the experiments, the frequency of the oscillatory sluice gate was set at a frequency in the range from 1.9 to 4.1 Hz to satisfy the assumption in the theory that there is no phase lag between the sluice gate motion and the discharge fluctuation. The submergence of the gate-opening, d_0, was between 7 and 85 mm in order to satisfy the assumption in the theory that the depth of the reservoir relative to the gate submergence is very large, approaching infinity. The basic Froude number F_0 was adjustable over the range from 0.3 to 2.4.

The experimental values for surface wave amplitude and phase at a fixed x-location are shown in Figure 31 as functions of F and in Figure 32 as functions of distance from the gate. The variation the fluctuating hydrodynamic pressure amplitude and phase at a fixed x-location near the gate are shown in Figures 33 as a functions of F and in Figure 34 as functions of submergence depth. The agreement between the experimental data points and the theoretical predictions (solid lines) is uniformly very good. This close agreement validates the application of Rayleigh's potential flow dissipative wave theory for

Figure 30. Schematic of the apparatus for confirmation experiments: ① sluice gate; ② reservoir wall; ③ reservoir; ④inlet tank; ⑤grids; ⑥crank mechanism; ⑦ DC motor; ⑧ pressure taps; ⑨ control valve; ⑩ pump; ⑪ pipe line; ⑫ discharge storage tank; ⑬ wave absorber; ⑭ quiescent free surface

After Ishii et al. (1977b) © 1977 JSME. Used with permission.

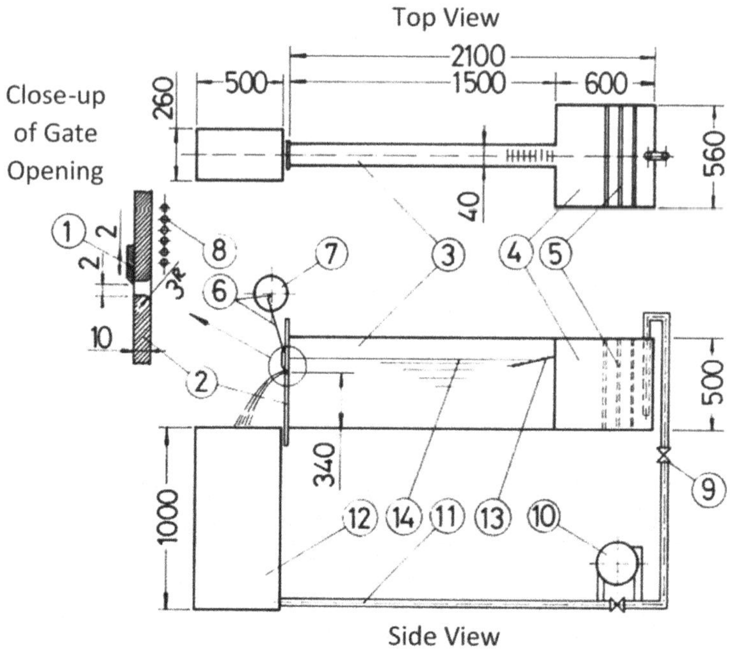

Figure 31. Measured surface wave (a) amplitude $\eta_{0\infty}$ (F, x′), and (b) phase $\beta_{0\infty}$ (F, x′) both as functions of Froude number at various downstream x-locations compared with theoretical predictions (solid lines); dimensionless distance $x' \equiv x/\lambda$ and λ is the dimensionless wavelength $\lambda = 2\pi / F^2 = 2\pi g / (d_0 \Omega^2)$

After Ishii et al. (1977b) © 1977 JSME. Used with permission.

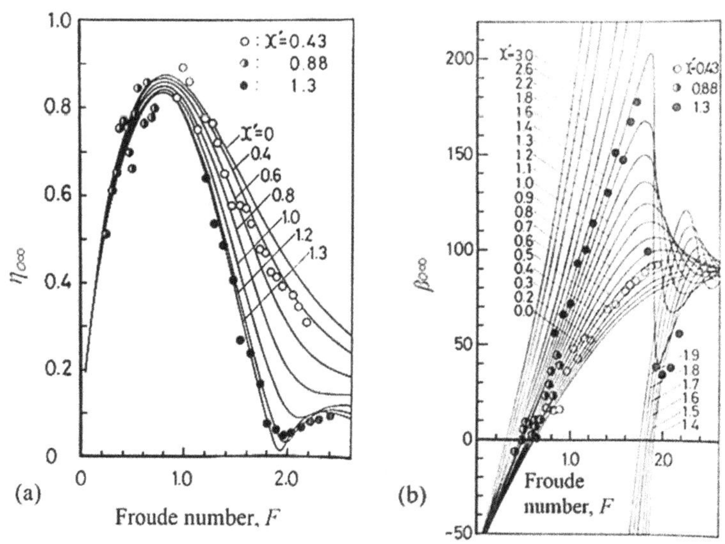

Figure 32. Measured variation of surface wave (a) amplitude $\eta_{0\infty}$ (F, x'), and (b) phase $\beta_{0\infty}$ (F, x') both as functions of distance from the gate, $x' \equiv x/\lambda$ where λ is the dimensionless wavelength $\lambda = 2\pi / F^2 = 2\pi g / \left(d_0 \Omega^2\right)$, and the Froude number compared with theoretical predictions (solid lines)

After Ishii et al. (1977b) © 1977 JSME. Used with permission.

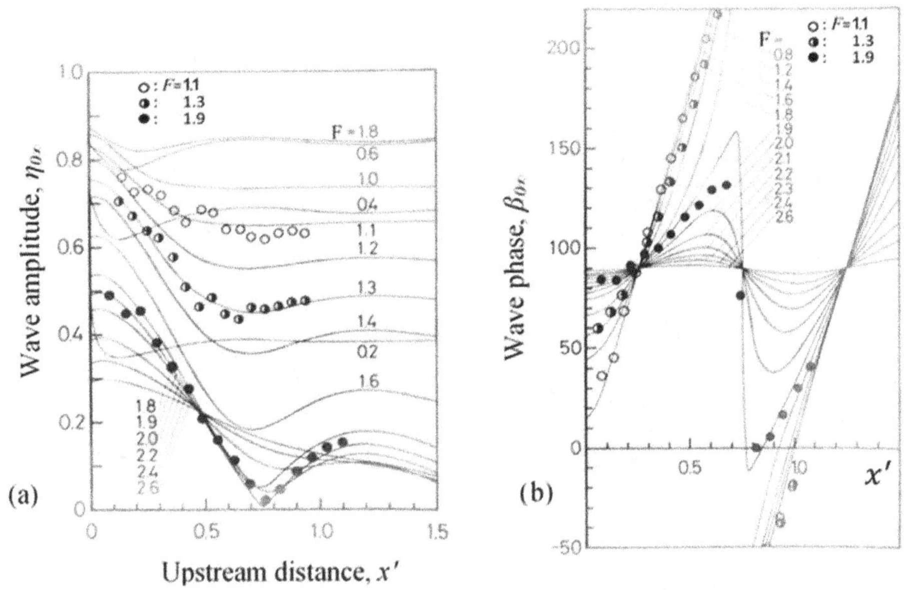

predicting oscillating pressures due to fluctuating discharge resulting from small amplitudes vertical gate vibrations at small Froude numbers.

A second set of experiments was undertaken in a new model gate apparatus, as shown in Figure 35, to confirm the validity of the theoretical analyses at higher Froude numbers. A special device was constructed to produce fluctuating discharge from the gate. A thin plate was amounted vertically at one end of a horizontal arm. The thin plate was placed near the gate lip along the upstream side of the skinplate. The rotation of the motor was converted into a small-amplitude oscillations of pinned horizontal arm. At the opposite end of the arm, the thin plate underwent small amplitude vertical oscillations. The gate itself was not vibrated making it possible to measure the flow-rate variation pressure with high accuracy, even at higher frequencies of forced vibration. The higher frequency vibration yielded higher Froude numbers.

The span length and height of the gate were 100 mm and 1300 mm, respectively. A displacement pickup was placed above the rotating arm near the thin plate to measure variations in the discharge opening.

The model gate was fixed in a 102 mm wide and 1200 mm deep water channel. The spaces between the sides of the gate and the water channel walls were filled with thin rubber tubes except along the channel walls near the sides of the thin oscillating plate. A plate glass window was fitted into both walls of the channel near to the base of the gate to allow for the measurement of discharge flow velocity in the gate opening using an LDV. The mean water depth above the lower edge of the thin plate, d_0, was 375 mm. The mean height of the gate opening, B, was 5 mm, and the vibration amplitude of the thin plate was 0.45 mm. The gate had no downstream submergence.

Figure 33. Measured composite resultant flow-rate variation pressure (a) amplitude $p_{r0\infty}(F, y)$, and phase $\beta_{r\infty}(F, y)$ at an x-location close to the gate surface as functions of Froude number for submergence depths compared with theoretical predictions (solid lines); distance $x' \equiv x/\lambda$ where λ is the dimensionless wavelength $\lambda = 2\pi / F^2 = 2\pi g/\left(d_0\Omega^2\right)$

After Ishii et al. (1977b) © 1977 JSME. Used with permission.

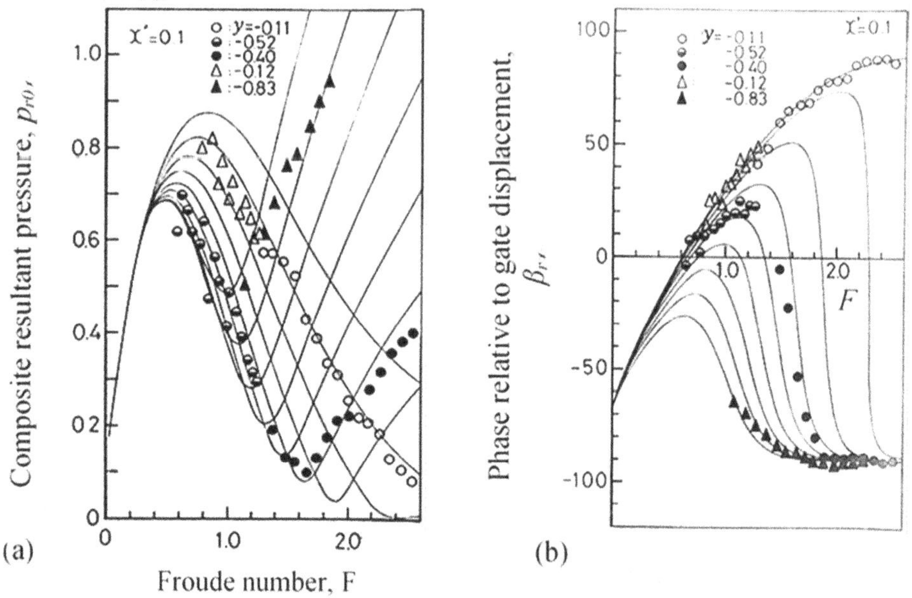

Figure 34. Measured composite resultant flow-rate variation pressure (a) amplitude $p_{r0\infty}(F, y)$, and phase $\beta_{r\infty}(F, y)$ at an x-location close to the gate surface as functions of submergence depth for several Froude numbers compared with theoretical predictions (solid lines); distance $x' \equiv x/\lambda$ where λ is the dimensionless wavelength $\lambda = 2\pi / F^2 = 2\pi g/\left(d_0\Omega^2\right)$

After Ishii et al. (1977b) © 1977 JSME. Used with permission.

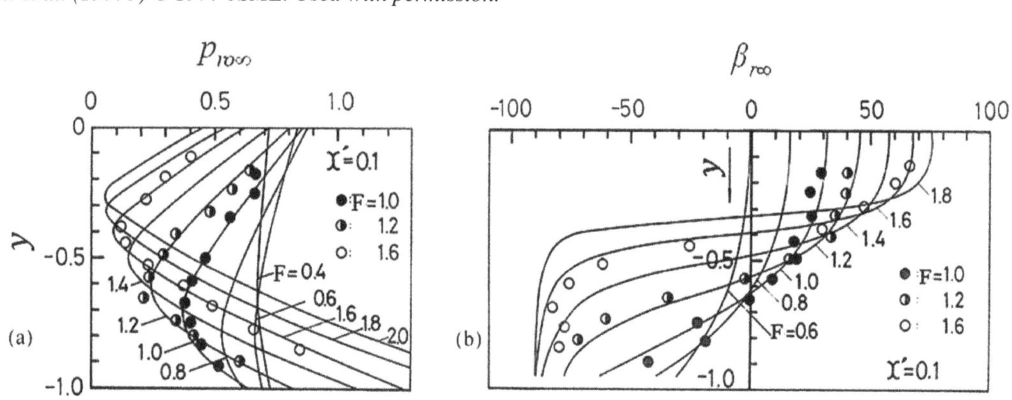

Figure 35. Schematic of vertical straight gate apparatus for confirmation experiments at higher Froude numbers

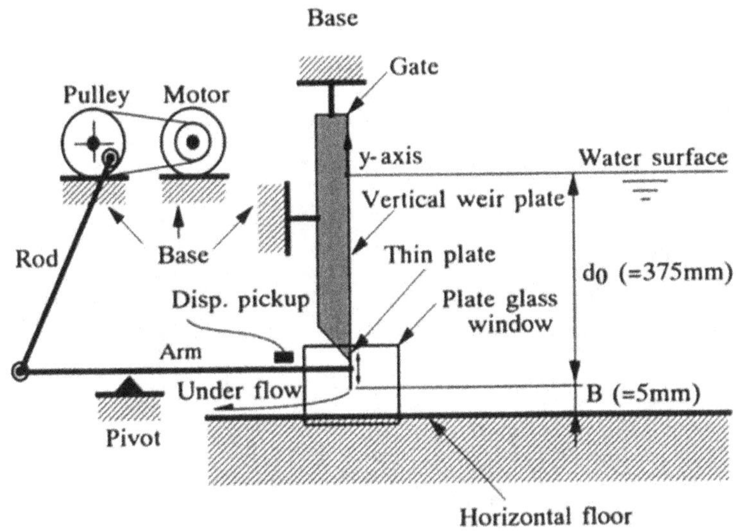

The lower edge of the model gate was forced to vibrate at frequencies ranging from 2 to 8 Hz, producing Froude numbers from 2.3 to 9.8. The fluctuating pressure was measured and processed to yield the pressure amplitude P_{r0} and the phase β_r relative to the displacement of the thin plate. The measured amplitude and phase data for the fluctuating pressure are plotted in Figures 36a and 36b, respectively Before plotting, the amplitude data was non-dimensionalized. The values in Figure 36a are the non-dimensional pressure amplitudes, p_{r0}.

The non-dimensional form of the flow-rate variation pressure was previously defined in Equation 123a, in which the representative value for the normalization is the product of $\sqrt{2}c_f$ and the water head $\rho g d_0$ (which corresponds to the amplitude Y_0 in the variation in the height of gate opening). However, few studies of measured values of the instantaneous flow-rate coefficient have been reported. In order to examine the values for the present case, the discharge flow velocity under the gate was measured using an LDV as the lower edge was vibrated. The amplitude of flow rate variations was calculated from the velocity data. The amplitude of flow-rate variation was calculated from the flow velocity data. The amplitude of flow-rate variations were divided by the flow-rate change induced by the quasi-static variation in the height of the gate opening, and this calculation yielded the instantaneous flow rate coefficient. As a result, the instantaneous flow rate coefficient was approximately 0.63 for the Froude numbers analyzed. All of the amplitude data shown in Figure 35a was normalized using the experimental data.

The non-dimensional amplitude p_{r0} increased gradually with the Froude number F, and rapidly increased with the depth below the water surface. Figure 35b shows the phase-lag of the fluctuating pressure. The phase-lag β_r maintained a constant value of 90 degrees except near the surface of water.

The theoretical resultant pressure amplitude, calculated from Equation 129b, and the phase-lag, calculated from Equation 129c, are both shown by solid lines in Figure 36. The theoretical values for both fluctuating pressure amplitude and phase-lag relative to gate motion associated with the flow-rate variations show good agreement with the experimental values.

Figure 36. Measured vertical distribution of flow-rate fluctuation pressure (a) amplitude p_{r0}, and (b) phase-lag, β_r relative to the gate displacement, both as functions of submergence y for Froude numbers from 2.3 to 9.8

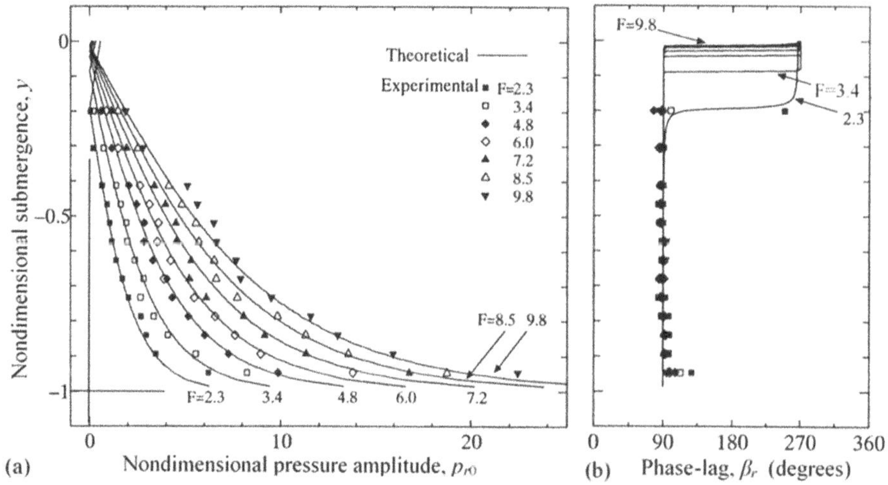

CONCLUSION

In this chapter we derived equations for determining the surface condition and pressure distributions in a reservoir with a vibrating gate inducing dispersive surface waves. The problem was reduced to an initial boundary value problem to solve the Laplace equation, Equation 8, with the initial conditions given in Equation 34, the surface condition given in Equation 33, and the boundary conditions for up-stream infinity and for infinite or finite reservoir depth given in Equation 35. This problem can be solved using mathematical techniques, such as Fourier integration and Fourier series transform techniques, Laplace transforms, complex integration techniques, the Cauchy integral theorem, and so on. With the known solution for the velocity potential, the elevation of dispersive water waves, $\tilde{\eta}\left(X;T\right)$, and the dispersive pressure fluctuation in the reservoir field, ΔP, can be determined using Equations 30 and 26, respectively.

Rayleigh's dissipative wave theory was then applied to a reservoir bounded by a vertical sluice gate with a small gate opening undergoing small amplitude vertical vibrations. More complicated realistic gate geometries will be considered in subsequent chapters: Chapter 5 for streamwise weir plate motion, Chapter 7 for long-span and flat-bottomed gates, Chapter 8 for eccentricity instability of Tainter gates, and Chapters 9 through 12 for Tainter gates.

The theoretical development was carried out for both finite and infinite reservoirs. The effects of finite reservoir depth on the free surface elevation are significant only for low Froude numbers, less than about 2, while the effects on the flow-rate variation pressure are quite significant for all Froude number. The potential flow solutions, once known, can be differentiated with respect to time to produce expressions for fluctuating pressure throughout the flow domain. The most significant result is that the fluctuating pressure field was shown to consist of two components, one in phase with the acceleration of the vertically vibrating gate, and the other in phase with the velocity of the vibrating gate.

The component of fluctuating pressure in phase with the gate's acceleration can be interpreted as an added mass term, represented by a dimensionless added mass coefficient. The added mass coefficient has been shown to be negligible except for a very limited region for F_0 less than about 2.0.

The component of fluctuating pressure in phase with the gate's velocity can be interpreted as a fluid damping term, or alternatively as a fluid excitation term when it takes on negative values. The fluid excitation is represented by a dimensionless fluid excitation coefficient. The fluid excitation coefficient was shown to approach a constant value for F_0 greater than about 10 for each reservoir depth ratio, and approaches a constant value of about 0.4 for an infinite depth reservoir.

The experimental studies cited and reproduced in Figures 31 to 34 and Figure 36 provide strong experimental support for the use of Rayleigh's dissipative wave theory to model the pressure loading on the gate surface due to small amplitude vertical vibrations of a sluice gate with free discharge.

REFERENCES

Atkinson, F. V. (1949). LXI. On Sommerfelds radiation condition. *Philosophical Magazine*, *40*(305), 645–651. doi:10.1080/14786444908561291

Baker, B. B., & Copson, E. T. (1939). *The Mathematical Theory of Huygens*. Oxford, UK: Clarendon.

Imaichi, K., & Ishii, N. (1977). Instability of an idealized Tainter gate system without damping caused by surface waves on the back water of dam. *Bulletin of the JSME*, *20*(146), 963–970. doi:10.1299/jsme1958.20.963

Ishii, N. (1992). Flow-induced vibration of long-span gates Part I: Model development. *Journal of Fluids and Structures*, *6*(5), 539–562. doi:10.1016/0889-9746(92)90018-X

Ishii, N., Imaichi, K., & Hirose, A. (1977a). Instability of elastically suspended Tainter gate system caused by surface waves on the reservoir of a dam. *Journal of Fluids Engineering*, *99*(4), 699–708. doi:10.1115/1.3448888

Ishii, N., Imaichi, K., & Takemoto, S. (1976). On waves generated at the back-water of dam by the motion of Tainter gate-system (in Japanese). *Trans. JSME*, *42*(364), 3929–2941. doi:10.1299/kikai1938.42.3929

Ishii, N., Imaichi, K., & Takemoto, S. (1977b). Water waves and structural loads caused by a periodic change of discharge from a sluice-gate. *Bulletin of the JSME*, *20*(146), 998–1007. doi:10.1299/jsme1958.20.998

Lamb, H. (1932). *Hydrodynamics*. New York, NY: Dover.

Petrikat, K. (1980). Seal vibration. In E. Naudascher & D. Rockwell (Eds.), *Practical Experiences with Flow-Induced Vibrations, IUTAM/IAHR Symposium, Karlsruhe, Germany*, (pp. 476-497). Heidelberg: Springer.

Rayleigh, L. (1883). The form of standing waves on the surface of running water. *Proceedings of the London Mathematical Society*, *1*(1), 69–78. doi:10.1112/plms/s1-15.1.69

Stoker, J. J. (1957). *Water Waves*. New York, NY: Interscience.

APPENDIX: NOMENCLATURE

A: Arbitrary small height of gate opening (m)
A: Arbitrary function
A, A_∞ : Amplitude of progressive wave
B, B_∞ : Amplitude of standing wave
c_f: Discharge coefficient for sluice gate
d_0: Depth of gate opening
d_r: Depth of upstream reservoir
dX, dY: Horizontal and vertical small dimensions of differential fluid element
dT: Small time (s)
f_B : Rayleigh's dissipative body force vector [N/kg]
F: Froude number
F_0: Basic Froude number.
g: Acceleration of gravity
i: Imaginary unit
$\vec{i}, \vec{j}$: Unit vectors in X and Y directions
j: Integer variable for Fourier integral transform
$M(t)$: Mass in differential fluid element
P: Fluid pressure (Pa)
p_r , P_r : Flow rate variation pressure [p_r is the dimensionless expression of P_r (Pa)]
$p_{r\infty}$: Flow rate variation pressure for reservoir of infinite depth
p_{r0}, $p_{r0\infty}$: Resultant amplitude of flow rate variation pressure
p_{rp} , $p_{rp\infty}$: Amplitude of progressive component of flow rate variation pressure
p_{rs}, $p_{rs\infty}$: Amplitude of standing component of flow rate variation pressure
q: Amplitude of fluctuating discharge $\Delta Q(T)$
Q_0: Steady discharge
s, S: Variable for Fourier integral transform [s is the dimensionless expression of S(1/m)]
S_0 : Constant in general solution $\bar{\Phi}$
S_0', S_j : Singular points for integrals in complex plane
t, T: Time [t is the dimensionless expression of T(s)]
u, v: Flow velocity in Cartesian coordinate directions X and Y
V: Fluid velocity vector (m/s)
W, W': Function in solution of velocity potential Φ
x, X: Horizontal coordinate [x is the dimensionless expression of X(m)]
x_G: Dimensionless expression of gate displacement of X_G
X_G: Downstream side displacement of long-span gate
$\overline{X_G}$: Amplitude of gate displacement
y, Y: Vertical coordinate [y is the dimensionless expression of Y(m)]
Y_0: Amplitude of sluice gate oscillation

Y_G: Downward displacement of sluice gate or gate opening change
β, β_∞ : Phase lag of free surface elevation
β^*: Reservoir to gate submergence depth ratio
$\beta_{rp}, \beta_{r\infty}$: Phase lag of resultant flow-rate variation pressure wave
δ: Modified amplitude of gate oscillation
ΔP_r: Flow rate variation pressure fluctuation from equilibrium state
ΔQ: Discharge variation
$\eta, \tilde{\eta}, \eta_\infty$: Elevation of water surface
$\eta_0, \eta_{0\infty}$: Amplitude of free surface elevation
$\eta_0, \tilde{\eta}_0$: Real coordinate of S_0' [η_0 is the dimensionless expression of $\tilde{\eta}_0$ (1/m)]
$\eta_j, \tilde{\eta}_j$: Real coordinate of S_j [η_j is the dimensionless expression of $\tilde{\eta}_j$ (1/m)]
Θ: Angle of weir plate relative to the horizontal dam sill
Θ_1, Θ_2: Rotation angles of dX and dY
λ: Dimensionless wavelength
$\mu, \tilde{\mu}$: Coefficient of Rayleigh's dissipative body forces [μ is the dimensionless expression of $\tilde{\mu}$ 1/s)]
ξ: Rotation of differential fluid element
$\xi_0, \tilde{\xi}_0$: Imaginary coordinate of S_0' [ξ_0 is the dimensionless expression of $\tilde{\xi}_0$ (1/m)]
$\xi_j, \tilde{\xi}_j$: Imaginary coordinate of S_j [ξ_j is the dimensionless expression of $\tilde{\xi}_j$ (1/m)]
ρ: Density (specific mass) [kg/m^3]
ζ: Variable for Laplace transform
$\bar{\Phi}$: Fourier integral cosine transform of Φ
Φ_0, Φ_1, Φ_2 : Phase function of S
$\bar{\Phi}^{(p)}$: Particular solution
Ψ: Stream function
Ψ_1, Ψ_2: Functions of S
Ω: In-water vibration frequency (1/s)
Ω_a: In-air vibration frequency (1/s)

Chapter 5
Streamwise Gate Vibration:
Added Mass and Damping

ABSTRACT

Streamwise vibrations of gates due to the bending flexibility of the skinplate of Tainter gates or the weir plate of long-span gates result in pushing-and-drawing of the water in the reservoir. During each cycle of vibration, the gate's motion must accelerate and then decelerate the water mass in contact with the vibrating gate surface, resulting in a substantial added mass effect. From simple single degree-of-freedom mass-spring-damper vibration theory, one understands that the effect of added mass is to lower the frequency of gate vibration. In addition to the push-and-draw effect, streamwise motion can also result in discharge fluctuation for inclined gates, providing a source of gate excitation. Rayleigh's wave theory analysis from the previous chapter is applied to provide an analysis framework for determining the magnitude of wave radiation damping and to calculate the added mass.

INTRODUCTION

Since a hydraulic gate must hold back a relatively dense liquid (the density of water is approximately 1000 times that of sea level air), it will be tensioned by a large hydrostatic or hydrodynamic force. One might be tempted to think that such a loaded gate could not possibly vibrate with the mass of water pushing against it. Under certain conditions, however, Tainter gates and long-span gates may easily and spontaneously vibrate despite the water load. The gates possess a degree-of-freedom in the streamwise direction due to the flexibility of the skinplate for Tainter gates or of the weir plate for long-span gates. The bending of these plates will push away and then draw back the water behind the gate in each cycle of vibration. During each cycle of vibration, the gate's motion must also accelerate and then decelerate the water in contact with the surface of the vibrating skinplate or weir plate. The mass of water that moves with the structure is called the added mass. From simple single degree-of-freedom mass-spring-damper vibration theory, one understands that the effect of added mass is to lower the frequency of gate vibration since the natural frequency is inversely proportional to the square root of the total system mass.

DOI: 10.4018/978-1-5225-3079-4.ch005

In addition to the added mass effect, we will briefly consider the viscous damping of the water on the gate vibration. Since the magnitude of viscosity is quite small, and the frequency of vibration is also relatively low, viscous damping will remain very small. Another form of damping is radiation damping due to the generation of surface waves. However, we will show that wave radiation damping becomes negligibly small for all practical gate applications.

The low levels of viscous and wave radiation damping permit the excitation of very quickly growing, spontaneous vibrations if a mechanism exists to supply energy to the gate motion. We will examine the energy source, and then show that – as with vertical vibrations of the weir plate – streamwise gate motion can result in fluctuations in the flow rate beneath the gate.

The focus of this chapter is on applying Rayleigh's wave theory from Chapter 4 to small-amplitude streamwise weir plate vibrations to determine the magnitude of wave radiation damping and to calculate the added mass. This problem is, as in Chapter 4, amenable to analytic solution.

The theoretical development in this chapter was previously presented in Ishii (1992).

THEORETICAL ANALYSIS OF PUSH-AND-DRAW PRESSURE

In Chapter 4, Rayleigh's wave theory was used to analyze the effect of small amplitude vertical vibrations of a vertical weir plate on the flow field. In this chapter, the same framework will be used to consider the hydrodynamic effects of small amplitude horizontal vibrations of the vertical weir plate.

The simplifications of the bounding geometry and boundary conditions that we will use for the present theoretical analysis of streamwise weir plate vibration are similar to those in the previous chapter. Figure 1 shows the simplified geometry in which the curved skinplate and the dam wall of the Tainter gate (see Figure 1 in Chapter 4) are replaced with a vertical planar weir plate oriented parallel to the *Y*-axis, while the inclined weir plate of the long-span gate (see Figure 2 in Chapter 4) is assumed to stand vertically along the *Y*-axis. The still water surface is used to define the *X*-axis. The analysis is done for a finite depth reservoir, which can then be extended to infinite depth reservoirs as well.

Boundary and Initial Conditions

In an operating gate, the hydraulic force exerted on the gate is in equilibrium with the restoring force due to the bending deformation of the gate. The gate will undergo bending vibrations about this equilibrium state. Here we assume the gate vibration is in the streamwise direction with an amplitude of $X_c(T)$, as shown in Figure 1. (Please note that in the definition of $X_C(T)$ in this chapter, the positive direction is upstream and is opposite that defined for $X_G(T)$ in Figure 27 in Chapter 4.)

The flow in the upstream channel is affected by the push-and-draw effect of the gate on the water. This fluid-structure interaction can be represented by the following boundary condition:

$$\left[\Phi_X\right]_{X=0} = \dot{X}_C\left(T\right) \text{for} - d_0 \le Y \le 0 \tag{1}$$

as shown in Figure 1, where Φ is the velocity potential, the subscript *X* represents the partial derivative with respect to *X*, and the over dot represents the derivative with respect to time *T*. In general, the amplitude of the lowest frequency gate vibration mode varies in the transverse (*Z*) direction, in the form of

Figure 1. Schematic side view of the simplified reservoir flow field and boundary conditions for push-and-draw pressure; the displacement is defined as positive when the gate movement is in the upstream direction
After Ishii (1992).

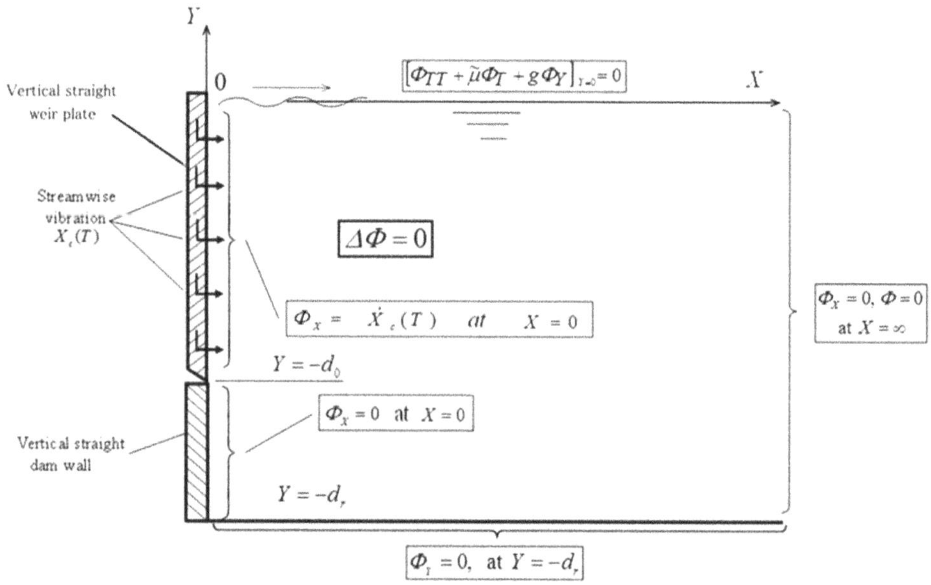

a half-wavelength sine wave (as portrayed in Figures 1 and 2 in Chapter 4). In the actual gate, the induced water waves do not travel as plane waves, but as two-dimensional waves. However, since the span is much larger than both the gate height, H, and the depth, d_0, these waves can be approximated as plane waves. Thus, the flow field in the upstream channel is regarded as two-dimensional in the vertical (X, Y) –plane. The X-axis is taken horizontally along the undisturbed free surface in the upstream direction and the Y-axis is vertically upward along the gate, as shown in Figure1. If the water is assumed incompressible and the flow is irrotational, the flow-field variation can be described by the Laplace equation

$$\Phi_{XX} + \Phi_{YY} = 0\left(\text{also written } \nabla^2\Phi = 0 \text{ or } \Delta\Phi = 0\right) \tag{2}$$

where the subscripts represent partial derivatives. The boundary conditions for the flow field are shown in Figure 1. If the mean gate opening is small relative to the gate submergence depth, d_0, the steady flow in the upstream channel is negligible. In this case, the boundary condition for the free surface is given by the following expression:

$$\left[\Phi_{TT} + \tilde{\mu}\Phi_T + g\Phi_Y\right]_{Y=0} = 0 \tag{3}$$

The boundary conditions on the fixed wall beneath the gate and the upstream-channel floor are:

$$\Phi_X\left(X,Y;T\right)=0 \text{ for X} = 0 \text{ and } -d_r \le Y \le -d_0 \tag{4a}$$

$$\Phi_Y\left(X,Y;T\right)=0 \text{ for } 0 \le X < \infty \text{ and } Y=-d_r \tag{4b}$$

These expressions state that the fluid velocity normal to each fixed boundary is zero. On the far-upstream boundary ($X = \infty$), the following boundedness conditions, from Stoker (1957), are imposed:

$$\Phi\left(\infty,Y;T\right)=0 \text{ and } \Phi_X\left(\infty,Y;T\right)=0 \tag{5}$$

The initial conditions prescribed for the velocity potential at time $T = 0$ are

$$\Phi\left(X,0;0\right)=0 \text{ and } \Phi_T\left(X,0;0\right)=0 \tag{6}$$

These conditions state that the free surface is initially at rest in its horizontal equilibrium position. Furthermore, at time $T = 0$, the following initial conditions are prescribed for the gate displacement $X_c(T)$:

$$X_c\left(0\right)=0 \text{ and } \dot{X}_c\left(0\right)=0 \tag{7}$$

The problem has thus been reduced to an initial and boundary value problem. We can solve the Laplace equation, Equation 2, subject to the boundary conditions given in Equations 1, 3, 4, and 5, and the initial conditions in Equations 6 and 7.

Solution to the Laplace Equation

To solve the initial and boundary value problem for the Laplace equation, the following Fourier integral transform (Sneddon, 1951) on X will be applied to Equation 2:

$$\bar{\Phi}\left(S,Y;T\right)=\sqrt{2/\pi}\int_0^{\infty}\Phi\left(X,Y;T\right)\cos SX\mathrm{d}X \tag{8}$$

The resulting equation is:

$$\bar{\Phi}_{YY}\left(S,Y;T\right)-S^2\bar{\Phi}\left(S,Y;T\right)=\sqrt{2/\pi}\Phi_X\left(0,Y;T\right) \tag{9}$$

in which $\bar{\Phi}$ is the Fourier cosine transform of Φ, and the boundary condition at $X = \infty$ in Equation 5 has been applied. The general solution for Equation 9 is given by a superposition of the particular solution $\bar{\Phi}^{(p)}(S,Y;T)$ and the homogeneous solution $A*\left(S;T\right)\cosh S\left(Y+d_r\right)$, as follows:

$$\bar{\Phi}\left(S,Y;T\right)=\bar{\Phi}^{(p)}\left(S,Y;T\right)+A*\left(S;T\right)\cosh S\left(Y+d_r\right) \tag{10}$$

in which the homogeneous solution, in terms of the hyperbolic cosine, automatically satisfies the boundary condition on the upstream-channel floor.

The particular solution $\bar{\Phi}^{(p)}(S,Y;T)$ satisfies the following equation:

$$\bar{\Phi}_{YY}^{(p)}-S^2\bar{\Phi}^{(p)}=\sqrt{2/\pi}\,\Phi_X(0,Y;T) \tag{11}$$

The finite depth of the upstream channel is given by d_r. In order to derive the particular solution, therefore, the following Fourier transform with respect to Y is applied to Equation 11:

$$\tilde{\bar{\Phi}}^{(p)}\left(S,n;T\right)\equiv\int_{-d_r}^{0}\bar{\Phi}^{(p)}\left(S,Y;T\right)\cos\frac{n\pi Y}{d_r}\,dY \tag{12}$$

The resulting transformed equation is

$$\tilde{\bar{\Phi}}^{(p)}\left(S,n;T\right)=-\sqrt{2/\pi}\cdot\frac{1}{S^2+\left(n\pi/d_r\right)^2}\dot{X}_c\cdot\frac{d_r}{n\pi}\cdot\sin\frac{n\pi d_0}{d_r} \tag{13}$$

The boundary conditions, Equation 1 and the first expression in Equation 4, were used in applying the Fourier transforms. Now the inverse Fourier transform with respect to n is applied to Equation 13:

$$\bar{\Phi}^{(p)}\left(S,Y;T\right)=\frac{1}{d_r}\tilde{\bar{\Phi}}^{(p)}\left(S,0;T\right)+\frac{2}{d_r}\sum_{n=1}^{\infty}\tilde{\bar{\Phi}}^{(p)}\left(S,n;T\right)\cos\frac{n\pi Y}{d_r} \tag{14}$$

with the following result:

$$\bar{\Phi}^{(p)}\left(S,Y;T\right)=-\sqrt{2/\pi}\left\{\frac{d_o/d_r}{S^2}+\frac{2}{\pi}\sum_{n=1}^{\infty}\frac{1}{n}\frac{\sin n\pi d_0/d_r}{S^2+\left(n\pi/d_r\right)^2}\cos\frac{n\pi Y}{d_r}\right\}\dot{X}_c \tag{15}$$

This equation satisfies the second boundary condition given in Equation 4.

If the Fourier integral transform in Equation 8 is applied to the free surface condition given in Equation 3, and then the general solution given by Equation 10 is substituted into the resulting equation, the following equation for the function $A*\left(S;T\right)$ can be derived (Stoker, 1957):

$$A^*_{TT}\left(S;T\right)+\tilde{\mu}A^*_T\left(S;T\right)+gS\tanh Sd_r\cdot A^*\left(S;T\right)=-\frac{1}{\cosh Sd_r}\left\{\bar{\Phi}^{(P)}_{TT}\left(S,0;T\right)+\tilde{\mu}\bar{\Phi}^{(P)}_T\left(S,0;T\right)\right\} \tag{16}$$

Substitution of the particular solution, $\bar{\Phi}^{(p)}$ given by Equation 15, into Equation 16 results in the following form:

$$A^*_{TT}\left(S;T\right)+\mu' A^*_T\left(S;T\right)+gS\tanh Sd_r A^*\left(S;T\right)$$

$$-\frac{\sqrt{2/\pi}}{\cosh Sd_r}\left\{\frac{d_0/d_r}{S^2}+\frac{2}{\pi}\sum_{n=1}^{\infty}\frac{1}{n}\frac{\cos\left(n\pi d_0/d_r\right)}{S^2+\left(n\pi/d_r\right)^2}\cos\left(\frac{n\pi Y}{d_r}\right)\right\}\left(\dddot{X}_c+\tilde{\mu}\ddot{X}_c\right) \tag{17}$$

The Fourier integral transform is applied to the initial condition, Equation 6, yielding

$$\bar{\Phi}\left(S,Y;0\right)=\bar{\Phi}_T\left(S,Y;0\right)=0 \tag{18}$$

Substitution of the general solution, Equation 10, into the initial condition, Equation 18, results in

$$\begin{aligned}&\bar{\Phi}^{(P)}\left(S,Y;0\right)+A^*\left(S;0\right)\cosh S\left(Y+d_r\right)=0 \text{ and}\\&\bar{\Phi}^{(P)}_T\left(S,Y;0\right)+A^*_T\left(S;0\right)\cosh S\left(Y+d_r\right)=0\end{aligned} \tag{19}$$

which, along with the initial conditions on the gate displacement, Equation 7, can be used to provide the initial conditions for the function $A^*\left(S;T\right)$:

$$A^*\left(S;0\right)=A^*_T\left(S;0\right)=0 \tag{20}$$

Applying the following Laplace transforms to Equation 17 with initial conditions given by Equations 20 and 7,

$$\overline{A^*}\left(S;\zeta\right)=\int_0^\infty e^{-\zeta T}A^*\left(S;T\right)dT \text{ and } \bar{\mathrm{X}}_c(\zeta)=\int_0^\infty e^{-\zeta T}X_c\left(T\right)dT \tag{21}$$

yields the following:

$$\overline{A^*}\left(S;\zeta\right)=\frac{\sqrt{2/\pi}}{\cosh Sd_r}\left\{\frac{\beta}{S^2}+\frac{2}{\pi}\sum_{n=1}^{\infty}\frac{1}{n}\cdot\frac{\sin n\pi\beta}{S^2+\left(n\pi/d_r\right)^2}\right\}\frac{\left(\zeta^3+\tilde{\mu}\zeta^2\right)}{\zeta^2+\tilde{\mu}\zeta+gS\tanh Sd_r}\bar{X}_c\left(\zeta\right) \tag{22}$$

The inverse Laplace transform of Equation 22 can be reduced to the form:

$$A^*\left(S;T\right)=\frac{\sqrt{2/\pi}}{\cos Sd_r}\left\{\left[\frac{d_0/d_r}{S^2}+\frac{2}{\pi}\sum_{n=1}^{\infty}\frac{1}{n}\frac{\sin n\pi d_0/d_r}{S^2+\left(n\pi/d_r\right)^2}\right]\right.$$
$$\left.\times\left\{\dot{X}_c\left(T\right)-gS\tanh Sd_r\int_0^T X_c\left(\tilde{\tau}\right)e^{-\tilde{\mu}\left(T-\tau'\right)/2}\cos\Phi_0\left(S\right)\left(T-\tilde{\tau}\right)\mathrm{d}\tilde{\tau}\right\}\right] \tag{23}$$

The small frictional coefficient $\tilde{\mu}\,(>0)$ was introduced in Equation 23 to attenuate the effect of the initial conditions with time.

The phase function, $\Phi_0\left(S\right)$, is given by

$$\Phi_0\left(S\right)=\sqrt{gS\tanh Sd_r-\tilde{\mu}^2/4} \tag{24}$$

Substitution of the particular solution, $\bar{\Phi}^{(p)}$, given by Equation 15, and the arbitrary function $A^*\left(S;T\right)$, given by Equation 23, into the general solution, Equation 10, results in the general solution of Equation 9 in the following form:

$$\bar{\Phi}\left(S,Y;T\right)=-\sqrt{\frac{2}{\pi}}\cdot\dot{X}_c\left(T\right)\left\{\frac{\left(d_0/d_r\right)}{S^2}+\frac{2}{\pi}\sum_{n=1}^{\infty}\frac{1}{n}\cdot\frac{\sin n\pi\left(d_0/d_r\right)}{S^2+\left(n\pi/d_r\right)^2}\cdot\cos\frac{n\pi Y}{d_r}\right\}$$
$$+\sqrt{\frac{2}{\pi}}\cdot\frac{\cosh S\left(Y+d_r\right)}{\cosh Sd_r}\left\{\frac{\left(d_0/d_r\right)}{S^2}+\frac{\pi}{2}\sum_{n=1}^{\infty}\frac{1}{n}\cdot\frac{\sin n\pi\left(d_0/d_r\right)}{S^2\cdot\left(n\pi/d_r\right)^2}\right\}$$
$$\times\left\{\dot{X}_c\left(T\right)-gS\tanh Sd_r\int_0^T X_c\left(\tilde{\tau}\right)e^{\tilde{\mu}\left(T-\tilde{\varsigma}\right)/2}\cos\Phi_0\left(S\right)\left(T-\tilde{\tau}\right)d\tilde{\tau}\right\} \tag{25}$$

The following inverse Fourier transform with respect to S:

$$\Phi\left(X,Y;T\right)=\sqrt{2/\pi}\int_0^{\infty}\bar{\Phi}\left(S,Y;T\right)\cos SXdS \tag{26}$$

when applied to Equation 25, yields the velocity potential $\Phi(X, Y; T)$:

$$\Phi\left(X,Y;T\right)=\frac{2}{\pi}\dot{X}_c(T)\left[-\int_0^\infty\left[\frac{\left(d_0/d_r\right)}{S^2}+\frac{2}{\pi}\sum_{n=1}^{\infty}\frac{1}{n}\frac{\sin n\pi\left(d_0/d_r\right)}{S^2+(\frac{n\pi}{d_r})^2}\cos\frac{n\pi Y}{d_r}\right\}\cos SXdS\right.$$

$$\left.+\int_0^\infty\left\{\frac{\left(d_0/d_r\right)}{S^2}+\frac{2}{\pi}\sum_{n=1}^{\infty}\frac{1}{n}\frac{\sin n\pi\left(d_0/d_r\right)}{S^2+(\frac{n\pi}{d_r})^2}\right\}\frac{\cosh S(Y+d_r)}{\cosh Sd_r}\cos SX\,dS\right] \tag{27}$$

$$-\frac{2}{\pi}g\int_0^\infty\left\{\frac{\left(d_0/d_r\right)}{S^2}+\frac{2}{\pi}\sum_{n=1}^{\infty}\frac{1}{n}\frac{\sin n\pi\left(d_0/d_r\right)}{S^2+\left(\frac{n\pi}{d_r}\right)^2}\right\}\frac{\cosh S(Y+d_r)}{\cosh Sd_r}S\tanh Sd_r\cos SXdS$$

$$\times\int_0^T X_c(\tilde{\tau})e^{-\frac{\tilde{\mu}}{2}(T-\tilde{\tau})}\cos\Phi_0(S)(T-\tilde{\tau})d\tilde{\tau}$$

The first and second integrals on S in the first term of Equation 27 can be reduced to the following expressions:

$$\text{First Integral}=-\frac{\beta}{\varepsilon}+\frac{\pi\beta}{2}X-\frac{d_r}{\pi}\sum_{n=1}^{\infty}\frac{1}{n^2}e^{-\frac{n\pi}{d_r}X}\sin n\pi\beta\cdot\cos\frac{n\pi Y}{d_r}$$

$$\text{Second Integral}=\left(\frac{1}{\varepsilon}-\frac{\pi}{2}X\right)\beta+\frac{d_r}{\pi}\sum_{n=1}^{\infty}\frac{1}{n^2}e^{-\frac{n\pi}{d_r}X}\sin n\pi\beta\frac{\cos\frac{n\pi}{d_r}(Y+d_r)}{\cos n\pi}$$

where ε represents a small value. As a result, the two integrations in the first term of Equation 27 cancel each other, resulting in the following expression for $\Phi\ (X,Y;T)$:

$$\Phi\left(X,Y;T\right)=-\frac{2}{\pi}g\int_0^\infty\left\{\frac{d_0/d_r}{S^2}+\frac{2}{\pi}\sum_{n=1}^{\infty}\frac{1}{n}\frac{\sin n\pi d_0/d_r}{S^2+\left(n\pi/d_r\right)^2}\right\}\frac{\cosh S\left(Y+d_r\right)}{\cosh Sd_r} \tag{28}$$

$$\times S\tanh Sd_r\cos SXdS\int_0^T X_c\left(\tilde{\tau}\right)e^{-\mu'(T-\tau')/2}\cos\Phi_0\left(S\right)\left(T-\tilde{\tau}\right)d\tilde{\tau}$$

PUSH-AND-DRAW PRESSURE

The pressure fluctuations in the upstream channel are generated by the push-and-draw effect, and will be called the *push-and-draw pressure*, denoted as $P_b(X, Y; T)$. This pressure is given by Equation 26 in Chapter 4 resulting in the following expression:

$$P_b\left(X,Y;T\right)=\frac{2}{\pi}\rho g\int_0^{\infty}\left\{\frac{d_0/d_r}{S^2}+\frac{2}{\pi}\sum_{n=1}^{\infty}\frac{1}{n}\frac{\sin n\pi d_0/d_r}{S^2+\left(n\pi/d_r\right)^2}\right\}\frac{\cosh\left(Y+d_r\right)}{\cosh Sd_r}$$
$$\times S\tanh Sd_r\cos SX\,\mathrm{d}S\left[X_c\left(T\right)-\int_0^T X_c\left(\tilde{\tau}\right)e^{-\mu'\left(T-\tau'\right)/2}\sin\Phi_0\left(S\right)\left(T-\tilde{\tau}\right)d\tilde{\tau}\right] \tag{29}$$

The following time derivative was used to obtain Equation 29:

$$\frac{d}{dT}\int_0^T X_c\left(\tilde{\tau}\right)e^{-\frac{\tilde{\mu}}{2}\left(T-\tilde{\tau}\right)}\cos\Phi_0\left(S\right)\left(T-\tilde{\tau}\right)d\tilde{\tau}$$
$$=X_c\left(T\right)-\Phi_0\left(S\right)\int_0^T X_c\left(\tilde{\tau}\right)e^{-\frac{\tilde{\mu}}{2}\left(T-\tilde{\tau}\right)}\sin\Phi_0\left(S\right)\left(T-\tilde{\tau}\right)d\tilde{\tau}$$

As was done in Chapter 4, dimensionless variables are introduced to represent the analytical results from Equation 29 in dimensionless form:

$$x\equiv\frac{X}{d_0},\quad y\equiv\frac{Y}{d_0},\quad s\equiv d_0S,\quad x_c\equiv\frac{X_c}{\bar{X}_c},\quad t\equiv\Omega_0T,\quad \tau\equiv\Omega_0\tilde{\tau},\mu\equiv\frac{\tilde{\mu}}{\Omega_0} \tag{30}$$

The representative length for the flow field is the submergence depth, d_0; the representative velocity, $d_0\Omega_0$, is the product of the depth d_0 and the circular natural vibration frequency in air, Ω_0; the representative amplitude of gate vibration in the middle of the gate, $\bar{X}_c$, is used to normalize the vibration amplitude.

The ratio of the upstream-channel depth, d_r, to the submergence depth, d_0, is represented by:

$$\beta^*\equiv d_r/d_0 \tag{31}$$

The dimensionless form $p_b(x, y; t)$ of the push-and-draw pressures $P_b(X, Y; T)$ is defined as:

$$p_b\equiv\frac{P_b/\rho g}{\bar{X}_c} \tag{32}$$

which represents the ratio of the fluctuating water head to the amplitude of the horizontal gate vibration, $\bar{X}_c$.

With the aid of Equations 30 to 32, the hydrodynamic pressure P_b given by Equation 29 can be re-written in the following dimensionless form:

$$p_b(x,y;t) = \int_0^{\infty} A_b(x,y;t)\,ds \left[x_c(t) - \frac{\varphi_0(s)}{F_0} \int_0^t x_c(\tau) e^{-\mu(t-\tau)/2} \sin\left\{ \frac{\varphi_0(s)}{F_0}(t-\tau) \right\} d\tau \right] \tag{33}$$

where A_b and $\varphi_0(s)$ are given by the following expressions:

$$A_b(x,y;t) = \frac{2}{\pi} \left\{ \frac{1}{\beta^* s^2} + \frac{2}{\pi} \sum_{n=1}^{\infty} \frac{\sin n\pi/\beta^*}{s^2 + (n\pi/\beta^*)^2} \right\} \frac{\cosh s(y+\beta^*)}{\cosh s\beta^*} s \tanh s\beta^* \cos sx \tag{34}$$

$$\varphi_0(s) \equiv \frac{\Phi_0(S)}{\sqrt{g/d_0}} = \sqrt{s \tanh s\beta^* - \frac{1}{4}\mu^2 F_0^2} \tag{35}$$

in which F_0 is the basic Froude number, defined as

$$F_0 \equiv \sqrt{d_0/g}\,\Omega_0 \tag{36}$$

As previously introduced in Chapter 4, this basic Froude number, F_0, characterizes the fluctuating flow field associated with a gate vibration at a circular in-air natural frequency, Ω_0.

The integral term in the push-and-draw pressure $p_b(x,y;t)$, given by Equation 33, represents the added mass and the damping due to fluid motion.

Assume that the gate undergoes a harmonic vibration $x_c(t)$:

$$x_c(t) = e^{i\Omega T} = e^{i(F/F_0)t} \tag{37}$$

where $i = \sqrt{-1}$, and Ω is the vibration frequency of the gate in flowing water, which is smaller than the in-air natural vibration frequency, Ω_0, due to added-mass effects.

F is the Froude number corresponding to the in-water frequency, defined as

$$F \equiv \sqrt{d_0/g}\,\Omega \tag{38}$$

It is of the same form as the basic Froude number, F_0, and characterizes the fluctuating flow field scaled with the in-water vibration frequency. The ratio of F/F_0 can be found from Equations 36 and 38 to be the ratio of the in-water to the in-air gate vibration frequencies:

$$F/F_0 = \Omega/\Omega_0 \tag{39}$$

If the gate vibration is given by Equation 37, the definite integral term in Equation 33 becomes

$$\int_0^t x_c(\tau) e^{-\mu(t-\tau)/2} \sin\left\{\frac{\varphi_0(s)}{F_0}(t-\tau)\right\} d\tau = \frac{F_0\varphi_0(s)}{W(s)} e^{i(F/F_0)t} \tag{40}$$

where the function, $W(s)$, is given by

$$W(s) = s\tanh s\beta^* - \left(F^2 - i\mu F F_0\right) \tag{41}$$

Substituting Equation 40 into Equation 33, the reduced fluid pressure $p_b(x, y; t)$ may be rewritten as

$$p_b(x, y; t) = -F^2 \int_0^\infty \frac{A_b(x, y; s)}{W(s)} ds e^{i(F/F_0)t} \tag{42}$$

neglecting the μ^2 terms.

The definite integral term in Equation 42 can be solved by introducing the integral contours in the complex s-plane shown in Figure 2. The function $A_b(x, y, s)$ in the integrand, which is given by Equation 34, has singular points at the origin $s = 0$ and at the locations given by $s = \pm jn\pi / \beta^*$ (where $n = 1, 2, ...$) on the imaginary axis. These singular points are indicated by the small open circles in Figure 2. Furthermore, the function $1/W(s)$ has singular points at the locations given by $s_0' = (\xi_0 + i\eta_0)$ near the real axis and at the points given by $s_j = (\xi_j + i\eta_j)$ (where $j = 1, 2, ...$) near the imaginary axis. These singular points are determined by the same non-dimensional transcendental equations given as Equations 109a and 109b in Chapter 4:

$$\xi_0 \tanh(\xi_0\beta^*) = F^2 \tag{43}$$

$$\eta_i \tan(\eta_i\beta^*) = -F^2 \tag{44}$$

Figure 2. Integral contours and singular points in the complex s-plane
From Ishii (1992). © 1992 Elsevier. Used with permission

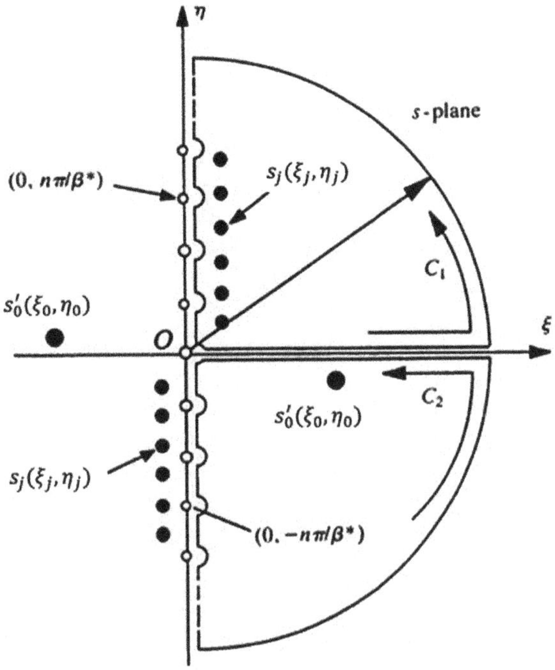

The values η_0 and ξ_j ($j = 1, 2, \ldots$) are small values depending upon the small positive value of μ; as given by Equation 100a in Chapter 4,

$$\eta_0 < 0 \text{ for } \xi_0 > 0 \text{ and } \eta_0 > 0 \text{ for } \xi_0 < 0;$$

as given by Equation 100b in Chapter 4,

$$\xi_j > 0 \text{ for } \eta_j > 0 \text{ and } \xi_j < 0 \text{ for } \eta_j < 0.$$

The solutions of these equations are shown in Figure 3. When $F^2\beta^*$ is comparatively small, the solution ξ_0 converges to zero and the solution $\eta_j\left(j = 1, 2, \ldots\right)$ converges to $j\pi / \beta^*\left(j = 1, 2, \ldots\right)$; when $F^2\beta^*$ becomes larger, ξ_0 converges to F^2 and $\eta_j\left(j = 1, 2, \ldots\right)$ converges to $(2j - 1)\pi / \beta^*\left(j = 1, 2, \ldots\right)$.

These singular points given by $s_0' = (\xi_0 + i\eta_0)$ and $s_j = (\xi_j + i\eta_j)$, ($j = 1, 2, \ldots$), are denoted by the small black dots in Figure 2. Solving the complex integral along the integral contours denoted by C_1 and C_2 in Figure 2, the definite integral along the real axis from 0 to ∞ is obtained as the summation of a residue value at each singular point enclosed in each integral contour.

The reduced fluid pressures $p_b\left(x, y; t\right)$ can then be written as follows:

Figure 3. Solution of the transcendental equations: (a) $\xi_0\beta^*$ tanh $\xi_0\,\beta^* = F^2\beta^*$ and (b) $\eta_j\beta^*$tanh $\eta_j\,\beta^*$ $= - F^2\,\beta^*$
From Ishii (1992). © 1992 Elsevier. Used with permission.

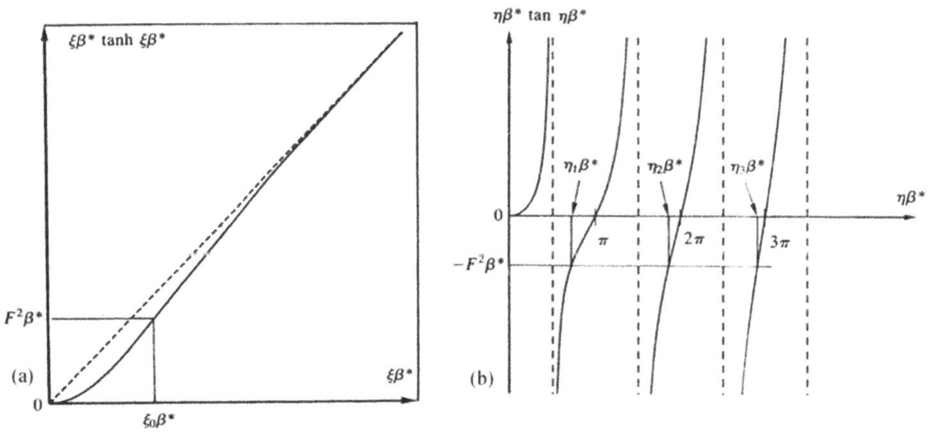

$$p_b(x,y;t) = 2F^4\left[\sum_{j=1}^{\infty}\left\{\frac{1}{\beta^*\eta_j^2}+\frac{2}{\pi}\sum_{n=1}^{\infty}\frac{1}{n}\frac{\sin n\pi/\beta^*}{\eta_j^2-\left(n\pi/\beta^*\right)^2}\right\}\frac{\cos\eta_j(y+\beta^*)}{\cos\eta_j\beta^*}\frac{\exp(-\eta_j x)}{W(i\eta_j)/i}\right.$$
$$+i\left\{\frac{1}{\beta^*\xi_0^2}+\frac{2}{\pi}\sum_{n=1}^{\infty}\frac{1}{n}\frac{\sin n\pi/\beta^*}{\xi_0^2+\left(n\pi/\beta^*\right)^2}\right\}\frac{\cosh\xi_0(y+\beta^*)}{\cosh\xi_0\beta^*} \tag{45}$$
$$\left.\times\frac{\exp\left\{-(\eta_0+i\xi_0)x\right\}}{W(\xi_0)}\right]e^{i(F/F_0)t}$$

where the functions $W(i\eta_j)/i$ and $W(\xi_0)$ are given by

$$W(i\eta_j)/i = \tan(\eta_j\beta^*)+\eta_j\beta^*/\cos^2(\eta_j\beta^*)$$
$$W(\xi_0) = \tanh(\xi_0\beta^*)+\xi_0\beta^*/\cosh^2(\xi_0\beta^*) \tag{46}$$

The real part of the harmonic gate vibration $x_c(t)$ given by Equation 37 is

$$x_c(t) = \cos\,(F/F_0)t \tag{47}$$

The fluid pressure corresponding to this gate vibration is given by the real part of Equation 45, that is

$$p_b(x,y;t) = p_{bs}(F,x,y)\cos\left(\frac{F}{F_0}t-\pi\right)+p_{bp}(F,x,y)\cos\left(\frac{F}{F_0}t-\xi_0 x+\pi/2\right) \tag{48}$$

The reduced push-and-draw pressure, $p_b\left(x, y; t\right)$, is composed of a standing-wave pressure component and a progressive-wave (or travelling-wave) component. Using the subscript "*s*" to denote the standing-wave pressure component and the subscript "*p*" the progressive-wave component, the amplitudes of these two pressure components are given by the following infinite-series summations:

$$p_{bs}(F,x,y) = -2F^4 \sum_{j=1}^{\infty} \left\{ \frac{1}{\beta^* \eta_j^2} + \frac{2}{\pi} \sum_{n=1}^{\infty} \frac{1}{n} \frac{\sin n\pi / \beta^*}{\eta_j^2 - \left(n\pi / \beta^*\right)^2} \right\} \frac{\cos \eta_j (y + \beta^*)}{\cos \eta_j \beta^*} \frac{\exp(-\eta_j x)}{W(i\eta_j) / i} \tag{49}$$

$$p_{bp}(F,x,y) = 2F^4 \left\{ \frac{1}{\beta^* \xi_0^2} + \frac{2}{\pi} \sum_{n=1}^{\infty} \frac{1}{n} \frac{\sin n\pi / \beta^*}{\xi_0^2 + \left(n\pi / \beta^*\right)^2} \right\} \frac{\cosh \xi_0 (y + \beta^*)}{\cosh \xi_0 \beta^*} \frac{\exp(-\eta_0 x)}{W(\xi_0)} \tag{50}$$

Upon substitution of $x = 0$ into Equation 48, the reduced push-and-draw pressure $p_b\left(0, y; t\right)$ acting on the gate surface is obtained.

The pressure amplitudes for the depth ratio $\beta^* = 1.0$ are shown in Figure 4. The vertical axis in these figures gives the dimensionless position *y* along the gate from the undisturbed water surface; the value *y* = -1 corresponds to the bottom of the gate. The abscissa value at *y* = 0 represents the reduced amplitude of the surface wave (Ishii & Imaichi, 1977a).

The major characteristics of the flow-induced pressure acting on the gate can be summarized as follows:

1. As the gate bottom (y = -1) is approached, the standing-wave pressure amplitude increases, as shown in Figure 4a.
2. As the water surface (*y* = 0) is approached, on the other hand, the progressive-wave (travelling-wave) pressure amplitude increases, as shown Figure 4b.
3. With increasing Froude number *F*, the standing-wave pressure amplitudes increase and the progressive-wave pressure amplitudes decrease, except near the surface (*y* = 0).
4. The maximum amplitude of the reduced standing-wave pressure appears at the gate bottom (y = -1) and reaches values of about 80 with *F* = 10 in Figure 4a. These maximum values are far larger than the maximum amplitudes of the reduced progressive pressure wave, which appears at the water surface (*y* = 0) and are less than about 2.0.
5. The progressive-wave pressure amplitude is always positive for all values of *F*, as shown in Figure 4b. In contrast, the standing-wave pressure amplitude takes on small negative values near the water surface (*y* = 0), as shown in the expanded view for small *y*-values in Figure 4a. A negative amplitude means that the standing-wave pressure is out-of-phase relative to that with a positive amplitude and dissipates energy from the gate.
6. Both the standing and progressive-wave amplitudes at the water surface (*y* = 0) tend toward zero as *F* increases. Resultant surface waves due to streamwise gate motion decrease in amplitude with increasing Froude number *F*.

Figure 4. Vertical distributions of the reduced push-and-draw pressure, $p_b(x, y; t)$, for $x = 0$ with a depth ratio $\beta^* = 1.0$: (a) amplitude $p_{bs}(F, 0, y)$ of the standing-wave pressure component; (b) amplitude p_{bp} (F, 0, y) of the progressive-wave pressure component
After Ishii (1992).

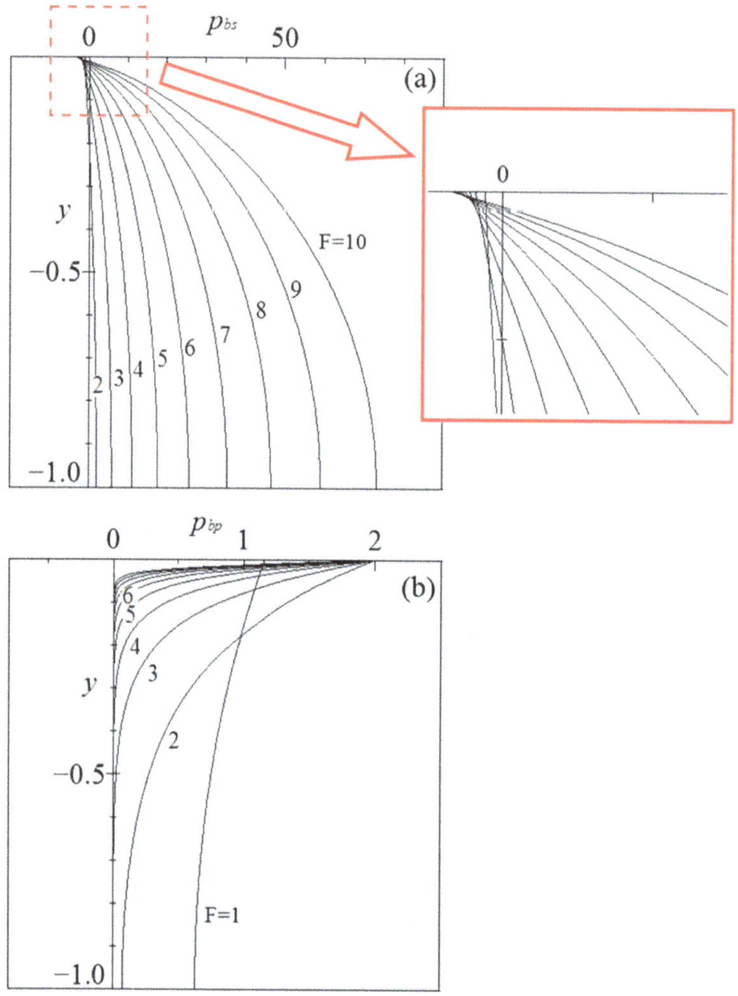

ADDED MASS EFFECTS AND SMALL WAVE-RADIATION DAMPING

We will now consider how the push-and-draw pressure, given in Equation 48, affects the vibration of the gate. The push-and-draw pressure acting on the gate (at $x = 0$) can be reduced to the following expression:

$$-p_b(0, y; t) = p_{bs}(F, 0, y)\cos\left(\frac{F}{F_0}t\right) + p_{bp}(F, 0, y)\cos\left(\frac{F}{F_0}t - \frac{\pi}{2}\right) \text{ or}$$
$$-p_b(0, y; t) = p_{bs}(F, 0, y)\cos\left(\frac{F}{F_0}t\right) + p_{bp}(F, 0, y)\sin\left(\frac{F}{F_0}t\right) \tag{51}$$

The gate vibration x_c is defined as positive when the gate moves in the upstream direction, and hence the hydrodynamic load on the gate in the positive direction of x_c becomes $-p_b$. The standing pressure amplitude $p_{bs,}$ is given in Equation 49 with $x = 0$ and the progressive pressure amplitude p_{bp} in Equation 50 with $x = 0$. The distribution of these pressure amplitudes along the gate is shown in Figure 4, in which the standing-wave pressure amplitude p_{rs}, shown in Figure 4a, takes on a large value that increases significantly with increasing Froude number *F*, while the progressive-wave pressure amplitude p_{bp}, shown in Figure 4b, takes on a small value that decreases rapidly with increasing Froude number *F*.

Equation 51 reveals to us that the standing wave component of the push-and-draw pressure, the first term on the right side, fluctuates in phase with the gate displacement and therefore acts as added mass in the gate vibration $x_c(t)$ given by Equation 47. The correlation of the standing-wave pressure component (dashed line with arrows) with the gate displacement (solid line) is shown schematically in Figure 5a, along with the gate velocity (dashed line). In the first quarter cycle when the gate moves downstream, the pressure load pushes against the gate movement, consuming energy from the gate vibration due to the negative value of the product of the pressure load and the vibration velocity. In the second quarter cycle, when the gate passes through the equilibrium point and is moving farther downstream, the pressure load on the gate acts in the same direction as the gate movement, supplying energy to the gate vibration due to the positive product of the pressure load and the vibration velocity. In the third quarter cycle, when the gate begins to move back upstream, the pressure load acts against the gate movement, consuming energy from the gate vibration due to the negative product of the pressure load and the vibration velocity. Finally, for the last quarter cycle when the gate passes through equilibrium and continues moving upstream, the pressure load acts on the gate in the same direction as the gate movement, supplying energy to the gate vibration due to the positive product of the pressure load and the vibration velocity. As a result, the net energy supply from the standing-wave pressure load to the gate vibration is zero for the full cycle. The standing-wave pressure load behaves like an inertial force on the gate vibration, which is called the added mass. This added mass effect on the gate vibration can be quite large since the standing-wave pressure amplitude is also very large, as shown in Figure 4a, and increases significantly with increasing Froude number, *F.*

Figure 5. Correlation of the push-and-draw pressure components with horizontal vibration displacement of the gate with positive displacement in the upstream direction: (a) standing-wave pressure component with zero net energy input and (b) progressive-wave component with negative net energy input to the gate

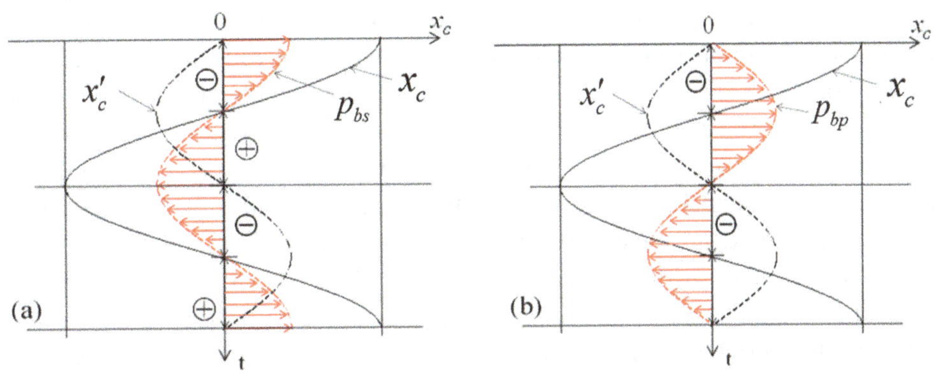

For a different perspective that can lead to better understanding, Equation 51 may be re-written as follows:

$$-p_b(0,y;t) = -p_{bs}(F,0,y)\left(\frac{F_0}{F}\right)^2 x_c'' - p_{bp}(F,0,y)\frac{F_0}{F}x_c' \tag{52}$$

where the cosine functions, $\cos\left[\left(F / F_0\right)t\right]$ and $\cos\left[\left(F / F_0\right)t - \pi / 2\right]$ in Equation 51 have been replaced by the acceleration term $-\left(F_0 / F\right)^2 x_c''$ and the velocity term $-\left(F_0 / F\right)x_c'$, respectively. The first term on the right hand side is the standing-wave pressure component which is proportional to the gate's acceleration, resulting in a significant added mass effect on the gate vibration. With this water added mass effect, the in-air natural vibration of the gate structure decreases substantially to a much lower in-water frequency.

In contrast, the second term on the right side of Equations 51 and 52 represents the progressive-wave pressure component, which fluctuates with a 90° phase lag relative to the gate displacement, as shown by a dashed line with arrows in Figure 5b. The product of the pressure load on the gate and the vibration velocity is consistently negative for the entire cycle, indicating that the progressive-wave pressure load always extracts energy from the gate vibration, damping the gate vibration. This damping will be termed wave-radiation damping. This wave-radiation damping is very small and decreases rapidly with increasing Froude number, proportional to the progressive-wave pressure amplitude shown in Figure 4b.

While these theoretical results make sense physically, experimental confirmation is essential before we attempt to apply these results to predict the behavior of large hydraulic gates regulating flow in rivers and reservoirs. Such experiments are discussed in the next section of this chapter.

EXPERIMENTAL CONFIRMATION OF THEORETICAL ANALYSIS

Figure 6 shows a schematic of the model gate set-up used to validate the analysis of the push-and-draw pressure when the gate oscillates horizontally. The span and height of the gate were 300 mm and 500 mm, respectively. The gate had a flat upstream surface.

The rigid gate model was attached vertically to the vibrating table that, in turn, was suspended by two stays from a fixed base (support structure). The rotary motion of the motor shaft was converted into small-amplitude linear parallel motion using a V-belt, pulley and connecting rod. A non-contact displacement pickup was placed on the downstream side of the gate to measure the horizontal vibration of the gate.

The apparatus was inserted into a 305 mm-wide, 450 mm-deep water channel that was approximately 5 m long. The channel floor was flat and horizontal. A small clearance was maintained between both sides of the gate and the channel wall, as well as between the lower edge of the gate and the channel floor, to avoid the gate contacting the channel wall and floor. Water was supplied at a steady rate from the upstream reservoir to maintain the mean water surface level. The mean water depth above the bottom of the gate, d_0, was 375 mm. The gate was not submerged on the downstream side.

The model gate shown in Figure 6 was forced to vibrate at frequencies ranging from 3 to 6 Hz and with a small amplitude of just less than 1.0 mm. The fluctuating pressure at the center of the vertical

Figure 6. Schematic of vertical flat gate model apparatus used in experiments to confirm the validity of the theoretical analysis of push-and-draw pressure fluctuations

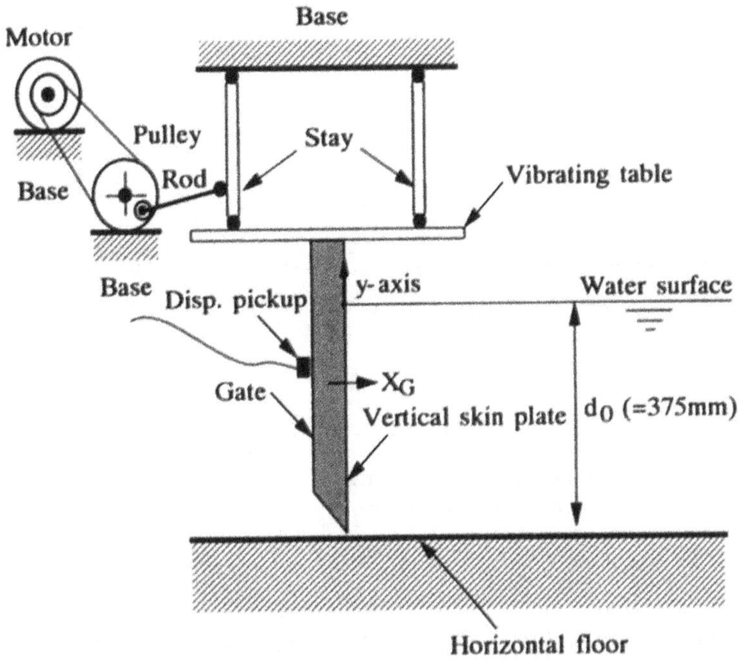

skinplate was measured using a small pressure transducer (manufactured by Toyoda Machine Works, Model PD104-0.1F).

The measured waveforms of the fluctuating pressure were analyzed using a spectrum analyzer to determine the pressure amplitude P_{b0} and phase β_b relative to the oscillating displacement of the gate in the horizontal direction. The measured hydrodynamic pressure on the skinplate was made dimensionless as p_{b0}, as in Equation 32. The measured results for p_{b0} and β_b are plotted as functions of water depth y from 0 to -1.0 in Figure 7. The Froude number F ranged from 3.4 to 7.2 in these measurements.

The solid lines in Figure 7 represent the theoretical results, calculated using Equation 51 in the following form:

$$p_b = p_{b0} \cos\left(\frac{F_0}{F} t - \beta_b\right) \tag{53a}$$

$$p_{b0} = \sqrt{p_{bs}^{\ 2} + p_{bp}^{\ 2}} \text{ and } \beta_b = \pi + \tan^{-1} \frac{p_{bp}}{p_{bs}} \tag{53b}$$

The theoretical results for the magnitude of the fluctuating pressure and the phase lag associated with the push-and-draw effect show excellent agreement with the experimental results in Figure 7. These

Figure 7. Measured amplitude and phase lag of push-and-draw pressure associated with small amplitude horizontal gate vibration pressure for Froude numbers from 3.4 to 7.2

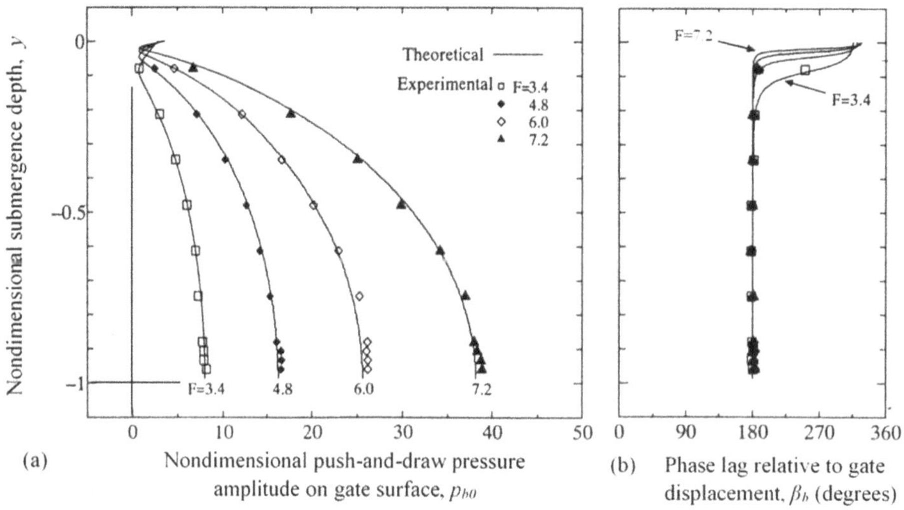

experimental data confirm the validity of the theoretical analyses developed in this chapter using the potential flow theory presented in Chapter 4 and based on the analysis by Rayleigh (1883).

DIMENSIONLESS ADDED MASS AND FLUID-DAMPING COEFFICIENTS

If we introduce ΔM_b and ΔC_b for the added mass and fluid-damping coefficients, respectively, caused by the push-and-draw effect of the gate on the water, their dimensionless representations Δm_b and Δc_b can be formulated as follows:

$$\Delta m_b \equiv \frac{\Delta M_b}{\rho d_0^2} \text{ and } \Delta c_b \equiv \frac{\Delta C_b}{\rho d_0 \sqrt{g d_0}} \tag{54}$$

The quantity ρd_0^2 is selected as the representative water mass. The quantity $\rho d_0^2 \sqrt{g d_0}$ in the second definition in Equation 54 represents the gravity force $\rho g d_0^2$ acting on the representative mass ρd_0^2 divided by the wave velocity $\sqrt{g d_0}$.

The dimensionless fluctuating water load acting on the gate, f_G, can be calculated by integrating the pressure $p_b(0, y; t)$ over the surface of the submerged gate from $y = -1$ to $y = 0$.

$$f_G \equiv \frac{F_G}{\rho g d_0 X_0} = \int_{-1}^{0} p_b(0, y; t) dy \tag{55}$$

The symbol F_G in Equation 55 is the fluctuating water load acting on the gate per unit span. Substituting Equation 52 into Equation 54 permits the fluctuating water load to be separated into two terms, one proportional to the gate's acceleration and one proportional to the gate's velocity:

$$\frac{f_G}{F_0^2} = \Delta m_b x_c'' + \frac{\Delta c_b}{F_0} x_c' \tag{56}$$

In Equation 56, the reduced added mass Δm_b and the reduced fluid-damping coefficient Δc_b are given by

$$\Delta m_b = \frac{1}{F^2}\int_{-1}^{0} p_{bs}\left(F,0,y\right)dy \text{ and } \Delta c_b = \frac{1}{F}\int_{-1}^{0} p_{bp}\left(F,0,y\right)dy \tag{57}$$

Substituting the standing-wave and progressive-wave pressure amplitude, given by Equations 49 and 50, and integrating with respect to y from -1 to 0, Δm_b and Δc_b can be reduced to

$$\Delta m_b = F^2 B_{ba}(F,\beta^*) \text{ and } \Delta c_b = F^3 B_{bv}(F,\beta^*) \tag{58}$$

where the functions $B_{ba}(F, \beta^*)$ and $B_{bv}(F, \beta^*)$, related to the push-and-draw effect of the gate on the water, are given by the following series summations:

$$B_{ba}(F,\beta^*) = 2\sum_{j=1}^{\infty}\left\{\frac{1}{\beta^*\eta_j^2} + \frac{2}{\pi}\sum_{n=1}^{\infty}\frac{2}{n}\frac{\sin n\pi/\beta^*}{\eta_j^2 - \left(n\pi/\beta^*\right)^2}\right\}\frac{\sin\eta_j\beta^* - \sin\eta_j(\beta^*-1)}{\eta_j\cos\eta_j\beta^* W(i\eta_j)/i} \tag{59}$$

$$B_{bv}(F,\beta^*) = 2\sum_{j=1}^{\infty}\left\{\frac{1}{\beta^*\xi_0^2} + \frac{2}{\pi}\sum_{n=1}^{\infty}\frac{2}{n}\frac{\sin n\pi/\beta^*}{\xi_0^2 + \left(n\pi/\beta^*\right)^2}\right\}\frac{\sinh\xi_0\beta^* - \sinh\xi_0(\beta^*-1)}{\xi_0\cosh\xi_0\beta^* W(\xi_0)} \tag{60}$$

The expressions for the reduced hydrodynamic coefficients, given by Equation 56, are functions of the Froude number, F, describing the fluctuating flow field, and the depth ratio, β^*. The function $B_{bv}(F, \beta^*)$ given by Equation 60 includes an infinite-series summation of terms containing the integer n. For increasingly larger n values, this infinite series converges rapidly to zero, with each term inversely proportional to the second or third power of n. Therefore, the series can be calculated numerically with high accuracy. The function $B_{ba}(F, \beta^*)$ given by Equation 59 includes double infinite-series summations on the integers n and j, the denominator of which includes the term $\left\{\eta_j^2 - \left(n\pi/\beta^*\right)^2\right\}$. As shown in Figure 3b, η_j is given by $(j\pi-\Delta)/\beta^*$, where $0 \leq \Delta < \pi/2$ (Δ converges to zero as $F^2\beta^* \to 0$ and to $\pi/2$ as $F^2\beta^* \to \infty$). If $j = n$, the denominator $\left\{\eta_j^2 - \left(n\pi/\beta^*\right)^2\right\}$ is at a minimum and thus the series term is at its maximum.

Figure 8. Calculated values of the reduced added mass, Δm_b, and the reduced fluid-damping coefficient, Δc_b, associated with the push-and-draw effect on the water due to streamwise vibration of the gate for various depth ratios, $1/\beta^*$

From Ishii (1992). © 1992 Elsevier. Used with permission.

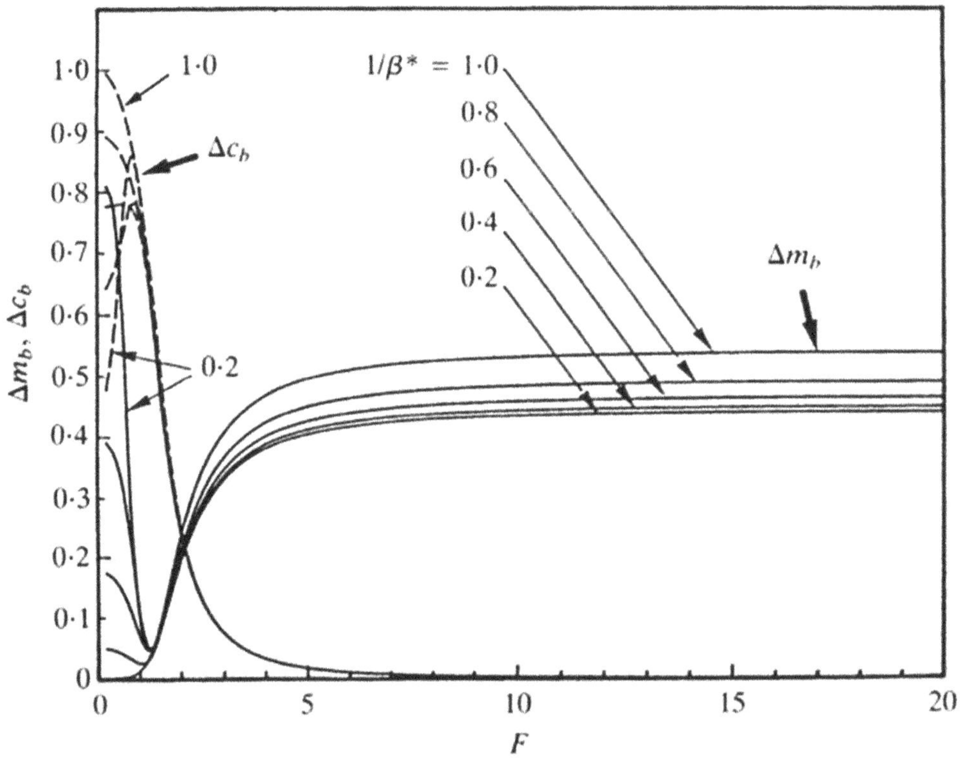

For the numerical calculation of $B_{ba}(F, \beta^*)$, therefore, it is important to sum correctly the double infinite-series terms around $j = n$. If $F^2\beta^* \to 0$ due to a small value of F, Δ converges to zero, and the double infinite-series diverges to ∞ for $j = n$. In this case, $B_{ba}(F, \beta^*)$ cannot be calculated accurately.

For a given Froude number, F, first the solutions ξ_0 and η_j of the transcendental equations, Equations 43 and 44, were calculated numerically, and then the functions $B_{ba}(F, \beta^*)$ and $B_{bv}(F, \beta^*)$ were computed. Substituting these calculated values into Equation 58, one obtains the reduced added mass and the reduced damping coefficients. The calculated results for Δm_b and Δc_b are shown in Figure 8, where values of $1/\beta^*$ are the third parameter. The gate is positioned over a horizontal concrete floor in most cases. Therefore, the submergence depth, d_0, is nearly equal to the upstream-channel depth, d_r, for small gate openings; hence, the depth ratio, β^*, is about 1.0. For this particular value of β^*, with increasing F, Δc_b decreases rapidly and is nearly zero for $F > 8.0$. In contrast, Δm_b increases rapidly from zero and approaches constant values. If an error of $\pm 1\%$ is permissible, Δm_b can be approximated as 0.54 for $F >$ 11.0. Δm_b and Δc_b were examined in a physical model by Ishii (1990) and the theoretical results presented in this chapter generally agreed well with the experiments.

As the depth ratio β^* increases ($1/\beta^*$ decreases), that is, as the upstream-channel depth becomes larger than the submergence depth, the flow disturbances caused by the gate vibration (that is, the flow-rate variation and the push-and-draw effect on the water) spread over a larger portion of the flow field, and thus the feedback effect of the disturbances to the gate vibration decreases. From this perspective, a smaller reduced added mass Δm_b is consistent with increasing β^* (decreasing $1/\beta^*$) for F larger than 1.0. All curves for F smaller than 1.0 change rapidly and are very complicated. However, the range $F < 1.0$ is less significant than $F > 1.0$ for two reasons: (i) in most cases, the Froude number, F, takes values far larger than 1.0, since gates are frequently operated with large submergence depths, d_0; (ii) if the depth d_0 is small, the Froude number F can be less than 1.0, but water mass also becomes small. Physically, this means that since the depth d_0 is small, only a small area of the gate can be in contact with the water. Therefore, since the area over which the pressure acts is small, no significant energy transfer from the gate to the water or from the water to the gate can occur.

The push-and-draw pressure results in a large added mass that can drastically reduce the in-air vibration frequency to a much lower in-water frequency. In addition, the push-and-draw pressure results in comparatively small wave-radiation damping, despite the large mass of water loading the gate. These two key factors increase the likelihood that even a massive, rigid hydraulic gate structure can become susceptible to quickly-growing, spontaneous vibration, if a mechanism exists to supply energy to the gate motion. Such a mechanism is the pressure fluctuation associated with flow-rate variation, as detailed in Chapter 4.

SUMMARY OF FLOW-INDUCED PRESSURE LOADING ON GATES DUE TO DISCHARGE FLUCTUATIONS AND STREAMWISE VIBRATIONS

In the previous sections, we derived expressions for push-and-draw pressure due to streamwise gate vibration and related the added mass, fluid excitation and fluid damping to the standing and progressive-wave pressure components. In the previous chapter, we analyzed the effect of vertical vibrations accompanied by fluctuating discharge. As a result, we have expressions for reduced added mass Δm_r and the fluid-excitation coefficient Δc_r due to flow-rate fluctuations:

$$\Delta m_r / k = \frac{1}{F^2}\int_{-1}^{0} p_{rp}\left(F,0,y\right)dy \text{ and } \Delta c_r / k = \frac{1}{F}\int_{-1}^{0} p_{rs}\left(F,0,y\right)dy \tag{61a}$$

These expressions were previously given in Equation 67 in Chapter 4. We also have expressions for the reduced added mass Δm_b and the fluid-damping coefficient Δc_b due to the push-and-draw pressure in terms of definite integrals of the respective pressure component amplitudes (previously shown as Equation 57):

$$\Delta m_b = \frac{1}{F^2}\int_{-1}^{0} p_{bs}\left(F,0,y\right)dy \text{ and } \Delta c_b = \frac{1}{F}\int_{-1}^{0} p_{bp}\left(F,0,y\right)dy \tag{61b}$$

From these relations, we reach the following conclusions:

1. The progressive-wave pressure component of $p_r(0, y; t)$, which is the first term on the right of Equation 96a in Chapter 4, depends on the vibratory acceleration. The corresponding amplitude p_{rp} $(F, 0, y)$ is, therefore, associated with the reduced added mass, $\Delta m_r/k$, as shown by the first of

the expressions in Equation 61a above. As the Froude number, F, increases, p_{rp} $(F, 0, y)$ decreases rapidly to a small value, as shown in Figure 18 in Chapter 4, and its definite integral representing the resultant fluid force or added mass also decreases. Since the definite-integral value is divided by F^2, the value of $\Delta m_r/k$ decreases rapidly to zero as F increases to 2 as shown by the solid lines in Figure 25 in Chapter 4.

2. The standing pressure-wave component of $p_r(0, y; t)$, which is the second term on the right of Equation 96a in Chapter 4, depends on the vibratory velocity and is negative if the corresponding amplitude $p_{rs}(F, 0, y)$ takes a positive value. The pressure, therefore, acts in the direction of gate movement, which means that $p_{rs}(F, 0, y)$ is associated with the reduced fluid-excitation coefficient, $\Delta c_r/k$ (see the second of the expressions in Equation 61a above). As the Froude number, F, increases, the amplitude $p_{rs}(F, 0, y)$ increases, as shown in Figure 19 in Chapter 4 and its definite integral, which represents the resultant fluid force that transfers energy from the flow to the gate, also increases. Since the definite-integral value is divided by F, $\Delta c_r/k$ approaches a constant value of 0.74 for F larger than about 8.0, as shown by the dashed lines in Figure 19 of Chapter 4.
3. The standing pressure-wave component of p_b $(0, y; t)$ due to streamwise vibration of the gate, which is the first term on the right-hand side of Equation 48, depends on the vibratory acceleration. Thus, as shown by the first expression in Equation 61b, the corresponding amplitude p_{bs} $(F, 0, y)$ is associated with the reduced added mass Δm_b. As the Froude number, F, increases, p_{bs} $(F, 0, y)$ increases, as shown in Figure 4a, and its definite integral representing the resultant fluid force or added mass also increases. Since the definite-integral value is divided by F^2, however, Δm_b approaches a constant value of 0.54 for F values larger than about 11.0, as shown by the solid lines in Figure 8.
4. The progressive pressure-wave component of $p_b(0, y; t)$ due to streamwise vibration of the gate, which is the second term on the right-hand side of Equation 48, depends on the vibratory velocity and is positive if the corresponding amplitude $p_{bp}(F, 0, y)$ takes a positive value. Hence, the pressure acts against the gate movement and transforms vibration energy of the gate into progressive pressure-wave energy. Consequently, as shown by the second expression in Equation 61b, $p_{bp}(F, 0, y)$ is associated with the reduced fluid-damping coefficient, Δc_b. As the Froude number, F, increases, $p_{bp}(F, 0, y)$ decreases rapidly to a small value, as shown in Figure 4b, and its definite integral, which represents the resultant fluid force producing fluid damping, also decreases. Since the definite-integral value is divided by F, Δc_b decreases rapidly to zero as F approaches 8.0 and greater values, as shown by the dotted lines in Figure 5.

POTENTIAL FOR SELF-EXCITED VIBRATION

The reduced fluid-damping coefficient Δc_b, shown in Figure 8, represents the energy dissipated into the water reservoir due to pressure waves induced by the push-and-draw effect of the gate movement. The reduced fluid-excitation coefficient Δc_r, shown by the dashed lines in Figure 25 in Chapter 4, represents the energy supplied by the flow to the gate due to pressure waves induced by the flow-rate variation. Therefore, these two terms together reveal the basic self-excitation characteristics of long-span or Tainter gates. The characteristics of the lines in Figure 5 indicate that the gate's push-and-draw effect on the water produces a large added mass, but small vibration-energy dissipation if the Froude number F exceeds about 8.0. On the other hand, for F exceeding about 1.0, the reduced fluid-excitation coefficient caused by the flow-rate variation increases rapidly and approaches a constant value for F values of about 5 or greater. The constant value depends on the gate inclination factor, k, and the depth ratio, β^* (see dashed lines in Figure 5).

Based on the analysis in Chapters 4 and 5, one can conclude that self-excitation is most likely for cases with F greater than about 8.0 and with large values of k and small values of β^*. As F decreases to values less than 8.0, the reduced fluid-damping coefficient increases (see dashed lines in Figure 5), and the reduced fluid-excitation coefficient becomes small (see dashed lines in Figure 25 in Chapter 4). For self-excited vibration to occur, the fluid excitation has to exceed both the mechanical and the fluid damping. Consequently, for $F < 8.0$ there is a reduced tendency toward self-excitation in spite of persisting flow-rate variations.

CONCLUSION

As a first step in the analysis of flow-induced vibrations of long-span gates or Tainter gates, we analyzed the flow field for a reservoir with a gate at small gate openings. We calculated the hydrodynamic pressure caused by small-amplitude gate vibrations at a small gate opening.

Our study showed that the basic parameters determining the vibration characteristics are the damping ratio of the vibrating system, the basic Froude number (containing the in-air natural frequency) F_0, the water-to-gate mass ratio, α_m, the ratio of upstream-channel depth to submergence depth of the gate bottom, β^*, and the gate inclination factor, k.

For an assumed vertical or streamwise, small-amplitude, oscillatory movement of the gate, the fluid excitation, the fluid damping and the added mass were derived in a dimensionless form, leading to the following conclusions. The flow-rate variation associated with vertical gate vibration (see Chapter 4) induces both standing and travelling (progressive) pressure waves. The standing-wave component is in phase with the gate velocity, and results in fluid excitation; the progressive-wave component is in phase with the gate acceleration and results in added mass. In this chapter, we showed that the push-and-draw pressure associated with streamwise vibration of the gate also induces standing and progressive pressure waves. The first component is in phase with the gate acceleration and results in added mass; the second component is in opposition to the gate velocity, that is, it is 180°out of phase with the gate, and results in fluid damping. The basic parameters determining the hydrodynamic coefficients are the Froude number (containing the in-water vibration frequency) F, α_m and β^*.

The changing values of fluid damping and excitation with F determines the self-excitation characteristics of a long-span gate. For F larger than about 8.0, the energy dissipation due to pressure waves induced by the push-and-draw effect is zero and the energy transfer from the flow to the vibrating gate due to the pressure waves caused by the flow-rate variation is large. Furthermore, the energy transfer becomes larger for larger values of k and smaller values of β^*. Therefore, a self-excited vibration is most likely for large F, large k, and small β^* values. The exciting force is mainly produced by the energy supply due to flow-rate variation while the vibration frequency is mainly determined by the added mass caused by the push-and-draw effects of the gate. If F is smaller than about 8.0, the long-span gate can also become unstable, but the self-excitation tendency is reduced, since the energy dissipation caused by the push-and-draw effect increases and the energy supply caused by the flow-rate variation decreases.

Based on the theoretical results presented in Chapters 4 and 5, the dimensionless equation of motion for the gate vibration, which includes integral fluid terms, can be reduced to a simple second-order differential equation. The vibration frequency and the fluid excitation, therefore, can be calculated for a variety of initial and boundary conditions. A more detailed and quantitative discussion of the vibration characteristics, gate instability and design criteria for safe gate operation will be presented in subsequent chapters (Chapter 7 for long-span gates and Chapters 8 to 12 for Tainter gates).

In Chapter 7, the development of Figure 8 showing the reduced added mass, Δm_b, and the reduced fluid-damping coefficient, Δc_b, associated with the push-and-draw effect as a function of F will be repeated and then extended to include a similar second figure showing the reduced added mass Δm_r, and the reduced fluid-excitation coefficient, Δc_r, associated with the flow-rate variation pressure. These coefficients will then be used in analyzing flow-induced vibrations of long-span gates. The repetition is intentional and is done to help the reader who does not read linearly as well as reinforcing the concepts before applying then in Chapter 7.

REFERENCES

Ishii, N. (1992). Flow-induced vibration of long-span gates Part I: Model development. *Journal of Fluids and Structures*, *6*(5), 539–562. doi:10.1016/0889-9746(92)90018-X

Ishii, N., Imaichi, K., & Takemoto, S. (1977). Water waves and structural loads caused by a periodic change of discharge from a sluice gates. *Bulletin of the Japan Society of Mechanical Engineers*, *20*(6), 998–1007. doi:10.1299/jsme1958.20.998

Rayleigh, L. (1883). The form of standing waves on the surface of running water. *Proceedings of the London Mathematical Society*, *1*(1), 69–78. doi:10.1112/plms/s1-15.1.69

Sneddon, I. N. (1951). *Fourier Transforms*. New York, NY: McGraw-Hill.

Stoker, J. J. (1957). *Water Waves*. New York, NY: Interscience.

APPENDIX: NOMENCLATURE

Table 1.

α_m	water to gate mass ratio, $\alpha_m \equiv \rho d_0^2/(\rho_* hH)$
β^*	ratio of upstream-channel depth to submergence depth of the gate bottom, called depth ratio in brief, $\beta^* \equiv d_r/d_0$
ΔQ	increase of flow rate under (m^2/s)
Δc_b, ΔC_b	fluid-damping coefficients caused by push-and-draw effect of gate on water [Δc_b is the dimensionless expression of ΔC_b (Pa s)]
Δc_r, ΔC_r	fluid-excitation coefficients caused by flow rate variation [Δc_r is the dimensionless expression of ΔC_r (Pa s)
Δm_b, ΔM_b	added mass caused by push-and-draw effect of gate on water [Δm_b is the dimensionless expression of ΔM_b (kg/m)]
Δm_r, ΔM_r	added mass caused by flow-rate variation [Δm_r is the dimensionless expression of ΔM_r (kg/m)]
ε	small positive constant
ζ_0	mechanical damping ratio
ζ_f	damping ratio due to fluid motion
η	imaginary part of s
η_0	small positive constant
η_j	solution of a transcendental equation
Θ	angle between gate and upstream-channel floor (deg)
μ, μ'	positive small constant representing small dissipative forces in water, $\mu \equiv \mu'/\omega_0$
ξ	real part of s
ξ_0	solution of a transcendental equation
ξ_f	excitation ratio due to fluid motion
ξ_j	small positive constant
ρ, ρ^*	specific mass of water and gate (kg/m^3)
τ, τ'	integral variable, $\tau \equiv \omega_0 \tau'$
Φ, $\bar{\Phi}$, Φ_b, Φ_r, Φ_p	velocity potential (m^2/s)
ϕ_0, Φ_0	phase function, $\phi_0 \equiv \sqrt{g/d_0}\Phi_0$
ω	vibration frequency (rad/s)
ω_0	natural vibration frequency in air (rad/s)

Section 2
Gate Excitation Mechanisms

Chapter 6
Overflow Gates:
Nappe Oscillations and Vibrations

ABSTRACT

This chapter focusses on instabilities associated with overflowing gates -- nappe oscillations and flap gate vibrations. A detailed mapping of potential initiating and sustaining energy sources is developed to help understand the mechanism of nappe oscillations. The precise mechanism of nappe oscillations remains an unsolved problem in fluid mechanics, even though we understand the process well enough to provide effective countermeasures to prevent nappe oscillations. We do not yet have the ability to definitively predict the onset criteria for nappe oscillations nor can we predict which mode of nappe oscillations will predominate and what produces changing modes with changing flow conditions. A method of determining added mass and in-water vibration frequency of flap gates is presented at the conclusion of the chapter.

INTRODUCTION

Decisions concerning the selection of gate types at any particular site are complex, and must take into account technical economic, environmental, and even aesthetic requirements. Erbisti (2004) lists specific factors to consider:

- Reliability of gate operation
- Weight of the structure
- Simplicity of operation
- Ease of maintenance
- Complexity of ancillary structures
- Magnitude and direction of hydraulic loads
- Required capacity of hoist motor
- Challenges of transport and on-site erection

DOI: 10.4018/978-1-5225-3079-4.ch006

Erbisti also notes, however, that there is no established routine method, or flow chart with a checklist, that leads consistently to the same choice. Experience and the expertise of the designers are often deciding factors in gate selection. Previous selection of a particular gate in a similar situation carries great weight in the decision.

Generally, gates specifically designed for overflow include flap, bear trap, drum and sector gates, as well as vertical-lift, fixed-wheel gates with hook-type crest structures (also called leaf gates). Inflatable rubber gates constitute another class of gates designed for overflow. Certain other gate types, such as long-span gates with a flap, Tainter/radial gates (called segment gates by Erbisti), as well as roller gates, all may be configured for overflow.

There are usually very good reasons for selecting an overflow gate as the control structure. Certainly, the gate must deliver the specified discharge capacity and regulate water surface levels. Almost as important, for most overflow gates, is the ability to pass floating debris and ice. Furthermore, the thin water curtain passing over the weir or gate, called the nappe, must be aerated to avoid flow-induced vibration and noise. Aeration can be achieved either by installing venting ducts or by adding spoilers (also called splitters, nappe breakers, or flow dividers) along the gate crest. The addition of spoilers in the design stage or as a retrofit when aeration is insufficient to attenuate flow-induced vibrations and/or tonal noise resulting from nappe oscillations carries an added maintenance cost. Although properly designed spoilers effectively attenuate nappe oscillations, they require cleaning to avoid accumulation of debris on the spoilers. Decisions to use spoilers instead of venting ducts may be based on construction costs. Depending on the amount of floating debris and how often debris has to be cleared from the spoilers, however, the cost of gate maintenance and debris removal may be higher than initially expected.

In other cases, certain gates are designed for simultaneous over- and underflow. Even when such gates have been specifically designed for either *just* overflow or *just* underflow, they may be used, sometimes unavoidably, for *simultaneous* over-and underflow. Details of gates with simultaneous over- and underflow can be found in Chapter 7 – Flow-Induced Vibration of Long-Span Gates.

In this chapter, we will consider flow-induced vibrations of overflow structures. We will review the mechanism of nappe oscillation and present several nappe oscillation countermeasures. Because flap gates have a propensity to vibrate, we will review several studies on this topic. Finally, we will present a method of estimating added mass for overflow gates, which provides a means of calculating natural frequencies of overflow structures.

OVERFLOW GATES

Flap Gates

Figure 1 shows schematics of several types of flap gate. Bottom-hinged flap gates (Figure 1a) are commonly used in irrigation systems, as weir or crest gates, and occasionally mounted atop vertical-lift or Tainter gates to permit passing of surface debris. A torque-tube flap gate (Figure 1b) consists of a tube with a rigidly connected gate structure. The actuator (often a hydraulic piston) drives the torque-tube to position the gate. Another variant of the flap gate is the fish-belly gate (Figure 1c), which also has a driving apparatus (not shown) similar to that of the torque tube gate. The "fish-belly" shape on the downstream side of the gate provides greater rigidity. The bear-trap gate in Figure 1d allows the gate leaf

Figure 1. Schematic diagrams of flap gates: (a) bottom-hinged, (b) torque-tube, (c) fish-belly, and (d) bear-trap

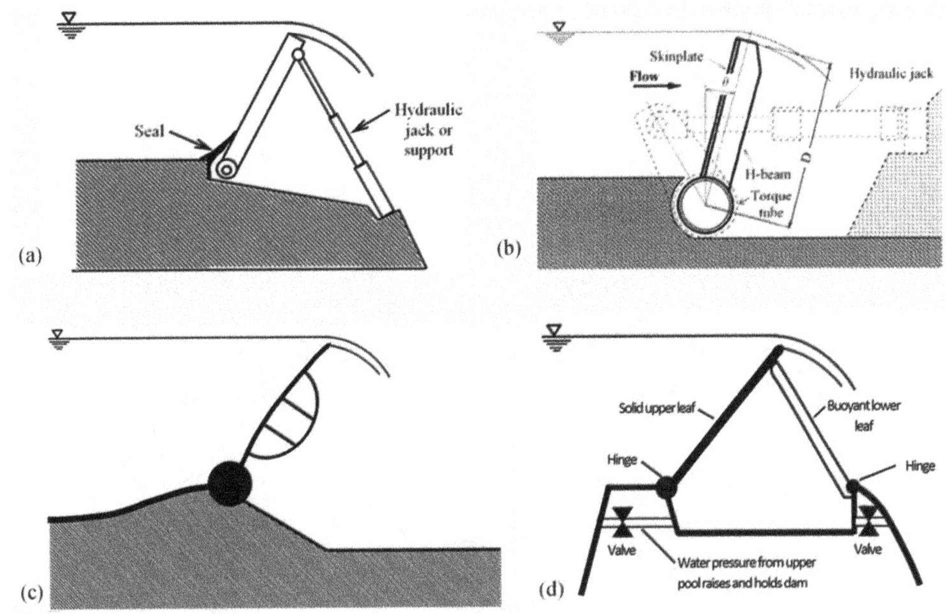

on the downstream side of the gate to fold under and be covered by the upstream gate leaf when it is in its horizontal position. The bear-trap gate is actuated via inflow from the upstream reservoir.

Drum, Sector, and Roller Gates

Drum and horizontal-axis sector gates are gate structures that can be retracted into a cavity on the gate crest or the river bed to permit unimpeded discharge. One fundamental distinction between a drum gate (Figure 2a) and a sector gate (Figure 2b) is the location of the mounting pin about which the gate rotates. In drum gates the rotation is about an upstream location, while the sector gate rotates about a downstream point. Drum gates are closed, buoyant volumes, while sector gates are not typically closed structures. Often horizontal axis sector gates have open bottom surfaces.

Leaf and Shell Gates

Multi-sectioned vertical leaf gates can be used as overflow gates, when they have a hook-type crest or a crest-mounted flap (Figures 3a). These leaf gates can also be used as underflow gates by raising the top leaf above the free surface and then raising the bottom leaf to permit underflow. Figure 3b shows a shell gate outfitted with a fish-belly flap gate to regulate overflow, In Figure 3c, a shell gate is in use as an underflow gate. Shell gate are sometimes used with simultaneous over- and underflow, as shown in Figure 3d. Unless the gate has been specifically designed for this purpose, simultaneous over-and underflow should be avoided, if possible. As previously mentioned in this chapter, simultaneous over- and underflow and the associated flow-induced vibration issues will be considered in Chapter 7 – Flow-Induced Vibration of Long Span Gates.

Figure 2. Schematic diagrams of (a) a drum gate, (b) a sector gate, and (c) a roller gate
After USGS (2013).

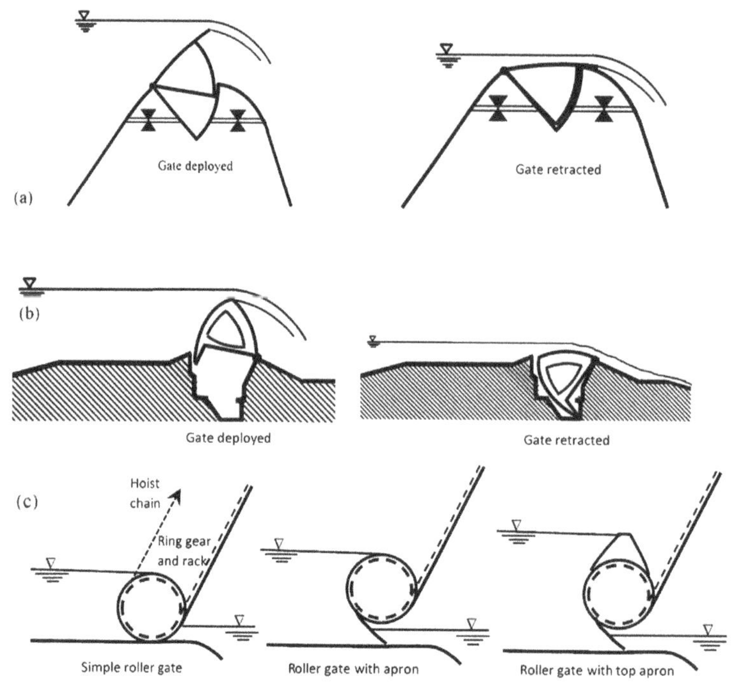

Figure 3. Schematic of leaf gates (a) overflow with hook-type crest and sidewall deflectors, (b) shell-type gate with only underflow, (c) shell-type gate with crest-mounted flap (d) long-span shell-type gate with simultaneous over- and underflow (flow right to left)

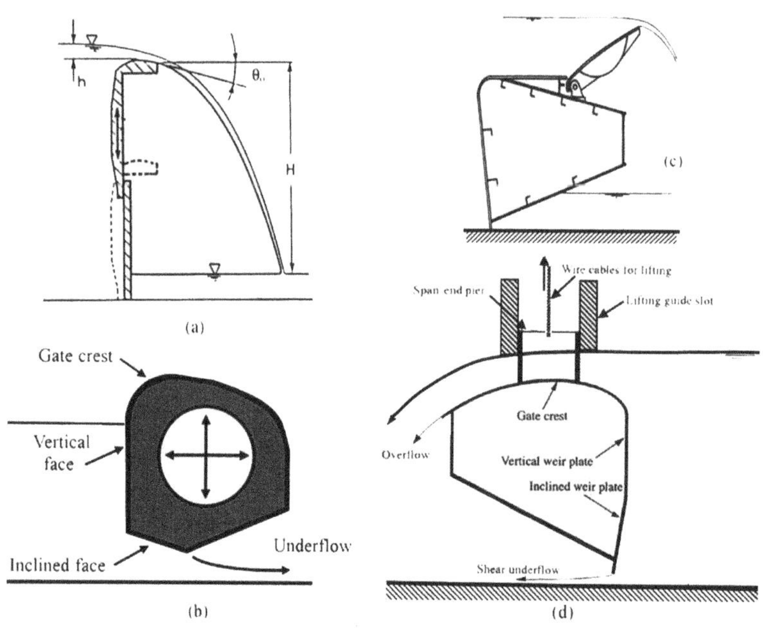

Fixed Wheel Gates

Fixed wheel gates have wheels mounted on each side of the rectangular gate structure. The wheels roll in tracks to permit the gate to be more easily raised with less friction than an equivalent slide gate without wheels resting on similar tracks. Vertical left wheel gate can be design to pass overflows when needed, as can be seen in the two examples in Figure 4.

Inflatable Weirs

A relatively recent development in surface water regulation is the inflatable weir, shown in Figure 5a. These weirs consist of flexible containment bladders that can be inflated with either air or water to provide a seasonal upstream reservoir for irrigation or recreation. Occasionally they are used to provide added capacity to an existing reservoir. The very first versions of inflatable dams did not have rear-mounted deflectors and were subject to flow-induced vibration. The addition of rear-mounted deflectors attenuated flow-induced vibration for air-inflated weirs, but Gebhardt et al. (2010) found that vibration may still be an issue for water-inflated membrane weirs. Another area of concern for inflatable weirs is abrasion of the deflated bags caused by flow-induced vibration, resulting in the need to replace the bags more frequently than was originally anticipated. Inflatable bladders have also been adapted for use as a raising and lowering mechanism for flap gates, as shown schematically in Figure 5b.

Figure 4. Schematic of a fixed-wheel gates for underflow and/or overflow
After Lewin (1995).

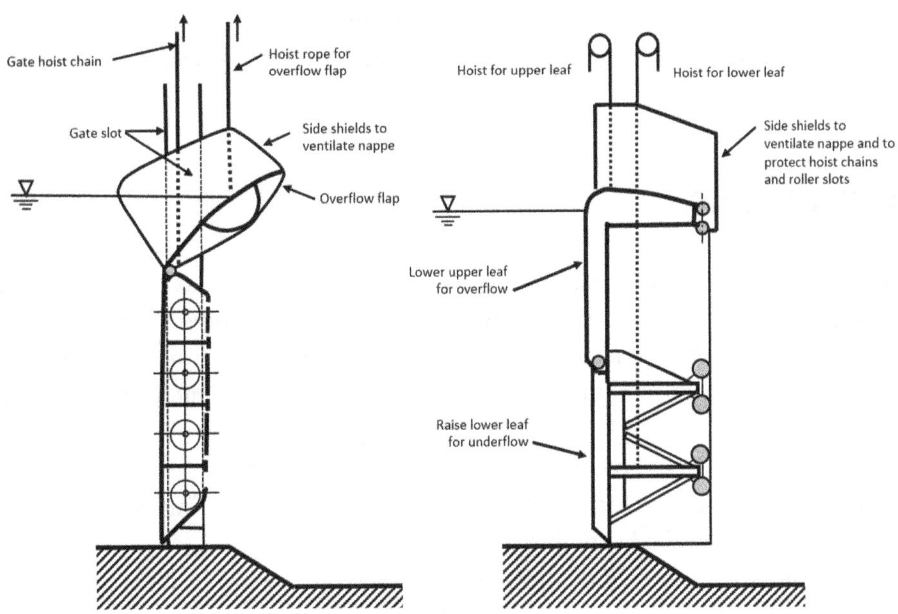

Figure 5. Schematic of (a) inflatable weir, and (b) inflatable air bladder for gate regulation beneath a flap gate with overflow

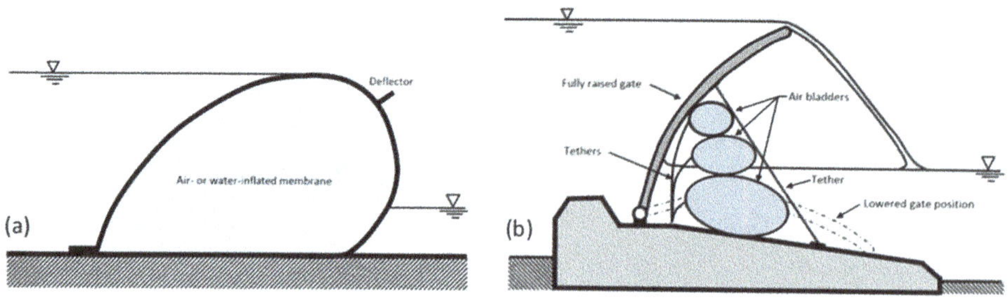

Nappe Aeration

The nappe is the thin water curtain that falls from the crest of an overflow gate or weir. The air cavity between the nappe and the downstream face of the gate is susceptible to fluctuating air pressures, which cause the nappe to oscillate and to radiate noise. In this situation, the air cavity behaves as a Helmholtz resonator, as described in Chapter 2– Vibrations: With Resonator and Stability Concepts. The nappe oscillation mechanism will be discussed in detail in the next section of this chapter.

Aside from the noise, nappe oscillations are undesirable because they may cause structural damage to the weir or gate. To prevent nappe oscillation, almost all overflow structures are equipped with ducts or other provisions to aerate the nappe. The required volumetric rate of air flow in the ducts is usually specified as a factor times the volumetric water flow rate. Gates that are equipped with side deflectors (see Figure 4 for example) that guarantee sufficient space between the nappe and the sidewall of the piers may still need additional aeration through ducting to prevent nappe oscillation. Long gates and weirs are susceptible to transverse acoustic standing waves beneath the nappe, which may serve to initiate or sustain nappe oscillations.

NAPPE OSCILLATIONS

Nappe oscillations can be seen and heard at many gates and weirs, as well as in water features and fountains around the world. Figure 6a shows oscillation of a water feature at the Radhuset in Oslo, Norway. Despite what might appear to be adequate aeration, with open areas on both sides of the nappe, the nappe undergoes oscillation. Similarly, in Figure 6b, the nappe at a fountain in Dunedin, New Zealand, comes over a relatively rough surface that generates longitudinal vorticity (the vertical white aerated lines on the back side of the nappe). Even though increased roughness of the brink structure is one nappe oscillation countermeasure, the nappe on this fountain still oscillates, producing an easily heard tone that some find annoying and others find soothing.

Other examples of nappe oscillations include the parabolic nappe of a laboratory-scale horizontal overflow (Figure 7), on a relatively low flap gate (Figure 8), and on leaf gates with relatively large fall heights (Figure 9).

Figure 6. Water features and fountains with nappe oscillations (a) at the Radhuset in Oslo, Norway, and (b) at a fountain in the Octagon in the center of Dunedin, New Zealand.
Photographs by C. Knisely.

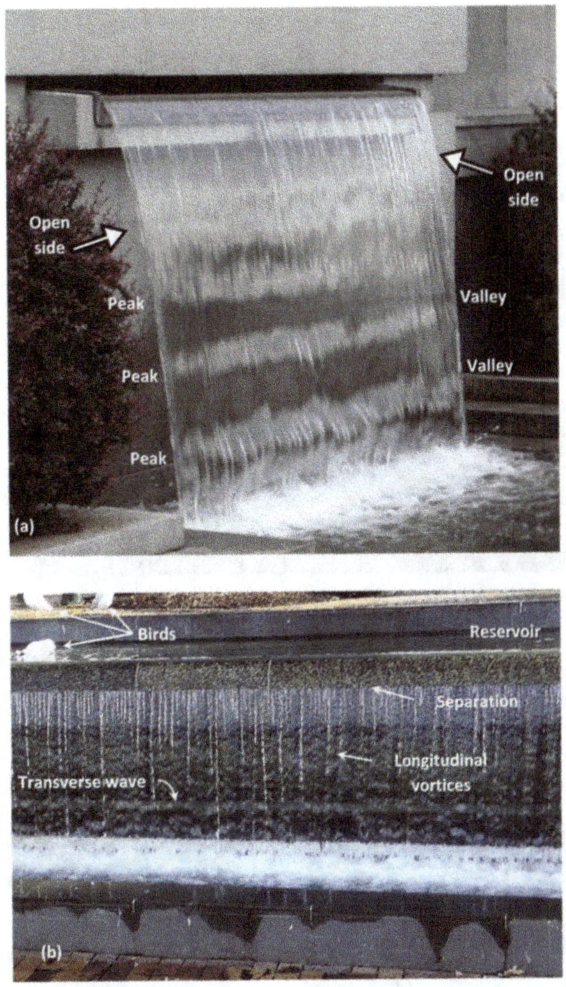

Figure 7 shows a frame from a video of the bottom portion of a laboratory-scale parabolic nappe with a 1.275 m fall height, 0.014 m brink depth and a 1 m span across the channel with oscillations at 19.8 Hz. (The vertical black rectangular shape in the bottom left is a measuring scale attached to the Plexiglas wall.) This laboratory structure was used in the experiments described later in this chapter to obtain data referred to as the Kyoto data set.

Figure 8 shows frames from a video of the nappe over a full-scale flap gate on the Shorenji Dam in Mie Prefecture, Japan. A case study on this gate will be presented later in this chapter. The downstream fall height was 1.09 m, and the overflow depth was approximately 0.017 m. In Figure 8a, spoilers are spaced approximately 0.40 m apart. Despite what appears to be substantial aeration (the estimated open area is approximately 17% of the total nappe surface), the nappe is still unstable and oscillates at about 22.3 Hz.

Figure 7. Extracted video frame of the bottom portion of a parabolic laboratory nappe with a fall height of 1.275 m. Horizontal lines were added to clearly show the nodal lines observed through a Plexiglas wall. Frequency of oscillation was approximately 19.8 Hz.
Photograph by C. Knisely.

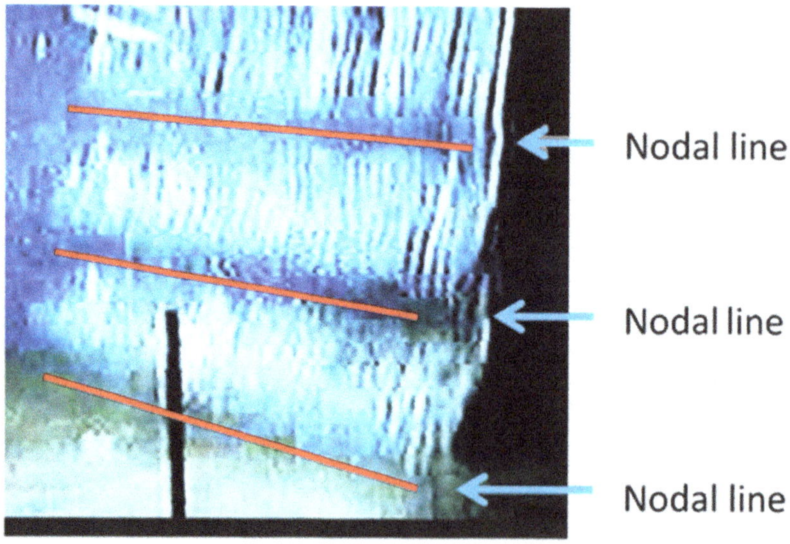

Figure 8. Flow over a flap gate (1.09 m fall height), showing (a) nappe oscillations despite aeration provided by crest-mounted spoilers, and (b) oscillations of the unaerated nappe after removal of the spoilers with coherent oscillations across the entire 14.8 m span of the weir.
Photographs by C. Knisely.

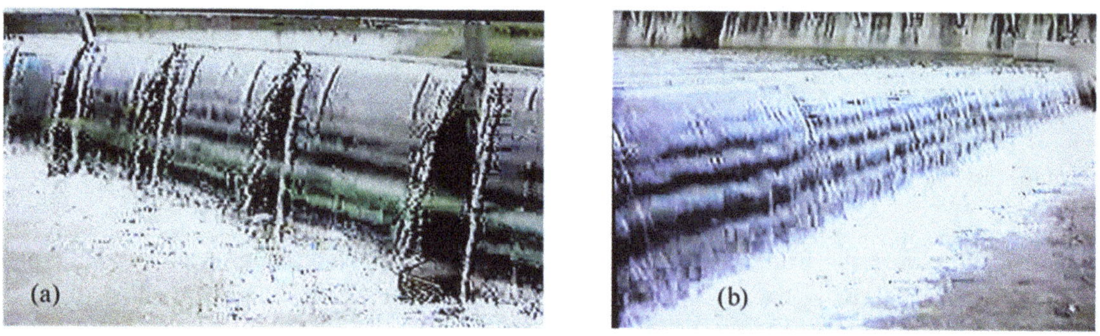

Figure 8b shows the same flap gate after the spoilers were removed. The small vertical striations result from the bases of the spoilers, which had not been ground flush with the gate overflow surface. The oscillations are quite strong, with the wave crests extending uniformly across the span of the gate. The nappe oscillations are coupled with flap gate vibrations, a mechanism to which we will return later in this chapter.

Figure 9a is a video frame from a study of leaf gate nappe oscillations. The gate span was approximately 10.8 m with a fall height of 4.6 m, an overflow depth of 0.15 m, and oscillations in the 16 Hz

Figure 9. (a) Frame from a video of nappe oscillations in the flow over a gate on the Seta River in Japan (10 m span and 4.6 m fall height), and (b) still photo from above the nappe showing crests of nappe oscillations as white line on the right side of the picture
Photograph by C. Knisely.

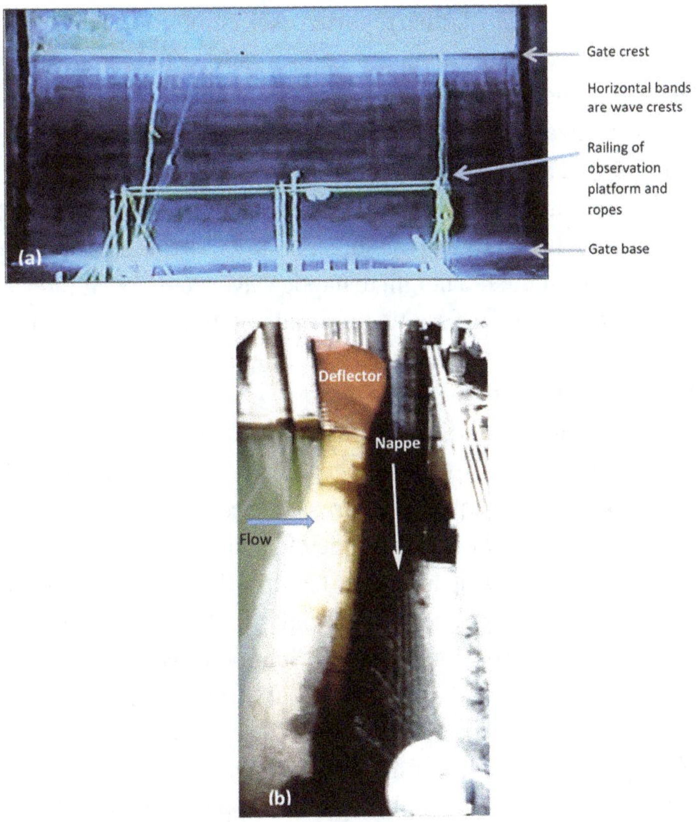

⅓-octave band. Detailed data for this oscillating nappe will be discussed in the section on Parabolic Nappes on Gates and Weirs later in this chapter. Figure 9b is a still photo looking down on the nappe with white lines of nappe crests visible on the right side of the photo.

Investigations of nappe oscillations have spanned at least eighty years and include studies by Petrikat (1941, 1958), Leo (1954), Hagerty & Shea (1955), Schwartz & Nutt (1963), Schwartz (1964, 1966), Partenscky & Khloeung (1967), Binnie (1972, 1974), Cremer & Ising (1972), Treiber (1974), and Kolkman (1974). More recent studies by Nakagawa et al. (1989), Knisely (1989, 1994), Casperson (1993, 1994, and 1996), Schmid & Henningson (2002), Sato et al. (2007), Crookston & Tullis (2013), Crookston et al. (2014), and Anderson (2014) have provided further insight into nappe oscillations.

The following observations are generally applicable to nappe oscillations:

- A closed air volume (or at least restricted airflow in and out of the air cavity), trapped between the overflowing nappe and the gate/weir structure, must be present
- Nappe oscillations always have multiple modes of oscillation
- The brink depth must be sufficient that the nappe does not tear apart
- When the nappe becomes too thick, nappe oscillations cease

The importance of a closed air volume was demonstrated by Binnie (1972) who, in experiments with a vertical water curtain discharged into a vacuum, found no signs of nappe instability. However, as shown in Figure 6a, there can still be oscillations even when the cavity is not completely closed. Falvey (2002) and Anderson (2014) both report oscillations in the nappe of a relatively large aerated labyrinth weir structures. Recently, Crookston et al. (2014) also reported oscillations on a labyrinth weir with open ends.

Nappe oscillations always have multiple modes of oscillation. This multi-mode characteristic had also been observed with impinging shear layer instabilities in cavities, edgetones, hole tones and other similar geometries. Indeed, Schwartz (1964) was one of the first to recognize the apparent similarity between these two phenomena. Thirty years later, Casperson (1993a, 1993b, and 1994) analyzed the nappe instability phenomenon as a Kelvin-Helmholtz shear layer instability (see Chapter 3 for more on shear layer instability). Although his analysis was substantially more eloquent, Casperson arrived at the same criteria concerning the ratio of the fall height to nappe wave length postulated by Schwartz (1966b).

After the nappe reaches a certain thickness that depends on fall height, and perhaps the span of the weir and secondarily on other geometrical parameters, the nappe simply has too much mass to be driven by the relatively small pressure fluctuations.

De Luca & Costa (1997) presented a spatial stability analysis of a falling liquid sheet. Schmid & Henningson (2002) also considered a vertically falling water curtain, but coupled it with a closed air volume to form a system with its own unique frequency characteristics, that is, as an eigenvalue problem. By finding functions that meet the boundary conditions they formulated, they were able to solve this eigenvalue problem computationally. They compared their results with experimental instability data to find that their lowest predicted instability frequency, when multiplied by the fall time, was 40% higher than the measured values. The meaning of this mismatch for Schmid & Henningson was that the instability occurs not because of a resonance at a single frequency, but due to the combination of two consecutive eigenvalues that differ in frequency by $2\pi/t_{fall}$. Girfoglio et al. (2013) and De Rosa et al. (2014) also contributed analyses of the global dynamics of nappe oscillation.

Several theories have been put forth to explain why the nappe oscillates. To test the hypothesis that the instability is caused by a difference in the relative velocity of the air and water streams, we can make use of White's (2006) sample calculation of the inviscid Kelvin-Helmholtz instability of an air-water interface. Including the effect of surface tension, Kelvin-Helmholtz instability requires an estimated relative velocity difference of about 6.7 m/s in the two streams. Even accounting for an estimated error of up to 50%, the required velocity difference would be greater than 3 m/s for the onset of these relatively short-wavelength waves (estimated 0.017 m wavelength). Longer wavelengths would require greater velocity differences. However, in practice, the initial brink velocity seldom reaches or exceeds 3 m/s at overflow depths for which nappe oscillations occur; the remainder of the required velocity difference would need to result from the wind speed, which is seldom steady. Such a situation would suggest a greater intermittency of nappe oscillations (mimicking those in naturally occurring wind gusts) than is observed in nature. These considerations suggest that the Kelvin-Helmholtz mechanism could serve to *initiate* nappe instability, but other mechanisms involving the air volume beneath the nappe would have to *sustain* the nappe oscillation.

Another reason why the nappe oscillates was proposed by Sato et al. (2007), Crookston and Tullis (2012), Tullis et al. (2014), and Anderson (2014). They argue that instantaneous perturbations to the initial motion of the nappe immediately after separation can serve as the starting point for oscillation. In this mechanism, each particle in the nappe follows a substantially parabolic trajectory and is subjected to oscillating pressures over the course of the fall. Schwartz (1964) computed such trajectories and showed that the pathlines of particles leaving the crest at successively later times gives the *appearance* of oscillation, but none of the particle streaklines oscillate with amplitudes observed in the composite nappe. Schwartz's calculations therefore appear to contradict the instantaneous perturbations theory.

As mentioned previously, nappe oscillations and impinging shear layer instabilities in cavities are similar in that both have multiple modes of oscillation. However, a fundamental difference in the mechanism of oscillation is the role of viscosity and vorticity. Due to the influence of viscosity, vortices are generated. In impinging shear layer instabilities, the vorticity distribution plays a significant role in the generation of the upstream influence of the impingement region (see Tinar and Rockwell (2014), for example). The induced feedback perturbation of an impinging shear layer can be understood in terms of the Biot-Savart induction law, which tells us that the time rate of change in the vorticity at the impingement edge results in induced velocity and pressure perturbations at the point of separation from the upstream corner. The presence of the impingement edge provides an organized feedback signal at the upstream separation edge that is undetectable in similar shear layers without impingement (Powell, 1961; Rockwell and Knisely, 1979a, 1979b; Ziada and Rockwell 1982, among others).

It is possible to show that an inviscid weir flow undergoing small harmonic perturbations in either the magnitude or direction of the brink velocity as the flow separates from the surface will produce an apparent oscillation of the nappe. The feedback in the nappe oscillation process does not necessarily rely on changes in vorticity as the nappe enters the downstream pool, but rather on the time-dependent volume changes of the air cavity beneath the nappe and the ensuing pressure fluctuations in that air space.

A schematic representation of possible mechanisms for the excitation of nappe oscillations is shown in Figure 10. Each of the oscillators -- inviscid perturbations in brink velocity vector, structural vibrations, Kelvin-Helmholtz interface instability, Helmholtz resonance, and transverse acoustic wave resonance – can initiate and amplify an initial disturbance, and each is capable of contributing to sustained oscillations, suggesting a single theoretical formulation that can predict all possible oscillations may not be possible. Each of the excitation mechanisms can function individually or together with other mechanisms. Therefore, it may be necessary to consider several potentially concurrent paths in order to predict the occurrence and behavior of nappe oscillations. The coupling of each oscillator to the nappe and possibly to other oscillators is through the fluctuating pressure in the space between the nappe and the weir/gate.

While the general mechanism of nappe oscillation is fairly well understood, the details of the mechanism remain less well understood. The lack of detailed knowledge limits our predictive abilities. For example, for a given overflow geometry, the available literature provides no guidelines for the *a priori* estimation of the onset of nappe oscillations, the range of water depths and fall heights over which oscillations will definitely occur, the specific frequency among the many possible frequencies at which the nappe will oscillate, or the optimal spacing and geometry for spoilers.

Determining the most effective position and geometry of spoilers for a given gate may involve a trial-and-error procedure on the completed full-scale structure. Spoilers are usually installed after the fact, being added to an existing structure to prevent unwanted nappe oscillations. The spoilers are generally spaced between ⅓ and ⅔ of the fall height. They often have a triangular shape with the apex pointing upstream (see Ogihara & Ueda, 1979, for example). Optimal configurations will have maximal suppression effectiveness with minimal weight, cost, and distortion of the gate discharge. These configurations must be determined for each gate or weir that experiences nappe oscillation. The countermeasures that work for one gate do not necessarily work for other gates, even those with substantially similar geometries.

We will review a selected number of empirical and theoretical studies in the following section to provide a basis for further discussion of the distinction between parabolic arc nappes and nappes due to

Figure 10. Possible mechanisms for nappe oscillations

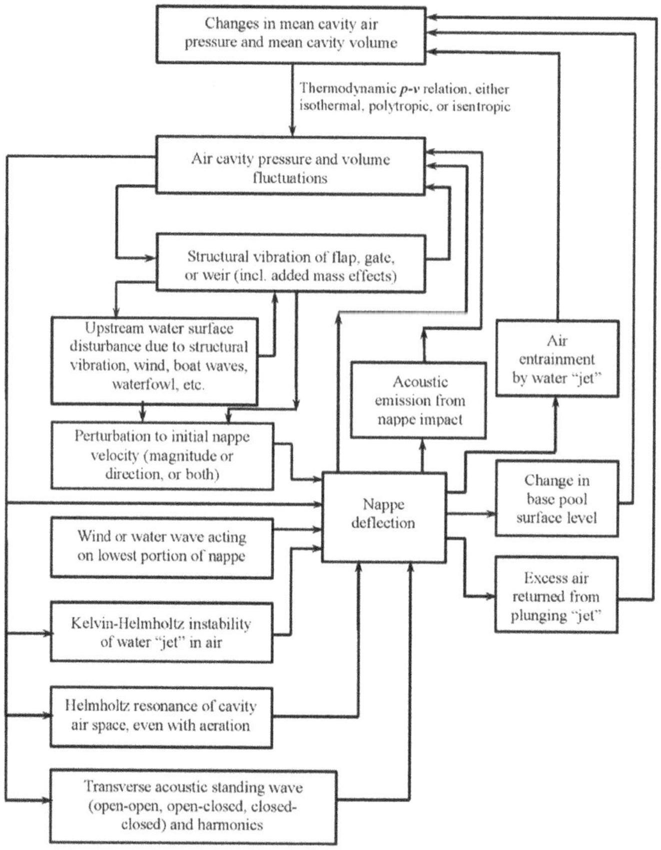

vertically falling water sheets with enclosed air volumes on one side of the nappe. Although both types of nappes undergo oscillations, the flow over a weir following a parabolic trajectory is in a different range of the characteristic dimensionless parameter than those for vertical water curtains. The difference may become a significant issue in trying to predict the frequency and mode at the onset of instability for the weir flows based on laboratory studies of vertical water curtains.

Early Studies on Nappe Oscillations

Many of the early studies on nappe oscillation were carried out on full-scale weirs or were done on laboratory models of weirs. In both cases the nappe geometry was parabolic. The data from the early work appears sound. More recent studies have involved vertical water curtains with restricted small-scale spans that add damping to the systems as the nappe moves back and forth between two comparatively closely spaced vertical walls. The non-dimensionalization of the early data differs from that of more recent work requiring some recalculation of the original data to make reasonable comparisons. The range of the dimensionless independent parameter (abscissa) and the non-dimensional frequencies of oscillation (ordinate) differ substantially between nappes with parabolic trajectories and vertically falling water curtains. Schwartz (1966a) provides concise summaries of the much of the early work on nappe oscillations. His review includes early German work from the 1930s onward (Fuhrmann, 1934; Müller, 1937; Fischer, 1938; Seifert 1940; Peters, 1940; Petrikat, 1958) and the studies by Pariset (1955) in France. Early theories about why nappes oscillate involved surface tension forces, rolling waves similar to ocean waves, resonances in the trapped air space, and analogies to sound tubes, radio transmitters and reed pipes. Most of these early theories were shown to be incomplete, unrealistic, or not supported by empirical data.

A relatively early laboratory test on a scaled model of a flap gate was undertaken by Leo (1954). His observations on nappe oscillations over a stationary flap provide further insight into nappe behavior. First, Leo suggested that the frequency of nappe oscillations could be estimated from the following expression, with less than 5% error:

$$f\sqrt{H} = 1.85N \text{ for } N = 2, 3, 4... \tag{1a}$$

or, after multiplication by $\sqrt{2 / g}$, the left side is approximately nappe frequency multiplied by the fall time $t_H = \sqrt{2H / g}$ of the overflow (neglecting any initial vertical velocity component), and Equation 1a becomes

$$ft_H = 0.835N \text{ for } N = 2, 3, 4... \tag{1b}$$

While Leo's nappe data with N = 4 does indeed agree with Equation 1a, two of his published data points (in his Figure 33) lie midway between N = 2 and N = 3, deviating by more than 20% from the predicted values. This discrepancy suggests that the criterion in Equation 1a lacks universality.

Secondly, Leo analyzed the effect of small perturbations at the separation point on the trajectory of fluid particles in the nappe. However, Leo did not take into account the effects of the oscillating pressure on the particle trajectories, as pointed out by Schwartz (1966a). It can be shown that perturbations to the horizontal velocity component can readily lead to volume fluctuations and sustained oscillation similar to those due to vertical perturbations. Leo considered only the vertical perturbations.

Thirdly, Leo evaluated the role of surface tension as a restoring force in the dynamics of the nappe. His conclusion was that, for the wavelengths he considered, surface tension provided a significant restoring force only for nappe thicknesses on the order of 0.01 mm.

As the nappe falls, gravitational acceleration plays a varied role depending on the initial velocity, initial flow direction and the fall height. For large fall heights with small initial velocities, the nappe accelerates as it falls, requiring the nappe thickness to decrease to satisfy continuity. The thinner nappe is more responsive to the pressure fluctuations, providing an increasingly large apparent amplitude of vibration near the bottom of the nappe. The stretching of the nappe from a linear to a convoluted form also requires the nappe thickness to decrease further. The frequency-wavelength relationship would suggest that if the apparent convective velocity of the nappe oscillation increases, then the wavelength would also increase. This deduction is confirmed by the increasing spacing between the "wave crests" in Figures 6, 8 and 9.

Surface tension acts on both the inner and outer nappe surface. The effect of surface tension depends on the thickness of the nappe and as well as on the radius of curvature of the nappe. Rayleigh formulated a critical wavelength for the undulations in a descending water jet that governs the breakup of the jet. Further study of the effects of surface tension on the nappe might be warranted because these thin regions of the nappe would appear to be a major difference between nappes with long fall time sand those with short fall times. The thinned region related near and between the "hairpin" turns in the nappe profile may have some role to play in the mode and frequency selection. Including the role of surface tension driven acceleration of the nappe in the formulation of a criterion for the self-selection of a preferred frequency of oscillation could prove fruitful. One might speculatively imagine a minimum wavelength near the bottom of a nappe for which the tight radius of curvature for the "stretched" nappe near the peaks and valleys reduces the restoring force of surface tension, permitting a small break in the nappe. The cavity pressure build up is temporarily bounded, and the nappe oscillation then "steps down" to the next available lower frequency with longer wavelength.

In later experimental studies on parabolic laboratory nappes, Schwartz (1964, 1966a) generated trajectories of nappe particles that were subjected to fluctuating pressure from the trapped air space. He showed that the pathline of particles that left the weir brink at successive time intervals produces an apparently oscillatory nappe profile even though no particle was undergoing oscillations with the same amplitude as the nappe. Schwartz later (1966b) suggested from his observations, and analogous to impinging shear layers, that the nappe should consist of an integer N plus ¼ wavelengths to sustain the oscillation. He also noted that the last quarter wavelength on the nappe near the tailwater moves in opposition to the pressure, that is, the last quarter wavelength is at its maximum distance from the base of the weir when the fluctuating pressure is at its minimum. Conversely, later in the cycle, the last quarter wavelength is closest to the weir when the fluctuating pressure is at its maximum. Sato et al. (2007) confirmed the anti-phase relationship between bottom-end nappe displacement and pressure observed by Schwartz.

From photos of oscillating nappes, Schwartz (1964, 1966a) also observed that the nappe had a ribbed appearance. He called the vertically-aligned, thickened portions of the nappe "ropes" with axes parallel to the flow direction, suggesting a three-dimensional influence on the nappe development.

Schwartz also observed that the amplitude of the trajectories of individual particles varied inversely in a nonlinear fashion with the frequency for a fixed amplitude of pressure fluctuation. Conversely, this inverse relationship means that small amplitude nappe displacements at a high frequency can produce the same fluctuating pressure levels as larger amplitude nappe displacements at a lower frequency.

The final observation that Schwartz deemed the most significant is that the maximum trajectory displacement occurs for particles that depart the separation point at the instant when the fluctuating pressure is zero. Later this phase relationship will be applied in Treiber's analysis of pressure feedback effects on the nappe oscillation.

A correlation for the onset of nappe oscillations in the flow over weirs was formulated by Partenscky & Khloeung (1967). Their correlation, using their notation, included the brink depth (e_0), the fall height (H), the depth of the downstream channel flow (d), and the Froude number, $\mathrm{Fr} = V_0 / \sqrt{g e_0}$, for the flow, in which V_0 is the initial nappe velocity at separation. Figure 11 shows the set-up and nomenclature employed by these authors.

Partenscky's empirical onset criterion was that H/e_0 must equal or exceed $H/e_{0,\,\mathrm{crit}}$ and is given in terms of the Froude number in the following inequality:

$$\frac{H}{e_0} \geq 21\mathrm{Fr}^{7.50} \ \text{for } d = 0 \quad \text{or} \quad \frac{H}{e_0} \geq 21\mathrm{Fr}^{8.10} \ \text{for } d \to \infty \tag{2}$$

Figure 11. Experimental set-up and nomenclature
After Partenscky & Khloeung (1967).

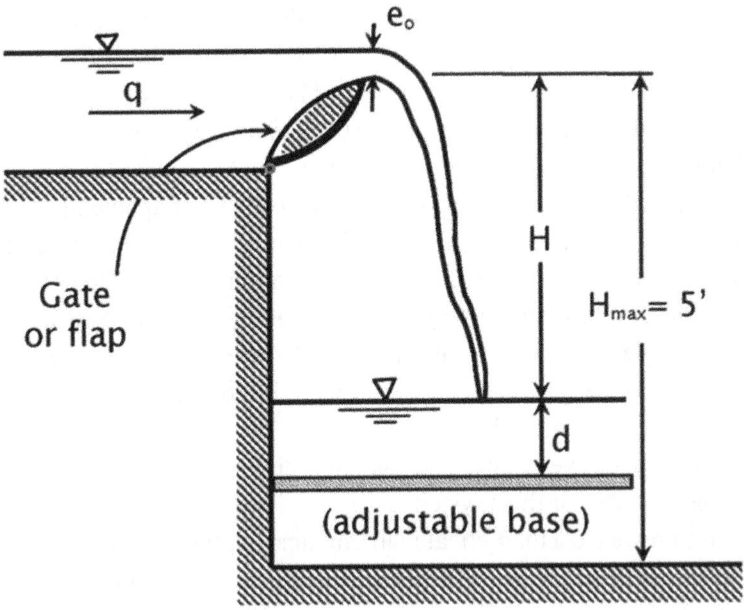

We will return to this criterion when we compare Partenscky's predicted onset criterion with experimental data in the literature.

Partenscky & Khloeung also provided an empirical correlation for the nappe oscillation frequency (f) in terms of a Strouhal number, $\mathrm{St} = fH / V_0$, as follows:

$$\mathrm{St} = 181\left(\frac{H}{e_0}\right)^{-0.275} \text{ for } \left(\frac{H}{e_0}\right) < 170 \tag{3a}$$

$$\mathrm{St} = 44 \text{ for } \left(\frac{H}{e_0}\right) \geq 170 \tag{3b}$$

Finally, Partenscky & Khloeung formed a non-dimensional pressure coefficient, $C_P \equiv \Delta P / \left(\frac{1}{2}\rho V_0^2\right)$, in which ΔP is the fluctuating pressure amplitudes and $\frac{1}{2}\rho V_0^2$ is the dynamic pressure at the nappe origin. They correlated the pressure coefficient with the ratio of fall height to brink depth:

$$C_P = \frac{1}{10}\left(28 - 5\ln\left(\frac{H}{e_0}\right)\right) \tag{4}$$

Cremer & Ising (1972) modeled nappe oscillations using an energy analysis of vertically falling water sheet with an enclosed air space on one side. A sketch of their set-up and terminology is given in Figure 12. They developed equations for the trajectory for a vertically falling freestream subjected to a fluctuating pressure in the enclosed air volume. The stream trajectory they calculated agreed well with photographs of the oscillating stream in their experimental apparatus. They then integrated the transverse (horizontal) displacement multiplied by the fluctuating pressure to determine the net energy transfer. They found that the fall height needed to be an integer number of wavelengths (in contrast to the observations of N + ¼ wavelengths by Schwartz) in order for the nappe to oscillate. The nappe entered the pool at the base of the tank perpendicular to the free surface, a condition which their theory predicted would yield the lowest possible energy loss. The average nappe excursion combined with the compressibility of the air volume permitted calculation of the fluctuating pressure. Although their physical model was too rigid, they were able to use the Nyquist-Strecker stability formulation to show that the system was unstable.

Cremer & Ising's math model yielded results that were in good agreement with their own vertical water curtain data. Predictions from their theory, however, should not be expected to agree precisely with data from parabolic nappes over weirs. The span of their physical model was only 0.10 m and may have been a factor in the stiffness of their system.

Kolkman (1974) also presented a theoretical mathematical analysis and experimental data on a vertical water curtain with an enclosed air space. Kolkman also found the data from his experimental model suggested it was much stiffer than full-scale weirs.

Figure 12. Apparatus with nomenclature
After Cremer & Ising (1972).

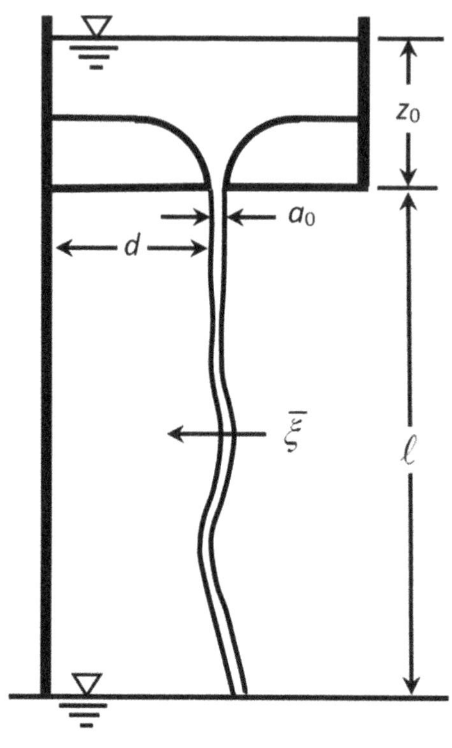

Treiber (1974) presented a slightly different formulation of the nappe energy analysis, similar to that of Cremer & Ising, but different in that Treiber assumed the flow stream was generated from an overflowing weir at some initial angle relative to horizontal and at some initial velocity that was determined by the head on the weir.

Treiber's analysis assumes a fluctuating pressure in the trapped air cavity:

$$\Delta p_1 = A \sin \omega t \tag{5}$$

The x- and y-coordinates, as defined in Figure 13, of a nappe element at some time t after separation and with no fluctuating air pressure can be determined by

$$\bar{x}\left(t\right) = \left(U_0 \cos \theta_0\right) t \tag{6a}$$

$$\bar{y}\left(t\right) = \left(U_0 \sin \theta_0\right) t + \frac{1}{2} g t^2 \tag{6b}$$

When the nappe element strikes the tailwater, it has traversed the distance H in the y-direction, permitting the fall time t_H to be defined from the solution of Equation 6b.

Figure 13. Geometry and notation
After Treiber (1974).

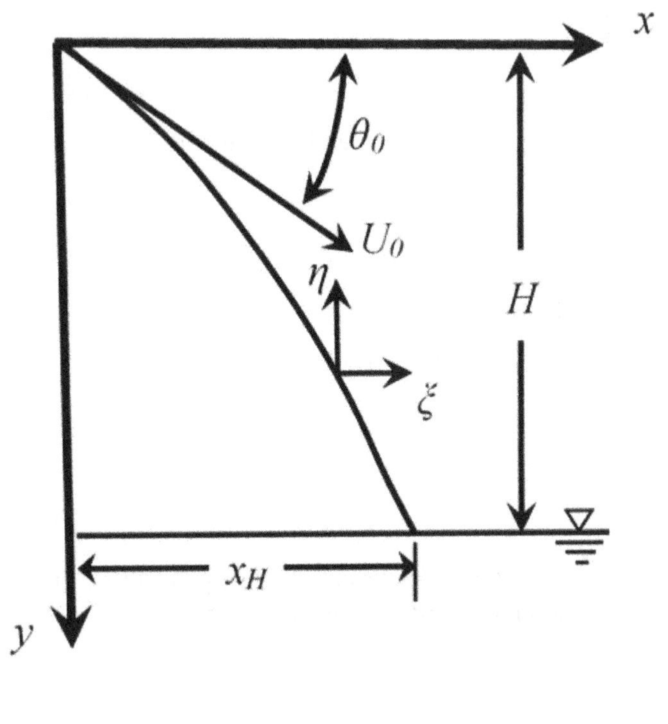

$$\bar{t}_H = \frac{-U_0 \sin\theta_0 + \sqrt{2gH + \left(U_0 \sin\theta_0\right)^2}}{g} \tag{7}$$

The values of x, y, and t_H in Equations 6 and 7 correspond to steady flow with no pressure forces acting on the nappe element. These values are designated with an overbar to indicate they are steady values.

A second set of Cartesian coordinates is then introduced. These coordinates are denoted as ξ and η to distinguish them clearly from the steady state coordinates (see Figure 13).

The fluctuating pressure, Δp_1, acts normal to the nappe surface and induces a normal acceleration denoted as $\ddot{n}$. On a unit nappe width basis, the force balance normal to the nappe surface requires

$$\ddot{n}dm = \Delta p_1 ds \tag{8}$$

in which dm is the differential nappe mass, and ds is the incremental nappe surface area per unit span on which the pressure Δp_1 acts. The nappe mass element dm per unit nappe width can be written in terms of the incremental nappe length, ds, as follows:

$$dm = \rho h ds \tag{9}$$

where the density of water is denoted by ρ and the symbol h represents the nappe thickness at any arbitrary location (x, y).

Continuity requires that the product of nappe thickness and instantaneous velocity remain constant, and thus

$$h = \frac{h_0 U_0}{u\left(x,y\right)} = \frac{q}{u\left(t\right)} \tag{10}$$

Correspondingly, the differential nappe mass can be written as

$$dm = \frac{\rho q ds}{u\left(t\right)} \tag{11}$$

and the corresponding expression for the normal acceleration as

$$\ddot{n} = \frac{\Delta p_1 u\left(t\right)}{\rho q} \tag{12}$$

Assuming the trajectories do not deviate substantially from the steady parabolic shape, Treiber decomposed the normal acceleration into ξ and η components:

$$\ddot{\xi}(t) = \frac{\Delta p_1 (U_0 \sin\theta_0 + gt)}{\rho q} \tag{13a}$$

$$\ddot{\eta}(t) = \frac{\Delta p_1 (U_0 \cos\theta_0)}{\rho q} \tag{13b}$$

For a nappe element leaving the weir at $t = 0$, the initial conditions on ξ and η are as follows:

$$\xi(0) = \dot{\xi}(0) = 0 \tag{14a}$$

$$\eta(0) = \dot{\eta}(0) = 0 \tag{14b}$$

In general, the pressure fluctuation can be at any arbitrary phase angle as the particle leaves the weir, so

$$\Delta p_1 = A \sin(\omega t + \beta) \tag{15}$$

Substituting the expression for Δp_1 given in Equation 15 into Equations 13a and b, integrating, and applying the initial conditions yields the following expressions for ξ and η:

$$\begin{aligned}\xi(t) = \frac{A}{\rho q}\Bigg[&\frac{-1}{\omega^2}(U_0 \sin\theta_0 + gt)\sin(\omega t + \beta) - \frac{2g}{\omega^3}\cos(\omega t + \beta) \\ &+ \frac{1}{\omega^2}(U_0 \sin\theta_0 - gt)(\sin\beta) + \left(\frac{2g}{\omega^3} + \frac{U_0 t \sin\theta_0}{\omega}\right)\cos\beta\Bigg]\end{aligned} \tag{16a}$$

$$\eta(t) = \frac{A}{\rho q}\left[\frac{U_0 \cos\theta_0}{\omega^2}\left(-\sin(\omega t + \beta) + \sin(\beta) + \omega t \cos(\beta)\right)\right] \tag{16b}$$

The nappe consists, of course, of fluid particles that leave the weir at different times. To find the instantaneous nappe profile at any arbitrary time τ, the phase angle of the pressure fluctuation is considered a variable. In general, fluctuating pressure phase angle at any time τ will be given as $\beta = \omega(\tau - t)$. Substituting this expression for β into Equations 16a and 16b results in equations for the coordinates of the instantaneous nappe profile at any arbitrary time $t = \tau$:

$$\xi(t,\tau) = \frac{A}{\rho q}\left[\frac{-1}{\omega^2}(U_0 \sin\theta_0 + gt)\sin(\omega\tau) - \frac{2g}{\omega^3}\cos(\omega\tau) + \frac{1}{\omega^2}(U_0 \sin\theta_0 - gt)\sin(\tau - t) + \left(\frac{2g}{\omega^3} + \frac{U_0 t \sin\theta_0}{\omega}\right)\cos\omega(\tau - t)\right] \tag{17a}$$

$$\eta(t) = \frac{A}{\rho q}\left[\frac{U_0 \cos\theta_0}{\omega^2}\Big(-\sin(\omega\tau) + \sin\big(\omega(\tau - t)\big) + \omega t\cos\big(\beta\omega(\tau - t)\big)\Big)\right] \tag{17b}$$

Integration of the nappe profile from the time of separation (t = 0) to the time of impact with the apron or the pool at the base of the weir ($t = t_H$) provides the volume deviation from the parabolic profile.

Treiber provides an expression for the volume fluctuation:

$$\Delta V(\tau) = \frac{-A}{\rho q}\frac{g^2 t_H}{\omega^3}\left[K_1 \sin\omega\tau + K_2 \cos\omega\tau\right] \tag{18}$$

where K_1 and K_2 are the following functions:

$$K_1 = \sin\omega t_H \left[1 - \left(\frac{2}{\omega t_H}\right)^2 - 2b\right] + \cos\omega t_H \left[\frac{4}{\omega t_H} + b\omega t_H\right] + \left[b + \frac{1}{3}\right]\omega t_H \tag{19a}$$

$$K_2 = \sin\omega t_H \left[-b\omega t_H - \frac{4}{\omega t_H}\right] + \cos\omega t_H \left[1 - \left(\frac{2}{\omega t_H}\right)^2 - 2b\right] + 1 + 2b + \left(\frac{2}{\omega t_H}\right)^2 \tag{19b}$$

The parameter b is given as follows:

$$b = \left(\frac{U_0}{g t_H}\right)^2 + \frac{U_0}{g t_H}\sin\theta_0 \tag{20}$$

With quite small pressure fluctuations at relatively low frequencies, one may assume isothermal compression in which $PV = P_0V_0$ = constant, or $P_0 dV + V_0 dP = 0$. With this assumption, the volume change in Equation 18 is accompanied by a pressure change given by

$$\Delta p_2(\tau) = \frac{P_0 A}{\rho q}\frac{g^2 t_H}{V_0 \omega^2}\left[K_1 \sin\omega\tau + K_2 \cos\omega\tau\right] \tag{21}$$

For the oscillation to be self-sustaining, this Δp_2 must be in phase with the initial pressure fluctuation Δp_1, and for the oscillations to grow, its amplitude must be greater than that of the initial pressure.

Since the initial pressure fluctuation had no cosine component, the value of K_2 must be set to zero. The zeros of this K_2 function provide a criterion for possible frequencies.

$$K_2 = 0 = \left[b\omega\, t_H + \frac{4}{\omega t_H}\right] \sin \omega\, t_H + \left[2b + \left(\frac{2}{\omega t_H}\right)^2 - 1\right] \cos \omega t_H - 1 - 2b - \left(\frac{2}{\omega t_H}\right)^2 \tag{22}$$

For $b \to 0$, the nappe has a long fall height with a relatively small initial velocity. As $b \to 0$, the roots of Equation 22 are given by

$$b \to 0, ft_H \to N + \frac{1}{2} \text{ with } N = 1,2,3\ldots \tag{23a}$$

As $b \to \infty$, the nappe has a very large initial velocity with a very small fall height; gravity does not affect the nappe velocity so the nappe thickness remains constant. As $b \to \infty$, one set of roots of Equation 22 is given by

$$b \to \infty, ft_H \to N \text{ with } N = 1,2,3\ldots \tag{23b}$$

A second set of roots can be found as $b \to \infty, ft_H \to N + \frac{1}{2}$. As Treiber notes, however, these roots do not agree with the trends of the available experimental data.

For values of b greater than 0.05 and modes N < 7, and for all values of b for $N \geq 7$, the roots of Equation 22 for the Nth mode can be given approximately by the following empirical curve fit:

$$f_N t_H = N + 0.25\left[1 + \tanh\left(\frac{\log b - \log\left(0.13 / N^{0.95}\right)}{\log\left(0.18 N^{0.003}\right)}\right)\right] \tag{24}$$

If the magnitude of Δp_2 is greater than that of the initially assumed pressure fluctuation, the nappe amplitude will grow, and the oscillations are self-sustaining. Growth of the nappe oscillation requires that

$$\frac{P_0 g^2 t_H^2 K_1}{\rho q V_0 \omega^3} > 1 \tag{25}$$

For vertical nappes and for values of b on the order of 0.1 or greater, Treiber's analysis provides the same frequency criteria as that of Cremer & Ising. However, Treiber's analysis is also applicable to parabolic nappes on weirs because he included nappe curvature in his solution and decomposed the normal acceleration into vertical and horizontal components.

To this point, Treiber's analysis is the most reasonable representation of oscillation of both vertical water curtains and parabolic nappes. Although Treiber's theoretical study does not address mode selection or provide guidance concerning the onset and termination of nappe oscillations in terms of the initial nappe parameters (h_0, U_0, and θ_0) and the fall height (H), it does provide a mathematical method for determining allowable frequencies. Furthermore, it provides a criterion for the growth of the oscillation. On the other hand, it does not provide a threshold criterion on how big the initial disturbance must be and it presupposes the existence of an initial fluctuating pressure.

Before we try to assess how well the theory matches experimental data, we will first re-frame Equations 2 and 3 into the form of $f_n t_H$ in terms of Treiber's b parameter (Equation 20). If we can find an appropriate non-dimensionalization scheme we should find consistency in data from very different geometries. Converting Partenscky's onset criterion to the Treiber format, we find it becomes

$$\left(f_n t_H\right)_{onset} = 624.18b^{0.8731} \tag{26}$$

and Partenscky's Strouhal criterion becomes

$$\left(f_n t_H\right)_{preferred} = 998.87b + 1.0686 \tag{27}$$

Parabolic Nappes on Gates and Weirs

There are a number of data sets in the literature for oscillating nappes with substantially parabolic trajectories. In addition to the aforementioned laboratory-scale studies by Leo (1954) and Partenscky & Khloeung (1967), Binnie (1972) reports on the instability of a horizontal slit jet with an enclosed air space beneath its parabolic trajectory. However, Binnie could independently vary the initial jet velocity and the initial transverse jet thickness, which sets his data apart from flows over weirs and gates. Treiber (1974) cites data from Kemp and Pullen (1961), found in the thesis by Schwartz (1966a). The Kemp and Pullen data was, at the time it was taken, possibly the only parabolic nappe data set in the English language literature.

At the Arai Weir near Kyoto, Japan, nappe oscillation studies were conducted on a full-scale double leaf, overflow gate (Nakagawa et al., 1988; Knisely, 1989). The 4 m fall height and 10.8 m span of the gate are the largest for which data are available in the open literature. This weir is shown schematically in Figure 14. Frequency and pressure fluctuation data were obtained in the enclosed air space and at three locations downstream of the nappe (Nakagawa et al., 1988). The data are reproduced in Figure 15, whereby the frequency data were normalized with the fall time, and the pressure data was scaled with the dynamic pressure corresponding to the average nappe velocity (H/t_H). The peak pressure fluctuation, when normalized by the characteristic dynamic pressure for the gate, $\frac{1}{2}\rho\left(H / t_H\right)^2$, reached a maximum value of about 0.73% (or approximately 65 Pa). Note also the close correspondence between the nappe oscillation frequency and the calculated frequency of a transverse acoustic standing wave calculated as an open-open mode with end corrections equal to $0.61r$, where r is the equivalent radius of the end area of the nappe ($r \approx \sqrt{\frac{2}{3} H x_H / \pi}$).

Figure 14. The Arai Weir on the Seta River, near Kyoto Japan showing (a) a bird's eye view of the Seta River and the Arai Weir, and (b) a side view of the leaf gate
From Knisely (1989). © 1989 Elsevier. Used with permission.

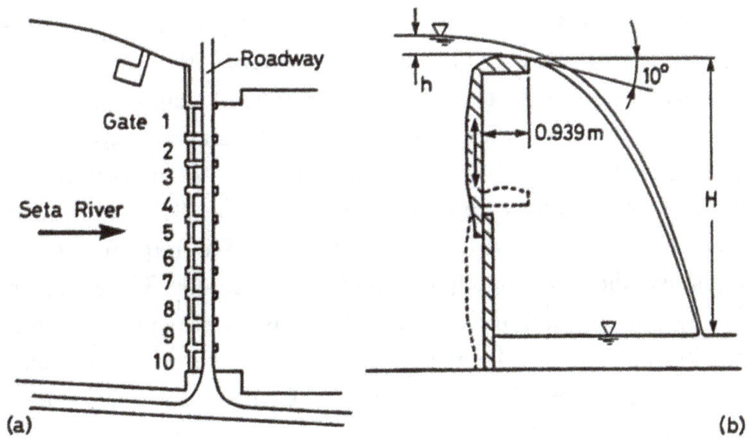

Figure 15. Normalized nappe oscillation data from the Arai Weir near Kyoto, Japan, showing $f_n t_H$ as a function of Treiber's b parameter. Pressure data are in percent of the characteristic dynamic pressure for the weir, given by $p_{rms}(f) / \left[\frac{1}{2}\rho\left(H / t_H\right)^2\right]$.

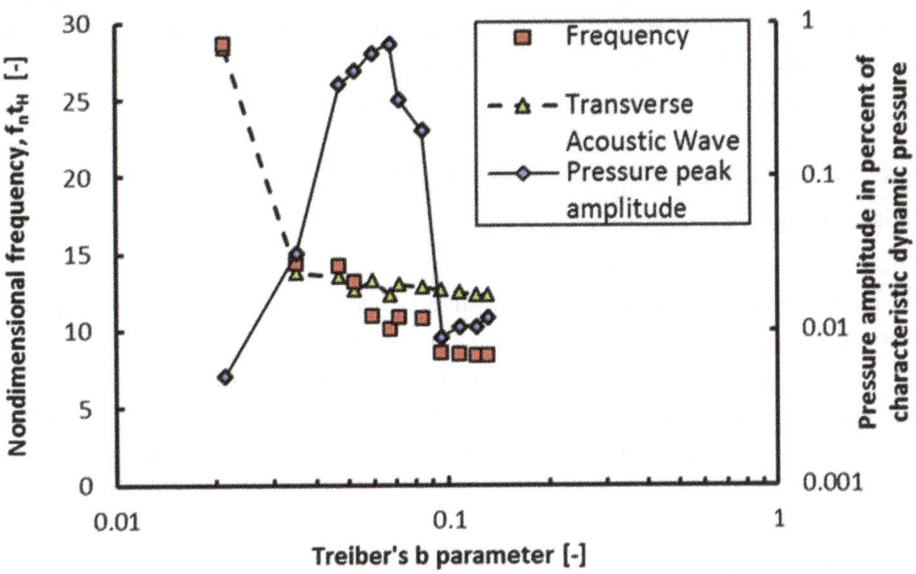

The first high frequency data point corresponds well with the second harmonic $\left(\lambda_a = S + 2\{0.61r\}\right)$, while the remainder of the frequency data for b < 0.1 agrees well with the estimated fundamental mode $\left(\frac{1}{2}\lambda_a = S + 2\{0.61r\}\right)$. For values of $b > 0.1$ at which the nappe frequency deviates more substantially from the acoustic mode, the pressure amplitude drops precipitously.

A logical conclusion from the data in Figure 15 is that the strong pressure peaks occur when the frequency of the pressure fluctuations corresponds to the frequency of the transverse standing acoustic wave. The oscillation is sustained by a standing transverse acoustic wave along the 10.8 m face of the gate. However, Nakagawa et al. suggest that the constant frequency value they calculated for the acoustic standing wave and the lack of pressure amplitude variation across the span of the gate together argue against the standing acoustic wave as the mechanism sustaining the nappe oscillation. A careful examination of their data indicates, however, that the end corrections for the open-open acoustic mode were neglected. By including approximate end corrections on each end of the open-open tube, the oscillation frequency is found to decrease slightly with increasing b value, as shown in Figure 15. Further, as the nappe frequency departs from the predicted acoustic standing wave frequency, the pressure fluctuation magnitude falls precipitously through more than an order of magnitude. The approximate end correction used was $0.61r$, where the equivalent radius, r, was defined as square root of the average vertical cross-sectional area beneath the nappe divided by π. The value of the end area used for end correction changes with the initial brink velocity as does the value of b. End correction should be used when calculating standing wave frequencies.

An important lesson to be learned from this study is that constant spanwise pressure distribution does not exclude the possibility that an acoustic wave sustains the nappe oscillation. Recall that the mechanisms in Figure 10 are not exclusive. Each of the mechanisms can act individually or in concert with each of the other mechanisms. Further, there is no provision that the action of one mechanism in any way precludes the existence of the other mechanisms.

A recent study on a labyrinth weir in North Carolina (USA) provides another example in which the transverse acoustic standing wave may be serving to fix the frequency of oscillation, while not necessarily being the only operable mechanism (Crookston et al., 2014). The data from this study are plotted in $f_n\, t_H$ vs. b format in Figure 16, with the following assumptions:

- The initial angle of the nappe was assumed to be 20°
- The initial velocity was assumed to be $1.72\sqrt{h_o g}$
- The speed of sound was assumed to be 343 m/s
- The end correction for the open tube was assumed to be $0.61r$, as was done with the Arai Weir.
- The open-open tube was assumed to support a ½-wavelength standing wave
- The open-closed tube was assumed to support a ¾-wavelength wave.
- Finally, the frequency data plotted by the triangle data points were assumed to be the center frequencies, 23.5 and 37.5 Hz, respectively, of the two ranges of 22 to 25 Hz and 35 to 40 Hz given by Crookston et al. (2014)

The prediction of the transverse standing wave frequency is complicated by a vertical step along the base of the labyrinth weir over which the nappe was formed. For this reason the data are presented for two fall heights; most likely, the actual frequency will lie between the two data sets. The data from the Arai Weir are also plotted for reference in Figure 16.

The results show that open area at the ends of the nappe does not necessarily supply sufficient aeration. The open area affects the frequency of the standing acoustic wave. Opening the nappe ends may attenuate the oscillations or it may serve to tune the standing wave frequency to more closely match the nappe frequency, and thereby amplify the pressure fluctuation in the air cavity under the nappe. We

Figure 16. Nappe oscillation frequencies for the labyrinth weir at the Linville Land Harbor Dam, North Carolina, USA
Extracted from values provided in Crookston et al. (2014).

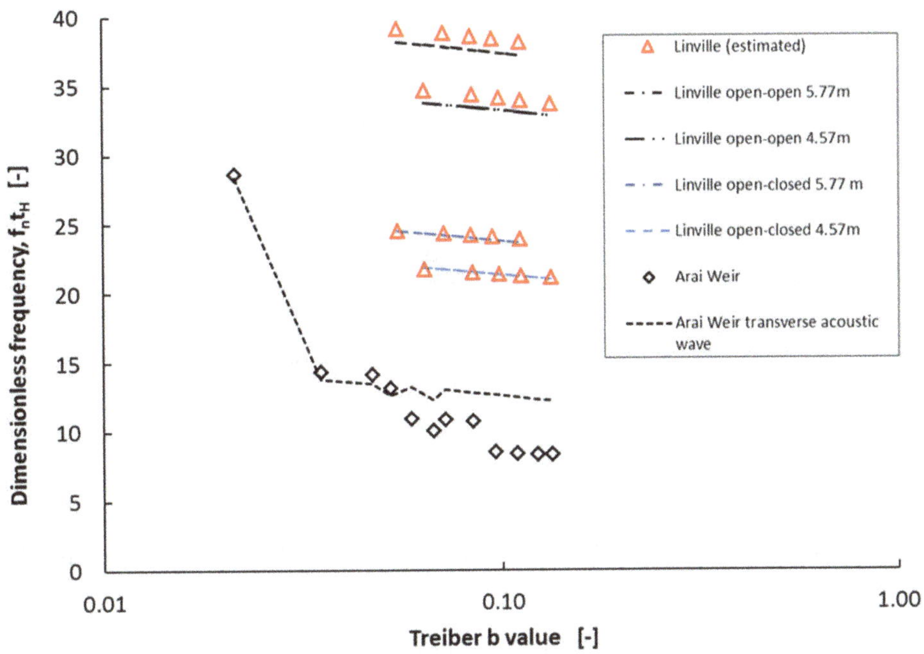

will offer a guideline later in this chapter concerning what percentage of the nappe surface area should be open to avoid nappe oscillation.

Anderson (2014) undertook a study of oscillating nappes on labyrinth weir models, and generated limited data. Without substantial assumptions or acquiring additional details, his data could not be included for comparison in Figure 16.

In the next section, we will present previously unpublished data on a lab-scale, parabolic weir, which was recorded by Mr. T. Kurimoto as part of his senior graduation thesis at Kyoto University in 1989.The study was on methods of attenuating nappe oscillation via aeration and via the use of compliant flaps suspended from the weir brink. Details of this study can be found in an internal report by Knisely et al. (1989), available from the author.

Kyoto Data Set From a Laboratory Scale Parabolic Nappe

Figure 17 shows the experimental set-up. Water was pumped from the reservoir to the horizontal channel, from which it was discharged as a parabolic nappe. The fall height was varied from 1.475 m to 1.175 m by regulating the depth of the downstream pool at the bottom of the nappe. The overflow depth was measured with a depth gage, and the discharge was measured with a V-notch weir downstream of the nappe. The discharge characteristics were well represented by the following discharge correlation:

$$q = K\sqrt{g}h_0^{3/2} \tag{28}$$

Figure 17. Set-up for laboratory study at Kyoto University, dimensions are in meters

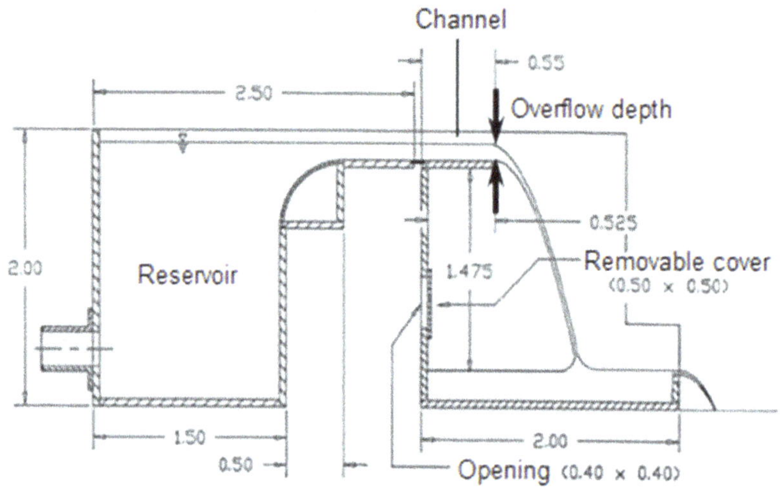

with K taking on an experimentally determined value of 1.790. The corresponding brink velocity is given by

$$U_0 = K\sqrt{h_0 g} \tag{29}$$

Table 1 provides an overview of the conditions studied. Initially the nappe oscillation characteristics with no countermeasure and no aeration were measured. Subsequently, nappe characteristics with increasing values of aeration were recorded in terms of R, the ratio of total area of the openings to the parabolic surface area of the nappe. The last set of tests was designed to study the efficacy of flexible flaps suspended from the brink for disrupting the oscillations (results previously reported in Knisely, 1994).

The parameters used to characterize the nappe oscillations are the same as in Figure 13. They include the initial velocity U_0, the initial direction of the nappe flow, θ_0, the fall height, H, and the location of impingement x_H. The time taken for the flow to fall the distance H, previously introduced as the fall time t_H, is given by

$$t_H = \frac{-U_0 \sin\theta_0 + \sqrt{2gH + \left(U_0 \sin\theta_0\right)^2}}{g} \tag{30}$$

This parameter, t_H, is used as a time scale.

The initial velocity, and the initial direction of the nappe can be combined with the fall time into the Treiber b parameter, given in Equation 20.

A small value of b represents an initially near zero nappe velocity with a large fall height in which the effect of gravity alone determines the water velocity at any point in the nappe. On the other hand, a

Table 1. Conditions for the Kyoto nappe studies

Fall Height (m)	Aeration (%)	Countermeasure	Comments
1.475	0	None	Reference condition
1.400	0	None	Reference condition
1.275	0	None	Reference condition
1.175	0	None	Reference condition
1.275	0.05	None	
1.275	0.2	None	
1.275	0.9	None	
1.275	2.0	None	
1.275	12.0	None	
1.275	0	1 Flap; 25 x 10 x 0.1 cm	Flex-flap
1.275	0	2 Flaps; 30 x 5 x 0.1 cm	
1.275	0	2 Flaps; 30 x 10 x 0.1 cm	Effective
1.275	0	3 Flaps; 30 x 10 x 0.1 cm	Effective
1.275	0	2 Flaps; 30 x 10 x 0.1 cm	With fan excitation
1.275	0	3 Flaps; 30 x 10 x 0.1 cm	With fan excitation

Includes entries from Knisely, 1994.

large b value is indicates a constant fall velocity through a relatively small fall height, with the velocity minimally affected by gravity.

The unaerated nappe data, shown in Figure 18, cover fall heights from 1.475 to 1.175 m. For the largest fall height, Figure 18a, the nappe oscillations are relatively strong and extend over a relatively wide range of b values from onset near $b = 0.01$ to termination near $b = 0.03$. Over the wide plateau of pressure amplitudes the fluctuations were between 200 and 250 Pa (or about 0.02 to 0.025 atmospheres, quite a small value). The peak amplitude of pressure fluctuations at higher b values was 333 Pa. As the fall height was decreased by raising the depth of the downstream pool, the broad b range of oscillations was split near the middle of the range into two separate ranges separated by a sharp valley (see Figure 18b). With a further decrease in the fall height, the upper range of oscillation was attenuated, while the lower range persisted (see Figures 18c and d). For these two lowest fall heights, distinct peaks in the pressure spectra in the upper b range of oscillation could still be clearly discerned, but the amplitude of these peaks was 2 to 3 orders on magnitude smaller than the corresponding peaks for the 1.475 m fall height. Similarly, more severe attenuation of the second range of nappe oscillation was also observed in the study of the effect of aeration, to be presented next.

In Figure 19 the composite sets of frequency and pressure data with a closed air cavity from the Kyoto study are plotted along with the Arai Weir data. The greater values of b for the Arai data are due to the greater overflow depth and correspondingly higher initial nappe velocity. The two data sets are qualitatively similar in overall characteristics, but differ in the details. The Arai Weir nappe oscillations have higher frequency nappe modes, but lower magnitudes of dimensionless pressure fluctuations.

The Partenscky onset criterion (Figure 19a) predicts the onset of nappe oscillation for the Kyoto data fairly well. The Arai Weir data are less well predicted. It seems the Partenscky criterion gives a

Figure 18. Variation of non-dimensional peak frequencies and dimensional amplitudes of air cavity pressure fluctuations (on log scale) as a function of Treiber's b parameter for fall heights of (a) 1.475 m, (b) 1.400 m, (c) 1.275 m, and (d) 1.175 m
All from the Kyoto data set.

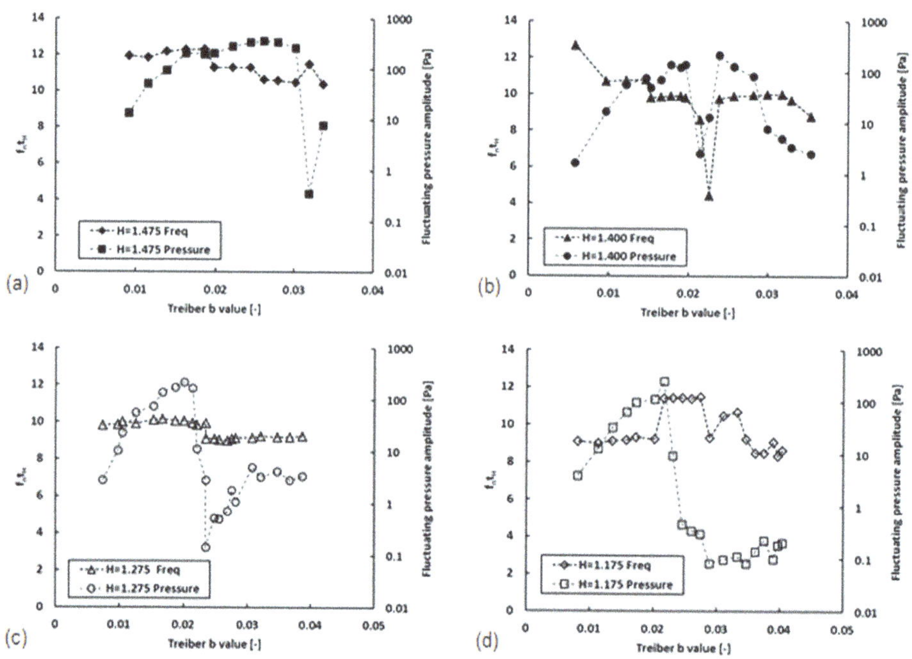

moderately accurate prediction of the onset for weir flows with close to horizontal overflow, but may need further refining for weirs with larger initial nappe inclinations and fall heights.

Aeration as a Countermeasure

The back wall of the Kyoto overflow chamber was opened by drilling successively more holes with increasing diameter, and ultimately removing the 0.5 x 0.5 m plate on the back wall beneath the overflow channel. The open area was very small relative to the nappe area. The first set of holes resulted in an open area that was 0.000246 times the nappe area. The open area ratio, R, the ratio of open area to nappe area, was used to characterize the amount of open area. The closed area ratio is defined as 1-R.

Figure 20 shows the frequency and pressure characteristics of nappe oscillations with increasing values of open area. In these tests, the fall height was held constant at 1.275 m. For 1.275 m fall height the upper range of nappe oscillation is already quite severely attenuated even when the air cavity was closed. As the open area increases, the nappe oscillation begins to decrease in both pressure amplitude and frequency at smaller *b* values. With an open area of about 12% of the nappe area, very small magnitude pressure fluctuations (on the order of 0.05% of the characteristic nappe dynamic pressure) are still present, as can be seen by the data in Figure 20b. The Arai Weir data shows pressure levels that are consistent with a nappe with an open area of somewhere between 1% and 10%. From reduced scale drawings of the gate in which the text was challenging to read, the estimated open area of the Arai Weir was between 9.6 to 12.5% of the nappe area, resulting from the space between the nappe and the piers.

Figure 19. Composite Kyoto data with closed air cavity with the Partenscky onset criterion. Both frequency and pressure data show qualitatively similar characteristics with those from the Arai Weir. (a) Frequency data show the tendency toward lower frequencies at larger b values. (b) Pressure amplitude data show a characteristic gradual rise and an abrupt decrease.

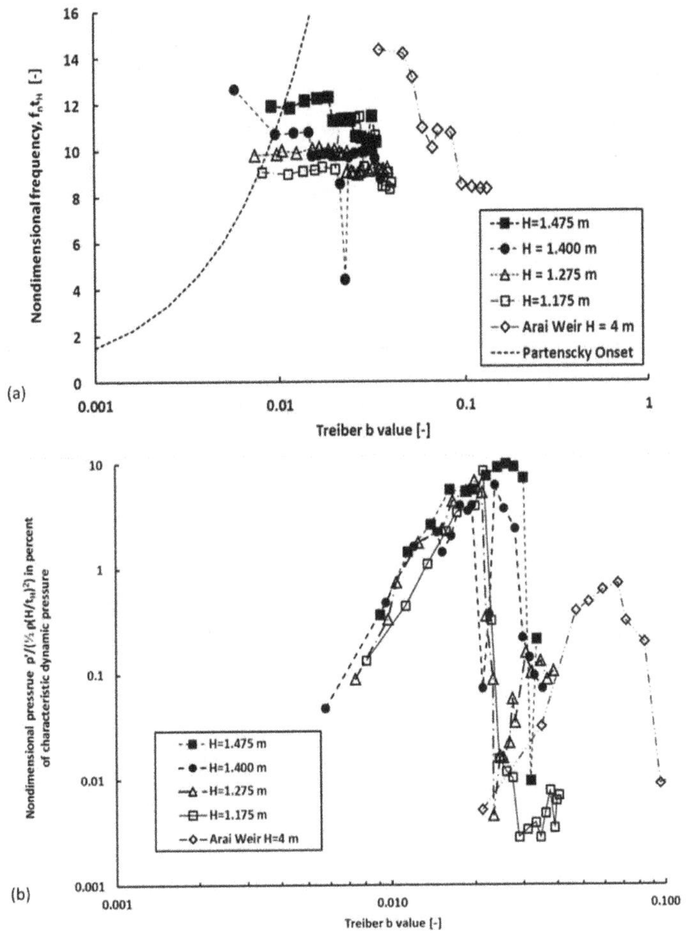

The frequency data becomes widely scattered with increasing aeration because there are many small frequency components competing for the available energy. The pressure spectra for the closed cavity and the cavity with increasing values of open area ratio, in Figure 21, show the decline from peak values about -22dB when R = 0 to peak values about -61 dB when R = 12%.

The non-dimensional peak pressure level for each closed area ratio, 1-R, is plotted for the Kyoto data in Figure 22. The area is estimated for the Arai Weir in this figure. The extrapolated curve fit to the data permits a somewhat speculative estimation of the required open area ratio to keep pressure fluctuation levels below a specified target level. The power law curve, fitted to the data, would suggest that to keep air cavity pressure fluctuations at or below 1/1000 of a percent of the nappe characteristic dynamic pressure, a closed area ratio of 75% (corresponding to an open area ratio of 25%) would be needed.

The corresponding required open area ratio would be about 25% of the nappe surface area. The assumption is that this area is directly opened to the atmosphere and would need to be adjusted to account

Figure 20. Composite data with aeration from Kyoto study: (a) frequency data, and (b) pressure data

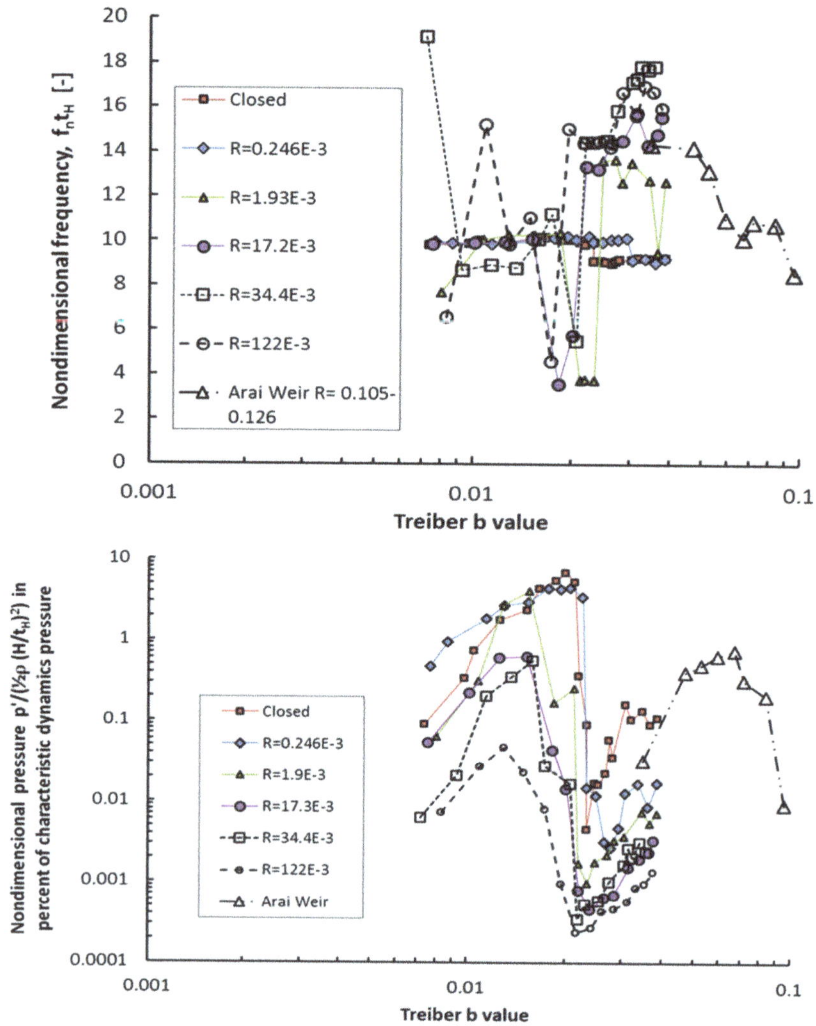

Figure 21. Fluctuating cavity air pressure spectra for closed and aerated cavities from the Kyoto data set with h_o = 0.015 m, for open area ratio values of (a) R = 0, (b) for R = 0.087, and (c) for R = 0.122

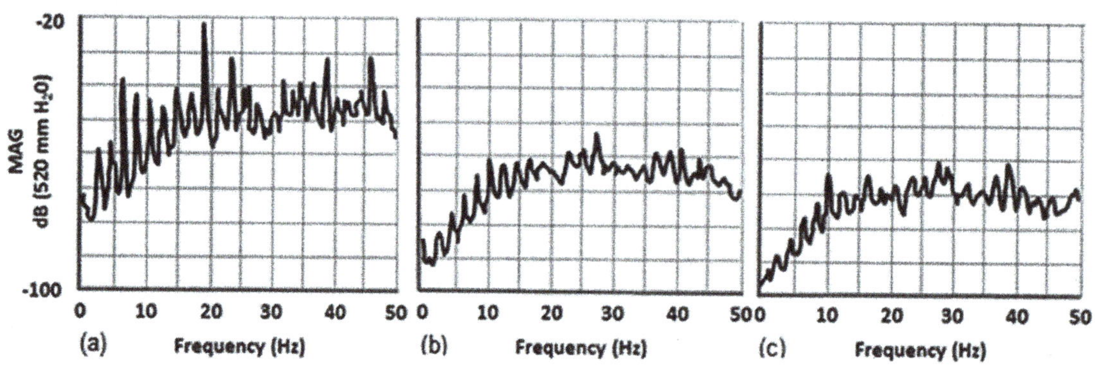

Figure 22. Estimate for required open area ratio to keep pressure fluctuation flow 0.001% of characteristic nappe dynamic pressure

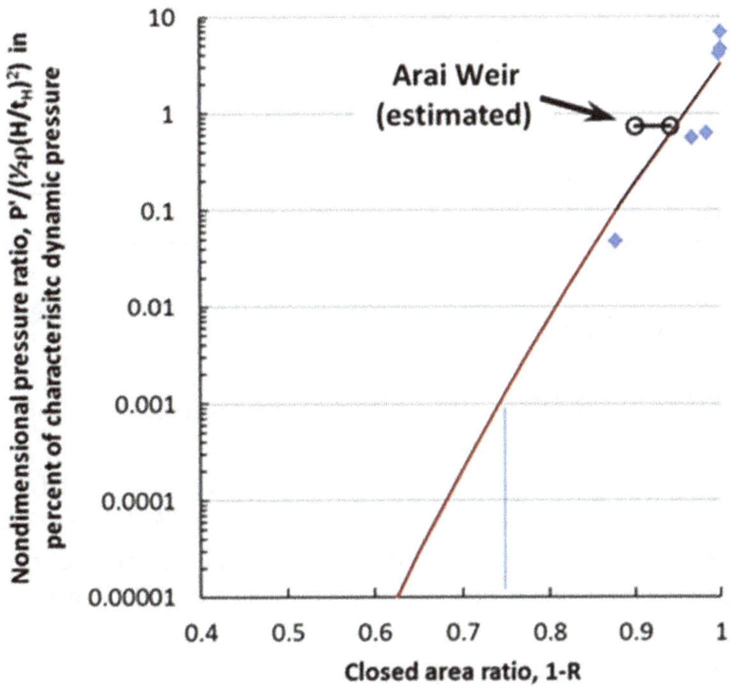

for pressure drop effects in ducts and piping. The 25% open area estimate would appear to be marginally consistent with the oscillating nappe previously shown in Figure 8a with an estimated 17% open area.

Note that opening 25% of the nappe surface may not be possible when a specified discharge capacity is required. Another option is to aerate the underside of the nappe through supply ducts. Lewin (1995) provides guidelines for nappe aeration in terms of required volumetric air flow rate per volumetric water flow rate:

$$Q_{air} = \alpha Q_{water} \tag{31}$$

Lewin (1995) provides a graph of α as a function of h_o/H for two ranges of brink Froude number. The two curves are well represented by a third order polynomial using the ratio of brink nappe thickness, h_o, to fall height, H, with the coefficients given in Table 2 for the two ranges of Froude number.

$$\alpha = a\left(\frac{h_o}{H}\right)^3 + b\left(\frac{h_o}{H}\right)^2 + c\left(\frac{h_o}{H}\right) + d \tag{32}$$

Table 2. Coefficients for use in Equation 32 for coefficient of air demand from curve fit to graph in Lewin (1995)

$Fr^2 = \frac{v_0^2}{gh_o}$ Range	Coefficient of $(h_o/H)^3$ a	Coefficient of $(h_o/H)^2$ b	Coefficient of (h_o/H) c	Constant d	Excel R^2 Value for Goodness of Fit
3-10	$1.3928\ 10^{-6}$	$-4.6021\ 10^{-4}$	$8.8930\ 10^{-2}$	$4.5667\ 10^{-2}$	0.99957
13-15	$1.52163\ 10^{-5}$	$-1.8202\ 10^{-3}$	$1.5782\ 10^{-1}$	$2.9529\ 10^{-2}$	0.99996

Labyrinth Weir Nappe Oscillations

To increase the discharge capacity of a weir, it can be folded across the flow channel to provide a much longer length of overflow as shown in the sketch in Figure 23.

Details concerning the design of labyrinth weirs can be found in Falvey (2002). Ongoing work on labyrinth weirs is being done by the Utah State University group under the direction of Prof. B. Tullis, (see Tullis et al., 2007; Crookston et al., 2014; and Anderson, 2014). Due to the convoluted structure of labyrinth weirs, adjacent nappes interact, complicating the analysis of these nappes. To determine the effect of weir geometry on the length of the interference region, Falvey (2002) provides an approximate curve fit for the length of the interference region, L_{de}, in terms of over flow depth, h, and the sidewall angle, α:

$$\frac{L_{de}}{h} = \exp\left(-0.052\alpha\right) \text{ for } 10° \le \alpha \le 90° \tag{33}$$

Figure 23. A labyrinth weir is folded across the discharge channel providing more overflow length and thereby providing greater discharge
After Paxson & Savage (2006).

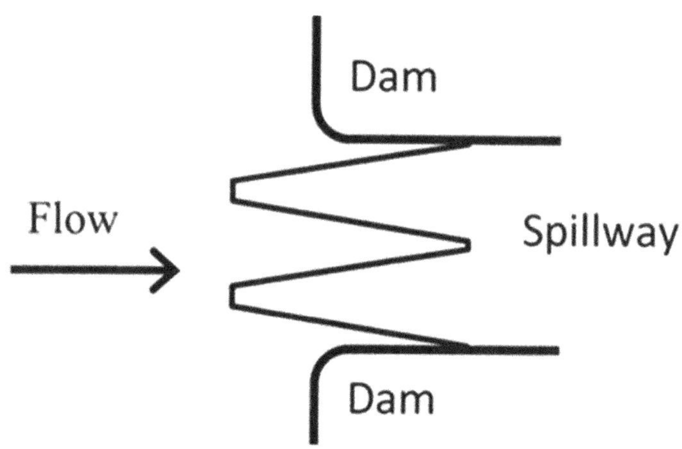

Falvey (2002) also discusses nappe oscillations in labyrinth weirs. He cites his work with Ervine (Ervine and Falvey, 1987), which found that the breakup length of the overflowing nappe was between $50h$ and $100h$. Their results suggest that for overflow depth to weir height, h/H, values less than 0.01, the nappe will break up and oscillations do not occur. The usual range of overflow depth to weir heights for which nappe oscillations can occur in labyrinth weirs is $0.01 \leq h / H \leq 0.06$.

One possible countermeasure to disrupt nappe oscillations is to roughen the crest. This approach was successful at the Avon weir in Australia (Metropolitan Water, Drainage and Sewerage Board, 1980). However, the disadvantage of the roughened crest was a 2% decrease in discharge.

Surging flow is another type of nappe instability that can occur on labyrinth weirs. This phenomenon results when the total upstream head on the weir, h_0, approaches $0.3H$, and air is evacuated from the space between the nappe and the weir. When the pressure in the air space becomes sufficiently low, air is entrained through the nappe and the pressure in the air pocket rises once again. The fluctuating pressure in the air pocket is accompanied by fluctuating discharge from the weir and fluctuations in the downstream water level. Falvey states that solutions that work to attenuate nappe oscillations do not necessarily work to prevent surging and vice versa.

Countermeasures for surging include splitters near the end of the weir crest and side wall aerators located high on the walls near the downstream side of the crest (Falvey, 2002). A common countermeasure for nappe oscillations is increased aeration, but Crookston et al. (2014) found persistent nappe oscillations at the labyrinth weir at Linville, despite what appeared to be sufficient aeration. Falvey (1980) and Anderson (2014) also reported nappe oscillations despite full aeration of the nappe. In these situations, the interaction of a standing acoustic wave may have supplied the fluctuating pressure to sustain the nappe oscillations (see "Early Studies on Nappe Oscillations" and Figure 16).

Other Countermeasures for Nappe Oscillation

In addition to roughening the weir/gate crest, aeration via splitters along the crest of the gate or weir, or installing aeration ducts, several other methods have been proposed for attenuating nappe oscillations. In the Kyoto nappe study, compliant flexible flaps, fastened to and hanging down from the horizontal weir surface, were found to be effective when

- Each flap was between 5 and 10% of the span (each flap was 5 to 10 cm wide for the 1 m span)
- The spacing was approximately 25 to 30% of the fall height apart (center-to-center distance)
- Each flap extended approximately 30% of the fall height beneath the weir surface.

Knisely (1994) provides evidence (see Figures 24 and 25) that in laboratory studies, the flaps were as effective as aeration in suppressing nappe oscillations. It is thought that these flaps draw down the nappe locally, producing a rippled or spanwise corrugated appearance across the nappe. The pressure pulsations in the air cavity cannot uniformly disturb the nappe in the regions near separation from the weir. This de-phasing of the excitation across the span of the structure is a well-known countermeasure in other types of vibratory systems, such as the helical strake countermeasure for vortex-induced vibrations.

Sato et al. (2007) showed that inserting box-like structures (baffles and baffle walls as they call them) near the base of the nappe can serve to dissipate nappe oscillations. This countermeasure may be

Figure 24. (a) Amplitude and (b) frequency of air pressure fluctuations in the cavity beneath a 1 m wide nappe with a fall height of 1.275 m, with 2 or 3 rubber flaps as countermeasure
After Knisely, 1994.

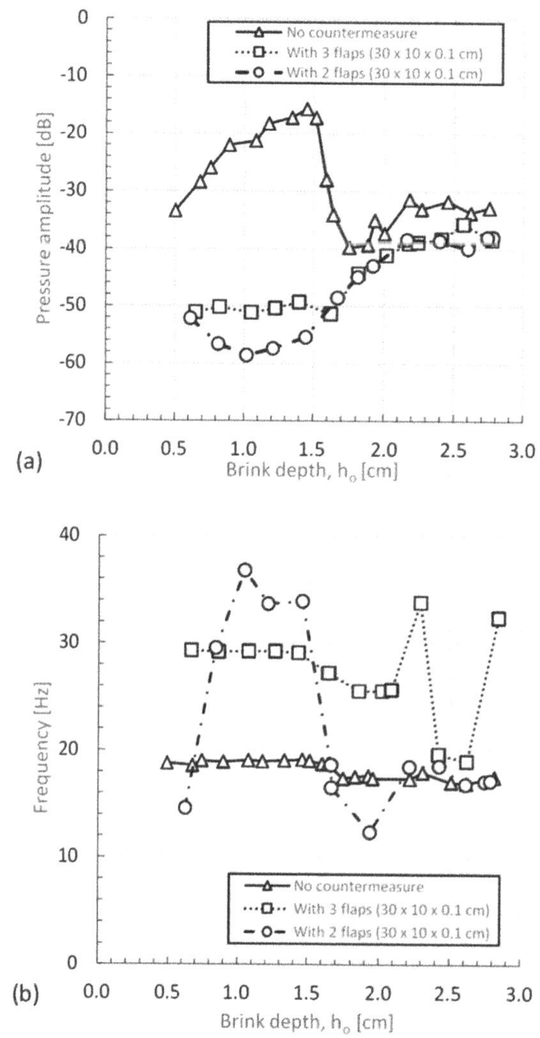

effective because of openings provided by the box-like structures which hinder the build-up of oscillatory pressure in the air cavity.

Anderson (2014) determined that roughening the weir crest prior to the point of separation can be effective. He also studied ways to disrupt the nappe near the impingement surface by using angled plates to deflect the nappe upon impingement.

When a transverse standing acoustic wave is participating in the nappe oscillation process, an effective countermeasure is to install multiple impervious, vertically-oriented barriers to reduce the wavelength of the standing wave. The shorter wavelength yields a higher acoustic frequency. By spacing the barriers judiciously, it is possible to decouple the acoustic mode by moving the acoustic frequency to a frequency beyond that at which the nappe can be excited. This solution is challenging when the gate is raised and lowered.

Figure 25. Spectra of pressure fluctuations in air cavity beneath a 1.0 m wide by 1.275 m high nappe with (a) no countermeasure, (b) two 30 x 10 x 0.1 cm flaps, and (c) three 30 x 10 x 0.1 cm flaps. All results with fan excitation directed at the base of the nappe
After Knisely, 1994.

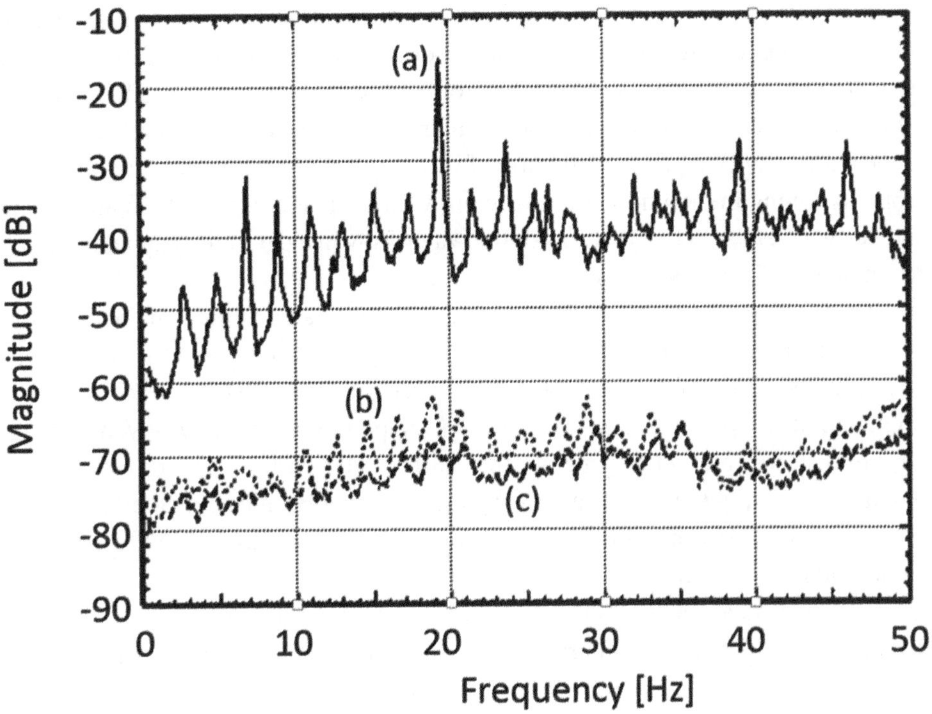

USBR Study of Nappe Oscillations on the Black Canyon Drum Gate

In 1939, the US Bureau of Reclamation investigated nappe oscillations on the drum gates at the Black Canyon Diversion Dam (see Glover et al., 1939 and Simmons, 1964). When the 64-ft span drum gate was between 30 and 70% open with overflow (0.22 to 1.45 feet of head), severe noise between 7 and 18 Hz, generated by the oscillating nappe, rattled windows and other building components in structures several hundred yards from the gate. In this particular case, the frequency of vibration increased with the increasing head on the gate. The report tied the nappe oscillation to the fluctuating pressure (maximum amplitude of pressure fluctuation was 0.06 psi) in the enclosed air volume. The cure for these oscillations, which were found to be of no danger to the gates or the weir, was to split the nappe with a single log at mid-span of the gates.

Similar nappe oscillations were found on the Prosser Creek Diversion Dam, with no gates on a fixed weir. Air vents with 15-in diameters were installed in the piers, about 200 ft apart. The aeration was insufficient to eliminate the nappe oscillations with flow depths of 0.5 to 0.7 ft of overflow depth. Subsequently, 6 x 8 x ½-in angle splitters were installed, extending horizontally from the crests on 33-ft centers with the apex of the angle pointed upward. With these splitters installed, the nappe was sufficiently aerated to attenuate nappe oscillations. Splitters with center spacings of less than 35 ft became the preferred countermeasure for nappe oscillations.

Vertical Water Curtains

In investigations of oscillations of water sheets, space and expense might dictate creating a vertical falling water curtain with an adjacent closed air space to simulate a naturally occurring parabolic nappe flowing over a gate or weir. A fundamental difference between the vertical water curtain and the parabolic nappe is the direction of the applied pressure loading and its effect on the trajectory of the fluid particles. For a vertical nappe, the pressure force acts in the horizontal direction. For a parabolic nappe, the pressure force acts in a nearly vertical direction (or perpendicular to the usually small overflow inclination) at the beginning of the nappe and changes the direction of its action to approach a horizontal pressure loading near the bottom, assuming the fall height is sufficiently large. The pressure fluctuation can cause both horizontal and vertical velocity perturbations of the water flow in a parabolic nappe, but only horizontal perturbations on an undeformed vertical water curtain.

Binnie (1971) investigated a vertically falling water sheet in a vacuum chamber and found that no nappe vibrations occurred. These results suggest that a second fluid (i.e., air) surrounding the water jet is required to induce the vibration. Binnie noted that the vibration is sensitive to the depth of the tailwater plunge pool, and also reported seeing instances of nappe vibration for fully vented nappes. Binnie used a planar, high-speed water jet to generate the falling water curtain. The resulting velocities were far higher than those typically found in overflow gates. Although Binnie accounted for the gravitational acceleration of the water sheet, he ignored any surface tension or skin friction effects on the closely spaced span end bounding vertical surfaces.

Frequency data for vertical water curtain oscillation are compared to those for parabolic nappe oscillation in Figure 26. Most vertical water curtain studies are undertaken at substantially higher velocities and shorter fall distances, yielding values of the *b* parameter that exceed those of parabolic nappes by more than an order of magnitude, some as much as three orders of magnitude higher. Also note that none of the "naturally" occurring nappe oscillations have less than approximately one complete wavelength (indicated by the parameter $f_n t_H$ values greater than 1.0) on the nappe, while there are laboratory data for vertical water curtains that can be excited with one-half wavelength on the vertical water curtain.

Vertical water curtain studies are of interest for determining characteristics of the Kelvin-Helmholtz instability and its coupling with an enclosed air volume. The need for higher velocities to excite the Kelvin-Helmholtz instability might suggest that the instability on parabolic nappes may occur due to one or more of the other mechanisms shown in Figure 10 serving to initiate the instability. Once initiated, the instability can then take on a Kelvin-Helmholtz character as it grows. As the instability grows, the pressure fluctuations also grow, and serve to initiate further cycles of the oscillation.

Vertical water curtain experiments have substantially higher Treiber *b* values than overflow nappes, indicating either very little time for gravitational acceleration to act, or such a high velocity that an acceleration due to gravity is negligibly small. In either case, the water velocity remains very close to constant for large *b* values. Constant fall velocity also implies constant nappe wavelength, perhaps in some way associated with the observation of less than one complete wavelength on the vertical water curtain.

Furthermore, the coupling between nappe thickness and initial velocity, inherent to free overfall, is missing in vertical water curtains since the slit width through which the water curtain passes can be adjusted independent of the head (provided a sufficient water flow rate is available). In free overfall, the brink depth and the initial overflow velocity are governed by a single parameter, the head above the gate.

Figure 26. Composite summary of nappe frequency data and vertical water curtain frequency data in the form of dimensionless frequency, $f_n\, t_H$, as a function of Treiber's b value. There is a mismatch between ranges of vertical water curtain instability (unfilled symbols) and parabolic nappe instability (filled symbols).

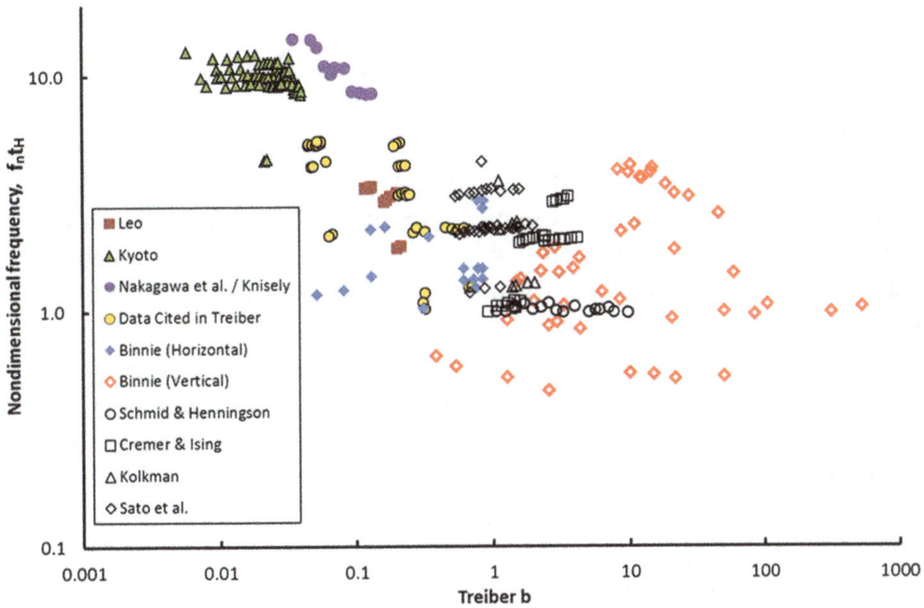

Another difference between the vertical water curtain with near constant thickness (large b values) and the parabolic nappes with lower *b* values is the occurrence of frequency jumps, indicating changes in the number of wavelengths on the nappe. Schwartz (1964) noted that the amplitude of the nappe response to a fluctuating pressure varied inversely with the frequency of pressure fluctuation. Conversely, it might then be argued that a small amplitude, high frequency disturbance can more easily sustain a given pressure fluctuation than a larger amplitude, lower frequency disturbance. The mass per wavelength of the higher frequency (shorter wavelength) disturbance is less, and especially near the bottom of the nappe, where the wavelength has been elongated and the nappe is thinner, a higher frequency pressure disturbance might be able to displace more easily the lower mass per wavelength nappe. It certainly appears that in "natural" nappes, the oscillations begin at relatively high frequencies and decrease in frequency within the same mode with increasing *b* parameter values. Most naturally occurring nappes appear to jump to at least one lower mode before the nappe becomes too thick for further excitation. Such frequency jump/mode change behavior is seldom observed in the available data for vertical water curtains.

Another factor to consider when relating vertical water curtain laboratory studies (and even parabolic nappe laboratory studies) with very small transverse spans to full-scale weirs is the effect of the sidewall damping on the nappe motion. Based on observation of the 1 m wide nappe in Kyoto, the sidewall damping appeared to deform the otherwise two-dimensional, oscillating nappe at distances of 3 to 5 cm on each spanwise end. The side wall friction may contribute to the excess rigidity commonly observed for with small-span models (see for example, Cremer & Ising, 1972 and Kolkman, 1974, among others).

VIBRATION OF INFLATABLE WEIRS

Despite substantial progress in the construction of stable inflatable weirs, some vibration issues persist. Chanson (1997) provides an overview of vibration mechanisms related to overflow over inflatable flexible-membrane weirs or dams (IFMDs, using Chanson's acronym) in both the inflated and the deflated condition.

Vibration and abrasion mechanisms for the deflated membrane include:

- Sub-atmospheric mean and fluctuating pressures due to a hydraulic jump causes the deflated membrane to lift up, which quickly leads to abrasion and failure
- Near-critical flow can occur for Froude numbers between 0.3 and 3.0. Unsteady undulations occur in the free surface, producing potentially sub-atmospheric pressures and transient flow separations near and over the deflated membrane. The membrane may fail due to abrasion (Chanson citing the work of Chervet, 1984).
- Wake-induced fluctuating pressures may also result in abrasive wear of the deflated membrane. The wake may result from flow past the anchoring hardware, over the membrane itself, or past any rear deflectors or fins if they are not properly recessed into the concrete. See Chanson (1997) for illustrations of this mechanism.
- Another source of abrasion of the membrane is from debris carried by the flow over the deflated membrane. Impacts of large pieces of debris -- such as logs, tree stumps, heavy branches, chunks of ice -- have the potential to puncture the membrane. Chanson suggests protective covering of the deflated membrane may be necessary when such debris-laden flow regularly passes over the deflated membrane.

For the inflated weir, vibration and abrasion mechanisms include:

- Sub-atmospheric pressures generated in the flow over air-inflated weirs may result in cavitation and separation on the downstream face of the inflated structure in overflow. Cavitation is the occurrence of vapor bubbles or pockets of vapor in otherwise liquid flows. The collapse of the vapor bubbles can lead to rapid erosion of the surface, and is accompanied by enhanced noise levels, vibration, and energy dissipation. Cavitation must be avoided in inflatable membrane structures. Chanson (1997) calls this general process *instability due to nappe adherence*; he cautions that predicting potential cavitation damage involves certain assumptions, but notes that there is an increased likelihood of cavitation with increasing structure radius. Thus, nappe adherence instability with potential cavitation and cavitation separation may not be seen in model studies, but may be found subsequently in the full-scale structure. The addition of a deflector on the rear surface is a recommended countermeasure for cavitation. Deflectors may, however, cause additional instabilities such as nappe oscillations.
- Near-critical flow instabilities as with the deflated membrane
- Debris-laden flows, as with the deflated membrane

Gebhardt et al. (2010), provide a different taxonomy of vibration mechanisms with 4 classifications:

1. **Nappe Vibrations:** Nappe oscillations and vibration of the inflatable dam structure occur when the inflatable structure is outfitted with a deflector. Nappe oscillations, as previously discussed in the present chapter, require the formation of an air cavity between the weir and the nappe. They generally occur with small overflow depths and low tailwater levels. Countermeasures for nappe oscillations include aeration of the air cavity and the addition of splitters along the crest of the weir or the end of the flap/deflector.
2. **Vibrations Due to Pressure Fluctuations:** This classification of vibration is the same as the *instability due to nappe adherence* coined by Chanson (1997) and described above.
3. **Vibrations Due to Uplift Forces:** With relatively high overflow and downstream water depth, the water-filled weir is elongated, and takes on a more streamlined shape. The lower pressure over the almost airfoil-like surface results in a suction force that raises the structure higher into the flow. The uplifted structure experiences great drag that pushes it back into the deformed, elongated shape. Once back in the elongated shape, the process begins again. This vibration mechanism occurs at lower frequencies than nappe adherence.
4. **Vibrations of the Deflated Membrane:** Vibrations of the deflated membrane include all of the mechanisms proposed by Chanson (1997), and include the hydraulic jump instability, the near-critical flow instability, and the wake flow instability.

Readers interested in further details concerning inflatable weirs are referred to Chanson (1997 and 1998), Gebhardt et al. (2010), and references cited therein.

FLAP GATE VIBRATION

Flap gates, especially those with elastic hoist cables or chains acting as spring elements, are extremely susceptible to flow-induced vibrations that are invariably initially coupled with nappe oscillations. Flap gates can vibrate in a variety of possible bending modes as well as in torsional modes. Leo (1954) noted that, in certain cases, the flap gate vibration could spontaneously jump from a bending mode to a torsional mode, but more commonly such a jump resulted from a large amplitude, externally applied excitation. As shown in Figure 10 nappe oscillations can respond to flap gate vibration, but the nappe oscillation could also result from a different excitation mechanism, and then serve to initiate flap gate vibration. Flap gate vibration can, however, persist at overflow depths at which the nappe oscillation mechanism is no longer functional due to the larger mass of water.

Leo's (1954) tests on models of flap gates provide a wealth of data that will be analyzed in the context of nappe oscillation mechanisms in the next section. We will then apply this knowledge better understand the vibration behavior of full-scale gates, using the Shorenji flap gate in Japan as an example.

Early Studies on Flap Gate Vibration

Leo (1954) tested a 1:10 scale model of the flap gate from the Dörverden Weir in Lower Saxony, Germany. Leo cited earlier works in his manuscript. Unfortunately, copies these earlier works could not be obtained in a timely fashion for inclusion in our manuscript.

The model had a 0.4 m long curved flap with a 6 m span and maximum fall height of 1.0 m. He tested the gate in a variety of configurations, both with and without open areas at the ends of the gate.

Estimates of the open-end area to nappe area for the data in Leo's report are between 2 to 4%. Based on the estimate in Figure 22, the reduction in peak pressure should be approximately 50%, while the data in Figure 27 show a peak amplitudes for data with both nappe ends open of about 70 to 80% of those found for the nappe with closed ends. The difference might well result from losses due to restrictions in the air passageway. The effective aeration of the open ends of the nappe appears to be about ½ of the physical open area. The other issue is that once the gate begins vibrating, it too adds to the volume change that drives the pressure fluctuations. The additional volume fluctuations due to gate motion could conceivably require more open area to avoid induced pressure fluctuations.

Leo found that the small open areas at the end of the flap gate had negligible effect on the vibration characteristics of the gate. Leo did note, however, slightly stronger pressure fluctuations with correspondingly higher vibration amplitudes, as well as the addition of two lower-frequency modes in the tests with closed air volumes, when compared to similar testing on the gates with open sides. Figure 27 is based on Leo's data, reworked to show his double amplitude (i.e, peak to peak) gate displacement as a function of the Treiber *b* parameter.

Leo suggested that three oscillators may be responsible for the vibration of the flap gate. The strongest and most persistent oscillator was the coupled system of flap gate and air volume; the second oscillator was also a coupled vibratory system, involving the deflection of the nappe mass coupled with the trapped

Figure 27. Comparison of flap gate double amplitude data (in mm/m) for flap gate model (6 m span) with open ends on nappe (unfilled symbols) and closed ends of nappe (filled symbols) in terms of the Treiber b parameter

Adapted from curve fits to data from Leo (1954).

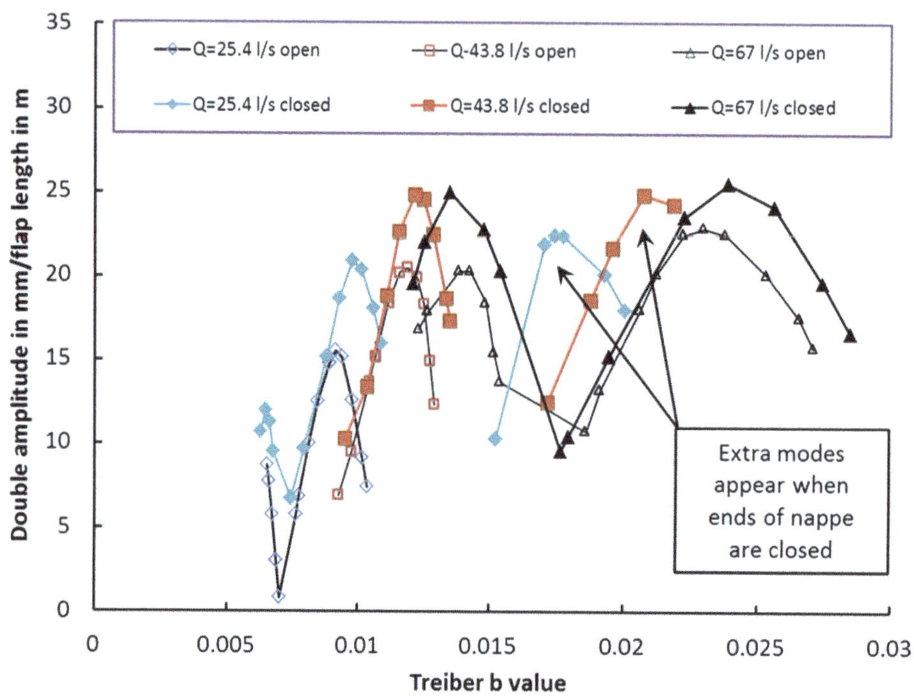

air volume acting as a spring. The third oscillator was a standing transverse acoustic wave beneath the 6 m span of the gate, serving to modulate the air pressure fluctuations in the restricted air cavity between the nappe and the gate, as shown by the fluctuating pressure signal in Figure 28.

Leo cautions against equating the nappe oscillation mechanism and the flap gate vibration mechanism, noting that the nappe oscillation occurs over a very limit range of overflow depths while the flap gate vibrations can occur over a very wide range of overflow depths, later quantified by Partenscky & Swain (1971), Kolkman (1984) and Půlpitel (1979). Partenscky & Swain suggested flap gate vibrations can occur for overflows up to 5% of the fall height; Kolkman (1984) cited personal experience with brink depths up to 10% of the fall height; and the data from Půlpitel show flap gate vibrations with brink depths of up to 15% of the fall height.

As expected from vibration theory, the flap gate elasticity and mass play a dominant role in determining the frequency of the flap gate-air cavity-nappe system, according to Leo. Another factor that must be determined is the added mass due to the water flowing over the gate. The added mass varies with the overflow depth. In most overflow gate problems, determining the added mass is usually a point of uncertainty requiring some form of experimental validation of assumed relationships. A method for computing the added mass will be presented later in this chapter.

From his experimental data, Leo showed that the frequency of the flap gate vibration for his model gate when multiplied by the square root of fall height scaled with the number of wavelengths on the nappe, as follows:

$$f_{fg}\sqrt{H} = 1.85N = 3.7, 5.55, 7.4, 9.25\ 11.1, 12.95, \ldots \quad (34)$$

where N is the number of wavelengths on the nappe. The $f_{fg}\sqrt{H}$ product when multiplied by $\sqrt{2/g}$ closely approximates the $f_n t_H$ parameter previously used in the analysis of nappe oscillations and graphs of nappe phenomena, such as Figure 18, 20 and 26. The dominance of the flap gate frequency may well make the corresponding nappe frequency values lie outside the N and N+½ limits in the previous dis-

Figure 28. (a) Flap gate tip displacement at 5.1 Hz, and (b) air cavity pressure fluctuation at 5.2 Hz, showing particularly strong modulation by standing acoustic wave at approximately 35 Hz
Data from Leo (1954).

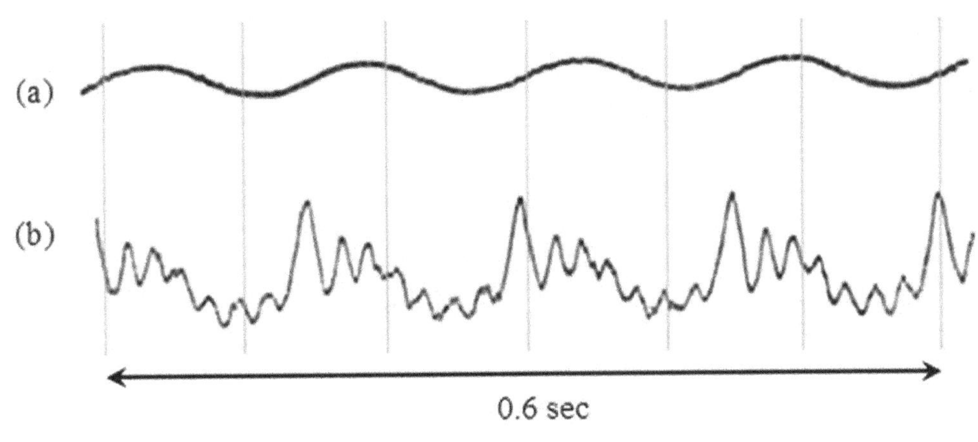

cussion of Treiber's analysis of nappe oscillation; however, Leo reported similar N and N+½ limits for his nappe data with a fixed gate. Another factor influencing the frequency of oscillation may be that of the standing acoustic wave noted in Leo's testing that could affect the frequency of nappe vibration both with a fixed gate and with a vibrating gate.

The fluctuating air cavity pressure links the nappe and gate motions and may bear evidence of the standing acoustic wave. Leo (1954) documented the air cavity pressure fluctuations with a vibrating flap gate and with the flap gate fixed. Figure 29 shows two sets of traces: the traces to the left, labeled (a), are the vibrating flap gate tip deflection and the corresponding cavity pressure fluctuation; those to the right, and labeled (b), are for a fixed, stationary flap gate under the same conditions. A very noticeable difference is the sharp peak in the cavity pressure fluctuation signal, the bottom signal in Figure 29a. The peak occurs in each cycle of the pressure fluctuation signal with a clearly defined phase relative to the gate amplitude signal. This peak in the fluctuating pressure is much weaker in the corresponding trace in Figure 29b, but the frequency of the cavity pressure fluctuation remains about the same, regardless of whether the gate is vibrating or fixed. Leo speculates that this larger peak in the pressure fluctuation indicates larger nappe amplitudes with the flap gate vibrating. The correspondingly larger nappe excursions in the lower reaches of the nappe produce a strong slapping sound as the nappe with these large convolutions slams into the surface of the plunge pool at the base of the nappe. This slapping sound commonly occurs with large amplitude nappe oscillations.

To test the hypothesis that the gate vibration is coupled with the cavity air pressure fluctuation, Leo installed a rigid plate (baffle) just beneath the flap gate such that it would not contact the gate when the gate vibrated (Figure 30). The small gap between the gate and the baffle was covered with a thin, freely rotating membrane that isolated the flap from the air cavity. Leo found that the flap gate did not vibrate under the same conditions where the gate had vibrated previously without the baffle. He concluded that isolating the flap gate from the air cavity beneath the nappe inhibited flap gate vibration.

The amplitude of gate vibration varies with fall height, flow rate and water brink depth (i.e., upstream head) over the gate, and the structural vibration mode of the flap gate. Plotting Leo's amplitude data for a fixed height with varied flow rate and, similarly, for a fixed flow rate with varied height as a function of Treiber's b parameter in Figure 31 demonstrates the similar characteristics of the two different sets of data, confirming the efficacy of the non-dimensionalization.

Figure 29. Pressure fluctuation (lower traces) in closed air cavity (a) with the flap gate vibrating, and (b) with the flap gate stationary
Using data from Leo, 1954.

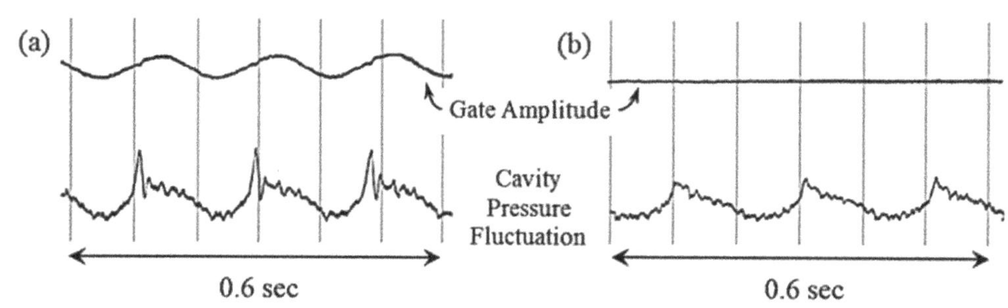

Figure 30. Isolation of the flap gate from the air cavity by a rigid baffle inhibits flap gate vibration
Adapted from Leo, 1954.

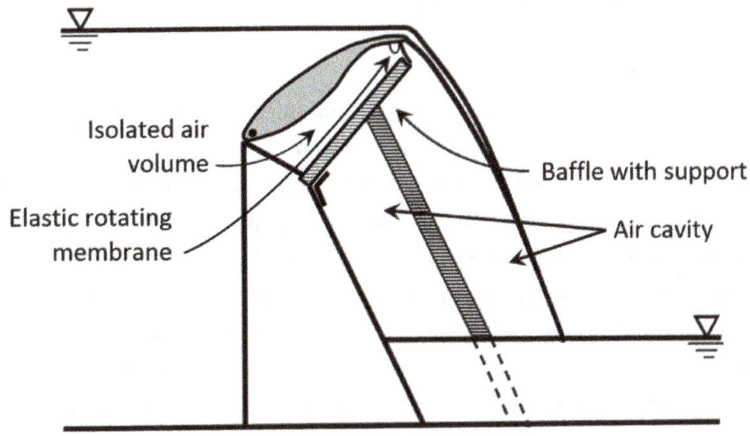

Figure 31. Equivalence of flap gate amplitude response for varying fall height (H) with fixed flow rate (Q) and varying flow rate with fixed fall height
Data from Leo (1954).

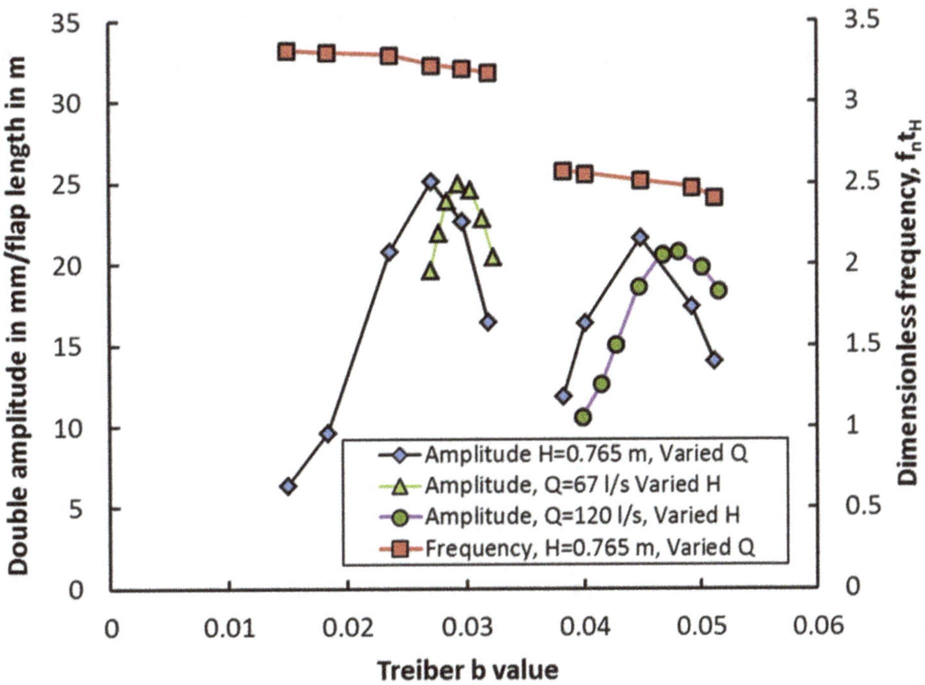

Leo (1954) also documented torsional vibration of the flap gate, in which the displacements of opposite ends of the 6 m gate were 180° out of phase. Figure 32 shows the two modes of torsional vibration recorded by Leo, with amplitudes of vibration comparable to those of the bending modes. Leo notes that the natural frequency in torsion was higher than that in bending for the flap gate he investigated, and resulted in the torsional modes occurring at fall heights that exceeded the range over which one bending mode vibration was found and before the next began, thus, in the "valleys" between the bending modes shown in Figure 27. On occasion, the bending and torsional modes could be excited concurrently, usually after the gate was manually displaced. On other occasions, and usually after manual displacement, the gate was observed to jump between torsional and bending modes. The frequencies for the bending modes in Figure 32 were estimated from the data scattered in Leo's report; they should be understood as being representative of the range of frequencies, instead of precisely measured data.

Leo recorded data from three starting flap positions. The first was when the flap was horizontal; the second was at the lowest position (below horizontal); and the third was at a high position. Unfortunately, the exact geometry corresponding to the low and high positions was not specified. Leo's observation was that the flap that was inclined in the lowest position was the most unstable, with maximum amplitudes of gate tip displacement roughly 1.6 to 1.8 times that of the horizontal gate. With the gate in the high position, the vibration amplitudes were somewhat reduced relative to the horizontal position.

In all cases, the gate vibration showed integer numbers of nappe wavelengths. Leo referred to these numbers as the modulation factor, the number of wavelengths on the nappe. In terms of the formulation presented by Treiber (1971), these modulation factors would be equivalent to integer mode numbers.

Partenscky & Swain's (1971) measurements on an elastically mounted flap gate model support Leo's observations that flap gate vibration frequency was the predominant system frequency. They found some

Figure 32. Torsional amplitude and estimated frequencies relative to bending amplitudes and estimated frequencies
Data from Leo (1954).

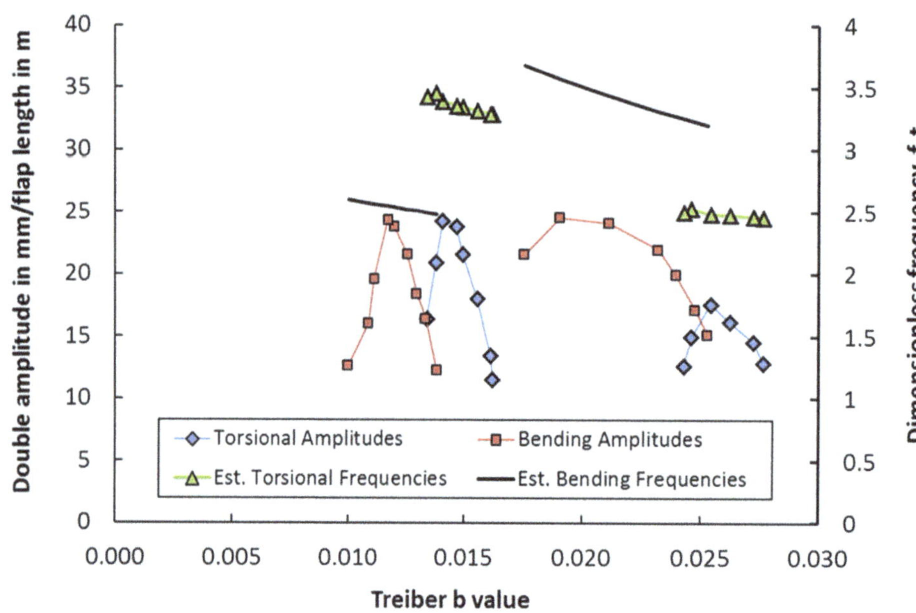

fall heights at which oscillations did not occur. Partenscky & Swain also noted strong modulation in the pressure fluctuation signal, suggesting a slight mismatch between the nappe oscillation frequency and the gate frequency or the possible influence of an acoustic mode.

Partenscky & Swain (1971) also formulated the equations of motion for an elastic flap gate that included the interaction effects of the three systems participating in flap gate vibration: the elastic flap gate, the closed air cavity, and the oscillating nappe. Using many input parameters from the model gate, the differential equations governing the vibration of this coupled system were derived and a matrix formulation of the solution was presented. Their set of three differential equations, representing the dynamic equilibrium of the flap, leads to a sixth order characteristic equation. This characteristic equation produced solutions in substantial agreement with the experimental values.

An overview of Partenscky & Swain's (1971) analysis follows.

The differential equations, representing the sum of the forces acting on the gate (ΣF_{HG}), the sum of the forces acting on the nappe (ΣF_{HN}), and the sum of the moments acting on the gate (ΣM_O), respectively, are:

$$\Sigma F_{HG} = m_1 \frac{d^2 x_1}{dt^2} + C_1 \left(\frac{dx_1}{dt} - \frac{dx_2}{dt} \right) + \left(K_1 - K_2 \right) x_1 - K_2 x_2 = 0 \tag{35}$$

$$\Sigma F_{HN} = m_2 \frac{d^2 x_2}{dt^2} + C_2 \left(\frac{dx_2}{dt} - \frac{dx_1}{dt} \right) + K_2 \left(x_2 - x_1 \right) + b_1 \frac{d\theta^*}{dt} - \eta_c b_2 \theta^* = 0 \tag{36}$$

$$\Sigma M_o = J \frac{d^2 \theta^*}{dt^2} + C_1 \frac{d\theta^*}{dt} + K_1 \theta^* - \eta_c a_1 \left(x_2 - x_1 \right) - a_2 \frac{d\theta^*}{dt} = 0 \tag{37}$$

The terms in these equations are defined in the appendix to this chapter (from Partenscky and Swain, 1971).

Partenscky & Swain (1971) assumed the form of the solutions to be

$$\theta^* = \theta_o e^{\lambda t}, x_1 = x_o e^{\lambda t}, x_2 = p_o e^{\lambda t} \tag{38}$$

in which $\lambda = \delta + i\,\omega$ and the values of θ_o, x_o, and p_o represent the maximum angular deflection the flap gate, the maximum translational deflection of the flap gate, and the maximum deflection of the nappe, respectively. With these assumed solutions, Equations 35 to 37 can be reduced to three algebraic equations:

$$\left(J\lambda^2 + C_1\lambda + K_1 - a_2\lambda \right) \theta_o - \eta_c a_1 p_{o+} \eta_c a_1 x_o = 0 \tag{39}$$

$$\left(b_1\lambda - \eta_c b_2 \right) \theta_o + \left(m_2\lambda^2 + C_2\lambda + K_2 \right) p_o - \left(C_2\lambda + K_2 \right) x_o = 0 \tag{40}$$

$$-\left(C_1\lambda + K_2\right)p_o + \left(C_1\lambda + K_1 + K_2 + m_1\lambda^2\right)x_o = 0 \tag{41}$$

Equations 39 to 40 can be written in matrix form and then reduced to a sixth order polynomial that can be factored into the product of three quadratics:

$$\left(\lambda^2 + A\lambda + B\right)\left(\lambda^2 + C\lambda + D\right)\left(\lambda^2 + E\lambda + F\right) = 0 \tag{42}$$

The values of A, B, C, D, E, and F are constants for a specified gate-nappe-air volume system.

The characteristic equation, Equation 42, yielded three negative roots and three positive roots for the particular flap gate considered by Partenscky & Swain (1971). The three negative roots cannot be realized physically. Two of the positive roots were in agreement with the observed system frequency. One of these two roots agreed well with the modulation frequency observed in the model study. The third positive root was left unexplained and its value was not reported.

The modulation observed by Partenscky & Swain may well result from a slight mismatch is the frequency of the flap gate and the nappe oscillation frequency driven by the wavelength criterion for the nappe. The modulation of one stronger signal by another weaker signal will appear as a modulation of the stronger dominant frequency, at a beat frequency given by the difference in frequency of the two signals. Spectral analysis of the signal will also reveal a higher frequency component at the sum of the two frequencies. The sum frequency is often hidden as noise in the composite signal and is not easily detected by visual inspection of the composite signal. Unfortunately, Partenscky & Swain did not provide the value for their third positive root to their characteristic equation. It might well have been the sum of the dominant flap gate frequency and the weaker nappe oscillation frequency.

To be able to predict the conditions under which flap gate vibrations will occur, analytical methods for calculating the hydrodynamic pressure on gates with underflow and gates with overflow are needed. In Chapters 4 and 5, we presented a method for gates with underflow that used potential theory for dissipative wave radiation. The theoretical predictions have been confirmed in laboratory-scale model experiments (see Chapters 7 and 11 for more model-scale results, in addition to those in Chapter 4 and 5), on full-scale long span gates (see Chapter 7), and on full-scale Tainter gate (see Chapters 12 and 14). However, until recently, there has been no confirmation of these analytical predictions on a full-scale gate with overflow. In the following section, we will present a method for calculating large magnitude hydrodynamic pressures acting on an overflow gate when it undergoes self-excited vibration. We will confirm the validity of the pressure estimation method on a full-scale gate and show how a method for predicting the added mass makes it possible to predict the in-water natural frequencies of overflow gates. The information presented in the following section was originally published in Anami et al. (2014).

Shorenji Flap Gate Case Study

The Shorenji Dam is located in Mie Prefecture, Japan, and regulates water flow for irrigation purposes. Anami et al. (2014) carried out field tests on the flap gate illustrated schematically in Figure 33. The gate has a height of 0.963 m, a span of 14.8 m and a mass of 3474 kg. The skinplate is attached rigidly to the torque-tube. The skinplate thickness is 9 mm, and the radius of the torque-tube is 156 mm. The distance from the torque-tube center to the top of the skinplate is 0.963 m. The skinplate is reinforced

Figure 33. Side view of 3474 kg flap gate with 14.8 m span located at the Shorenji Dam in Mie Prefecture, Japan
From Anami et al. (2014). © 2014 IJMERR. Used with permission.

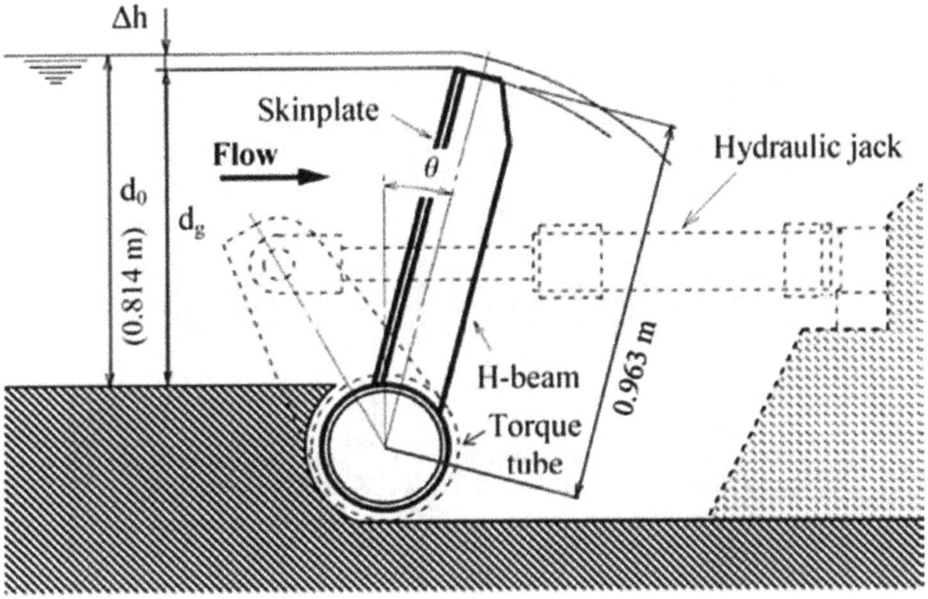

by 12 H-beams, 194 mm in height, as seen in the upstream view in Figure 34. The whole gate can be rotated around the torque-tube by a hydraulic jack (on one side only) shown as a dashed line in Figure 33. The inclination angle of the gate relative to the vertical direction, denoted by θ in Figure 33, can be varied from 15° to 90°.

In-Air Natural Vibration Characteristics

The dam consisted of two such gates, located side-by-side with a center pier. By lowering the one gate completely, the upstream reservoir was emptied and the other gate could be instrumented for in-air modal testing. The first series of tests employed an impact hammer and accelerometers. The gate was struck with an impact hammer on one of the H-beams near the spanwise center, indicated by the ⊗ mark in Figure 34. The acceleration response of the gate was measured by 6 servo-type mechanical accelerometers, which were sequentially positioned at the 36 points indicated by the small black circles on the H-beams in Figure 34. These points are denoted by a letter and a number according to the following scheme: A, B, and C represent locations top to bottom on a beam and 1, 2 ... 12 represent the H-beams left to right across the span. The accelerometer data was then processed using a digital computer to yield the modal frequencies, modal damping ratios and the mode shapes of the in-air natural vibration. Anami et al. (2014) provide the details of the data processing.

Figure 34 also shows the spoilers that were installed along the gate crest to attenuate self-excited vibration. It was not possible to arrange for their removal before the in-air modal testing, which required scaling of the measured in-air natural frequencies for the analysis of the gate without spoilers.

Figure 34. Upstream view of flap gate showing locations of accelerometers and horizontal impact location with a magnified view of the impact location
After Anami et al. (2014).

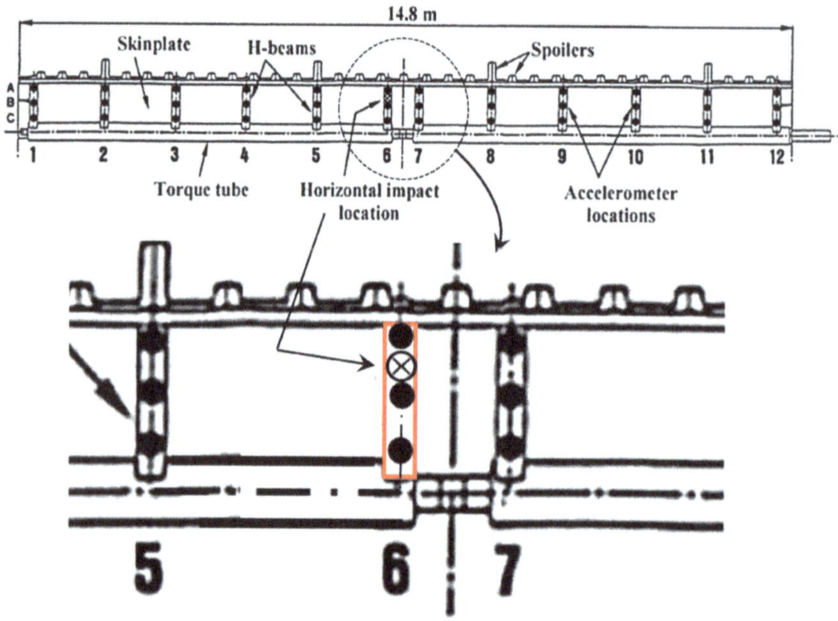

An overview of the frequency, the damping ratio, and the mode shape for these natural vibrations is shown in Table 3. The frequency of the fundamental mode is 30.5 Hz, and its mode shape suggests that it is essentially a uniform vibration across the entire span of the gate. A cross-sectional view of this fundamental mode is shown in Figure 35. This figure shows both the maximum upstream and downstream displacements of the skinplate. The numbers near the top and the letters on the right of the gate correspond to the measurement locations given in Figure 34. From this mode chart, it is evident that the gate is undergoing rotational vibration. The small black circles on the right side of Figure 35 show rough estimates of the centers of rotation. The bottom edge of the skinplate was taken as the zero reference location. The position of the center of rotation ranges from the center of the torque tube to its outer surface.

Table 3. Dynamic characteristics of the first three modes of the gate

Mode Name	Frequency f_a[Hz]	Damping Ratio ζ_a	Mode Shape
M_1	30.5	0.030	
M_2	39.5	0.019	
M_3	42.3	0.006	

From Anami et al. (2014).

Figure 35. Fundamental mode of natural vibrations of the Shorenji flap gate in air showing modal amplitude and height of rotational center at each spanwise measurement location
From Anami et al. (2014). © 2014 IJMERR. Used with permission.

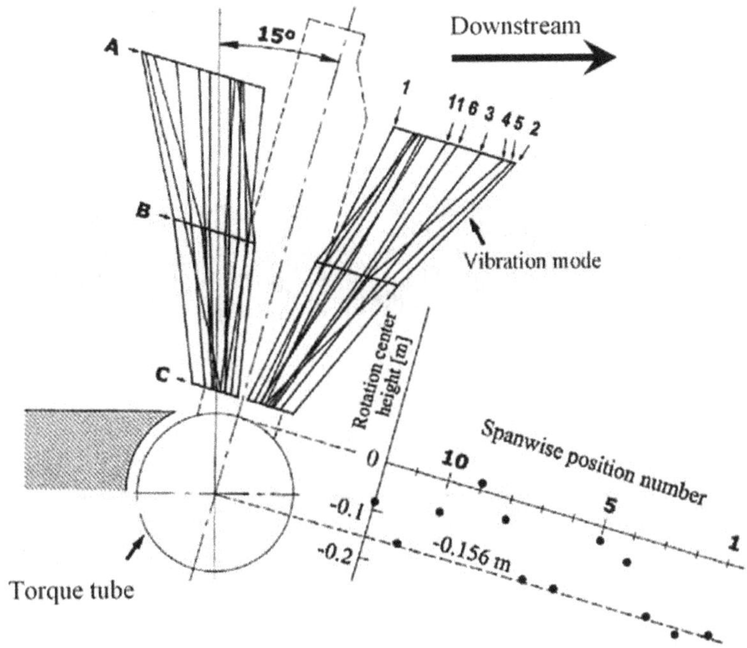

The second series of tests was designed to measure the in-water self-excited vibration of the gate. For these tests, the gate crest spoilers were removed. The upstream reservoir was restored, with upstream water depth maintained at 0.814 m. The bottom of the skinplate was aligned with the upstream riverbed, as shown in Figure 33, and water flowed over the instrumented test gate. Tests were carried out predominantly for a gate inclination angle of 17.2°, as well as for angles from 15° to about 33°. The accelerometers produced real-time records of the growth of the self-excited vibrations. These real-time waveforms could then be processed to yield the frequency of vibration and the excitation ratio (the negative of the damping ratio).

An acceleration waveform of the self-excited vibration with the gate at an inclination angle of 17.2° and an overflow depth of 16 mm (measured at Point 1A) is shown in Figure 36. The vibration grew rapidly and reached steady state after about 1 second. The excitation ratio ξ was 0.0072, and the frequency f_w was 22 Hz.

For gate inclination angles from 15° to about 33°, the excitation ratios and frequencies were measured at two locations. In Figure 37, the open circles represent the data from Point 1A, while the filled circles are data from Point 12A. Although there is scatter in the data, the excitation ratio gradually decreases as the gate inclination angle increases, most likely because of the increase in the overflow and the decrease in the volume of the air cavity formed on the downstream side of the gate. On the other hand, the frequency is locked onto a value of about 21.9 Hz (estimated initial Mode 11, jumping to Mode 10) for inclination angles less than about 22.2°. The frequency suddenly decreases to around 19.5 Hz (estimated Mode 9) once the inclination angle exceeds about 22.2°. With a continued increase in the inclination angle, the frequency gradually increases. The constant frequency at small inclination angles suggests a feedback mechanism most likely related to the oscillations of the nappe.

Figure 36. Waveform of streamwise in-water acceleration for a gate inclination angle of 17.2° and an overflow depth of 16 mm as measured at point 1A
From Anami et al. (2014). © 2014 IJMERR. Used with permission.

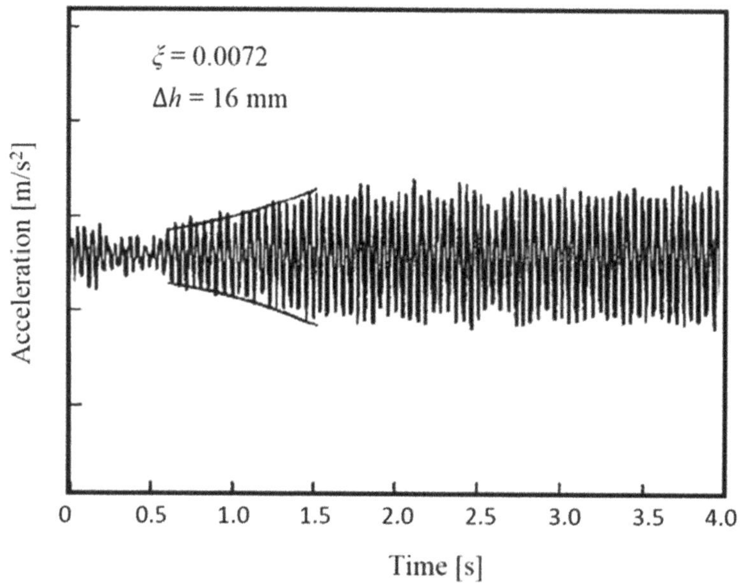

Figure 37. Variation of in-water self-excited vibration as a function of gate inclination angle. Vibration is characterized by (a) excitation ratio, and (b) frequency. Open circles are for Point 1A; filled circles are for Point 12A
From Anami et al. (2014). © 2014 IJMERR. Used with permission.

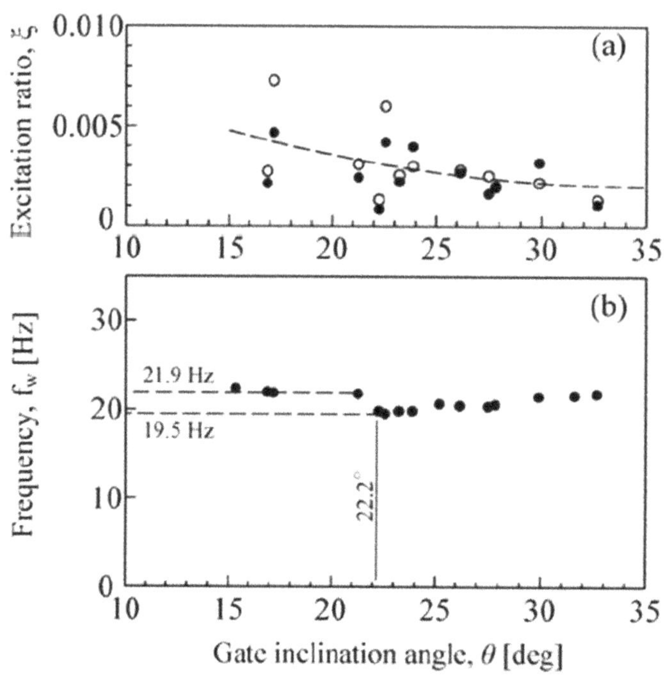

Recasting the frequency and amplitude data from the flap gate vibration in terms of Treiber's *b* value yields the characteristic behavior shown in Figure 38. Because the gate is supported by the hydraulic jack on one side only, the gate inclination angles on the left and right sides differ slightly, as does the fall height, *H*, and hence the fall time, t_H. Because of these differences, the Treiber *b* values on the left and right sides also differ slightly. The *b* values are higher than those from Leo (1954) because of the relatively smaller fall height. Note that despite the slight difference in inclination, fall height and fall times, the frequencies measured on the left and right sides are very consistent and essentially the same, within the experimental error. The flap gate vibration amplitudes are larger on the unsupported left side, than on the right side which is constrained by the hydraulic jack.

Since the lowest frequency natural vibration mode in air usually corresponds to the self-excited in-water mode, an analysis of the hydrodynamic pressure for this mode will be presented in the next section. Once the hydrodynamic pressure is known, the ratio of the frequency of self-excited in-water vibration to the frequency of the in-air natural vibration can be calculated. We will show that the theoretical calculations are in good agreement with the experimental data.

Hydrodynamic Pressure

Following a procedure similar to that developed in Chapters 4 and 5 for gates with underflow, the flow field for overflow gates will be simplified to permit calculation of the hydrodynamic pressure. We will consider the case in which the skinplate stands vertically and undergoes rotational vibration with a small amplitude, as shown on the left side of Figure 39. The dimensional coordinate *Y* corresponds to the vertical skinplate surface and is defined as positive upward from the mean water surface, which corresponds to the dimensional coordinate *X*. The water depth of the upstream reservoir is represented by d_0.

Figure 38. Measured characteristics of flap gate vibration on 0.963 m flap gate with maximum fall height of 1.275 m and overflow depths between 0.014 and 0.087 m at inclination angles from vertical ranging from 15 to 33° (in the downstream direction)

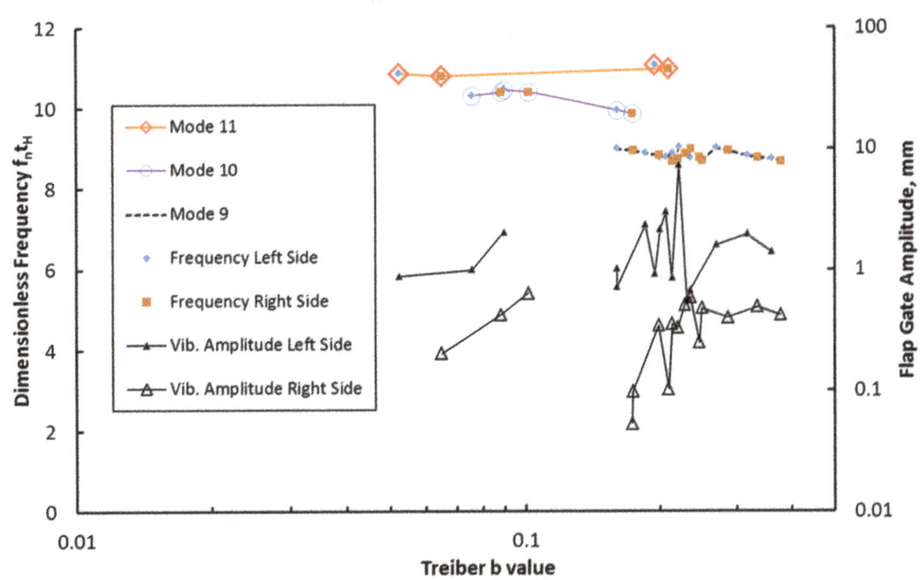

Figure 39. Vertical weir-plate model on the left, and calculated values of vibration-induced hydrodynamic pressure on the right
From Anami et al. (2012a), © 2012 Elsevier. Used with permission.

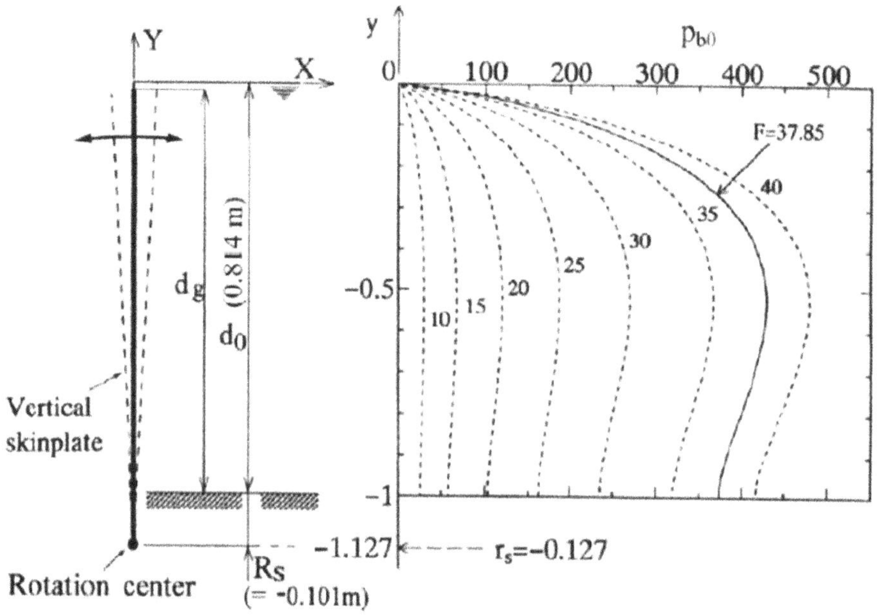

The top of the vertical skinplate above the riverbed is given as d_g; the height of the rotation center above the riverbed is given as R_s. As shown in Figure 39, the rotation center is lower than the upstream riverbed so that R_s takes on a negative value. The pressure change near the water surface is comparatively small. Therefore, the effect of the overflow water on the induced hydrodynamic pressure at larger depths is assumed to be insignificant. Instead, the overflow water depth was disregarded and it was replaced by a rigid wall vibrating with the vertical skinplate in the analytical model. The hydrodynamic pressure induced by the vertical skinplate vibration can be calculated by applying the potential-flow theory developed by Rayleigh (1945) and Lamb (1932) for dissipative wave-radiation problems. See Chapters 4 and 5 and the references cited therein for the details of the theory.

The vertical axis y in Figure 39 is the non-dimensional coordinate along the vertical skinplate:

$$y \equiv Y / d_0 \tag{43}$$

The value $y = -1$ corresponds, therefore, to the upstream riverbed. The abscissa is the non-dimensional pressure amplitude, defined by the following expression:

$$p_{b0} \equiv \frac{P_{b0} / \rho g}{\left(d_0 - R_s\right)\Psi_0} \tag{44}$$

where Ψ_0 is the amplitude of the rotation angle of the vertical skinplate. Thus, p_{b0} represents the ratio of the hydrodynamic pressure head corresponding to P_{b0} to the vibration amplitude at the top of the vertical skinplate.

An expression for hydrodynamic pressure induced by the rotational vibration of the vertical skinplate, P_b, as defined by Anami et al. (2012a), can be reduced to the non-dimensional pressure p_b acting on the vertical skinplate at $x = 0$. This non-dimensional pressure p_b is given by the following superposition of the acceleration and velocity terms:

$$p_b = p_{bs}\left(\frac{F_0}{F}\right)^2 \psi'' + p_{bp}\left(\frac{F_0}{F}\right)\psi' \tag{45}$$

where ψ represents the non-dimensional rotation angle and the prime represents the derivative relative to the non-dimensional time, t, given as follows:

$$\psi \equiv \frac{\Psi}{\Psi_0} \text{ and } t \equiv \frac{T}{1/\Omega_a} \tag{46}$$

The Froude number F and the basic Froude number F_0 are defined by

$$F \equiv \sqrt{\frac{d_0}{g}}\Omega_w \text{ and } F_0 \equiv \sqrt{\frac{d_0}{g}}\Omega_a \tag{47}$$

The symbols Ω_w and Ω_a are the in-water and in-air frequencies, respectively. Therefore, the ratio of the two Froude numbers, F/F_0, represents the in-water to in-air frequency ratio:

$$\frac{F}{F_0} = \frac{\Omega_w}{\Omega_a} \text{ or } \frac{F}{F_0} = \frac{f_w}{f_a} \tag{48}$$

The variables p_{bs} and p_{bp} in Equation 45 are the non-dimensional pressure amplitudes of the so-called "standing and progressive pressure-wave components," respectively, which are given in the following series summations:

$$p_{bs}(x,y) = -\frac{2F^4}{1-r_s}\sum_{j=1}^{\infty}\left[\frac{1}{\beta^*\eta_j^2}\left(r_s - \frac{1}{2}\right) + \frac{2}{\pi}\sum_{n=1}^{\infty}\frac{1}{n}\frac{1}{\eta_j^2 - \left(n\pi/\beta^*\right)^2} \times \left\{r_s \sin\frac{n\pi}{\beta^*} + \frac{\beta^*}{n\pi}\left(\cos\frac{n\pi}{\beta^*} - 1\right)\right\}\right] \times \frac{\cos\eta_j\left(y+\beta^*\right)}{\cos\eta_j\beta^*}\frac{1}{W'\left(i\eta_j\right)/i} \tag{49}$$

$$p_{bp}(x,y) = \frac{2F^4}{1-r_s}\left[\frac{1}{\beta^*\xi_0^2}\left(r_s - \frac{1}{2}\right) + \frac{2}{\pi}\sum_{n=1}^{\infty}\frac{1}{n}\frac{1}{\xi_0^2 + \left(n\pi / \beta^*\right)^2} \times \left\{r_s \sin\frac{n\pi}{\beta^*} + \frac{\beta^*}{n\pi}\left(\cos\frac{n\pi}{\beta^*} - 1\right)\right\}\right] \times \frac{\cos\xi_0\left(y+\beta^*\right)}{\cos\xi_0\beta^*}\frac{1}{W'\left(\xi_0\right)} \tag{50}$$

The symbol r_s represents the reduced rotation center height, defined by

$$r_s \equiv \frac{R_S}{d_0} \tag{51}$$

which takes on a negative value in the present case. The value was assumed to be the average of the experimental values shown in Figure 35 which was 0.101 m, giving a non-dimensional average height of the rotation center of r_s = -0.127 which is also indicated in Figure 35.

The variables ξ_0 and η_j (j=1,2,3…) in Equations 49 and 50 are the solutions of the following two transcendental equations:

$$\xi_0 \tanh\left(\xi_0\beta^*\right) = F^2 \text{ and } \eta_j \tan\left(\eta_j\beta^*\right) = -F^2 \text{ with j=1,2,3,}\cdots \tag{52}$$

In addition, the functions $W'\left(i\eta_j\right)/i$ and $W'(\xi_0)$ included in Equations 49 and 50 are given, respectively, by the following two expressions:

$$W'\left(i\eta_j\right)/i = \tan\eta_j\beta^* + \frac{\eta_j\beta^*}{\cos^2\eta_j\beta^*} \text{ and } W'\left(\xi_0\right) = \tanh\xi_0\beta^* + \frac{\xi_0\beta^*}{\cosh^2\xi_0\beta^*} \tag{53}$$

where β^* represents the ratio of the gate submergence depth d_0 to the upstream reservoir depth, which has been assumed to take on a value of 1.0 for the present case.

The resultant vector magnitude of hydrodynamic pressure $\left(p_{b0} \equiv \sqrt{p_{bp}^2 + p_{bs}^2}\right)$ of the standing and progressive pressure components can be calculated. The calculated amplitudes of the hydrodynamic pressure, p_{b0}, are shown on the right side of Figure 39 in terms of the Froude number, introduced in Equation 47. The Froude number represents the dynamic similarity of flow field and can also be written as follows:

$$F \equiv 2\pi f_w \times \sqrt{\frac{d_0}{g}} \tag{54}$$

Substituting the measured frequency of the self-excited vibrations, f_w, of about 20.9 Hz, as shown in Figure 37b, and the water depth, d_0, of 0.814 m, yields a Froude number of 37.9 for these tests. The non-dimensional pressure amplitude distribution for this Froude number is shown by the solid line in Figure 39. As previously mentioned, in these calculations the skinplate rotation center height R_s was assumed to take on the mean experimental value of –0.101 m, with a corresponding value of r_s = –0.127 (rounded to –0.13 in subsequent calculations).

The hydrodynamic pressure attains its maximum value of about 430 (corresponding to 0.43 m of water) near the center of the vertical skinplate when the top of the vertical skinplate vibrates with an amplitude of merely 1 mm.

Knowing pressure amplitude and phase relative to the gate motion permits the assessment of the equivalent added mass, knowing added mass permits the calculation of the in-water vibration frequency. Details follow in the next section. Additional details can be found Chapter 9 and in Anami et al. (2012b)

Added Mass and Frequency Ratio

The water-added moment of inertia, ΔI, can be calculated by integrating the hydrodynamic pressure moment around the rotation center over the vertical skinplate. As a result, one may introduce the non-dimensional added mass Δm_ψ, defined by

$$\Delta m_{\psi} \equiv \frac{\Delta I / \left(d_0 - R_S\right)^2}{\rho d_0^2 W_0} \tag{55}$$

The added moment of inertia was divided by the square of the rotation radius (d_0-R_S), to convert it into the equivalent mass. W_0 is the gate span of 14.8 m. The denominator is the representative water mass contained in the volume of $d_0^2 W_0$. The calculated results for the non-dimensional added mass as a function of the inclination angle are shown in Figure 40a, where the solid line is for the non-dimensional rotation center mean height of –0.13. In addition, the broken lines are for the r_s values of 0.0 and –0.19, which were the maximum and minimum values of the non-dimensional rotation center height r_s from the experimental data. An increase in the inclination angle is simulated by decreasing the height of the vertical skinplate top from the riverbed, d_g. As a result of this slight decrease in d_g, the added moment of inertia, ΔI, decreases slightly, with a corresponding slight decrease in the non-dimensional added mass.

The in-water frequency of the flap gate, f_w, is substantially reduced relative to its natural frequency in the air, f_{a1}, due to the added mass of the water. The ratio of these two frequencies can be calculated from the following expression (Anami & Ishii, 1998):

$$\frac{f_w}{f_{a1}} \equiv \frac{1}{\sqrt{1 + \alpha_{\psi} \Delta m_{\psi}}} \tag{56}$$

where α_ψ is the water to gate mass ratio, defined by

$$\alpha_{\psi} \equiv \frac{\rho d_0^2 W_0}{I_{\psi} / \left(d_0 - R_S\right)^2} \tag{57}$$

The numerator in the above expression is the representative water mass and the denominator is the equivalent vertical skinplate mass which can be calculated by dividing the moment of inertia of the vertical skinplate, I_{ψ}, by the square of the rotation radius, $(d_0 - R_s)^2$. The moment of inertia, I_{ψ}, for the Shorenji flap gate was 545.8 kg·m², resulting in a mass ratio of 15.1. The solid line (for $r_s = -0.13$) and broken lines (for $r_s = 0$ and -0.19) in Figure 40b show the calculated results for the frequency ratio. As the inclination angle increases, the frequency ratio shows a slight increasing tendency, caused by a slight decrease in the non-dimensional added mass.

Unfortunately, the spoilers shown in Figure 34 were still attached when the in-air vibration tests were made. Afterwards, they were removed for the in-water self-excited vibration tests. The measured in-air natural frequency of 30.5 Hz for the fundamental mode, shown in Table 3, was scaled to 34.1 Hz for the flap gate without spoilers. Dividing the measured frequency data shown in Figure 37b by the corrected

Figure 40. (a) The calculated added mass for the Shorenji flap gate. The solid line is for the non-dimensional rotation center mean height of –0.13. The broken lines are for the maximum and minimum values of r_s from the experimental data. (b) Calculated in-water to in-air frequency ratio for the Shorenji flap gate. The solid and broken lines correspond to the r_s values in (a). The black filled data circles are the experimental data.

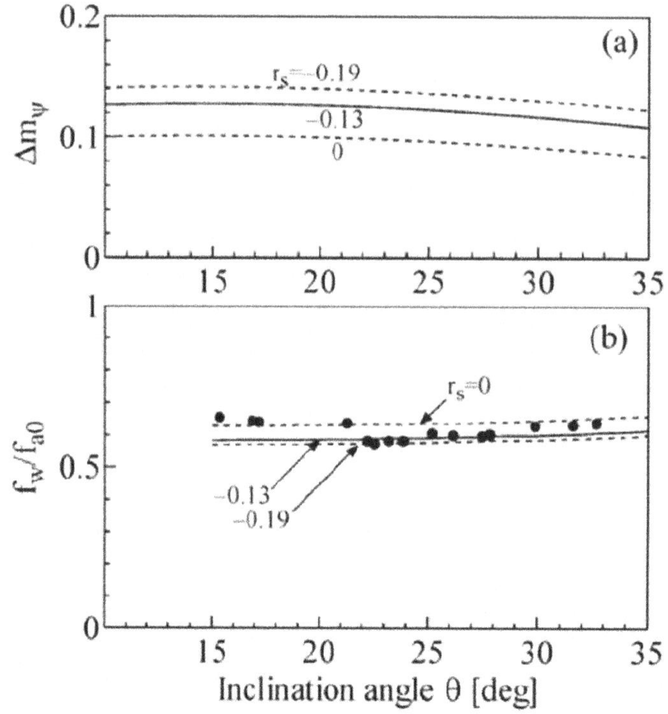

in-air natural frequency, the measured frequency ratio data are obtained, as shown in Figure 40b. The theoretical calculations accurately reflect the measured data.

Conclusions From the Shorenji Flap Gate Study

The analytical method for calculating the hydrodynamic pressure exerted on vibrating gates, based on the potential flow theory for the dissipative wave radiation problems, has been previously validated through laboratory-scale model gate testing on various gate geometries. In this case study, the analytical method was indirectly validated for a full-scale gate by using the measured frequency data for the self-excited vibrations of the flap gate at the Shorenji Dam in Japan. Thus, we suggest that the same analytical method may be used to predict the in-water vibration frequency and the induced hydrodynamic force for flap gates as well as for other types of gates, such as Tainter gates, long-span gates, and drum gates with overflow in full-scale applications.

CONCLUSION

More than half a century ago, Schwartz (1966) concluded that the nappe oscillation phenomena on overflow gates were so complex that none of the many proposed theories could adequately address all of the observations. Today we know more about nappe behavior and how to attenuate nappe oscillations through a host of countermeasures, from roughening the crest to installing splitters, suspended flaps and other devices to change the dynamics of the nappe.

Our knowledge still does not permit us to predict onset conditions for nappe oscillations, their initial frequency, or subsequent frequency behaviors with changing nappe parameters. The study of Partenscky & Swain (1971) presents a method of computing the dynamic behavior of a flap gate coupled to nappe oscillations. In addition, we have developed a method for estimating the effect of the water added mass on the dynamics of gates with overflow, permitting the prediction of the in-water natural frequencies of overflow gates.

REFERENCES

Anami, K., & Ishii, N. (1998). In-water streamwise vibration of Folsom Dam radial gates in California. In *Proceedings of 1998 ASME/JSME Joint Pressure Vessels and Piping Conference, PVP-Vol. 363,* (pp. 87-93). New York, NY: American Society of Mechanical Engineers.

Anami, K., Ishii, N., & Knisely, C. W. (2012a). Pressure induced by vertical planar and inclined curved weir-plates undergoing streamwise rotational vibration. *Journal of Fluids and Structures, 29*, 35–49. doi:10.1016/j.jfluidstructs.2011.11.007

Anami, K., Ishii, N., & Knisely, C. W. (2012b). Added mass and wave radiation damping for flow-induced rotational vibrations of skinplates of hydraulic gates. *Journal of Fluids and Structures, 35*, 213–228. doi:10.1016/j.jfluidstructs.2012.07.008

Anami, K., Ishii, N., Knisely, C. W., & Matsumoto, Y. (2014). Field test validation of analytical model for vibration characteristics of a flap gate undergoing self-excited vibration. *International Journal of Mechanical Engineering and Robotics Research*, *3*(4), 1.

Anderson, A. A. (2014). *Causes and Countermeasures for Nappe Oscillation: An Experimental Approach* (M.S. Thesis). Department of Civil and Environmental Engineering, Utah State University, Logan, UT.

Binnie, A. M. (1972). The stability of a falling sheet of water. *Proceedings of the Royal Society of London. Series A, Mathematical and Physical Sciences*, *326*(1565), 149–163. doi:10.1098/rspa.1972.0002

Binnie, A. M. (1974). Resonating waterfalls. *Proceedings of the Royal Society of London. Series A, Mathematical and Physical Sciences*, *339*(1619), 435–449. doi:10.1098/rspa.1974.0132

Casperson, L. W. (1991). Fluttering fountains. *Journal of Sound and Vibration*, *162*(2), 251–262. doi:10.1006/jsvi.1993.1117

Casperson, L. W. (1994). Fluttering fountains: Stability criteria. *Journal of Applied Physics*, *75*(10), 4892–4894. doi:10.1063/1.355775

Casperson, L. W. (1995). Fluttering fountains: Annular geometry. *Journal of Applied Physics*, *79*(3), 1275–1278. doi:10.1063/1.361022

Chanson, H. (1996). *Some hydraulic aspects during overflow above inflatable flexible membrane dam*. Report CH47/96, 3-3-3-9. Department of Civil Engineering, University of Queensland, Australia.

Chanson, H. (1997). *Hydraulics of open channel flow*. Kent, UK: Butterworth-Heinemann.

Chanson, H. (2002). *Hydraulics of stepped chutes and spillways*. Boca Raton, FL: CRC Press.

Cremer, L., & Ising, H. (1972). Selbsterregte Schwingungen von Überfallstrahlen. *Acta Acustica united with Acustica*, *27*(2), 94–107.

Crookston, B. M., Anderson, A., Shearin-Feimster, L., & Tullis, B. P. (2014). *Mitigation investigation of flow-induced vibrations at a rehabilitated spillway*. Paper presented at the 5th International Symposium on Hydraulic Structures. Hydraulic Structures and Society: Engineering Challenges and Extremes. Brisbane, Australia. doi:10.14264/uql.2014.30

Crookston, B. M., & Tullis, B. P. (2012). Hydraulic design and analysis of labyrinth weirs. II: Nappe aeration, instability, and vibration. *Journal of Irrigation and Drainage Engineering*, *139*(5), 371–377. doi:10.1061/(ASCE)IR.1943-4774.0000553

De Luca, L., & Costa, M. (1997). Instability of a spatially developing liquid sheet. *Journal of Fluid Mechanics*, *331*, 127–144. doi:10.1017/S0022112096003916

De Rosa, F., Girfoglio, M., & de Luca, L. (2014). Global dynamics analysis of nappe oscillation. *Physics of Fluids*, *26*(12), 122109. doi:10.1063/1.4904752

Erbisti, P. C. F. (2004). *Design of Hydraulic Gates*. Rotterdam: A.A. Balkema.

Falvey, H. T. (1980). Bureau of Reclamation experience with flow-induced vibrations. In E. Naudascher & D. Rockwell (Eds.), *Practical Experiences with Flow-Induced Vibrations, IUTAM/IAHR Symposium*, (pp. 386-398). Heidelberg, Germany: Springer. doi:10.1007/978-3-642-81528-7_42

Falvey, H. T. (2002). Nappe oscillation. In H. T. Falvey (Ed.), *Hydraulic Design of Labyrinth Weirs* (pp. 69–76). Reston, VA: ASCE Press. doi:10.1061/9780784406311.ch07

Fischer, H. (1938). Die Schwingungen an überstromten Wehen und ihre Beseitigung. *Zeitschrift des Vereins deutscher Ingenieure, 82*(1), 27 - 28.

Fuhrmann, O. (1934). *Schwingungsuntersuchungen an Überströmten Beweglichen Wehren* (Vol. 16). Berlin, Germany: Preußische Druckerei.

Gebhardt, M., Nestmann, F., Schweizerhof, K., & Kemnitz, B. (2010). On the causes of vibrations and the effects of countermeasures at water-filled inflatable dams. *Proceeding of 1st European IAHR Congress.*

Girfoglio, M., De Rosa, F., Coppola, G., & de Luca, L. (2013). Global eigenmodes of free-interface vertical liquid sheet flows. *Computational Methods in Multiphase Flow VII, 79*, 285–295. doi:10.2495/MPF130241

Glover, R. E., Thomas, C. W., & Hammett, T. F. (1939). *Report on vibration studies made at Black Canyon Dam, Boise project.* USBR HYD-58. Accessed 12 May 2015 at http://www.usbr.gov/pmts/hydraulics_lab/pubs/HYD/HYD-58.pdf

Hagerty, W. W., & Shea, J. F. (1955). A study of the stability of plane fluid sheets. *Journal of Applied Mechanics, 22*(3), 509–514.

Kemp, A. R., & Pullen, R. A. (1961). *Report on the Vibrating Nappe Phenomenon.* Johannesburg, South Africa: Department of Civil Engineering, University of the Witwatersrand.

Knisely, C. W. (1989). On the acoustics of nappe oscillation. In Č. Maksimović & M. Radojković (Eds.), *Proceedings of Hydrocomp 89* (pp. 510–519). Rotterdam: Elsevier.

Knisely, C. W. (1994). Nappe oscillation countermeasure. In *Contributed Papers in Fluids Engineering 1994: Proceedings of 1994 ASME Fluids Engineering Division Summer Meeting* (pp. 17-24). Lake Tahoe, NV: American Society of Mechanical Engineers.

Kolkman, P. A. (1974). Instability of a vertical water-curtain closing an air chamber. In E. Naudascher, (Ed.), *Flow-Induced Structural Vibrations, IUTAM/IAHR Symposium*, (pp 17-33). Heidelberg, Germany: Springer. doi:10.1007/978-3-642-48279-3_2

Kolkman, P. A. (1984). Gate Vibrations. In P. Novak (Ed.), *Developments in Hydraulic Engineering 2* (pp. 55–112). London, UK: Elsevier Applied Science.

Kyotoh, H. (2002). Incipient oscillations of a sheet of falling water and the instability mechanisms. *Journal of Hydroscience and Hydraulic Engineering, 20*(1), 77–93.

Leo, B. T. (1954). Selbstgesteuerte Schwingungen an überströmten Wehren. Veröffentlichungen der Forschungsanstalt für Schiffahrt, Wasser und Grundbau, No. 3. Berlin: Akademie.

Lewin, J. (1995). *Hydraulic Gates and Valves in Free Surface Flow and Submerged Outlets*. London, UK: Thomas Telford.

Metropolitan Water, Sewage and Drainage Board. (1980). Investigation into spillway discharge noise at Avon Dam. *ANCOLD Bulletin*, *57*, 31–36.

Müller, O. (1937). *Das bei Überfall schwingende Wehr als selbsterregtes, gekoppeltes System unter Berucksichtigung gewisser Analogien zum Röhrensender und zur Zungenpfeife. Mitteilung No.33*. Berlin: Preußische Versuchsanstalt für Wasserbau und Schiffsbau.

Ogihara, K., & Ueda, S. (1979). C11 Flap gate oscillation. In E. Naudascher & D. Rockwell (Eds.), *Practical Experiences with Flow-Induced Vibrations, IUTAM/IAHR Symposium*, (pp. 466-470). Heidelberg, Germany: Springer.

Pariset, E. (1955). Etude sur la vibration des lames déversantes. In *Proceedings of the 6th IAHR Congress*, (Vol. 3, pp. 1-15). The Hague,Netherlands

Partenscky, H. W. (1964). *Forces dynamiques agissant sur les portes des écluses lors de la dérivation partielle des crues* (PhD Dissertation). Université de Toulouse, Toulouse, France.

Partenscky, H. W., & Khloeung, S. I. (1967). *Oscillations de lames déversantes non-aerees*. Paper presented at the 12th IAHR Congress, Seminar paper S6, Ft. Collins, CO.

Partenscky, H. W., & Swain, A. (1971). *Theoretical study on flap gate oscillation*. Paper presented 14th IAHR Congress, Paper B25, Paris, France.

Paxson, G., & Savage, B. (2006). Labyrinth spillways: comparison of two popular USA design methods and consideration of non-standard approach conditions and geometries. *Proceedings of the. International Junior Researcher and Engineer Workshop on Hydraulic Structures,* 37-46.

Peters, K. (1940). *Flatterschwingungen von Überfallstrahlen und Fahnen. Mitteilungen des Hydraulischen Instituts der Technischen Hochschule München* (Vol. 10). Munich, Germany: R. Oldenbourg.

Petrikat, K. (1958). Vibration tests on weirs and bottom gates. Water Power, 10, 52-57, 99-104, 147-148, 190-197.

Půlpitel, L. (1979). Some experiences with curing flap gate vibration. In E. Naudascher & D. Rockwell (Eds.), *Practical Experiences with Flow-Induced Vibrations, IUTAM/IAHR Symposium*, (pp. 471-472). Heidelberg, Germany: Springer.

Rayleigh, L. (1878). On the instability of jets. *Proceedings of the London Mathematical Society*, *10*(1), 4–13. doi:10.1112/plms/s1-10.1.4

Sato, Y., Miura, S., Morii, S., & Ohkubo, S. (2007). Behavior of a falling water sheet. *Journal of Environmental Engineering*, *2*(2), 394–406. doi:10.1299/jee.2.394

Schmid, P. J., & Henningson, D. S. (2002). On the stability of a falling liquid curtain. *Journal of Fluid Mechanics*, *463*, 163–171. doi:10.1017/S002211200200873X

Schwartz, H. I. (1964). Nappe oscillation. *Journal of the Hydraulics Division*, *90*(6), 129–143.

Schwartz, H. I. (1966a). *Some Aspects of Nappe Oscillations* (PhD Dissertation). University of the Witwatersrand, Johannesburg, South Africa.

Schwartz, H. I. (1966b). Edgetones and nappe oscillation. *The Journal of the Acoustical Society of America, 39*(3), 579–582. doi:10.1121/1.1909930

Schwartz, H. I., & Nutt, L. P. (1963). Projected nappes subject to transverse pressure. *Journal of the Hydraulics Division, 89*(4), 97–104.

Seifert, R. (1940). Über Schwingungen von Wehren. *Zeitschrift des Vereins deutscher Ingenieure, 84*(7), 105-110.

Simmons, W. P. (1964). *Experiences of the Bureau of Reclamation with Flow-induced Vibrations*. USBR Report HYD-538. Retrieved May 12, 2015 from http://www.usbr.gov/pmts/hydraulics_lab/pubs/HYD/HYD-538.pdf

Sumi, T. (1990). Three-dimensional characteristic of nappe oscillation and the estimation of sound pressure levels. *Journal of Hydroscience and Hydraulic Engineering, 8*(1), 89–99.

Treiber, B. (1974). Theoretical study of nappe oscillation. In E. Naudascher (Ed.), *Flow-Induced Structural Vibrations, IUTAM/IAHR Symposium*, (pp. 34-46). Heidelberg, Germany: Springer.

Tullis, B., Young, J., & Chandler, M. (2007). Head-discharge relationships for submerged labyrinth weirs. *Journal of Hydraulic Engineering, 133*(3), 248–254. doi:10.1061/(ASCE)0733-9429(2007)133:3(248)

USGS. (2013). *Measurement and Computation of Streamflow: Volume 1. Measurement of Stage and Discharge; Volume 2. Computation of Discharge*. Water Supply Paper 2175, U.S. Geological Survey. Retrieved May 12, 2015 from http://pubs.usgs.gov/wsp/wsp2175/

Vrijer, A. (1979). Stability of vertically movable gates. In E. Naudascher & D. Rockwell (Eds.), *Practical Experiences with Flow-Induced Vibrations, IUTAM/IAHR Symposium*, (pp. 428-434). Heidelberg, Germany: Springer.

APPENDIX

Definition of terms in Partenscky & Swain (1971) equations of motion for the flap gate.

$J \cdot \frac{d^2\theta *}{dt^2}$ = Inertia moment of the flap and the mass of water participating in the oscillation;

$C_1 \cdot \frac{d\theta *}{dt}$ = Damping moment due to frictional effects;

$K_1\theta *$ = Restoring moment due to the structural elasticity of the gate system;

$\eta_c \cdot a_1 \cdot (x_1 - x_2)$ = Disturbing moment caused by the pressure variation in the enclosed air volume due to the relative displacement of the nappe with respect to the gate;

$a_2 \cdot \frac{d\theta *}{dt}$ = Vibration inducing moment caused by the pressure variation in the enclosed air volume due to the oscillation of the nappe

$m_1 \cdot \frac{d^2x_1}{dt^2}$ = Inertia force of the moveable flap gate;

$C_1 \cdot \left(\frac{dx_1}{dt} - \frac{dx_2}{dt}\right)$ = Damping force due to the relative motion of the flap gate and the nappe;

$(K_1 + K_2)x_1$ = Horizontal component of the total elastic force caused by the motion of the gate;

$K_2 \cdot x_2$ = Horizontal component of the force caused by the variation of the air pressure due to the motion of the nappe;

$m_2 \cdot \frac{d^2x_2}{dt^2}$ = Inertia force of the nappe;

$C_2 \cdot \left(\frac{dx_2}{dt} - \frac{dx_1}{dt}\right)$ = Damping force due to the relative motion of the flap gate and the nappe;

$K_2 \cdot (x_2 - x_1)$ = Restoring force caused by the elastic compression of the air volume due to the relative motion of the nappe with respect to the gate;

$b_1 \cdot \frac{d\theta *}{dt}$ = Vibration inducing force caused by the pressure variation in the enclosed air volume due to the oscillation of the gate;

$\eta_c \cdot b_2 \cdot \theta *$ = Exciting force caused by the pressure variation of the enclosed air volume due to the motion of the gate; in which: J = the moment of inertia of the flap gate system and the water mass participating in the oscillation about the axis of rotation;

$C_1 = \frac{\delta_1 \omega_1 J_1}{\pi}$ = Overall damping coefficient of the gate system; wherein δ_1 is the damping coefficient of the flap gate in air obtained from transient vibration tests; ω_1 is the natural angular frequency of the gate, and $J_1 = K_1/a_1$; $\theta * = \theta_o e^{\lambda t}$, where θ_o is the maximum angular deflection of the gate; $\lambda = \delta + i\omega$ in which δ is the total damping of the gate-air-nappe system and ω is the corresponding frequency of oscillation of the system;

$K_1 = \omega_1^2 m_1$ = Spring constant of the gate system (obtained from model tests); m_1 = mass of the gate system;

η_c = Coupling factor (assumed to be equal to 1.0);

$a_1 = \dfrac{L^2 B^2 P_m n_p H}{2V_m}$ = Disturbing force caused by the pressure variation in the air volume due to the relative displacement of the nappe with respect to the gate; where P_m is the average pressure below gate; $n_p \approx 1.1$= exponent for polytropic compression; V_m is the average volume of air enclosed;

η_c = Coupling factor (assumed to be equal to 1.0);

$a_2 = \dfrac{P_m L^2 B}{2\omega_1 \theta_o}$ = Factor which considers the vibration inducing moment caused by the undulation of the nappe;

$K_2 = \dfrac{B^2 H^2 P_m n_p}{V_m}$ = Elastic coefficient of the air enclosed beneath the nappe;

$C_2 = \dfrac{m_2 \delta_2 f_n}{\pi}$ = Damping coefficient of the nappe; where m_2 = mass of the nappe; δ_2 =damping of the nappe;

$b_1 = \dfrac{P_m BH}{\omega_1 \theta_o}$ = Factor which considers the vibration inducing force due to the oscillation of the gate;

$b_2 = \dfrac{L^2 B^2 P_m n_p H}{2V_m} = a_1$ = Disturbing force acting on the nappe due to the oscillation of the gate;

f_n = Frequency of oscillation of the enclosed air beneath the nappe.

Chapter 7
Flow-Induced Vibration of Long-Span Gates:
One and Two Degrees-of-Freedom

ABSTRACT

In this chapter the theoretical equations for fluctuating pressures due to vertical and streamwise gate motions developed in Chapters 4 and 5 are used to derive equations of motion for long-span gates with underflow, overflow and simultaneous over- and underflow. Theoretical development of analysis methods is supported by laboratory and full-scale measurements. Specifically, this chapter considers long-span gate instabilities including one degree-of-freedom vibration of gates with underflow and free discharge, one degree-of-freedom vibration of a gate with submerged discharge and vortex shedding excitation, a two degree-of-freedom vibration of long-span gates with only underflow, and two degrees-of-freedom vibration of long-span gates with simultaneous over and underflow. A method is developed to predict pressure loading on the crest of the gate with overflow.

INTRODUCTION

Gates that dam wide upstream reservoirs or wide rivers typically have spans that are much larger than the gate heights, making then relatively elastic in the streamwise direction. They readily undergo elastic streamwise bending. Long-span gates are commonly installed with the weir plate inclined toward the downstream side such that the larger upstream hydrostatic pressure will create a water tight seal as it presses down on the gate in its closed position. While the inclination of long-span gates in the downstream direction may, depending on the gate mounting system, assure a water tight seal, it also makes these gates press-shut devices. As discussed in Chapter 2, press-shut devices, while statically stable, are dynamically unstable. The dynamic instability of a downstream inclined long-span gate results from its streamwise bending displacement.

DOI: 10.4018/978-1-5225-3079-4.ch007

A conventional long-span gate with a weight of 49 tons, constructed on the Ara River, Japan, is shown in Figure 1. The weir plate is 28 m long and about 3 m high. It is inclined at 30° (relative to vertical) in the downstream direction. The hydraulic load on the weir plate is supported by the gate slots in the concrete piers at the ends of the gate span. The gate is raised along the slot. In contrast, a different type of long-span gate (new at the time of its design) located in Schinznach, close to Zurich, Switzerland, is shown in Figure 2. The weir plate is 22.5 m long and about 3.4 m high and is inclined at 30° in the downstream direction (the same as for the Ara River gate). The gate has a weight of 100 ton. In order to decrease the required force to raise the gate, the hydraulic load exerted on the weir plate is supported by a pair of trunnion pins located on the upstream side of the gate in the piers at the gate ends. The large hydraulic pressure on the gate bottom, creates a moment about the upstream trunnion pins and acts to raise the gate, rotating it about the trunnion pins. Further, the weir plate has a circular arc shape, which increases the hydraulic opening moment on the heavy gate. As a result, the required force needed to raise the gate in rotation about the trunnion pins is drastically reduced. The operation is easily accomplished and the gate opens smoothly despite its very substantial weight. Although the gate opens smoothly, the design makes the gate behave as a press-open, with inherent problems in sealing the gate when it is in its closed position. This type of gate is known as a reversed Tainter gate.

At small openings most, if not all, long-span gates are susceptible to serious gate vibrations as a result of flow beneath the gate, as reported by Petrikat (1980), Ishii et al. (1987), and Ishii (1990a; 1990b; 1990c). The long-span gate constructed on the Ara River, shown in Figure 1, also underwent similar vibrations at a frequency between 3 and 5 Hz, with a water depth of 2.7 m upstream and almost 0 m downstream. The vibration occurred at small gate openings from 8 to 25 mm and the vibration amplitude was about 4 mm at the spanwise center.

In 1977, the Schinznach Gate shown in Figure 2 experienced violent vibrations at a frequency of 4.1 Hz with an upstream water depth of 3.4 m and essentially 0 m downstream depth, as shown in Figure 3. The gate opening was 25 mm at each end and 8 mm in the middle of the gate. Due to streamwise vibrations of the gate skin-plate, surface waves with a straight front parallel to the gate took were generated, as shown in Figure 3a, at the onset of small amplitude vibrations. As the vibration amplitude increased, the waves were amplified, as shown in Figure 3b, and then violent broken waves were generated, as

Figure 1. Long-span gate on the Ara River, Japan, with a weight of 49 tons: (a) side view from downstream, (b) cross-sectional view

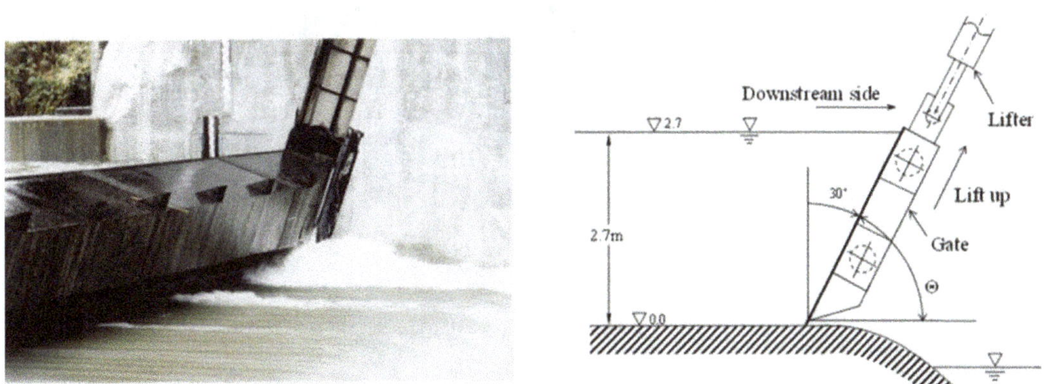

(a) (b)

Figure 2. Long-span gate at Schinznach, Switzerland, with a weight of 100 tons, showing (a) side view from downstream, (b) cross-sectional view
After Petrikat (1980).

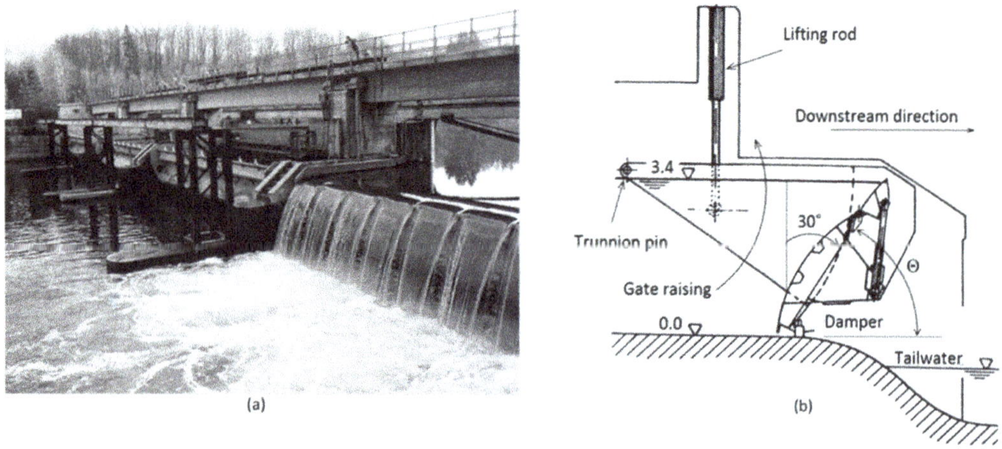

Figure 3. An example of streamwise vibration of the long-span gate at Schinznach, Switzerland: (a) small waves; (b) growing waves; (c) initial broken waves; (d) violent broken waves; (e) schematic view of vibration mode
Photos courtesy of P. Billeter.

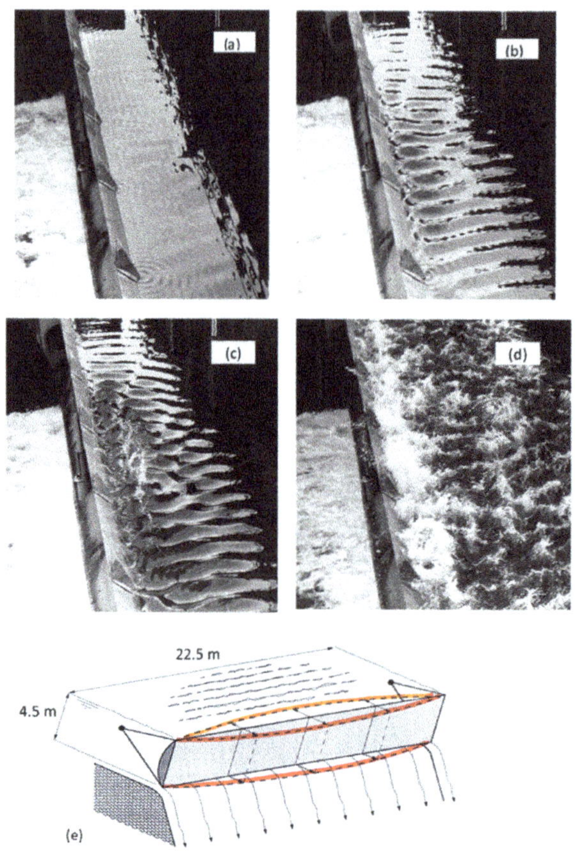

shown in Figures 3c and 3d. In order to prevent the violent vibration, a vibration damper was attached beneath the weir plate, as shown in Figure 2b.

The vibration mode of the gate, shown schematically in Figure 3e, exhibits a half-wavelength streamwise and parallel vibration mode shape with a node at each end. Since this gate behaves as press-shut device, it is dynamically stable when considering possible vibration of the whole gate about the trunnion pin. As for possible coupled-mode vibration of the streamwise bending vibration and the whole gate vibration around the trunnion pin, further discussion and research will be needed.

A common feature of both the Schinznach Gate and the Ara River Gate is their inclined skin-plates at about 30° in the downstream direction. The gate vibrations produce surface waves and fluctuating hydraulic pressures for two reasons. The first is due to the variation in flow rate caused by changes of the gate opening, and the second is due to the push-and-draw effect by the gate on the water. These two mechanisms for inducing pressure fluctuation were presented in Chapter 4 and 5, respectively. As was shown by Ishii et al. (1987, 1992), the dynamic stability of such gates depends on whether the hydraulic-pressure fluctuation, resulting from these effects, augments or damps the gate vibrations. In order to ascertain the gate stability, the hydraulic-pressure fluctuations resulting from the gate vibrations must, therefore, be determined first.

The hydrodynamic pressure caused by the changes of the gate opening is called the flow-rate variation pressure. Chapter 4 provided the theoretical development of the flow-rate-pressure. The pressure fluctuation due to the push-and-draw effect of the gate on the water is called the push-and-draw pressure. Chapter 5 provided the theoretical development if the push-and-draw pressure. These theoretical analyses of Chapter 4 and 5 rely on the potential-flow theory treating the wave-radiation problem as an initial-value and boundary-value problem, as developed by Rayleigh (1883, 1945) and Lamb (1904, 1932),

Since there is no downstream submergence on the gates at Schinznach and on the Ara River, and since the gates are of limited streamwise extent, vortices cannot form beneath these gates. Moreover, turbulence effects in the accelerated approach flow can be neglected. For these reasons application of potential-flow theory is appropriate. The analysis of gates vibration in the absence of vortex excitation using potential flow theory has been shown to provide reasonably accurate predictions of the flow-induced vibrations of Tainter gates (radial gates) in the studies of Imaichi & Ishii (1977), Ishii & Imaichi (1977, 1982), Ishii et al. (1977, 1980), Ishii et al. (1983) and Ishii & Naudascher (1984, 1992) and of long-span gates by Ishii & Knisely(1992a, 1992b) and Ishii et al. (1992, 1995).

In contrast to the gate vibrations with essentially 0 m downstream water depth, vibrations of submerged or flat-bottomed long-span gates, such as those studied by Thang et al (1987), Naudascher (1987), Ishii et al. (1989), Ishii & Knisely (1988;1989; 1990) are caused by vortex formation. The analysis of the vibration of such gates will require careful inclusion of the effects of vortex formation on the pressure loading on the gates. Gates with submerged discharge and long-span gates simultaneous over-and under flow will be considered as the last two topics in the present chapter.

For most of the present chapter, except for the section on Viscous Flow Effects, an analytical approach to the flow-induced vibration problem of long-span gates at small openings is presented, assuming small amplitude vibrations in the single half-wavelength bending mode shown in Figure 3c. First, the resultant hydrodynamic pressure is determined based the flow-rate variation pressure (from Chapter 4) and the push-and draw pressure (from Chapter 5).

If a long-span gate vibrates, the upstream water is pushed and drawn by the streamwise gate motion and a pressure fluctuation component proportional to the gate acceleration is induced, which produces an added-mass effect on the gate. At the same time, the back-and-forth motion of the water induces a pressure fluctuation component proportional to the gate velocity, which consumes some energy produced by the gate motion in the form of dissipative waves, thus producing fluid damping.

The flow-rate variation caused by the fluctuating gate opening induces a pressure fluctuation component proportional to the gate acceleration and leads to an additional added-mass effect on the gate. At the same time, the flow-rate variation pressure component proportional to the gate velocity supplies energy from the flowing fluid to the gate, resulting in fluid excitation.

As a result, the composite resultant hydrodynamic pressure on the gate is represented by a superposition of two pressure components, one proportional to gate acceleration and one proportional to gate velocity.

Subsequently, the equation of motion for the gate is derived, in which the reduced added mass, and the reduced fluid-excitation and damping coefficients are determined. The solutions for the vibration frequency and the net fluid-excitation ratio are all determined.

Finally, these theoretical results are confirmed by experiments using a simple two-dimensional model for the flow-induced flexural vibration of long-span gates. The agreement between the model test results and the theoretical predictions demonstrates that the use of Rayleigh's wave theory in the derivation of the hydrodynamic pressure is applicable to the analysis of the flow-induced vibration of long-span gates.

RESULTANT HYDRODYNAMIC PRESSURE INDUCED BY STREAMWISE GATE VIBRATION

In order to simplify the analysis of the unsteady flow field, a simplified gate model is assumed. The assumed model has a vertical skin-plate and an inclined floor under the gate. The floor is inclined at an angle Θ, as shown in Figure 4. The angle Θ represents the angle between the inclined weir plate and the horizontal floor shown in Figures 1(b) and 2(b). The variation of this angle Θ determines how much the flow rate varies due to the streamwise gate vibration. On an operating gate, the hydraulic force exerted on the gate is in equilibrium with the restoring force due to the bending deformation of the gate. The gate undergoes bending vibrations about this equilibrium state. The fundamental bending mode (Figure 4) of the vibrational displacement is given by the following expression:

$$X\left(Z;T\right) = X_c\left(T\right)\sin\left(\frac{Z}{l} + \frac{1}{2}\right)\pi \tag{1}$$

The span length l is equal to one-half of a wavelength. The amplitude in the middle of the gate is represented by X_c (T). The horizontal axis along the gate is represented by Z with its origin O in the middle of the gate. The gate movement in the upstream direction is defined as positive.

The gate-vibration amplitude varies as sin $(Z/l + 1/2)\pi$ in the Z direction, as given by Equation (1). Therefore, the induced water waves do not travel as plane waves, but as two-dimensional wave forms. However, since the span length, l, is much larger than the gate height, H, and the gate submergence depth, d_0, these waves can be approximated as plane waves. Thus, the flow field in the upstream channel is regarded as two-dimensional in the vertical (X, Y) - plane, with the X-axis taken horizontally along the undisturbed free surface in the upstream direction and the Y-axis vertically upward along the gate, as shown in Figure 4.

Figure 4. Overview of the model of a long-span gate with an inclined floor. The gate dams a wide upstream channel and undergoes streamwise vibration in the fundamental half-wavelength bending mode After Ishii (1992).

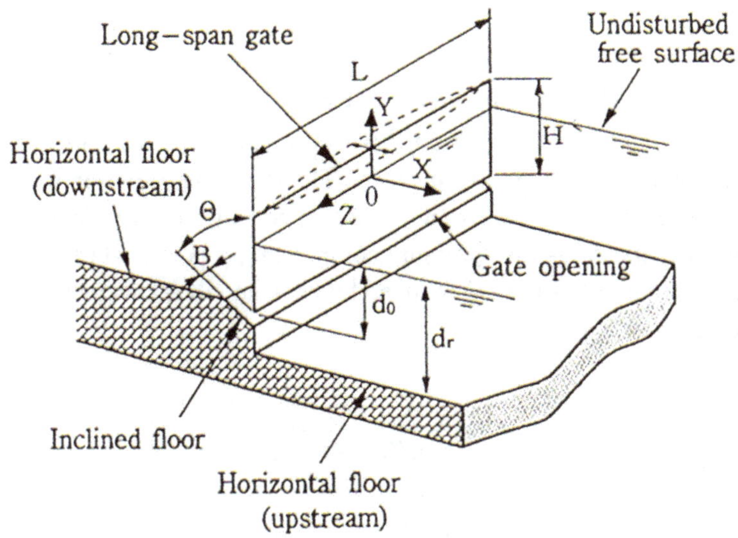

On the basis of this assumption, the flow-rate variation pressure P_r $(X, Y; T)$ induced by the gate vibration can be made dimensionless in the following expression, after the definition in Equation 131 in Chapter 4:

$$p_r(x,y;t) \equiv \frac{P_r(X,Y;T)}{\sqrt{2c_f}\rho g \bar{X}_c \sin(Z/l+1/2)\pi \cdot \cos\Theta} = \frac{1}{\sqrt{2c_f}\cos\Theta} \cdot \frac{P_r(X,Y;T)/\rho g}{\bar{X}_c \sin(Z/l+1/2)\pi} \tag{2}$$

where $\bar{X}_c$ represents the maximum amplitude in the middle of the gate. The dimensionless axes (x, y) and the dimensionless time t are all the same as defined in Equation 73 in Chapter 4:

$$x \equiv \frac{X}{d_0}, y \equiv \frac{Y}{d_0}, t \equiv \Omega_0 T$$

Now the dimensionless amplitude x_c (t) is introduced as

$$x_c(t) \equiv \frac{X_c(T)}{\bar{X}_c} \tag{3}$$

With this dimensionless amplitude x_c (t), the dimensionless flow-rate-variation pressure on the gate, p_r $(0, y; t)$ is given in the following form:

$$p_r(0,y;t) = \left(\frac{F_0}{F}\right)^2 p_{rp}\left(F,0,y,\beta^*\right)\cdot x_c'' - \left(\frac{F_0}{F}\right) p_{rs}\left(F,0,y,\beta^*\right)\cdot x_c' \tag{4}$$

which is the negative of Equation 134 in Chapter 4, since x_G introduced in Equation 132 in Chapter 4 was defined as positive in the downstream direction, but x_c in this chapter is defined as positive in the upstream direction. The Froude number F and the basic Froude number F_0 are defined in Equation 74 in Chapter 4 and in Equations 36 and 38 in Chapter 5. The flow-rate-variation pressure consists of two terms, the first term is the progressive-wave pressure component proportional to the gate acceleration and the second term is the standing-wave pressure component proportional to the gate velocity.

The push-and-draw pressure P_b $(X, Y; T)$ induced by the gate vibration can be non-dimensionalized, using Equation 32 in Chapter 5:

$$p_b(x,y;t) \equiv \frac{P_b(X,Y;T)}{\rho g\bar{X}_c \sin(Z/l+1/2)\pi} = \frac{P_r(X,Y;T)/\rho g}{\bar{X}_c \sin(Z/l+1/2)\pi} \tag{5}$$

With the dimensionless amplitude x_c (t), the reduced push-and-draw pressure on the gate, p_b $(0, y; t)$ is given as follows from Equation 52 in Chapter 5:

$$p_b(0,y;t) = p_{bs}(F,0,y)\left(\frac{F_0}{F}\right)^2 x_c'' + p_{bp}(F,0,y)\left(\frac{F_0}{F}\right) x_c' \tag{6}$$

The push-and-draw pressure also consists of two terms, the first term is the standing-wave pressure component proportional to the gate acceleration, and the second term is the progressive-wave pressure component proportional to the gate velocity. With an assumed small amplitude of vibration X_c, both the flow-rate-variation pressure P_r and the push-and-draw pressure P_b also are small. The resultant pressure $P(X, Y; T)$ can be given, therefore, by a simple superposition of these two pressures:

$$P(X,Y;T) = P_r(X,Y;T) + P_b(X,Y;T) \tag{7}$$

Now introduce the following definition for dimensionless resultant pressure p:

$$p\ (x,y;t) \equiv \frac{P(X,Y;T)}{\rho g\bar{X}_c \sin(Z/l+1/2)\pi} = \frac{P\ (X,Y;T)/\rho g}{\bar{X}_c \sin(Z/l+1/2)\pi} \tag{8}$$

With the definitions in Equations 2 and 5, in addition to Equation 8, the resultant pressure p can be reduced to

$$p(x,y;t) = kp_r(x,y;t) + p_b(x,y;t) \tag{9}$$

where k is defined as

$$k \equiv \sqrt{2}c_f \cos\Theta \tag{10}$$

This coefficient k includes the gate inclination angle Θ. Thus, k can be interpreted as the "gate inclination factor."

EQUATION OF MOTION

The gate mass per unit length is assumed to be $\rho_* hH = M / L$, where H is the gate height, h is the equivalent gate thickness, ρ_* is the gate density, M is the gate mass, and L is the gate span. The hydrodynamic pressure exerted on the gate is obtained by substituting $X = 0$ into P(X, Y; T). The resulting gate vibration excited by P(0, Y; T) is governed by the following equation of motion:

$$\rho_* hH\frac{\partial^2 X}{\partial T^2} + C\frac{\partial X}{\partial T} + EI\frac{\partial^4 X}{\partial Z^4} + \int_{-d_0}^{0} P\left(0,Y;T\right)dY = 0 \tag{11}$$

Since it was assumed that the gate vibrates in the single bending mode given by Equation 1, the above equation of motion can be reduced to the following expression:

$$\ddot{X}_c + 2\Omega_0\zeta_0\dot{X}_c + \Omega_0^2 X_c + \int_{-d_0}^{0} P\left(0,Y;T\right)dY \Big/ \left\{\rho_* hH \sin\left(Z/l + 1/2\right)\pi\right\} = 0 \tag{12}$$

where Ω_0 is the natural vibration frequency in air, and ζ_0 is the damping ratio.

By using the dimensionless values introduced in Equation 73 in Chapter 4 in addition to Equations 2, 5 and 8, one can write the above equation in the following form:

$$x_c'' + 2\zeta_0 x_c' + x_c + \frac{\alpha_m}{F_0}\int_{-1}^{0}\left\{kp_r\left(0,y;t\right) + p_b\left(0,y;t\right)\right\}dy = 0 \tag{13}$$

where the prime represents the derivative with respect to the dimensionless time t. The parameter α_m is defined as

$$\alpha_m = \rho d_0^2 / \left(\rho_* hH\right) = \rho d_0^2 / \left(M / L\right) \tag{14}$$

The physical meaning of the numerator of α_m can be understood with the help of Figure 5, where ρd_0^2 represents the mass of water in a volume of a square with sides equal to the gate submergence times a unit spanwise length. The denominator of α_m represents the gate mass per unit length in the Z direction. Thus, α_m represents the ratio of water-to-gate masses.

Figure 5. Representative mass of water, ρd_0^2, used in the formulation of the water-to-gate mass ratio and as a reference in the discussion of added mass
From Ishii (1992). © 1992 Elsevier. Used with permission.

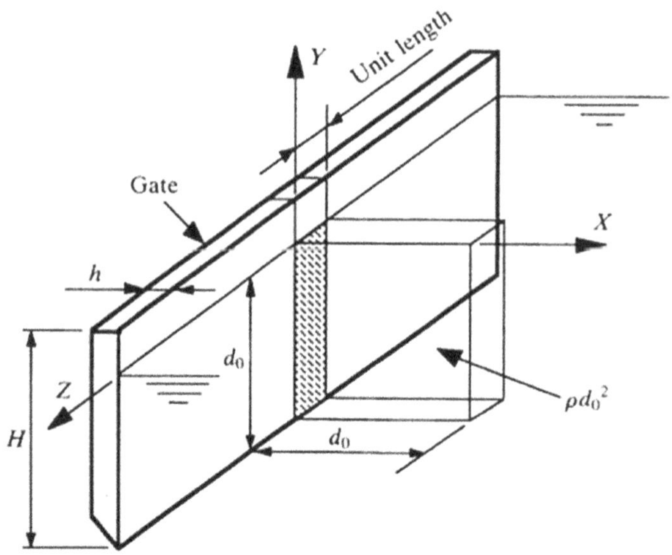

IDENTIFICATION OF ADDED MASS, FLUID EXCITATION AND FLUID DAMPING

Substitution of Equation 4 for the flow-rate variation pressure and Equation 6 for the push-and-draw pressure into Equations 12 and 13 leads to the equation of motion for the gate in the form of the following second-order homogeneous differential equation:

$$\left\{1+\alpha_m(\Delta m_r+\Delta m_b)\right\}x_c''+2(\zeta_0-\xi_r+\zeta_b)x_c'+x_c=0 \tag{15}$$

The second term in Equation 4, that is, the flow-rate-variation standing-wave pressure component results in a negative damping term, corresponding to the negative damping ratio represented by $-\xi_r$ in Equation 15. The symbol, ξ_r, used to represent this negative damping, is called the fluid-excitation ratio. A small fluid damping effect is also produced by the second term of Equation 6, that is, the push-and-draw progressive-wave pressure component, resulting in the fluid damping ratio represented by ζ_b. These excitation and damping ratios are defined in the following forms, respectively:

$$\xi_r \equiv \frac{\Delta C_r}{2\rho_* hH\Omega_0} \text{ and } \zeta_b \equiv \frac{\Delta C_b}{2\rho_* hH\Omega_0} \tag{16}$$

where ΔC_r is the fluid excitation coefficient caused by the flow-rate-variation standing-wave pressure component and ΔC_b is the fluid damping coefficient caused by the push-and-draw progressive-wave pressure component. Now another dimensionless representation of these fluid excitation and damping coefficients is introduced in the following forms, respectively:

$$\Delta c_r \equiv \frac{\Delta C_r}{\rho d_0 \sqrt{g d_0}} \text{ and } \Delta c_b \equiv \frac{\Delta C_b}{\rho d_0 \sqrt{g d_0}} \tag{17}$$

The representative quantity for this definition is $\rho d_0 \sqrt{g d_0}$, which represents the gravity force $\rho g d_0^2$ acting on the representative mass ρd_0^2 divided by the wave velocity $\sqrt{g d_0}$. As a result, the fluid-excitation ratio ξ_r and the fluid damping ratio ζ_b, defined by Equation 16, can be related to the reduced fluid coefficients, respectively, as follows:

$$\xi_r = \alpha_m \frac{\Delta c_r}{2F_0} \text{ and } \zeta_b = \alpha_m \frac{\Delta c_b}{2F_0} \tag{18}$$

With careful arrangement of all expressions when Equations 6 and 13 are substituted into Equation 4, one can derive the reduced fluid-excitation coefficient Δc_r and the reduced fluid damping coefficient Δc_b in integrals of the flow-rate-variation standing pressure component p_{rs} and of the push-and-draw progressive pressure component p_{bp}, respectively:

$$\Delta c_r = \frac{k}{F} \int_{-1}^{0} p_{rs}(F, 0, y) dy \text{ and } \Delta c_b = \frac{1}{F} \int_{-1}^{0} p_{bp}(F, 0, y) dy \tag{19}$$

The water added mass consists of two contributions: ΔM_r from the first term in Equation 4, resulting from the flow-rate-variation progressive-wave pressure component, and ΔM_b from the first term in Equation 6, due to the push-and-draw standing-wave pressure component. The effects of these two added mass terms are included in their non-dimensional forms, Δm_r and Δm_b, in the first term of Equation 15. The non-dimensional forms of these added mass coefficients are given as follows:

$$\Delta m_r \equiv \frac{\Delta M_r}{\rho d_0^{\,2}} \text{ and } \Delta m_b \equiv \frac{\Delta M_b}{\rho d_0^{\,2}} \tag{20}$$

The representative quantity for this definition, ρd_0^2, has been previously introduced with reference to Figure 5. With the careful arrangements of all expressions when Equations 6 and 13 are substituted into Equation 4, one can derive these reduced added mass terms in the following integral forms of the flow-rate-variation progressive-wave pressure component p_{rp} and of the push-and-draw standing-wave pressure component p_{bs}, respectively:

$$\Delta m_r = \frac{k}{F^2} \int_{-1}^{0} p_{rp}(F, 0, y) dy \text{ and } \Delta m_b = \frac{1}{F^2} \int_{-1}^{0} p_{bs}(F, 0, y) dy \tag{21}$$

The pressure amplitudes, p_{rs} and p_{bp} in Equations 19 as well as p_{rp} and p_{bs} in Equation 21, were previously derived in Equations 128b and 128c in Chapter 4 for the flow-rate variation pressure and in Equations 49 and 50 in Chapter 5 for the push-and-draw pressure. Integration of these pressure amplitudes over the gate submergence can be theoretically performed, resulting in the expressions given in Equations 22 and 23 that follow.

For the fluid excitation and damping ratio:

$$\Delta c_r / k = -F^2 \frac{2}{\beta^*} \sum_{j=1}^{\infty} \left\{ \frac{1}{\eta_j^2} + \sum_{n-1}^{\infty} \frac{\cos n\pi / \beta^*}{\eta_j^2 - \left(n\pi / \beta^* \right)^2} \right\} \frac{\sinh \eta_j \beta^* - \sin \eta_j (\beta^* - 1)}{\eta_j \cos(\eta_j \beta^*) W(i\eta_j) / i} \tag{22a}$$

$$\Delta c_b = 2F^3 \sum_{j=1}^{\infty} \left\{ \frac{1}{\beta^* \xi_0^2} + \frac{2}{\pi} \sum_{n=1}^{\infty} \frac{2}{n} \frac{\sin n\pi / \beta^*}{\xi_0^2 + \left(n\pi / \beta^* \right)^2} \right\} \frac{\sinh \xi_0 \beta^* - \sinh \xi_0 (\beta^* - 1)}{\xi_0 \cosh \xi_0 \beta^* W(\xi_0)} \tag{22b}$$

For the water added mass:

$$\Delta m_r / k = F \frac{2}{\beta^*} \left\{ \frac{1}{\xi_0^2} + \sum_{n=1}^{\infty} \frac{\cos n\pi / \beta^*}{\xi_0^2 + \left(n\pi / \beta^* \right)^2} \right\} \frac{\sinh \xi_0 \beta^* - \sinh \xi_0 (\beta^* - 1)}{\xi_0 \cosh(\xi_0 \beta^*) W(\xi_0)} \tag{23a}$$

$$\Delta m_b = 2F^2 \sum_{j=1}^{\infty} \left\{ \frac{1}{\beta^* \eta_j^2} + \frac{2}{\pi} \sum_{n=1}^{\infty} \frac{2}{n} \frac{\sin n\pi / \beta^*}{\eta_j^2 - \left(n\pi / \beta^* \right)^2} \right\} \frac{\sin \eta_j \beta^* - \sin \eta_j (\beta^* - 1)}{\eta_j \cos \eta_j \beta^* W(i\eta_j) / i} \tag{23b}$$

Characteristics of Added Mass, Fluid Excitation, and Fluid Damping

The functions Δc_b and $\Delta m_r / k$ given by Equations 22b and 23a are in terms of infinite-series summations with respect to the integer *n*. As *n* becomes larger, the infinite series converge rapidly to zero, in inverse proportion to the second or third power of *n*. Therefore, they can be calculated numerically with high accuracy. The functions $\Delta c_r / k$ and Δm_b given by 22a and 23b are in terms of double infinite-series summations with respect to the integers *n* and *j*, the denominator of which includes the term $\left\{ \eta_j^2 - \left(n\pi / \beta^* \right)^2 \right\}$. As shown in Figure 3b in Chapter 5, η_j is given by $(j\pi - \Delta)/\beta^*$, where $0 \le \Delta < \pi/2$ (Δ converges to zero as $F^2\beta^* \to 0$ and to $\pi/2$ as $F^2\beta^* \to \infty$). If $j = n$, the denominator $\left\{ \eta_j^2 - \left(n\pi / \beta^* \right)^2 \right\}$

is the smallest and thus the series term is largest. It is important for the numerical calculation of $\Delta c_r / k$ and Δm_b, therefore, to correctly sum the double infinite-series terms around $j = n$. If $F^2\beta^* \rightarrow 0$ because of small F, in particular, Δ converges to zero, and the double infinite-series diverges to ∞ for $j = n$. In this case, $\Delta c_r / k$ and Δm_b cannot be calculated accurately.

For a given Froude number, F, the solutions ξ_0 and η_j of the transcendental Equations 43 and 44 in Chapter 5 were first calculated numerically, and then the functions $\Delta c_r / k$, $\Delta m_r / k$, Δc_b and Δm_b were calculated. The calculated results Δm_r and Δc_r are shown in Figure 6, and those for Δm_b and Δc_b in Figure 7. In these figures, the value of $1/\beta^*$, the depth ratio, defined as the gate submergence d_0 to the reservoir depth d_r, is given as a parameter. The parameter β^* was previously introduced in Equation 106b in Chapter 4 as well as in Equation 31 in Chapter 5.

The gate is positioned over a horizontal concrete floor in most cases. Therefore, the submergence depth, d_0, is nearly equal to the upstream-channel depth, d_r, for small gate openings; hence, the depth ratio β^* is about 1.0. For this particular value of β^*, the reduced added mass, Δm_r, divided by k shows the same tendency as the reduced fluid-damping coefficient Δc_b. As F increases, $\Delta m_r / k$ and Δc_b both decrease asymptotically approaching zero, $\Delta m_r / k$ for $F > 2.0$ and Δc_b for $F > 8.0$. Further, the reduced fluid-excitation coefficient, Δc_r divided by k and the reduced added mass, Δm_b both increase rapidly from zero and approach constant values. If an error of $\pm 1\%$ is permissible, $\Delta c_r / k$ and Δm_b can be approximated by 0.74 and 0.54 for $F > 11.0$, respectively. From a study by Ishii (1990b) using a laboratory model, the measured values of Δm_b and Δc_b were found to agree well with the theoretical results presented in this chapter.

Figure 6. Calculated results for the reduced added mass Δm_r and the reduced fluid-excitation coefficient Δc_r associated with the flow-rate variations for various $1/\beta^*$ values: $-\Delta m_r / k$; $--\Delta c_r / k$

From Ishii (1992). © 1992 Elsevier. Used with permission.

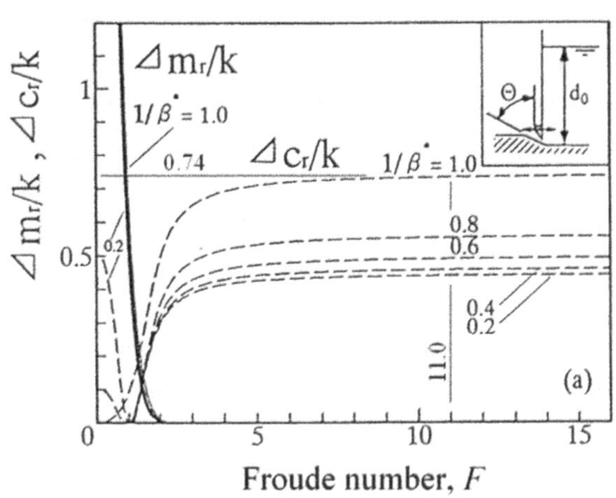

Figure 7. Calculated results for the reduced added mass, Δm_b, and the reduced fluid-damping coefficient, Δc_b, associated with the push-and-draw effect of the gate on the water for various $1/\beta^*$ values: $-\Delta m_b$; $--\ \Delta c_b$, same figure previously given as Figure 8 in Chapter 5
From Ishii (1992). © 1992 Elsevier. Used with permission.

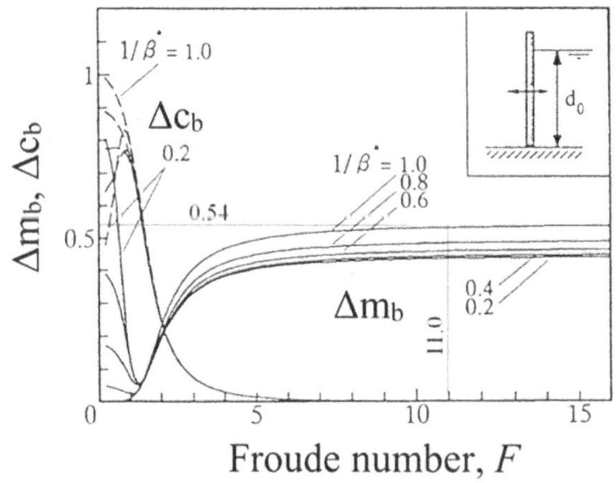

As the depth ratio, β^*, increases ($1/\beta^*$ decreases), meaning as the upstream-channel depth d_r becomes larger than the submergence depth d_0, the flow disturbances caused by the gate vibration (*e.g.*, the flow-rate variation and the push-and-draw effect on the water) spread over a larger region of the flow field, and thus the feedback effect of the disturbances to the gate vibration decreases. For this reason, the reduced fluid-excitation coefficient $\Delta c_r/k$ and the reduced added mass Δm_b both decrease with increasing β^* (decreasing $1/\beta^*$) for F larger than 2.0. All curves for F smaller than 2.0 change rapidly and are very complicated. However, this range of small Froude numbers, $F < 2.0$, is not as significant as that for $F > 2.0$ for two reasons:

1. In most practical applications, the Froude number, F, is far larger than 2.0, since gates are normally operated with large depths d_0
2. Even if the depth d_0 is small, the Froude number F may be less than 2.0, but the water mass becomes small so that the water-to-gate mass ratio, α_m, given by Equation 14 is also small. As is evident from Equation 18, the fluid-excitation and damping ratios are always small for $F < 1.0$, even though Δc_r and Δc_b may change rapidly. Physically, this means that since the area of the gate in contact with the water is small for small depths d_0, the amount of energy transferred from the gate to the water or from the water to the gate becomes insignificantly small.

With these calculated results for Δc_r and Δm_r for the flow-rate-variation pressure and for Δc_b and Δm_b for the push-and-draw pressure, the fluid-excitation ratio ξ_r and the fluid damping ratio ζ_b can be calculated using Equation 18.

VIBRATION FREQUENCY AND NET FLUID-EXCITATION RATIO

For comparatively small rates of change in the gate vibration amplitude, the solution of Equation 15 is given by

$$x_c = \exp\left\{(\xi + i)\left(\frac{\Omega}{\Omega_0}\right)t\right\} = \exp\left\{(\xi + i)\left(\frac{F}{F_0}\right)t\right\} \tag{24}$$

where ξ represents the resultant excitation ratio. Substitution of this expression in Equation 24 into Equation 15 results in an expression for the in-water to in-air vibration frequency ratio, Ω / Ω_0, that is, for the ratio of the Froude number to the basic Froude number F / F_0, given by

$$\frac{F}{F_0} = \sqrt{\frac{1 - \left\{1 + \alpha_m(\Delta m_r + \Delta m_b)\right\}\xi^2}{1 + \alpha_m(\Delta m_r + \Delta m_b)}} \tag{25a}$$

With $\xi \ll 1.0$, the above expression can be reduced to the following simpler form:

$$\frac{F}{F_0} = \frac{1}{\sqrt{1 + \alpha_m(\Delta m_r + \Delta m_b)}} \tag{25b}$$

The resultant excitation ratio ξ can be written in the following manner:

$$\xi = \frac{(\xi_r - \zeta_b) - \zeta_0}{\sqrt{1 + \alpha_m(\Delta m_r + \Delta m_b)}} = \frac{F}{F_0}\left\{(\xi_r - \zeta_b) - \zeta_0\right\} \tag{26}$$

in which $(\xi_r - \zeta_b)$ in the numerator is the *net fluid-excitation ratio* which represents the net excitation effect of the fluid motion on the gate, as given by

$$\xi_r - \zeta_b = \frac{\alpha_m}{2F_0}(\Delta c_r - \Delta c_b) \tag{27}$$

If this net fluid-excitation ratio exceeds the mechanical damping ratio due to friction and structural damping effects, which is equivalent to a mechanical damping effect small than the net fluid-excitation ratio,

$$\zeta_0 < (\xi_r - \zeta_b) \tag{28}$$

then the resultant excitation ratio ξ, given by Equation 26, is a positive number, and the gate system possesses a dynamic instability.

The reduced added mass Δm_r and Δm_b, included in the right hand side of Equation 25 are functions of the Froude number F, and thus an iterative calculation method can be applied in order to find the Froude number ratio F/F_0. By using the characteristics for the added mass Δm_r and Δm_b, shown in Figures 6 and 7, the Froude number ratio F/F_0, and thereby the in-water to in-air vibration frequency ratio Ω/Ω_0, can be calculated, as shown in Figure 8a. The parameter in Figure 8a is the water-to-gate mass ratio α_m, defined by Equation 14. As shown in Figure 6, with increasing values of Froude number F, the added mass Δm_r rapidly decreases and becomes nearly zero for Froude numbers, F, larger than about 2. Returning to Figure 8a for values of basic Froude number F_0 less than about 2.0, the frequency ratio F/F_0 is independent of the flow-rate variation factor k, and can be approximated by

Figure 8. Theoretical solutions for flow-induced vibration of long-span gates for $1/\beta^* = 1.0$: (a) In-water to in-air vibration frequency ratio F/F_0; (b) net fluid-excitation ratio $(\xi_r - \zeta_b)$ or stability criterion representing minimum mechanical damping ratio, Minimum(ζ_0), for dynamic stability

From Ishii & Nakata (1995). © 1995 A. A. Balkema. Used with permission.

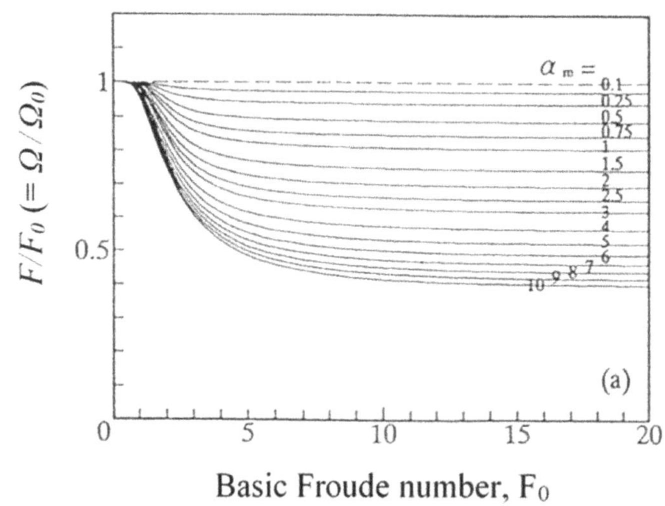

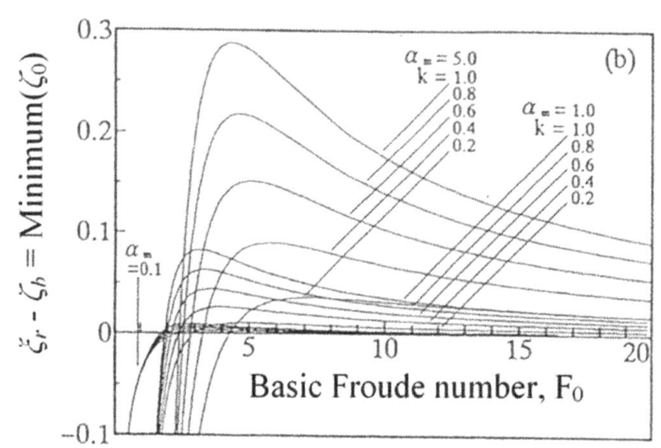

$$\frac{F}{F_0} = \frac{1}{\sqrt{1+\alpha_m \Delta m_b}} \qquad \text{for } F > 2.0 \tag{29}$$

From small values of F_0, the ratio F/F_0 gradually decreases from 1.0 and then approaches a constant value that itself decreases with increasing α_m values. Figure 7 shows that as the Froude number F increases to more than about 11.0, Δm_b converges to a value of 0.54 for $1/\beta^* = 1.0$. As a consequence, the constant value of F/F_0 for $F > 11.0$ and for $1/\beta^* = 1.0$ can be calculated as

$$\frac{F}{F_0} = \frac{1}{\sqrt{1+0.54\alpha_m}} \qquad \text{for } F > 11.0 \text{ and } 1/\beta^* = 1.0 \tag{30}$$

From Equation 30, one can easily understand that the in-air vibration frequency drastically decreases in water, due to the water added mass effect. The in-water vibration frequency decreases to 0.52 times the in-air vibration frequency for $\alpha_m = 5.0$ and to 0.4 times the in-air frequency for $\alpha_m = 10.0$.

By using the characteristics for the fluid-excitation coefficient Δc_r shown in Figure 6 and the fluid-damping coefficient Δc_b shown in Figure 7, the net fluid-excitation ratio $(\xi_r - \zeta_b)$ from Equation 27 can be calculated as a function of F_0, as shown in Figure 8b. The gate inclination factor k has been included as the parameter in Figure 8(b). The ordinate represents the minimum damping ratio, Minimum(ζ_0), necessary for gate stability. One may understand that the gate is dynamically stable, if the mechanical damping ratio ζ_0 caused by friction and structural damping falls beneath each characteristic curve. With increasing F_0, $(\xi_r - \zeta_b)$ rapidly increases, exhibiting its maximum value of 0.28 at $F_0 = 4.2$ for $\alpha_m = 5.0$ and k=1.0, and then gradually decreasing to zero. This maximum value of the net fluid-excitation ratio indicates that in the vicinity of a basic Froude number of 4.0, a very violent self-excited vibration, far beyond our initial expectation is induced, since the mechanical damping effect for gate bending vibration is generally very small.

On the whole, the value of $(\xi_r - \zeta_b)$ decreases with decreasing α_m and k. For $F > 8.0$, the fluid-damping coefficient Δc_b decreases to almost zero, as shown in Figure 7 and hence $(\xi_r - \zeta_b)$ can be calculated as

$$\frac{\xi_r - \zeta_b}{k} = \frac{\alpha_m}{2F_0}\left(\Delta c_r / k\right) = 0.37\frac{\alpha_m}{F_0} \qquad \text{for } F > 8.0 \tag{31}$$

where the function $\Delta c_r / k$ on the right hand side of Equation (31) is presented graphically in Figure 6.

EXPERIMENTAL VERIFICATION OF VIBRATION FREQUENCY AND FLUID-EXCITATION RATIOS

Experimental Setup

The behavior of a long-span gate vibrating streamwise in an upright state on a horizontal river bed with an inclined sill beneath the gate is examined using the laboratory model in Figure 9. The vibrating table is supported by two stays with length of 700 mm. With this parallel motion mechanism, the vertical gate table translates in the horizontal direction. For small amplitude vibrations, the motion of the gate induced the rotation of the stays is essentially linear parallel motion. The spanwise length of the model gate is 297 mm, and the model gate is installed in a uniform horizontal channel of 300 mm in width. This model gate is called a section model and corresponds to a cut-out portion of a long-span gate, such as that shown in Figure 4. The equivalent dynamic mass of the entire gate model, *M*, is 26.7 kg.

Piano wires, attached to the right and left sides of the vibrating table, are connected to the right arm and the left base, respectively. The arm is supported by a pin with coiled springs fitted on the top. By adjusting the vertical location of the spring attachment, the natural frequency of the model gate can be adjusted. The shape of the lower end of the gate is a sharp edge, and the floor just under the sharp lower end of gate is inclined, forming the gate angle Θ between the vertical gate and the inclined floor. This inclination angle Θ was fixed at two values of 30° and 45°. The entire model can be adjusted vertically, allowing for different gate openings, *B*, between the sharp lower end of the gate and the inclined floor. The depth of gate submergence on the upstream side, d_0, was fixed at 290 mm. Given values for the gate mass *M* and the gate submergence depth d_0, the water-to-gate mass ratio α_m defined by Equation 14 takes on a value of 0.95. The gate opening height *B* was fixed at 2 mm far smaller than the gate submergence d_0. Thereby the depth ratio β^* of the upstream channel depth to d_0 takes on a value of about 1.0.

Figure 9. Two-dimensional section model for flow-induced flexural vibration of long-span gate
From Ishii & Nakata (1995). © 1995 A. A. Balkema. Used with permission.

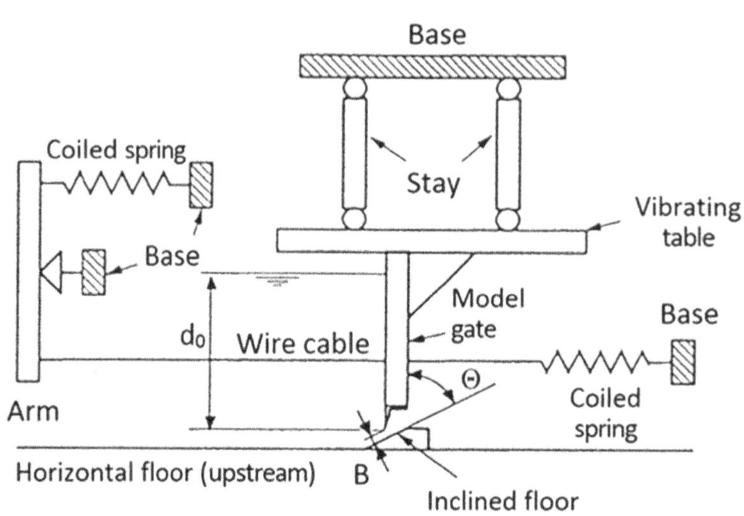

Results for In-Water Vibration Frequency, Excitation Ratio, and In-Air Damping Ratio

Samples of the self-excited vibration waveforms observed when the floor inclination angle Θ was set to 45° are shown in Figure 10. In all cases, the gate started to vibrate spontaneously with no additional external disturbances, and initially the vibration amplitude grew exponentially and then approached a constant value.

Special attention was paid to the exponentially increasing amplitude curve, to calculate the vibration frequency Ω and the excitation ratio ξ. The calculated data versus the in-air vibration frequency Ω_0 are shown in Figure 11, where the Figure 11(c) shows the in-air damping ratio due to mechanical damping effects in the present vibrating gate model. The in-air vibration frequency was varied from 2.09 Hz to 6.03 Hz.

Figure 10. Transient vibration waveforms of model gate in flowing water: (a) for Ω_0 = 2.09 Hz; (b) for Ω_0 = 4.41 Hz; (c) for Ω_0 = 6.03 Hz

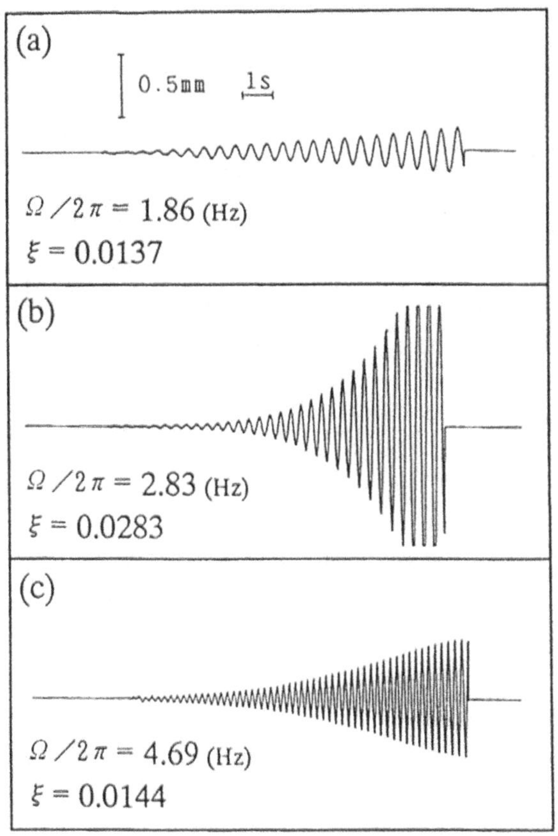

Comparison of Experimental With Theoretical Results

The abscissa Ω_0 in Figure 11 can be transformed to the basic Froude number as defined in Equation (36) in Chapter 7 to permit comparison of the experimental results with the theoretical predictions. The vibration frequency ratio data in Figure 11a can be re-formatted as shown in Figure 12a. The solid curve shows the theoretical predictions calculated from Equation 25b for a water-to-gate mass ratio α_m of 0.95. The data for both inclination angles agree well with the theoretical curves. Substituting the data shown in Figure 11 into Equation 2), the corresponding net fluid-excitation ratio $(\xi_r - \zeta_b)$ can be calculated and subsequently plotted in Figure 12b. The data for both inclination angles again agree very well with the corresponding theoretically calculated curves.

Figure 11. Experimental results for flow-induced vibration of model gate for d_0=290 mm, B=2 mm and α_m=0.95: (a) in-water to in-air vibration frequency ratio Ω/Ω_0; (b) excitation ratio ξ; (c) mechanical damping ratio ζ_0

From Ishii & Nakata (1995). © 1995 A. A. Balkema. Used with permission.

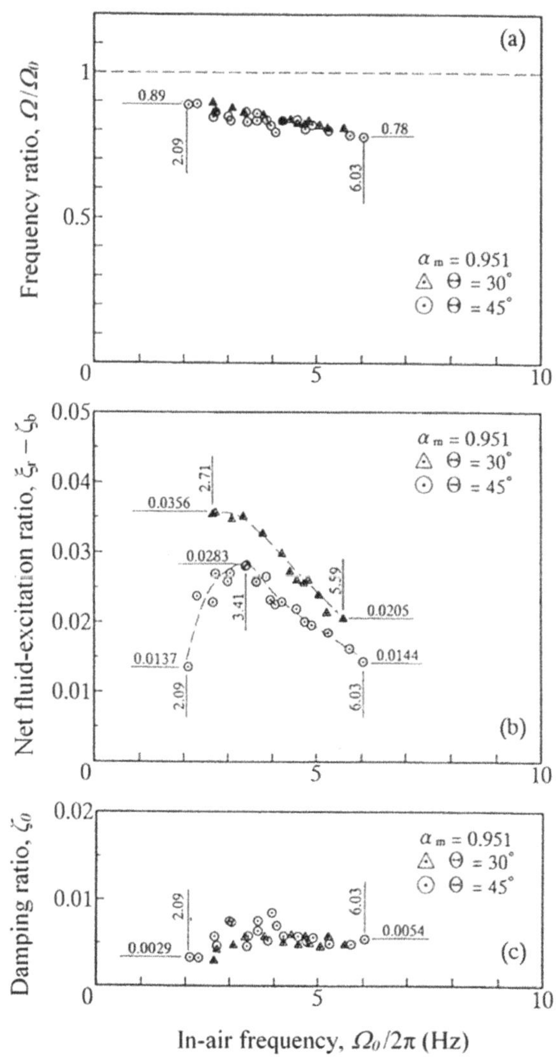

Figure 12. Comparison of experimental results with theoretical results as functions of the basic Froude number F_0 for (a) the vibration frequency ratio F/F_0, and (b) net fluid-excitation ratio $\xi_r - \zeta_b$
From Ishii & Nakata (1995). © 1995 A. A. Balkema. Used with permission.

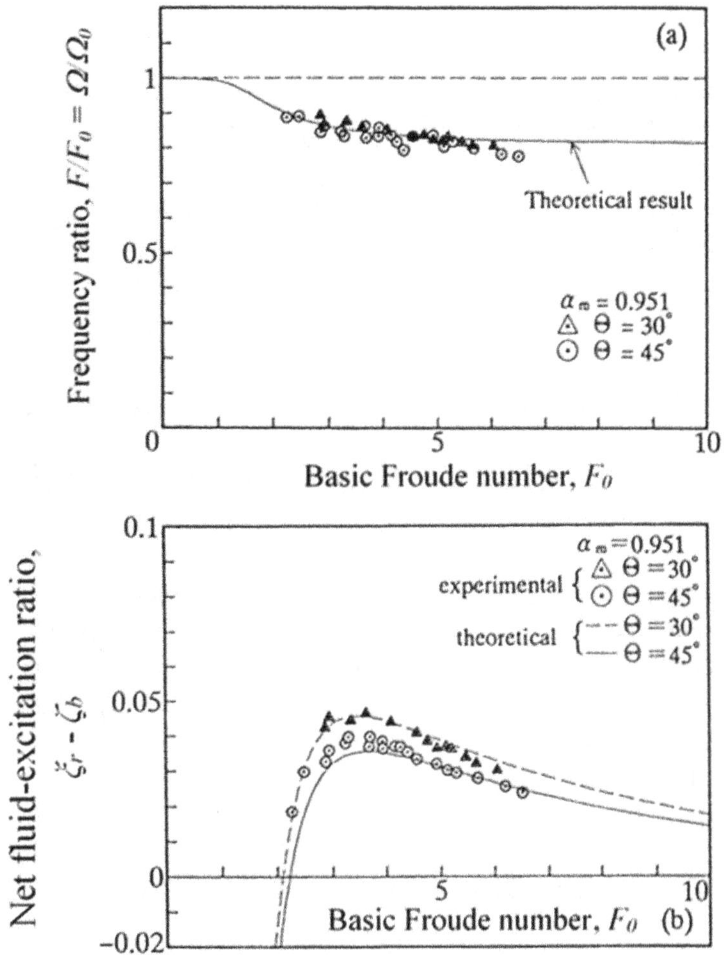

In order to draw the theoretical curves in Figure 12b, it was first necessary to determine the unsteady discharge coefficient c_f in the flow-rate variation factor k defined by Equation 10. For long-span gates, the vibration direction of gate is not usually perpendicular to the discharge flow, as was the case for the Tainter gate studied by Ishii & Naudascher (1992). Based on the difference in the vibration direction, it was suspected that c_f for long-span gates takes on values smaller than 1.0. To validate this suspicion, the gate was forced to vibrate with a small amplitude of 0.25 mm. The distribution of fluctuating flow velocity between the gate bottom and the inclined floor was carefully measured using a laser Doppler velocimeter. The gate submergence depth was 300 mm. The unsteady discharge coefficient c_f calculated from the measured discharge distribution is shown in Figure 13.The unsteady discharge coefficient c_f was approximately constant between 0.50 to 0.52 for vibration frequencies less than about 3 Hz, but subsequently decreases gradually with increasing vibration frequency.

Figure 13. The unsteady discharge coefficient, c_f, decreases with frequencies above 3 Hz
From Ishii & Nakata (1995). © 1995 A. A. Balkema. Used with permission.

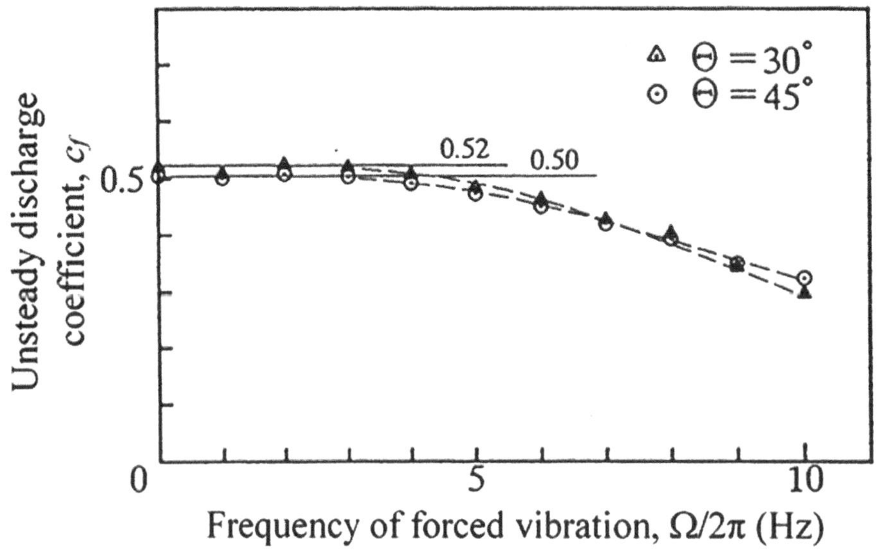

From the experimental data for unsteady discharge coefficient, the gate inclination factor k can be calculated and plotted in Figure 14, as a function of the Froude number F. Based on the characteristics of k, the net fluid-excitation $\xi_r - \zeta_b$ corresponding to the experimental data shown in Figure 12b can be calculated, and subsequently plotted as dashed curve for $\Theta = 30^o$ and the solid curve for $\Theta = 45^o$, in Figure 12. From these comparisons, one may conclude that the theoretical analysis can accurately predict the model test results for the net fluid-excitation ratio.

DYNAMIC INSTABILITY OF SCHINZNACH AND ARA RIVER LONG-SPAN GATES

As an example of application of the present theoretical analyses, the dynamic instability levels the 100-ton Schinznach and 49-ton Ara River long-span gates were analyzed. Table 1 shows the major input parameters for estimation and the resulting assessments. With given values of gate mass, spanwise length l, submergence depth d_0 and in-water vibration frequency Ω, the water-to-gate mass ratio α_m and the Froude number F can be calculated using Equation 14 and Equation 38 in Chapter 5, respectively, as shown in Table 1. The Froude number F is between 10 and 16.5, close enough to assume the asymptotic limit for values larger than 11.0. The upstream-channel to gate submergence depth ratio β^* is 1.0. The corresponding basic Froude number F_0 values can be calculated from Equation 30, resulting in values of 18 to 30. As a result, the net fluid-excitation ratio divided by the gate inclination factor, $(\xi_r - \zeta_b)/k$, can be calculated from Equation 31, since F is larger than 8.0. The unsteady discharge coefficient c_f varies with the vibration frequency, as shown in Figure 13, and takes on an average value of about 0.50 in the vibration frequency range of 3 Hz to 5 Hz. Now the gate inclination factor k can be calculated using Equation 10, and takes on a value of 0.35 for both gates with the gate inclination angle Θ of 60°. Finally, the net fluid-excitation ratio $(\xi_r - \zeta_b)$ is found to be 0.014 for the Schinznach gate and 0.030 to 0.018 for the Ara River gate.

Figure 14. Gate inclination factor k depends on both the gate inclination angle and the unsteady discharge coefficient
From Ishii & Nakata (1995). © 1995 A. A. Balkema. Used with permission.

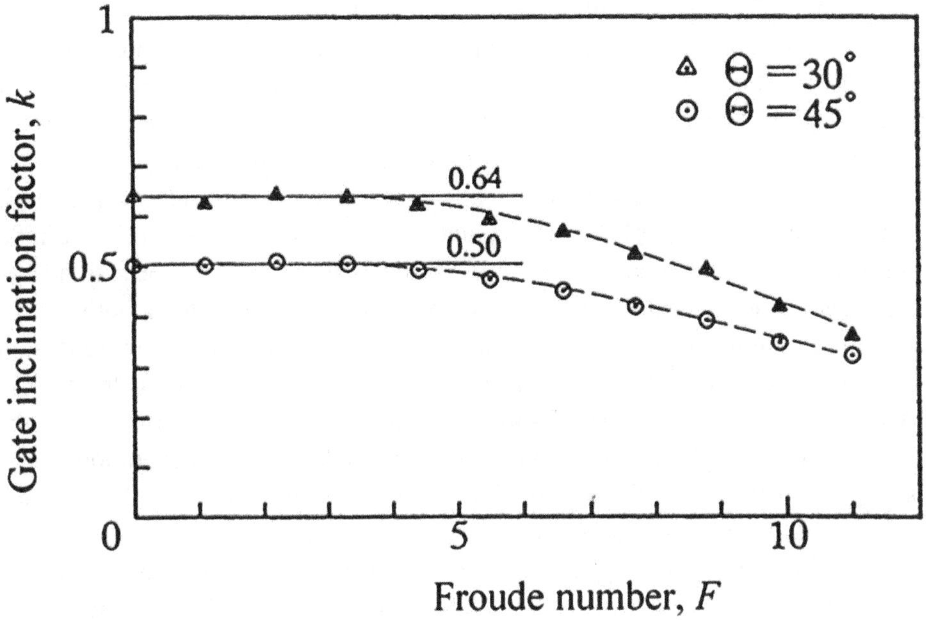

Table 1. Major parameters for gates and calculated values for dynamic instability level for 100-ton Schinznach and 49-ton Ara River long-span gates

Gate	l (m)	d_0 (m)	$\Omega/2\pi$ (Hz)	β^*	α_m	F	F_0	$\frac{\xi_r - \zeta_b}{k}$	Θ (deg)	c_f	k	$\xi_r - \zeta_b$
Schinznach (100 ton)	22.5	3.4	4.1	1.0	2.6	15.2	23.8	0.04	60	0.5	0.35	0.014
Ara-river (49 ton)	28	2.7	3 to 5		4.2	9.9 to 16.5	17.9 to 30.0	0.087 to 0.052				0.036 to 0.018

Long-span gates are generally lightly damped, with low-levels of mechanical damping in the low frequency bending modes. Thus, the criterion for dynamic instability in Equation 28 is easily satisfied. As a result, one may conclude that the Schinznach gate is susceptible to spontaneous vibration with the maximum excitation ratio of 0.014, while the Ara River gate has an even greater susceptibility to more violent spontaneous vibration with the maximum excitation ratio of 0.018 to 0.030, which is about twice as large as the maximum for the Schinznach gate.

VISCOUS FLOW EFFECTS ON GATES WITH UNDERFLOW

Long span gates are frequently used for flood control and in irrigation systems in wide, shallow rivers. Long–span gates can be plate-like structures, as shown in Figure 15a or shell-type structures, as shown in Figure 15b. Most long-span gates, regardless of the cross-sectional design, are supported by cable lift systems at each end of the span. Shell-type long-span gates consist of a long tube-like structure to provide sufficient bending rigidity for the hydrostatic loads in both the horizontal and vertical directions. Shell-type long-span gates are frequently used in irrigation schemes in Japan.

As the streamwise extent of the gate (gate thickness) increases, viscous effects present in the flow beneath the gate and cannot be modeled by the potential flow theory used to this point in this chapter. The flow analysis must include the viscous effects. Viscous effects also become a significant concern for thin plate-like gates when the gates are submerged on the downstream side. In both cases, gates with significant streamwise extent and gates with submerged discharge, the mean flow separation with the associated shedding of vorticity from solid surfaces requires inclusion of viscous terms, usually in the form of the two-dimensional, unsteady Navier-Stokes equations. Inclusion of turbulence in the flow will require a full three-dimensional unsteady Navier-Stokes formulation. The generation of a shear layer beneath the gate presents the possibility of shear layer interactions with the body, leading to instability-induced vibrations and movement-induced vibrations.

A common design feature of shell-type gates is an added flap gate on the crest of the main gate, as shown in Figure 16. As might be expected the fluid loading above the shell-structure and the added mass effects in horizontal direction serve to reduce the natural frequencies in the vertical and horizontal directions, respectively. A coalescence of the vertical and horizontal frequencies can lead to a coupled-mode instability. Some long-span shell-type gates are designed to be self-flooding, with holes along the downstream surface to permit the inflow of water to add mass to the structure. The mass entering through the inflow portals is another factor in reducing the natural frequencies of the structure.

All long-span gates must be able to pass the maximum design flood. To do so, some long-span shell-type gates may be designed for simultaneous over- and underflow. Long-span gates with simultaneous over- and underflow will be considered in the subsequent section of this chapter.

Figure 15. Long-span gates with underflow from left to right: (a) a long-span plate gate with a bending mode normal to the longer inclined plate axis acting as a press-shut device, and (b) a long-span shell-type gate that can, under certain circumstances, act as a coupled-mode press-shut device
After Ishii et al. (1994).

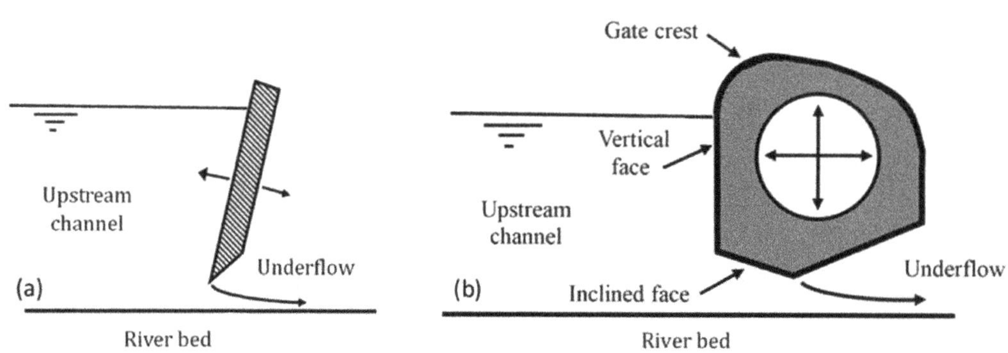

Figure 16. A long-span shell-type gate with horizontal and vertical degrees-of-freedom with underflow may become susceptible to coupled-mode vibration when vertical and horizontal frequencies approach one another due to fluid loading and added mass effects
From Nakata et al. (1996). © 1996 JSME. Used with permission.

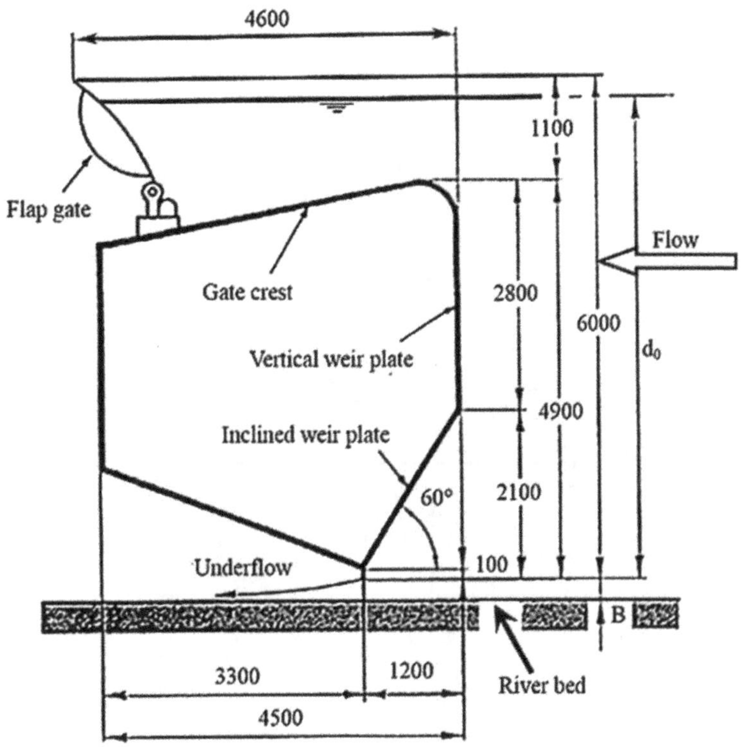

Shear Layer in Underflow Beneath a Gate

Local pressure gradients and the discharge beneath a gate with a small gate opening are very sensitive to miniscule variations in the gate opening. As discussed in Chapter 5, discharge fluctuations can induce both progressive-wave and standing-wave pressure fluctuations in the upstream reservoir, and by extension of the concepts, in the downstream reservoir for submerged gates as well. Pressure fluctuations lead to added mass effects, fluid damping effects, and fluid excitation effects in thin gates with free discharge. Thick gates, both inclined and flat-bottomed, and gates with submerged discharge are subject to multiple modes of vibration as documented by Billeter and Staubli (2000). Previously, transverse vibrations of gates with one vertical (transverse) degree-of-freedom in an underflow were investigated by Hardwick (1974), Kolkman (1976), Thang (1984, 1990), Thang & Naudascher (1986), among others. Streamwise gate vibrations associated with a horizontal (streamwise) single degree-of freedom have been studied by Jongeling (1988), Ishii & Knisely (1989, 1990), and Thang (1990). Systematic classifications of gate excitation and vibration mechanisms have been developed by Naudascher & Rockwell (1994) and Kolkmann & Jongeling (2007).

Figure 17. (a) Laboratory set-up for multiple mode testing of a submerged flat-bottomed gate with small gate openings (0.5 < s/d <1), and (b) mapping of possible modes and trajectories of gate vibration for both one degree-of-freedom (top and bottom rows) and two degrees-of-freedom (center rows); P.S. denotes press-shut; P.O. denotes press-open
After Billeter & Staubli (2000).

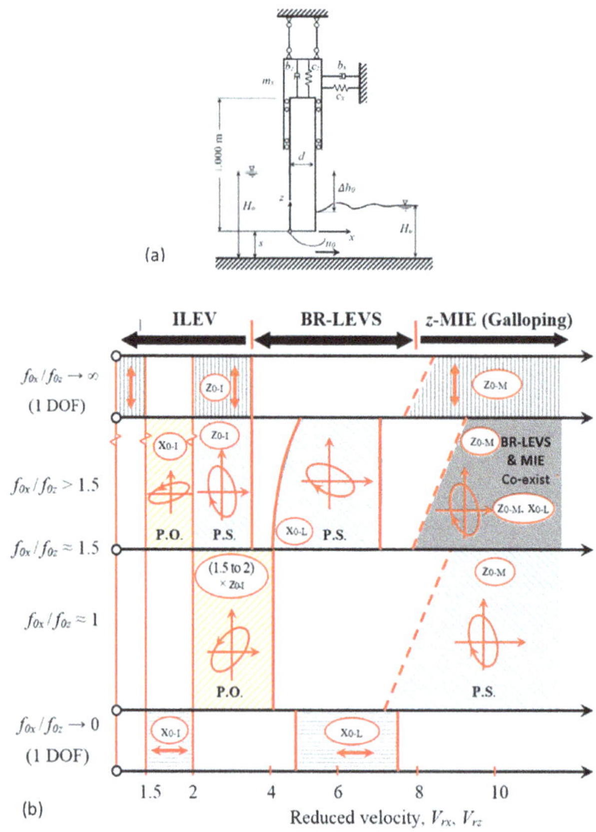

Figure 17a shows a schematic of the model flat-bottomed gate with two degrees-of-freedom used in a study by Billeter & Staubli (2000). They considered small gate openings, $0.5 < s/d < 1.0$, but did not specifically look at variations of gate behavior with s/d values. Similarly they did not vary the difference between the upstream and downstream gate submergence. They assert that the gate opening ratio and the submergence ratio will quantitatively affect their results, but do not alter the fundamental excitation mechanisms they investigated. Figure 17b is their mapping of the various modes and trajectories the gate underwent in their experiments. The excitation involved impinging leading edge vortices (ILEV), body resonant leading edge vortex shedding (BR-LEVS), and movement induced excitation (MIE - galloping). A short description of these various excitation mechanisms follows. In the mapping in Figure 17b, f_{x0}/f_{z0} is the ratio of predominant frequencies of vibration in the horizontal (x) and the vertical directions (z), and the reduced velocities along the abscissa are based on the respective frequencies with the gate width d as the length scale, so $V_{rx} = u_0/f_{x0}\, d$ and $V_{rz} = u_0/f_{z0}\, d$. In the mapping, the amplitudes for the two degrees-of-freedom vibrations are referenced to the corresponding amplitude for the single degree-of-freedom in the same reduced velocity range. Billeter & Staubli (2000) reported that the addition of a second degree-of-freedom extended the range of reduced velocities over which vibration occurred and resulted in larger amplitudes of vibration for $f_{x0}/f_{z0} = 1$, but otherwise did not affect the maximum amplitude of vibration substantially. For details on the evolution and interaction of these various modes, please see the original paper.

In Figure 17b, the one degree-of-freedom in the x-direction occurs over two reduced velocity ranges when motion in the z-direction is constrained. In the lower velocity range, $1.5 < V_{rx} < 2.0$, pressure fluctuations result from the impingement of the vortex shed from the leading edge on the gate bottom. The frequency of this mode is determined by the shear layer instability and the length from the leading edge to the impingement location. The net fluctuating pressure loading on the vertical faces was attributed to discharge fluctuation resulting from the varied discharge flow area associated with the passage of flow structures along the bottom of the gate, as suggested by Ishii & Knisely (1988; 1989; 1990). This lower velocity range is denoted as impinging leading edge vortex (ILEV). In the higher velocity range, $4.0 < V_{rx} < 8.0$, small-amplitude, horizontal gate motion imposes its frequency on the vortices shed from the leading edge, producing body-resonant leading edge vortex shedding (BR-LEVS) excitation. The shed vortices no longer impinge on the bottom surface of the gate, but rather are shed into the downstream reservoir behind the gate where they produce a fluctuating pressure loading on the downstream gate face.

One degree-of-freedom vibration in the z-direction (with the x-direction constrained) occurs over multiple velocity ranges. Two relatively low reduced velocity regions of vertical vibrations, $1.0 < V_{rz} < 1.5$ and $2.0 < V_{rz} < 3.8$, shown in Figure 17b, were identified by Billeter and Staubli (2000). The higher of these two ranges is driven by the ILEV mechanism, while to lower region is interpreted as a subharmonic response to the ILEV excitation. The higher reduced velocity single degree-of-freedom z-direction, $V_{rz} > 8$ to 10, is attributed to movement-induced excitation (MIE) akin to the galloping of rectangular cylinders in cross-flow.

The two degrees-of-freedom response depends strongly on the frequency ratio of the horizontal and vertical vibrations, fx_0/f_{z0}. When the two frequencies are substantially equal, the second row from the bottom in Figure 17b shows both modes can be driven by ILEV excitation with a maximum amplitude in the vertical direction at 1.5 to 2.0 times the amplitude of the one degree-of-freedom vertical vibration and over a range of V_{rz} that extends to higher velocities than for the one degree-of-freedom vertical vibration case. At reduced velocities greater than about 8, it appears that the MIE excitation drives the coupled vibration in the vertical direction and the x-component responds as driven vibration following the vertical forcing. When the ratio of frequencies exceeds 1.5, the third row from the bottom pertains and show that at low reduced velocities, the x-direction ILEV excited vibration serves as the driver of the coupled system. At slightly higher reduced velocities, the vertical response to the ILEV excitation serves as the driver for the coupled vibration. At a higher range of reduced velocities, $5 < V_{rx} < 7.5$, the x-direction response to the BR-LEVS excitation serves as the driver for the coupled vibration. At slightly higher reduced velocities, that is, at V_{rx} greater than about 8.0, the two modes each have their separate excitation mechanisms; the x-vibration is driven by the BR-LEVS excitation while the vertical mode is an MIE-driven vibration.

Instability of Full-Scale Gate at Very Small Gate Openings With Submerged Discharge

A set of Tainter gate field tests were conducted on a 77-ton Tainter gate, denoted as "Gate P," in October 2007. In the testing, small levels of damped vibrations of the Tainter gate in running water were triggered by breaking a small diameter steel rod attached between the bottom center of the skinplate and the surface of the dam crest (see Anami et al., 2013 for details of the testing method). From these tests, the gate's inherent in-water vibration frequencies and mode shapes were clearly identified. Using the measured dynamic characteristics of the gate as input into the theoretical analysis, the dynamic stability of the gate was carefully examined over a range of operating conditions to ascertain the gate's susceptibility to the coupled-mode dynamic instability (see Chapter 10 for the theoretical analysis of the self-excited coupled-mode instability mechanism). It was determined that the gate is stable at all operating conditions relative to the coupled-mode mechanism.

At small gate openings and with a 0.42 m downstream submergence, however, the gate experienced a strong instability-induced vibration due to the separated flow beneath the gate discharging into the downstream reservoir. This instability involves the generation of vortices, swirling regions of fluid, in the separated flow region just downstream of the gate.

A photograph of Gate P is shown in Figure 18. The configuration and major dimensions of the gate are shown in Figure 19. The skinplate has a height of 12.5 m and a span of about 15.24 m. The length of radial arm is 15.24 m. The whole gate mass is about 77 tons. The skinplate has 23 vertical I-beams and 3 horizontal girders, as shown in Figure 19. The radial arm structure has 3 struts strengthened with a truss structure. The radial arms are inclined about 16 degrees inward, toward the span center. The gate is raised by two massive chains attached, one to each spanwise end of the skinplate.

Instrumentation

As shown by the small circles in Figure 19, the vibrational acceleration of the gate was measured at 12 points on the vertical ribs along the skinplate, at 2 points on the radial arm and at 2 points on the hoist frame and hoisting chain. The measurement positions are denoted "A" to "F" along the skinplate bottom, "F1" to "F7" along the skinplate center, "R1" and "R2" along the radial arm and "H1" and "H2" for the

Figure 18. The 77-ton Gate P viewed from the downstream side with the gate closed showing downstream gate submergence

hoist frame and chain. On the skinplate, radial (essentially streamwise) accelerations were measured, as shown by the arrows with no designation near their tips. In addition, tangential (essentially vertical) accelerations were also measured at "A", "F2", "F5" and "F7," as shown by the arrows with a "T" designation near their tips. At position "F7," the spanwise acceleration was measured as well, as shown by arrows with "Sp" designations near their tips. On the radial arm, the tangential and spanwise accelerations were measured, as shown by arrows with "T" and "Sp" designations, respectively. On the hoist frame, the vertical acceleration was measured, as shown by arrows with a "T" designation. In addition, vertical and spanwise accelerations were measured on the hoisting chain, as shown by arrows with "T" and "Sp" designations, respectively. In total, 24 servo-type accelerometers (Lion Co.: LS-20C, LS-10C) were used. The accelerometers were attached to the gate, with strong magnetic bases.

In addition, a gate-opening gauge (Kyowa Co. DPT-05MDS) was attached with duct tape to the right side of the skinplate, as shown by the small square in Figure 19c. The gate-opening gauge has a thin wire cable, the tip of which was fastened with special adhesive to the concrete pier.

Time Traces

Vibratory accelerations of the gate, in response to the input energy from the breaking of the steel-rod or from the flowing water (ambient excitation), were recorded on computers through A/D converters.

Figure 19. Drawing of Gate P showing dimensions and positions of accelerometers and gate-opening gage (small square marker) from (a) the side view, (b) the top view, and (c) the view from downstream; all dimensions in mm

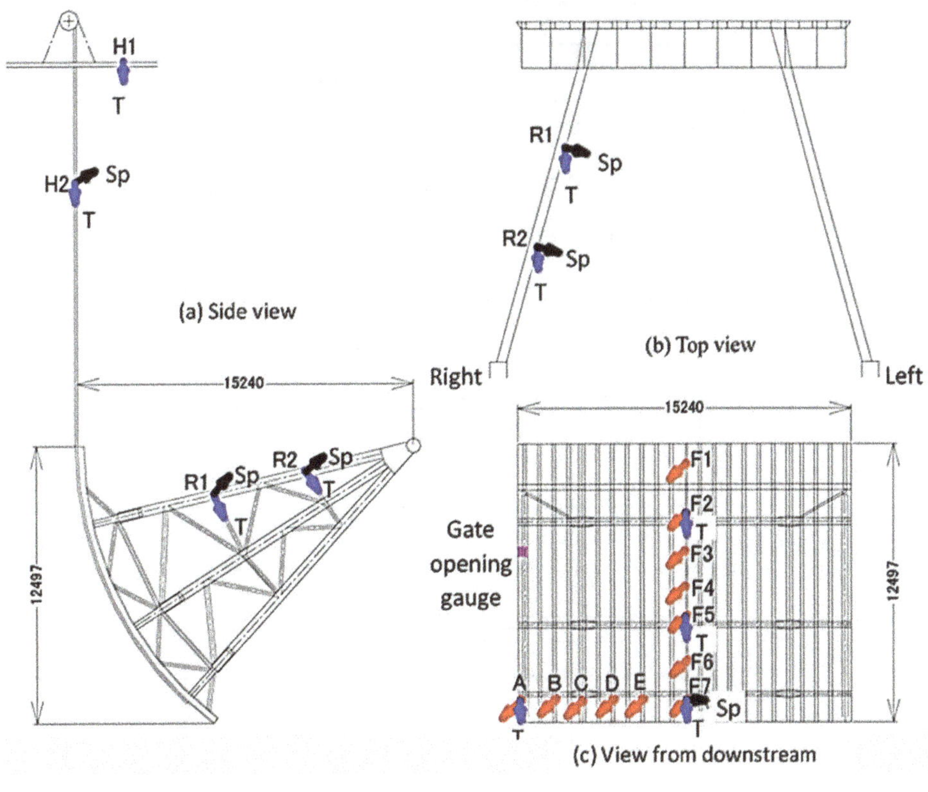

The data sampling rate was at 3 kHz or 1.5 kHz. Measured vibration accelerations for each of the tests were sequentially displayed as real time waveforms. The tests using ambient excitation with a very small gate opening of 35 mm produced the acceleration signals shown in Figures 20 (radial), 21 (radial), and 22 (tangential). The acceleration signals for this small opening exhibit large amplitude accelerations with the abscissa scale of 2×G (twice the acceleration of gravity). With a gate opening of 76 mm, the acceleration signals, shown in Figure 24, as an example, with an abscissa scale of 0.1×G (that is, 1/20 of the scale in Figures 20 to 22), show no conspicuous vibrations. Similar traces with no indication of vibration were recorded at all other accelerometer locations with gate openings larger than 76 mm.

Frequency Spectra

Measured vibration accelerations were analyzed using DADiSP/2000 WORKSHEET from Astrodesign, Inc. The Fourier transform was performed using this software to produce amplitude spectra, and compute the frequency, damping ratio and phase-lag to generate structural mode shapes.

Figure 20. Time traces of radial (streamwise) acceleration signals from 6 accelerometers positioned along the bottom of the skinplate with a gate opening of B = 35 mm

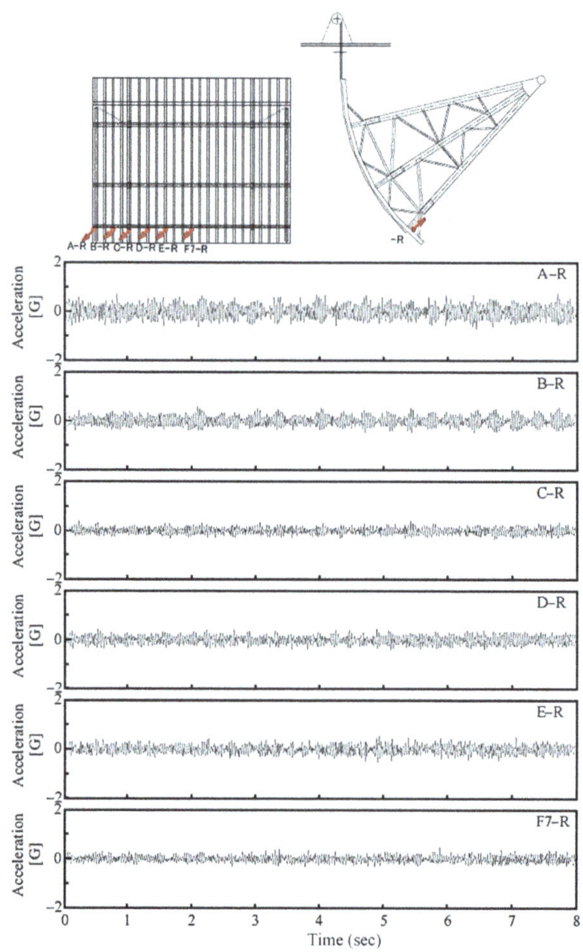

Figure 21. Time traces of radial (streamwise) acceleration signals from 7 accelerometers positioned on the spanwise centerline of the skinplate with a gate opening of B = 35 mm

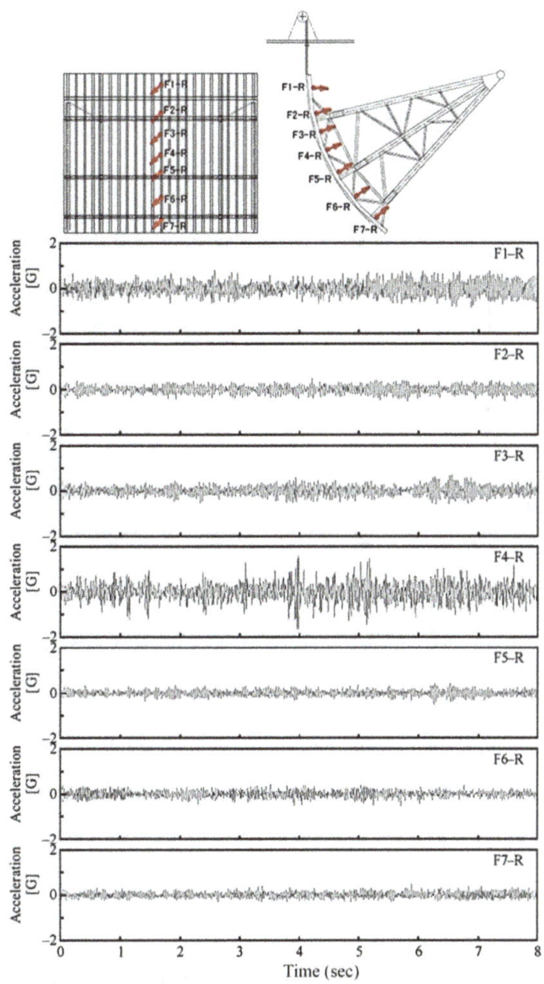

When the gate opening was very small, for example, $B = 35$ mm, high level vibrations occurred, as shown in the spectra in Figures 24 to 26 (also see Figures 20 to 22 for corresponding time traces). In these spectra several sharp peaks are present in the frequency range from 25 Hz to 35 Hz. From careful examination of the spectra in Figures 24 to 26, consistent peaks at 28.3 Hz were identified. These peaks at a relatively high frequency suggest possible instability-induced excitation, related to vortices in the separated flow from the bottom edge of the skinplate as it passes into the downstream reservoir. Such excitation would not be expected if the gate had free discharge with no downstream submergence.

The acceleration spectra from the ambient excitation with a small gate opening ($B = 35$ mm) produced a clearly identifiable and consistent vibration mode at a relatively high frequency, as shown in Figure 27. Along the vertical center line of the skinplate, a two-wavelength mode of streamwise bending vibration is seen in Figure 27a, accompanied by a one-wavelength mode of spanwise bending vibration of the radial arms. An additional one-wavelength streamwise bending mode was identified along the

Figure 22. Time traces of tangential (vertical) acceleration signals from 6 accelerometers on skinplate and radial arms with a gate opening of B = 35 mm

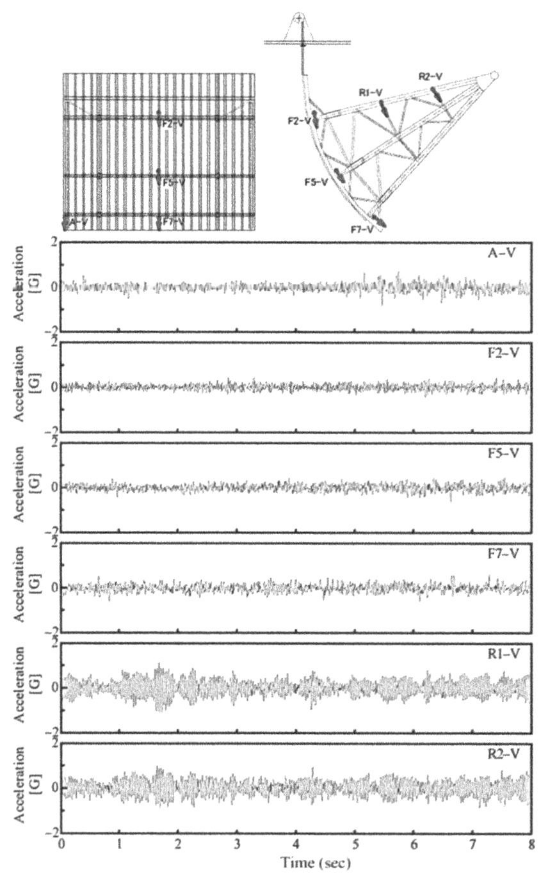

bottom of the skinplate, that is, in the spanwise direction, in Figure 27c.The deformation distributions with multiple nodes along the gate centerline as well as in the spanwise direction would reduce the integrated effect of this vibratory deformation on the gate's moment of inertia about the trunnion pin. Thus, the vibration mode at 28.3 Hz is not expected to result in any significant associated susceptibility to the coupled mode instability.

The vibration at 28.3 Hz is a concern for the gate owner and should receive attention. Analysis that follows strongly suggests that this vibration results from instability-induced excitation, produced by vortex formation due to the separation of the flow from the bottom edge of the skinplate and passing into the downstream reservoir. This mechanism of vibration is distinctly different from the coupled-mode instability. The frequency map in Figure 17 shows the occurrence of instability-induced excitation, denoted as ILEV in Figure 17, occurring over the range of reduced velocities from 1 to 4, roughly. A review of the substantial literature on IIE and the details of this excitation mechanism can be found in Naudascher & Rockwell (1994).

The spectra in Figure 28 are typical of the spectra obtained from all acceleration traces with gate openings larger than about 76 mm. No consistent structural mode shape could be identified by these very low level and somewhat random acceleration signals.

Figure 23. Time traces of radial (streamwise) acceleration signals from 6 accelerometers positioned along the bottom of the of the skinplate with a gate opening of B = 76 mm showing negligible vibrations

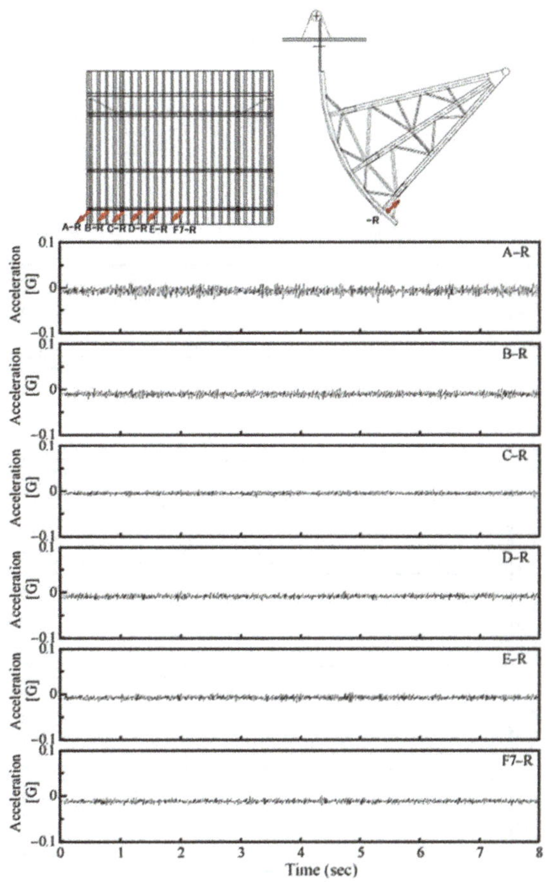

Analysis of the Mechanism

The downstream side of Gate P was submerged to a depth of about 420 mm. With ambient excitation at very small gate openings of 31 mm to 35 mm, a relatively high level, noisy vibration occurred. The vibration frequency was determined to be 28.3 Hz and the acceleration reached amplitudes of about 0.2×G at F1-R (in the streamwise direction at the top center of the skinplate) and 0.5×G at R1-S (in the spanwise direction on the radial arm). The suspected source of this vibration is flow separation and vortex interactions with the gate in the submerged downstream region. To confirm this suspicion, a detailed examination of this vibration was undertaken to ascertain whether it was a movement-induced vibration or an instability-induced vibration.

A flat-bottom gate with submergence on the downstream side, illustrated in Figure 29a, is susceptible to instability-induced-excitation resulting in self-sustained vibrations (Naudascher & Rockwell, 1994). The onset conditions for this self-excited vibration have been empirically formulated in terms of two non-dimensional parameters: the reduced velocity, V_r, and the ratio of the gate opening B to the streamwise gate width S. The reduced velocity V_r is defined by

Figure 24. Amplitude spectra of streamwise accelerations corresponding to the traces in Figure 20 with accelerometers along the bottom of the skinplate with a gate opening of B = 35 mm

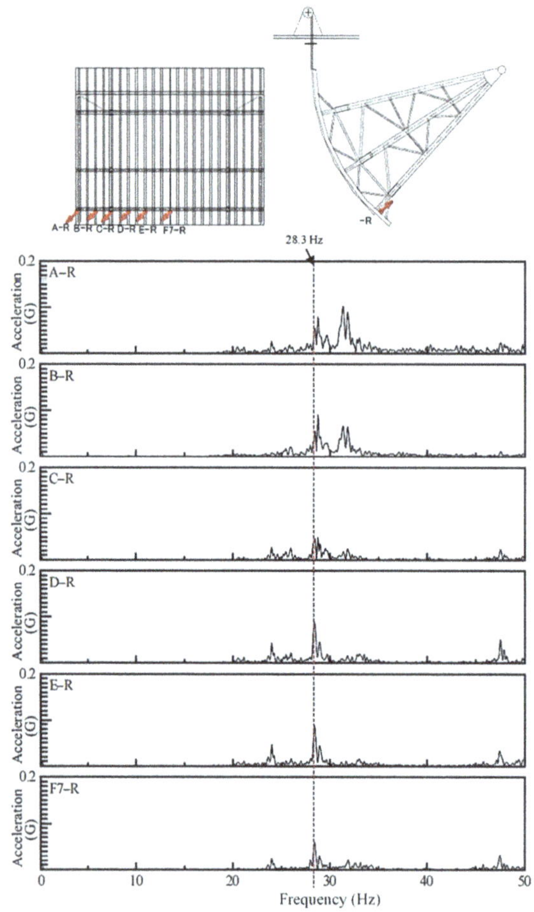

$$V_r = V / \left(f \cdot S\right) \tag{32}$$

where V represents the jet-flow velocity under the gate, f represents the vibration frequency in Hz and S represents the streamwise width of gate bottom.

Table 2 presents the parameter ranges for the two excitation mechanisms being considered. The first row is for instability-induced vibrations that occur at a reduced velocities, V_r, between 2 and 4 for gate opening ratios, B/S, smaller than 1.0. The second row is for movement-induced vibrations that typically occur at reduced velocities, V_r, between 20 and 40 for gate opening ratios between 5 and 10.

The jet flow velocity V can be calculated as

$$V = \sqrt{2g\left(d_0 - h\right)} \tag{33}$$

Figure 25. Amplitude spectra of streamwise accelerations corresponding to the traces in Figure 21 with accelerometers along the vertical centerline of the skinplate with a gate opening of B = 35 mm

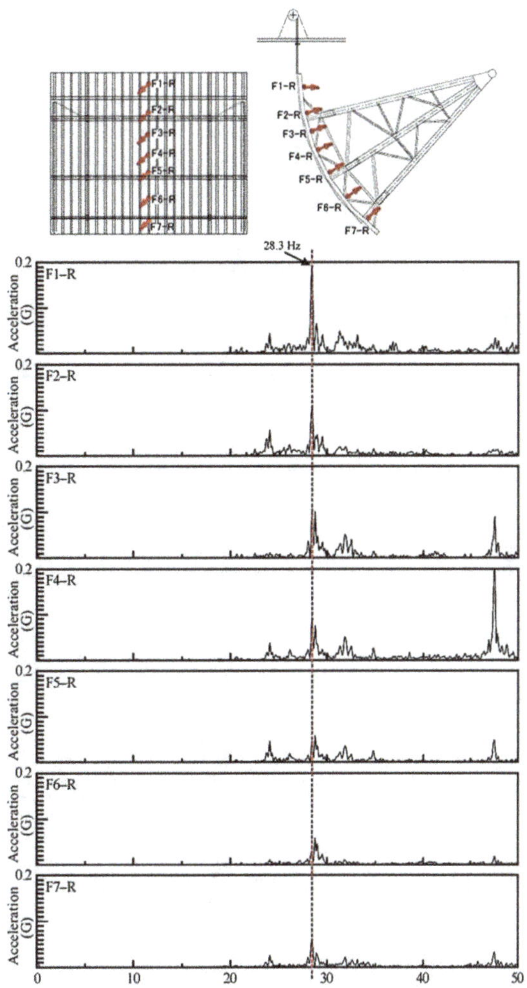

and equals 15.2 m/s with an upstream gate submergence d_0 of 12.02 m and the downstream gate submergence h of 0.42 m.

Gate P does not, however, have a simple rectangular flat bottom. As shown in Figure 29b, the gate bottom progressing downstream from the upstream face consists of a relatively short flat bottom region with a streamwise length of 115 mm followed by a region inclined at 45° with a horizontal length of 260 mm. The corresponding characteristic length, S, is then bounded as

$$0.115\text{m} \le S \le \sim 0.318\text{m} \tag{34}$$

With a gate opening of B = 35 mm, the gate opening ratio B/S is substantially less than 1.0:

$$\sim 0.11 \le B / S \le 0.30 \tag{35}$$

Figure 26. Amplitude spectra of tangential accelerations corresponding to the traces in Figure 22 with accelerometers on the skinplate and radial arms with a gate opening of B = 35 mm

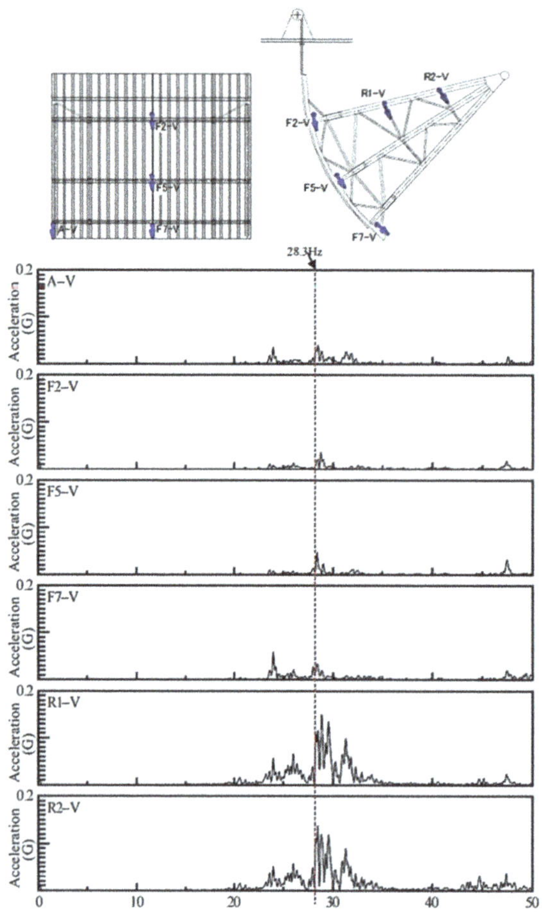

This *B/S* range would suggest the mechanism is instability-induced excitation if the reduced velocity is in the appropriate range.

The value of the reduced velocity V_r for Gate P can be estimated using the measured value of vibration frequency of 28.3 Hz in Equation 33:

$$V_r = V / \left(f \cdot S\right) = \frac{15.2\text{m/s}}{28.3\text{Hz} \times \left(0.115 \text{ to} \sim 0.318\right)\text{m}} = 4.7 \text{ to} \sim 1.7 \tag{36}$$

This V_r range closely agrees with the value shown in the mapping of Figure 17 and similar sources of empirically determined parameter ranges for vibration due to instability-induced excitation (Naudascher & Rockwell, 1994).

The measured relatively strong and noisy vibration at a frequency of 28.3 Hz with ambient excitation with a small gate opening is an instability-induced vibration caused by the flow instability in the separated flow from the bottom edge of the skinplate passing into the downstream reservoir.

Figure 27. Mode shape at a relatively high frequency of 28.3 Hz (a) side view and (b) top view

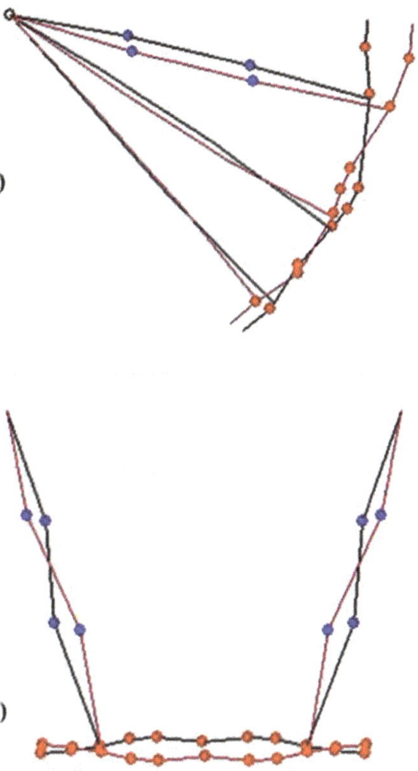

(a)

(b)

In order to prevent this vibration, the downstream submergence would need to be eliminated to permit free discharge on the downstream side of the gate. To do this, the downstream river bed should be maintained to remove debris and sediment build-up such that the downstream side of the gate always has free, non-submerged discharge. While this remediation may be expensive, or there may be environmental objections, the removal of rocks and accumulated sediments from the immediately downstream reaches of the river, if permitted, is expected to eliminate this vibration problem.

If such measures are not possible and if free discharge cannot be maintained, then the existence of this instability for small gate openings would dictate a policy of traversing the gate as quickly as possible through the range of small openings.

Fluctuating Discharge Beneath a Submerged Streamwise Vibrating Gate

A very significant component in the streamwise excitation of a one or two degree-of-freedom gate is the periodic fluctuation in the effective flow area beneath the gate. A flat-bottomed gate vibrating horizontally in a horizontal channel has no geometric change in the available flow area due to the gate vibration. Thus, one might naively suspect that streamwise motion of the gate would not be accompanied by discharge fluctuations, and that the gate would be dynamically stable. Contrary to this expectation and, especially for gates with submerged discharge, gate motion does cause discharge fluctuations (Ishii & Knisely, 1988, 1989, and 1990). The discharge fluctuations appear to result from, or at least be ac-

Figure 28. Amplitude spectra of streamwise accelerations corresponding to the traces in Figure 23 with accelerometers along the bottom of the skinplate with a gate opening of B = 76 mm

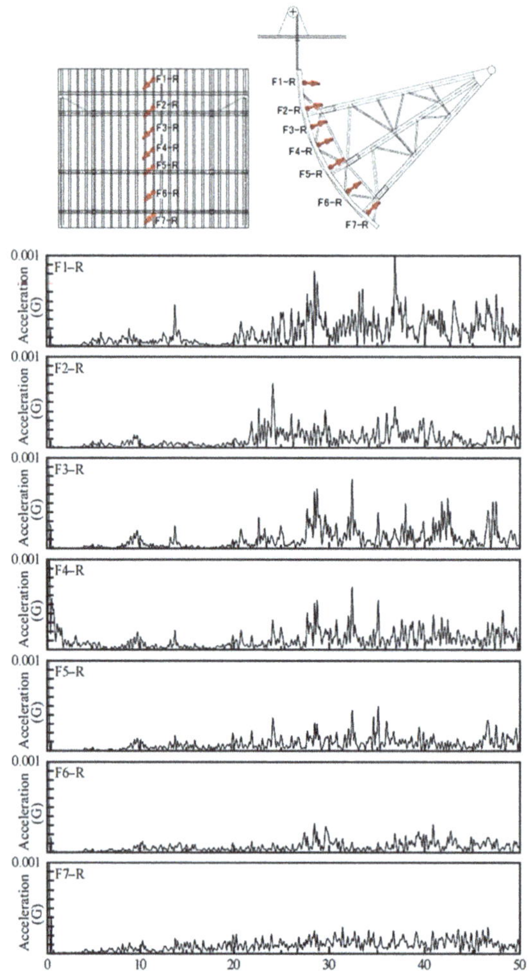

companied by, large scale shear layer vortices shed from the leading edge of the gate and the resultant flapping motion of the shear layer.

The large scale vortices have been shown to result in pressure loadings on the upstream and downstream gate faces which sustain the vibration. Over a limited range of reduced velocities and gate openings, these large scale shear layer vortices respond quite readily to the gate motion, resulting in a violent vertical oscillations of the separated shear layer. The periodic pinching down of the separated shear layer results in discharge fluctuations that are not related to any geometrical variation in the opening height, but rather are induced by the hydrodynamics of the shear layer.

Ishii and Knisely (1988, 1990) undertook hot-film anemometer measurements in the shear layer, flow visualization, and pressure measurement on a flat-bottomed gate forced to vibrate in the streamwise direction. A schematic diagram of their apparatus is shown in Figure 30b, along with a list of relevant parameter values (Figure 30a) and the definitions of non-dimensional parameters (Figure 30c). The parameters in this experiment were sized to produce an unstable shear layer based on the stability criteria

Figure 29. (a) A flat-bottomed gate, and (b) schematic of the bottom of Gate P with a more complicated bottom shape

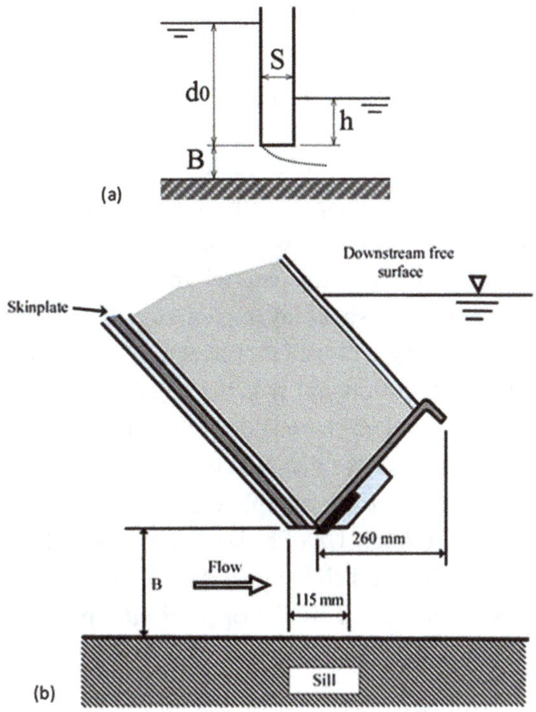

Table 2. Empirical parameter ranges for instability-induced and movement-induced vibrations for a submerged, vertical flat-bottomed gate

Excitation Mechanism	*B/S*	V_r
Instability-Induced Vibration	$(B/S) < 1.0$	$2 < V_r < 4$
Movement-Induced Vibration	$5 < (B/S) < 10$	$20 < V_r < 40$

Figure 30. Measurement of shear layer fluctuation showing (a) values of parameters; (b) schematic diagram of the submerged flat-bottomed gate apparatus, and (c) definition of dimensionless parameters
From Ishii & Knisely (1990) © 1990 JSME. Used with permission.

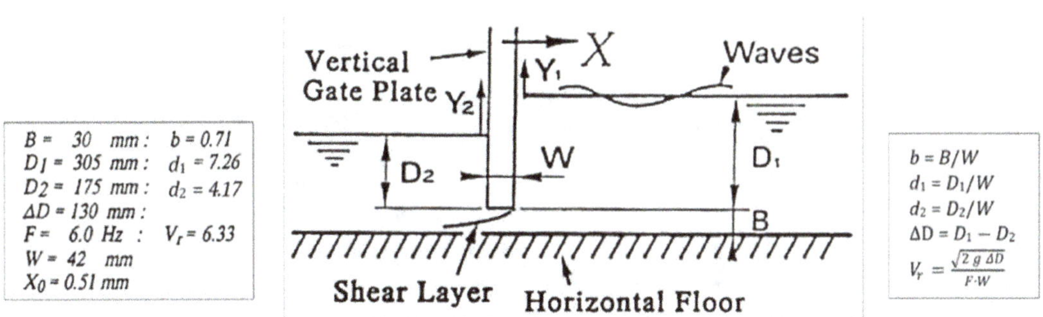

of Thang & Naudascher (1986). The gate thickness was 42 mm and the gate opening was 30 mm. The gate was driven at 6.0 Hz with an amplitude of $X_0 = 0.51$ mm.

An estimate of the magnitude of the fluctuating discharge coefficient, Δc_d, and its phase, ϕ_d, relative to the gate motion was measured using a hot-film anemometer beneath the leading edge of the separated shear layer. The anemometer was sequentially placed at various vertical locations in the gate opening.

The digitization of the flow visualization photographs at multiple streamwise and vertical locations permitted an estimation of the maximum and minimum excursions of the shear layer. The estimated excursions of the shear are shown in Figure 31, where the distance H is measured from the bottom of the channel and is non-dimensionalized with the gate width as $h = H/W$.

Figure 31a shows the non-dimensional shear layer excursion. The phase of the shear layer oscillation is also plotted in Figure 31b, suggesting a constant convective speed of approximately 0.511•($2\pi FW$), in which F is the frequency of gate vibration and W is the streamwise extent of the gate (see Figure 30). For more on convective speed and wavelength considerations, please see the original paper.

These excursions were extracted from dye visualization in the water flow beneath the gate. The dye visualization photographs in Figure 32 show a complete cycle of gate motion, starting at the maximum upstream gate location and stepping though 1/8 of a cycle between each photograph. The white inserted rectangle marks the gate cross-section and the sinusoidal trace in the white area shows the cycle of vibration with the current location indicated by the open circle on this sinusoid. Careful study of the

Figure 31. (a) Shear layer excursion about its mean position, and (b) shear layer maximum and minimum positions relative to the channel bottom
From Ishii & Knisely (1990). © 1990 JSME. Used with permission.

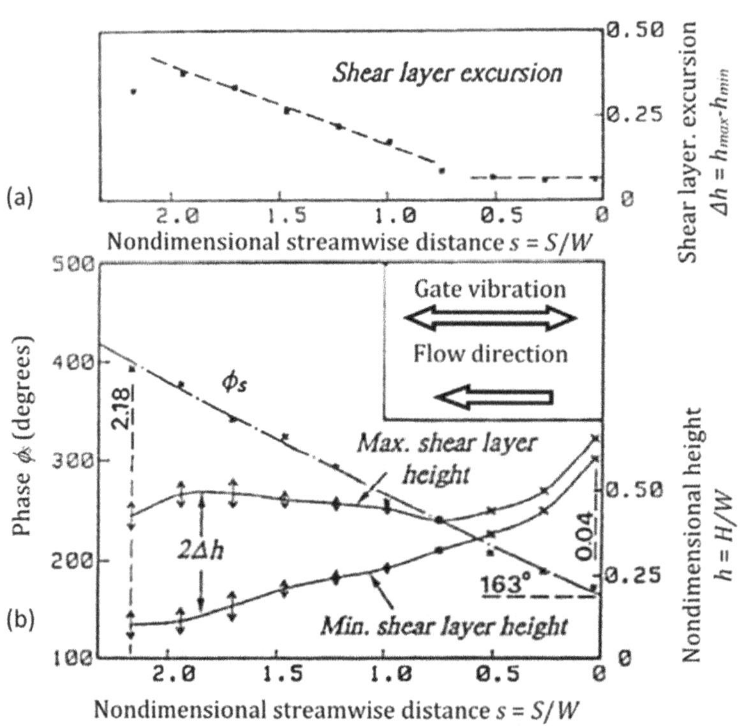

Figure 32. Dye visualization of the shear layer beneath a submerged flat-bottomed gate undergoing forced vibration at V_r = 6.33 with a reduced gate opening of b =0.71 and a reduced amplitude x_0 = X_0/W of 0.00238

From Ishii & Knisely (1990). © 1990 JSME. Used with permission.

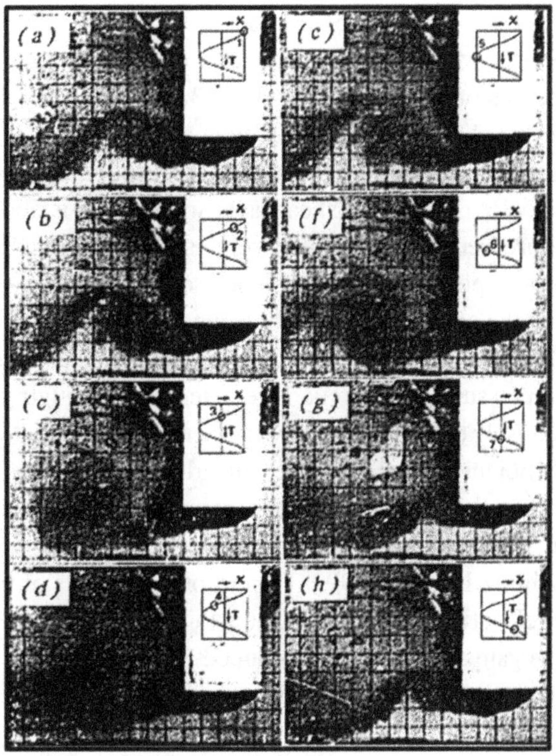

photographs reveals that as the gate moves downstream from its maximum upstream position, as in Figures 32a to 32d, the shear layer in the downstream region is moving downward, essentially in phase with the gate velocity. During the second half cycle, the gate moves upstream and the shear layer beneath the downstream gate corner is drawn upwards to pinch off the previous large vortex and generate a new smaller vortex that can be clearly identified in Figures 32g and 32h.

The measured fluctuating discharge coefficient was found to be 0.0056 for all vertical locations directly beneath the gate's leading edge with a phase angle of 295° relative to the gate displacement.

In addition to the quantification of shear layer fluctuation using hot film measurements and flow visualization, pressure measurements on the vertical gate faces were recorded and analyzed. The fluctuating pressure amplitude and phase relative to gate displacement are shown in Figure 33. The pressure amplitudes on the upstream gate face (ΔP_1) and on the downstream gate face (ΔP_2) are normalized as:

$$\Delta p_1 = \Delta P_1 / \left(2\rho g X_0\right) \text{ and } \Delta p_2 = \Delta P_2 / \left(2\rho g X_0\right) \quad (37)$$

The component of pressure in phase with the gate velocity provides a measure of the energy transfer from the flow to the gate. From the integration of the pressure amplitude times the sine of the phase angle over each of the gate surfaces, Ishii & Knisely concluded that energy transfer to the moving gate from the fluid occurs in roughly equal measures on the downstream face and on the upstream face. Based on these observations and measurements, one should expect to find streamwise gate vibrations over a certain range of reduced velocities for which this energy transfer occurs.

Ishii & Knisely (1989) considered how the magnitude and phase of the discharge fluctuations vary with the reduced velocity. Using the same apparatus as in their 1988 paper, they varied the oscillation frequency to change the reduced velocity. At frequencies of 3.60, 5.32, 5.72, 6.00 and 7.40 Hz and corresponding reduced velocities of 10.55, 7.14, 6.64, 6.33 and 5.13, they determined the wavelength, wavenumber and convective speed of the shear layer fluctuations. Figure 34 shows the wavelength and convective speed increase and the wavenumber decreases correspondingly with increasing reduced velocity. When the reduced velocity is 5.13, the shear layer wavelength is about 2.19 and the shear layer excursion was at its largest observed value. Speculatively, Ishii &Knisely (1989) suggest the maximum shear layer excursion might be expected as the wavelength of the shear layer approaches twice the gate width, further suggesting that the pressure fluctuations due to gate motion and those due to convection of shear layer vortices may be mutually reinforcing. The maximum shear layer excursion, however, does not correspond to maximum pressure excitation on the gate due to changing phase relations.

From their data, Ishii & Knisely (1989) note that with increasing reduced velocity from 5.13 to 10.55, the fluctuating discharge coefficient decreases close to linearly from 0.60% to 0.48%. Similarly, the phase-lag of the fluctuating discharge relative to gate displacement also shows a linear decrease from 309° to 228°. The flapping motion of the shear layer was shown to be consistent with the discharge fluctuations.

Continuing their work, Ishii & Knisely (1992a) report on the effects of the gate opening height, the upstream-downstream water elevation differences, water-to-gate mass ratio, and gate width on the measured excitation acting on the gate as functions of reduced velocity.

The findings in Ishii & Knisely (1992a) are summarized as follows:

1. For a very lightly damped or undamped gate, self-excited streamwise vibrations may be expected over the range of reduced velocities $V_r = \sqrt{2\rho\Delta D} \,/\, (FW)$ (where ΔD is the upstream-downstream elevation difference, F is the gate vibration frequency in Hz, and W is the streamwise gate thickness) from 4.5 to 7.0 for reduced gate openings ($s = S/W$, where S is the dimensional gate opening) from 0.52 to 1.2 (see Figure 35a).

Figure 33. Fluctuating pressure magnitudes and phases on (a) the downstream face, and (c) on the upstream face of a submerged flat-bottomed gate undergoing forced vibration, as shown in (b)
From Ishii & Knisely (1990). © 1990 JSME. Used with permission.

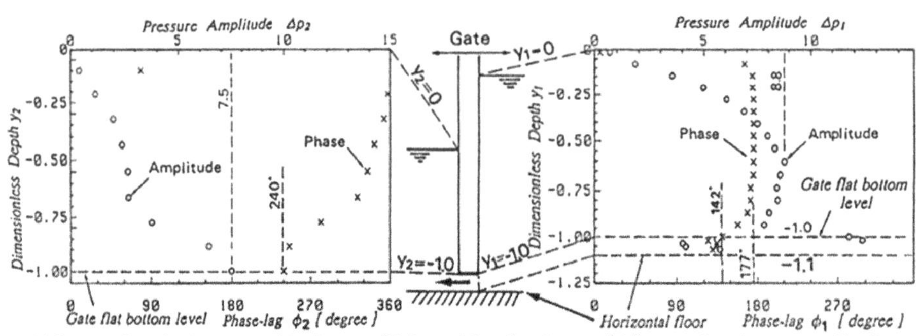

(a) Pressure along the downstream side of gate (b) Gate and flow domain (c) Pressure along the upstream side of gate

Figure 34. Non-dimensional wavelength, wave number, and convective speed of the shear layer oscillations depend on the reduced velocity in experiments using forced gate vibration at a small amplitude
From Ishii & Knisely (1989) © 1989 Elsevier. Used with permission.

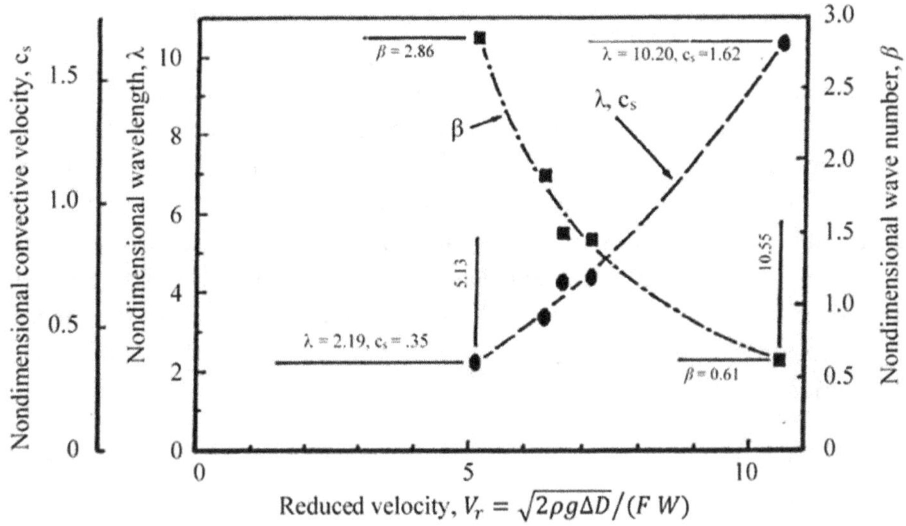

Figure 35. (a) Amplitude, (b) fluid excitation ratio, and (c) added mass for streamwise self-excited vibration of a submerged flat-bottomed gate due to shear layer vortices
From Ishii & Knisely (1992a). © 1992 JSME. Used with permission.

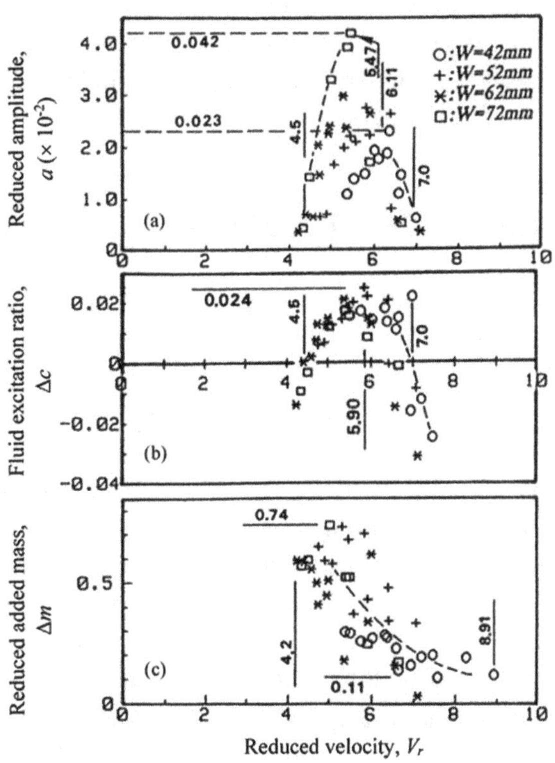

2. Fluid excitation (negative fluid damping) reaches a peak value of about 0.024 with a gate opening of about 0.71 and a reduced velocity between 5.34 and 5.9 (see Figure 35b).
3. The reduced added-mass (the added mass divided by the representative mass of water) decreases with increasing reduced velocity, from about 0.74 at $V_r = 4.2$ to about 0.11 at $V_r = 8.91$ (see Figure 35c).
4. The maximum amplitude occurs at different reduced velocities and reduced gate openings from those at which the fluid excitation ratio reaches its maximum. Larger gate widths resulted in larger amplitudes. Correct prediction of amplitudes will require knowledge of spring nonlinearities, damping nonlinearities, and amplitude effects on added mass.
5. A design criterion for dynamic stability of long span gates with one degree-of-freedom in the streamwise direction is that the available mechanical and structural damping ratio must be greater than the product of the maximum possible fluid excitation ratio, Δc, times the water-to-gate mass ratio, α_m. The maximum value of the fluid-excitation ratio was shown to be about 0.024. Therefore, stability criterion is given as

$$\zeta_a > 0.024\alpha_m \text{ or } \mathrm{Sc} > 0.30 \quad (38)$$

Sc is the Scruton number, a mass damping parameter used frequently in one of several forms. Here the Scruton number is given as $Sc = 4\pi\zeta_a / \alpha_m$. The water-to-gate mass ratio is known in the design stage. Sufficient damping must be available in the gate structure to meet the stability criterion or an additional damping element should be installed to meet the criterion in Equation 38.

The cited works to this point have provided an understanding of the shear layer generated discharge fluctuations that can drive a single degree-of-freedom system to self-excited streamwise vibrations. Most of the specific studies have considered submerged discharge, but similar shear layer dynamics can affect thick gates even when they have free discharge.

GATES WITH TWO DEGREES-OF FREEDOM IN UNDERFLOW

Now we will consider long-span gates with two degrees-of-freedom with underflow only. Specifically shell-type gates as shown in Figure 16 with an inclined weir plate, both with and without a crest-mounted flap gate, are known to be susceptible to coupled mode vibration. Laboratory investigations by Ishii & Knisely (1992b) and Nakata el al. (1996) were undertaken to explore the mechanism of two degree-of-freedom vibration of gates with underflow

Ishii & Knisely (1992b) examined the case of a plate gate with an inclined bottom surface, or from a different perspective, when the upstream gate surface includes an inclined face, as shown in Figure 36. The main objective of their work was to explore the fundamental mechanism of two degree-of-freedom self-excited gate vibration. The thickness of the gate, *W*, was 42 mm, the span of the gate, *L*, was 295 mm and the mass of the gate, *M*, was 9.12 kg. The bottom of the gate was inclined at an angle, Θ, of 17.5° (relative to the horizontal channel bottom). The corner formed by the vertical face and the inclined bottom was rounded with a radius of about 5 mm to minimize flow separation. The water depth above the gate bottom, d_0, was maintained at 375 mm and the gate opening, *B*, was fixed at 5 mm. The coiled springs and levers, shown in Figure 36, simulated the gate elasticity. The restoring force of these springs was transmitted to the gate by the levers. Varying the position of attachment of the springs to the levers permitted fine adjustment of the gate natural frequencies. The coiled spring on the downstream side of the gate was needed to provide in-air counter tension allowing the measurement of the horizontal gate frequency and damping ratio in air. The horizontal vibration frequency in air, Ω_{ax}, was fixed at 11.32 Hz (71.1 rad/s) with an in-air damping ratio, ζ_{ax}, of 0.0025.

The parameters governing dynamic similarity in this experiment, and as previously introduced in Chapters 4 and 5, were chosen to be the basic Froude number and the water-to-gate mass ratio, following Ishii et al. (1987). The basic Froude number, $F_{0x} = \Omega_{ax}\sqrt{d_0 / g}$, is completely determined by the upstream gate submergence d_0 and the vibration frequency in air, Ω_{ax}. The water-to-gate mass ratio, $\alpha_m = \rho d_0^2 / (M / L)$, depends on the upstream gate submergence and the gate mass per unit length. When the values of these two non-dimensional parameters are held constant, the ratio of the in-water vibration frequency to the in-air vibration frequency is also constant. Thus, an alternate Froude number based on submergence and in-water frequency is also constant. Discussion of the details of this interpretation of the Froude numbers introduced in this paragraph for the gate vibration problem can be found in Ishii et al. (1987). The basic Froude number for this set of experiments was 13.9. The water-to-gate mass ratio was fixed at 4.55

The experiment consisted of selecting an in-air natural frequency in the range $3.5 < \Omega_{ay} < 12.9$, in the vertical (y) direction, measuring the in-air damping, ζ_{ay}, in the y-direction, fixing the upstream water depth and gate opening, and then recording the resulting vibration time history.

Horizontal Vibration Drives Vertical Vibration

Figure 37 shows the resulting measurements of (a) in-air damping ratios for y-direction vibrations, (b) the in-water vibration frequencies in the x- and y-directions, and (c) the excitation ratios in the x- and

Figure 36. Two degrees-of-freedom test apparatus for plate gates with inclined upstream face
From Ishii & Knisely (1992b). © 1992 Elsevier. Used with permission.

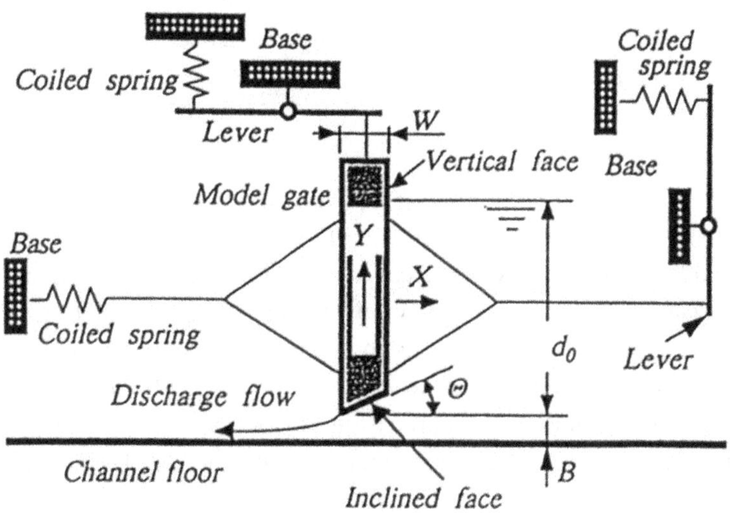

Figure 37. Measurements of (a) y-direction in-air damping, (b) in-water vibration frequencies in the x- and y-directions, and (c) fluid excitation ratios in the x- and y-directions, as functions of Ω_{ay}, the in-air y-direction vibration frequency, with F_0 =13.9, Ω_{ax} =11.32 Hz, and B/d_0 = 0.0133
From Ishii & Knisely (1992b). © 1992 Elsevier. Used with permission.

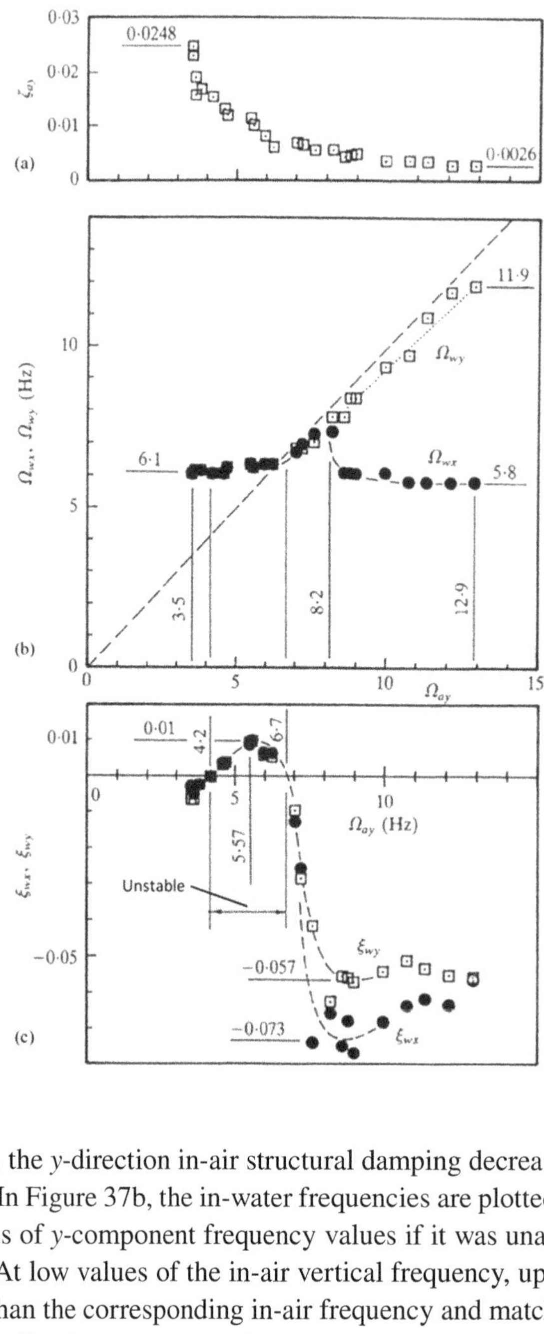

y-directions. In Figure 37a, the y-direction in-air structural damping decreases almost exponentially as Ω_{ay} increases from 3.5 Hz. In Figure 37b, the in-water frequencies are plotted. The diagonal dashed line indicates the expected locus of y-component frequency values if it was unaffected by the water on the upstream side of the gate. At low values of the in-air vertical frequency, up to about Ω_{ay} = 6.7, the in-water frequency is higher than the corresponding in-air frequency and matches the in water x-direction frequency. The y-direction vibration appears to be synchronized with the x-direction vibration at the in-water x-direction frequency. In Figure 37c, the excitation ratio values in both directions are positive, and almost equal, over the range of vertical frequencies from 4.2 to 6.7 Hz, indicating self-excitation of both degrees-of-freedom. The x- and y-frequencies are nearly constant at 6.1 to 6.3 Hz over the range of self-excitation. When the in-air vertical vibration frequencies are above 6.7 Hz, the excitation ratios become negative, indicating damping of the vibrations. When the damping becomes negative, the vibration frequencies increase, suggesting the vertical frequency is attempting to entrain the horizontal, but the vibration cannot be sustained. Beyond a vertical in-air frequency of 8.2 Hz, the two frequencies diverge, each with a separate value. The horizontal frequency remains approximately constant at about 5.8 Hz, while the vertical in-water frequency increases almost linearly with the vertical in-air frequency.

When self-excited vibrations occur, the vertical vibrations appear at the frequency of horizontal vibrations, suggesting that the main vibration is in the horizontal direction. From the analysis in Chapter 5 and the above analysis of a long-span gate undergoing streamwise vibration, the fluid force resulting from the horizontal vibration can be calculated. It consists of two terms: one proportional to the gate velocity and the other proportional to gate acceleration. The proportionality coefficients are called fluid excitation and added mass coefficients, and are denoted as ΔC_x and ΔM_x, or in non-dimensional form as the reduced fluid excitation coefficient, Δc_x, and the reduced added mass coefficient, Δm_x. With known values of the in-water to in-air horizontal frequencies ($\gamma_{xx} = \Omega_{wx}/\Omega_{ax}$), the in-air damping ratio, ζ_{ax}, and the in-water excitation, ξ_{wx}, the evaluation of the reduced fluid excitation coefficient and the reduced added mass coefficient from the theoretical equations below following from Ishii et al. (1987) and Ishii (1992):

$$\Delta c_x = \frac{2}{\alpha_m} \frac{F_{0x}}{\sqrt{1-\zeta_{ax}^2}} \left[\zeta_{ax} + \left(\frac{\xi_{wx}}{\gamma_{xx}} \right) \sqrt{\frac{1-\xi_{wx}^2}{1-\zeta_{ax}^2}} \right] \tag{39}$$

$$\Delta m_x = \frac{1}{\alpha_m} \left[\left(\frac{1-\xi_{wx}^2}{1-\zeta_{ax}^2} \right) \left(\frac{1}{\gamma_{xx}^2} \right) - 1 \right] \tag{40}$$

Note that the fluid excitation coefficient, Δc_x, does not include any contribution from structural damping.

Using Equations 39 and 40, the measured data in Figure 37 were converted to these two dimensionless forms and plotted versus γ_{xy}, the ratio of the in-water horizontal frequency to the in-air vertical frequency, in Figure 38b. Plotted in this manner, small values of Ω_{ay} correspond to large γ_{xy} and large Ω_{ay} to small γ_{xy} on the left of the abscissa. Note that the region of positive reduced fluid excitation ratio extend from $\gamma_{xy} = 0.97$ to $\gamma_{xy} = 1.5$, with a maximum value of 0.13 at $\gamma_{xy} = 1.13$. Over the range of positive reduced excitation coefficients, the vertical and horizontal vibrations, in Figure 38a, are synchronized. Note also the added coefficient mass remains relatively constant, between 0.48 and 0.55.

Figure 38a shows the non-dimensional frequency characteristics. The following frequency ratios are defined:

- $\gamma_{wyx} = \Omega_{wy}/\Omega_{wx}$: The ratio of in-water vertical frequency to in-water horizontal frequency
- $\gamma_{yy} = \Omega_{wy}/\Omega_{ay}$: The ratio of in-water vertical frequency to in-air vertical frequency
- $\gamma_{xx} = \Omega_{wx}/\Omega_{ax}$: The ratio of in-water horizontal frequency to in-air horizontal frequency, corresponding to the forcing frequency/system frequency in SDOF forced vibration theory
- γ_{nyy}: The ratio Ω_{wy}/Ω_{ay} *at the onset of self-excitation*, called the inherent frequency ratio (not plotted explicitly in Figure 38a), to be used subsequently in the discussion of a stability criterion

Figure 38. Measured data in non-dimensional form as functions of γ_{xy}, the in-water x-direction to in-air y-direction frequency ratio, showing (a) frequency ratios; (b) fluid excitation and added mass coefficients calculated from Equations 34 and 35, and (c) measured phase lag of the y-displacement relative to the x-displacement

From Ishii & Knisely (1992b). © 1992 Elsevier. Used with permission.

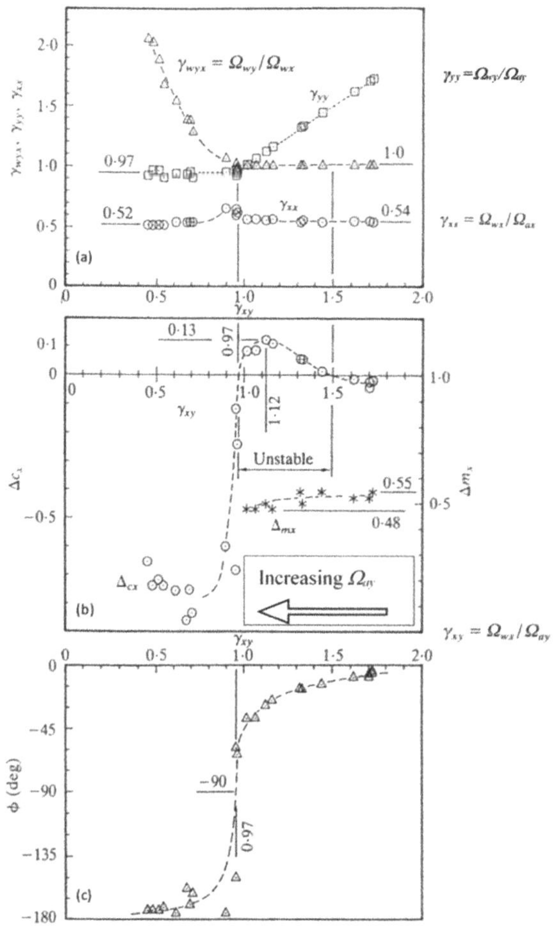

The first three are plotted as functions of γ_{xy}, assuming the horizontal frequency is serving to force the vertical motion.

When self-excited vibrations do not occur, the vertical vibration frequency does not correspond to the horizontal vibration frequency, as shown by the values of $\gamma_{wyx} = \Omega_{wy}/\Omega_{wx}$ different from 1.0 in Figure 38a. However, just after externally displacing and releasing the gate, it vibrates at large amplitudes and the vertical vibration frequency is the same as the horizontal vibration frequency, permitting the measurement of the phase shift of the vertical vibration relative to the horizontal vibration. At large amplitudes, the two vibrations can be regarded as having a common frequency, with a phase shift in the vertical vibration.

$$X(t) = \bar{X}\cos(\Omega_{wx}T) \text{ and } Y(t) = \bar{Y}\cos(\Omega_{wx}T - \phi) \tag{41}$$

The measured values of the phase lag, ϕ, shown in Figure 38c, might suggest that the y-displacement leads the x-displacement, however, the actual phase lag can be represented as the given value $+2\pi$, to avoid issues of causality. Although the y-displacement always leads the x-displacement, the magnitude of the lead decreases from180° at small γ_{xy} to 90° at $\gamma_{xy} = 0.97$ (the onset of self-excited vibrations) and approaches zero at higher values of γ_{xy}. This phase characteristic, while not identical with, is nonetheless very similar to that of an SDOF forced oscillator (compare with Figure 1 in Chapter 2), supporting the suggestion that the vertical vibration might be a response vibration to the forcing of the main horizontal vibration.

Both the vertical and the horizontal vibrations are driven by the pressure fluctuations in the reservoir. From the analysis in Chapter 5, we know that, at Froude numbers in the range of these experiments, the pressure due to the push-and-draw effect from the x-direction displacement in Equation 41 can be written as

$$p_b = P_b / \left(\rho g \bar{X}\right) = p_{bs} \cos\left(\Omega_{wx} T - \pi\right) \tag{42}$$

where the negligibly small progressive-wave pressure component need not be considered at the basic Froude number ($F_{ox} = 13.9$) in this test. Since the horizontal motion leads the fluid pressure by 180°, the phase relationship between the vertical motion and the pressure due to horizontal motion can be written as $\phi_p = \phi - \pi$. The vertical motion can then be represented by the following expression:

$$Y\left(t\right) = \bar{Y} \cos\left[\left(\Omega_{wx} T - \pi\right) - \phi_p\right] \tag{43}$$

The phase between the pressure fluctuation and the y-direction motion, resulting from the pressure acting on the inclined face is plotted in Figure 39. The shape is the same as that in Figure 1 of Chapter 2, with phase values ranging from zero to 180° at large $\gamma_{xy.}$ The phase is 90° when $\gamma_{xy} = \gamma_{nyy} = 0.97$ (under the conditions for this set of data), the natural frequency ratio of Ω_{wy} to Ω_{ay} at the onset of self-excitation.

The annotations on the plot in Figure 39 of "Press-open "and "Press-shut" result from examination of the trajectories generated from the experimentally measured displacements in the x- and y- directions, shown in Figure 40.

In Figure 40, positive X values are for displacement in the upstream direction while positive Y values are upward displacement of the gate. The trajectories (a) to (d) in Figure 40 correspond to frequency ratios (γ_{xy} given in the figure) at which self-sustained vibrations do not occur. In these cases, the gate is given an initial displacement and the subsequent trajectories of the damped free vibration are recorded. The trajectories (e) and (f) in Figure 40 were recorded for frequency ratios at which self-sustained vibrations occur. The data in trajectory (e) correspond to the maximum value of the reduced fluid excitation coefficient occurring at $\gamma_{xy} = 1.12$. The embedded vertical and horizontal time traces show how rapidly the vibration amplitudes grow. Trajectory (f) is for $\gamma_{xy} = 1.32$ with a lower, but still positive, reduced fluid excitation coefficient. In these two cases only a limited number of cycles are needed to show the relatively flat elliptical trajectories with clockwise progression, indicating energy transfer from the flowing water to the gate motion.

Two important observations can be made concerning these trajectories. The first is that in cases where the vibration is quickly damped in trajectories (a) to (e), the gate moves upward and downstream

Figure 39. Phase difference between fluctuating pressure resulting from x-component displacement and the y-component response to the pressure loading on the inclined face
From Ishii & Knisely (1992b). © 1992 Elsevier. Used with permission.

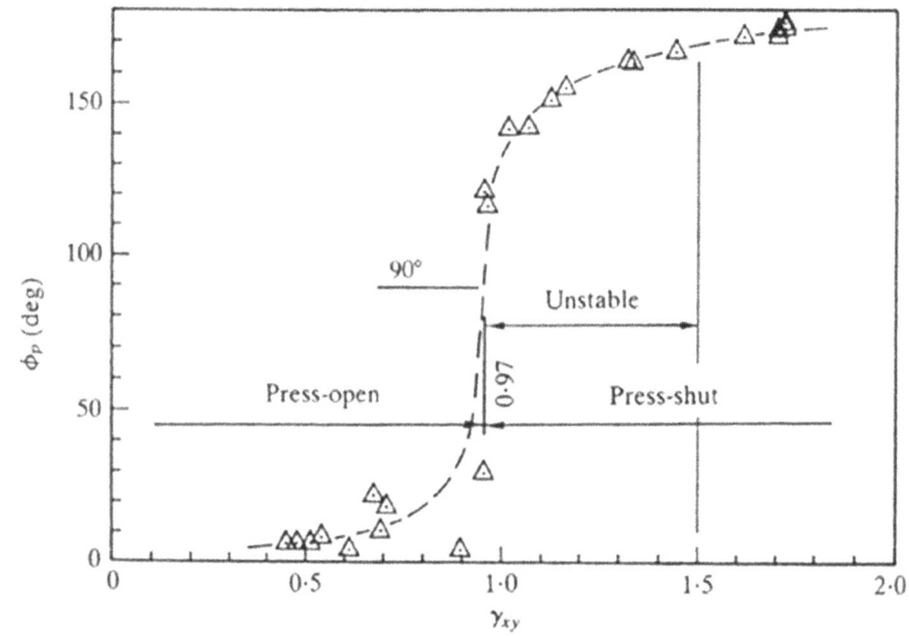

(to the left on the page). Such motion is exactly what is expected of a press-open device. The second observation is that for cases in which self-sustained vibrations occur, trajectories (e) and (f), the gate motion is downward and downstream, indicative of a press-shut device. In trajectory (d), even though the vibration is slowly damped, the trajectory approaches that of a press-shut device. The frequency ratio in this case, the γ_{xy} value of 0.96 is just slightly less than the onset ratio and the coupling of the vertical and horizontal motion through the fluid force is not quite sufficient to produce self-sustained vibrations.

Trajectory (e) in Figure 40 corresponds to the largest observed vibration amplitudes for a 17.5°weir plate inclination angle. The symbol ε represents the ratio of vertical to horizontal amplitude. The value of ε in trajectory (e) is 0.98, indicating that the mean inclination of the gate motion is approximately 45° from the horizontal axis. Trajectory (f) with a slightly larger frequency ratio, γ_{xy} =1.32 has a smaller reduced fluid excitation coefficient (see Figure 38b) and the amplitude of vibration is reduced. The amplitude ratio, ε, decreases to 0.67 indicating a greater amplitude reduction in the vertical direction than in the horizontal direction. Thus, the mean inclination of the gate motion is also reduced.

The classification of the gate as a press-open or press-shut device, depending on the ratio γ_{xy} helps explain an additional aspect of Figure 38a. On the left side of the unstable range, the vertical and horizontal frequencies are not related. Over this range of γ_{xy}, the vertical and horizontal motions are so quickly damped that there is no chance for synchronization, resulting in an extremely stable gate. However, for conditions corresponding to the region to the right of the unstable range in Figure 38(a), the gate still behaves as a press-shut device and has a strong tendency toward dynamic instability. Here the motions are weakly damped and synchronization of the vertical and horizontal frequencies does occur before the vibrations are eventually damped.

Figure 40. Measured trajectories of model gate displacement indicative of press-open and press-shut behavior for (a) γ_{xy} = 0.45 (Ω_{ay} = 12.9 Hz); (b) γ_{xy} = 0.68 (Ω_{ay} = 8.98 Hz); (c) γ_{xy} = 0.90 (Ω_{ay} = 8.20 Hz); (d) γ_{xy} = 0.96 (Ω_{ay} = 7.03 Hz), ϕ = -58.7°, ε = 1.38; (e) γ_{xy} = 1.12 (Ω_{ay} = 5.57 Hz), ϕ = -28.8°, ε = 0.98, with vertical and horizontal displacement time traces; (f) γ_{xy} = 1.32 (Ω_{ay} = 4.59 Hz), ϕ = -16.6°, ε = 0.67

From Ishii & Knisely (1992b). © 1992 Elsevier. Used with permission.

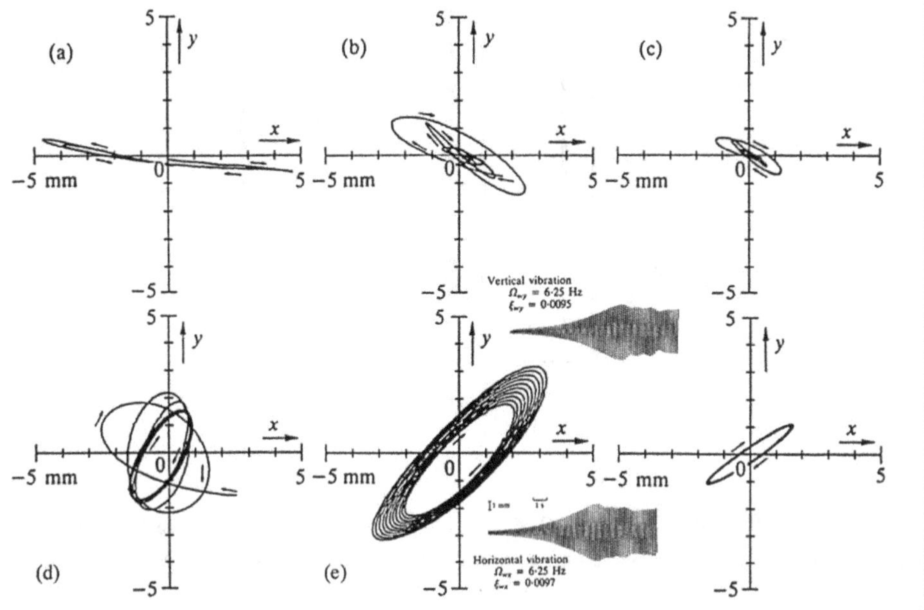

Similar data from a subsequent study examining the effects of the lower weir plate inclination angle on the gate dynamics and instability were published in Nakata et al. (1996). The laboratory apparatus was the same as that of Ishii & Knisely (1992b), shown in Figure 36a and the same procedure was followed. The tested gates had inclinations from 17.5° to 65°, as shown in Figure 41. All gate models had a fixed height with a fixed streamwise gate thickness of 42 mm, a span of 295 mm. and a gate mass of 9.12 kg. In all cases, the gates had free discharge with no downstream submergence. The gate upstream submergence was 375 mm and the mean gate opening was set at 5 mm. The in-air horizontal frequency was held constant at Ω_{ax} = 11.32 Hz (71.1 rad/s) with an in-air damping ratio of ζ_{ax} = 0.0025 in the x-direction. The in-air natural frequency in the vertical direction could be varied between $3 < \Omega_{ay} < 13$ Hz. The basic Froude number was held constant at $F_{0x} = \Omega_{ax}\sqrt{d_0 / g}$ =13.9 and the water-to-gate mass ratio, $\alpha_m = \rho d_0^2 / (M / L)$ was also kept constant at 4.55.

For the gate with an inclined weir plate angle, Θ =17.5°, the most unstable self-excited vibration appeared when Ω_{ay} was 5.6 Hz. The corresponding measured vibration waveforms and gate trajectories are presented at the top of Figure 42. The in-water vibration frequency, Ω_w = 6.25 Hz, and the maximum excitation ratio, $\xi_{w,\,max}$ = 0.010, were extracted by digitizing the measured vibration waveforms and analyzing the data using a computer. For the measured gate displacement trajectory, on the right side in Figure 42, the positive X is in the upstream direction while positive Y is upward. The relatively flat

Figure 41. Gate models used to ascertain the effect of inclination angle of the lower portion of upstream weir plate weir on two degrees-of-freedom vibration
From Nakata et al. (1996). © 1996 JSME. Used with permission.

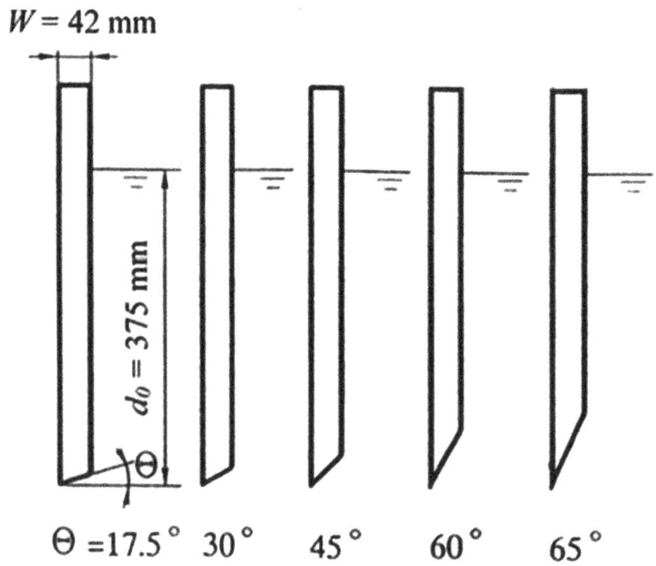

elliptical shape of this trajectory is traversed in the clockwise direction and is indicative of press-shut behavior, corresponding to net energy transfer from the flowing water to the gate as discussed above relative to Figure 40.

In addition to the data for the 17.5° weir plate, measured vibration waveforms and gate trajectories under the most unstable conditions for weir plate inclination angles of 30°, 45°, 60° and 65° are presented sequentially from top to bottom in Figure 42. The gate trajectories for all inclination angles are elliptical and are traversed in the clockwise direction. The mean inclination of these elliptical traverses becomes steeper for increasing weir plate inclination angles. For increasing weir plate inclination angles, the excitation ratio $\xi_{w,max}$ (corresponding to maximally unstable self-excited vibrations) initially increases, reaches its maximum value of 0.032 for an inclination angle of 60°, and then subsequently decreases to 0.024 at 65°.

Using the values for α_m, ζ_{ax}, ξ_{wx}, γ_{xx}, and F_{0x} in Equation 39, the reduced fluid excitation coefficient in the x-direction, Δc_x, was calculated for the maximally unstable self-excited vibrations for each weir plate inclination angle. A similar expression, replacing the x-subscripted variables with the corresponding y-subscripted variables was used to calculate the reduced fluid excitation coefficient in the y-direction, Δc_y, for the maximally unstable self-excited vibrations. As shown in Figure 43, the reduced fluid excitation ratio in the x-direction, Δc_x, initially increases with increasing inclination angle, reaching a maximum value of about 0.38 for an inclination angle of 60°, before decreasing with the increase of inclination angle to 65°. The y-direction reduced fluid excitation coefficient, Δc_y, increases monotonically with increasing inclination angle. The third curve in Figure 43 is the vector sum of the x- and y-components in the direction of inclination of the corresponding trajectory in Figure 42. Given the relative strength of the x-direction component, this third curve follows the same trend as the x-direction reduced fluid excitation coefficient.

From Figures 42 and 43, the gate model with the 60° weir plate inclination angle is known to have the largest reduced fluid excitation coefficient. The recorded data for the gate with the 60° inclination over the range of in-air vertical vibration frequencies are plotted in Figure 44 as raw data and in Figure 45 as non-dimensional parameters to facilitate comparison with the 17.5°inclination data in Figures 37 and 38. In Figure 44a, the measured in-air vertical damping is plotted. The in-air damping is needed in the subsequent computation of the reduced fluid excitation coefficient. The in-water vibration frequencies in the *x*- and *y*-directions, plotted versus the in-air vertical frequency in Figure 44b, coalesce during self-excited vibrations from Ω_{ay} = 4.3 to 7.3 Hz. The excitation ratios in both the *x*- and *y*-directions, plotted in Figure 44c, are both positive over this same range of in-air vertical frequencies, as expected for self-excited gate vibrations. When the in-air vertical vibration frequency Ω_{ay} was 8.0 Hz, just slightly greater than 7.3 Hz, the horizontal and vertical vibrations were still synchronized with each other, as shown in the Figure 44b, but even with an initial displacement, the gate vibration was damped, thus indicating negative values of the excitation ratio, as shown in Figure 44c. When Ω_{ay} was greater than 8.0 Hz and after an initial displacement, the gate exhibited uncoupled damped vibrations at distinct frequencies in the horizontal and vertical directions. These distinct frequencies are called the *inherent natural frequencies in water*, denoted as Ω_{nwx} and Ω_{nwy}. In this regime, the in-water vertical vibration followed the in-air value, with a slight frequency reduction due to the added mass, resulting in the ratio

Figure 42. Horizontal and vertical waveforms and vibration trajectory for each gate model under conditions corresponding to its maximum instability
From Nakata et al. (1996). © 1996 JSME. Used with permission.

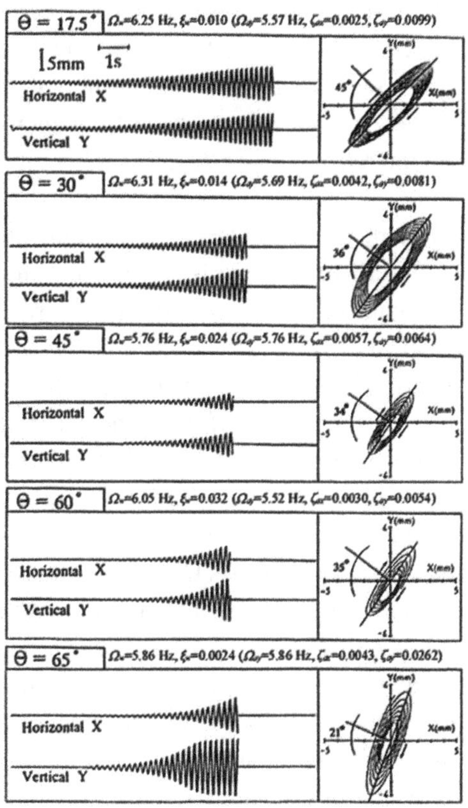

Figure 43. Maximum fluid excitation coefficients in the horizontal and vertical directions; vector sum in the direction of trajectory inclination in Figure 42 increases with increasing inclination
From Nakata et al. (1996). © 1996 JSME. Used with permission.

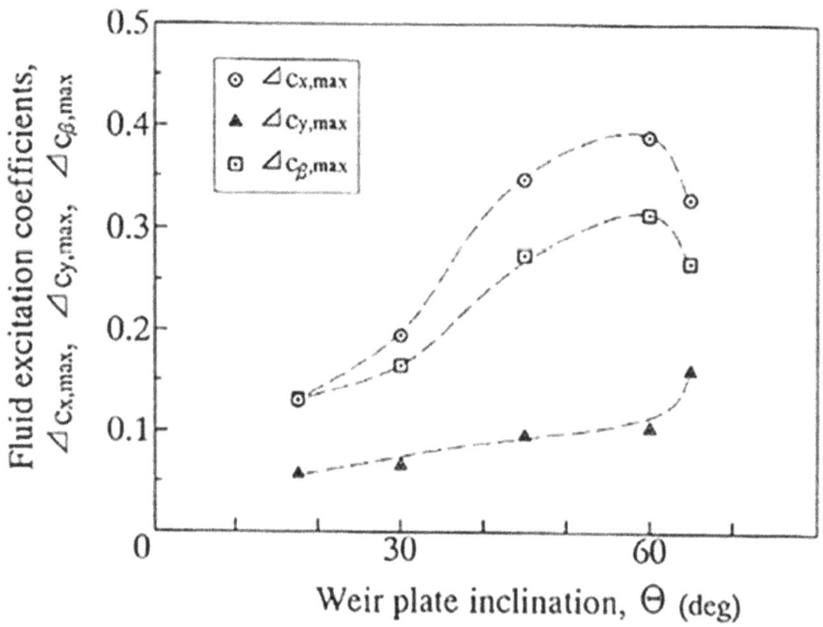

of the inherent y-direction natural frequency in water to the in-air y-direction natural frequency being very close to constant. This ratio is denoted as $\gamma_{nyy} \equiv \Omega_{nwy}/\Omega_{ay} \approx 0.98$ for these model tests. The in-water horizontal vibration frequency in this region was close to constant at about $\Omega_{nwx} = 5.3$ Hz, permitting the definition of $\gamma_{nxy} \equiv \Omega_{nwx}/\Omega_{nwy}$, called the inherent in-water frequency ratio. For comparison with the plots in Figure 38 and 39 that used the ratio $\gamma_{xy} \equiv \Omega_{wx}/\Omega_{ay}$, this inherent in-water frequency ratio, γ_{nxy}, is *approximately* equal to $\gamma_{nyy}\,\gamma_{xy}$.

The frequency ratio graphs in Figure 45a with $\Theta = 60°$ look quite similar to those in Figure 38 with $\Theta = 17.5°$. One effect of the larger inclination angle is the larger values of Δc_x in Figure 45b relative to those in Figure 38b, which also means the termination ratio, γ_{term} will also be higher (apparently not reached in the testing by Nakata et al., 1996). A second effect of the larger inclination of the weir plate is the deviation of the phase angle for values of γ_{nxy} for values greater than the onset ratio.

In addition to the 60° weir plate data and a repetition of the earlier 17.5°weir plate data, data were also obtained for inclined weir plate angles of 30°, 45°, and 65°. The series of non-dimensional data from 17.5° to 65°weir plate inclinations are plotted in Figure 46 as a function of the inherent in-water frequency ratio, $\gamma_{nxy} \equiv \Omega_{nwx}/\Omega_{nwy}$. For all weir plate inclination angles, the gate is most unstable when the inherent in-water frequency ratio was near 1.0. As the weir plate inclination increases, the unstable region becomes wider, and $\Delta c_{x,max}$ increases, reaching its maximum of 0.40 at Θ=60°, which is about 3 times larger than the value of 0.13 recorded at Θ=17.5°. Another aspect of the influence of the vertical component on the coupled mode vibration, seen in Figure 46, is that when Θ =17.5°, the region of coupled mode vibration is essentially locked to the horizontal frequency component. With increasing weir plate angles, the extent of this locking to the horizontal frequency decreases, and when Θ reaches 60 to 65°, the departure of the coupled mode frequency from the horizontal frequency occurs over a large

Figure 44. Gate self-excitation characteristics for Θ =60° as functions of in-air vertical frequency showing (a) the in-air damping in the vertical direction, (b) the in-water frequencies in the horizontal and vertical directions, and (c) the excitation ratio in the horizontal and vertical directions
From Nakata et al. (1996). © 1996 JSME. Used with permission.

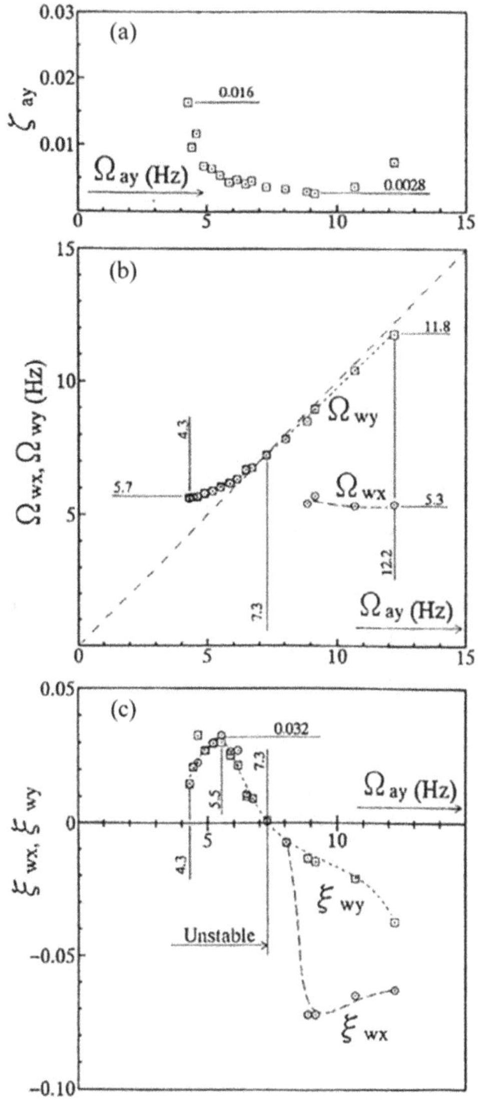

portion of the region of self-excited vibration. Both of these phenomena, that is, the increasing vertical fluid excitation and the changing nature of frequency locking, will require more extensive investigation before any firm conclusions can be drawn.

All reduced fluid excitation coefficient data are plotted together as a functions of the inherent in-water frequency ratio, γ_{nxy}, in Figure 47. From this plot, the expanded width of the unstable region and the concurrent growth of the peak reduced fluid excitation coefficient are obvious up to the 60° weir plate inclination. A drastic change is apparent as the angle Θ changes from 60° to 65°, suggesting further

Figure 45. Non-dimensional vibration characteristics for Θ = 60°, the inclination angle with the maximum excitation ratio showing (a) frequency ratios, (b) reduced fluid excitation coefficient, and (c) phase angle of vertical displacement relative to the horizontal displacement
From Nakata et al. 1996. © 1996 JSME. Used with permission.

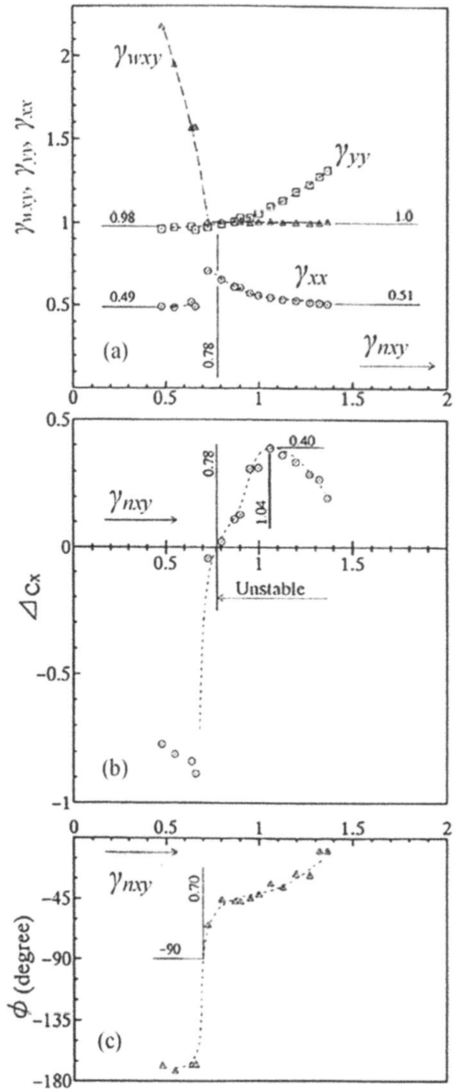

clarification is required before a complete understanding of the effects of the weir plate inclination can be obtained.

Comparison With One Degree-of-Freedom Theoretical Predictions

Assuming that the modeling of the two degree-of-freedom vibration as a main vibration (here the x-direction) driving the vibration in the second coordinate direction is a sufficiently accurate approximation, the question arises as to whether the one degree-of-freedom analysis from earlier in this chapter

Figure 46. Measured data in for $\Theta = 17.5°, 30°, 45°, 60°$ and $65°$ as functions of γ_{nxy}, the inherent in-water frequency ratio, showing (a) frequency ratios; and (b) x-direction reduced fluid excitation coefficients

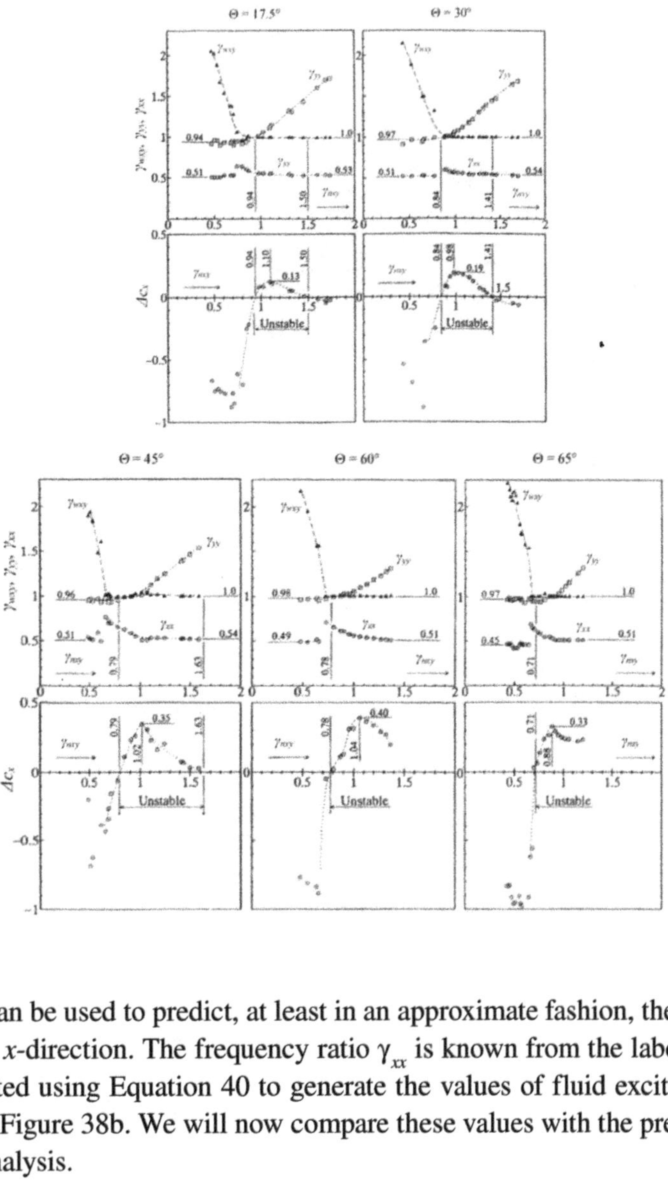

and from Chapter 5 can be used to predict, at least in an approximate fashion, the characteristics of the main vibration in the x-direction. The frequency ratio γ_{xx} is known from the laboratory measurements and Δm_x was calculated using Equation 40 to generate the values of fluid excitation and added mass coefficient plotted in Figure 38b. We will now compare these values with the predictions from the one degree-of-freedom analysis.

The one degree-of-freedom analysis begins by using Equation 30 to calculate the ratio of $\gamma_{xx} = F_x/F_{ox}$ using the value of the water-to-gate mass ratio, $\alpha_m = 4.55$ based on the experiment, with $1/\beta^* \approx 1.0$, and initially assuming the added mass coefficient could be approximated as the asymptotic value $\Delta m_b = 0.54$. From this initial calculation of $F_x = \gamma_{xx} F_{0x}$, the value of Δm_b was calculated using the Equation 23b or approximated from Figure 7. Now the new value of Δm_b was used to refine the estimate of F_x. The result of this iterative process is plotted in Figure 48, showing a frequency ratio of $\gamma_{xx} = 0.54$. Using the values on the graph (in-water x-direction frequency of about 6.1 Hz) and the in-air frequency of 11.32 Hz, yields the calculated frequency ratio value of 0.539, which is in excellent agreement with the measured value.

Figure 47. Effect of inclination angle on reduced fluid damping as a function the γ_{nxy}
From Nakata et al. (1996). © 1996 JSME. Used with permission.

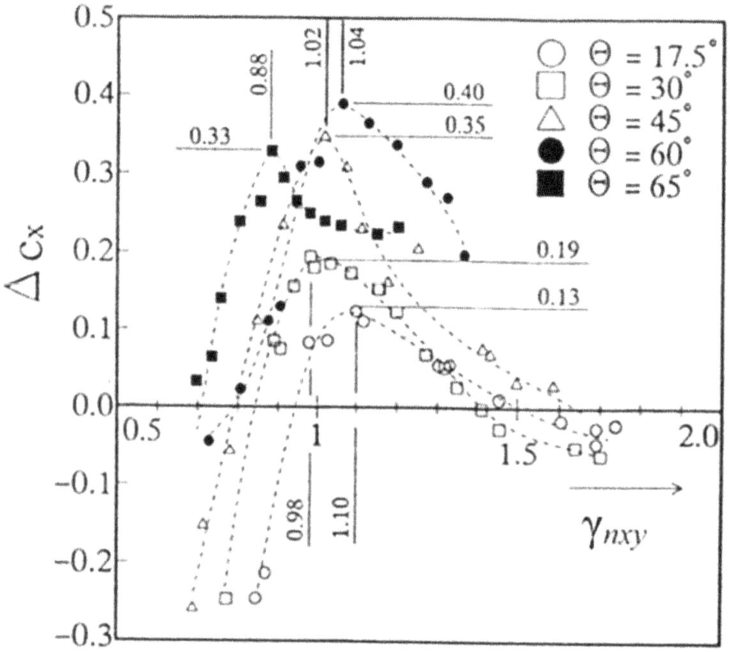

Figure 48. Calculation of frequency ratio using one degree-of-freedom analysis from Chapter 5
After Ishii & Knisely (1992b). © 1992 Elsevier. Used with permission.

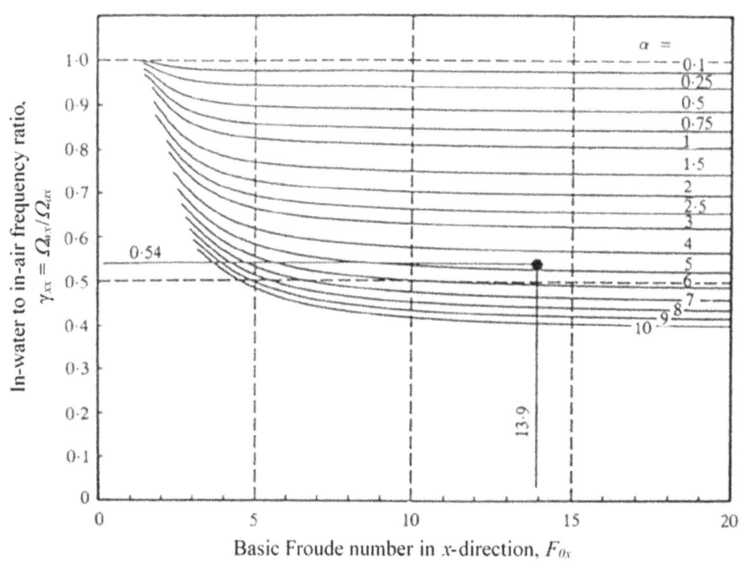

Next this value of $\gamma_{xx} = 0.54$ was used to calculate the Froude number in the x-direction as $F_x = \gamma_{xx} F_{ox} = (0.54)\times(13.9) = 7.51$. From this Froude number, the reduced added mass can be found as shown in Figure 49. The calculated value of Δm_x =0.52 agrees substantially with measured values ranging between 0.48 and 0.55 in Figure 38b. Figure 49 also shows that the added fluid damping due to flow rate variation is negligible at $F_x = 7.51$.

Having found that the one degree-of-freedom analysis provides accurate results, we will now calculate the reduced push-and-draw pressure acting on the upstream face of the gate. The resulting calculated distribution of the magnitude dimensionless pressure, p_{bs}, along the gate is shown in Figure 50. The ordinate y is the distance below the free surface divided by the upstream gate submergence d_0 and is positive upwards. The free surface is at y = 0 and the bottom of the gate is at y = -1. The pressure magnitude is negligibly small near the surface, but becomes progressively larger near the bottom of the gate. The pressure magnitude increases with increasing Froude number, F_x. Since the pressure represented in this figure is 180° out of phase with the gate motion, it appears as an added mass in the x-direction force balance on the gate.

The pressure magnitude for the constant value of $F_x = 7.51$ examined in Ishii & Knisely (1992b) is plotted as the dashed line in Figure 50. The pressure magnitude is about 45 near the lower end of the gate, indicating that the actual fluid pressure is about 45 times the characteristic water pressure, $\rho g\bar{X}$, corresponding to the gate vibration amplitude $\bar{X}$. For shell-type long-span gates or other gates with inclined upstream surfaces, this large pressure acts normal to the inclined gate surface and may produce vertical vibration.

Figure 49. Reduced added mass calculation for values in the experiment
From Ishii & Knisely (1992b). © 1992 Elsevier. Used with permission.

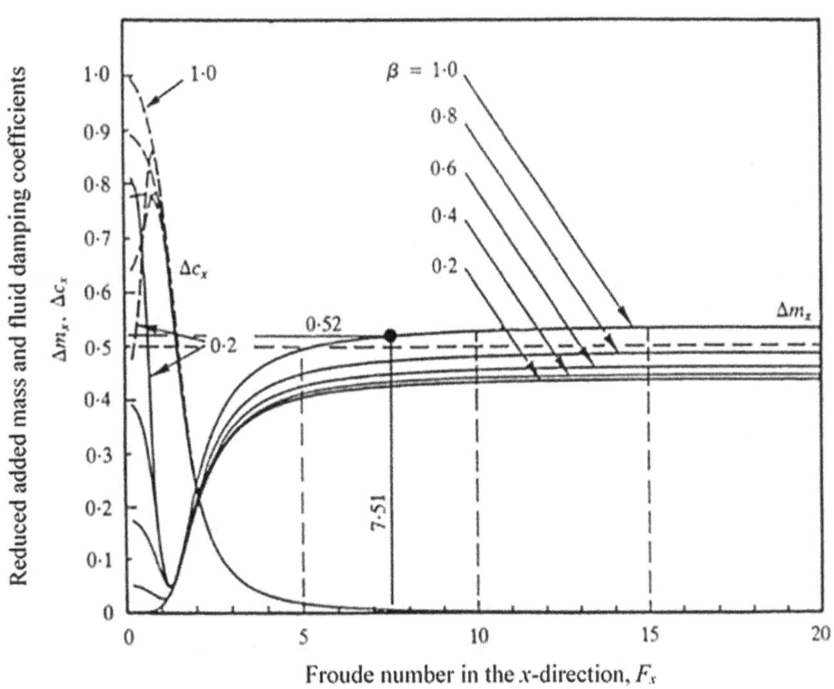

Figure 50. Calculated non-dimensional push-and-draw pressure acting on the gate from the one degree-of-freedom theory (see Chapter 5 for details)
After Ishii & Knisely (1992b). © 1992 Elsevier. Used with permission.

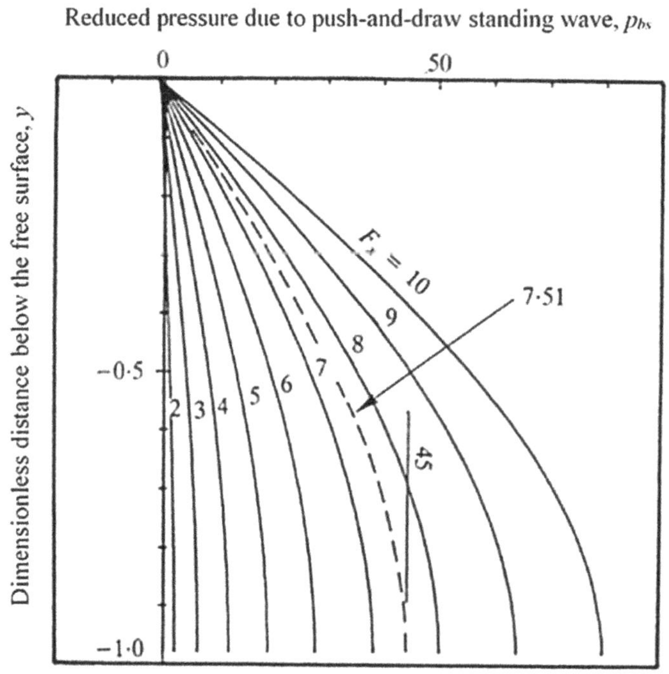

Vertical Vibratory Response and Energy Input

The vertical motion can be considered to be a forced response to the fluctuating pressure acting on the inclined lower gate surface. This assumption is supported by the characteristic phase distribution shown in Figures 26c and 27. The gate motion can be represented by Equation 43. The distribution of ϕ_p shown in Figure 39 is the same as that for forced vibration in Figure 1 of Chapter 2, except the value of γ_{xy}, where the phase lag passes through 90° is not 1.0, but rather 0.97. This is because the natural vibration frequency in water, Ω_{nwy}, is 0.97 times the in-air frequency, as shown in Figure 38a, due to added mass effects. The frequency ratio, γ_{xy}, can be re-written in terms of the in-water frequency and the inherent frequency ratio. γ_{nyy}, as follows:

$$\gamma_{xy} = \frac{\Omega_{wx}}{\Omega_{nwy}}\frac{\Omega_{nwy}}{\Omega_{ay}} = \frac{\Omega_{wx}}{\Omega_{nwy}}\gamma_{nyy} \tag{44}$$

The horizontal in-water frequency, Ω_{wx}, serves as the forcing frequency for the vertical vibrations, and the natural vertical in-water frequency, Ω_{nwy}, acts as the system frequency. Resonance occurs when the forcing and system frequencies match, that is, the ratio $\Omega_{wx}/\Omega_{nwy} = 1.0$ in Equation 44. At resonance, therefore, $\gamma_{xy} = \gamma_{nyy}$ and the phase angle ϕ_p, is 90°. The vertical vibration is therefore considered the forced response to pressure fluctuations generated by the horizontal gate motion.

By eliminating the time parameter from Equations 41 for $X(T)$ and 43 for $Y(T)$, an equation for Y as a function of X can be obtained, as follows:

$$Y = \bar{Y}\left(\frac{X}{\bar{X}}\cos\phi + \delta\sin\phi\sqrt{1-\frac{X}{\bar{X}}}\right) \tag{45}$$

where $\delta = +1$ when $0 < \Omega_{wx}T < \pi$, that is, when $X / \bar{X}$ is decreasing from +1 to -1, and $\delta = -1$ when $\pi < \Omega_{wx}T < 2\pi$, that is when $X / \bar{X}$ is increasing from -1 to +1. Equation 45 describes the elliptical trajectories of gate displacement. Certain special cases are worth considering in detail:

1. When the frequency ratio $\gamma_{xy} = \gamma_{nyy}$ (= 0.97, for $\Theta = 17.5°$), the phase ϕ in Equation 45 is -90° (see Figure 38c) and the phase between the fluctuating pressure induced by the horizontal movement and the vertical motion, ϕ_p, is 90° (see Figure 39). This 90° phase relation indicates that maximum energy transfers from the pressure fluctuations to the vertical motion and the amplitude ratio $\varepsilon = \bar{Y} / \bar{X}$ will be a maximum. Assuming maximum energy transfer occurs and $\varepsilon > 1.0$, then the trajectory described by Equation 45 is an ellipse with the major axis along the Y-axis since cos $\phi = 0$. Furthermore, since sin ϕ is negative, the ellipse is traversed in the clockwise direction. The trajectory would resemble that of trajectory (d) in Figure 40.
2. When the frequency ratio γ_{xy} is slightly greater than γ_{nyy}, the phase ϕ increases from -90° toward 0°, and the effectiveness of the energy transfer from pressure fluctuations to vertical gate vibration decreases and ε descreases correspondingly. Such behavior is evident in the transition from trajectory (d) to (e) and then to (f) in Figure 40. This argument does not preclude an increase in the Y-amplitude, but merely states that the ratio of the Y-amplitude to the X-amplitude must decrease. The X-amplitude is expected to increase until a maximum value of Δc_x is reached. Furthermore, since the magnitude of sin ϕ decreases as ϕ ranges from -90° to 0°, the eccentricity of the elliptical orbit decreases. In addition, the inclination of the major axis decreases from being perpendicular to the X-axis to being along the X-axis, which indicates a transition from a neutral device to a press-shut device. These tendencies are confirmed by examining the trajectories found in trajectories (e) and (f) in Figure 40.
3. When γ_{xy} increases further, the Y-direction motion becomes extremely small and the gate vibrates essentially only in the horizontal direction.
4. When the frequency ratio $\gamma_{xy} < \gamma_{nyy}$, the major axis of the ellipse has a negative slope, indicating the gate behaves as a press-open device. This observation is confirmed by referring to trajectories (a) to (c) in Figure 40.

From these preceding four points, the inherent vibration ratio, γ_{nyy}, is seen to play a major role in determining the onset of vibration.

When the vertical vibration begins, one must consider the effect of the flow rate variation beneath the gate due to the changing discharge area. The theory of flow-rate-variation pressure was developed in Chapter 4, where it was shown that the principal component of pressure caused by flow rate variation, P_r, is a term that lags the vertical motion given in Equation 43 by 90°. Symbolically,

$$p_r = P_r / \left(\rho g \bar{Y}\right) = p_{rs} \cos\left\{\Omega_{wx} T - \left(\frac{3\pi}{2} + \phi_p\right)\right\} \tag{46}$$

or, after the addition of 2π to the phase,

$$p_r = p_{rs} \cos\left\{\Omega_{wx} T - \left(\phi_p - \frac{\pi}{2}\right)\right\} \tag{47}$$

An additional fluctuating progressive-wave pressure component which lags the vertical gate motion is also induced, but at the Froude number in this case negligible (see Chapter 4 for more details).

From Equation 128c in Chapter 4, the vertical distribution of non-dimensional pressure amplitude, p_{rs}, acting on the upstream face of the gate can be calculated and is presented graphically in Figure 51. Near the water surface, the amplitude is negligibly small, but as the water depth increases, the amplitude increases almost exponentially. The pressure distribution corresponding to a Froude number of F_x = 7.51, plotted as the dashed line, has a maximum non-dimensional pressure amplitude of about 20 near the gate opening.

Figure 51. Reduced pressure amplitude acting on gate surface due to flow-rate-variation pressure resulting from vertical vibrations
After Ishii & Knisely (1992b). © 1992 Elsevier. Used with permission.

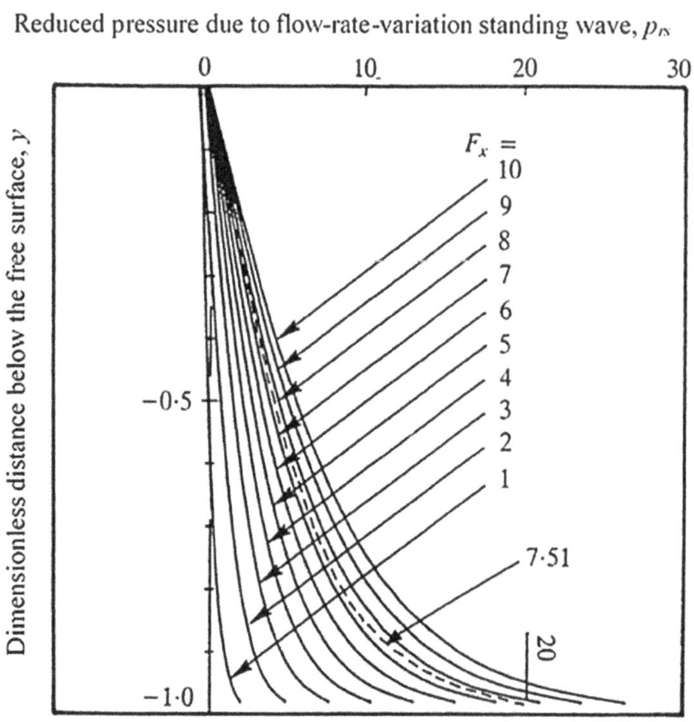

Closed Energy Cycle for Two Degrees-of-Freedom

Coupled-mode vibrations of a two degree-of-freedom gate with only underflow with free discharge result in a fluctuating pressure with two distinct components. The first is due to the horizontal motion and lags the horizontal gate motion by180°; the second is due to discharge fluctuations associated with vertical vibrations and lags the first by an additional 90°.

Over one cycle of vibration,

1. The fluid pressure due to horizontal motion is 180° out of phase with the horizontal motion and neither supplies nor consumes energy for horizontal vibrations
2. The pressure due to horizontal motion supplies energy to the vertical vibrations where the energy supplied per cycle is proportional to -sin ϕ and is always positive since -180° < ϕ < 0°
3. The pressure fluctuation due to fluctuating discharge resulting from vertical motion either supplies energy to the vibration or extracts energy from the horizontal vibration depending on the sign of -cos ϕ. Energy is supplied to the horizontal motion when the gate acts like a press shut device [that is, when 90° < ϕ_p < 180° and -cos ϕ is negative (recall $\phi_p = \phi + \pi$] and vibratory energy is extracted from the gate vibration when the gate acts like a press-open device (0° < ϕ_p < 90° and -cos ϕ is positive)
4. The fluid pressure due to fluctuating discharge resulting from vertical motion consumes vertical vibratory energy since the pressure lags the vertical displacement by 90°, so that when the vertical gate velocity reaches its downward maximum, the fluctuating pressure acts upward on the gate bottom, directly opposing the vertical gate motion.

These four listed points can be interpreted schematically as the closed cycle energy flow diagram shown in Figure 52. When horizontal gate motion occurs, a pressure fluctuation is induced, which, in turn, drives the vertical vibration. The discharge fluctuations resulting from the vertical motion generate new pressure fluctuations which amplify the original horizontal motion. For this mechanism to function, the gate must behave as a press-shut device and the energy supplied to the vertical vibration in item (2) above must exceed the energy consumed by the pressure resulting from discharge fluctuations, item (4) above. Since the non-dimensional pressure amplitude near the gate opening due to horizontal motion (about 45 for the present Froude number) is greater than the corresponding value due to vertical motion (about 20 for the present Froude number), the gate can be expected to vibrate as long as ϕ is between -90° and about -15°. These values of ϕ correspond to values of γ_{xy} between the onset ratio, 0.97, and the termination ratio, 1.5 (for the Θ = 17.5° inclined weir plate). Beyond a value of γ_{xy}= 1.5, the energy supplied to the gate by item (2) is less than that removed by item (4), and even though the gate still acts as a press-shut device, there can be no vibration. Clearly, this last statement must be adjusted as the inclination angles increase and the instability region expands.

Figure 52. Closed energy cycle for coupled-mode vibration of shell-type gates with inclined weir plates in underflow with free discharge
From Ishii & Knisely (1992b). © 1992 Elsevier. Used with permission.

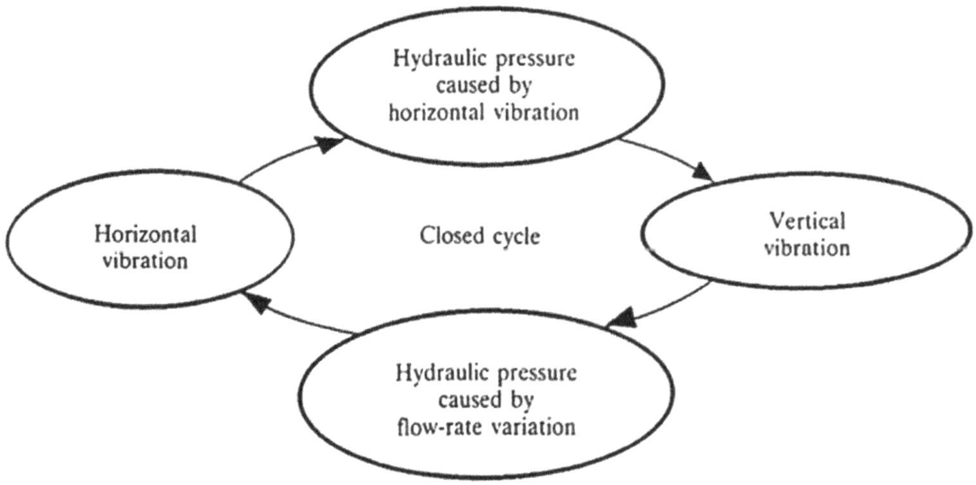

Stability Criterion

Frequency Criterion

If the ratio of horizontal in-water to vertical in-air frequencies, γ_{xy}, of a shell-type long-span gate (or other gate with inclined upstream surface) is between the onset ratio, γ_{nyy}, and the termination ratio, γ_{term}, the possibility of self-excited vibrations is high, unless the gate is equipped with a special damping device. Therefore, dynamic stability is ensured if $\gamma_{xy} < \gamma_{nyy}$ or if $\gamma_{xy} > \gamma_{term}$, which leads to the following frequency criteria for dynamic stability:

$$\Omega_{nwy} > \Omega_{wx} \text{ or } \Omega_{ay} < \Omega_{wx} / \gamma_{term} \tag{48}$$

For dynamic stability of shell-type long-span gates, the in-water vertical natural frequency should be greater than the in-water horizontal frequency, or the in-air vertical frequency should be less than the in-water natural frequency divided by the termination ratio (about 1.5 for $\Theta = 17.5°$ to about 2.0, or more, for $\Theta = 60°$). The region of instability in terms of γ_{nxy}, is plotted for each inclination angle in Figure 53a. The lower bound is the onset ratio and the upper bound is the termination ratio, used in Equation 48. The size of the filled circle indicates the relative strength of the instability, located near $\gamma_{nxy} = 1$ for all values of Θ.

Figure 53. Effect of inclination angle on (a) stability criterion and (b) on the maximum reduced fluid excitation coefficient
From Nakata et al. (1996). © 1996 JSME. Used with permission.

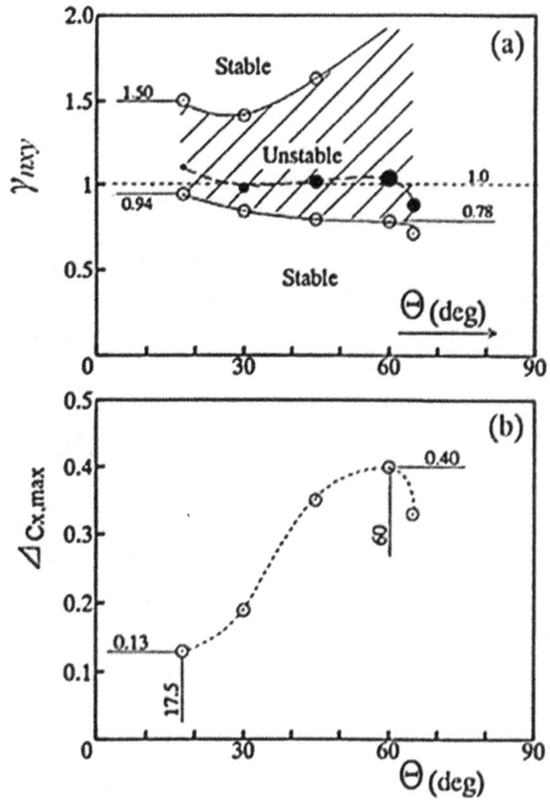

Damping Criteria

From the reduced excitation coefficient formulation in Equation 39, a criterion can be developed for the required damping needed to prevent vibrations. The damping level is sufficient if the excitation ratio in water in the horizontal direction, ξ_{wx}, is less than zero. For stable operation of the gate this criterion requires

$$\zeta_{ax} > \left(\Delta c_x / 2F_{0x}\right)\alpha_m \text{ or } Sc > 2\pi\Delta c_x / F_{0x} \tag{49}$$

where Sc is the Scruton number characterizing the mass-damping behavior of the shell-type gate.

To prevent vibration, the in-air damping ratio in the horizontal direction, which corresponds to the damping provided by a damper, must be larger than the combination of Δc_x, F_{ox} and α_m given on the right side of the first expression in Equation 49, or the Scruton number must be larger than the combination of Δc_x and F_{ox} given on the right side of the second expression in Equation 49. When the value of $(\Delta c_x/2F_{ox})$ is known, the damping ratio, ζ_{ax}, can be calculated as a function of the mass ratio, α_m. The difficulty involved in this calculation is that the value of Δc_x varies with the basic Froude number, F_{ox}, and with the angle of inclination, Θ, of the upstream gate surface, as shown in Figure 53b. The maximum values of Δc_x are presented in Figure 53b as a function of the inclination angle, but at a fixed F_{0x} of 13.9. In experiments to date the basic Froude number has been fixed, and thus a damping criterion for all possible Froude numbers cannot be obtained. However, for a gate model with $\Theta = 17.5°$ operating at a basic Froude number of $F_{ox} = 13.9$, the damping criterion of Equation 49 reduces to

$$\zeta_{ax} < 0.0047\alpha_m \text{ or } Sc > 0.059 \quad (50)$$

On the basis of the foregoing analysis, the shell-type gate shown in Figure 16, which is in common use in Japan with a weir plate inclination angle of 60°, is most likely not an optimal design as it may be susceptible to the coupled mode vibration observed in the studies reviewed. All of the model tests, however, were performed for a fixed value of basic Froude number, which controls the dynamic similarity of the fluctuating flow that accompanies coupled-mode gate vibration.

LONG-SPAN GATES WITH SIMULTANEOUS OVER- AND UNDERFLOW

Long span gates are frequently used for flood control and in irrigation systems in wide, shallow rivers. In Japan, especially, they are often of a shell-type construction, consisting of a very long tube-like structure, supported on a cable lift system at each end of the span, as shown in Figure 54. Some long-span gate are designed to be self-flooding, with holes along the downstream surface to permitting the inflow of water to add mass to the structure. All long-span gates must be able to pass flood level water, and may be designed for simultaneous over- and underflow.

Initially, a review of the somewhat limited literature concerning gates with simultaneous over- and underflow will be presented. Subsequently, the dynamics of long-span shell-type gates will be considered.

The self-weight of the gate results in a mean vertical deformation of the gate, even when suspended in air. The gate will also have a streamwise mean deformation due to the mean loading of the water at differing levels on the on the upstream and downstream faces. In addition to the static deflections, the gate can also experience dynamic deformation, undergoing beam bending in both the vertical and horizontal directions.

Usually, a minimum of three modes of vibration must be considered. The first mode is the gate mass suspended on elastic cables acting as a vertical rigid-body mass-spring-damper system. While this mode has been observed on one occasion for one full scale gate (see Ishii et al., 1995), it has not occurred while the gate was instrumented, and data substantiating the occurrence of this vibration mode are not available.

The second mode is a streamwise bending mode, usually the lowest frequency half wavelength mode. The third mode is a vertical bending mode, again usually the lowest frequency half-wavelength bending mode. A fourth mode, the lowest frequency torsional mode of the shell-type long-span gates, is at a substantially higher frequency and very small amplitudes due to the high torsional rigidity, and is seldom a significant contributor to long-span gate instability.

The instability of a long-span gate with a flap gate, but with no overflow (see Figure16) requires consideration of the role of the fluid loading above the main gate, and its effect on the in-water frequency of the vertical mode. It may be that without the added water load the horizontal and vertical modes are at substantially different frequencies; the fluid loading can lower the vertical mode frequency to the point that it matches that of the horizontal mode, permitting the coupling of the two modes.

Figure 54. Simultaneous over- and underflow (flow from right to left) past the cross-section of a full-scale long-span gate showing one span-end pier with lifting cable and guide slot. Accelerometer locations are marked as black squares. Length dimensions are in mm.
From Ishii et al. (1995). © 1995 Elsevier. Used with permission.

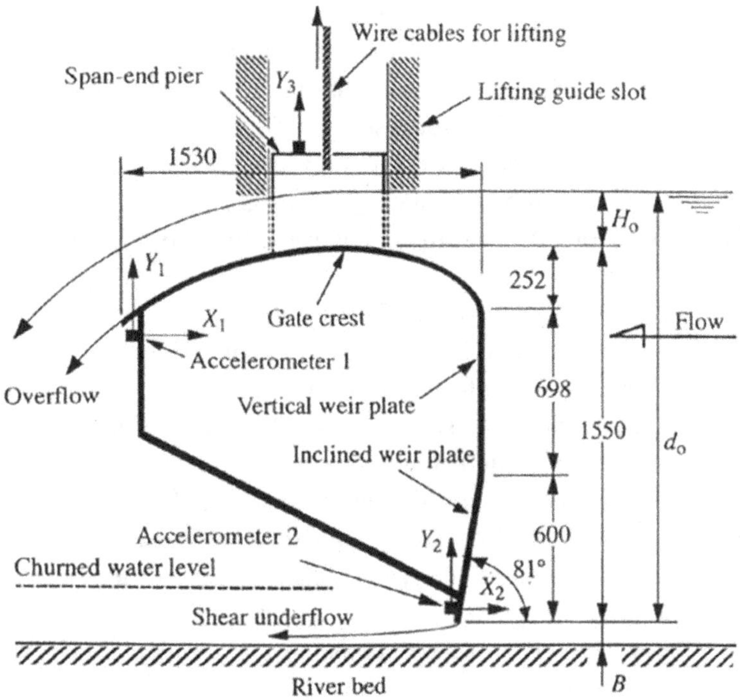

Even with just underflow, as portrayed in Figure 15, the horizontal and vertical bending modes of shell-type long-span gates can become coupled, under certain conditions. The analysis of this coupled-mode mechanism was the topic of the preceding section of this chapter and is summarized in the schematic of the closed energy cycle in Figure 52. A small horizontal perturbation results in a fluctuating pressure load on the inclined portion of the upstream weir plate. The pressure loading can drive a small vertical response vibration. The motion of the vertical vibration changes the gate opening and results in a discharge fluctuation that produces pressure loading that, in turn, drives the horizontal vibration, that then subsequently enhances the vertical motion, and the coupled vibration quickly grows.

For gates with simultaneous over and underflow, a similar decomposition of the vibration into a main vibration in one direction and a response vibration in the other will be explored in this portion of the chapter. The instability analysis of a gate, such as that if Figure 54, with simultaneous over- and underflow, however, is complicated by the addition of vertical and horizontal added masses. With appropriate approximate scaling of the effective gate height, the method developed for calculating the vibration characteristics long-span gates with only underflow will be adapted for the case of long-span, shell-type gates with simultaneous over- and underflow in this section of the chapter.

Literature Concerning Gates With Simultaneous Over-and Underflow

Early Works

As early as 1953, Petrikat (1953) reported the destruction of a laboratory model of a Tainter gate when it was exposed to simultaneous over- and underflow. He considered the vibration analogous to that of an elastic or elastically mounted cylinder excited by the shedding of alternating vortices in the form of a Kármán vortex street. A few years later, he suggested (Petrikat, 1958) that the vibrations due to simultaneous over- and underflow could be attenuated or eliminated by disrupting the original flow pattern, as shown in Figure 55a, by adding a row of spoilers (0.3 m high, 0.3 m long, spaced 0.7 m apart, at the end of the flap surface inclined at 45° downstream relative to the flap surface) in conjunction with sill modifications, as shown in Figure 55b.

Naudascher (1959; 1961) measured force coefficients on stationary rigid gate models with simultaneous over- and underflow. The measured fluctuating force coefficients were then assumed to act as a forcing function on elastic gate models with an assumed vertical degree-of-freedom. He suggested gate and sill geometries to attenuate vibrations. He also noted the effects of aeration and the downstream water surface level on the reduced vibration amplitudes.

Naudascher (1961) noted the need to consider the individual and interactive roles of (1) the upper jet flow over the gate crest (referred to also as nappe previously considered in Chapter 6); (2) the lower jet formed in the underflow region beneath the gate; (3) the space between jets downstream of the gate; and (4) the elasticity of the gate.

Naudascher (1961) followed the lead of Petrikat and again considered the vibration due to the simultaneous over- and underflow past a Tainter, roller, or vertical lift leaf gate with a vertical degree of freedom to be analogous to Kármán vortex shedding from a cylindrical body. He considered the vibration as a vortex-street-like oscillation of the wake mass downstream of a flat plate with negligible streamwise dimension and formulated a non-dimensional parameter similar to the Strouhal number for vortex shedding from cylinders.

Figure 55. Segment weir with simultaneous over- and underflow: (a) flow pattern resulting in vibration, and, (b) a cure for the resulting vibrations
After Petrikat (1958).

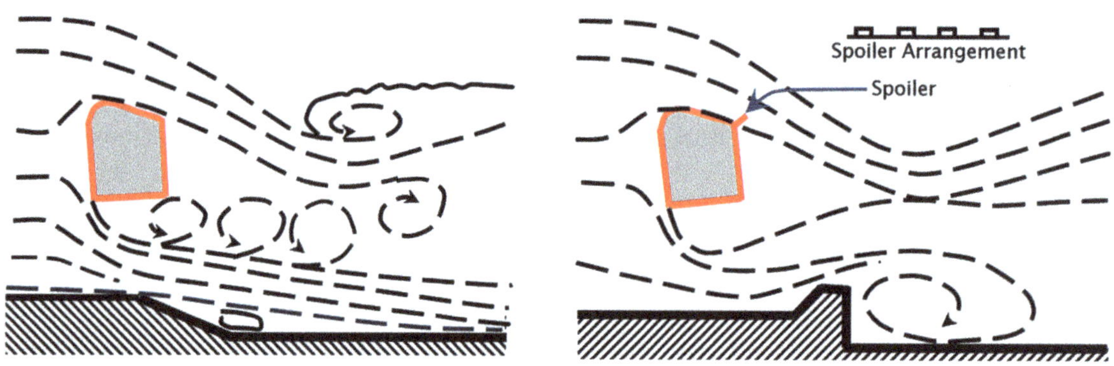

More recently, Husain (1999) provided further clarification of the hydrodynamics of gates with simultaneous over- and underflow. He tried to decompose the flow by considering separately the underflow and the overflow and then attempting to find an interaction factor that would permit the combination of the separate loading for the case of simultaneous over- and underflow. He found the fluctuating pressure coefficient and the fluctuating force on the gate bottom to be much larger on a vibrating gate than on a stationary gate. Similarly he found amplification of the fluctuating pressure and force coefficients on the gate crest with vibration, relative to those on the stationary gate. Finally, he concluded that there is a further magnification of coupling and interaction when the dominant frequency of the fluctuating forces is close to the gate natural frequency.

Castro and Bustamante (2002) attempted to analyze the discharge by determining fictitious heights for both the underflow and the overflow. They assumed the underflow was emanating from a fictitious tunnel of some fictitious height that could supply pressurized water to be discharge beneath the gate. The overflow was from another fictitious height reservoir. The goal was to determine the streamline separating the overflow from the underflow. Their laboratory results validated their proposed theoretical scheme. They did not consider vibration of the gate.

Laboratory Study of Simultaneous Over-and Underflow

Motivated by problems with the vibration of installed full-scale long-span gates when exposed to simultaneous over and underflow, Ishii et al. (1994) undertook a systematic laboratory study of two degree-of-freedom model gates with simultaneous over- and underflow.

For shell-type gates with inclined upstream weir plates, such as that shown in Figure 54, any pressure fluctuation will act with a vertical component on both the inclined weir plate and the gate crest; both surfaces have a significant influence on the observed coupled mode self- excited vibrations. The individual roles of the inclined weir plate and the gate crest were delineated in two separate experiments: the first on a gate model with an inclined weir plate, but no overflow (see the preceding section "Two Degree-of-Freedom Gates with Underflow") and the second on a gate model with overflow, but with no inclined weir plate.

In the preceding section of this chapter the mechanism of coupled-mode self-excited vibration caused by the inclined weir plate was analyzed using the theoretical results for the inclined long-span gate with slender streamwise dimension. The second source of vertical loading on a gate with simultaneous over- and is due to the pressure acting on the gate crest. Measurement and analysis of the pressure acting on a laboratory-scale gate crest will be developed in the following section of the chapter.

Gate Model Parameters

The gate height, H, was 330 mm, the gate thickness, W_0, was 70 mm and the transverse gate span, L, 294 mm. The mass of the entire gate was 11.97 kg. The gate crest corresponded to the Creager nappe inner curve for 33 mm of overflow (Creager & Justin 1927). The actual overflow, H_0, was comparatively large, 50 mm, yielding an upstream submergence, d_0 (see Figure 56), of 380 mm. The gate opening height, B, was maintained at a small value of 3 mm. To aerate effectively the closed cavity formed by the large overflow and small underflow, a vertical spoiler was installed on the channel floor. The spoiler had an isosceles triangle cross-section with a base width of 7 cm and a side length of 10 cm; the apex of the triangle pointed upstream. The spoiler was 400 mm high, extending from a short circular pin base on

Figure 56. Schematic of gate model with two degree-of-freedom, a small gate opening for underflow, and relatively large overflow passing over the gate crest
From Ishii et al (1994). © 1994 Elsevier. Used with permission.

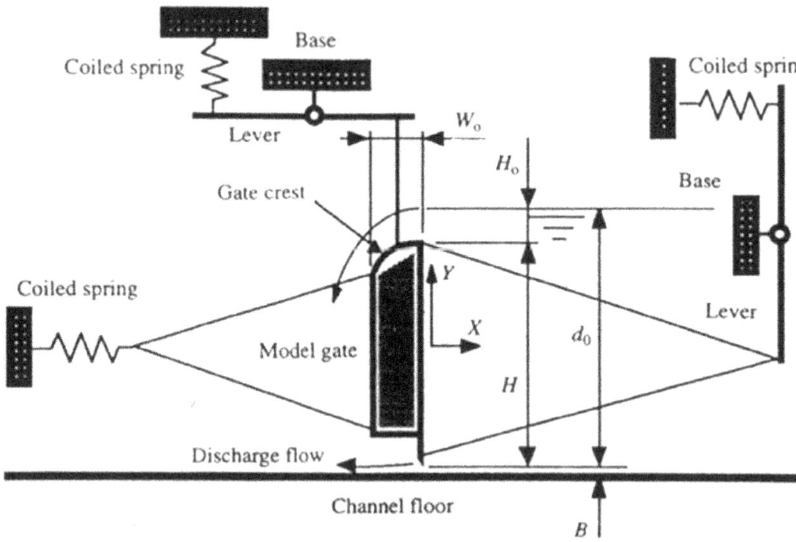

the channel floor to a second support point above the upstream water level. It was located along the gate centerline approximately 5 to 10 cm downstream of the gate.

The model gate was supported by coiled springs to simulate the gate elasticity. The two coiled springs (one on each side of the channel) on the upstream side were connected to four points on the upstream face of the model gate by means of two levers. The position of attachment of these springs to the levers could be varied to give a fine adjustment of the gate natural frequencies. The coiled springs on the downstream side controlled the horizontal gate frequency and kept the in-air damping ratio low. The coiled spring and lever arrangement shown at the top of Figure 56 simulated the vertical elasticity of the gate. With this arrangement, the model gate could undergo streamwise and vertical vibrations, but is constrained from any appreciable torsional motion.

Streamwise gate motion produces fluctuating overflow and corresponding hydrodynamic pressure fluctuations on the gate crest. The fluctuating hydrodynamic load on the gate crest forces the gate to vibrate in the vertical direction at the in-water frequency of streamwise gate vibrations. Considering the gate as a simple mechanical oscillator, it is known that the gate response in the vertical direction due to forcing at the in-water streamwise frequency depends primarily on the ratio of the forcing frequency (i.e., the in-water streamwise frequency) to the in-water natural frequency of vertical vibration. Recalling that there is a 180° phase change as the forcing frequency progresses from frequencies below to frequencies above resonance, the resulting vertical gate motion is expected show such a phase reversal relative to the horizontal motion. As was found for two degrees-of-freedom gates with underflow, the gate is expected to behave as a dynamically stable press-open device below resonance; at and above resonance it is expected to do so as a dynamically unstable press-shut device.

The in-air frequency of streamwise vibration, Ω_{ax}, was fixed at 9.17 Hz and the in-air vertical frequency, Ω_{ay}, varied over the range $3.79 < \Omega_{ay} < 8.26$ Hz. The in-air damping ratio of the horizontal vibration, ζ_{ax}, was as small as possible and had a value of 0.0031.

The fundamental dimensionless quantities dominating the dynamic similarity of the fluctuating flow and the vibrations of the long-span gate with a small underflow but no overflow are a basic Froude number $F_{0x} = \Omega_{ax}\sqrt{d_0 / g}$ (from Equation 74 in Chapter 4) and the water-to-gate mass ratio,

$$\alpha_m = \rho d_0^2 / \left(\rho_* hH\right) = \rho d_0^2 / \left(M / L\right),$$

as defined in Equation 14.

When these two quantities are held constant, the in-water to in-air horizontal frequency ratio is also kept constant. In addition, an alternate Froude number, based on the in-water frequency, also remains constant. The values of F_{ox} and α_m for the model experiments reported herein were 11.3 and 3.55, respectively. The overflow depth, $H_0 = 50$ mm, and the gate opening, $d_0 = 3$ mm, were also held constant.

Measured Data

For a selected value of vertical frequency, the in-air damping was measured by displacing the gate vertically and releasing it with no water in the channel. The damping was measured by the decay of the vibration amplitude. With the channel filled with water to the above specifications, the gate was again given a displacement, if it did not undergo self-excited vibration. The in-water frequencies and damping/excitation ratios were measured in both the vertical and horizontal directions. The measured values are shown in Figure 57. The in-air damping in Figure 57a is qualitatively similar to that in Figure 44a, but almost twice as large. The in-water horizontal and vertical vibration frequencies in Figure 57b and their corresponding excitation ratios ξ_{wx} and ξ_{wy}, in Figure 57c were extracted from the x- and y-direction displacement traces. Over the region $4.65 \leq \Omega_{ay} \leq 5.46$ Hz, the excitation ratios takes on positive values, indicating exponential growth of perturbations over this region. Over two different ranges of in-air natural vertical frequencies, Ω_{ay}, given by $\Omega_{ay} < 4.65$ Hz and $\Omega_{ay} >$ about 6 Hz, gate vibration was damped even when given an initial displacement. In these dynamically stable regions, the in-water horizontal and vertical frequencies are not the same.

The in-water horizontal frequency, Ω_{wx} remains approximately constant between about 5.10 to 5.13 Hz and is about equal to the horizontal natural frequency in still water, Ω_{nwx}, but is significantly less than the in-air horizontal frequency of 9.17 Hz, due to the large added mass effect. The vertical in-water frequency, on the other hand, is close to the corresponding in-air frequency, except over the limited range of $4.65 \leq \Omega_{ay} \leq 6$ Hz, where the vertical in-water frequency coalesces with the horizontal in-water frequency. In this range, the motion in the two directions is coupled and the gate undergoes spontaneous self-sustained vibrations. The gate motion traces out elliptical orbits. The excitation ratio reaches its maximum value of about 0.017, as a result of resonance when Ω_{ay} takes on a value of 4.97 Hz.

In Figure 58, the data were non-dimensionalized as was done for the gate with inclined weir plate in underflow. The abscissa in Figure 58, γ_{xy}, is the ratio of the in-water streamwise frequency to the in-air vertical frequency. Recall that the values of γ_{xy} are inversely proportional to the in-air vertical frequency, Ω_{ay}.

The reduced fluid excitation coefficient, Δc_x, was computed using Equation 39, presented earlier in this chapter. Over a limited range of $0.92 < \gamma_{xy} < 1.05$, the non-dimensional fluid excitation coefficient, Δc_x, takes on a positive value and reaches its maximum value of 019 at $\gamma_{xy} = 0.99$. For frequency ratios in this range of positive Δc_x the gate will undergo self-excited vibration, if the mechanical damping in

Figure 57. Measured values of (a) in-air damping ratio for vertical vibrations, (b) in-water vibration frequencies, and (c) excitation ratios of model gate, as functions of the in-air vertical frequency, Ω_{ay}
From Ishii et al (1994).

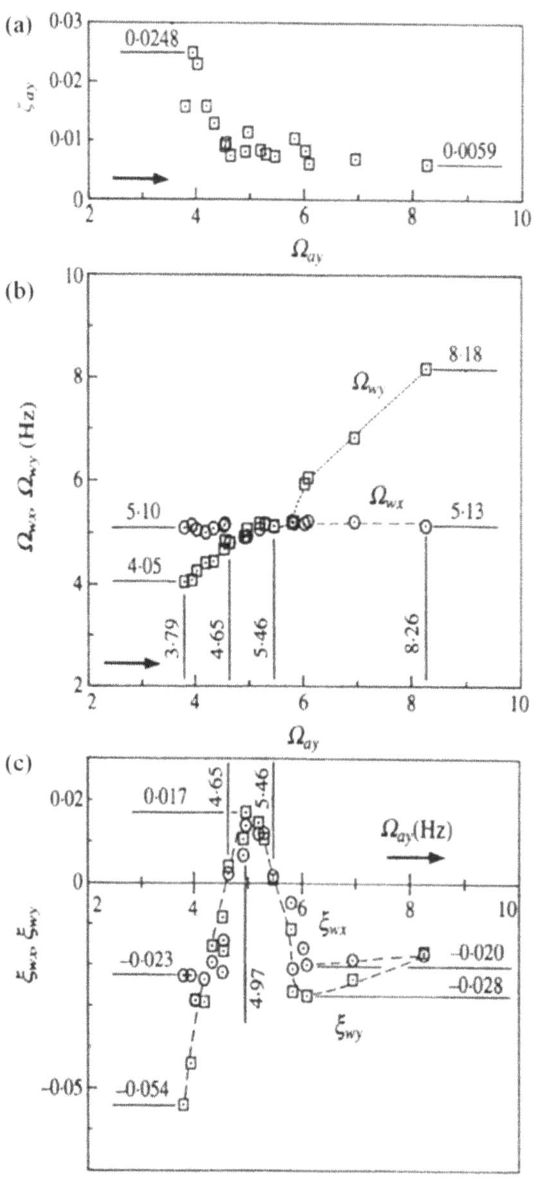

the vibrating system is less than the negative damping supplied by the fluid flow. Recall the reduced excitation coefficient represents the fluid excitation only and does not include structural damping.

The reduced added mass coefficient, Δm_x, computed using Equation 40 and shown in Figure 58c ranges from about 0.59 to 0.67. For the frequency ratio, γ_{xy}, between 0.99 and 1.03, the horizontal frequency decreases somewhat, apparently being entrained by the lower vertical frequency. In this range, therefore, a correct value for Δm_x, could not be calculated and three data points were omitted.

Figure 58. Vibration frequency ratios, reduced fluid excitation coefficient, reduced added mass coefficient and phase lag of vertical motion relative to horizontal motion, shown as functions of the ratio of in-water streamwise to in-air vertical frequencies, γ_{xy}
From Ishii et al. (1994). © 1994 Elsevier. Used with permission.

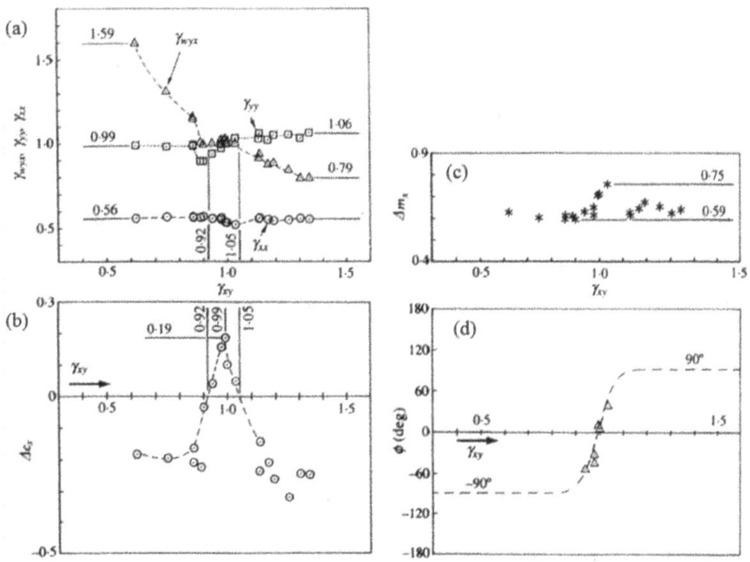

The phase, ϕ, in Figure 58d is the phase-lag of the vertical vibration relative to the horizontal vibration. This phase-lag changes from -90 to 90° for increasing values of γ_{xy}, and passes through zero when the frequency ratio, γ_{xy}, equals the inherent vertical frequency ratio, γ_{nyy}.

The data patterns in Figure 58 are reminiscent of very similar patterns in the data for the two degree-of-freedom gate with inclined weir plate in Figure 45 and 46. Many of the previously discussed inferences about the vertical vibration being the response to the forced load imparted on the gate surface due the fluctuating pressure generated by the streamwise gate vibration can also be made for the current model, a vertical gate with simultaneous over- and underflow with no inclined gate face.

Figure 59 shows the characteristic trajectories (with the upstream direction to the right), that have been previously shown to be associated with a press-shut device undergoing self-excited vibrations (see Figures 40 and 42). Trajectories (b) to (f) correspond to the unstable region for frequency ratios, γ_{xy}, from 0.94 to 1.03 and clearly exhibit press-shut behavior with downward gate motion while moving downstream and upward gate motion while moving upstream For $\gamma_{xy} = 0.86$, the gate moves essentially in only the horizontal direction, as shown in trajectory (a). Trajectories (g) and (h) exhibit random motion for corresponding values of $\gamma_{xy} = 1.13$ and $\gamma_{xy} = 1.17$, respectively. No instance of press-open gate behavior was observed, despite the anticipation of such from the fundamental dynamics of mechanical vibrations and the previous experience with gate having inclined upstream weir plates.

In order to establish the relationship between horizontal gate motion and the induced pressure on the gate crest, the gate model was forced to vibrate with a horizontal amplitude of 1.5 mm at a frequency of 5.1 Hz, corresponding to the in-water natural horizontal frequency. The fluctuating pressure on the gate crest was monitored using a semiconductor type pressure transducer (see Ishii et al., 1994 for details). The pressure transducer was positioned successively at locations along the gate centerline with 5 mm

Figure 59. Trajectories of gate motion which are indicative of press-shut behavior
From Ishii et al. (1994). © 1994 Elsevier. Used with permission.

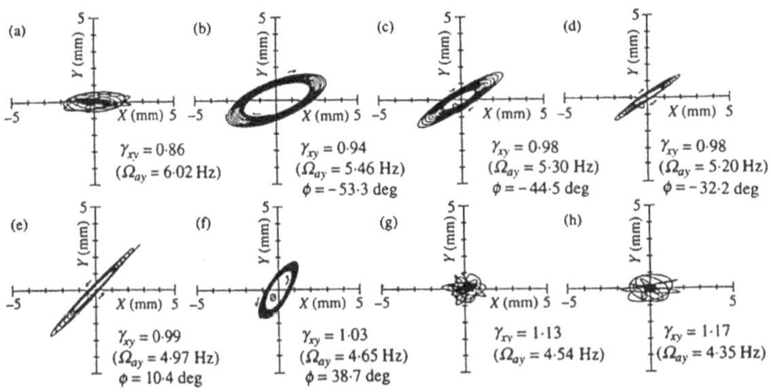

spacing between each location. Since the gate vibration is limited to very small amplitudes in the X-direction, these pressure measurements provide an accurate estimate of the fluctuating pressure acting on the gate crest.

Assuming the upstream direction is positive and the horizontal motion is again given by Equation 41 with a change in the amplitude notation, that is

$$X(t) = X_0 \cos\left(\Omega_{wx} T\right) \tag{51}$$

the induced fluctuating pressure acting on the gate crest can be written as

$$P_y(t) = P_{y0} \cos\left(_{wx} T - _0\right) \tag{52}$$

The measured pressure amplitude, P_0, and the phase lag, ϕ_0, relative to horizontal gate displacement are plotted in Figure 60, where the abscissa is the horizontal distance downstream of the vertical gate face. The pressure amplitude decreases at increasing downstream distances, while the phase lag, a most significant parameter in the discussion of energy transfer and coupling, takes on a value of 120° at the most upstream location, gradually decreasing with increasing downstream distance and increases slightly near the downstream gate face. The mean value of the pressure phase lag is about 90°. This indicates that the pressure reaches its maximum value as the gate passes through equilibrium while moving in the downstream direction. A rough estimate of the pressure magnitude is about 59 Pa, yielding a per-unit-displacement pressure amplitude of 39 Pa/mm of horizontal gate motion.

Vertical Response Characteristics

The pressure exerted on the gate crest acts downward and is the driving force for the vertical vibrations. However, the positive Y-axis is upwards. Therefore, the phase between the vertical displacement and the horizontal displacement is the sum of the phase shift of the forcing function (i.e., the pressure) and the mechanical response of the structure. Thus, $\phi = \phi_0 + \pi + \phi_p$ where ϕ_p is the phase response of the

Figure 60. Distributions of amplitude and phase lag of fluctuating pressure acting on gate crest during forced horizontal vibrations as functions of distance downstream from the vertical gate face
From Ishii et al. (1994). © 1994 Elsevier. Used with permission.

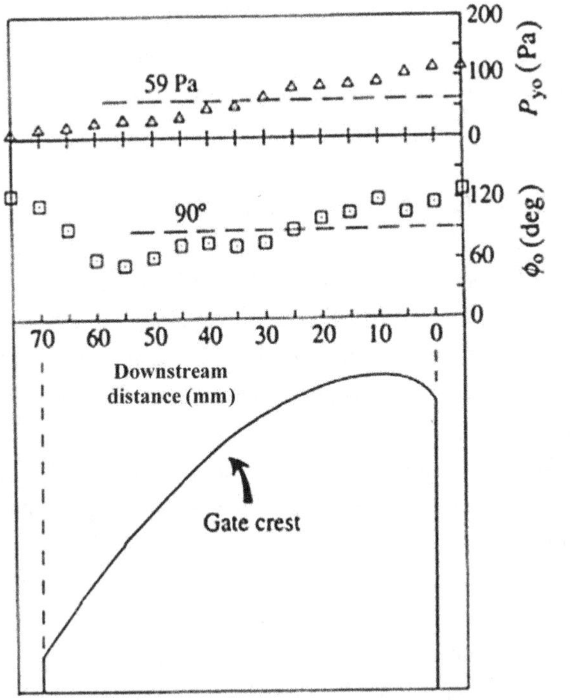

classical forced oscillator. Substituting this expression for the phase angle into Equation 51, re-written here with a different amplitude notation

$$Y(T) = Y_0 \cos\left(\Omega_{wx} T - \phi\right) \tag{53}$$

yields the following expression for the vertical displacement:

$$Y(T) = Y_0 \cos\left(\Omega_{wx} T - \phi_0 - \pi - \phi_p\right) \tag{54}$$

in which the phase angle is measured relative to the horizontal displacement. Substituting the measured value of π/2 for the pressure phase lag, ϕ_0, into Equation 54 yields the following expression:

$$Y(T) = Y_0 \cos\left[\Omega_{wx} T - \left(\phi_p - \frac{\pi}{2}\right)\right] \tag{55}$$

The phase lag of $Y(T)$ relative to $X(T)$ is the phase lag of the forced oscillator minus $\pi/2$, that is $\phi = \phi_p - \pi / 2$. The value of ϕ is expected to range from -90° to 90°, corroborating the experimental data in Figure 58d.

Eliminating the time dependence in the $X(T)$ and $Y(T)$ in Equations 51 and 54 permits the derivation of an equation for the trajectory of the gate motion. From this equation one can conclude that the mean slope of the trajectory is always positive or, providing theoretical support for the absence of press-open devices that would have a negative mean slope.

Field Study Simultaneous Over- and Underflow

Full-scale field measurements were undertaken in May 1988 on a shell-type long-span gate used for irrigation purposes, previously illustrated in Figure 54 and repeated here as Figure 61a for convenience The gate is 1.55 m high with a span length of about 25 m. The gate crest is smoothly curved for overflow and the front surface consists of vertical and inclined weir plates. The height to (streamwise) width ratio of the shell section is about 1.0.

The gate was designed to be lifted above the river when not in use. When increased upstream water levels were required, the gate could be lowered at a relatively low but constant rate to rest directly on the river bed, damming the river. The upstream water level then rose gradually and ultimately overflowed the crest of the gate. The overflow height reached about 25 to 30 cm, yielding a maximum upstream water level of about 1.8 m to provide water for irrigation. When resting on the river bed, the gate experienced no flow-induced vibration. When the irrigation task was completed, the gate was raised at a constant rate to return to the elevated, stowed position. No vibration was encountered during routine raising of the gate.

When the gate is in service and the river discharge exceeds the design capacity of the overflow gate, the upstream water level will rise above the retaining walls, leading to flooding of the surrounding region.

Under such circumstances, the gate owner insisted on using the gate for flood control as well as for its design use. By raising the gate and regulating the gate opening, discharge capacity is augmented and the gate can be used to control the upstream water level, despite its initial design for fixed position operation. The gate experienced severe self-excited vibrations at small gate openings. The vibrations were observed to involve coupling between the horizontal and vertical directions. The nappe was sufficiently aerated, as shown in Figures 51b and 51c, and played no role in the vibrations.

With the gate opening set to 5 to 7 cm, the upstream water level could be adequately controlled, but severe self-excited vibrations occurred. Gate vibrations at various underflow gate openings and overflow depths were carefully measured using accelerometers, with special attention being paid to excitation from an initial quiescent state to a stable vibrational state with a steady amplitude of vibration.

The vibrations of the gate in flowing water depend fundamentally on its in-air vibration characteristics, such as the natural frequencies, damping ratios and vibration mode shapes. Experimental modal analysis was undertaken for both horizontal and vertical bending modes with the gate in its elevated position. The added mass and the reduced fluid excitation coefficients were extracted from the experimental data for in-water and in-air vibration frequencies, damping ratios, and the excitation ratios.

Application of the theory from the preceding portion of this chapter to the full-scale gate permitted the computation of the theoretical values for the in-water vibration frequency, the added mass, and the fluid excitation coefficient for a given value of the in-air vibration frequency. The theoretically predicted results were in good agreement with the experimental results.

Figure 61. Full-scale gate with simultaneous over- and underflow (a) drawing of gate cross-section showing accelerometer locations as black squares with length dimensions in mm; (b) overflow across entire span (c) effective spoilers aerating the nappe
From Ishii et al., (1995).

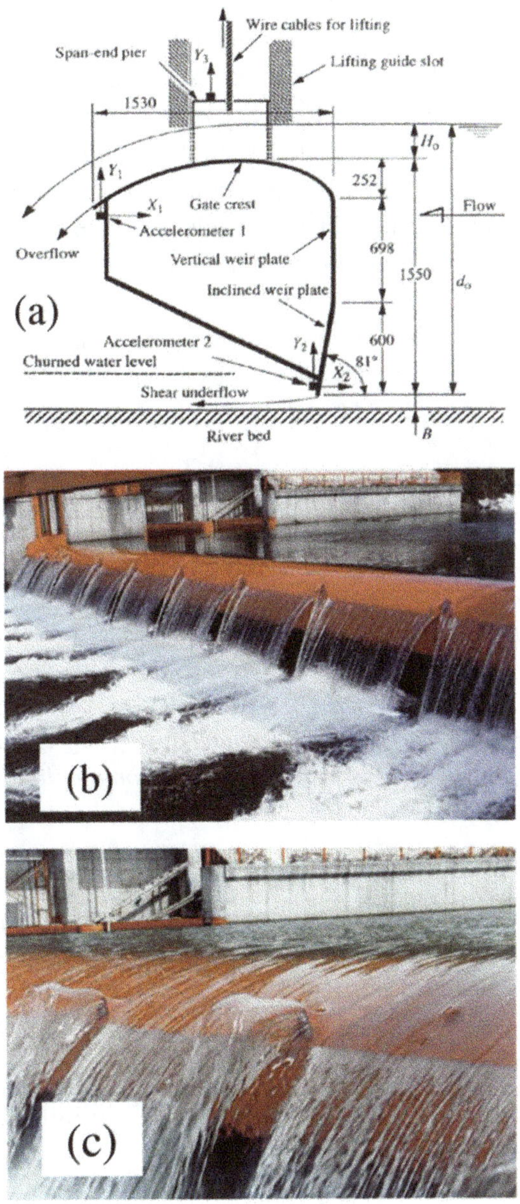

Measured Data on Gate Dynamics

Gate vibrations were monitored at two points at the spanwise center of the gate. The first point was located just under the tail of the gate crest, and the second point just behind the gate lip, as a shown in Figure 61a. Integration of the accelerations provided the horizontal and vertical vibration displacements,

X_1 and X_2, and Y_1 and Y_2. In addition, the vertical vibrational displacement, Y_3 at the end of the gate span in the gate slot was also recorded to examine the spanwise mode of vertical vibration. Samples of the X_1 and Y_1 displacements are shown in Figure 62.

The in-water vibration frequency and the excitation ratio (defined as the logarithmic increment divided by 2π) were extracted by digitizing the waveforms for computer analysis of the data. The in-water vibration frequencies in the horizontal and vertical directions, Ω_{wx} and Ω_{wy}, were both 3.32 Hz, and the excitation ratios in the horizontal and vertical directions, ξ_{wx} and ξ_{wy} were 0.015 and 0.019, respectively.

The maximum amplitudes, X_{max} and Y_{max}, of the horizontal and vertical motion are of 13.8 and 5.1 mm, respectively. These amplitude data are plotted as circles in Figure 63, corresponding to $B = 5$ cm on the horizontal axis which represents the gate opening height. The measured height of overflow, H_0 (see Figure 61a), was 25.0 cm, and is denoted by a triangle in Figure 63. The impingement of the overflow impeded the shallow underflow, resulting in an average depth of the churning downstream water of about 20 cm, just behind the gate lip.

When the gate opening height was increased from 5 to 6 cm and subsequently to 7 cm, the maximum vibration amplitudes decreased rapidly, but the gate still experienced self-excited vibration. The in-water vibration frequency was maintained at 3.32 Hz. When the gate opening height was set at 8 cm, however, no vibrations occurred.

The gate trajectory, shown in Figure 64, was obtained y plotting the Y displacement versus the X displacement. Such trajectories are typical of gates behaving as coupled-mode press-shut devices. This type of vibration involves the coupling of the horizontal and vertical bending vibrations through the hydrodynamic forces acting on the gate crest and on the inclined weir plate in the vertical direction and on the vertical and inclined weir plates in the horizontal direction.

Figure 62. Samples of horizontal and vertical displacement during self-excited coupled-mode vibrations exhibiting initial exponential growth followed by steady vibration amplitudes
From Ishii et al. (1995). © 1995 Elsevier. Used with permission.

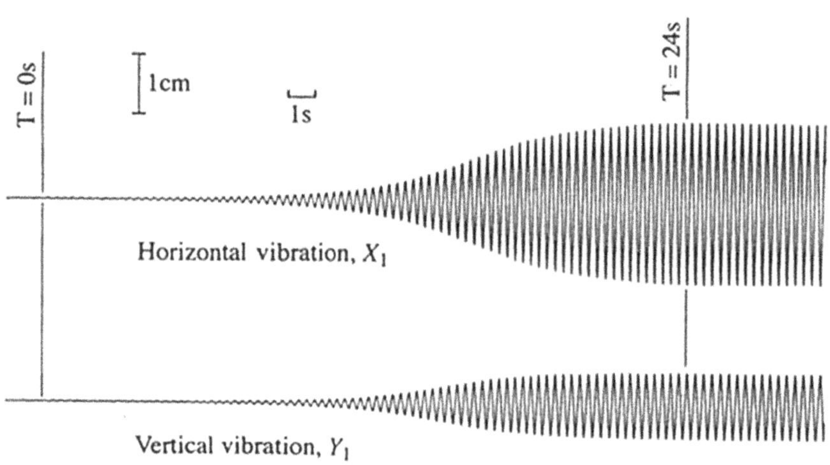

Figure 63. Maximum amplitudes of vibration in both the horizontal and vertical directions, and height of overflow as functions of gate opening height
From Ishii et al (1995). © 1995 Elsevier. Used with permission.

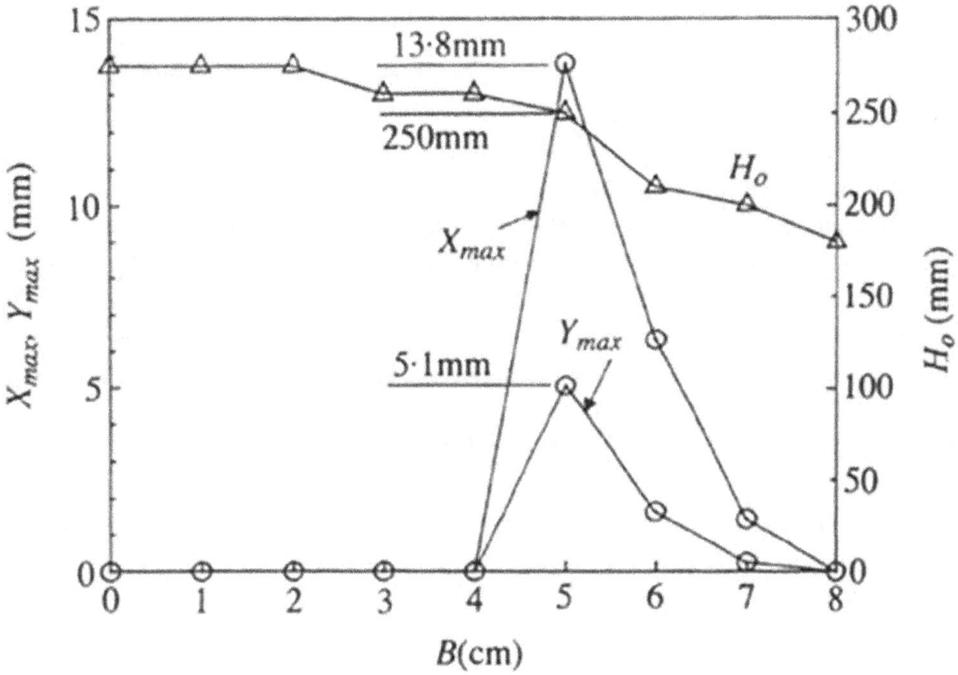

Figure 64. Trajectory of gate during coupled-mode vibration, showing an inclined elliptical trajectory traversed in the clockwise direction
From Ishii et al. (1995). © 1995 Elsevier. Used with permission.

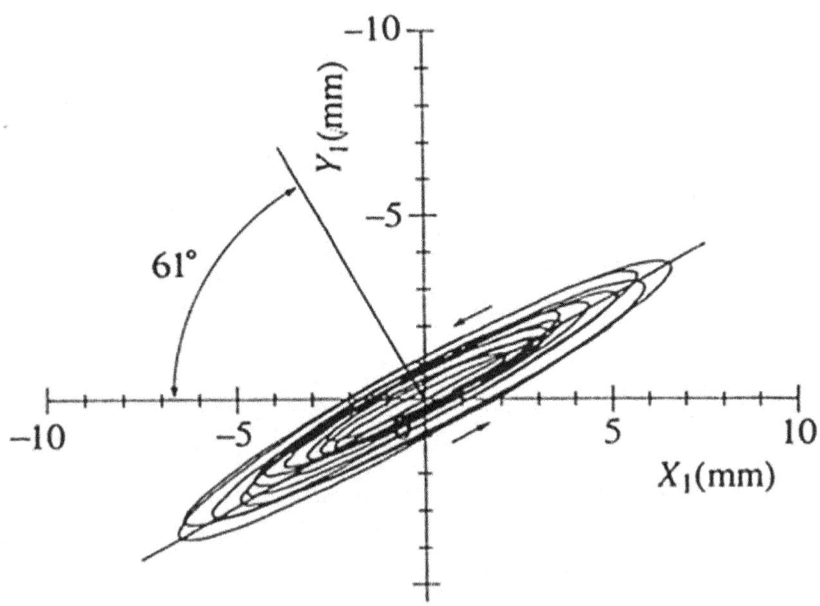

In-Air Mode Shapes and Frequencies

With the gate lifted from the water, in-air modal analysis was undertaken, with detail given in Ishii et al. (1995). The fundamental vibration modes are mode 1V at a frequency of 4.69 Hz and mode 1H at 6.45 or 6.46 Hz. These two mode shapes are essentially half-wavelength bending modes in the horizontal and vertical directions, though mode 1V has a small rotational component about the spanwise axis. The damping ratio is 0.027 to 0.042 for mode 1V and 0.004 to 0.002 for mode 1H. The full-scale gate was so designed that the bending rigidity in the horizontal direction is slightly larger than in the vertical direction. This slightly higher horizontal rigidity is reflected in the higher natural frequency of the half-wave mode in the horizontal direction (6.46 Hz) relative to the vertical mode (4.69 Hz).

The second vibration modes are mode 2V at a frequency of 16.2 Hz and mode 2H at 20.7 Hz. Both are essentially the full-wave bending mode in their respective direction. Mode 2V appears only in the vertical transfer function and mode 2H only in the horizontal transfer function. The damping ratio was 0.014 for both modes. Additional details of the modal analysis of the gate can be found in Ishii et al. (1995).

Added Mass and Reduced Fluid Excitation Coefficients

The added mass coefficient, ΔM_x and the reduced fluid excitation coefficient, ΔC_x, can be calculated by the dimensional equations similar to the non-dimensional expressions used previously and given in Equations 39 and 40. The dimensional forms are:

$$\Delta M_x = M\left[\left(\frac{1-\xi_{wx}^2}{1-\zeta_{ax}^2}\right)\left(\frac{\Omega_{ax}}{\Omega_{wx}}\right)^2 - 1\right] \tag{56}$$

$$\Delta c_x = \frac{2M\Omega_{ax}}{\sqrt{1-\zeta_{ax}^2}}\left[\zeta_{ax} + \xi_{wx}\left(\frac{\Omega_{ax}}{\Omega_{wx}}\right)\sqrt{\frac{1-\xi_{wx}^2}{1-\zeta_{ax}^2}}\right] \tag{57}$$

The in-air vibration frequency, Ω_{ax}, will be taken to be the circular frequency 29.47 rad/s (4.69 Hz) for the lowest-frequency mode 1V. The gate mass, M, is 26,492 kg (mass per unit span is 1,075.2 kg/m). Substituting the experimental data: Ω_w = 20.86 rad/s (3.22 Hz), ξ_{wx} = 0.015, and ζ_{ax} =0.027 into Equations (56) and (57), results in added mass and fluid excitation coefficient (per unit span) of values of 1,071 kg/m and 2,876 Pa s, respectively. These values are shown in Table 3, for later comparison with theoretical results.

Theoretical Calculations

The theoretical analysis developed in Chapter 5 and the present chapter was used to generate the plot of in-water to in-air vibration frequency ratio shown in Figure 65a. The abscissa is the basic Froude number, given as $F_0 = \Omega_{ax}\sqrt{d_0 / g}$ = (29.47 rad/s)(1.8 m/ 9.81 m/s^2)$^{1/2}$ = 12.6 in this full-scale test.

Table 3. Comparison of theoretical values with experimental data for the full-size gate

Parameter	Experiment	Theory Using Equivalent Gate Height, H_e	Theory Using Physical Gate Height
Ω_{ax} (Hz)	4.69	—	—
Ω_{ax} (rad/s)	29.5		
Ω_{wx} (Hz)	3.32	3.30	3.16
Ω_{wx} (rad/s)	20.9	19.9	19.85
ΔM_x kg	26,389	25,980	30,783
ΔC_x N/(m/s)	70,867	65,537	74,463

Before using the plot in Figure 65a to determine the Froude number ratio, $F/F_0 = \Omega_{wx}/\Omega_{ax}$, the value of the water-to-gate mass ratio, $\alpha_m = \rho d_0^2 / \left(M / L\right)$, must be computed. The numerator ρd_0^2 represents the added mass. Since the height of the present full-scale gate with overflow is smaller than the upstream submergence, d_0, the definition of α_m must be corrected slightly. Further, a correction in the added mass due to the smooth curve of the gate crest must be taken into consideration. However, the exact evaluation of the effect of the smooth curve on the reduction of the added mass appears to be exceedingly difficult. As a crude approximation, it is assumed here that half of the gate crest height is effective in producing added mass. Accordingly, the effective height of gate, H_e, is assumed to be given by H_e = 1.55 - 0.252/2 = 1.424 m.

Replacing d_0 by H_e, the approximate water-to-gate mass ratio becomes $\alpha_m = \rho H_e^2 / \left(M / L\right)$, which has a value of 1.886 for this full-scale gate. When calculating the basic Froude number, F_0, as was done above, the upstream submergence, d_0, measured from the gate lip to the free surface as shown in Figure 61a, must still be used since it governs the water velocity.

Using the calculated values of α_m, and F_0, the corresponding frequency ratio, $F/F_0 = \Omega_{wx}/\Omega_{ax}$, is 0.704 from Figure 65a. Alternatively, for Froude numbers greater than 10, the value $F/F_0 = \Omega_{wx}/\Omega_{ax}$, can be calculated by $F/F_0 = \Omega_{wx}/\Omega_{ax} = 1 / \sqrt{1 + 0.54\alpha_m}$ = 0.704 for α_m =1.886. The corresponding calculated in-water vibration frequency, Ω_{wx} is 20.75 rad/s (3.30 Hz), which corresponds almost exactly to the measured value. The value of F is 8.89.

If one chooses to ignore the added mass reduction due to crest curvature and uses the physical gate height of 1.55 m as the length scale in calculating α_m, the resulting calculated in-water frequency is 3.16 Hz, which is 4.8% lower than measured. Results obtained using the physical gate height as the length scale are given in the right most column of Table 3. Since these values for frequency and added mass depart further from the measured values than those obtained using the equivalent gate height, H_e, all other results are based on using the equivalent height as the length scale.

The added mass of the gate as well as the fluid damping and excitation coefficients can be determined from the graphs in Figures 65b and 65c. The reduced added mass, Δm_b, and the reduced fluid damping coefficient, Δc_b, associated with the push-and-draw effect of the gate on the water are shown in Figure 65b, while Figure 65c shows the reduced added mass, Δm_r, and the reduced fluid excitation coefficient, Δc_r, associated with the flow-rate variation,. The parameter $1/\beta^*$ is the ratio of the submergence depth, d_0, to the upstream-channel depth. The present full-scale gate is over a horizontal bed and is operated at small openings; accordingly, the $1/\beta^*$ parameter takes on a value insignificantly different from 1.0. At

Figure 65. (a) In-water to in-air vibration frequency ratio calculated from a theoretical study of a longitudinal slender long-span gate undergoing streamwise vibration, (b) the reduced added mass, Δm_b, and the reduced fluid damping coefficient, Δc_b, associated with the push-and-draw effect of the gate on the water, and (c) the reduced added mass, Δm_r, and the reduced fluid excitation coefficient, Δc_r, associated with flow-rate variations
From Ishii et al.(1995). © 1995 Elsevier. Used with permission.

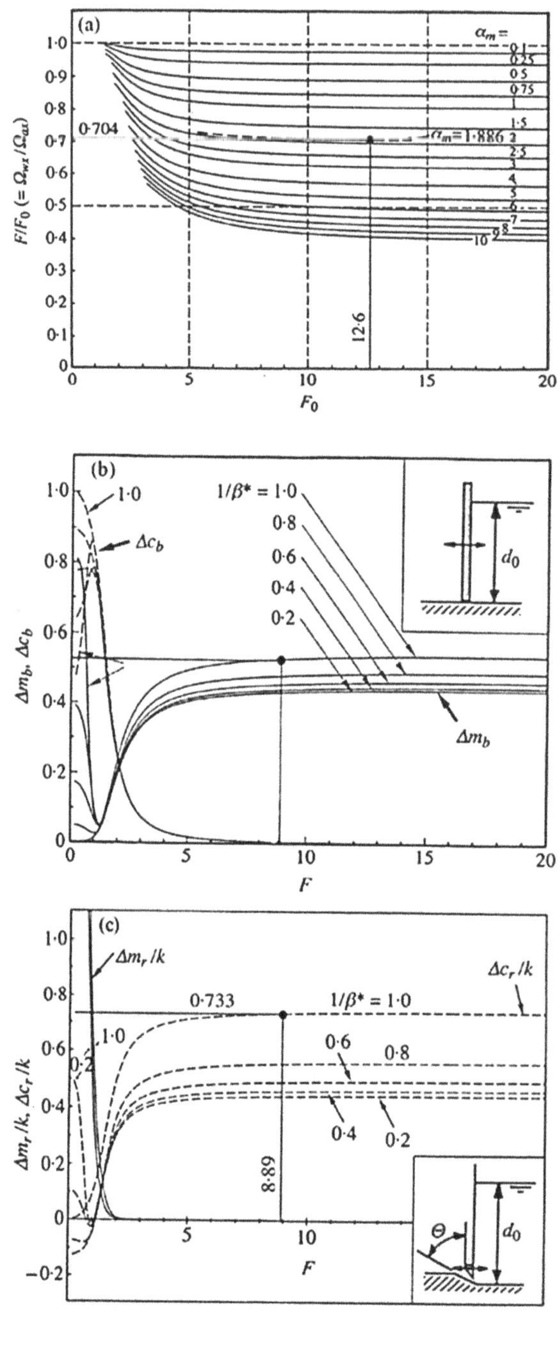

a Froude number F = 8.89, the values for the added mass, $\Delta m_b = 0.52$ and the fluid excitation coefficient fluid $\Delta c_r/k = 0.733$. The other two coefficients are zero. Recall that k *is* the inclination factor, defined by $k = \sqrt{2}c_f \cos\Theta$ where Θ is the inclination angle of the bed just under the gate, measured from vertical, as illustrated in Figure 65c. It directly affects the flow-rate variation due to streamwise gate vibration. As shown by the gate motion trajectories in Figure 64, the inclination angle of the present full-scale gate is $\Theta = 61°$. Assuming that $c_f = 1.0$, k takes on a value of 0.686, and $\Delta c_r = 0.50$.

The effective gate height, H_e, is used as the length scale in the expressions for the resultant reduced added mass (for horizontal motion), Δm_b, and the resultant reduced excitation coefficient Δc_r. The resulting expressions are

$$\Delta m_x = \Delta m_b + \Delta m_r = \frac{\Delta M / L}{\rho H_e^2} \text{ and } \Delta c_x = \Delta c_r - \Delta c_b = \frac{\Delta C / L}{\rho H_e \sqrt{g H_e}} \tag{58}$$

With Δm_r and Δc_b equal to zero, the resulting values of ΔM and ΔC, can be determined by substituting the appropriate remaining values. The added mass $\Delta M = 25{,}980$ kg and the fluid excitation coefficient $\Delta C = 65{,}537$ N s/m, both of which corresponds quite well to their respective experimental values in Table 3. The added mass is almost as large as the gate mass (26,492 kg).

Using the physical gate height as the length scale leads to less accurate results for added mass and approximately the same accuracy for the fluid excitation coefficient, as is also shown in Table 3.

Dynamic Fluid Loading

The dynamic design of gates requires at least a moderately accurate estimate of the dynamic fluid loading on the gate. If the hydrodynamic pressure amplitude is given by $P(Y)$, where Y represents the upward axis along the vertical weir plate, measured from the water surface, then the fluid force amplitude acting on the whole gate can be calculated by

$$F_{Load} = \int_{-L/2}^{L/2} dZ \int_{-d_0}^{-H_0} P(Y)\, dY \tag{59}$$

where Z is the horizontal axis along the gate span with its origin at the spanwise midpoint. The non-dimensional hydrodynamic pressure due to gate vibration consists of the flow-rate variation pressure component and the push-and-draw pressure component. The magnitude of the sum of these two pressure components along vertical location corresponding to the vertical weir plate from the gate bottom to the water surface, as function a of the Froude number, F, is plotted in Figure 66.

From the plotted results, the resultant pressure be calculated as shown in Figure 66 for $F = 8.89$. The abscissa, p, is a non-dimensional form of the pressure amplitude, $P(Y)$:

$$p(y) = \frac{P(Y)}{\rho g X_c \sin\left(\frac{Z}{L} + \frac{1}{2}\right)\pi} \tag{60}$$

The non-dimensional vertical coordinate is scaled using the gate submergence d_0, that is $y = Y/d_0$. The value $y = 0$ corresponds to the free surface of the water and $y = -1$ to the bottom of the gate. The non-dimensional value corresponding to vertical positions of the top of the gate is $y = -0.14$ and is shown by the horizontal line in Figure 66.

The fluid loading in Equation 59 can then be rewritten in terms of the non-dimensional pressure in Equation 60 as

$$\frac{F_{Load}}{X_c} = \frac{2}{\pi}\rho g L d_0 \int_{-1}^{-0.14} p(y)\, dy \tag{61}$$

The pressure curve $F = 8.89$ Froude number, shown as the dashed curve in Figure 66, increases as the non-dimensional depth, y, increases. Although this theoretical result is rigorously valid only for a longitudinally slender long-span gate with underflow, and with no overflow, it is reasonable to assume that this pressure distribution along the gate will be roughly the same for the present gate with both under- and overflow. The good agreement between other theoretical and experimental results in the preceding section suggests that this rough assumption will not lead to significant error. With $F = 8.89$, the integral in Equation 61takes on a value of 38.7, which when substituted, along with $\rho = 1000$ kg/m^3, $g = 9.8$ m/s^2, $L= 24.64$ m and $d_0 = 1.88$ m, into Equation 61 yields a load per unit mid-span streamwise gate deflection of $F_{Load}/X_c = 110$ kN/cm.

The maximum amplitude of the mid-span gate deflection, X_c, was 1.38 cm, as shown in Figure 63. The corresponding predicted maximum dynamic fluid loading, F_{Load}, is 151 kN. The static fluid loading on the gate was 370 kN, giving a maximum total fluid loading of 521 kN, or about 1.4 times the static loading.

Mechanism of Self-Excited Coupled-Mode Vibration

The mechanism for sustaining the oscillation is as follows: when the gate moves in the downstream direction, it also moves downward; the flow rate under the gate decreases causing the deceleration of the upstream approach flow; due to this deceleration, the hydrodynamic pressure in the upstream channel increases, and the increased hydrodynamic pressure pushes the gate in the downstream direction, amplifying the original downstream motion of the gate. Analogous, but reversed behavior occurs as the gate moves upward and upstream simultaneously, amplifying the original upstream motion. The vertical vibration responds as a forced vibration due to the fluctuating pressure loading on the inclined weir plate and the gate crest.

Second, it was shown that the theory developed for the self-excited vibrations of a longitudinally slender long-span gate installed over an inclined bed, although not rigorously valid, contained enough of the essential physics that it could be applied to the case of simultaneous over- and underflow. This theory can predict the in-water vibration frequency, the added mass, and the fluid excitation coefficient for a given value of the in-air vibration frequency. The theoretically predicted results are in good agreement with the experimental results.

Figure 66. Resultant fluctuating pressure amplitude in a vertical plane corresponding to the vertical weir plate of the gate extending from the gate bottom to the water surface
From Ishii et al. (1995). © 1995 Elsevier. Used with permission.

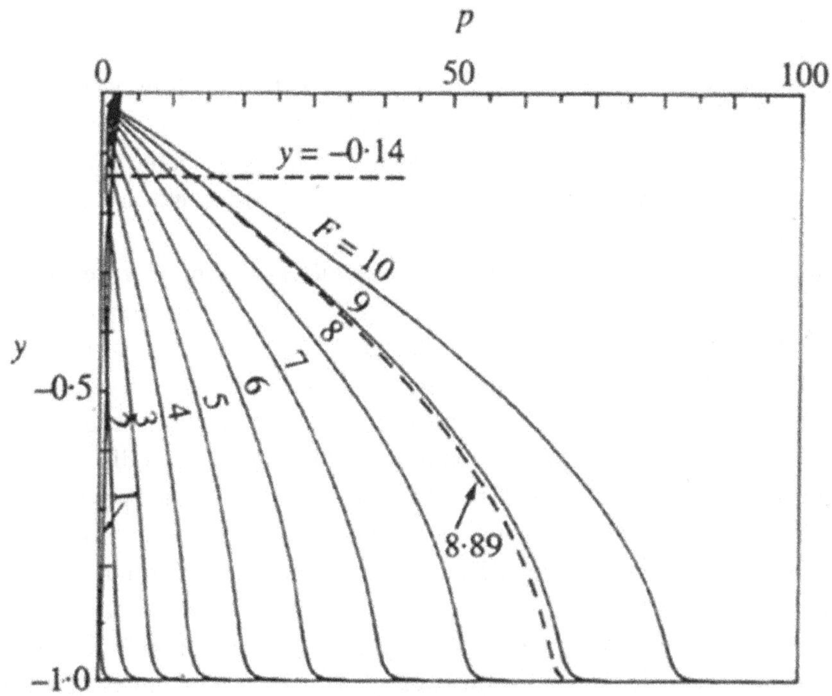

The fundamental reason for the good agreement of experimental and theoretical results can be inferred from Figure 66, in which the resultant pressure amplitude distribution along the weir plate is shown. The pressure amplitude in the region near the water surface, $0 > y > -0.14$, corresponding to the overflow height, is fairly small compared to the pressure amplitude over the rest of the gate.

Thus, the predominant loading on the gate is due to the underflow, and one may conclude that the previously developed theory for a slender gate with no overflow can be used to provide an excellent approximation of the behavior of the present gate, despite geometrical differences and the observed need of overflow to excite vibrations.

Further, the dynamic fluid loading on the full-scale gate due to gate vibration was calculated. Such dynamic loading calculations are essential to the dynamic design of the gate. To the authors' knowledge there is no method currently available for predicting this dynamic loading on the gate and often, in practice, it is neglected. In this study, the hydrodynamic force on the full-scale gate per unit vibratory displacement of gate was calculated, as well as the actual hydrodynamic force exerted on the full-scale gate when vibrating at its maximum amplitude.

The dimensionless parameters associated with shear-layer excited vibration were computed, but did not match those of the present full-scale gate vibration. The observed self-excited vibrations of the present full-scale gate were not due to the dynamic characteristics of the shear layer in the underflow.

CONCLUSION

A method for the analysis of the susceptibility of long-span gates with free discharge and inclined in the downstream direction to flow-induced vibrations has been presented. The hydrodynamic pressure caused by small-amplitude gate vibrations at a small gate opening was calculated, and the equation of motion for the gate vibration was derived and solved. The theoretical analyses were validated by experiments on a two-dimensional model gate. The derived analysis method was then applied to compute the dynamic instability and susceptibility to violent spontaneous vibrations at the 100-ton Schinznach Gate in Switzerland and the 49-ton Ara River Gate in Japan, both of which underwent violent spontaneous vibration in the past.

The present theoretical analyses show that the basic parameters determining the vibration characteristics are the damping ratio of the vibrating system, the basic Froude number (containing the in-air natural frequency) F_0, the water-to-gate mass ratio, α_m, the ratio of upstream-channel depth to submergence depth of the gate bottom, β^*, and the gate inclination factor, k.

Given an initial streamwise oscillatory movement of the gate, the fluid excitation, the fluid damping and the added mass were derived in a dimensionless form, leading to the following conclusions. The flow-rate variation induces both standing and travelling (progressive) pressure waves. The standing-wave component is 180° out of phase with the gate velocity and results in fluid excitation; the progressive-wave component is in phase with the gate acceleration and results in added mass. The push-and-draw effect of the gate on the water also induces standing and progressive pressure waves. The standing wave component is in phase with the gate acceleration and, again, results in added mass; the progressive wave component is in phase with the gate velocity and results in fluid damping. The basic parameters determining the hydrodynamic coefficients are the Froude number (containing the in-water vibration frequency) F, α_m, β^* and k.

The equation of streamwise motion for long-span gates at small gate openings, reduced to a simple second-order linear differential equation, was solved to provide insight into the fundamental characteristics of the in-water to in-air vibration frequency ratio and the net fluid-excitation ratio. The vibration frequency ratio takes on a value naturally smaller than 1.0, due to added mass effects of fluid motion. As F_0 increases, the vibration frequency ratio initially decreases, but then approaches a constant value. The value of the constant, itself, decreases with increasing α_m. For instance, the frequency ratio approaches 0.52 for $\alpha_m = 5.0$ and 0.40 for $\alpha_m = 10.0$. The net fluid-excitation ratio has a significant physical meaning. If the net fluid-excitation ratio exceeds the mechanical damping ratio, the gate will always become dynamically unstable. For small values of F_0, the net fluid-excitation ratio takes on a negative value, indicating a stable situation for the gate. As F_0 increases, however, the net fluid-excitation ratio increases rapidly to a positive maximum value at about $F_0 = 4.0$, and then decreases gradually approaching zero for very large values of F_0. The gate instability level increases with increasing α_m or k. It is significant to note that even if F_0 becomes far larger than 4.0, the net fluid-excitation ratio remains positive. From this characteristic, one may understand that for small mechanical damping ratios, the gate is unstable over a wide range of F_0.

Flow-induced vibration tests for a small, two-dimensional model gate were performed. The vibration frequency ratio and the net fluid-excitation ratio were calculated from the exponentially growing transient wave forms. Meanwhile, based on the unsteady discharge coefficient calculated from the discharge flow velocity measured by a laser-Doppler velocimeter, the net fluid-excitation ratio corresponding to the model test results was theoretically calculated. The theoretically predicted results agree well with the experimental data.

The flow-induced flexural vibration of long-span gates with free discharge is independent of any vortex formation beneath the gate that appears with submerged discharge.. Since the gates have no downstream submergence and are of limited streamwise thickness with no flat bottom regions, the use of potential flow analysis permitted a full theoretical solution that is in good agreement with the experimental results. This application of potential flow analysis is indeed a rare case of a successful theoretical analysis of flow-induced vibrations.

Gates with submerged discharge are susceptible to instability induced excitation (IIE) as the separated shear layer impinges on other gate structures or discharges into the submerged downstream reservoir. An example of full-scale gate undergoing IIE was presented showing how to estimate the values for parameters associated with IIE. Another effect of submerged discharge is that it can result in another source of discharge fluctuations. The excursions of the separated shear layer beneath a flat bottomed gate were documented and shown to be related to variations in the flow rate beneath the gate.

Even with free discharge, long-span gates with two degrees-of-freedom and with an inclined portion of their upstream weir plates can undergo self-excited coupled-mode vibrations. The results of the experimental study showed that over a certain range of the ratio $\gamma_{xy} = \Omega_{wx}/\Omega_{ay}$, the gate behaved as a press-shut device that was statically stable, but dynamically unstable. By combining the experimental results with theoretical analysis, the dynamic instability was shown to be driven by the horizontal motion inducing a fluctuating pressure that, in turn, produced vertical motion. The vertical motion caused a discharge fluctuation, which produced an additional fluctuating pressure that reinforced the initial horizontal motion.

In addition, onset and termination criteria for the gate motion in terms of the frequency ratio γ_{xy}, as well as criteria for the dynamic stability of the gate in terms of frequency and damping were presented. The region of instability as well as the intensity of the instability were shown to depend on the inclination angle of the weir plate. Similarly, the parameters in the criteria for stability need to vary with the inclination angle.The basic mechanism presented herein is expected to function in any shell-type long-span gate, but the stability criteria, especially the damping criterion, need to be examined for different gate geometries and Froude numbers.

A two-dimensional gate model with only a vertical upstream weir plate with no inclined lower portion was constructed to investigate the coupling of the crest loading with the horizontal vibration. Dynamic characterization of the resulting vibration, as was done in the preceding section, is presented in terms of the frequency of vibration, the excitation ratio (the negative of the damping factor) and the trajectories of gate motion. The trajectories of gate motion suggest that, due to the coupling of vertical and streamwise vibratory motion, the gate behaves as a press-shut device, that is, as a device that tends to move toward reduced discharge with increasing upstream potential. In addition, other dynamic parameters including the ratio of horizontal to vertical natural frequencies, the ratio of in-water to in-air frequency, a reduced fluid excitation coefficient, a reduced added mass and the phase difference between vertical and horizontal vibratory motions are presented.

In an opportunity to confirm the theoretical results from the laboratory, a field test was undertaken on a gate for which no vibrations were observed when the gate was subjected to underflow alone. Self-sustained vibrations were observed only with simultaneous over-and underflow over a narrow range of gate openings. From this field test, one might conclude that the overflow provided the primary excitation. However, such a conclusion is not truly justifiable. The underflow alone could have provided the bulk of the negative damping required to excite the vibration, but fell just short of overcoming the mechanical damping. Only a small amount of additional excitation due to overflow would have then been necessary to produce self-sustained vibration. Thus, the overflow mechanism could potentially provide only a small fraction of the total fluid excitation, and yet still be essential for sustaining the vibration. Further work on the interaction of the underflow and overflow mechanisms is clearly required before one can accurately assess the amount of excitation attributable to each surface for any flow configurations.

The method of analysis developed in the present chapter will be applied to the self-excited coupled-mode instability of Tainter gates in Chapter 10.

REFERENCES

Anami, K., Ishii, N., Knisely, C. W., Tsuji, T., & Oku, T. (2013). Steel-rod breaking excitation method to identify dynamic instability of full-scale Tainter gates. *Hydro Review Worldwide*, *21*(6), 34-39. Available at http://www.hydroworld.com/articles /print/volume-21/issue-6/articles/civil-structures/method-for-identifying-dynamic-instability.html

Billeter, P., & Staubli, T. (2000). Flow-induced multiple-mode vibrations of gates with submerged discharge. *Journal of Fluids and Structures*, *14*(3), 323–338. doi:10.1006/jfls.1999.0274

Castro Delgado, M., & Bustamante, X. H. (2002). Experimental analysis of combined overflow and underflow through a vertical plane gate [in Spanish]. *Hydraulic Engineering in Mexico*, *XVII*(1), 83–91.

Creager, W. P., & Justin, J. D. (1927). *Hydro-Electric Handbook*. New York: John Wiley & Sons.

Husain, D. (1999). Dynamic study of simultaneous underflow-overflow past low head sluice gates. *JKAU. Engineering and Science*, *11*(1), 51–67.

Imaichi, K., & Ishii, N. (1977). Instability of an idealized Tainter gate system without damping caused by surface waves on the backwater of dam. *Bulletin of the Japan Society of Mechanical Engineers*, *20*(146), 963–970. doi:10.1299/jsme1958.20.963

Ishii, N. (1990a). *Flow-induced vibration of long span gates (added mass and fluid damping)*. Presented at IAHR Symposium 1990, Belgrade, Yugoslavia.

Ishii, N. (1990b) Flow-induced vibration of long-span gates (Verification of added mass and fluid damping). *JSME International Journal, Series II (Fluid Engineering, Heat Transfer, Power, Combustion, and Thermophysical Properties)*, *33*, 642-648.

Ishii, N. (1990c). Flow-induced vibration of long-span gate (4th report, vibration frequency ratio and fluid damping ratio). Transactions of the Japan Society of Mechanical Engineers (Series. C), 56(531), 2887-2892.

Ishii, N. (1992). Flow-induced vibration of long-span gates part I: Model development. *Journal of Fluids and Structures, 6*(5), 539–562. doi:10.1016/0889-9746(92)90018-X

Ishii, N., & Imaichi, K. (1982). Instability of elastically suspended Tainter gate system caused by surface waves on the reservoir of a dam (Approximate solution) [in Japanese]. *Transactions of the Japan Society of Mechanical Engineers, Series B, 48*(428), 640–647. doi:10.1299/kikaib.48.640

Ishii, N., Imaichi, K., & Hirose, A. (1977). Instability of elastically suspended Tainter gate system caused by surface waves on the reservoir of a dam. *Journal of Fluids Engineering, 99*(4), 699–709. doi:10.1115/1.3448888

Ishii, N., Imaichi, K., & Hirose, A. (1980). Dynamic instability of Tainter gates. In E. Naudascher & D. Rockwell (Eds.), *Practical Experiences with Flow-Induced Vibrations* (pp. 452–460). Berlin: Springer. doi:10.1007/978-3-642-81528-7_49

Ishii, N., Imaichi, K., & Takemoto, S. (1977). Water waves and structural loads caused by a periodic change of discharge from a sluice gate. *Bulletin of the Japan Society of Mechanical Engineers, 20*(6), 998–1007. doi:10.1299/jsme1958.20.998

Ishii, N., Imaichi, K., & Yamasaki, M. (1983). Flow-induced vibration of single arm gates. *Proceedings 20th IAHR Congress*, 317-324.

Ishii, N., & Knisely, C. W. (1988). Shear layer behavior under a streamwise vibrating gate and its significance to gate vibration. In *Proceedings of the 2nd International Symposium on Fluid Control, Measurement, Mechanics and Flow Visualization*, (pp. 45-49). Bedford, UK: Stephens & Associates.

Ishii, N., & Knisely, C. W. (1989). Shear layer dynamics and unsteady discharge beneath a long-span gate undergoing streamwise vibrations. In Č. Maksimović & M. Radojković (Eds.), *Proceedings of Hydrocomp 89* (pp. 500–509). Rotterdam: Elsevier.

Ishii, N., & Knisely, C. W. (1990). Shear layer behaviour under a streamwise vibrating gate and its significance to gate vibration. *JSME International Journal. Series 3, Vibration, Control Engineering, Engineering for Industry, 33*(2), 131–138. doi:10.1299/jsmec1988.33.131

Ishii, N., & Knisely, C. W. (1992a). Flow-induced vibration of long-span gates due to shed vortices (Vibration criteria, level of fluid excitation and added mass). *JSME International Journal. Series 3, Vibration, Control Engineering, Engineering for Industry, 35*(1), 1–8. doi:10.1299/jsmec1988.35.1

Ishii, N., & Knisely, C. W. (1992b). Flow-induced vibration of shell-type long-span gates. *Journal of Fluids and Structures, 6*(6), 681–703. doi:10.1016/0889-9746(92)90003-L

Ishii, N., Knisely, C. W., & Nakata, A. (1992). Coupled-mode vibration of gates with simultaneous over- and underflow. In *Symposium on Flow-Induced Vibration and Noise-Volume 6* (pp. 193-205). New York, NY: ASME.

Ishii, N., Knisely, C. W., & Nakata, A. (1994). Coupled-mode vibration of gates with simultaneous over- and underflow. *Journal of Fluids and Structures, 8*(5), 455–469. doi:10.1006/jfls.1994.1022

Ishii, N., Knisely, C. W., & Nakata, A. (1995). Field study of a long-span shell-type gate undergoing flow-induced vibration. *Journal of Fluids and Structures, 9*(1), 19–41. doi:10.1006/jfls.1995.1002

Ishii, N., Knisely, C. W., & Sanagi, T. (1989). Flow-induced vibration of long-span gates caused by shedding vortices (1st Report, vibration criteria, level of fluid excitation and added mass) [in Japanese]. *Transactions of the Japan Society of Mechanical Engineers, Series B, 55*(519), 2724–2730. doi:10.1299/kikaic.55.2724

Ishii, N., & Nakata, A. (1995). Flow-induced flexural vibration of long-span gates (vibration frequency and fluid-excitation ratio). In P. Bearman (Ed.), *Flow-Induced Vibration* (pp. 139–148). Rotterdam: A. A. Balkema.

Ishii, N., & Naudascher, E. (1984). Field tests on natural vibration modes of a Tainter gate. In K. V. H. Smith (Ed.), *Channels and Channel Control Structures* (pp. 209–222). Southampton, UK: Springer. doi:10.1007/978-3-662-11300-4_17

Ishii, N., & Naudascher, E. (1992). A design criterion for dynamic stability of Tainter gates. *Journal of Fluids and Structures, 6*(1), 67–84. doi:10.1016/0889-9746(92)90056-9

Ishii, N., Naudascher, E., & Thang, N. (1987). A flow-induced vibration of long-span gates. *Proceedings of the International Conference on Flow-Induced Vibrations, Bowness-on-Windermere, BHRA*, 293-303.

Jongeling, T. H. G. (1988). Flow-induced self-excited in-flow vibrations of gate plates. *Journal of Fluids and Structures, 2*(6), 541–566. doi:10.1016/S0889-9746(88)80022-7

Kolkman, P. A., & Jongeling, T. H. G. (2007). Dynamic Behaviour of Hydraulic Structures. Delft, The Netherlands: Deltares (formerly Delft HydraulicsWL).

Lamb, H. (1904). On deep water waves. *Proceedings of London Mathematical Society, 2,* 371-400.

Lamb, H. (1932). *Hydrodynamics*. Mineola, NY: Dover.

Nakata, A., Ishii, N., Knisely, C. W., Irie, K., & Moriya, Y. (1996). Effects of weir plate inclination angle on flow-induced vibrations of long-span, shell-type gates. *JSME International Journal. Series C, Dynamics, Control, Robotics, Design and Manufacturing, 39*(2), 203–211. doi:10.1299/jsmec1993.39.203

Naudascher, E. (1959). Beitrag zur Untersuchung der schwingungserregenden Kräfte an gleichzeitig über-und unterströmten Wehrverschlüssen. *Technische Mitteilungen Krupp, 17*(5).

Naudascher, E. (1961) Vibration of gates during overflow and underflow. *Journal of the Hydraulics Division, 87*(5), 63-86.

Naudascher, E. (1987). Flow-induced streamwise vibrations of structures. *Journal of Fluids and Structures, 1*(3), 265–298. doi:10.1016/0889-9746(87)90243-X

Naudascher, E., & Rockwell, D. (1994). *Flow-Induced Vibrations: An Engineering Guide; IAHR Hydraulic Structures Design Manual*. Rotterdam: A.A. Balkema.

Petrikat, K. (1953). Die schwingungsanfachenden Kraefte in Wehrbau. *MAN-Forschungsheft, 4*, 69.

Petrikat, K. (1958). Vibration tests on weirs and bottom gates. *Water Power, 190*, 52–57.

Petrikat, K. (1980). Seal vibration. In E. Naudascher & D. Rockwell (Eds.), *Practical Experiences with Flow-Induced Vibrations* (pp. 476–497). Berlin: Springer. doi:10.1007/978-3-642-81528-7_54

Rayleigh, J. W. S. (1883). The form of standing waves on the surface of running water. *Proceedings of the London Mathematical Society*, *15*(1), 69–78. doi:10.1112/plms/s1-15.1.69

Rayleigh, J. W. S. (1945). *The Theory of Sound*. Mineola, NY: Dover.

Thang, N. D., Lin, Q. H., & Naudascher, E. (1987). Flexural streamwise vibration of gate plates under vortex action. In *Proceedings International Conference on Flow Induced Vibrations* (pp. 171-181). London, UK: British Hydraulic Research Association.

Thang, N. D., & Naudascher, E. (1986). Self-excited vibrations of vertical-lift gates. *Journal of Hydraulic Research*, *24*(5), 391–404. doi:10.1080/00221688609499316

APPENDIX: NOMENCLATURE

B: Gate openings
c_f: Unsteady discharge coefficient
C: Damping coefficient
d: Gate width of flat-bottomed gate model with two degrees-of-freedom
d_0: Gate submergence depth
d_r: Upstream-channel depth
E: Young's modulus
f: Vibration frequency (Hz)
f_{x0}: Predominant frequency of vibration in the horizontal (x) direction
f_{z0}: Predominant frequency of vibration in the vertical (z) direction
F, F_x: Froude number
F_0, F_{0x}: Basic Froude number
F_{Load}: The dynamic fluid loading on the gate
g: Acceleration of gravity
h: Equivalent gate thickness
h: Downstream gate submergence (2nd meaning, read in context)
H: The gate height,
H_0: Height of overflow
H_e: The effective height of gate
I: Moment of inertia of area
j: Integer number
k: Gate inclination factor
l : Span length, half a wavelength
L: Gate span length
M: The gate mass
n: Integer number
O: Origin of horizontal axis, at the middle of the gate.
p, P: Resultant pressure [p is the dimensionless expression of P]
P_0: Measured pressure amplitude
p_b, P_b: Resultant push-and-draw pressure [p_b is the dimensionless expression of P_b]
p_{bp}: Amplitude of progressive-wave component of push-and-draw pressure
p_{bs}: Amplitude of standing-wave component of push-and-draw pressure
p_r, P_r: Resultant flow-rate-variation pressure [p_r is the dimensionless expression of P_r]
p_{rp}: Amplitude of progressive-wave component of flow-rate-variation pressure
p_{rs}: Amplitude of standing-wave component of flow-rate-variation pressure
P_y: Fluctuating pressure acting on the gate crest
s: Gate openings at flat-bottomed gate model with two degrees-of-freedom
s: Reduced gate openings ($s=S/W$) (2nd meaning, read in context)
S: Streamwise width of gate bottom
S: Dimensional gate opening (2nd meaning, read in contex)

Sc: The Scruton number
t, T: Time [t is the dimensionless expression of T]
u_0: Flow velocity under the gate
V: Jet-flow velocity under the gate
V_{rx}: Reduced velocity single degree-of-freedom x-direction
V_{rz}: Reduced velocity single degree-of-freedom z-direction
W: Streamwise gate thickness
W_0: Gate thickness
x, X: Horizontal coordinate along undisturbed free surface [x is the dimensionless expression of X(m)]
$\bar{X}$: Gate vibration amplitude in horizontal direction
X_0: Amplitude of gate vibration
x_c, X_c: Amplitude in the middle of the gate [x_c is the dimensionless expression of X_c (m)]
$\bar{X}_c$: Maximum amplitude in the middle of the gate
X_{max}: Maximum amplitudes of the horizontal motion
y, Y: Vertical coordinate [y is the dimensionless expression of Y (m)]
$\bar{Y}$: Gate vibration amplitude in vertical direction
Y_{max}: The maximum amplitudes of the vertical motion
Z: Horizontal coordinate along the gate [z is the dimensionless expression of Z (m)]
α_m: Water-to-gate mass ratio
β^*: Reservoir to gate submergence depth ratio
δ: Factor that determines the thickness of the elliptical trajectory of the gate displacement
ΔC: Fluid excitation ratio
Δc_r, ΔC_r: Fluid excitation coefficient caused by the flow-rate-variation standing-wave pressure component [Δc_r is the dimensionless expression of ΔC_r]
Δc_b, ΔC_b: Fluid damping coefficient caused by the push-and-draw progressive-wave pressure component [Δc_b is the dimensionless expression of ΔC_b]
Δc_d: Magnitude of the fluctuating discharge coefficient
Δc_x, ΔC_x: Fluid excitation coefficient [Δc_x is the dimensionless expression of ΔC_x]
ΔD: Upstream-downstream elevation difference
ΔM: Added mass
Δm_b, ΔM_b: Water added mass from the push-and-draw standing-wave pressure component [Δm_b is the dimensionless expression of ΔM_b]
Δm_r, ΔM_r: Water added mass from the flow-rate-variation progressive-wave pressure component [Δm_r is the dimensionless expression of ΔM_r]
Δm_x, ΔM_x: Added mass coefficient [Δm_x is the dimensionless expression of ΔM_x]
Δp_1, ΔP_1: Pressure amplitude on the upstream gate face [Δp_1 is the dimensionless expression of ΔP_1]
Δp_2, ΔP_2: Pressure amplitude on the downstream gate face [Δp_2 is the dimensionless expression of ΔP_2]
ε: Represents the ratio of vertical to horizontal amplitude.
ϕ: Measured values of the phase lag
ϕ_0: Pressure phase lag
ϕ_d: Phase lag of fluctuating discharge coefficient
ϕ_p: Phase angle relationship between the vertical motion and the pressure due to horizontal motion
γ_{nxy}: Inherent in-water frequency ratio ($\gamma_{nxy} = \Omega_{nwx}/\Omega_{nwy}$)

γ_{nyy}: Inherent frequency ratio *at the onset of self-excitation* ($\gamma_{nyy} = \Omega_{wy}/\Omega_{ay}$)
γ_{term}: Termination ratio
γ_{wyx}: Ratio of in-water vertical frequency to in-water horizontal frequency ($\gamma_{wyx} = \Omega_{wy}/\Omega_{wx}$)
γ_{xx}: Ratio of in-water to in-air horizontal frequencies ($\gamma_{xx} = \Omega_{wx}/\Omega_{ax}$)
γ_{xy}: Ratio of the in-water horizontal frequency to the in-air vertical frequency ($\gamma_{xy} = \Omega_{wx}/\Omega_{ay}$)
γ_{yy}: Ratio of in-water vertical frequency to in-air vertical frequency ($\gamma_{yy} = \Omega_{wy}/\Omega_{ay}$)
Θ: Gate inclination angle
ρ: Density of water
ρ_*: Density of gate
Ω: In-water vibration frequency
Ω_0: The natural vibration frequency in air
Ω_{ax}: In-air vibration frequency
Ω_{ay}: The in-air vertical frequency
Ω_{nwx}: *Inherent natural frequency* in still water in horizontal vibration
Ω_{nwy}: *Inherent natural frequency* in still water in vertical vibration
Ω_w: In-water vibration frequency
Ω_{wx}: In-water vibration frequency in horizontal direction
Ω_{wy}: In-water vibration frequency in vertical direction
ξ: Resultant excitation ratio
ξ_r: Fluid-excitation ratio
ξ_0 and η_j: Solutions of transcendental equations
$\xi_{w,\,max}$: Maximum excitation ratio
ξ_{wx}: Excitation ratio in water in horizontal direction
ξ_{wy}: Excitation ratio in water in vertical direction
ζ_0: Damping ratio
ζ_a: Damping ratio
ζ_{ax}: In-air damping ratio, in the x-direction
ζ_{ay}: In-air damping, in the y-direction
ζ_b: Fluid damping ratio

Chapter 8
Eccentricity Instability:
A Mechanism for Tainter (Radial) Gate Vibration

ABSTRACT

A significant vibration mode of Tainter gates is the rigid-body vibration of the gate around the trunnion pin. Any such rigid-body vibration of a Tainter gate is accompanied by periodic changes in the flow rate through the gate opening which, in turn, results in fluctuating loads on the gate. The induced loads are then fed back to excite or damp the initiating rigid-body vibration. The placement of the center of the circular-arc skinplate eccentric to and beneath the trunnion pin center results in the gate behaving as a press-shut device, with a strong likelihood of susceptibility to an intense dynamic instability. In this chapter the theory behind this eccentricity instability is developed and its potential applicability to the Wachi Dam failure examined with the conclusion that eccentricity instability was most likely not the cause of the Wachi failure.

INTRODUCTION

Many Tainter gates have been installed and continue to operate safely without destructive dynamic instabilities. Figure 1 shows the massive Tainter gates at the Itaipu Dam in Brazil as an example of gates that serve their as-designed function without dynamic instabilities. As discussed in Chapter 1, however, Itaipu Dam did have one isolated incident, speculatively related to vibrations, resulting in the failure of a rod in a hydraulic cylinder (see Cassidy, 2000). Figure 2 shows a second example of apparently stable gates at the Benmore Dam in New Zealand. There have been few notable gate failures among the many successful installations. Gate accidents since 1960 are documented in Table 1 in Chapter 1. The lessons we need to learn are from the instances when large Tainter gates do experience destructive failure.

Among the reported accidents is the Wachi Dam gate failure in Japan on July 2, 1967, when a 37-ton Tainter gate failed during pre-discharge tests (Yano, 1968). Additional details and analysis of the Wachi Dam gate failure are presented in Chapter 12. After the Wachi Dam gate failure, one of the co-authors

DOI: 10.4018/978-1-5225-3079-4.ch008

Figure 1. Massive gates (20 x 21.34 m, each gate weighing 3185 kN or 74 tons) at Itaipu Dam in Brazil exemplify the many Tainter gates installed worldwide that are apparently stable and have not experienced recurrent dynamic instabilities.

Figure 2. Tainter gates (10.6 x 11.9 m) at the Benmore Dam in New Zealand are another example of apparently stable gates with no reported vibration problems.
Photograph by C. W. Knisely.

of this book (Professor N. Ishii) initiated an investigation of the flow-induced vibration of Tainter gates. He identified an intense self-excited vibration, which he called eccentricity instability, which occurs under certain conditions (Ishii & Imaichi, 1976; Imaichi & Ishii, 1977, Ishii & Imaichi, 1977; Ishii et al., 1980; Ishii & Imaichi, 1982; Ishii & Naudascher 1992; Ishii, 1995). This instability is restricted to Tainter gates in which the center of the circular-arc skinplate is offset relative to the trunnion pin center.

As shown in Figure 3, Tainter gates are elastically suspended with long wire cables or chains and may, therefore, vibrate easily around the trunnion pin due to external disturbances acting on the gate. Field tests by Ishii & Naudascher (1984) on the natural vibration modes of a Tainter gate in Japan revealed that the rigid-body vibration of the gate around the trunnion pin is a significant vibration mode. Any such rigid-body vibration of a Tainter gate is accompanied by periodic changes in the flow rate through the gate opening which, in turn, results in fluctuating loads on the gate. The induced loads are then fed back to excite or damp the initiating rigid-body vibration. The placement of the center of the circular-arc skinplate beneath the trunnion pin center results in the gate behaving as a press-shut device, with a strong likelihood of susceptibility to an intense dynamic instability. On the other hand, when the center of the circular-arc skinplate is above the trunnion pin center, the gate behaves as a press-open device, which makes sealing the gate difficult, but which usually makes the gate immune to dynamic instabilities.

In this chapter we develop the theoretical basis for predicting whether an eccentric Tainter gate is susceptible to eccentricity instability, a dangerous type of flow-induced vibration. As a first step, we will use the analysis developed in Chapter 4 to determine the fluctuating hydrodynamic loads on the gate induced by rigid-body, small-amplitude rotary vibrations at small gate openings bounded by an infinite depth upstream reservoir. Next we will derive the exact equation of motion for a Tainter gate undergoing small-amplitude vibrations. This equation of motion turns out to be a homogeneous differential equation, the exact solution of which can be obtained. From the solution we are able to calculate

Figure 3. Eccentric Tainter gate on the dam crest of an infinite reservoir, with flow from right to left
From Ishii & Naudascher (1992). © 1992 Elsevier. Used with permission.

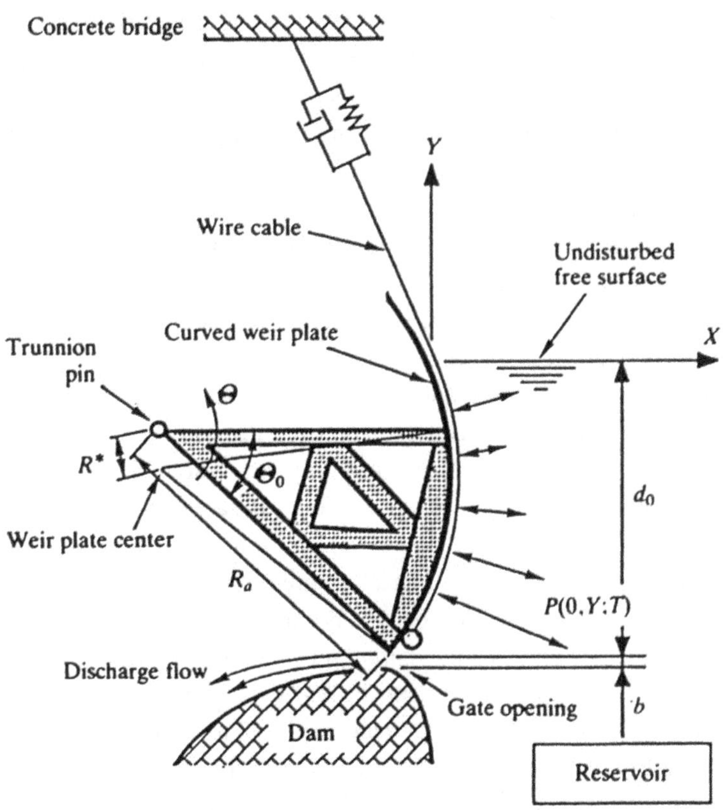

the excitation ratio and the non-dimensional vibration frequency. With known values of the excitation ratio and the frequency of vibration, the major characteristics of the flow-induced vibration of eccentric Tainter gates can be determined.

We will then present a design criterion for the dynamic stability of Tainter gates, based on the theoretical analysis, applicable to any vibration frequency. The equivalent damping ratio is given as a simple function of the modified water-to-gate mass ratio. Using this design criterion, the mass ratio necessary for dynamic stability can be calculated for any equivalent damping ratio and, thus, the permissible eccentricity of the trunnion pin or the minimal moment of inertia of the gate can be determined.

Finally, the validity of the analysis will be determine by comparing the calculated results with the experimental results reported by Ishii et al. (1980).

HYDRODYNAMIC PRESSURE INDUCED BY RIGID-BODY ROTARY OSCILLATION

The two-dimensional Tainter gate system to be considered is shown schematically in Figure 3. The upstream reservoir depth is assumed infinite. The skinplate has a circular-arc form and is supported by stays that are connected to two trunnion pins fixed on the concrete piers. The wire cables clamped to the lower end of the skinplate can be wound or unwound by a winch on the bridge, rotating the gate about the trunnion pins.

Tainter gates are normally designed so that the center of the circular-arc skinplate coincides with the trunnion pin axis when there is no hydraulic load on the gate. When the gate is installed, however, the hydraulic force exerted on the skinplate induces a structural deformation that generally causes an eccentricity about the trunnion pin. In addition, the gate may be designed for enhanced static stability by positioning the center of the skinplate below the center of the trunnion pin. In the equilibrium state, the moment about the trunnion pin due to the weight of the Tainter gate and the hydraulic pressure exerted on the skinplate is equal to that of the lifting force of the wire cables. Since the wire cables are fairly long (usually 15 to 25 m) and operated at about 10% of their yield load, they are moderately elastic. In this way, the Tainter gate and the wire cables form a rotary vibration system around the trunnion pin. A Tainter gate with an eccentric trunnion pin will start to oscillate about its equilibrium state whenever the gate is suddenly moved up or down or when the water level in the reservoir suddenly changes. As a result of this rotary oscillation, the gate opening, b, will change periodically, producing discharge fluctuations that result in a flow-rate variation hydrodynamic pressure fluctuation. For eccentric Tainter gates, the rigid-body rotary oscillation is accompanied by a small streamwise movement of the circular-arc skinplate, which induces a push-and-draw hydrodynamic pressure fluctuation. This streamwise vibration, however, has a negligibly small amplitude for small values of gate eccentricity.

In Chapter 4, we derived the theoretical flow-rate variation hydrodynamic pressure when the skinplate at the reservoir outlet and the dam crest beneath it are both vertical. In contrast, Tainter gates have circular-arc skinplates and the dam sill under the gate is curved. Despite the differences in the shape and orientation of the boundaries, the analysis in Chapter 4 can be applied to determine a rough theoretical hydrodynamic pressure for the curved geometries. Experimental testing on model gates then allows us to refine the theoretical calculations to take into account the boundary and orientation differences.

The angle Θ shown in Figure 3 denotes the angular oscillatory deflection of the whole gate, defined to be positive in the counter-clockwise direction. This oscillatory gate motion results in the upward displacement of the lower edge of the skinplate, which is given by $R_a\Theta$, where R_a represents the distance of the lower edge of the skinplate from the trunnion pin center. The corresponding increase in vertical gate opening is given by $R_a\Theta\cos\Theta_0$, which corresponds to the negative of the downward vibration Y_G in Chapter 4:

$$R_a \cos\Theta_0 \cdot \Theta = -Y_G\left(T\right) \tag{1}$$

where Θ_0 represents the elevation angle of the trunnion pin from the lower edge of the skinplate, as shown in Figure 3. The dimensionless form of the oscillatory angle Θ is now defined as follows:

$$\theta \equiv \frac{\Theta}{\bar{\Theta}} \tag{2}$$

where $\bar{\Theta}$ represents the amplitude of the oscillatory angle. Consequently, Equation 1 can be re-written as follows:

$$Y_G\left(T\right) = -R_a\bar{\Theta}\cos\Theta_0 \cdot \theta \tag{3}$$

which indicates that $R_a\bar{\Theta}\cos\Theta_0$ corresponds to Y_0 in Equations 36 and 123a in Chapter 4. Therefore, the flow-rate variation pressure P_r for the present geometry can be reduced to the following dimensionless form, p_r:

$$p_r\left(x, y; t\right) \equiv \frac{P_r(0, Y; T)}{\sqrt{2c_f}\rho g R_a\bar{\Theta}\cos\Theta_0} \text{ or } \frac{1}{\sqrt{2c_f}} \cdot \frac{P_r(0, Y; T) / \rho g}{R_a\bar{\Theta}\cos\Theta_0} \tag{4}$$

where Y_0 in Equation 123a in Chapter 4 was replaced by $R_a\bar{\Theta}\cos\Theta_0$. From this definition for dimensionless pressure, the flow-rate variation pressure p_r is given by the negative of Equation 134 in Chapter 4. In the present geometry, the reservoir depth is assumed infinite and hence the flow-rate variation pressure $P_{r\infty}$ for the infinite reservoir depth is reduced to the following dimensionless form $p_{r\infty}$:

$$p_{r\infty}(0, y; t) = \left(\frac{F_0}{F}\right)^2 p_{rp\infty}(F, 0, y) \cdot \theta'' - \left(\frac{F_0}{F}\right) p_{rs\infty}(F, 0, y) \cdot \theta' \tag{5}$$

in which $p_{rp\infty}$ and $p_{rs\infty}$ are given by Equations 125b and 125c, respectively, in Chapter 4 for an infinite depth reservoir. The dimensionless depth $y = Y/d_0$ and time $t = \Omega_0 T$ are defined in Equation 73 in Chapter 4. The Froude number $F = \sqrt{d_0 / g\Omega}$ and the basic Froude number $F_0 = \sqrt{d_0 / g_0\Omega_a}$ are defined in Equation 74 in Chapter 4. The prime in Equation 5 represents a derivative with respect to the dimensionless time t.

EQUATION OF MOTION

The moment due to the transient hydrodynamic pressure $P_{r\infty}(0,Y;T)$ on the skinplate rotates the Tainter gate about the trunnion pin. The distance between the resultant vector of the transient hydrodynamic pressure exerted on the skinplate and the trunnion pin axis is assumed to be an approximately constant value, R^*. It follows, then, that the hydrodynamic pressure moment M_p (per unit span) about the trunnion pin center, which turns the Tainter gate in the counter-clockwise direction (see Figure 3), is given by

$$M_p = -c_g R^* \int_{-d_0}^{0} P_{r\infty}(0,Y;T)dY \tag{6}$$

For the limits of the above definite integral, it is assumed that both the amplitude of the surface waves and the change of the gate opening are very small compared to the gate submergence depth d_o. The curved skinplate has been replaced by a vertical flat plate to simplify the flow boundary contour. Therefore, when evaluating the turning moment acting on the Tainter gate, the area of the vertical flat plate has to be corrected to some extent. The coefficient c_g represents the correction factor for the area of the skinplate, which is given by the ratio of the area of the curved skinplate to that of its projection in the vertical plane. Using Equations 4 and 5, the hydrodynamic pressure moment M_p can be reduced to the following expression:

$$M_p = -\frac{1}{2}k_\theta \frac{\rho d_0^{\,2}}{1/R_a^{\,2}} \cdot \frac{1}{d_0/g}\bar{\Theta}\left\{\left(\frac{F_0}{F}\right)^2 \int_{-1}^{0} p_{rp\infty}(F,0,y)dy \cdot \theta'' - \left(\frac{F_0}{F}\right)\int_{-1}^{0} p_{rs\infty}(F,0,y)dy \cdot \theta'\right\} \tag{7}$$

in which k_θ represents an eccentricity factor defined as follows:

$$k_\theta = 2\sqrt{2}c_f c_g \frac{R^*}{R_a}\cos\Theta_0 \tag{8}$$

which takes on a positive value when the skinplate center is located beneath the trunnion-pin axis ($R^* > 0$) and a negative value when these positions are reversed ($R^* < 0$). The absolute value of k_θ increases with the absolute value of the eccentricity, $|R^*|$. The equation of motion for small oscillations of the Tainter gate system about its equilibrium position is given in the following basic form:

$$-I_\theta\ddot{\Theta} - C_\theta\dot{\Theta} + K_\theta\Theta + M_p = 0 \tag{9}$$

or

$$\ddot{\Theta} + 2\Omega_0\zeta_0\dot{\Theta} + \Omega_0^{\,2}\Theta - M_p / I_\theta = 0 \tag{10}$$

where I_0 represents the moment of inertia (per unit spanwise length) about the trunnion pin center, Ω_0 is the natural vibration frequency in air, and ζ_0 is the mechanical damping ratio. Substituting Equation 7 into Equation 10 and reducing it to the non-dimensional form with Equation 73 in Chapter 4 and Equation 2 above, the equation of motion can be re-written in the following form:

$$\left\{1+\frac{k_\theta \alpha_{m\theta}}{2F^2}\int_{-1}^{0} p_{rp\infty}(F,0,y)dy\right\}\theta''+2\left\{\zeta_0-\frac{k_\theta \alpha_{m\theta}}{4F_0 F}\int_{-1}^{0} p_{rs\infty}(F,0,y)dy\right\}\theta'+\theta=0 \tag{11}$$

where $\alpha_{m\theta}$ represents the (torsional) water-to-gate mass ratio, defined by

$$\alpha_{m\theta}\equiv\frac{\rho d_0^{\,2}}{I_\theta / R_a^{\,2}} \tag{12}$$

The physical meaning of the numerator on the right side of Equation 12, ρd_0^2, was explained in Figure 5 in Chapter 7, and represents the water mass in a volume of the unit span times a square with a side equal to the gate submergence. The denominator in Equation 12, I_θ / R_a^2, represents the equivalent mass of the gate per unit span. Large values of the mass ratio $\alpha_{m\theta}$ are indicative of large gate submergence depths or of a lightweight gate construction.

The equation of motion, Equation 11, can be reduced to the form of a homogeneous second-order differential equation:

$$\left\{1+\alpha_{m\theta}\Delta m_\theta\right\}\theta''+2(\zeta_0-\xi_{r\theta})\,\theta'+\theta=0 \tag{13}$$

The flow-rate-variation standing-wave pressure component $p_{rs\infty}(F,0,y)$, given by Equation 125c in Chapter 4, results in a negative damping term, which is reflected in Equation 13 as the negative damping ratio $-\xi_{r\theta}$ in the second term. The magnitude of this damping ratio $\xi_{r\theta}$ is called the fluid-excitation ratio. This excitation ratio is defined in a form similar to that for a damping ratio:

$$\xi_{r\theta}\equiv\frac{\Delta C_{r\theta}}{2I_\theta \Omega_0} \tag{14}$$

where $\Delta C_{r\theta}$ is the fluid excitation coefficient caused by the flow-rate variation standing-wave pressure component. The dimensionless form of this fluid excitation coefficient can be written in the following form:

$$\Delta c_{r\theta}\equiv\frac{\Delta C_{r\theta}}{\rho d_0\sqrt{g d_0}R_a^{\,2}} \tag{15}$$

The quantity in the denominator, $\rho d_0^2 \sqrt{g d_0} R_a^2$, represents the gravity force moment $\rho g d_0^2 R_a$ acting on the representative mass ρd_0^2 divided by the angular wave velocity $\sqrt{g d_0} / R_a$. As a result, the fluid-excitation ratio $\xi_{r\theta}$ defined by Equation 14 can be related to the reduced fluid damping coefficient $\Delta c_{r\theta}$ as follows:

$$\xi_{r\theta} = \alpha_{m\theta} \frac{\Delta c_{r\theta}}{2F_0} \tag{16}$$

which corresponds to a torsional form of Equation 18 in Chapter 7.

The flow-rate-variation progressive-wave pressure component $p_{rp\infty}$ $(F, 0, y)$, given by Equation 125b in Chapter 4, results in the added mass in the first term of Equation 13. The added mass is defined as

$$\Delta m_{r\theta} \equiv \frac{\Delta I_\theta / R_a^{\,2}}{\rho d_0^{\,2}} \tag{17}$$

which corresponds to a torsional form of Equation 20 in Chapter 7.

From a term-by-term comparison of Equations 11 and 13, the reduced added mass coefficient, $\Delta m_{r\theta}$, and the reduced fluid-excitation coefficient, $\Delta c_{r\theta}$, emerge in their respective integral forms of the flow-rate variation progressive-wave pressure component $p_{rp\infty}$ and of the standing-wave component, $p_{rs\infty}$:

$$\Delta m_{r\theta} = \frac{k_\theta}{2F^2} \int_{-1}^{0} p_{rp\infty}(F, 0, y) dy \tag{18}$$

$$\Delta c_{r\theta} = \frac{k_\theta}{2F} \int_{-1}^{0} p_{rs\infty}(F, 0, y) dy \tag{19}$$

The integral of $p_{rp\infty}$ in Equation 18 can be evaluated from $y = -1$ to 0 following a procedure similar to that in Chapter 4. After evaluating the integral, $\Delta m_{r\theta}$ can be reduced to the following expression:

$$\Delta m_{r\theta} / k_\theta = 2 \frac{1}{F^3} e^{-F^2} \left(1 - e^{-F^2}\right) \tag{20}$$

The integral of $p_{rs\infty}$ in Equation 19, while more complicated, can be evaluated from $y = -1$ to 0, again following the procedure in Chapter 4. The integration yields the following expression:

$$\Delta c_{r\theta} / k_\theta = \frac{1}{\pi} f_{nq}(F) \tag{22}$$

where the function $f_{nq}(F)$ is defined as follows:

$$f_{nq}(F) = \left(1 - \frac{1}{F^2}\right)\ln 2 - \int_0^\infty \frac{\sin s - \sin 2s}{s^2 + F^4}\,ds + \frac{1}{F^2}\int_0^\infty \frac{s(\cos s - \cos 2s)}{s^2 + F^4}\,ds \tag{23}$$

Equation 23 is a significant characteristic function for determining the dynamic instability of eccentric Tainter gates. The characteristics of this function are shown in Figure 4. The function $f_{nq}(F)$ reaches its minimum of about 0.03 when F is about 1.0; it increases rapidly for values of F below 1.0, and it increases to ln 2 as F increases. If F is close to 1.0, the water waves on the free surface become resonant with the gate motion in a manner analogous to that of a forced mass-spring-viscous damper system (Ishii & Imaichi, 1977); when resonance occurs, the waves effectively carry away the energy supplied by the gate motion to the fluid. Thus, $f_{nq}(F)$ assumes a negative value for $f \approx 1.0$.

EXCITATION RATIO AND VIBRATION FREQUENCY

If we assume that the damping and excitation ratios ξ, ζ_0 and $\xi_{r\theta}$ are small, the solution of Equation 13 can be written in the following form:

$$\theta = \exp\left\{(\xi + i)\frac{\Omega}{\Omega_0}t\right\} = \exp\left\{(\xi + i)\frac{F}{F_0}t\right\} \tag{24}$$

where ξ represents the resultant excitation ratio. Substitution of this expression into Equation 13 results in the following characteristic equation:

Figure 4. Evaluation of the integral function f_{nq} in Equation 23
After Ishii & Naudascher (1992). © 1992 Elsevier. Used with permission.

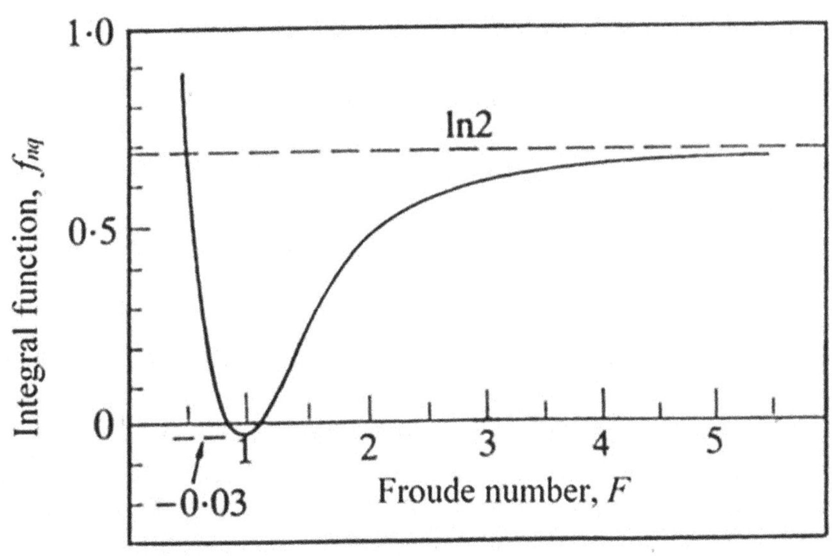

$$\left(1+\alpha_{m\theta}\Delta m_{r\theta}\right)\left(\xi+i\right)^2\left(\frac{F}{F_0}\right)^2+2\left(\zeta_0-\xi_{r\theta}\right)\left(\xi+i\right)\left(\frac{F}{F_0}\right)+1=0 \tag{25}$$

The real part of this equation can be solved for F/F_0, the ratio of the Froude number to the basic Froude number. After linearization (neglecting higher order products of ξ, ζ_0 and $\xi_{r\theta}$ that are negligibly small for the assumed small values of the damping and excitation ratios), the ratio F/F_0 is given by:

$$\frac{F}{F_0}=\frac{1}{\sqrt{1+\alpha_{m\theta}\Delta m_{r\theta}}} \tag{26}$$

Substitution of Equation 20 into Equation 26 results in the following equation:

$$\frac{F}{F_0}=\frac{1}{\sqrt{1+2k_\theta\alpha_{m\theta}e^{-F^2}\left(1-e^{-F^2}\right)/F^3}} \tag{27}$$

where F/F_0 is equivalent to the in-water to in-air vibration frequency ratio, Ω/Ω_0.

The imaginary part of Equation 25 can be solved for the resultant excitation ratio, ξ, which is given by

$$\xi=\frac{\xi_{r\theta}-\zeta_0}{\sqrt{1+\alpha_{m\theta}\Delta m_{r\theta}}}=\frac{F}{F_0}\left(\xi_{r\theta}-\zeta_0\right) \tag{28}$$

Substitution of Equations 16 and 22 into Equation 28 results in the following equation:

$$\xi=\left(\frac{F}{F_0}\right)\left(\frac{k_\theta\alpha_{m\theta}}{2\pi F_0}f_{nq}(F)-\zeta_0\right) \tag{29}$$

The three fundamental parameters determining the excitation ratio ξ and the non-dimensional vibration frequency F/F_0 are:

1. The basic Froude number F_0
2. The modified mass ratio, defined as the product of the eccentricity factor k_θ and the mass ratio $\alpha_{m\theta}$
3. The equivalent viscous damping ratio ζ_0

To generate numerical values for the excitation ratio, ξ, and the non-dimensional vibration frequency, F/F_0, as functions of the three parameters listed above, F/F_0 must be calculated first since it is needed to calculate ξ. To do so, we assume a value of $k_\theta\,\alpha_{m\theta}=1.5$ and input it into Equation 27. The challenge is that Equation 27 cannot be solved explicitly to compute the F value corresponding to a chosen F_0. Equation 27 can, however, be explicitly solved for F_0 as a function of F. The calculated values of F/F_0 are plotted as a function of the basic Froude number F_0 in Figure 5a. When higher order damping/

excitation terms are retained, Equation 26 becomes more complicated (see Ishii & Naudascher, 1992), and the effects of ζ_0 on the values of F/F_0 are indiscernible.

As just noted, the influence of ζ_0 on F/F_0 is rather small, in fact, almost indiscernible on the scale of Figure 5a. This result can be expected from elementary vibration theory. As F_0 becomes smaller, the graphed value of F/F_0 decreases until it reaches 0.58 at $F_0 = 0.8$. The gate vibration, $\theta(t)$, induces a fluid force on the gate surface composed of two components (Ishii & Imaichi 1977): one is the acceleration-based component which is out-of-phase by π relative to $\theta(t)$ and the other is the velocity-based component which is out-of-phase $\pm\pi / 2$ relative to $\theta(t)$. Since this theoretical analysis assumes that the vibrations have a small amplitude, the horizontal gate motion due to a small trunnion-pin eccentricity has been neglected, and only pressures induced by flow-rate variations are included. Therefore, this analysis assumes no added mass moving with the gate. However, the acceleration component in the fluid force acts as an added mass, whose effect is termed the *equivalent added mass effect*. As F_0 becomes smaller, the equivalent added mass increases and, thus, the gate vibration frequency decreases due to the increased equivalent added mass with decreasing F_0, as shown in Figure 5a.

Once the value of F/F_0 is known, the value of ξ can be calculated from Equation 29 for a range of equivalent damping ratios, with results shown in Figure 5b. If $F_0 < 0.8$, ξ exceeds 0.1, as seen in Figure 5b. This result contradicts the assumption that the excitation ratio is comparatively small; that is, an accurate evaluation is not possible in this region. For this reason, the calculated results are not shown for $F_0 < 0.8$. Near $F_0 = 1.14$, the excitation ratio takes a negative minimum value, independent of the ζ_0 value. If F_0 is about 1.14, the non-dimensional vibration frequency F/F_0 is about 0.836 (Figure 5a), and hence the Froude number, F, is about 0.95 (from Equation 20). As explained in the discussion of Figure 4, when the Froude number, F, is about 1.0, the function $f_{nq}(F)$ becomes negative due to a resonance effect of the water waves on the free surface. If the function $f_{nq}(F)$ is negative, however, the excitation ratio ξ must also be negative, as is evident from Equation 29. The minimum value of ξ decreases as ζ_0 increases, as is also evident from Equation 29. For $F_0 > 1.14$, ξ increases gradually up to a maximum near $F_0 = 2.12$. If F_0 exceeds about 2.0, F/F_0 is very close to 1.0 (Figure 5a). Substituting $F/F_0 = 1.0$ into Equation 29, ξ can be expressed as

$$\xi = \frac{k_\theta \alpha_{m\theta}}{2\pi F_0} f_{nq}(F) - \zeta_0 \tag{30}$$

For $F_0 > 2.0$, the curves in Figure 5b closely parallel one another, and the distance between any two curves is close to the difference of their respective ζ_0 values. As $F_0 \to \infty$, the function $f_{nq}(F) / F_0$ in the first term on the right of Equation 30 approaches zero. Consequently, the curve for each ζ_0 value in Figure 5b approaches the negative value of its corresponding ζ_0 as $F_0 \to \infty$. Finally, the maximum value of $f_{nq}(F) / F_0$ can be found to be 0.235 when $F_0 = 2.12$; the maximal value of ξ at $F_0 = 2.12$ is given by

$$\max(\xi) = 0.037 k_\theta \alpha_{m\theta} - \zeta_0 \tag{31}$$

Substitution of $k_\theta\, \alpha_{m\theta} = 1.5$ into Equation 31, yields $\max(\xi) = 0.0561 - \zeta_0$. Thus, for F_0 larger than about 1.0, the excitation ξ will assume negative values (indicating damping of the gate oscillation) if the equivalent viscous damping ratio $\zeta_0 > 0.0561$ provided $k_\theta\, \alpha_{m\theta} = 1.5$.

Figure 5. Calculated results for (a) frequency ratio and (b) resultant excitation ratio for modified mass ratio $k_\theta \alpha_{m\theta} = 1.5$ and the equivalent viscous damping ratio from zero to 9%
From Ishii & Naudascher (1992). © 1992 Elsevier. Used with permission.

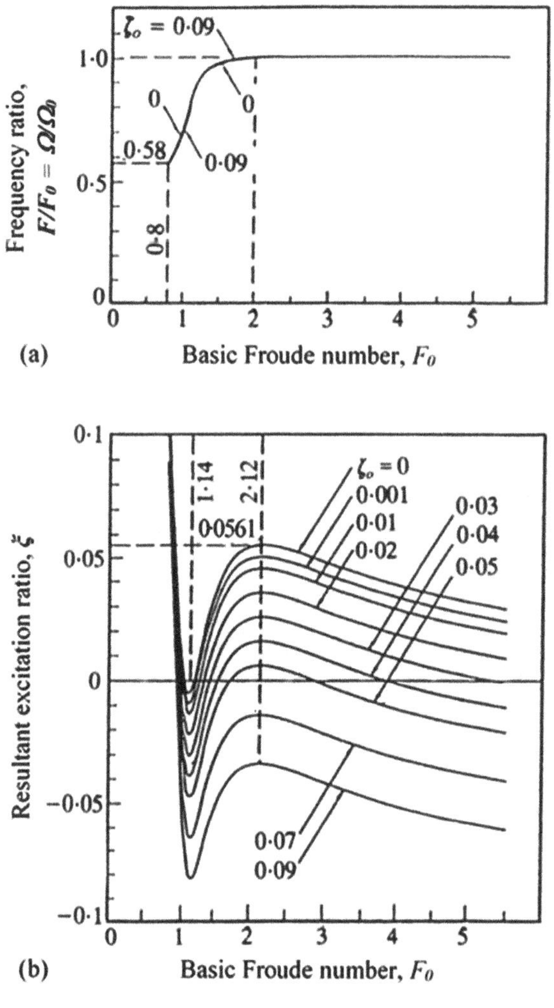

Figures 6a and 6b show how the modified mass ratio affects the values of F/F_0 and ξ. The equivalent viscous damping, ζ_o, is held constant at 0.04. As $k_\theta \alpha_{m\theta}$ increases, the effect of the equivalent added mass increases, so that F/F_0 decreases. However, this tendency occurs only if F_0 is less than about 2.0; if F_0 is larger than 2.0, then F/F_0 is 1.0 for all $k_\theta \alpha_{m\theta}$ values. Therefore, the characteristics of ξ for $F_0 >$ 2.0 may also be represented by Equation 30, and ξ still possesses a maximum value at $F_0 = 2.12$ which, from Equation 31, is given as

$$\max(\xi) = 0.037 k_\theta \alpha_{m\theta} - 0.04 \tag{32}$$

Equation 32 shows that the maximum value of ξ increases linearly with $k_\theta \alpha_{m\theta}$. For F_0 values greater than 1.14, ξ takes on negative values as long as $k_\theta \alpha_{m\theta}$ is less than 1.07. From the relationship in Equa-

Figure 6. Effect of modified mass ratio, $k_\theta \alpha_{m\theta}$, on (a) the frequency ratio, and (b) the resultant excitation ratio for an equivalent damping ratio $\zeta_0 = 0.04$
After Ishii & Naudascher (1992). © 1992 Elsevier. Used with permission.

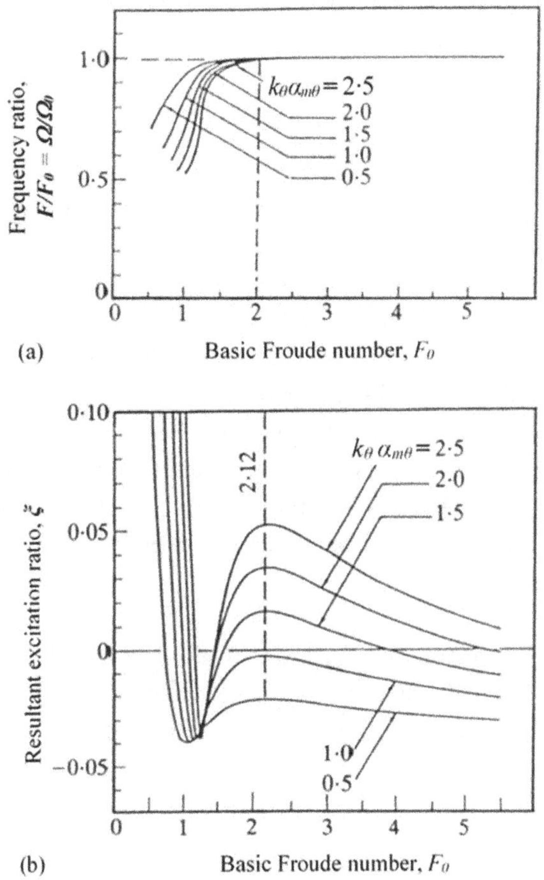

tion 32, the upper limit on the values of $k_\theta \alpha_{m\theta}$ for which gate stability is maintained can be determined for any given equivalent damping ratio ζ_o. Finally, as $F_0 \rightarrow \infty$, the function $f_{nq}(F) / F_0$ in Equation 30 approaches zero, and all curves in Figure 6b approach the negative of the ζ_o value, in this example $\xi \rightarrow -0.04$ as $F_0 \rightarrow \infty$.

A DESIGN CRITERION FOR DYNAMIC STABILITY OF ECCENTRIC GATES

As shown in Figures 5b and 6b, for values of the basic Froude number $F_0 > 1.14$, the excitation ratio ξ is positive when the equivalent damping ratio ζ_0 is comparatively small. The maximum value of the excitation ratio ξ is given by Equation 31, and a design criterion for dynamic stability of eccentric Tainter gates can be developed from this relationship. If the maximum value of the excitation ratio ξ is negative, then the Tainter gate is dynamically stable for all $F_0 > 1.14$. The design criterion for dynamic stability for $F_0 > 1.14$ is given by the following relationship:

$$\zeta_0 > 0.0374 k_\theta \alpha_{m\theta} \text{ or } \mathrm{Sc} > 0.47 \tag{33}$$

where Sc is the Scruton number characterizing the mass-damping conditions of a Tainter gate:

$$\mathrm{Sc} \equiv \frac{4\pi\zeta_0}{k_\theta \alpha_{m\theta}} \tag{34}$$

In practice, the equivalent damping ratio can readily be obtained by impact testing (see, for example, Ishii & Naudascher, 1984).

The positive values of the excitation ratio ξ in the range $F_0 < 1.14$ are of little practical significance. For a large prototype Tainter gate, d_o is at least 10 m, so $F_0 < 1.14$ would correspond to natural vibration frequencies, $\Omega_o/2\pi$, of about 0.16 Hz or less, substantially below common prototype frequencies. Indeed, it was found from field tests by Ishii & Naudascher (1984) that the natural rotational frequency of a typical Tainter gate in air was about 9 Hz. On the other hand, if the submergence depth, d_o, is small, F_0 may be less than 1.14 even for comparatively large natural vibration frequencies. In such cases, however, $k_\theta \alpha_{m\theta}$ also becomes small, and the unstable region for $F_0 < 1.14$ decreases, as seen from Figure 6b.

Based on the design criterion given in Equation 33, two cures for dynamic instability of eccentric Tainter gates can be proposed:

1. Increase the damping for a given modified mass ratio $k_\theta \alpha_{m\theta}$ so that the design criterion given in Equation 33 is satisfied. The dynamic instability of the collapsed 12 m high by 9 m wide Tainter gate at the Wachi Dam in Japan was examined by Ishii et al. (1980) using the following parameters:

$$k' \equiv 2\sqrt{2} c_f c_g \rho \frac{R^* R_a^3}{I_\varsigma} \sin^2 \Theta_0 \cos \Theta_0 \quad \text{and} \quad R_d' \equiv \frac{d_0}{R_a \sin \Theta_0} \tag{35}$$

The design factor, k', which depends on the moment of inertia and the size of the Tainter gate, was 0.21 for the collapsed gate; the non-dimensional submergence depth R_d' was 1.62; and the modified mass ratio $k_\theta \alpha_{m\theta}$ was 0.55. Note that $k_\theta \alpha_{m\theta}$ can be written as follows:

$$k_\theta \alpha_{m\theta} = k' R_d'^2 \tag{36}$$

According to Equation 33, an equivalent damping ratio ζ_0 larger than 0.021 would have been necessary to overcome the instability. Under normal circumstances, mechanical friction due to the side seals (rubber against concrete over the full height of the gate on both sides) would provide such damping. For the gate in question, however, one might speculate that the friction may have been reduced by excessive wear on the side seals due to a prolonged, asymmetrical lifting of the gate by the wire cables on each side of the gate. However, we suggest that an additional, and more likely, instability mechanism was responsible for the Wachi Dam gate failure, as discussed in Chapter 12.

2. Reduce the eccentricity from the trunnion pin or increase the moment of inertia of the gate to provide dynamic stability. Recall that the modified mass ratio $k_\theta\,\alpha_{m\theta}$, is proportional to the ratio of trunnion-pin eccentricity to the moment of inertia for the gate as defined in Equations 8 and 12. The maximum value of the modified mass ratio $k_\theta\,\alpha_{m\theta}$ for which the gate remains stable can be calculated using the equivalent damping ratio determined, for example, by impact testing (Ishii & Naudascher, 1984) and the design criterion in Equation 33.

COMPARISON OF THEORY WITH EXPERIMENTAL RESULTS

To confirm the approximate solutions obtained in the previous sections, the calculated results were compared with the experimental results reported by Ishii et al. (1980). The Tainter gate model used (Figure 3) had the following dimensions: the lower-stay inclination angle, $\Theta_0 = 39.2°$; the lower-stay length, $R_a = 138$ mm; the skinplate radius, 135 mm; the skinplate span, 40 mm. The moment of inertia about the trunnion pin per unit width was 0.0575 kg m. The center of the skinplate was situated beneath the trunnion pin with an adjustable eccentricity, R^*. The gate was placed in a water channel, 1,500 mm long, 500 mm high, and 40 mm deep. The clearance between the sides of the gate and the water-channel walls was set to about 0.5 mm. Since the theoretical solutions of the vibration frequency ratio and the excitation ratio assumed small ratios b/d_0, only experimental results for the smallest gate opening corresponding to $b/d_0 = 0.062$ are considered, although other gate openings were also investigated (Ishii et al., 1980). The gate was most unstable at small gate openings. As the gate opening increased, in general, the gate became more stable. The experimental data presented here represents the worst case scenario among the cases tested. It can be argued that, as the gate opening is increased, the mean flow velocity in the reservoir increases and the fluid pressure phase-lag relative to the gate vibration increases, and thus the energy supplied from the fluid motion to the vibrating gate decreases.

The results for the non-dimensional vibration frequency, F/F_0, are shown in Figure 7 for a non-dimensional gate opening $b/d_0 = 0.062$. The unfilled circles represent the experimental data for a design factor $k' = 0.76$ and a non-dimensional submergence depth $R'_d = 0.93$, resulting in a modified mass ratio $k_\theta\alpha_{m\theta}$ of 0.66. The equivalent damping ratio ζ_0 was about 0.018 (Ishii et al., 1980).

The theoretical curve for $k_\theta\,\alpha_{m\theta} = 0.66$ and $\zeta_0 = 0.018$ is shown by the solid line in Figure 7. For small basic Froude numbers, F_0, the experimental values of F/F_0 lie slightly below the theoretical curve. The deviation, however, remains within 5%. Within this 5% tolerance, the frequency prediction closely agrees with the experimental data.

A similar comparison of the theoretical and experimental values of the excitation ratio, ξ, is shown in Figure 8, again for $b/d_0 = 0.062$. The open circles represent the experimental values for a design factor $k' = 1.22$ and a non-dimensional submergence depth $R'_d = 1.4$. The modified mass ratio $k_\theta\,\alpha_{m\theta}$ was 2.39 and the measured equivalent damping ratio, ζ_0, about 0.037 (Ishii et al., 1980). The theoretical curve for $k_\theta\,\alpha_{m\theta} = 2.39$ and $\zeta_0 = 0.037$ is shown by the solid line. The experimental data lie above the theoretical curve.

Note that the dashed curve plotted through the experimental data is approximately parallel to the theoretical curve. As explained in connection with Figure 5b, the theoretical curves for $F_0 > 2.0$ are parallel to each other, and the distance between any two curves is close to the difference between their respective equivalent damping ratios, ζ_0. Therefore, it may be plausible that the difference between the experimental and the theoretical data is due to an error in the complicated experimental evaluation of the

Figure 7. Experimental measurements of frequency ratio as a function of basic Froude number compared with theoretical prediction for $k_\theta \alpha_{m\theta} = 0.66$
From Ishii & Naudascher (1992). © 1992 Elsevier. Used with permission.

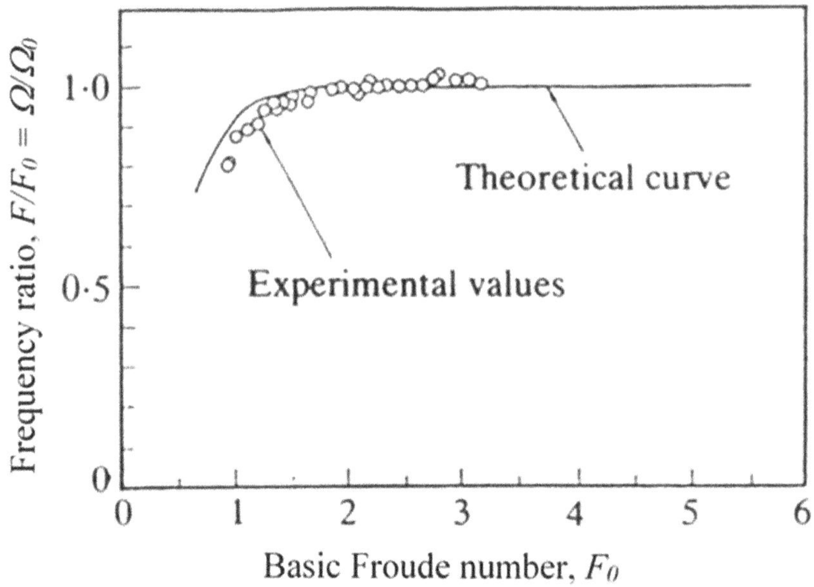

Figure 8. Experimental measurements of excitation ratio as a function of basic Froude number compared with theoretical prediction for $k_\theta \alpha_{m\theta} = 2.39$
From Ishii & Naudascher (1992). © 1992 Elsevier. Used with permission.

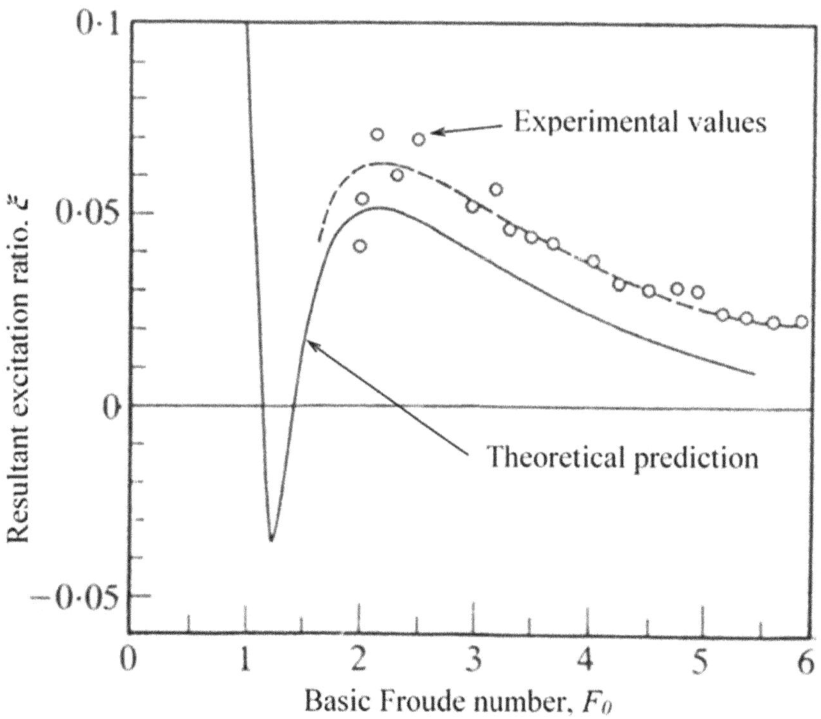

equivalent damping ratio, ζ_0. For the model gate, damping was due to Coulomb friction at the trunnion pin, internal friction in the coiled spring, and hydraulic friction due to water flow in the gap between the gate and the channel walls. The evaluation procedure included establishing a steady-state vibration and then calculating ζ_0 based on the assumption of an equivalent viscous damper in the vibration system. It is conceivable that this procedure may have led to a relatively large error. As shown by the dashed line in Figure 6, the experimental data fit better with the curve for $\zeta_0 = 0.025$, so the possible error in the measured damping might have been of the order of 30%. The reason that this potentially incorrect damping did not substantially affect the frequency ratios in Figure 7 is that the non-dimensional vibration frequency F/F_0 was found to be unaffected by the damping ratio, ζ_0, as shown in Figure 5a.

CONCLUSION

Eccentricity instability is a type of self-excited flow-induced vibration that can affect eccentric Tainter gates. The characteristics of eccentricity instability about the trunnion pin at small gate openings were determined by an approximate solution of the equation of motion. We can draw the following conclusions from the theoretical results and their experimental verification.

1. The instability and vibration characteristics of eccentric Tainter gates depend, essentially, on three parameters:
    a. The basic Froude number, F_0, which may also be interpreted as an inverse reduced velocity. The basic Froude number is composed of the submergence depth and the natural frequency of the vibration system in air.
    b. The modified mass ratio, $k_\theta\, \alpha_{m\theta}$, which relates the moments of inertia of water and gate.
    c. The equivalent damping ratio, ζ_0, which represents the damping effects on the vibration system.
2. The vibration frequency of the Tainter gate is almost independent of ζ_0. For F_0 greater than about 2.0, the vibration frequency is equal to the in-air vibration frequency, independent of the value of $k_\theta\, \alpha_{m\theta}$. For $F_0 < 2.0$, the vibration frequency gradually decreases from that in air as F_0 decreases or as $k_\theta\, \alpha_{m\theta}$ increases. This frequency decrease is due to an equivalent added mass effect.
3. The dynamic instability increases as $k_\theta\, \alpha_{m\theta}$ increases or as ζ_0 decreases. The excitation ratio of the Tainter gate system takes negative values only in a limited range around $F_0 = 1.14$; for $F_0 > 1.14$ and $F_0 < 1.14$, it assumes positive values. The basic Froude number of full-scale Tainter gates, in general, is much larger than 1.14. In the range of significance for full-scale gates, in other words, when $F_0 > 1.14$, the excitation ratio reaches a maximum value given by $\xi = 0.0374 k_\theta\, \alpha_{m\theta} - \zeta_0$ when $F_0 = 2.12$.
4. To ensure the dynamic stability of eccentric Tainter gates for $F_0 > 1.14$, the Scruton number, *Sc*, must be larger than 0.47, regardless of the natural vibration frequency. From this design criterion, the amount of damping and the magnitude of the moment of gate inertia necessary for dynamic stability can be determined, as can the allowable eccentricity of the skinplate center from the trunnion pin.

REFERENCES

Imaichi, K., & Ishii, N. (1977). Instability of an idealized Tainter gate system without damping caused by surface waves on the back water of dam. *Bulletin of the JSME, 20*(146), 963–970. doi:10.1299/jsme1958.20.963

Ishii, N., & Imaichi, K. (1982). Instability of elastically suspended Tainter gate system caused by surface waves on the reservoir of a dam (Approximate Solution) [in Japanese]. *Transactions of the Japan Society of Mechanical Engineers, Series B, 48*(428), 640–647. doi:10.1299/kikaib.48.640

Ishii, N., Imaichi, K., & Hirose, A. (1977). Instability of elastically suspended Tainter gate system caused by surface waves on the reservoir of a dam. *Journal of Fluids Engineering, 99*(4), 699–708. doi:10.1115/1.3448888

Ishii, N., Imaichi, K., & Hirose, A. (1980). Dynamic instability of Tainter gates. In E. Naudascher & D. Rockwell (Eds.), *Practical Experiences with Flow-Induced Vibrations* (pp. 452–460). Berlin: Springer. doi:10.1007/978-3-642-81528-7_49

Ishii, N., Imaichi, K., & Takemoto, S. (1977). Water waves and structural loads caused by a periodic change of discharge from a sluice-gate. *Bulletin of the JSME, 20*(146), 998–1007. doi:10.1299/jsme1958.20.998

Ishii, N., Knisely, C. W., & Nakata, A. (1995). Field study of a long-span shell-type gate undergoing flow-induced vibrations. *Journal of Fluids and Structures, 9*(1), 19–41. doi:10.1006/jfls.1995.1002

Ishii, N., & Naudascher, E. (1984). Field tests on natural vibration modes of a Tainter gate. In *Channels and Channel Control Structures* (pp. 209–222). Berlin: Springer. doi:10.1007/978-3-662-11300-4_17

Ishii, N., & Naudascher, E. (1992). A design criterion for dynamic stability of Tainter gates. *Journal of Fluids and Structures, 6*(1), 67–84. doi:10.1016/0889-9746(92)90056-9

Yano, K. (1968). On the event of the gate destruction. *Disaster Prevention Research Institute Annuals of Kyoto University*, 1-17.

APPENDIX: NOMENCLATURE

b: Mean gate opening height

c_f : Discharge coefficient

c_g : Area coefficient

d_0: Submergence depth of skinplate lower edge

$f_{nq}(F)$: Characteristic function in determining the dynamic instability

F : Froude number $\left(=\sqrt{d_0 / g}\,\Omega\right)$

F_0 : basic Froude number $\left(=\sqrt{d_0 / g}\,\Omega_a\right)$

g: Acceleration of gravity

i: Imaginary unit

I_θ: Moment of (per unit spanwise length) anout trunnion pin of gate

k_θ : Eccentricity factor

k' : Design factor

M_p: Hydrodynamic pressure moment (per unit spanwise length)

p_r, P_r : Flow-rate variation pressure [p_r is the dimensionless expression of P_r (Pa)]

$p_{r\infty}$: Flow-rate variation pressure for infinite reservoir depth

$p_{rp\infty}$: Flow-rate-variation progressive-wave pressure component

$p_{rs\infty}$: Flow-rate-variation standing-wave pressure component

R_a: Distance of skinplate lower edge from trunnion pin center

R^*: Distance of hydrodynamic pressure resultant vector on skinplate from trunnion pin

R_d' : Non-dimensional submergence depth

s: Variable for Fourier integral transform

Sc: Scruton number

t: Time

x, X: Horizontal coordinate along undisturbed free surface [x is the dimensionless expression of X (m)]

y, Y: Vertical coordinate [y is the dimensionless expression of Y (m)]

Y_0: Amplitude of skinplate oscillation

Y_G: Downward vibration of skinplate

$\alpha_{m\theta}$: (Torsional) water-to-gate mass ratio

$\Delta c_{r\theta}$, $\Delta C_{r\theta}$: Fluid excitation coefficient [$\Delta c_{r\theta}$ is the dimensionless expression of $\Delta C_{r\theta}$]

$\Delta m_{r\theta}$: Reduced added mass

θ, Θ: Oscillatory angle around trunnion pin of Tainter gate [θ is the dimensionless expression of Θ (degree)]

Θ_0: Elevation angle of trunnion pin from skinplate lower edge

$\bar{\Theta}$: Amplitude of oscillatory angle

ζ_0: Equivalent viscous damping ratio

ξ : Resultant excitation ratio
$\xi_{r\theta}$: Fluid-excitation ratio
ρ: Specific mass (density) [kg/m^3]
Ω: In-water vibration frequency
Ω_a: In-air vibration frequency

Chapter 9
Pressure on a Submerged Plate Undergoing Streamwise Rotation:
Theory and Validating Experiments

ABSTRACT

In this chapter theoretical expressions are developed for the pressure loading on a vertical flat plate bounding an upstream reservoir with the plate undergoing rotational streamwise oscillation in close proximity to a bounding surface beneath the plate. The flow field in the reservoir can be decomposed into two parts. The first component of the flow field is due to only the rotational vibratory motion of the rigid weir plate (about some point on the weir plate), while the gate remains entirely closed with no discharge. The second component of the flow field results from the up-and-down vibration of the weir plate. Data from model scale testing in the laboratory and field tests on full-scale Tainter gates show excellent to good agreement with the theoretical predictions, validating the use of the theory.

INTRODUCTION

The geometry of a Tainter gate on a dam crest in complicated, involving a curved gate surface and a curved gate crest. Simplifying these surfaces by modeling them as planar surfaces with vertical and horizontal faces (in the case of the upstream reservoir), a solution can be found for the pressure loading on the gate. In this chapter, we develop the theoretical framework for using analyses similar to those in Chapters 4, 5, and 7 to determine fluctuating pressures due to push-and draw vibrations and due to discharge fluctuations, in-water to in-air frequency ratios, and in-water damping. Data from model scale testing in the laboratory and field tests on full-scale Tainter gates show excellent to good agreement with the theoretical predictions, validating the use of the theory.

DOI: 10.4018/978-1-5225-3079-4.ch009

VERTICAL, FLAT, AND RIGID WEIR PLATE

Boundary and Initial Conditions

A Tainter gate with a circular-arc skinplate, is typically installed on a dam crest with a curved profile, such as the Creager crest (see Creager & Justin, 1927). If one insists on an exact representation of such curved boundaries, analytical modeling of the flow field is essentially impossible for this geometry. To make the problem tractable, the skinplate was replaced by a vertical flat, rigid weir plate and the curved dam crest was replaced by a horizontal, flat bed with a vertical step down at the gate location to allow for the greater depth of the reservoir, as shown in Figure 1. This sudden step down represents the sudden increase in reservoir depth on the upstream side of the dam crest. If the analysis is restricted to small vibrations of the weir plate, it should be possible to calculate the hydrodynamic pressure with sufficient accuracy even with such coarse representation of the gate geometry. The accuracy of this approximate method will be confirmed by model experiments. The water surface was treated as a free surface. By simplifying the boundaries in this manner, a theoretical solution for the flow field is now possible.

The following analysis is based on the assumption that the flow field in the reservoir can be decomposed into two parts, as explained in detail in Anami and Ishii (1998), Anami et al. (2000a). The first component of the flow field is due to only the rotational vibratory motion of the rigid weir plate (about some point on the weir plate), while the gate remains entirely closed with no discharge, as shown in Figure 1a. The second component of the flow field results from the up-and-down vibration of the weir plate, as shown in Figure 1b. The weir plate is further assumed to be driven simultaneously with the weir plate rotation. The flow field fluctuation is driven by the sum of two gate motions. Further, viscous effects will be assumed negligible, permitting this flow field to be modeled as a potential flow with a velocity potential Φ that satisfies the Laplace equation. The potential flow assumption also allows the superposition of the effects of the two gate motions. If the weir plate rotational vibration is assumed to be of small enough amplitude, the flow field can be treated two-dimensionally, because the span of the weir plate is assumed to be large.

Figure 1. Vertical flat-and-rigid weir plate and two-dimensional flow field; (a) Streamwise vibration inducing push-and-draw pressure P_b, and (b) vertical vibration inducing flow-rate variation pressure P_r
From Anami & Ishii (2001).

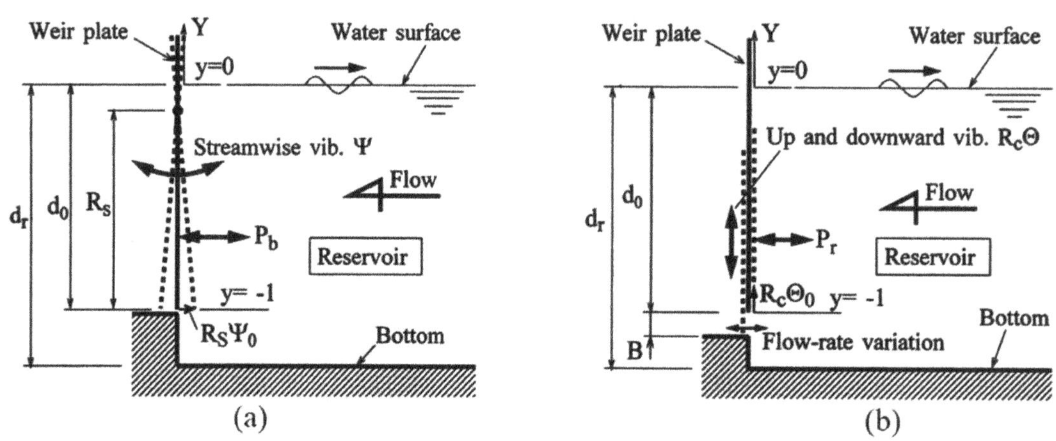

The Laplace equation is given by the following two-dimensional expression:

$$\Phi_{XX} + \Phi_{YY} = 0 \qquad \left(0 \le X < \infty, -d_r < Y \le 0\right) \tag{1}$$

where the X-axis is taken to be positive along the still water surface in the upstream direction, and the Y-axis is taken as positive upward along the vertical weir plate. The subscript on Φ indicates partial differentiation. The velocity potential Φ is assumed to be a function of X, Y, and time T.

The surface velocity of the weir plate due to streamwise vibrations must coincide with the fluid velocity. The movement of the rigid weir plate performing the counter-clockwise rotation Ψ can be equated to the fluid velocity at X=0:

$$\Phi_X\left(0, Y; T\right) = -\left\{-Y - \left(d_0 - R_S\right)\right\}\Psi_T(T) = \left(d_0 - R_S + Y\right)\Psi_T(T) \qquad \left(-d_0 \le Y < 0\right) \tag{2}$$

where d_0 is the submergence depth of the weir plate, and R_S is the rotation center height of the weir plate measured from its lower end, as shown in Figure 1a.

When a Tainter gate rotates as a rigid body counterclockwise through an angle Θ, the gate opening is given by $R_c\Theta$, where R_c is the radius of the bottom end of the weir plate. It is assumed that the discharge flow ΔQ per unit width through the gate opening is given by the product of the average flow velocity and the gate opening:

$$\Delta Q = c_f\sqrt{2gd_0} \cdot R_c\Theta = q\Theta \tag{3}$$

where

$$q = c_f\sqrt{2gd_0} \cdot R_c \tag{4}$$

and c_f is the instantaneous flow-rate variation coefficient.

Variation in the flow rate drives the fluctuations in the reservoir flow field, thus generating waves on the water surface. Assuming small amplitude vertical vibrations of the weir plate (representing opening area variation due to the rotational vibration of the whole-body gate about the trunnion pin), the resulting waves also have small amplitudes. Therefore, the boundary condition on the water surface is given in the following expression:

$$\left[\Phi_{TT} + \tilde{\mu}\Phi_T + g\Phi_Y\right]_{Y=0} = 0 \tag{5}$$

where the second term introduces a damping body force ($\tilde{\mu}\rho \times$ velocity) proportional to the flow velocity, as was done in Chapters 4 and 5, to approximate the small fluid damping effect (see Rayleigh, 1945; Lamb, 1932; and Stoker, 1957) for more details. The frictional coefficient $\tilde{\mu}$ has the units of 1/s.

On the vertical wall beneath the weir plate and on the reservoir bed, the normal flow velocity on each boundary is zero, resulting in the following boundary conditions:

$$\left.\begin{array}{ll} \Phi_X\left(0,Y;T\right)=0 & \left(-d_r\leq Y<0\right) \\ \Phi_Y\left(X,-d_r;T\right)=0 & \left(0\leq X<\infty\right) \end{array}\right\} \tag{6}$$

The upstream extent of the reservoir is assumed to be infinite and contain no obstacles. The flow field fluctuation induced by the weir plate vibration propagates upstream. However, due to the small damping effect in the fluid, the fluctuation amplitude is attenuated with distance and the fluctuation cannot be transmitted infinitely far upstream. In order to represent this practical consideration, the following boundary conditions are imposed on the velocity potential Φ and on the flow velocity Φ_X in the far upstream region ($X=\infty$):

$$\Phi\left(\infty,Y;T\right)=0,\qquad \Phi_X\left(\infty,Y;T\right)=0 \tag{7}$$

Furthermore, in order to treat the present flow field as an initial-value problem, it is assumed that at, and prior to, the initial time ($T\leq 0$) there is no disturbance in the reservoir flow field:

$$\Phi\left(X,Y;0\right)=0\,,\qquad \Phi_T\left(X,0;0\right)=0 \tag{8}$$

For the weir plate vibration, the following initial conditions are assumed:

$$\Psi(0)=0,\qquad \dot{\Psi}(0)=0,\qquad \Theta\left(0\right)=0,\qquad \dot{\Theta}\left(0\right)=0 \tag{9}$$

With this formulation, the problem is reduced to a Neumann-type boundary- and-initial-value problem of the equation, Equation 1. The solutions of this problem can be analytically derived using Fourier integrals along with Fourier and Laplace transforms.

Hydrodynamic Pressures Induced by Vibration of a Weir Plate

The reservoir extends infinitely in the upstream direction (i.e., the X-direction). Therefore, the Fourier integral transform with respect to X:

$$\bar{\Phi}\left(S,Y;T\right)=\sqrt{\frac{2}{\pi}}\int_0^\infty \Phi\left(X,Y;T\right)\cos SXdX \tag{10}$$

can be applied to the velocity potential Φ, reducing Equation 1, the Laplace equation, to the following form:

$$\bar{\Phi}_{YY}\left(S,Y;T\right)-S^2\bar{\Phi}\left(S,Y;T\right)=\sqrt{\frac{2}{\pi}}\Phi_X\left(0,Y;T\right) \tag{11}$$

where the boundary condition given in Equation 7 was used. The flow velocity at X=0 appears on the right-hand side of the above expression.

The general solution to Equation 11 is given by the superposition of the particular solution $\bar{\Phi}^{(p)}$ and the homogenous solution $A^* \cosh S\left(Y + d_r\right)$:

$$\bar{\Phi}\left(S,Y;T\right) = \bar{\Phi}^{(p)}\left(S,Y;T\right) + A^*\left(S;T\right)\cosh S\left(Y + d_r\right) \tag{12}$$

where the coefficient $A*$ is a function of the integral variable S and time T.

The particular solution $\bar{\Phi}^{(p)}$ satisfies the following boundary condition from Equation 6:

$$\bar{\Phi}_Y^{(p)}\left(S,-d_r;T\right) = 0 \tag{13}$$

The speed conditions are imposed at Y= 0 as follows:

$$\bar{\Phi}_Y^{(p)}\left(S,0;T\right) = 0 \tag{14}$$

The following expression will be obtained when the general solution, Equation 12, is substituted into Equation 11:

$$\bar{\Phi}_{YY}\left(S,Y;T\right) - S^2\bar{\Phi}\left(S,Y;T\right) = \sqrt{\frac{2}{\pi}}\Phi_X\left(0,Y;T\right) \tag{15}$$

In the Y-direction, the reservoir depth is bounded, and after applying the following Fourier transform,

$$\tilde{\bar{\Phi}}^{(p)}\left(S,n;T\right) \equiv \int_{-d_r}^{0} \bar{\Phi}\left(S,Y;T\right)\cos\frac{n\pi Y}{d_r}dY \; (n = 0,\, 1,\, 2\ldots) \tag{16}$$

to Equation 12, the particular solution is obtained.

The following expression results from taking the Fourier transform of Equation 15 and then applying the conditions of Equations 13 and 14:

$$\begin{aligned}\tilde{\bar{\Phi}}^{(p)}\left(S,n;T\right) = & -\sqrt{\frac{2}{\pi}}\frac{d_r}{n\pi}\frac{\Psi_T\left(T\right)}{S^2 + \left(n\pi / d_r\right)^2}\left\{R_S \sin\frac{n\pi}{\beta^*} - \frac{d_r}{n\pi}\left(1 - \cos\frac{n\pi}{\beta^*}\right)\right\} \\ & + \sqrt{\frac{2}{\pi}}\frac{\Theta\left(T\right)}{S^2 + \left(n\pi / d_r\right)^2} q \cos n\pi / \beta^*\end{aligned} \tag{17}$$

When n = 0, Equation 17 can be reduced to the following expression:

$$\bar{\tilde{\Phi}}^{(p)}\left(S,0;T\right)=-\sqrt{\frac{2}{\pi}}\frac{\Psi_T\left(T\right)}{S^2}\left(R_S d_0-\frac{1}{2}d_0^2\right)+\sqrt{\frac{2}{\pi}}\frac{q}{S^2}\Theta\left(T\right) \tag{18}$$

In obtaining these expressions, the boundary condition of Equation 3 is treated in the following manner. The integration of the flow velocity $\Phi_X\left(0,Y;T\right)$ over a small region of 2δ in extent at the bottom end of the weir plate is assumed to be equal to the flow-rate variation $\Delta Q(T)$, that is:

$$\Delta Q\left(T\right)\left(=q\Theta_T\right)=-\lim_{\delta\to 0}\int_{-d_0-\delta}^{-d_0+\delta}\Phi_X\left(0,Y;T\right)dY \tag{19}$$

The flow velocity near the discharge opening changes so that the volumetric discharge is equal to ΔQ. Furthermore, it is assumed that this condition also applies in the limit as Y approaches $-d_0$ by permitting the height of the discharge opening $2\delta\to 0$.

The parameter β^* in Equations 17 and 18 is the ratio of the upstream reservoir depth d_r to the gate submergence depth d_0:

$$\beta^*\equiv\frac{d_r}{d_0} \tag{20}$$

When $\beta^*=1.0$ the reservoir depth is equal to the gate submergence depth, and when $\beta^*>1.0$ the reservoir is deeper than gate submergence depth.

The inverse Fourier transform

$$\bar{\Phi}\left(S,Y;T\right)\equiv\frac{1}{d_r}\bar{\tilde{\Phi}}\left(S,0;T\right)+\frac{2}{d_r}\sum_{n=1}^{\infty}\bar{\tilde{\Phi}}\left(S,n;T\right)\cos\frac{n\pi Y}{d_r}\ (n=0,\ 1,\ 2....) \tag{21}$$

is applied for any value of n value to give the particular solution $\bar{\Phi}^{(p)}$ in the following equation:

$$\begin{aligned}\bar{\Phi}^{(p)}\left(S,Y;T\right)=&-\sqrt{\frac{2}{\pi}}\Psi_T\left(T\right)\left[\frac{1}{\beta^*S^2}\left(R_S-\frac{1}{2}d_0\right)\right.\\&\left.+\frac{2}{\pi}\sum_{n=1}^{\infty}\frac{1}{n}\frac{1}{S^2+\left(n\pi/d_r\right)^2}\left\{R_S\sin\frac{n\pi}{\beta^*}+\frac{d_r}{n\pi}\left(\cos\frac{n\pi}{\beta^*}-1\right)\right\}\cos\frac{n\pi Y}{d_r}\right]\\&+\sqrt{\frac{2}{\pi}}\Theta\left(T\right)\frac{q}{d_r}\left\{\frac{1}{S^2}+2\sum_{n=1}^{\infty}\frac{\cos n\pi/\beta^*}{S^2+\left(n\pi/d_r\right)^2}\cos\frac{n\pi Y}{d_r}\right\}\end{aligned} \tag{22}$$

To the arbitrary function $A^*(S;T)$, the general solution, Equation 12, and the boundary condition of the free surface, Equation 5, are used to obtain the following:

$$\begin{aligned} &A_{TT}^{*}\left(S;T\right)+\tilde{\mu}A_{T}^{*}\left(S;T\right)+gS\tanh Sd_{r}A^{*}\left(S;T\right) \\ &=-\frac{1}{\cosh Sd_{r}}\left\{\bar{\Phi}_{TT}^{(p)}\left(S,0;T\right)+\tilde{\mu}\bar{\Phi}_{T}^{(p)}\left(S,0;T\right)\right\} \end{aligned} \tag{23}$$

The subsequent substitution of the particular solution, Equation 22, into Equation 23 results in an equation for determining the arbitrary function:

$$\begin{aligned} &A_{TT}^{*}\left(S;T\right)+\tilde{\mu}A_{T}^{*}\left(S;T\right)+gS\tanh Sd_{r}A^{*}\left(S;T\right) \\ &=\sqrt{\frac{2}{\pi}}\frac{1}{\cosh Sd_{r}}\left[\left(\Psi_{TTT}\left(T\right)+\mu\Psi_{TT}\left(T\right)\right)\left\{\frac{1}{\beta^{*}S^{2}}\left(R_{S}-\frac{1}{2}d_{0}\right)\right.\right. \\ &\left.\left.+\frac{2}{\pi}\sum_{n=1}^{\infty}\frac{1}{n}\frac{1}{S^{2}+\left(n\pi/d_{r}\right)^{2}}\left\{R_{S}\sin\frac{n\pi}{\beta^{*}}+\frac{d_{r}}{n\pi}\left(\cos\frac{n\pi}{\beta^{*}}-1\right)\right\}\right\}\right. \\ &\left.-\frac{q}{d_{r}}\left\{\frac{1}{S^{2}}+2\sum_{n=1}^{\infty}\frac{\cos n\pi/\beta^{*}}{S^{2}+\left(n\pi/d_{r}\right)^{2}}\right\}\left(\Theta_{TT}\left(T\right)+\mu\Theta_{T}\left(T\right)\right)\right] \end{aligned} \tag{24}$$

With the fluid assumed to be quiescent at and before the initial time and the free surface assumed similarly quiescent, the initial conditions on $\bar{\Phi}$ are as follows from Equation 8:

$$\bar{\Phi}\left(S,Y;0\right)=0,\bar{\Phi}_{T}\left(S,Y;0\right)=0 \tag{25}$$

The following equations can be obtained by substituting Equation 14 into these initial conditions, Equation 25:

$$\bar{\Phi}^{(p)}\left(S,Y;0\right)+A^{*}\left(S;0\right)\cosh S\left(Y+d_{r}\right)=0$$

$$\bar{\Phi}_{T}^{(p)}\left(S,Y;0\right)+A_{T}^{*}\left(S;0\right)\cosh S\left(Y+d_{r}\right)=0 \tag{26}$$

By substituting particular solution, Equation 22, into Equations 26, the initial conditions on the water surface in terms of the function $A^*(S;T)$ can be obtained:

$$A^{*}\left(S;0\right)=0 \text{ and } A_{T}^{*}\left(S;0\right)=0 \tag{27}$$

Applying the Laplace transform and the inverse transform based on the convolution theorem to the result, the arbitrary function $A^*(S;T)$ satisfying the initial condition, Equation 27, is obtained in the form:

$$
\begin{aligned}
A^*(S;T) = \sqrt{\frac{2}{\pi}} \frac{1}{\cosh Sd_r} & \left[\left[\frac{1}{\beta^* S^2} \left(R_S - \frac{1}{2} d_0 \right) \right. \right. \\
& \left. \left. + \frac{2}{\pi} \sum_{n=1}^{\infty} \frac{1}{n} \frac{1}{S^2 + \left(n\pi / d_r \right)^2} \left\{ R_S \sin \frac{n\pi}{\beta^*} + \frac{d_r}{n\pi} \left(\cos \frac{n\pi}{\beta^*} - 1 \right) \right\} \right] \right. \\
& \times \left\{ \Psi_T(T) - gS \tanh Sd_r \int_0^T \exp\left\{ -\tilde{\mu} \left(T - \tilde{\tau} \right) / 2 \right\} \cos \Phi_0(S) \left(T - \tilde{\tau} \right) \Psi\left(\tilde{\tau} \right) d\tilde{\tau} \right\} \\
& - \frac{q}{d_r} \left\{ \frac{1}{S^2} + 2 \sum_{n=1}^{\infty} \frac{\cos n\pi / \beta^*}{S^2 + \left(n\pi / d_r \right)^2} \right\} \\
& \left. \times \left\{ \Theta(T) - gS \tanh Sd_r \int_0^T \frac{\exp\left\{ -\tilde{\mu} \left(T - \tilde{\tau} \right) / 2 \right\}}{\Phi_0(S)} \sin \Phi_0(S) \left(T - \tilde{\tau} \right) \Theta\left(\tilde{\tau} \right) d\tilde{\tau} \right\} \right]
\end{aligned} \tag{28}
$$

In Equation 2, the effect of the initial condition dissipates with increasing time due to the introduction of the small frictional coefficient $\tilde{\mu}$. The phase function $\Phi_0(S)$ is given in the following expression:

$$
\Phi_0(S) = \sqrt{gS \tanh Sd_r - \tilde{\mu}^2 / 4} \tag{29}
$$

Substituting the resulting particular solution $\overline{\Phi}^{(p)}$ and the arbitrary function $A^*(S;T)$, the general solution $\overline{\Phi}$ is obtained:

$$
\begin{aligned}
& \overline{\Phi}(S, Y; T) = \\
& -\sqrt{\frac{2}{\pi}} \Psi_T(T) \left[\frac{1}{\beta^* S^2} \left(R_S - \frac{1}{2} d_0 \right) + \frac{2}{\pi} \sum_{n=1}^{\infty} \frac{1}{n} \frac{1}{S^2 + \left(n\pi / d_r \right)^2} \left\{ \begin{aligned} & R_S \sin \frac{n\pi}{\beta^*} \\ & + \frac{d_r}{n\pi} \left(\cos \frac{n\pi}{\beta^*} - 1 \right) \end{aligned} \right\} \cos \frac{n\pi Y}{d_r} \right] \\
& + \sqrt{\frac{2}{\pi}} \frac{\cosh S\left(Y + d_r \right)}{\cosh Sd_r} \left[\frac{1}{\beta^* S^2} \left(R_S - \frac{1}{2} d_0 \right) + \frac{2}{\pi} \sum_{n=1}^{\infty} \frac{1}{n} \frac{1}{S^2 + \left(n\pi / d_r \right)^2} \left\{ \begin{aligned} & R_S \sin \frac{n\pi}{\beta^*} \\ & + \frac{d_r}{n\pi} \left(\cos \frac{n\pi}{\beta^*} - 1 \right) \end{aligned} \right\} \right] \\
& \times \left\{ \Psi_T(T) - gS \tanh Sd_r \int_0^T \exp\left\{ -\tilde{\mu} \left(T - \tilde{\tau} \right) / 2 \right\} \cos \Phi_0(S) \left(T - \tilde{\tau} \right) \Psi\left(\tilde{\tau} \right) d\tilde{\tau} \right\} \\
& + \sqrt{\frac{2}{\pi}} \Theta(T) \frac{q}{d_r} \left\{ \frac{1}{S^2} + 2 \sum_{n=1}^{\infty} \frac{\cos n\pi / \beta^*}{S^2 + \left(n\pi / d_r \right)^2} \cos \frac{n\pi Y}{d_r} \right\} \\
& - \sqrt{\frac{2}{\pi}} \frac{\cosh S\left(Y + d_r \right)}{\cosh Sd_r} \frac{q}{d_r} \left\{ \frac{1}{S^2} + 2 \sum_{n=1}^{\infty} \frac{\cos n\pi / \beta^*}{S^2 + \left(n\pi / d_r \right)^2} \right\} \\
& \times \left\{ \Theta(T) - gS \tanh Sd_r \int_0^T \frac{\exp\left\{ -\tilde{\mu} \left(T - \tilde{\tau} \right) / 2 \right\}}{\Phi_0(S)} \sin \Phi_0(S) \left(T - \tilde{\tau} \right) \Theta\left(\tilde{\tau} \right) d\tilde{\tau} \right\}
\end{aligned} \tag{30}
$$

The first and second terms result from the streamwise rotational vibration; the first term is the particular solution and the second term is the arbitrary function. The third and fourth terms result from the flow-rate variation; the third term is the particular solution and the fourth term is the arbitrary function.

In order to derive the velocity potential $\Phi(X,Y;T)$, the following inverse Fourier transform in S is applied:

$$\Phi\left(X,Y;T\right) \equiv \sqrt{\frac{2}{\pi}}\int_0^\infty \bar{\Phi}\left(S,Y;T\right)\cos SXdS \tag{31}$$

The velocity potential can also be written in the form of a superposition of the velocity potentials $\Phi^{(b)}$ and $\Phi^{(r)}$:

$$\Phi\left(X,Y;T\right) = \Phi^{(b)}\left(X,Y;T\right) + \Phi^{(r)}\left(X,Y;T\right) \tag{32}$$

where $\Phi^{(b)}$ represents the velocity potential induced by the push-and-draw effect of the gate on the water,

$$\begin{aligned}
\Phi^{(b)}\left(X,Y;T\right) = &-\frac{2}{\pi}\Psi_T\left(T\right)\int_0^\infty\left[\frac{1}{\beta^* S^2}\left(R_S-\frac{1}{2}d_0\right)\right.\\
&\left.+\frac{2}{\pi}\sum_{n=1}^{\infty}\frac{1}{n}\frac{1}{S^2+\left(n\pi/d_r\right)^2}\left\{R_S\sin\frac{n\pi}{\beta^*}+\frac{d_r}{n\pi}\left(\cos\frac{n\pi}{\beta^*}-1\right)\right\}\cos\frac{n\pi Y}{d_r}\right]\cos SXdS\\
&+\frac{2}{\pi}\Psi_T\left(T\right)\int_0^\infty\frac{\cosh S\left(Y+d_r\right)}{\cosh Sd_r}\left[\frac{1}{\beta^* S^2}\left(R_S-\frac{1}{2}d_0\right)\right.\\
&\left.+\frac{2}{\pi}\sum_{n=1}^{\infty}\frac{1}{n}\frac{1}{S^2+\left(n\pi/d_r\right)^2}\left\{R_S\sin\frac{n\pi}{\beta^*}+\frac{d_r}{n\pi}\left(\cos\frac{n\pi}{\beta^*}-1\right)\right\}\right]\cos SX\,dS\\
&-\frac{2}{\pi}\int_0^\infty\frac{\cosh S\left(Y+d_r\right)}{\cosh Sd_r}\left[\frac{1}{\beta^* S^2}\left(R_S-\frac{1}{2}d_0\right)\right.\\
&\left.+\frac{2}{\pi}\sum_{n=1}^{\infty}\frac{1}{n}\frac{1}{S^2+\left(n\pi/d_r\right)^2}\left\{R_S\sin\frac{n\pi}{\beta^*}+\frac{d_r}{n\pi}\left(\cos\frac{n\pi}{\beta^*}-1\right)\right\}\right]\cos SX\,dS\\
&\times gS\tanh Sd_r\int_0^T\exp\left\{-\tilde{\mu}\left(T-\tilde{\tau}\right)/2\right\}\cos\Phi_0\left(S\right)\left(T-\tilde{\tau}\right)\Psi\left(\tilde{\tau}\right)d\tilde{\tau}
\end{aligned} \tag{33}$$

and $\Phi^{(r)}$ represents the velocity potential induced by the flow-rate variation:

$$\Phi^{(r)}\left(X,Y;T\right)=\frac{2}{\pi}\Theta\left(T\right)\frac{q}{d_r}\int_0^{\infty}\left\{\frac{1}{S^2}+2\sum_{n=1}^{\infty}\frac{\cos n\pi/\beta^*}{S^2+\left(n\pi/d_r\right)^2}\cos\frac{n\pi Y}{d_r}\right\}\cos SXdS$$
$$-\frac{2}{\pi}\frac{q}{d_r}\Theta\left(T\right)\int_0^{\infty}\frac{\cosh S\left(Y+d_r\right)}{\cosh Sd_r}\left\{\frac{1}{S^2}+2\sum_{n=1}^{\infty}\frac{\cos n\pi/\beta^*}{S^2+\left(n\pi/d_r\right)^2}\right\}\cos SXdS \qquad (34)$$
$$+\frac{2}{\pi}\frac{q}{d_r}\int_0^{\infty}\frac{\cosh S\left(Y+d_r\right)}{\cosh Sd_r}\left\{\frac{1}{S^2}+2\sum_{n=1}^{\infty}\frac{\cos n\pi/\beta^*}{S^2+\left(n\pi/d_r\right)^2}\right\}\frac{gS\tanh Sd_r}{\Phi_0\left(S\right)}\cos SXdS$$
$$\times\int_0^{T}\exp\left\{-\tilde{\mu}\left(T-\tilde{\tau}\right)/2\right\}\sin\Phi_0\left(S\right)\left(T-\tilde{\tau}\right)\Theta\left(\tilde{\tau}\right)d\tilde{\tau}$$

The integration over the infinite range of S is indicated by the limits in these integral expressions. The integrations of the first and second terms of Equation 33 are as follows:

First term:

$$\int_0^{\infty}\left[\frac{1}{\beta^*S^2}\left(R_S-\frac{1}{2}d_0\right)+\frac{2}{\pi}\sum_{n=1}^{\infty}\frac{1}{n}\frac{1}{S^2+\left(n\pi/d_r\right)^2}\times\left\{R_S\sin\frac{n\pi}{\beta^*}+\frac{d_r}{n\pi}\left(\cos\frac{n\pi}{\beta^*}-1\right)\right\}\cos\frac{n\pi Y}{d_r}\right]\cos SXdS$$
$$=\frac{1}{\beta^*}\left(R_S-\frac{1}{2}d_0\right)\left(\frac{1}{\varepsilon}-\frac{\pi}{2}X\right)+\sum_{n=1}^{\infty}\frac{1}{n}\frac{d_r}{n\pi}\exp\left(-\frac{n\pi X}{d_r}\right) \qquad (35)$$
$$\times\left\{R_S\sin\frac{n\pi}{\beta^*}+\frac{d_r}{n\pi}\left(\cos\frac{n\pi}{\beta^*}-1\right)\right\}\cos\frac{n\pi Y}{d_r}$$

Second term:

$$\int_0^{\infty}\frac{\cosh S\left(Y+d_r\right)}{\cosh Sd_r}\left[\frac{1}{\beta^*S^2}\left(R_S-\frac{1}{2}d_0\right)\right.$$
$$\left.+\frac{2}{\pi}\sum_{n=1}^{\infty}\frac{1}{n}\frac{1}{S^2+\left(n\pi/d_r\right)^2}\times\left\{R_S\sin\frac{n\pi}{\beta^*}+\frac{d_r}{n\pi}\left(\cos\frac{n\pi}{\beta^*}-1\right)\right\}\right]\cos SXdS$$
$$=\frac{1}{\beta^*}\left(R_S-\frac{1}{2}d_0\right)\left(\frac{2}{\varepsilon}-\pi X\right)+\frac{d_r}{\pi}\sum_{n=1}^{\infty}\frac{1}{n^2}\left\{R_S\sin\frac{n\pi}{\beta^*}+\frac{d_r}{n\pi}\left(\cos\frac{n\pi}{\beta^*}-1\right)\right\} \qquad (36)$$
$$\times\exp\left(-\frac{n\pi X}{d_r}\right)\frac{\cos n\pi/d_r\left(Y+d_r\right)}{\cos n\pi}$$

Similarly, the integrations of the first and second terms of Equation 34 are as follows:

First term:

$$\int_0^\infty \left\{ \frac{1}{S^2} + 2\sum_{n=1}^{\infty} \frac{\cos n\pi / \beta^*}{S^2 + \left(n\pi / d_r\right)^2} \cos \frac{n\pi Y}{d_r} \right\} \cos SXdS$$
$$= \frac{1}{\varepsilon} - \frac{\pi}{2} X + \sum_{n=1}^{\infty} \frac{d_r}{n} \exp\left(-\frac{n\pi X}{d_r}\right) \cos \frac{n\pi}{\beta^*} \cdot \cos \frac{n\pi Y}{d_r} \tag{37}$$

Second term:

$$\int_0^\infty \frac{\cosh S\left(Y + d_r\right)}{\cosh Sd_r} \left\{ \frac{1}{S^2} + 2\sum_{n=1}^{\infty} \frac{\cos n\pi / \beta^*}{S^2 + \left(n\pi / d_r\right)^2} \right\} \cos SXdS$$
$$= \frac{1}{\varepsilon} - \frac{\pi}{2} X + \sum_{n=1}^{\infty} \frac{d_r}{n} \exp\left(-\frac{n\pi X}{d_r}\right) \cos \frac{n\pi}{\beta^*} \cdot \cos \frac{n\pi / d_r \left(Y + d_r\right)}{\cos n\pi} \tag{38}$$

in which ε is a small constant. Cauchy's integral theorem was used in obtaining Equations 36 and 38. In the process of determining Equations 33 and 34, the term $1/\varepsilon$ contained in Equations 35 to 38 is offset. Therefore, the general solution of the potential $\Phi^{(b)}$ and $\Phi^{(r)}$ can be written as follows:

$$\Phi^{(b)}\left(X,Y;T\right) = -\frac{2}{\pi} g \int_0^\infty \frac{\cosh S\left(Y + d_r\right)}{\cosh Sd_r} \left[\frac{1}{\beta^* S^2} \left(R_S - \frac{1}{2} d_0\right) \right.$$
$$\left. + \frac{2}{\pi} \sum_{n=1}^{\infty} \frac{1}{n} \frac{1}{S^2 + \left(n\pi / d_r\right)^2} \left\{ R_S \sin \frac{n\pi}{\beta^*} + \frac{d_r}{n\pi} \left(\cos \frac{n\pi}{\beta^*} - 1 \right) \right\} \right]$$
$$\times \cos SXS \tanh Sd_r dS \times \int_0^T \Psi\left(\tilde{\tau}\right) \exp\left\{-\tilde{\mu}\left(T - \tilde{\tau}\right)/2\right\} \cos \Phi_0\left(S\right)\left(T - \tilde{\tau}\right) d\tilde{\tau} \tag{39}$$

and

$$\Phi^{(r)}\left(X,Y;T\right) = \frac{2}{\pi} \frac{q}{d_r} g \int_0^\infty \frac{\cosh S\left(Y + d_r\right)}{\cosh Sd_r} \left\{ \frac{1}{S^2} + 2\sum_{n=1}^{\infty} \frac{\cos n\pi / \beta^*}{S^2 + \left(n\pi / d_r\right)^2} \right\} \frac{S \tanh Sd_r}{\Phi_0\left(S\right)} \cos SXdS$$
$$\times \int_0^T \exp\left\{-\tilde{\mu}\left(T - \tilde{\tau}\right)/2\right\} \sin \Phi_0\left(S\right)\left(T - \tilde{\tau}\right) \Theta\left(\tilde{\tau}\right) d\tilde{\tau} \tag{40}$$

The fluctuating hydrodynamic pressure, $P(X,Y;T)$, exerted on the weir plate, is given by the following equation:

$$P\left(X,Y;T\right) = -\rho \Phi_T\left(X,Y;T\right). \tag{41}$$

Substituting Equations 32, 39, and 40 into Equation 41, yields the following expressions for the fluctuating pressure, $P_b(X,Y;T)$, caused by the push-and-draw effect (from Equation 39) and the fluctuating pressure, $P_r(X,Y;T)$, caused by the flow-rate variation (from Equation 40):

$$P_b\left(X,Y;T\right)=\frac{2}{\pi}\rho g\int_0^{\infty}\frac{\cosh S\left(Y+d_r\right)}{\cosh Sd_r}\left[\frac{1}{\beta^* S^2}\left(R_S-\frac{1}{2}d_0\right)\right.$$
$$\left.+\frac{2}{\pi}\sum_{n=1}^{\infty}\frac{1}{n}\frac{1}{S^2+\left(n\pi/d_r\right)^2}\left\{R_S\sin\frac{n\pi}{\beta^*}+\frac{d_r}{n\pi}\left(\cos\frac{n\pi}{\beta^*}-1\right)\right\}\right] \tag{42}$$
$$\times\left\{\Psi\left(T\right)-\Phi_0\left(S\right)\int_0^T\Psi\left(\tilde{\tau}\right)\exp\left\{-\tilde{\mu}\left(T-\tilde{\tau}\right)/2\right\}\sin\Phi_0\left(S\right)\left(T-\tilde{\tau}\right)d\tilde{\tau}\right\}\times\cos SXS\tan Sd_r dS$$

and

$$P_r\left(X,Y;T\right)=-\frac{2}{\pi}\frac{q}{d_r}\rho g\int_0^{\infty}\frac{\cosh S\left(Y+d_r\right)}{\cosh Sd_r}\left\{\frac{1}{S^2}+2\sum_{n=1}^{\infty}\frac{\cos n\pi/\beta^*}{S^2+\left(n\pi/d_r\right)^2}\right\}\frac{S\tanh Sd_r}{\Phi_0\left(S\right)}\cos SXdS$$
$$\times\Phi_0\left(S\right)\int_0^T\Theta\left(\tilde{\tau}\right)\exp\left\{-\tilde{\mu}\left(T-\tilde{\tau}\right)/2\right\}\cos\Phi_0\left(S\right)\left(T-\tilde{\tau}\right)d\tilde{\tau} \tag{43}$$

In the remainder of this chapter, and as in Chapters 4, 5, and 7, the fluctuating pressure in Equation 42 is referred to as "push-and-draw pressure" and that in Equation 43 as "flow-rate variation pressure".

Now, in order to express the hydrodynamic pressure in terms of non-dimensional groups, the following dimensionless variables are introduced:

$$\begin{aligned}
&x\equiv X/d_0, \quad y\equiv Y/d_0, \quad s\equiv d_0S, \quad t\equiv\Omega_aT\\
&\tau\equiv\Omega_a\tilde{\tau}, \quad \mu\equiv\tilde{\mu}/\Omega_a, \quad r_s\equiv R_S/d_0\\
&\psi\left(t\right)\equiv\Psi\left(T\right)/\Psi_0, \quad \theta\left(t\right)\equiv\Theta\left(T\right)/\Theta_0, \quad \varphi_0\equiv\Phi_0/\sqrt{d_0/g}
\end{aligned} \tag{44}$$

where the weir plate submergence depth d_0 was selected as the representative length of the flow field, and the period $1/\Omega_a$ of the in-air natural vibration of the weir plate was selected as the representative time. With these parameters, the non-dimensional form of the hydrodynamic pressures are given by

$$p_b\equiv\frac{P_b/\rho g}{R_S\Psi_0} \quad \text{and} \quad p_r\equiv\frac{P_r/\rho g}{\sqrt{2c_f}R_c\Theta_0} \tag{45}$$

The numerators on the right hand side of both Equations 45 are the water head corresponding to the hydrodynamic pressure. The denominator in each case is the displacement amplitude of the lower end of the weir plate, since $R_S\Psi_0$ and $R_c\Theta_0$ represent the amplitude of the weir plate. Introducing this non-dimensionalization into Equations 42 and 43 yields the following non-dimensional forms:

$$p_b\left(x,y;t\right)=\frac{2}{\pi}\rho g\int_0^\infty \frac{\cosh s\left(y+\beta^*\right)}{\cosh s\beta^*}\left[\frac{1}{\beta^* s^2}\left(1-\frac{1}{2r_s}\right)\right.$$
$$\left.+\frac{2}{\pi}\sum_{n=1}^{\infty}\frac{1}{n}\frac{1}{s^2+\left(n\pi/\beta^*\right)^2}\left\{\sin\frac{n\pi}{\beta^*}+\frac{\beta^*}{n\pi}\frac{1}{r_s}\left(\cos\frac{n\pi}{\beta^*}-1\right)\right\}\right]\times s\tanh s\beta^*\cos sxds \tag{46}$$
$$\times\left\{\psi\left(t\right)-\frac{\varphi_0\left(s\right)}{F_0}\int_0^t\psi\left(\tau\right)\exp\left\{-\mu\left(t-\tau\right)/2\right\}\sin\frac{\varphi_0\left(s\right)}{F_0}\left(t-\tau\right)d\tau\right\}$$

and

$$p_r\left(x,y;t\right)=-\frac{2}{\pi}\frac{1}{\beta^*}\int_0^\infty\frac{\cosh s\left(y+\beta^*\right)}{\cosh s\beta^*}\left\{\frac{1}{s^2}+2\sum_{n=1}^{\infty}\frac{\cos n\pi/\beta^*}{s^2+\left(n\pi/\beta^*\right)^2}\right\}s\tanh s\beta^*\cos sxds$$
$$\times\frac{1}{F_0}\int_0^t\theta\left(\tau\right)\exp\left\{-\mu\left(t-\tau\right)/2\right\}\cos\frac{\varphi_0\left(s\right)}{F_0}\left(t-\tau\right)d\tau \tag{47}$$

where $\varphi_0(s)$ and F_0 are defined by the following expressions, respectively:

$$\varphi_0\left(s\right)\equiv\sqrt{s\tanh s\beta^*-\frac{\tilde{\mu}^2F_0^2}{4}}\quad\text{and}\quad F_0\equiv\sqrt{\frac{d_0}{g}}\Omega_a \tag{48}$$

F_0 is termed the basic Froude number, which represents the square root of the ratio of the inertia force $\rho(d_0\Omega_a)^2/d_0$ to the gravity force ρg.

Periodic Solution of Hydrodynamic Pressure

In order to calculate the hydrodynamic pressure when the weir plate undergoes periodic vibration at a frequency of Ω_w[rad/s], the dimensionless angular displacement of the skinplate, $\psi(t)$ and $\theta(t)$, are denoted by the following complex forms:

$$\psi\left(t\right)\equiv\exp\left\{i\Omega_w T\right\}=\exp\left\{i\left(F/F_0\right)t\right\}\quad\text{and}\quad\theta\left(t\right)\equiv\exp\left\{i\Omega_w T\right\}=\exp\left\{i\left(F/F_0\right)t\right\} \tag{49}$$

where the Froude number representing the dynamic similarity of the flow field, F, is again defined by

$$F\equiv\sqrt{\frac{d_0}{g}}\Omega_w \tag{50}$$

In this definition, the product of the representative length d_0 and the circular vibration frequency Ω_w is selected as the characteristic velocity.

When the periodic vibration of the weir plate is given in the form of Equation 49, the integral with respect to τ in Equations 46 and 47 can be reduced to the following forms:

$$\int_0^t \psi(\tau)\exp\{-\mu(t-\tau)/2\}\sin\frac{\varphi_0(s)}{F_0}(t-\tau)\,d\tau = \frac{F_0\varphi_0(s)}{W(s)}\exp\{i(F/F_0)t\} \tag{51}$$

and

$$\int_0^t \theta(\tau)\exp\{-\mu(t-\tau)/2\}\cos\frac{\varphi_0(s)}{F_0}(t-\tau)\,d\tau = \frac{i\,FF_0}{W(s)}\exp\{i(F/F_0)t\} \tag{52}$$

where the function $W(s)$ is given by:

$$W(s) = s\tanh s\beta^* - F^2\{1 - i(F_0/F)\mu\} \tag{53}$$

Substituting the integrated results into Equations 47 and 48, and expanding the integral on s to a complex integral, the dimensionless pressures can be shown to be of the following forms:

$$\begin{aligned} p_b(x,y;t) = {} & 2F^4\exp\left(i\frac{F}{F_0}t\right)\left[\sum_{j=1}^{\infty}\left\{\frac{1}{\beta^*\eta_j^2}\left(1-\frac{1}{2r_s}\right)\right.\right. \\ & \left.\left. +\frac{2}{\pi}\sum_{n=1}^{\infty}\frac{1}{n}\frac{1}{\eta_j^2-(n\pi/\beta^*)^2}\left\{\sin\frac{n\pi}{\beta^*}+\frac{\beta^*}{n\pi}\frac{1}{r_s}\left(\cos\frac{n\pi}{\beta^*}-1\right)\right\}\right\}\times\frac{\cos\eta_j(y+\beta^*)}{\cos\eta_j\beta^*}\frac{\exp(-\eta_j x)}{W'(i\eta_j)/i}\right. \\ & +i\left\{\frac{1}{\beta^*\xi_0^2}\left(1-\frac{1}{2r_s}\right)+\frac{2}{\pi}\sum_{n=1}^{\infty}\frac{1}{n}\frac{1}{\xi_0^2+(n\pi/\beta^*)^2}\right. \\ & \left.\left.\times\left\{\sin\frac{n\pi}{\beta^*}+\frac{\beta^*}{n\pi}\frac{1}{r_s}\left(\cos\frac{n\pi}{\beta^*}-1\right)\right\}\right\}\times\frac{\cosh\xi_0(y+\beta^*)}{\cosh\xi_0\beta^*}\frac{\exp(-i\xi_0 x)}{W'(\xi_0)}\right] \end{aligned} \tag{54}$$

and

$$p_r\left(x, y; t\right) = 2\frac{F^3}{\beta^*}\exp\left(i\frac{F}{F_0}t\right)\left[i\sum_{j=1}^{\infty}\left\{\frac{1}{\eta_j^2} + 2\sum_{n=1}^{\infty}\frac{\cos n\pi / \beta^*}{\eta_j^2 - \left(n\pi / \beta^*\right)^2}\right\}\frac{\cosh \eta_j\left(y + \beta^*\right)}{\cosh \eta_j\beta^*}\frac{\exp\left(-\eta_j x\right)}{W'\left(i\eta_j\right) / i}\right.$$
$$\left. - \left\{\frac{1}{\xi_0^2} + \frac{2}{\pi}\sum_{n=1}^{\infty}\frac{1}{n}\frac{1}{\xi_0^2 + \left(n\pi / \beta^*\right)^2}\right\}\frac{\cosh \xi_0\left(y + \beta^*\right)}{\cosh \xi_0\beta^*}\frac{\exp\left(-i\xi_0 x\right)}{W'\left(\xi_0\right)}\right] \quad (55)$$

where Cauchy's integral theorem has been applied. Additionally, the terms with the small frictional coefficient μ were disregarded at this stage. ζ_0 and η_j (j=1, 2, 3…), included in Equations 54 and 55, are the solutions to the following transcendental equations:

$$\xi_0 \tanh\left(\xi_0\beta^*\right) = F^2 \quad \text{and} \quad \eta_j \tan\left(\eta_j\beta^*\right) = -F^2\,(j = 1, 2, 3,\ldots) \quad (56)$$

As shown in the graphical representation of the solution in Figure 2a, the solution ξ_0 approaches 0, when $F^2\beta^*$ is comparatively small, and approaches F^2, as $F^2\beta^*$ increases. As shown in Figure 2b, the solution η_j (j = 1, 2, 3…) approaches $j\pi$, as the order of j increases. With these definitions for ξ_0 and η_j, the functions $W'(i\eta_j)$ and $W'(\xi_0)$ included in Equations 54 and 55, are given, respectively, by the following expressions:

$$W'\left(i\eta_j\right) / i = \tan \eta_j\beta^* + \frac{\eta_j\beta^*}{\cos^2 \eta_j\beta^*}$$

Figure 2. Solutions to the transcendental equations; (a) ξ_0 tanh ($\xi_0\beta$) = F^2, and (b) η_j tan ($\eta_j\beta$*) = -F^2*
From Anami et al. (2000a). © 2000 JSME. Used with permission.

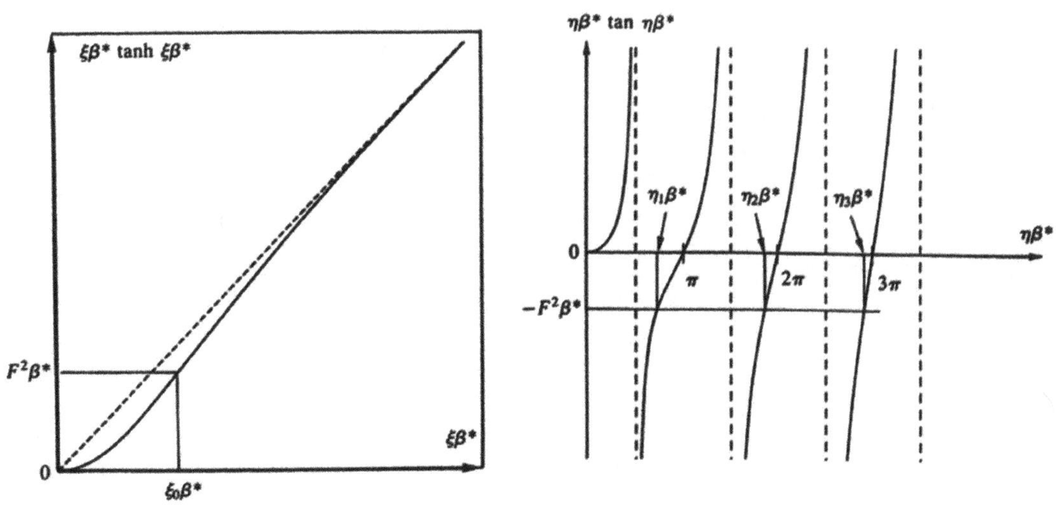

$$W'\left(\xi_0\right) = \tanh \xi_0 \beta^* + \frac{\xi_0 \beta^*}{\cosh^2 \xi_0 \beta^*} \tag{57}$$

When the gate motion is given by the real part of Equation 49, that is:

$$\psi\left(t\right) = \cos \frac{F}{F_0} t \quad \text{and} \quad \theta\left(t\right) = \cos \frac{F}{F_0} t \tag{58}$$

the real parts of Equations 54 and 55 are the resulting periodic hydrodynamic pressure and are represented by the following expressions:

$$p_b\left(x, y; t\right) = p_{bs}\left(x, y\right) \cos\left(\frac{F}{F_0} t - \pi\right) + p_{bp}\left(x, y\right) \cos\left(\frac{F}{F_0} t - \xi_0 x + \frac{\pi}{2}\right) \tag{59}$$

$$p_r\left(x, y; t\right) = p_{rs}\left(x, y\right) \cos\left(\frac{F}{F_0} t - \frac{\pi}{2}\right) + p_{rp}\left(x, y\right) \cos\left(\frac{F}{F_0} t - \xi_0 x - \pi\right) \tag{60}$$

where the first term represents the standing-wave pressure component, and the second term the progressive-wave pressure component. The pressure amplitudes $p_{bs}(x, y)$, $p_{bp}(y)$, $p_{rs}(x, y)$ and $p_{rp}(y)$ are given in the following series summations:

$$\begin{aligned} p_{bs}\left(x, y\right) = -2F^4 \sum_{j=1}^{\infty} \left\{ \frac{1}{\beta^* \eta_j^2} \left(1 - \frac{1}{2r_s}\right) + \frac{2}{\pi} \sum_{n=1}^{\infty} \frac{1}{n} \frac{1}{\eta_j^2 - \left(n\pi / \beta^*\right)^2} \right. \\ \left. \times \left\{ \sin \frac{n\pi}{\beta^*} + \frac{\beta^*}{n\pi} \frac{1}{r_s} \left(\cos \frac{n\pi}{\beta^*} - 1\right) \right\} \right\} \frac{\cos \eta_j \left(y + \beta^*\right)}{\cos \eta_j \beta^*} \frac{\exp\left(-\eta_j x\right)}{W'\left(i\eta_j\right) / i} \end{aligned} \tag{61}$$

$$\begin{aligned} p_{bp}\left(y\right) = 2F^4 \left\{ \frac{1}{\beta^* \xi_0^2} \left(1 - \frac{1}{2r_s}\right) + \frac{2}{\pi} \sum_{n=1}^{\infty} \frac{1}{n} \frac{1}{\xi_0^2 + \left(n\pi / \beta^*\right)^2} \right. \\ \left. \times \left\{ \sin \frac{n\pi}{\beta^*} + \frac{\beta^*}{n\pi} \frac{1}{r_s} \left(\cos \frac{n\pi}{\beta^*} - 1\right) \right\} \right\} \frac{\cosh \xi_0 \left(y + \beta^*\right)}{\cosh \xi_0 \beta^*} \frac{1}{W'\left(\xi_0\right)} \end{aligned} \tag{62}$$

$$p_{rs}\left(x,y\right)=-2\frac{F^3}{\beta^*}\sum_{j=1}^{\infty}\left\{\frac{1}{\eta_j^2}+2\sum_{n=1}^{\infty}\frac{\cos n\pi/\beta^*}{\eta_j^2-\left(n\pi/\beta^*\right)^2}\right\}\frac{\cosh\eta_j\left(y+\beta^*\right)}{\cosh\eta_j\beta^*}\frac{\exp\left(-\eta_j x\right)}{W'\left(i\eta_j\right)/i} \tag{63}$$

$$p_{rp}\left(y\right)=2\frac{F^3}{\beta^*}\left\{\frac{1}{\xi_0^2}+\frac{2}{\pi}\sum_{n=1}^{\infty}\frac{1}{n}\frac{1}{\xi_0^2+\left(n\pi/\beta^*\right)^2}\right\}\frac{\cosh\xi_0\left(y+\beta^*\right)}{\cosh\xi_0\beta^*}\frac{1}{W'\left(\xi_0\right)} \tag{64}$$

Comparing Equations (59) and (60) with the weir plate vibration (Equation 58), the first term in Equation 59, the standing-wave pressure component of p_b, is seen to have a 180° phase-lag, resulting in an added mass effect on the weir plate vibration, while the second term in Equation 59, the progressive-wave pressure component of p_b, is seen to have a 90° phase-shift at $x = 0$, resulting in a damping effect on the weir plate vibration. Moreover, the first term in Equation 60, the standing-wave pressure component of p_r has a -90° phase-lag, resulting in excitation while the second term in Equation 60, the progressive-wave pressure component of p_r, has a 180° phase shift at $x = 0$, again resulting in an added mass effect.

Hydrodynamic pressure measured in experiments is the result of superposition of these terms. The following expressions are obtained as the results of superposition of Equations 59 and 60:

$$p_b = p_{b0}\cos\left(\frac{F}{F_0}t-\varphi_b\right) \tag{65}$$

$$p_r = p_{r0}\cos\left(\frac{F}{F_0}t-\varphi_r\right) \tag{66}$$

where the pressure amplitude, p_{b0} and p_{r0}, and phase-lag, φ_b and φ_r, are obtained by the following expressions, respectively.

$$p_{b0}=\sqrt{p_{bs}^2+p_{bp}^2}\quad\text{and}\quad\varphi_b=\tan^{-1}\frac{p_{bp}}{p_{bs}}$$

$$p_{r0}=\sqrt{p_{rs}^2+p_{rp}^2}\quad\text{and}\quad\varphi_r=-\tan^{-1}\frac{p_{rs}}{p_{rp}} \tag{67}$$

Major Characteristics of Push-and-Draw Pressure

The push-and-draw hydrodynamic pressure acting on the weir plate can be calculated by substituting $x = 0$ into Equation 59. Figure 3 shows the resulting pressure for a depth ratio $\beta^*=1.0$, and a reduced rotation center height $r_s=1.0$, which means the rotation center is located at the water surface. The parameter is the Froude number F which ranged from 5 to 35. The amplitude of the progressive-wave pressure component, p_{bp}, given by Equation 62, is shown in Figure 3a, and that of the standing-wave pressure component, p_{bs}, given by Equation 61, is shown in Figure 3b. The resultant vector magnitude of the hydrodynamic pressure of the progressive and standing pressure components, p_{b0}, given by Equation 67, is shown in Figure 3c. The corresponding phase lag φ_b relative to the counter-clockwise vibration of the weir plate is shown in Figure 3d. In all of these figures, the ordinate y represents the reduced depth along the weir plate. The standing pressure component increases rapidly as the y position approaches the lower end of the weir plate. On the other hand, progressive pressure amplitude p_{bp} is quite small compared to the standing pressure amplitude. Moreover, the small amplitude appears only near the water surface. It decreases rapidly and is very close to zero at greater depths. Even for a Froude number F of 5, the influence of progressive pressure extends only to $y = -0.2$. When the Froude number is smaller than 5, the progressive pressure component becomes larger than the standing pressure component. Since, however, the Froude number F generally takes on comparatively large values for vibrations of gates, the calculated results for small Froude number are not shown.

Since the progressive pressure amplitude is small, the resultant pressure amplitude p_{b0} exhibits much the same distribution as the standing pressure amplitude, as shown in Figure 3c. The phase-lag φ_b shown in Figure 3d is 180° except near the water surface. Since standing pressure amplitude p_{bs} of Figure 3b is negative near the water surface, phase-lag ϕ_b at the water surface ($y = 0$) approaches 360°. The 180° phase lag of the push-and-draw pressure indicates that it reaches its maximum value when the weir plate is at its most downstream position. Therefore, since the push-and-draw pressure acts in opposition to the movement of the weir plate, it acts effectively as an added mass for the vibration of the weir plate.

The magnitude of the added mass become more extreme as the Froude number F increases. The standing pressure amplitude p_{bs} becomes even larger, and progressive pressure amplitude p_{bp} becomes extremely small. Therefore, the resultant pressure amplitude p_{b0} becomes identical to the standing pressure amplitude p_{bs}, and phase-lag φ_b is 180° over essentially the entire domain ($y = -0.03$ to $y = -1.0$).

Another sample calculation is shown in Figure 4, for a Froude number $F = 35$ and a depth ratio $\beta^*=1.0$ (corresponding to the conditions at the Folsom Dam gate). The reduced height r_s is shown as a parameter, since it depends upon the structural dynamics of the gate. The hydrodynamic pressure for $r_s \to \infty$ has been calculated by Ishii (1992) and is shown as the dashed line in Figure 4.

The progressive-wave pressure component appears only near the water surface, and its amplitude is small enough to neglect, relative to that of the standing pressure-wave component due to the relatively high Froude number $F = 35$ used in these calculations. Therefore, one may conclude that the progressive pressure-wave component itself is of no practical significance in this situation. In other words, the damping effect due to fluid motion is very small, since the progressive pressure wave can dissipate very little energy at such small amplitudes as it propagates in the upstream direction. It should be noted, however, that the progressive pressure takes on larger amplitudes, at smaller values of Froude number.

The standing pressure amplitude, on the other hand, takes on quite a large value, far larger than that of the progressive pressure. Independent of the parameter r_s, the standing pressure amplitude near the water surface becomes very small, thus suggesting that almost no standing waves will appear on the

Figure 3. Hydrodynamic pressure induced by a rotating flat-and-rigid weir plate, for r_s=1.0 and β=1.0; (a) amplitude of the progressive-wave pressure component, (b) amplitude of the standing-wave pressure component, (c) resultant pressure amplitude, and (d) phase-lag relative to the gate rotation*
After Anami & Ishii (2001).

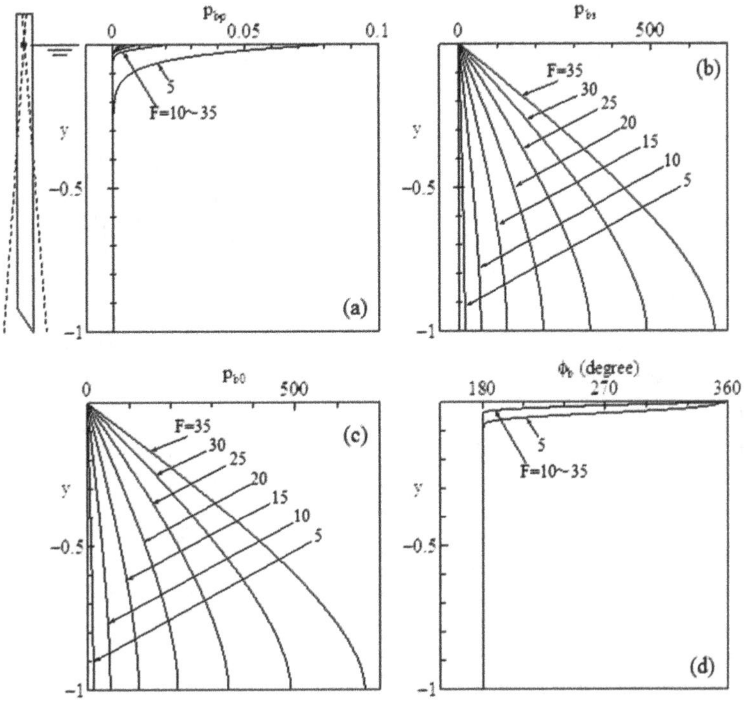

water surface. However, a large standing pressure arises in the water beneath the free surface. When the rotation center is above the water surface (r_s > 1.0), the standing pressure rapidly increases in amplitude with increasing submergence, reaching its maximum value at the lower end of the weir plate. The amplitude of this pressure maximum gradually decreases with lowered rotation center position. When the rotation center is just at the water surface level (r_s = 1.0), the maximum amplitude is 73% (the value is 664) of the value corresponding to uniform parallel translation ($r_s = \infty$; shown by dotted line). When the rotation center lies beneath the water surface (r_s < 1.0), the standing pressure initially decreases with increasing submergence (yielding negative values), and subsequently increases, again exhibiting a maximum amplitude at the lower end of the weir plate. This maximum value continues to decrease with a further lowering of the rotation center position. From the first term on the right hand side of Equation 57, the negative amplitude of the standing pressure suggests that the phase-lag between the pressure and the weir plate motion becomes zero. Thus, the standing pressure fluctuates in phase with the counter-clockwise vibration of the weir plate. Therefore, with a submerged rotation center, the phase of the standing pressure above and below the rotation center will differ by 180°.

The resultant vector magnitude of hydrodynamic pressure p_{b0} of the progressive and standing pressure components is shown in Figure 4c. The pressure amplitude increases continuously with depth when the rotation center is above the water surface (r_s > 1.0). When the rotation center is located beneath the surface (r_s < 1.0), the amplitude first increases, then decreases to zero, and then increases again. The depth at which the pressure amplitude goes to zero does not agree with the rotation center position; the

Figure 4. Hydrodynamic pressure induced by rotating flat-and-rigid weir plate, for $r_s > 0.5$, $F = 35$ and $\beta^* = 1.0$; (a) the amplitude of the progressive-wave pressure component, (b) the amplitude of the standing-wave pressure component, (c) the resultant pressure amplitude, and (d) the phase-lag

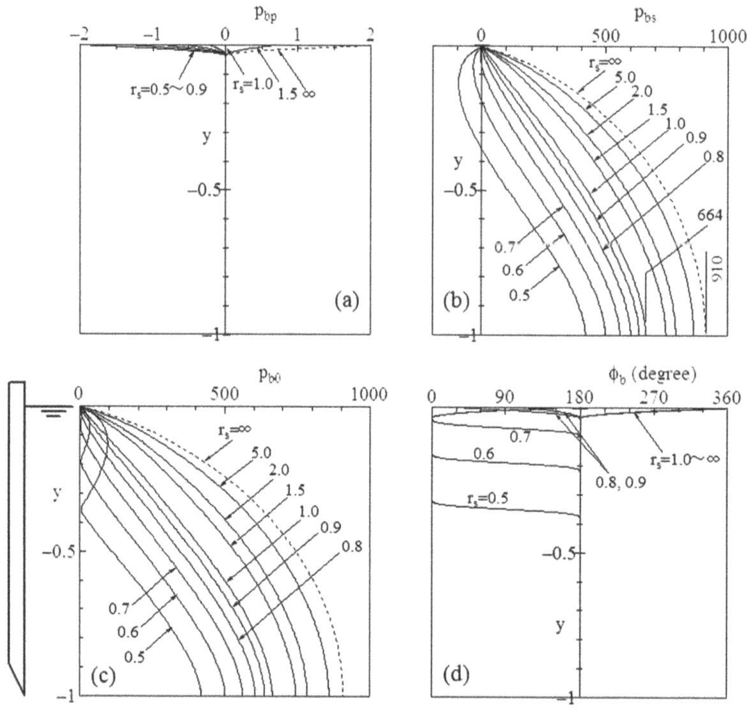

zero pressure location appears at a location slightly above that of the rotation center. For instance, the rotation center for $r_s = 0.5$ is located at $y = -0.5$, while the depth for zero pressure amplitude is at $y = -0.36$. This difference is the influence of the very large pressure over the lower portion of the weir plate. The phase relative to the counter-clockwise movement of the weir plate is always out-of-phase (180°) for $r_s > 1.0$, while for $r_s < 1.0$, it is out-of-phase (180°) on the lower portion of the weir plate and is in-phase (0°) on the upper portion of the weir plate. These phase-lag characteristics indicate that the hydrodynamic pressure amplitude reaches its maximum, when the flow near the weir plate is most decelerated.

The general shape of the hydrodynamic pressure distribution along the weir plate is relatively unchanged by changes in the Froude number F, as shown in Figure 3. The scaling of changes in the distribution with F can be characterized by considering the maximum pressure amplitude at the bottom edge of the weir plate. This maximum pressure amplitude is plotted as a function of the Froude number F in Figure 5, where the water depth ratio β^* was fixed at 1.0. The parameter in Figure 5 is the reduced rotation center height r_s. The pressure amplitude increases approximately quadratically with Froude number F. For an r_s value of 1.0, for instance, the reduced pressure amplitude was 217 at $F = 20$. With a doubling of the Froude number to $F = 40$, the pressure amplitude has grown by almost four-fold, taking on a value of 868. With a reduced pressure amplitude of 1360 at F=50, a slight vibration amplitude of 1 mm at the bottom end of the weir plate will result in a hydrodynamic pressure head that is 1360 times this displacement, that is, 1.36 m water head pressure. As the rotation center position moves to higher locations, the maximum pressure amplitude increases correspondingly.

Figure 5. Effect of Froude number upon the maximum hydrodynamic pressure at the lower end of the weir plate; water depth ratio β=1.0*
From Anami et al (2000a). © 2000 JSME. Used with permission.

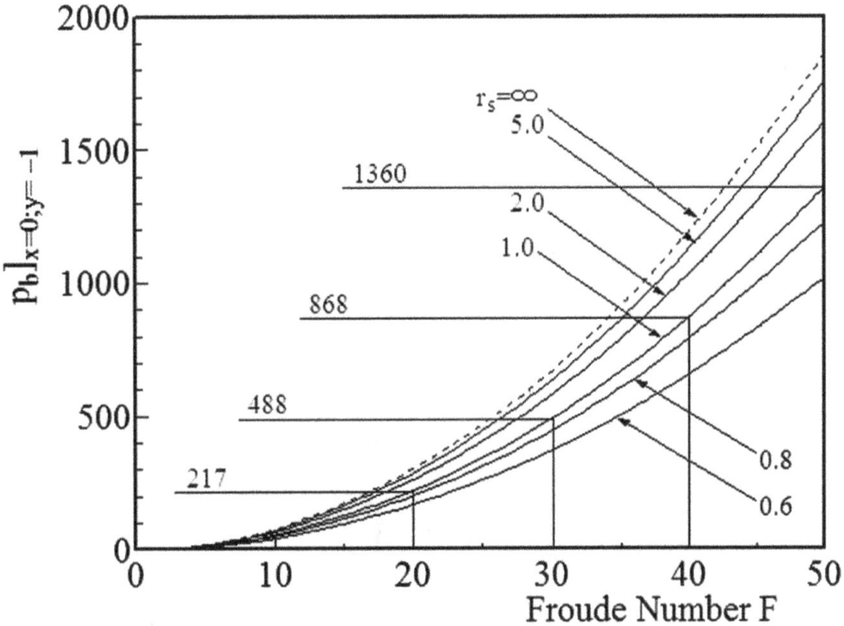

The effect of the reservoir depth on the hydrodynamic pressure is shown in Figure 6, where the Froude number F was fixed at 30 and the reduced rotation center height r_s at 1.0. The parameter in the figure is the reduced reservoir depth as defined by the ratio of the reservoir depth to the submergence depth of weir plate bottom edge, β^*. The hydrodynamic pressure consistently decreases, distinctly near the weir plate lower end, with increasing β^*. This is because the spreading of flow field into the reservoir increases with increasing reservoir depth, thus weakening the pressure build-up effect due to the weir plate vibration. For the same reason, the depth at which the pressure maximum occurs also moves upward for the weir plate bottom edge with increasing reservoir depth, as is shown by the dashed line in Figure 6. The maximum pressure of 362 at $\beta^*=2.0$ is about 74% of the pressure maximum (488) corresponding to a reservoir with $\beta^*=1.0$.

Major Characteristics of Flow-Rate Variation Pressure

The flow-rate variation pressure acting on the weir plate can be calculated by substituting $x = 0$ into Equation 60. Figure 7 shows the result for the depth ratio $\beta^*=1.0$. The parameter in Figure 7 is the Froude number F, which varies from 5 to 35. The amplitude of the progressive pressure-wave component, p_{rp}, is shown in Figure 7a, and that of the standing-wave pressure component, p_{rs}, is shown in Figure 7b. The resultant vector magnitude of hydrodynamic pressure of the progressive and standing pressure components, p_{r0}, is shown in Figure 7c. The corresponding phase lag ϕ_r relative to the downward vibration of the weir plate is shown in Figure 7d.

The progressive pressure amplitude p_{rp} takes on a value even smaller than the push-and-draw pressure. A very small pressure amplitude appears only near the water surface. The standing pressure amplitude

Figure 6. Effect of the depth ratio on the hydrodynamic pressure: Froude number F=30; reduced rotation center height r_s=1.0
From Anami et al. (2000a). © 2000 JSME. Used with permission.

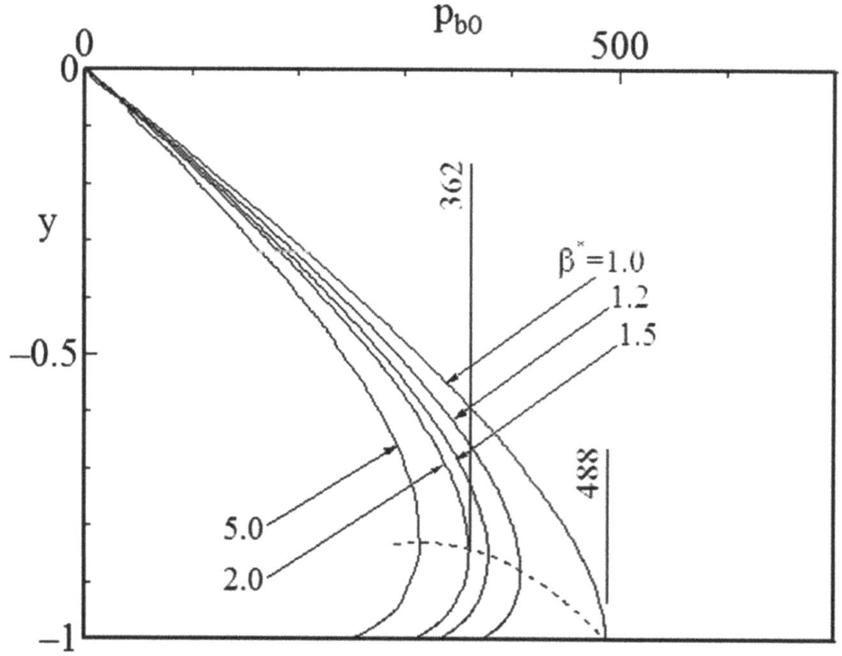

p_{rs} rapidly increases toward the bottom end of the weir plate, although it is almost zero near the water surface. Since the progressive pressure amplitude is small, the resultant hydrodynamic pressure p_{r0} has almost the same distribution as the standing pressure amplitude.

The phase-lag is 90° except near the water surface. Since the progressive pressure amplitude p_{rs} takes on a negative value near the water surface, the phase-lag φ_b at the water surface (y=0) approaches 270°. The 90° phase-lag shows that the flow-rate variation pressure reaches its maximum value, when the weir plate descends and passes the mid-point. Although the amplitude is quite small relative to the push-and-draw pressure, it has a great significance, none the less. If the weir plate is moving in the downstream direction (in addition to the up and down motion) at the time it reaches the midpoint of downward motion, this flow-rate variation pressure will supply energy to the horizontal movement of the weir plate, and the weir plate amplitude increases. In contrast, if the weir plate is moving in the upstream direction at the instant it reaches the midpoint of the downward motion, the flow-rate variation pressure will consume energy from the horizontal weir plate movement, damping the horizontal vibration.

Added Mass Coefficient and Added Damping Coefficient

The two-dimensional model of a weir plate and a flow field to which the following discussion pertains was previously shown in Figure 1. The discharge opening is in the closed state and the vertical rigid weir plate performs a rotational vibration in the streamwise direction. The counterclockwise rotation angle is represented by Ψ. The push-and-draw pressure arises in the flow field of upstream reservoir. The hydrodynamic pressure acting on the weir plate is represented by P_b. The Y-axis is taken as positive upward

Figure 7. Hydrodynamic pressure induced by up and down vibrational weir plate, for β=1.0; (a) amplitude of the progressive pressure-wave component, (b) amplitude of the standing pressure-wave component, (c) resultant pressure amplitude, and (d) phase-lag*
From Anami & Ishii (2001).

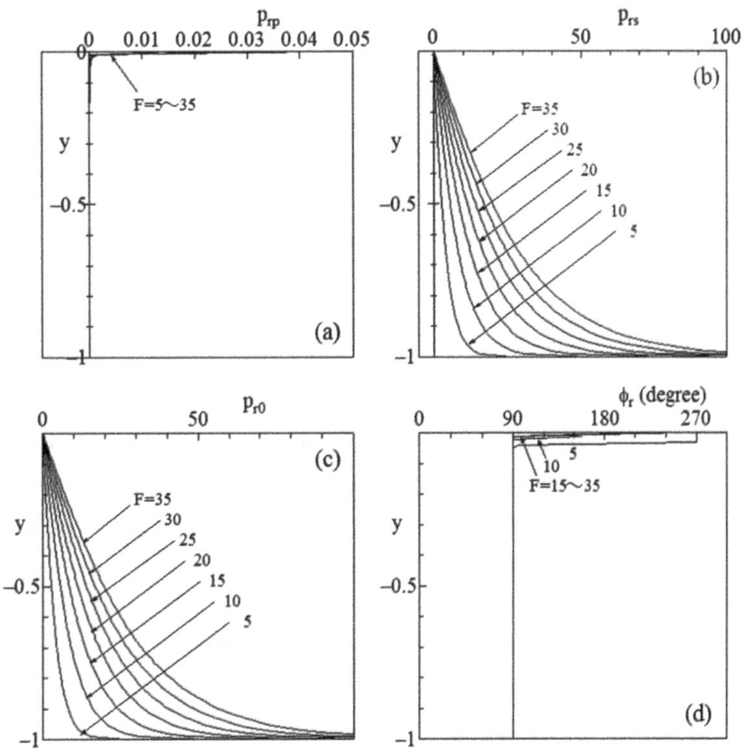

along the vertical weir plate, from the still water level. The weir plate submergence depth is represented by d_0 and the rotation center height from the weir plate bottom by R_S, resulting in the distance from the center of hydrodynamic pressure to the rotation center of $\{-Y-(d_0-R_S)\}$. Therefore, the equation of motion of the vertical rigid weir plate performing rotational vibration with a small amplitude can be written as:

$$I_\psi \ddot{\Psi} + C_\psi \dot{\Psi} + K_\psi \Psi + W_0 \int_{-d_0}^{0} P_b \left\{-Y - \left(d_0 - R_S\right)\right\} dY = 0 \tag{68}$$

Dividing this whole equation by the moment of inertia around the rotation center, I_ψ, the following equation can be derived:

$$\ddot{\Psi} + 2\zeta_a \Omega_a \dot{\Psi} + \Omega_a^2 \Psi + \frac{W_0}{I_\psi} \int_{-d_0}^{0} P_b \left\{-Y - \left(d_0 - R_S\right)\right\} dY = 0 \tag{69}$$

where Ω_a and ζ_a represent the in-air natural vibration frequency and in-air damping ratio, respectively. W_0 is the span of the weir plate.

Using the dimensionless variables defined in Equation 44, the equation of motion, Equation 69, can be reduced to the following dimensionless form:

$$\psi'' + 2\zeta_a\psi' + \psi + \frac{\alpha_\psi}{F_0^2}\int_{-1}^{0}\left(1 - \frac{1+y}{r_s}\right)p_b dy = 0 \tag{70}$$

where the prime represents differentiation with respect to dimensionless time t. F_0 is the basic Froude number representing the dynamic similarity of the flow field, when $d_0\Omega_a$ is selected as the representative speed. The parameter α_ψ represents the water-to-gate mass ratio, defined by

$$\alpha_\psi \equiv \frac{\rho d_0^2 W_0}{I_\psi / R_S^2} \tag{71}$$

in which $\rho d_0^2 W_0$ in the numerator is the representative mass of water in a rectangular parallelepiped given by the product of the square of one side d_0 and the span W_0. The term I_ψ / R_S^2 in the denominator is the equivalent mass determined by the moment of inertia of the weir plate around its rotation center, I_ψ, divided by the square of rotation center height R_S. Thus, larger values of this water-to-gate mass ratio correspond to lighter weight designs in which the water mass is larger than the weir plate mass.

Assuming the weir plate performs periodic rotational vibration given by

$$\Psi = \Psi_0 \cos\Omega_w T \qquad \text{or} \qquad \psi = \psi_0 \cos\frac{F}{F_0}t \tag{72}$$

the reduced hydrodynamic pressure p_b acting on the weir plate can be obtained by setting $x = 0$ in Equation 59:

$$p_b(0, y; t) = p_{bs}(0, y)\cos\left(\frac{F}{F_0}t - \pi\right) + p_{bp}(y)\cos\left(\frac{F}{F_0}t + \frac{\pi}{2}\right) \tag{73}$$

(F/F_0) represents the in-water to in-air vibration frequency ratio, as explained in Anami et al. (1999, 2000b), Ishii (1990a, 1990b), where F_0 and F were previously defined in Equations 48 and 50, respectively:

$$\frac{F}{F_0} = \frac{\sqrt{d_0 / g}\,\Omega_w}{\sqrt{d_0 / g}\,\Omega_a} = \frac{\Omega_w}{\Omega_a} \tag{74}$$

The cosine functions $\cos\{(F/F_0)t-\pi\}$ and $\cos\{(F/F_0)t + \pi/2\}$ in Equation 73 can be replaced by an acceleration term $(F_0/F)^2\psi''$ and a velocity term $(F_0/F)\psi'$, respectively, permitting the hydrodynamic pressure p_b to be reduced to

$$p_b\left(0,y;t\right)=p_{bs}\left(0,y\right)\left(\frac{F_0}{F}\right)^2\psi''+p_{bp}\left(y\right)\left(\frac{F_0}{F}\right)\psi' \tag{75}$$

Substituting this expression into Equation 70, the equation of motion can be further reduced to

$$\begin{aligned}&\left\{1+\alpha_\psi\frac{1}{F^2}\int_{-1}^{0}\left(1-\frac{1+y}{r_s}\right)p_{bs}\left(0,y\right)dy\right\}\psi''\\&+\left\{2\zeta_a+\frac{\alpha_\psi}{F_0}\frac{1}{F}\int_{-1}^{0}\left(1-\frac{1+y}{r_s}\right)p_{bp}\left(y\right)dy\right\}\psi'+\psi=0\end{aligned} \tag{76}$$

However, it is also possible to re-write the equation of motion, Equation 70, in a different form by denoting the added moment-of-inertia by ΔI and the wave-radiation-damping moment coefficient by ΔC, as given by Equation 77 below.

$$\left(1+\frac{\Delta I}{I_\psi}\right)\psi''+2\left(\zeta_a+\frac{\Delta C}{2\Omega_a}\right)\psi'+\psi=0 \tag{77}$$

Since the acceleration and velocity terms in Equation 77 must be in agreement with the corresponding terms in Equation 76, ΔI and ΔC must then be given by the following expressions:

$$\Delta I=\rho d_0^2W_0R_S^2\frac{1}{F^2}\int_{-1}^{0}\left(1-\frac{1+y}{r_s}\right)p_{bs}\left(0,y\right)dy \tag{78}$$

$$\Delta C=\rho d_0^2W_0R_S^2\frac{1}{FF_0}\int_{-1}^{0}\left(1-\frac{1+y}{r_s}\right)p_{bp}\left(y\right)dy \tag{79}$$

As a result, the dimensional parameters in the above expressions can be selected as representative scaling quantities in reducing ΔI and ΔC to their non-dimensional forms. After substitution, ΔI and ΔC can be non-dimensionalized as

$$\Delta m_\psi\equiv\frac{\Delta I/R_S^2}{\rho d_0^2W_0}\qquad\text{and}\qquad\Delta c_\psi\equiv\frac{\Delta C/R_S^2}{\rho g d_0^2W_0/\sqrt{gd_0}} \tag{80}$$

where the definitions are arranged to help understand of their physical meanings. In the first definition, the equivalent added mass $\Delta I/R_S^2$ is divided by the representative water mass $\rho d_0^2W_0$. Therefore, Δm_ψ is called the reduced equivalent added mass. In the second definition, the equivalent damping coefficient

$\Delta C/R_S^2$ is divided by the ratio of the gravity force acting on the representative water mass $\rho g d_0^2 W_0$ to the velocity of a long wave $\sqrt{gd_0}$. Thus, Δc_ψ is called the reduced wave-radiation damping coefficient. Finally, Δm_ψ and Δc_ψ are given in the form of the following integral expressions for the standing and progressive pressure amplitudes:

$$\begin{aligned} \Delta m_\psi\left(F\right) &= \frac{1}{F^2}\int_{-1}^{0}\left(1-\frac{1+y}{r_s}\right)p_{bs}\left(0,y\right)dy \\ \Delta c_\psi\left(F\right) &= \frac{1}{F}\int_{-1}^{0}\left(1-\frac{1+y}{r_s}\right)p_{bp}\left(y\right)dy \end{aligned} \tag{81}$$

Both Δm_ψ and Δc_ψ are functions of the Froude number F. As indicated in the second expression, the wave-radiation damping coefficient results from the progressive wave component. Therefore, the interpretation of the wave-radiation damping coefficient, introduced here, is that it represents the energy dissipation by progressive-waves traveling in the upstream direction.

The calculated reduced equivalent added mass and the reduced wave-radiation damping coefficient for a representative sample case are shown in Figure 8. Given the Froude number F, the solutions ξ_0 and η_j (j = 1, 2,...) of the transcendental Equations 56 can be calculated by Newton's method. Then, the series summations in Equations 61 and 62 were calculated, and Δm_ψ and Δc_ψ given by Equation 81 could then be calculated. Calculations were made for $d_0 = d_r$, that is, depth ratio $\beta^* = 1.0$, since full-scale weir plates are installed, in many cases, over a river bed with no step changes in the sill, or on a dam crest. The abscissa is the Froude number F and the parameter is the reduced rotation center height r_s.

The equivalent added mass Δm_ψ takes on a small value when the Froude number is small, and gradually increases with increasing F, and then approaches an asymptotic constant value for Froude numbers larger than about 10. The asymptotic value decreases with decreasing reduced rotation center height r_s. When $r_s = \infty$, the rotation center is infinitely above the water surface, and the weir plate undergoes a uniform parallel vibration in the streamwise direction. Such a case was studied by Ishii (1990a, 1990b, 1992) and the equivalent added mass for uniform streamwise vibration of the weir plate is shown by the dashed line which asymptotes to the value of 0.54 for large Froude numbers. In the case when $r_s = 1.0$, the rotation center is at the water surface level, the asymptotic value decreases to 0.21. This value is only about 39% of the value 0.54 when the weir plate performs uniform parallel vibration. Thus, the reduced rotation center height has a significant influence on the reduced equivalent added mass.

In contrast to the added mass characteristic, the wave-radiation damping coefficient Δc_ψ takes on a large value when the Froude number is small, and rapidly decreases with increasing the Froude number and approaches zero for the Froude numbers larger than about 10, for all rotation center heights. This characteristic can be explained when one recalls that the wave-radiation damping coefficient is induced by the progressive pressure waves, as demonstrated in the second expression in Equation 78. The reduced wave velocity, c, of the progressive pressure-wave is given by

$$c\left(\equiv\frac{C}{d_0\Omega_a}\right)=\frac{1}{\xi_0}\cdot\frac{F}{F_0}\to\frac{1}{F_0F} \tag{82}$$

Figure 8. Reduced equivalent added mass Δm_{ψ} and reduced wave-radiation damping coefficient Δc_{ψ}, calculated for a vertical, flat, rigid weir plate undergoing rotational vibration
Anami & Ishii (2001).

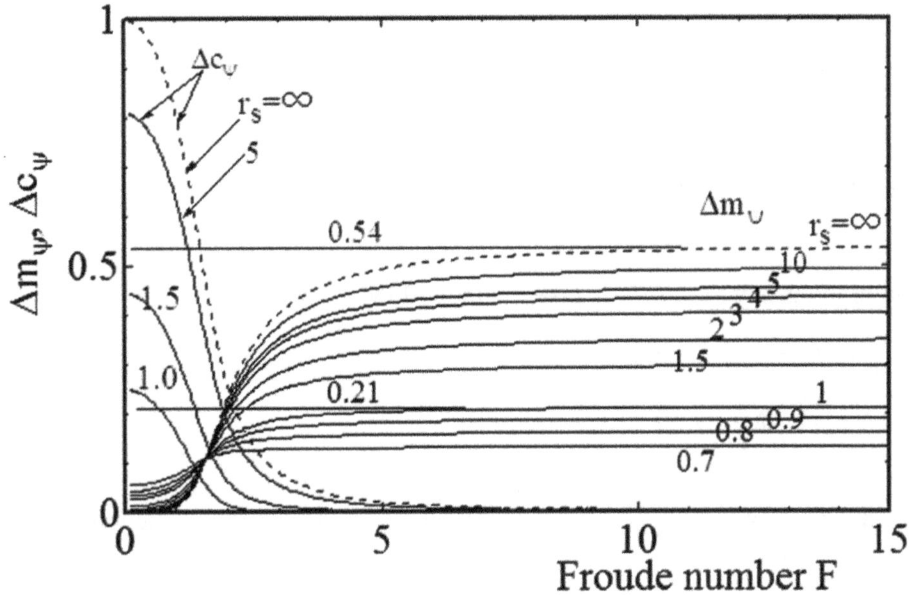

and becomes smaller with increasing Froude numbers. Thus, the energy supplied to the fluid by the weir plate motion cannot be effectively dissipated in the form of progressive-wave pressures in the upstream direction. The energy dissipation with wave velocity decreasing as in Equation 82 is further divided by the Froude number F, resulting in the previously noted significant decrease in wave-radiation damping coefficient for increasing Froude number.

In-Water to In-Air Frequency Ratio

In contrast to the progressive wave characteristics, when F becomes moderately large, the standing pressure-wave component of the hydrodynamic pressure is proportional to the square of F. This is because the smaller the progressive-wave velocity becomes, the more effectively the standing wave pressure component can store the energy transferred to the fluid by the weir plate motion. From the first expression in Equation 78, the ΔI term (which subsequently will form the equivalent added mass $\Delta I/R_S^2$) is divided by the square of F yielding an equivalent added mass that asymptotically approaches a constant value.

Substituting the expressions in Equation 81 into the equation of motion, Equation 76, yields the following equation:

$$\left(1+\alpha_{\psi}\Delta m_{\psi}\right)\psi''+2\left(\zeta_a+\zeta_f\right)\psi'+\psi=0 \tag{83}$$

where ζ_f is the wave-radiation damping ratio, defined by

$$\zeta_f\left(F\right) \equiv \alpha_\psi \frac{\Delta c_\psi\left(F\right)}{2F_0} \tag{84}$$

This wave-radiation damping ratio is a function of the Froude number. Here assume the solution of the equation of motion, Equation 83, is given by

$$\psi\left(t\right) = \exp\left\{-\Omega_w\zeta_w + i\Omega_w\right\}T = \exp\left(-\zeta_w + i\right)\left(\frac{F}{F_0}\right)t \tag{85}$$

The vibration frequency ratio F/F_0 is given in the following form:

$$\frac{F}{F_0}\left(=\frac{\Omega_w}{\Omega_a}\right) = \sqrt{\frac{1}{1+\alpha_\psi \Delta m_\psi\left(F\right)} - \zeta_w^{\ 2}} \tag{86a}$$

Assuming the damping ratio ζ_w is sufficiently small ($\zeta_w << 1.0$), F/F_0 can be reduced to

$$\frac{F}{F_0} = \frac{1}{\sqrt{1+\alpha_\psi \Delta m_\psi\left(F\right)}} \tag{86b}$$

In addition, the damping ratio ζ_w can be given by

$$\zeta_w = \frac{\zeta_a + \zeta_f}{\sqrt{1+\alpha_\psi \Delta m_\psi\left(F\right)}} \tag{87}$$

Given the basic Froude number F_0, the Froude number F can be numerically calculated from Equation 86, while the damping ratio ζ_w can be calculated from Equation 87.

The standing pressure-wave component $p_{bs}(0,y)$ in the equivalent added mass Δm_ψ in the first expression of Equation 81 is a quite complicated series summation of the Froude number, as shown in Equation 61. Because of this complicated summation, it is difficult to obtain the Froude number F for a given basic Froude number F_0 in numerical calculations to obtain the solution of Equation 86. However, conversely, it is comparatively easy to calculate the basic Froude number F_0 from Equation 86, for a given Froude number F. Of course, then the specified Froude number F is the solution obtained for the basic Froude number F_0 and in this manner the corresponding pairs of values of F and F_0 are obtained. Thus, the calculated results for the vibration frequency ratio F/F_0 as a function of F_0 are shown in Figure 9a, for $r_s = 1.0$ indicating that the rotation center height of weir plate is at the still water surface level. Figure 9b shows the calculated results for the wave-radiation damping ratio ζ_f due to fluid motion. The depth ratio β^* of the reservoir to the gate submergence was fixed at 1.0. The abscissa shows the basic Froude number F_0, and the parameter is the water-to-gate mass ratio α_ψ.

Figure 9. Characteristics of in-water to in-air frequency ratio and fluid damping ratio for $\beta^*=1.0$, $r_s=1.0$: (a) in-water to in-air vibration frequency ratio, (b) fluid damping ratio
From Anami & Ishii (2001).

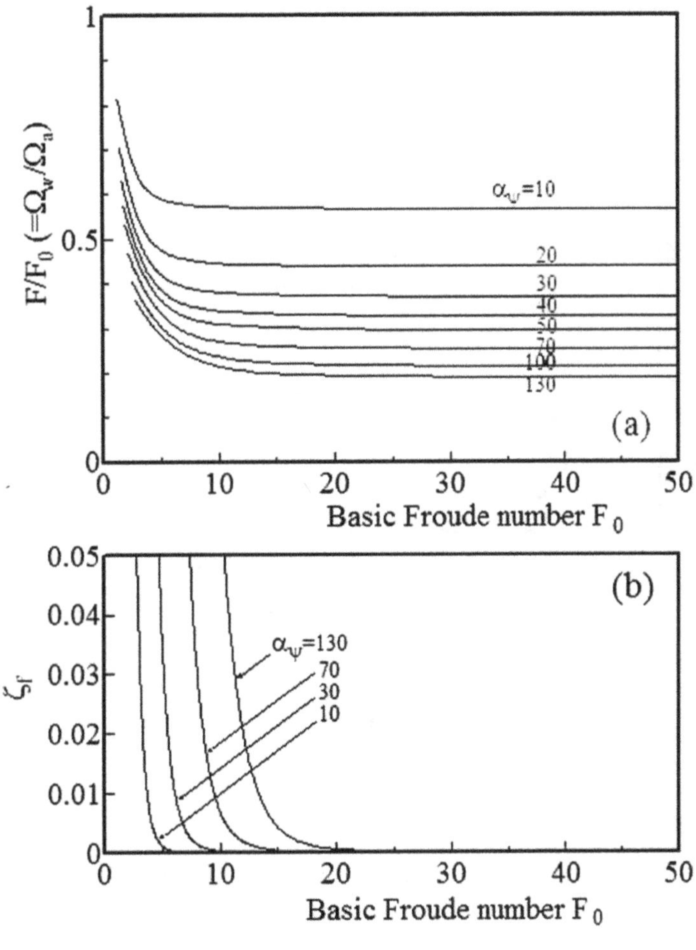

As the basic Froude number F_0 increases, the vibration frequency ratio F/F_0 rapidly decreases, approaching a constant value for basic Froude numbers F_0 larger than about 20. The asymptotic value decreases with increasing water-to-gate mass ratio. Therefore, for increasingly lightweight design (high water-to-gate mass ratio), the in-air vibration frequency is increasingly reduced when the gate is placed in flowing water.

As shown in Figure 9b, the wave-radiation damping ratio ζ_f takes on a comparatively large value for small basic Froude numbers F_0. As the basic Froude number F_0 increases, however, ζ_f rapidly decreases. If the basic Froude number F_0 is larger than about 20, ζ_f is not significantly different from zero, even if the mass ratio is 130. The significant inference of this result is that there is no damping effect due to wave-radiation upon the weir plate vibration, when the basic Froude number is larger than about 20.

For the basic Froude number F_0 larger than about 20, the vibration frequency ratio F/F_0 approaches a constant value. These asymptotic values of F/F_0 can be plotted, as in Figure 10, with the water-to-gate mass ratio α_ψ on the abscissa and with the reduced rotation center height r_s as the parameter. As the mass ratio α_ψ increases, the asymptotic vibration frequency ratio F/F_0 gradually decreases. In addition, for the same mass ratio, the vibration frequency ratio decreases with increasing reduced rotation center height.

Figure 10. Characteristics of in-water to in-air frequency ratio for $F_0 > 20$
From Anami & Ishii (2001).

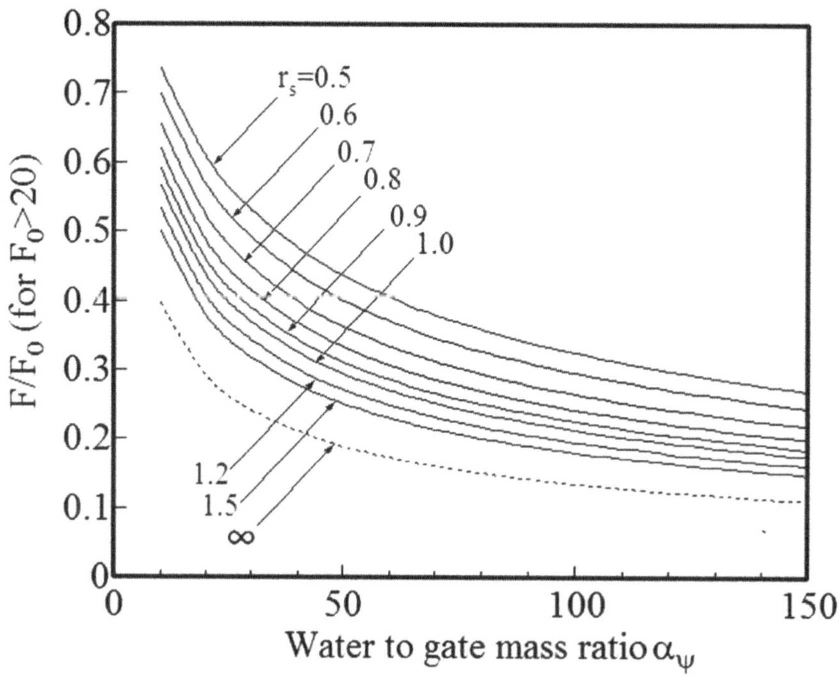

INCLINED CIRCULAR-ARC SKINPLATE

Empirical Extension of Theory

In practice, Tainter gates have circular arc skinplates, and are usually installed inclined in the downstream, direction. In this section, we examine what corrections are needed to apply the theoretical analysis for the vertical, flat weir plate to the inclined circular-arc skinplate.

The two-dimensional model of the inclined circular-arc skinplate and flow field is shown in Figure 11, where the elastically supported circular-arc rigid skinplate performs a small rotational vibration around the rotation center P on the skinplate. The flow field is assumed to extend infinitely in the upstream direction. In order to simplify the flow field, the curved dam crest is replaced by a flat floor which extends infinitely in the upstream direction.

The circular-arc skinplate is inclined in the downstream direction. This inclination is denoted by Θ_s which is the angle from a vertical line to the chord drawn between the water surface position O and lower end Q of the circular-arc skinplate. This inclination angle Θ_s is given by the following expression:

$$\Theta_S = \frac{\Phi_O - \Phi_Q}{2} \tag{88}$$

where Φ_O is the elevation angle of point O and Φ_Q the depression angle of point Q, looking from the trunnion center (also assumed to be skinplate center), O_T.

Figure 11. Vibrating circular-arc rigid skinplate and two-dimensional flow field
From Anami & Ishii (2001).

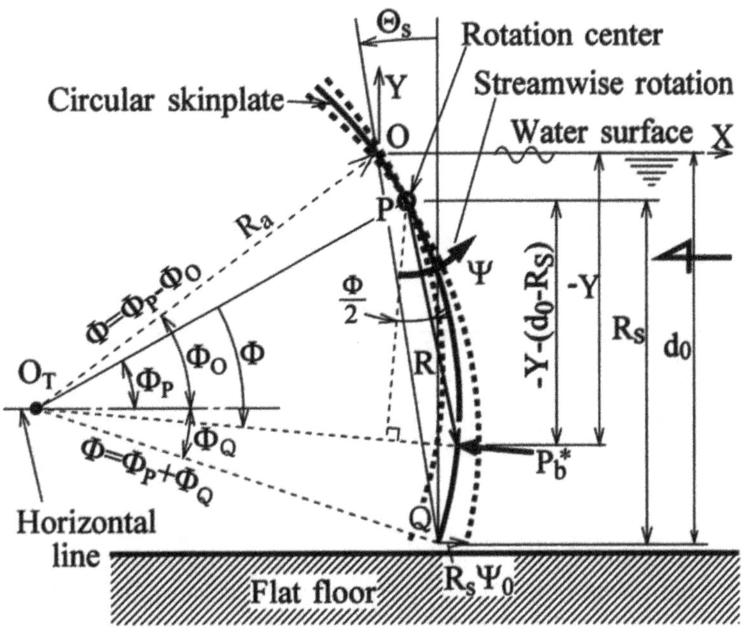

The counter-clockwise rotation angle of the circular-arc skinplate about point P is represented by Ψ. Assuming a small amplitude rotational vibration, the flow-rate variation beneath the gate due to this rotational vibration is assumed negligible. Thus, only the push-and-draw pressure arises in the upstream flow field. The push-and-draw pressure is represented by $P_b{}^*$, shown by the vector acting on the skinplate. The location of the center of pressure of $P_b{}^*$ on the skinplate is denoted by the clockwise rotation angle Φ about the trunnion center (point O_T) and measured from the radial line passing through the rotation center P to the radial line passing through the center of pressure of $P_b{}^*$. The chord length from the skinplate rotation center (point P) to the center of pressure is denoted by R. A relationship can be developed between Φ and R, and is given by

$$2R_a \sin\frac{\Phi}{2} = R \text{ or } R_a \cos\frac{\Phi}{2} d\Phi = dR \tag{89}$$

where R_a is the radius of the circular-arc skinplate.

The span of the two-dimensional model shown in Figure 11 is represented by W_0. The hydrodynamic force acting on a small portion $R_a d\Phi W_0$ of the circular-arc skinplate is given by $R_a d\Phi W_0 P_b^*$. In addition, the hydrodynamic force vector is located at a distance, given by $R\cos(\Phi / 2)$, from the rotation center P. The moment around the rotation center, due to the hydrodynamic force can be obtained by multiplying the force by the distance through which it acts. The equation of motion of a small rotational vibration of the circular-arc skinplate can then be derived as follows:

$$\ddot{\Psi} + 2\zeta_a \Omega_a \dot{\Psi} + \Omega_a^2 \Psi + \frac{W_0}{I_\psi} \int_\Phi P_b^* R R_a \cos\frac{\Phi}{2} d\Phi = 0 \tag{90}$$

This equation of motion can be re-arranged using Equation 89 for the angle Φ and the chord length R, as follows:

$$\ddot{\Psi} + 2\zeta_a \Omega_a \dot{\Psi} + \Omega_a^2 \Psi + \frac{W_0}{I_\psi} \int_R P_b^* R dR = 0 \tag{91}$$

Finally, the equation of motion can be reduced to the following non-dimensional form:

$$\psi'' + 2\zeta_a \psi' + \psi + \frac{\alpha_\psi}{r_s F_0^2} \int_r p_b^* r dr = 0 \tag{92}$$

Now the integral term in Equation 92 must be converted into a usable form that can be integrated. To do this, the chord length R from the rotation center P to the center of the hydrodynamic pressure is replaced by its vertical projected length $\{-Y-(d_0-R_S)\}$, where Y is the vertical coordinate taken as positive upward from the water surface, as shown in Figure 11. This vertical projected length $\{-Y-(d_0-R_S)\}$ can be given as a function of the angle Φ:

$$-Y - d_0 + R_s = R_a \sin\Phi_P + R_a \sin(\Phi - \Phi_P) \tag{93a}$$

or, after non-dimensionalization

$$-y - 1 + r_s = r_a \left\{\sin\Phi_P + \sin(\Phi - \Phi_P)\right\} \tag{93b}$$

where Φ_P is the elevation angle of the rotation center P, and y and r_a are the dimensionless equivalents of the coordinate Y and the circular-arc skinplate radius R_a, respectively, defined by

$$y \equiv Y / d_0 \quad \text{and} \quad r_a \equiv R_a / d_0 \tag{94}$$

The chord length R of the first expression in Equation 89 can be reduced to

$$r\left(\equiv R_a / d_0\right) = 2r_a \sin\frac{\Phi}{2} \tag{95}$$

Let Δ be the "vertical projection error," defined as the difference between the reduced chord length r and its reduced projected length $\{r_s\text{-}1\text{-}y\}$, and given as follows:

$$\Delta \equiv r - \left(r_s - 1 - y\right) \quad \text{or} \quad r = r_s - 1 - y + \Delta \tag{96}$$

Using Equations 93 and 95, Δ can be given alternatively by

$$\Delta = r_a \left\{ 2\sin\frac{\Phi}{2} - \sin\Phi_P - \sin(\Phi - \Phi_P) \right\} \tag{97}$$

Since the angle Φ is related to y by Equation 93, this vertical projection error is a function of y. As a result, using Equation 96, the equation of motion, Equation 92, can be re-arranged as follows:

$$\psi'' + 2\zeta_a \psi' + \psi - \frac{\alpha_\psi}{r_s F_0^2} \int_0^{-1} p_b^* \left(r_s - 1 - y + \Delta \right) \left(1 - \frac{d\Delta}{dx} \right) dy = 0 \tag{98}$$

Hence,

$$\psi'' + 2\zeta_a \psi' + \psi + \frac{\alpha_\psi}{F_0^2} \int_{-1}^{0} p_b^* \left(1 - \frac{1 + y - \Delta}{r_s} \right) \left(1 - \frac{d\Delta}{dx} \right) dy = 0 \tag{99}$$

where the integral term in Equation 92 has been converted to a definite integral with respect to the vertical coordinate y. The differential dΔ/dy can be replaced with

$$\frac{d\Delta}{dy} = \frac{d\Delta}{d\Phi} \cdot \frac{d\Phi}{dy} \tag{100}$$

where $d\Delta/d\Phi$ can be found from Equation 97:

$$\frac{d\Delta}{d\Phi} = r_a \left\{ \cos\frac{\Phi}{2} - \cos(\Phi - \Phi_P) \right\} \tag{101}$$

In addition, $d\Phi/dy$ can be determined from Equation 93b:

$$\frac{d\Phi}{dy} = \frac{-1}{r_a \cos(\Phi - \Phi_P)} \tag{102}$$

Finally, $d\Delta/dy$ can be given by

$$\frac{d\Delta}{dy} = 1 - \frac{\cos(\Phi / 2)}{\cos(\Phi - \Phi_P)} \tag{103}$$

If the push-and-draw pressure p_b^* can be given as a function of y, the definite integral in Equation 96 can then be evaluated. For this purpose, assume that p_b^* for the circular-arc skinplate can be approximated by a correction factor times the push-and-draw pressure for the vertically projected flat weir plate of the circular-arc skinplate. Otherwise, it would be exceedingly difficult to complete rigorously the theoretical analysis of push-and-draw pressure induced by the rotational vibration of the circular-arc skinplate, strictly adhering to a curved flow boundary such as the curved skinplate. Therefore, if the above approximation is acceptable, it is possible to solve for the gate motion due to the push-and-draw pressure. One might readily object to the above approximation as inappropriate for the present analysis, due to both the circular-arc geometry of the skinplate and its inclination in the downstream direction. However, experimental results (to be presented subsequently) suggest the approximation is completely appropriate since the distribution of push-and-draw pressure acting on the circular-arc skinplate shows a very similar tendency to the pressure distribution on a vertical, flat, and rigid weir plate. The correction factor can, and will, be determined through experiments.

Adopting the proposed approximate treatment of the push-and-draw pressure, the push-and-draw pressure p_b^* acting on the circular-arc skinplate can be given by the product of the pressure correction factor δ_p and the hydrodynamic pressure on a vertical, flat, and rigid weir plate:

$$p_b\left(0,y;t\right)=\delta_p\left\{p_{bs}\left(0,y\right)\left(\frac{F_0}{F}\right)^2\psi''+p_{bp}\left(y\right)\left(\frac{F_0}{F}\right)\psi'\right\} \tag{104a}$$

or

$$\frac{p_b\left(0,y;t\right)}{\delta_p}=\left\{p_{bs}\left(0,y\right)\left(\frac{F_0}{F}\right)^2\psi''+p_{bp}\left(y\right)\left(\frac{F_0}{F}\right)\psi'\right\} \tag{104b}$$

The right-hand side (RHS) of Equation 104b is the push-and-draw pressure acting on a vertical, flat, and rigid weir plate. The first term on the RHS is the acceleration term and the second the velocity term. The coefficient $p_{bs}(0, y)$ is the amplitude of the standing-wave pressure component, and $p_{bp}(y)$ is the amplitude of the progressive-wave component. The standing-wave pressure is, in general, a function of both the non-dimensional vertical coordinate, y, and the non-dimensional horizontal coordinate x $(=X/d_0)$ from the skinplate surface along the still water surface, but when evaluated at $x=0$, it corresponds to the pressure on the skinplate. The progressive-wave component is a function of the non-dimensional y coordinate. The Froude number F, reduced rotation center height r_s, and depth ratio β^* $(=d_r/d_0)$ are involved as parameters; the depth ratio β^* is 1.0 in this case, since the Tainter gate is installed over a flat floor.

Now, Equation 104 is substituted into the equation of motion, Equation 92, yielding

$$(1+\delta_p\alpha_\psi\Delta m_\psi^*)\psi''+2(\zeta_a+\zeta_f)\psi'+\psi=0 \tag{105}$$

where ζ_f is the modified wave-radiation damping ratio, defined by

$$\zeta_f(F) \equiv \delta_p \alpha_\psi \frac{\Delta c_\psi^*(F)}{2F_0} \tag{106}$$

The variables $\Delta m_\psi{}^*$ and $\Delta c_\psi{}^*$ represent the equivalent added mass and wave-radiation damping coefficient, respectively. Both are dimensionless quantities given by the following expressions:

$$\Delta m_\psi^*\left(F\right) = \frac{1}{F^2}\int_{-1}^{0}\left(1 - \frac{1+y-\Delta}{r_s}\right)\left(1 - \frac{d\Delta}{dy}\right) p_{bs}\left(0, y\right) dy$$

$$\Delta c_\psi^*\left(F\right) = \frac{1}{F}\int_{-1}^{0}\left(1 - \frac{1+y-\Delta}{r_s}\right)\left(1 - \frac{d\Delta}{dy}\right) p_{bp}\left(y\right) dy \tag{107}$$

Both $\Delta m_\psi{}^*$ and $\Delta c_\psi{}^*$ are functions of the Froude number F. The equivalent added mass is due to the standing-wave pressure component $p_{bs}(0, y)$ and the wave-radiation damping coefficient is due to the progressive-wave pressure component $p_{bp}(y)$.

For the Folsom Dam Tainter gate, the gate submergence depth d_0 was 13.26 m at the time of failure. The radius of the skinplate is 14.33 m. In addition, from experimental modal analysis, the rotation center height R_S was determined to be 9.6 m, which has an elevation angle Φ_P of 17°. With these values, the reduced rotation center height r_s is 0.72, and the reduced radius of the circular-arc skinplate r_a is 1.08. The elevation angle of the water surface Φ_O was 37.5° and the depression angle of the lower end of the skinplate Φ_Q was 18.5°. The vertical projection error Δ can be calculated from these conditions, as shown by the solid line in Figure 12. Recall that $y = 0$ represents the water surface and $y = -1.0$, the lower end of the skinplate. The vertical projection error Δ is almost zero everywhere except near the water surface. It takes on a small negative value near the water surface, but this small non-zero value causes little problem in the analysis because of the small hydrodynamic pressure near the water surface. Since the rotation center for the Folsom Dam Tainter gate is just slightly beneath the water surface level and the vertical projection error Δ is negligible, it follows that $d\Delta/dy$ can also be neglected. As a result, the two expressions in Equation 107 become identical with the corresponding expressions for a vertical, flat, and rigid weir plate.

The solution of the equation of motion, Equation 105, can be easily obtained, assuming that the solution is of the form $exp\{(-\zeta_w + i)\times(F/F_0)\times t\}$. The in-water to in-air vibration frequency ratio and the in-water damping ratio are given by

$$\frac{F}{F_0}\left(= \frac{\Omega_w}{\Omega_a}\right) = \sqrt{\frac{1}{1+\delta_p\alpha_\psi\Delta m_\psi^*\left(F\right)} - \zeta_w^2} \approx \frac{1}{\sqrt{1+\delta_p\alpha_\psi\Delta m_\psi^*\left(F\right)}} \tag{108}$$

$$\zeta_w = \frac{\zeta_a + \zeta_f}{\sqrt{1+\delta_p\alpha_\psi\Delta m_\psi^*\left(F\right)}} \tag{109}$$

Figure 12. Vertical projection error, Δ, of the chord length r relative to the projected reduced chord length
From Anami & Ishii (2001).

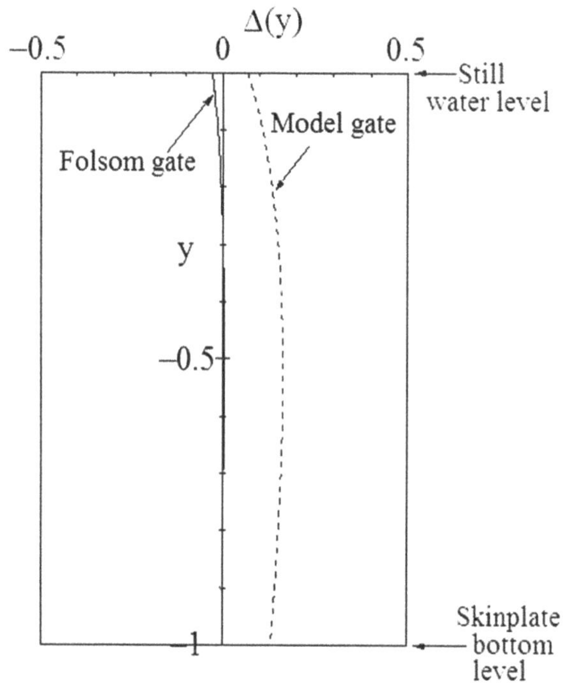

If the pressure correction factor δ_p is known, the vibration frequency ratio and damping ratio can be calculated using the above expressions, Equations 108 and 109, for a specified value of Froude number F. From the given value of F and the calculated frequency ratio, the value of the basic Froude number F_0 can be determined. Therefore, given a value of the basic Froude number F_0, the Froude number can be numerically determined from Equation 108, and then the damping ratio can be calculated from Equation 109.

Validation by Model Experiments

In order to determine the push-and-draw pressure induced by a rotational vibration of an inclined circular-arc skinplate, forced vibration tests were conducted. The first set of tests were done with a circular-arc skinplate model on a curved crest spillway model (called the practical reservoir). Subsequent tests were done using a circular-arc skinplate model above a flat horizontal bed and then an inclined flat plate skinplate above a flat bed. Finally free-vibration tests were undertaken with the circular-arc skinplate model above an inclined channel section beneath the gate model to minimize the contribution of flow rate variation as the gate was allowed to vibrate.

Pressure Correction Factors for Practical Reservoir Flow Field

To make the problem of pressure loading on a circular-arc skinplate undergoing rotational streamwise vibration on a spillway crest mathematically tractable, the circular-arc skinplate was replaced by a vertical, flat, rigid weir plate undergoing streamwise rotational vibration, and the curved dam crest against which the gate closes was modeled as a horizontal surface with a vertical step down at the gate position to allow for the greater depth of the reservoir. To determine how applicable the results from this analysis might be, a series of laboratory test were undertaken using a circular arc skin plate above a spillway model with a Creager crest (Creager & Justin, 1927) and a step down on the upstream side, as shown in Figure 13.

The two-dimensional experimental setup of the approximately 1/38-scale Folsom Dam model is shown in Figure 13, where the circular-arc skinplate has a radius of 375 mm and the rotation center "P" has a height of 279 mm from the skinplate lower end. The width of the experimental setup is 300 mm. The submergence depth d_0 is fixed at 290 mm, and the upstream reservoir depth is 840 mm. Thus, the values of the reduced reservoir depth, β_0* (see the following paragraph), and the reduced rotation center height, r_s, are 2.90 and 0.97, respectively. The skinplate was forced to perform streamwise rotational vibrations with a small displacement amplitude of 0.3 mm at the bottom end of the skinplate. The push-and-draw hydrodynamic pressure on the skinplate was measured over the skinplate surface, with a high sensitivity pressure transducer, for a variety of forced vibration frequencies Ω_w, overhang distances from the skinplate lower end, L_0, and skinplate inclination angles Θ_s. The small, highly sensitive transducer had an adapter with a 0.5 mm-diameter opening, and could be positioned perpendicular to the streamline along the skinplate, 5 mm away from its surface. With the small vibration amplitudes in this study, this simple pressure measurement method provided accurate pressure data, as was verified by calibrating against data from a flush-mounted transducer.

Figure 13. Approximately 1/38-scaled experimental setup of an inclined circular-arc skinplate forced to undergo streamwise rotational vibration, and the Creager curve dam crest with the vertical step down on the upstream side
From Anami et al. (2005). © 2005 ASME. Used with permission.

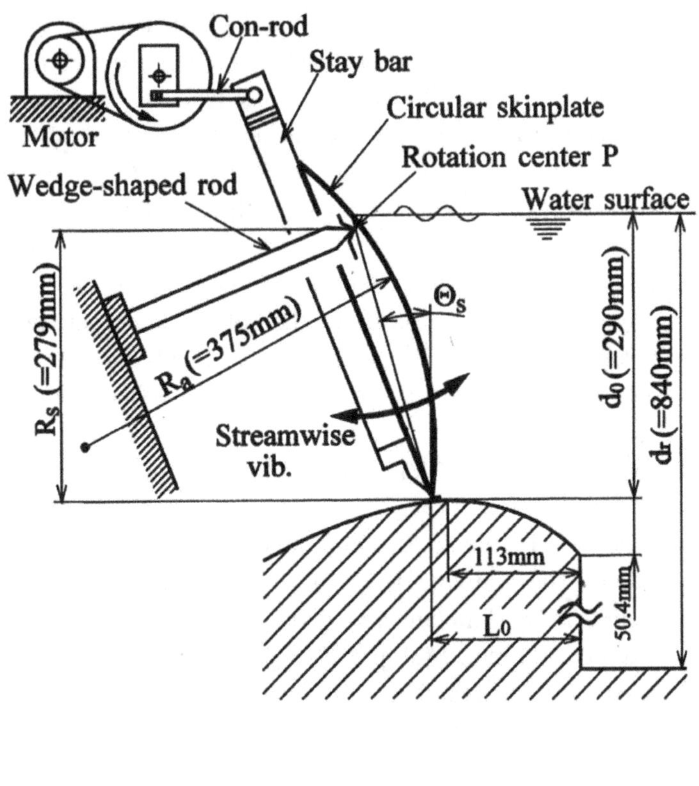

As the Froude number F increases, the standing pressure amplitude p_{bs} increases as F^2, resulting in a significant added mass effect, while the progressive component p_{bp} decreases rapidly, indicating no damping effect due to fluid motion in the limit of large F. It is noted here that the numerical value of β^* in the corresponding theoretical model differs from the apparent value (denoted β_0^*) in the physical model due to the influence of the dam crest upstream of the actual gate. The corresponding theoretical value of β^* must be determined as an empirical function of the parameter $\gamma \equiv L_0/d_0$, the reduced overhang length (i.e., the non-dimensional distance to the vertical step down).

The push-and-draw hydrodynamic pressure on the skinplate was measured for various values of both the reduced overhang length, γ, and the downstream inclination angle, Θ_s. Measured data were compared with corresponding theoretical values calculated for the simplified flow field model, to determine empirical values of two pressure correction factors, the pressure coefficient δ_p, introduced in Equation 104a, and the reduced reservoir depth, β^*, introduced in the preceding paragraph.

Measured magnitudes of the push-and-draw pressure acting on the circular-arc skinplate are plotted in Figure 14, where corresponding theoretical results for the simplified reservoir flow field are shown as solid lines. The theoretical calculations used *empirically determined values of reduced reservoir depth β^* and pressure coefficient δ_p* that when substituted into Equation 59 produced good agreement between the measured data and the theoretical curves as shown in Figure 14a. The effect of the reduced overhang length, $\gamma \equiv L_0/d_0$, on the pressure magnitude is discernible in Figure 14a, where the Froude number F was fixed at 7.0 (forcing frequency of 6.47 Hz). As γ decreases, the pressure magnitudes consistently decrease, distinctively near the bottom end of the skinplate. This decrease is due to the increased flow field spreading with decreased L_0. The empirical values of the reduced reservoir depth β^* and pressure coefficient δ_p producing good agreement with the data in Figure 14a are plotted in Figure 15a as functions of reduced overhang length. For γ values larger than about 0.5, β^* and δ_p take on constant values of 1.0 and 0.83, respectively. As γ decreases, the corresponding value of β^* to be used in the theoretical analysis increases, approaching its apparent value, $\beta_0^*(= 2.87)$ for a sudden step down just beneath the gate, i.e., when $\gamma = 0$, while δ_p decreases gradually to about 0.73 when $\gamma = 0$.

The effects of skinplate inclination angle Θ_s and Froude number F are shown in Figure 14b, where γ was fixed at 0.5 for model experiments and β^* at 1.0 for theoretical calculations. The measured data for four different skinplate inclination angles were reduced using the pressure coefficient δ_p vs. Θ_s, shown in Figure 15b. The empirical values of δ_p for each inclination angle Θ_s were independent of Froude number F. Good agreement between the measured data and the theoretical curve was obtained at each Froude number F.

From the empirical pressure correction factors shown in Figure 15, one may conclude that the excessive push-and-draw hydrodynamic pressure exerted on the failed Tainter gate at Folsom Dam with $\Theta_s = 9.5°$ and $\gamma = 0.5$ can be well approximated by the theoretical calculations for the simplified reservoir flow field, using a pressure coefficient δ_p of 1.0 and a reduced reservoir depth β^* of 1.0.

Inclined Circular-Arc Skinplate Above Horizontal Bed

The experimental setup is shown in Figure 16, in which an inclined circular-arc skinplate undergoes the streamwise rotational vibration. The skinplate, made of acrylic resin, has a radius R_a of 375 mm and a width W_0 of 298 mm. The circular-arc skinplate is inclined in the downstream direction and is installed in a flat channel with a width of 300 mm. Multiple V-shaped grooves were made on the back side of the skinplate and served to support the skinplate with a wedge-shaped rod at the rotation center P, as

Figure 14. Measured amplitudes of push-and-draw hydrodynamic pressure on the circular-arc skinplate undergoing periodic streamwise rotational vibration at reduced rotation center height r_s = 0.97 and apparent reduced reservoir depth, β=2.90, and corresponding simulations by theoretical analysis for simplified reservoir flow field; (a) pressure magnitudes at F = 7.0 and Θ_s = 16°, portraying the effects of reduced overhang length, and (b) pressure magnitudes at γ = 0.5, showing the dependence on skinplate inclination angle Θ_s and Froude number F*

From Anami et al. (2005). © 2005 ASME. Used with permission.

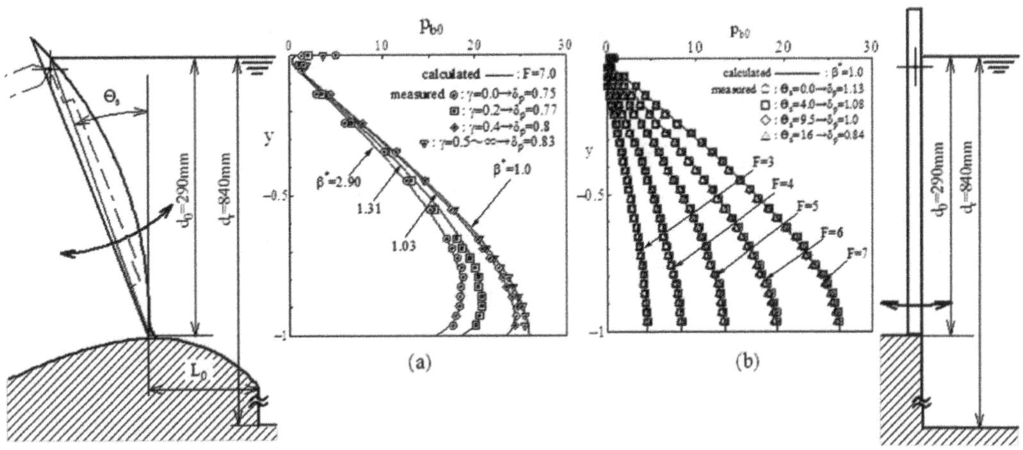

close to the skinplate surface as possible. A wedge-shaped rod was attached at each side of the skinplate. The upper end stay bars, attached to the back surface of the skinplate, were forced to vibrate, forcing the circular-arc skinplate to undergo rotational vibration around of the rotation center P. The discharge opening under the skinplate was adjusted to be as small as possible. In addition, a thin rubber sheet is attached to seal completely the discharge flow. This arrangement was intended to permit the measurement of only the push-and-draw pressure with no contribution from the flow-rate variation pressure. A clearance of about 1 mm was maintained at each side of the skinplate, to which a 3 mm diameter rubber tube was attached on the upstream side to seal completely any leakage flow.

The submergence depth d_0 was fixed at 290 mm, which was the same value as in the experiments to verify the push-and-draw pressure due to the streamwise rotational vibration of a vertical, flat, and rigid weir plate. In the present case, the central angle of the submerged portion of the circular-arc skinplate is 48°, which is close to the 56° for the failed Folsom Tainter gate. The elevation angle of water surface level O is 40°, and the depression angle of the lower end Q is 8°, and the inclination angle Θ_s of the chord $\overline{OQ}$ is 16°.

Just as in the experiments with the vertical weir plate, the larger the forced vibration amplitude of the skinplate, the higher the accuracy of the pressure measurements. In such a case, however, breaking waves on the free surface are easily generated, leading to greater errors in the pressure data. Thus, since the forced vibration frequency cannot be raised sufficiently, the Froude number is restricted to relatively small values. To obtain as large a Froude number as possible, the vibration amplitude was kept at a small value, while the measurement accuracy was enhanced as much as possible by utilizing ensemble averaging. In the present experiments, the amplitude of the lower end of the skinplate was adjusted from 0.52 mm to 0.55 mm, and the forced vibration frequency was varied from 2.78 Hz to 6.48 Hz. The rotation

Figure 15. Empirical data for pressure coefficient and reduced reservoir depth for the push-and-draw hydrodynamic pressure induced by streamwise rotational vibration of the circular-arc skinplate over Creager curve dam crest with a vertical step down on the upstream side; (a) δ_p and β^* at $F=7.0$ and $\Theta_s = 16^o$ showing dependence on reduced overhang length, $\gamma \equiv L_0/d_0$, and (b) δ_p dependence on skinplate inclination angle Θ_s at a fixed value of $\gamma = 0.5$

From Anami et al. (2005). © 2005 ASME. Used with permission.

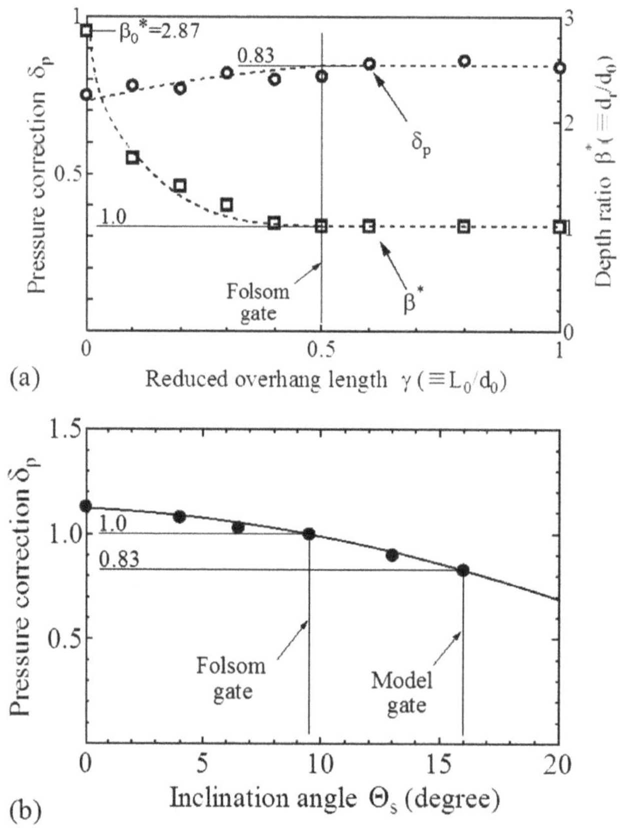

center P was fixed at a height of R_s = 280 mm, which was very close to the level of the water surface, resulting in a reduced rotation center height r_s of 0.97.

In the present experiments, the forcing frequencies corresponded to Froude numbers F from 3 to 7. The calculated results for the push-and-draw pressure for the vertical weir plate for each Froude number with a reduced rotation center height r_s of 0.97 are shown by the solid lines in Figure 17. Figure 17a shows the pressure amplitude and Figure 17b shows the phase-lag.

The push-and-draw pressure amplitude P_{b0}*, measured by experiments, can be reduced to a corresponding dimensionless form:

$$p_{b0}^* = \frac{P_{b0}^* / \rho g}{R_s \Psi_0} \tag{110}$$

Figure 16. Experimental setup of inclined circular-arc skinplate undergoing rotational vibration
From Anami & Ishii (2001).

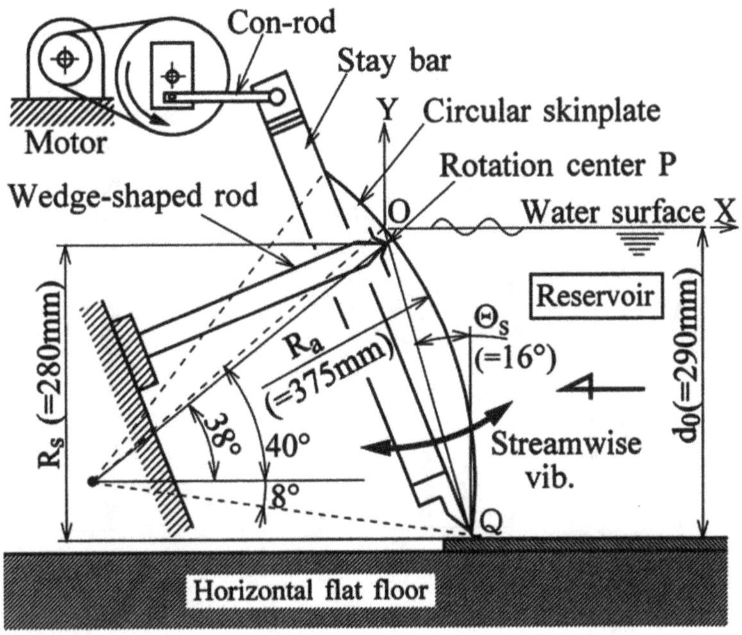

where the denominator is the streamwise vibration displacement amplitude of the lower end of the skinplate. The measured data of p_{b0}^*, plotted in Figure 17a, are divided by the pressure correction coefficient $\delta_p = 0.83$ to provide an excellent correlation with the theoretical values. All of the experimental data, both amplitude and phase, are in excellent agreement with the corresponding theoretical predictions. Therefore, one may conclude that the push-and-draw pressure on the circular-arc skinplate can be correctly calculated, based on the push-and-draw pressure on a vertical, flat, and rigid weir plate, with the introduction of an empirical value for the pressure correction coefficient, δ_p. In the present model test configuration, $\delta_p = 0.83$.

The physical basis for the perhaps unexpected validity of the above approximation will be sought in the following paragraph. Measured data for the push-and-draw pressure induced by the parallel vibration of a flat weir plate inclined at 10° are plotted in Figure 18, where the theoretical curves calculated for the projected vertical, flat weir plate are shown by the dotted lines for comparison. The measured data are distinctively smaller than the corresponding theoretical curve, near the bottom end of the weir plate. This is due to the increased flow field spreading with increased inclination of the weir plate. In contrast, as the present experimental setup in Figure 16 shows, even though the skinplate chord is inclined in the downstream direction, the lower end of the circular arc skinplate surface, which induces the largest pressure amplitudes, is almost vertical, looking more like a vertical flat weir plate than an inclined flat plate. Therefore, the flow field spreading found for the inclined flat plate does not occur, permitting the approximation of the push-and-draw pressure on the circular-arc skinplate as that on a vertical, flat weir plate with little error.

Figure 17. Measured push-and-draw pressure p_{b0}^*/δ_p on a circular-arc skinplate inclined at 16°, undergoing rotational vibration with r_s = 0.97 showing (a) amplitude, and (b) phase-lag
From Anami & Ishii (2001).

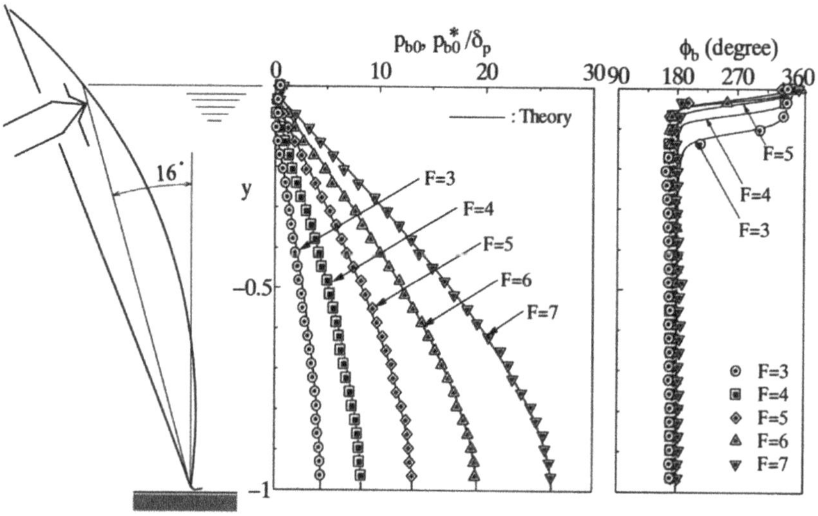

Figure 18. Push-and-draw pressure induced by 10° inclined flat weir plate, undergoing streamwise parallel vibration
From Anami et al. (2003). © 2003 JSCE. Used with permission.

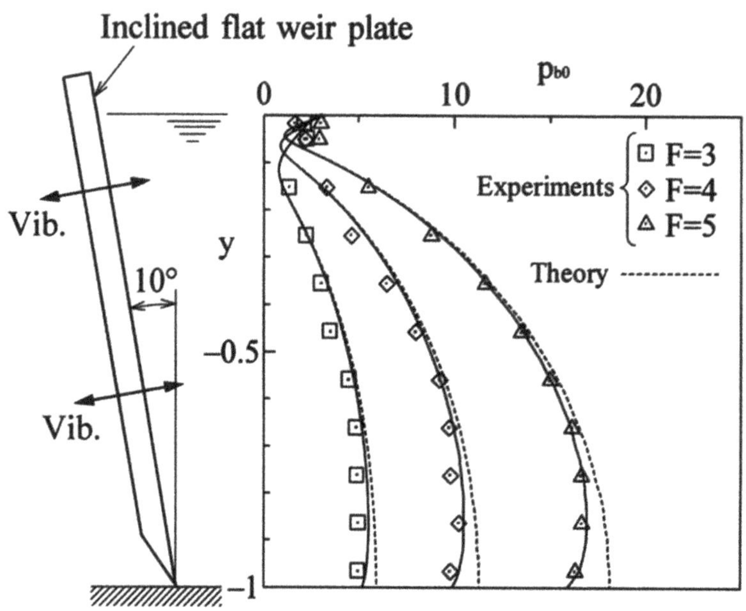

Streamwise Vibration of Inclined Flat Weir Plate

Similar measurements of the push-and-draw pressure were made for a variety of inclination angles Θ_s. In all cases, use of the theoretical calculations for the projected vertical, flat, and rigid weir plate to determine correctly the push-and-draw pressure on the circular-arc skinplate was validated. The empirical values of the pressure correction coefficient δ_p, from Figure 15b, are re-plotted in Figure 19 for convenience. The pressure correction coefficient δ_p gradually increases with decreasing inclination angle Θ_s. As Θ_s decreases, the lower portion of skinplate becomes increasingly inclined in the upstream direction, thus increasing the enclosed effect of the flow field. As a result, the push-and-draw pressure increases, thus increasing the pressure correction coefficient to values greater than 1.0, as shown in Figure 19. For the present model gate with $\Theta_s = 16°$, the pressure correction coefficient takes on a value of $\delta_p = 0.83$, while for the Folsom Dam Tainter gate with $\Theta_s = 9.5°$ the pressure correction coefficient takes on a value of $\delta_p = 1.0$.

The assumption that the push-and-draw pressure acting on the inclined circular-arc skinplate can be empirically evaluated by using the theoretical analysis for the vertical, flat, and rigid weir plate has been validated. Then, in order to confirm further the theoretical analysis of flow-induced vibration of the skinplate using empirically evaluated pressure, the free-vibration tests with an elastically-supported circular-arc skinplate were conducted.

Figure 19. Pressure correction coefficient δ_p for evaluating push-and-draw pressure, for inclined circular-arc skinplate
From Anami & Ishii (2001)

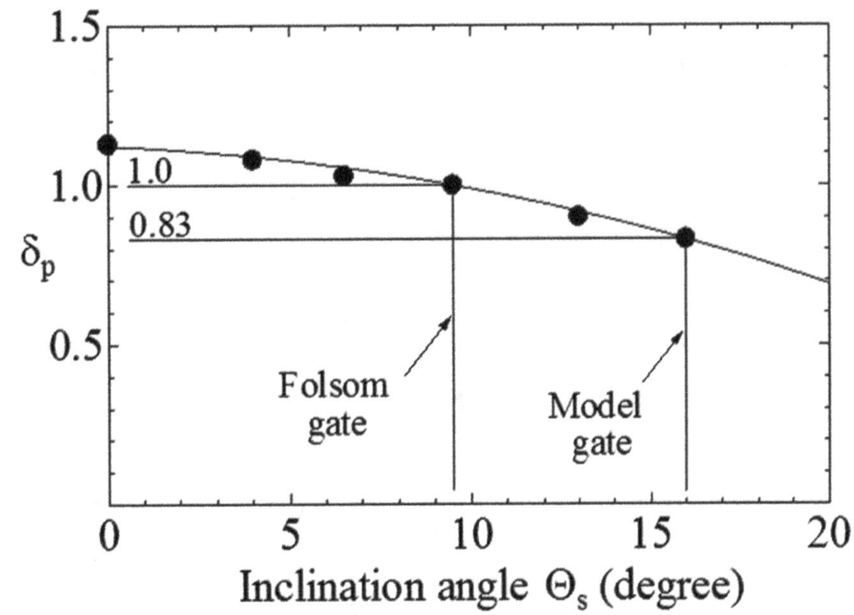

Free Vibration With an Inclined Bed

The experimental setup of the inclined circular-arc skinplate undergoing the rotational vibration is shown in Figure 20. In the present experiments, however, the radial arms are tightly fixed on the base to allow only the skinplate streamwise rotational vibration. The skinplate rotation center P has a height R_S of 515 mm from the bottom end of the skinplate. In order to adjust the natural vibration frequency of the skinplate, a coil spring was attached to the back of the skinplate. The circular-arc skinplate which has a radius R_a of 455 mm was inserted in the 300 mm-width channel. The mass of the skinplate was 15 kg. In the model experiment, mechanical friction was kept as small as possible, since the model was light-weight. To maintain the lowest possible mechanical friction, the skinplate could not be permitted to contact the channel walls. Therefore, the width of skinplate was chosen to be 290 mm, with 5 mm clearances on each side of the skinplate. The floor just under the skinplate is inclined 22° in the downstream direction. This is to minimize any flow-rate variation pressure by keeping the gate motion parallel to the bounding surface. Therefore, only the push-and-draw pressure arises in the upstream reservoir. The average discharge opening was set at 3 mm.

The in-air and in-water free vibration tests were conducted for the three cases shown in Table 1. In Case 1, the gate submergence depth d_0 was 420 mm, and the moment of inertia I_ψ was 0.944 kg·m^2. The in-air natural vibration frequency was varied from 5.21 Hz to 12.49 Hz. Examples of measured time histories are shown in Figure 21. Figure 21a shows the time trace at a relatively low frequency, while Figure 21b shows a similar trace for a relatively higher frequency. The upper time history in both cases is the in-air free vibration waveform, and lower one is the in-water free vibration waveform. In both in-air and in-water testing, the gate was manually displaced with a maximum amplitude of 1 mm and then the small amplitude skinplate vibrations were recorded. The resulting data show an in-water to in-air

Figure 20. Experimental setup of inclined circular-arc skinplate undergoing rigid streamwise rotational vibration
From Anami & Ishii (2001)

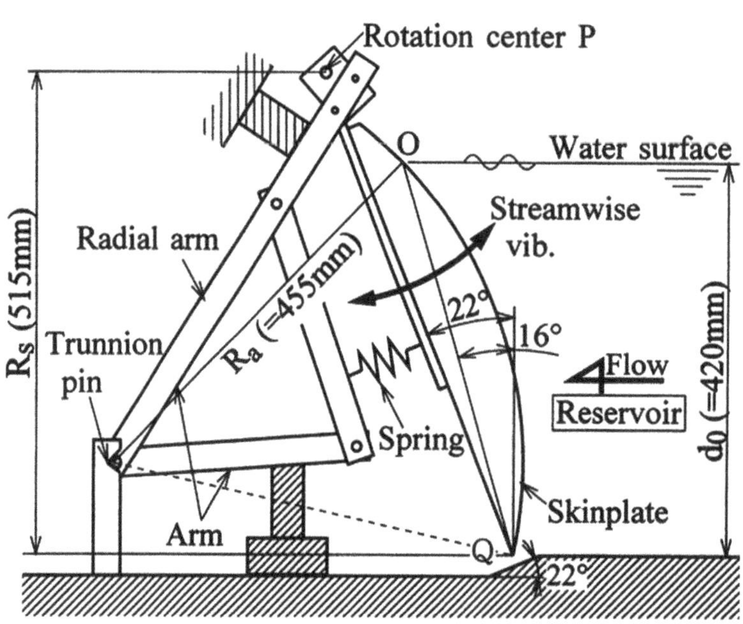

Table 1. Specifications of model tests for damped free vibrations, both in-air and in-water

Test name	Water depth d_0 (mm)	Moment of inertia I_ψ (kgm^2)	Rotation center height R_S (mm)	Water to gate mass ratio, α_ψ	Rotatio n center reduced height, r_s	Range of in-air natural vibration frequency Ω_a (Hz)
Case 1		0.944		14.9		5.21~12.49
Case 2	420	1.74	515	8.07	1.23	4.79~12.15
Case 3		2.38		5.89		3.53~10.14

From Anami & Ishii (2001).

vibration frequency ratio Ω_w/Ω_a of 0.49 for Figure 21a, and 0.48 for Figure 21b. Thus, even when the tested vibration frequency was different, the vibration frequency ratio takes on almost the same value. The value of the in-water damping ratio is much larger than the in-air damping ratio.

Similar experiments were conducted in detail for Case 1. In addition, as shown in Table 1 for Cases 2 and 3, the gate submergence depth d_0 was maintained at 420 mm, the moment of inertia I_ψ of the skin-plate was increased, and detailed experiments were conducted. With these values, the water-to-gate mass ratio α_ψ decreases from 14.9, to 8.07 and then to 5.89, as shown in Table 1. All test results are plotted in Figure 22. Figure 22a shows the vibration frequency ratio Ω_w/Ω_a, for all cases, and Figure 22b shows the in-air and in-water damping ratios ζ_a and ζ_w only for Case 1. The test results corresponding to the time histories in Figure 21 are denoted by the "a" and "b" designations in Figure 22b. The abscissa is the basic Froude number F_0. In these experiments, F_0 varied from 4.59 to 16.25.

For the circular-arc skinplate shown in Figure 20, the vertical projection error Δ from Equation 97 takes on the value previously shown in Figure 12 with a dashed line, with a maximum value just slightly larger than 0.1 The reduced rotation center height r_s was 1.23, as shown in Table 1. The equivalent added mass Δm_ψ and the wave-radiation damping coefficient Δc_ψ can then be calculated as a function of Froude number F from Equation 107 with these empirical data. The skinplate inclination angle Θ_s was 16°, and the corresponding pressure correction coefficient δ_p is 0.83, as shown in Figure 19. Furthermore, as shown in Table 1, the mass ratio α_ψ is known. Therefore, the in-air to in-water vibration frequency ratio Ω_w/Ω_a can be calculated using Equation 108, and the in-water damping ratio ζ_w can be calculated using Equation 109. When calculating ζ_w, the average value 0.004 of the experimental data was used for the in-air damping ratio ζ_a. The theoretically calculated results are shown in Figure 22 by the solid lines.

As shown in Figure 22a, the experimental data and the theoretical predictions of the in-air to in-water vibration frequency ratio are in good agreement. On the other hand, when considering the damping ratio data in Figure 22b, the measured data shown by the filled circles are significantly greater than the theoretically calculated values, shown by the solid line. This difference is believed to result from the comparatively large 5 mm leakage clearance at each side of the skinplate. The damping caused by the leakages from the clearances is quite large. Such additional and large damping effects are unavoidable in the present small-scale model tests. Of course, in practical full-scale Tainter gates, since the damping

Figure 21. Test records of free vibrations of Case 1 for the circular skinplate, (a) time traces in-air and in-water at a low frequency, and (b) time traces in-air and in-water at a higher frequency
From Anami & Ishii (2001).

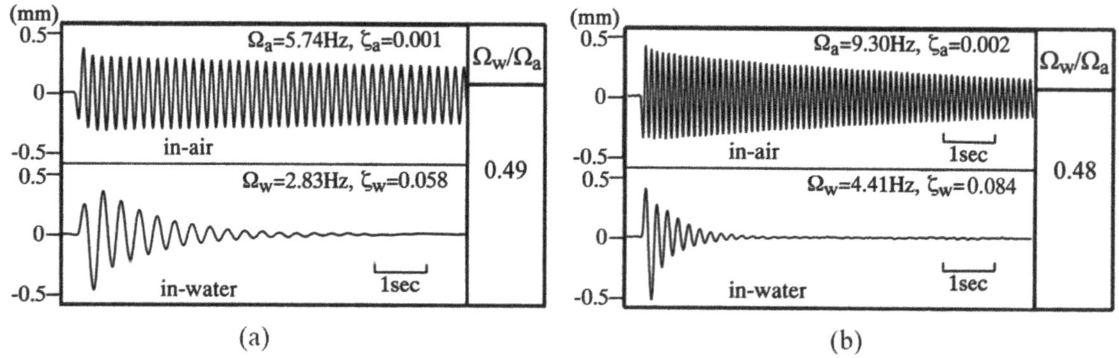

Figure 22. Free vibration test results of the circular skinplate plotted over basic Froude number F_0 with data for (a) vibration frequency ratio Ω_w/Ω_a, and (b) damping ratio ζ_a in-air and ζ_w in-water for Case 1
From Anami & Ishii (2001).

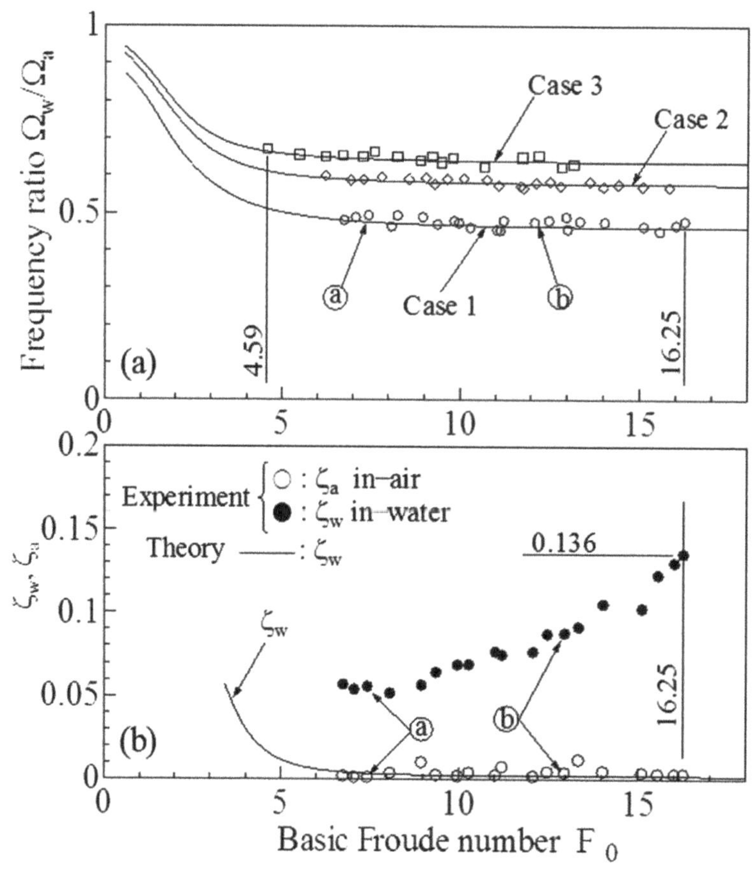

effect by such leakage does not arise at all, the theoretical results and measured data are expected to show good agreement for the damping ratio also. In any event, the damping ratio ζ_w does not have a significant influence on the vibration frequency ratio, following from Equation 108. Of greatest importance is that the calculated and measured results for the vibration frequency ratio exhibit good agreement with each other, as shown in Figure 22a.

Validation on Full-Scale Gates

Field vibration tests were undertaken on a 29-ton Tainter gate (hereafter called Tainter gate "A") and a 34-ton Tainter gate (hereafter called Tainter gate "B") in Japan. The in-air and in-water vibration characteristics were measured for gate "A." In addition to the in-air and in-water vibration characteristics for gate "B," the hydrodynamic pressure was also measured. The validation of the hydrodynamic pressure predictions was done using the measured data from gate "B". The validation of vibration frequency ratio and wave radiation damping coefficient were accomplished using the measured data from both gate "A" and "B." More details concerning the field tests on these gates can be found in Chapter 14-- Full-Scale Gate Measurements.

A photograph of Tainter gate "B," taken from downstream, is shown in Figure 23. The skinplate has a height of 9.8 m, a radius of 11.0 m, and a span of 9.0 m.

In the field vibration tests of Tainter gate "B", the hydrodynamic pressure due to the gate vibration could be simultaneously measured, in addition to the vibration response. A pressure tap was drilled in the skinplate to attach a pressure transducer. The pressure tap is located 700 mm from the bottom of the skinplate.

Figure 24 shows the lower portion of the skinplate. Figure 24a is an enlarged view of the pressure transducer attached to the skinplate, and Figure 24b is a sketch of the pressure transducer configuration.

The measured hydrodynamic pressure is shown in Figure 25. The simultaneously recorded vibration waveforms at the bottom center of the skinplate are also shown. As shown in the frequency spectra,

Figure 23. Thirty-four ton Tainter gate "B" in Japan, as viewed from downstream

Figure 24. Lower portion of the skinplate of Tainter gate "B" showing (a) enlarged view of the pressure transducer attached to skinplate, and (b) the installed configuration of the pressure transducer on the skinplate

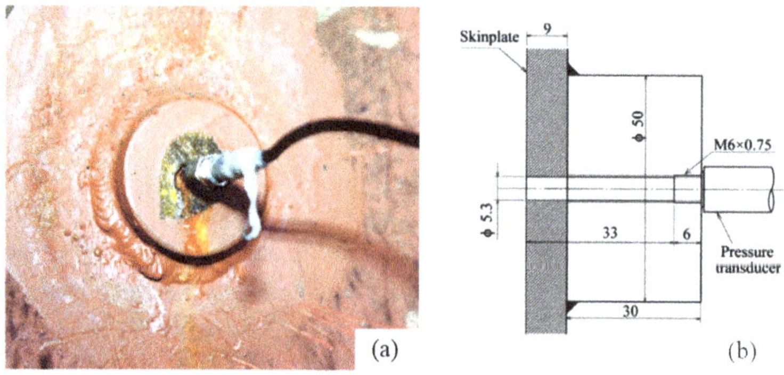

Figure 25. Representative samples of vibration accelerations and hydrodynamic pressure fluctuation measured on Tainter gate "B" showing (a) a real-time wave form, and (b) the corresponding frequency spectra
From Ishii & Anami (2001).

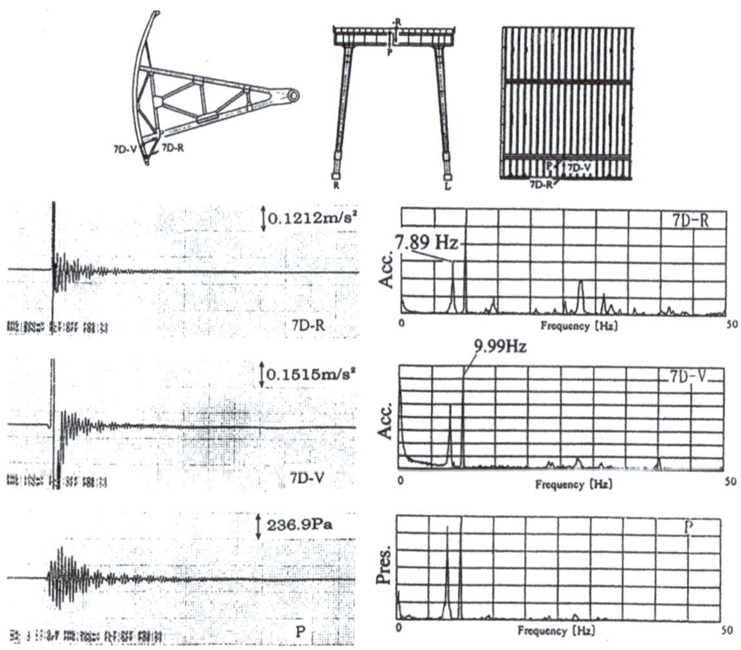

the two sharp peaks in the pressure spectrum of pressure agree well with the vertical natural vibration frequency at 7.89 Hz and with the streamwise natural vibration frequency at 9.99 Hz, Therefore, these hydrodynamic pressure fluctuations clearly appear to be associated with skinplate vibrations.

As shown in Figure 25, two significant spectral peaks appear at frequencies of 7.35 Hz and 9.99 Hz. The skinplate was vibrating in both the tangential (vertical) direction (*V*) and in the streamwise (radial) direction (R). Therefore, the measured hydrodynamic pressure amplitude P_0 is the resultant of the push-and-draw pressure amplitude P_{b0}* caused by the streamwise vibration and the flow-rate-variation pressure amplitude P_{r0} caused by the tangential vibration.

The push-and-draw and flow-rate-variation pressures have a 90° phase difference. Therefore, the resultant pressure amplitude P_0 is given by

$$P_0 = \sqrt{\left(P_{b0}^*\right)^2 + \left(P_{r0}\right)^2} \tag{111}$$

Using the previously introduced non-dimensionalization for amplitudes P_{b0}* and P_{r0}, the non-dimensional forms can be written as

$$p_{b0}^* = \frac{P_{b0}^* / \rho g}{X_0} \quad \text{and} \quad p_{r0} = \frac{P_{r0} / \rho g}{\sqrt{2}\, c_f Y_0} \tag{112}$$

X_0 and Y_0 are the amplitudes of the streamwise and tangential vibrations of the bottom end of the skinplate. With these definitions, $p_{b0}*/\delta_p$ and p_{r0} can be theoretically calculated using Equations 61 to 64.

The theoretical results for the full-scale Tainter gate "B" are compared with the experimental measurements in Figure 26. For the streamwise vibration, the Froude number F was 61 and the average of reduced rotation center height r_s was 0.55. With these values, the push-and-draw pressure amplitude p_{b0} was calculated, as shown by the solid line in Figure 26a. Since the range of r_s was from 0.53 to 0.61, calculated results are also shown by dashed lines for the maximum and minimum r_s values. For the tangential vibration, the Froude number was 45. With this value, the flow-rate-variation pressure p_{r0} was calculated, as shown in Figure 26b.

Figure 26. Hydrodynamic pressure acting on the lower center lower portion of the skinplate of Tainter gate "B" for two conditions: (a) $p_{b0}*/\delta_p$ with $F = 61$ and $r_s = 0.55$, and (b) p_{r0} with $F = 45$
From Ishii & Anami (2001).

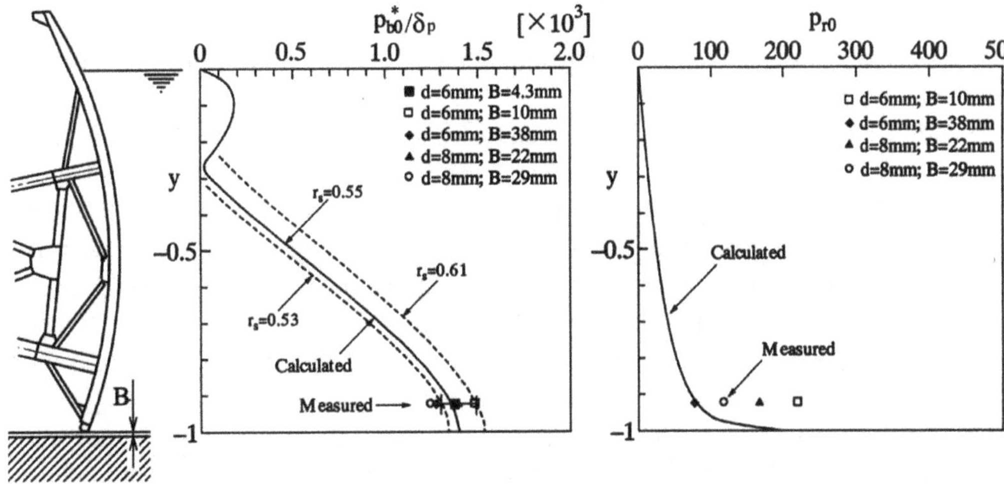

To determine the values of the experimental data plotted in Figure 26, the push-and-draw and flow-rate variation pressures had to be extracted from the measured resultant pressure P_0 for the vibration amplitudes X_0 and Y_0.

The leading vibration for 9.99 Hz is the streamwise rotational vibration. Even if the flow-rate variation pressure is completely ignored, it has little effect on the push-and-draw pressure data, since the flow-rate variation pressure is about 1/10 of the push-and-draw pressure. The skinplate of Tainter gate "B" has an inclination angle Θ_s of 1°, for which δ_p takes 1.1 from the empirical values presented in Figure 19. Therefore, the measured data at 9.99 Hz were reduced to non-dimensional data, using a pressure correction coefficient δ_p of 1.1. The plotted measured data are in good agreement with the theoretical results. In the theoretical analyses, it was assumed that the streamwise lower-mode bending vibration of the skinplate could be treated as a rotational rigid vibration. Since the theoretical predictions show good agreement with the field test results, one may therefore conclude that such a simplified treatment of the bending vibration is valid in practical applications, as long as the vibration amplitude is considered small.

The leading vibration for 7.35 Hz is the tangential vibration. It was quite difficult to extract the small flow-rate-variation pressure from the measured resultant pressure data. In this study, however, the theoretical push-and-draw pressure has been verified. Therefore, the flow rate variation pressure was calculated by subtracting the calculated push-and-draw pressure from the resultant pressure for 7.35 Hz. The push-and-draw pressure was theoretically calculated for conditions corresponding to the vertical vibration, that is, for a Froude number $F = 45$, a reduced rotation center height $r_s = 0.55$ and a reduced measurement location, resulting in $p_{b0}^*/\delta_p = 774$. With this value, the push-and-draw pressure P_{b0}^* can be calculated by

$$P_{b0}^* = p_{b0}^* \rho g X_0 \tag{113}$$

where the streamwise vibration amplitude X_0 was assumed to be that at measurement location 7D on the bottom end of the skinplate. Then, the flow-rate variation pressure amplitude P_{r0} can be extracted from the resultant pressure P_0 using Equation 111. The data are ultimately reduced to non-dimensional form using the second expression of Equation 112. The non-dimensional flow-rate variation pressure is plotted in Figure 26b. Despite some scatter in the measured data, the data show substantial agreement with the theoretical results, as seen in this figure.

The present experimental results are extremely valuable since they represent the first occasion in which the theoretical analysis of the hydrodynamic pressure acting on the gate was validated on an installed, full-scale Tainter gate.

The calculation of the vertical projection error, Δ, for Tainter gates "A" and "B" was done in the same way as for the Folsom Tainter gate. The calculated results in Figure 27show that the vertical projection error for these two full-scale gates is also almost zero, as was previously demonstrated in Figure 12 for the Folsom Dam gate. In the case of the Folsom Dam gate, the skinplate rotation center was located near the water surface. In contrast, the rotation center is near the middle of the submergence depth for Tainter gate "B" and even lower for Tainter gate "A." In any case, as shown in Figure 27, the vertical projection error Δ is close enough to zero to be neglected. Therefore, one can safely assume that Δ always approaches zero for any large-sized installed Tainter gate.

As a result, when Δ is negligible, the expressions for reduced added mass and reduced wave radiation damping, given in Equations (107) revert to those for the flat plate gate given in Equations (81).

Figure 27. Correction of vertical projection, Δ, of reduced chord length r relative to projected reduced chord

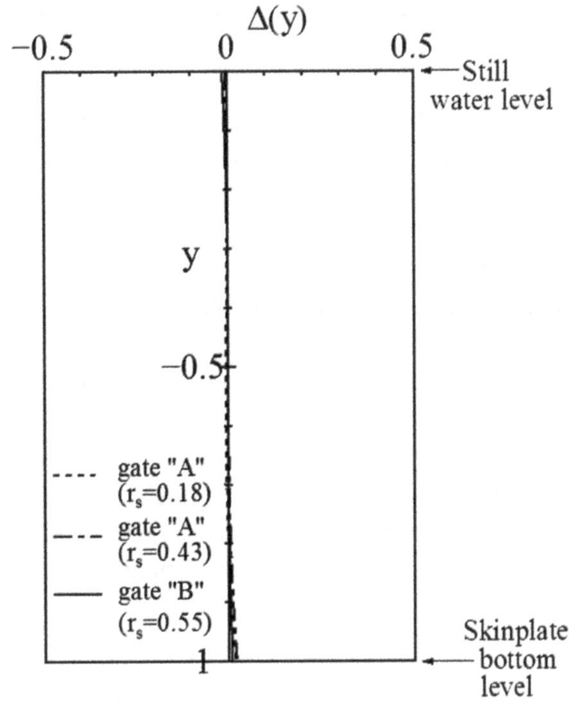

Further, the in-water to in-air vibration frequency ratio $\Omega_{w\psi}/\Omega_{a\psi}$, the in-water damping ratio $\zeta_{w\psi}$ and the wave-radiation damping ratio $\zeta_{f\psi}$ can be calculated using

$$\frac{\Omega_{w\psi}}{\Omega_{a\psi}} = \frac{1}{\sqrt{1+\delta_p \alpha_\psi \Delta m_\psi (F)}} \tag{114}$$

$$\zeta_{w\psi} = \frac{\zeta_a + \zeta_{f\psi}}{\sqrt{1+\delta_p \alpha_\psi \Delta m_\psi (F)}} \tag{115}$$

$$\zeta_{f\psi}\left(F\right) \equiv \delta_{\mathrm{p}} \alpha_\psi \frac{\Delta c_\psi \left(F\right)}{2F_0} \tag{116}$$

These expressions are the same as those previously given for the vertical flat weir plate, Equations 86b, 87 and 84, respectively, when the pressure correction factor, $\delta_p \rightarrow 1.0$.

The major characteristics of $\Omega_{w\psi}/\Omega_{a\psi}$ and $\zeta_{f\psi}$ of calculations for Tainter gates "A" and "B" are shown by the solid lines in Figure 28. The abscissa represents the basic Froude number F_0. The parameters are the water-to-gate mass ratio α_ψ and the reduced rotation center height r_s. These parameters, α_ψ and r_s, are 0.32 and 0.18 for Mode M_{x1} of Tainter gate "A", 26.6 and 0.43 for Mode M_{x2} of Tainter gate "A", and 101 and 0.55 for Mode M_x of Tainter gate "B".

The basic Froude number F_0 is relatively large for full-scale Tainter gates. F_0 is 133 and 137 for Modes M_{x1} and M_{x2} of Tainter gate "A", respectively, and is 185 for Mode M_x of Tainter gate "B". The measured Ω_w/Ω_a data are plotted against these large Froude numbers in Figure 28a. The field vibration tests results are in close agreement with the calculated results indicated by solid lines.

The theoretical and experiment results for the model gate (Figure 22) are also shown in the same figure. Until now, the theoretical analysis had only been validated for basic Froude numbers less than 16.25. With these unique field measurements, the theoretical analysis has now been validated for large basic Froude numbers as well. In addition, the present theoretical analysis still assumed that the streamwise lower-mode bending vibration of the skinplate could be replaced by a rotational vibration of a flat and rigid weir plate. Since the experimental results are in such close agreement with the theoretical predictions, this assumption retains its validity even for the full-scale Tainter gates.

Regarding Tainter gate "B", the in-air and in-water damping ratios were determined from the results of the steel bar breaking experiments. The measured in-air damping ratio ζ_a was 0.018 and the in-water

Figure 28. In-water to in-air vibration frequency ratio and wave-radiation damping ratio of the skinplate streamwise vibration

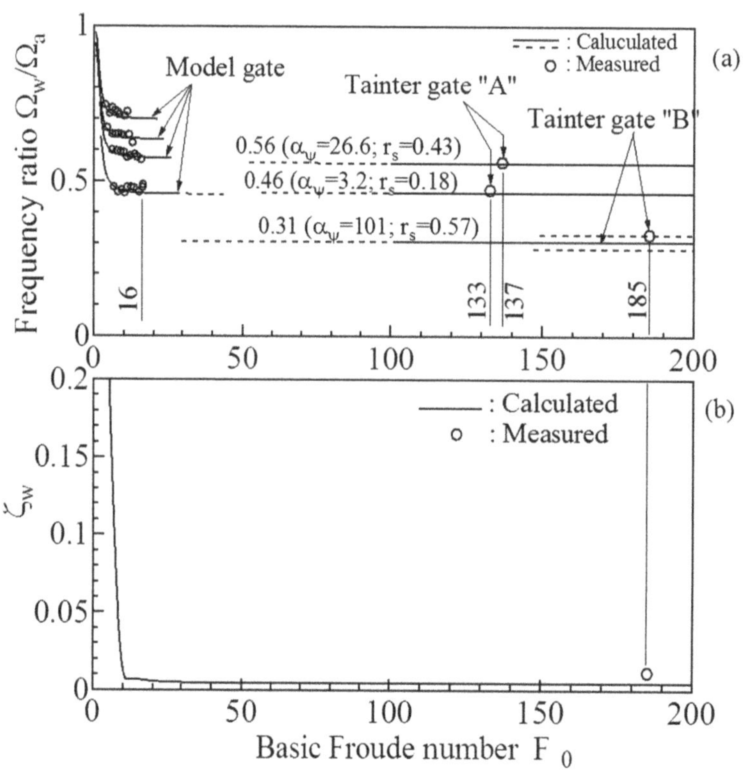

damping ratio ζ_w was 0.011. Using these damping values, the wave-radiation damping ratio was calculated as follows:

$$\zeta_{f\psi} = \frac{\zeta_w}{1/\sqrt{1+\delta_p \alpha_\psi \Delta m_\psi}} - \zeta_a = \frac{0.011}{0.031} - 0.018 = 0.017$$

This data point is plotted in Figure 28b. The measured value is a slightly larger than the theoretically predicted value of almost zero. This difference is believed to be due to the large frictional forces at the rubber seals attached to each side of the skinplate. These frictional forces do not appear in the in-air experiments. Considering this damping from these seals, the difference of about 0.017 seems quite reasonable. In any event, one may conclude that in full-scale Tainter gates, the effect of wave-radiation damping is very small, as theoretically predicted.

CONCLUSION

The potential flow theory from Chapter 4, 5 and 7 was extended to permit the calculation of pressure loading on circular-arc Tainter gates. The theory modelled the gate as undergoing combined rotation in the streamwise direction with up-and down vertical vibration. The pressure loadings again consisted of components, one associated with the push-and-draw effect due to the streamwise rotation and the other due to the flow-rate variation resulting from the vertical vibration.

An empirical correction factor was introduced to account for the curvature of the skinplate. The correction was incorporated into the theory. Values for the correction factor we determined through model scale and full-scale testing.

The results from the analysis in this chapter, supported by data from experimental testing, permits the very accurate prediction of the pressure loading on a curved Tainter gate. The ratio of in-water to in-air frequencies can also be determined as can the fluid damping. The analysis in this chapter also sets the stage for development of the theory of self-excited coupled-mode instability of Tainter gates in the next chapter.

REFERENCES

Anami, K., & Ishii, N. (1998). In-water streamwise vibration of Folsom Dam radial gates in California, In *Proceedings of 1998 ASME/JSME Joint Pressure Vessels and Piping Conference*, (Vol. 363, pp. 87-93). New York, NY: American Society of Mechanical Engineers.

Anami, K., & Ishii, N. (2001). Analyses of FIV of Folsom Dam Tainter gates and verifications by model tests. In *Proceedings of Commemorative International Symposium for the 40th Anniversary for the Foundation of the Osaka Electro-Communication University*, (*Vol. 2*, pp.165-181). Osaka: The Osaka Electro-Communication University.

Anami, K., Ishii, N., Knisely, C. W., & Matsumoto, Y. (1999). In-water natural vibration frequency of a flap gate and its theoretical calculation. *Proceedings of JSME Dynamics and Design Conference '99*, B, 265-268. (in Japanese)

Anami, K., Ishii, N., Knisely, C. W., & Matsumoto, Y. (2000b). Measured and calculated vibration characteristics of a full-scale flap gate undergoing self-excited vibration. In M. Souli (Ed.), *Proceedings of 2000 ASME Pressure Vessels and Piping Conference*, (Vol. 414-1, pp. 23-29). New York, NY: American Society of Mechanical Engineers.

Anami, K., Ishii, N., Knisely, C. W., & Oku, T. (2005). Hydrodynamic pressure load on Folsom Dam Tainter gate at onset of failure due to flow-induced vibrations. In *Symposium on Flow-Induced Vibration 2005, Proc. of ASME Pressure Vessels and Piping Conference*, (Vol. 4, pp. 557-564). New York, NY: American Society of Mechanical Engineers. doi:10.1115/PVP2005-71444

Anami, K., Ishii, N., & Takano, W. (2003). Empirical evaluation method of hydrodynamic pressures induced by vibrations of inclined circular-arc skinplate of Tainter gates. [in Japanese]. *Journal of Structural Engineering*, *49A*, 645–651.

Anami, K., Ishii, N., & Yamasaki, M. (2000a). Hydrodynamic pressure induced by rotating skinplate of Folsom Dam Tainter gate [in Japanese]. *Transactions of the Japan Society of Mechanical Engineers, Series B*, *66*(652), 3116–3123. doi:10.1299/kikaib.66.652_3116

Creager, W. P., & Justin, J. D. (1927). *Hydro-Electric Handbook.* John Wiley & Sons.

Ishii, N. (1990a). Flow-induced vibration of long-span gates (verification of added mass and fluid damping). *JSME International Journal, Series II*, *33*(4), 642–648.

Ishii, N. (1990b). Flow-induced vibration of long span gates (4th report: Vibration frequency ratio and fluid damping ratio). *Transactions of the Japan Society of Mechanical Engineers*, *56*(531), C, 49–54.

Ishii, N. (1992). Flow-induced vibration of long-span gates (Part I: Model development). *Journal of Fluids and Structures*, *6*(5), 539–562. doi:10.1016/0889-9746(92)90018-X

Ishii & Anami. (2001). Application of experiences and of recent findings with Tainter gates in Japan on the Folsom Dam Gate Stability. In *Proceedings of Commemorative International Symposium for the 40th Anniversary for the Foundation of the Osaka Electro-Communication University*, (*Vol. 2*, pp.165-181). Osaka, Japan: The Osaka Electro-Communication University.

Lamb, H. (1932). *Hydrodynamics.* Cambridge University Press.

Rayleigh, J. W. S. (1945). *The Theory of Sound.* New York: Dover.

Stoker, J. J. (1957). *Water Waves.* New York: Interscience Publications.

APPENDIX: NOMENCLATURE

A^*: Coefficient of homogenous solution
B: Gate opening
c, C: Wave velocity of progressive pressure-wave [c is the dimensionless expression of C]
c_f: Instantaneous flow-rate variation coefficient
d_0: Gate submergence depth
d_r: Depth of the reservoir
F: Froude number
F_0: Basic Froude number
g: Acceleration of gravity
i: Imaginary unit
I_ψ: Moment of inertia of the weir plate around its rotation center
j, n: Positive integer
p_b, P_b: Push-and-draw pressure [p_b is the dimensionless expression of P_b]
p_{b0}: Amplitude of push-and-draw pressure
p_{b0}^*, P_{b0}^*: Amplitude of push-and-draw pressure acting on the circular-arc skinplate [p_{b0}^* is the dimensionless expression of P_{b0}^*]
p_{bp}: Amplitude of progressive pressure-wave component of p_b
p_{bs}: Amplitude of standing pressure-wave component of p_b
p_r, P_r: Flow-rate variation pressure [p_r is the dimensionless expression of P_r]
p_{r0}, P_{r0}: Amplitude of flow-rate-variation pressure [p_{r0} is the dimensionless expression of P_{r0}]
p_{rp}: Amplitude of progressive pressure-wave component of p_r
p_{rs}: Amplitude of standing pressure-wave component of p_r
r_s, R_S: Height of skinplate rotation center [r_s is the dimensionless expression of R_s]
R_c: Radius of the bottom end of the weir plate
s, S: Integral variable
t, T: Time t is the dimensionless expression of T]
W: Function of the integral variable, s
W_0: Spanwise length
x, X: Horizontal coordinate along undisturbed free surface [x is the dimensionless expression of X(m)]
y, Y: Vertical coordinate [y is the dimensionless expression of Y (m)]
z, Z: Horizontal coordinate along vertical weir-plate [z is the dimensionless expression of Z (m)]
α_ψ: Water-to-gate mass ratio
β^*: Reduced reservoir depth, $\beta^* = d_r/d_0$
Δ: Vertical projection error
δ_p: Pressure correction coefficient
ΔC: Wave radiation damping coefficient
Δc_ψ: Reduced wave radiation damping coefficient
ΔI: Added moment of inertia
Δm_ψ: Reduced equivalent added mass
ΔQ: Discharge flow per unit width through the gate opening

ϕ, Φ: Velocity potential [ϕ is the dimensionless expression of Φ]
$\Phi^{(b)}$: Velocity potential induced by the push-and-draw variation
$\Phi^{(r)}$: Velocity potential induced by the flow-rate variation
ϕ_0, Φ_0: Phase function [ϕ_0 is the dimensionless expression of Φ_0]
ϕ_b: Phase-lag
η_j,ξ_0: Solution of a transcendental equation
$\tilde{\eta}$: Positive small constant representing small dissipative forces in water
ρ: Density of water
θ, Θ: Counterclockwise rotation angle of whole rigid gate [θ is the dimensionless expression of Θ]
Θ_0: Angular amplitude of rotational vibration of whole rigid gate
Θ_s: Gate inclination angle
ρ: Density of water
$\tilde{\mu}$: Positive small constant representing small dissipative force in water
ψ,Ψ: Counter-clockwise rotation angle of skinplate [ψ is the dimensionless expression of Ψ]
Ψ_0: Angular amplitude of rotational vibration of skinplate
Ω_a: Natural vibration frequency in air
Ω_w: Vibration frequency in water
ζ_a: In-air damping ratio
ζ_f: Wave radiation damping ratio
ζ_w: In-water streamwise damping ratio

Chapter 10
Theory of Self-Excited Coupled-Mode Vibration of Tainter Gates:
A Concern for Gate Designers

ABSTRACT

In general, any mechanism that produces an unbalanced moment may also serve to initiate a rigid body rotation of a Tainter gate about the trunnion pin. From the modal analysis testing on an intact Folsom Dam Tainter gate, and an understanding of the concepts of flow-rate variation pressure and push-and-draw pressure presented in Chapters 4 and 5, respectively, a conceptual model of the vibration mechanism can be formulated. The whole gate rotation induces a flow-rate variation pressure and a coupled inertia torque on the skinplate, as presented in Chapter 4. Both the flow-rate-variation pressure and the inertia torque excite the skinplate to rotate in a bending mode shape about a horizontal nodal line. In the present chapter we will develop the theory behind such an instability mechanism, called the self-excited coupled-mode instability, culminating in the graphical representation of the Folsom Dam gate instability in terms of a dynamic stability criterion diagram under the conditions at which failure occurred.

INTRODUCTION

In this book Tainter gates references can be found in most every of chapter. One special type of instability, eccentricity instability was presented in Chapter 8. The unbalanced moment acting on the gate due to the eccentricity is an essential feature of the eccentricity instability mechanism. In general, any mechanism that produces an unbalanced moment may also serve to initiate a rigid body rotation of the Tainter gate about the trunnion pin. In the present chapter we will present the theory behind another instability mechanism, called the self-excited coupled-mode instability.

DOI: 10.4018/978-1-5225-3079-4.ch010

Tainter gates, or radial gates as they are also known, are frequently used as crest gates on dam spillways for water level regulation. The design of these Tainter gates is such that the resultant hydraulic load due to pressure exerted on the skinplate usually passes through the trunnion pin. In this manner, the hydraulic load is usually borne by the trunnion pin as shown in Figure 1. Mechanical friction in Tainter gates is much lower than in other types of gates, such as vertical lift gates. Further, a portion of the gate weight is also carried by the trunnion pin, permitting the use of a relatively a small capacity hoisting motor and smooth operation of the gate. For this reason, Tainter gates are well suited for larger installations, such as the crest gates on the Itaipu Dam in Brazil, as was shown in Figure 1 in Chapter 8, as well as many other sites world-wide. Tainter gates are also installed on the dam crest at the Folsom Dam on the American River near Sacramento, California. A drawing of a Tainter gate from the Folsom Dam is shown in Figure 1. The gate has a height of 15.5 m, a width of 12.8 m with a skinplate radius of 14.33 m. The whole gate mass is 87.03×10^3 kg.

According to an eyewitness account, one of these massive Tainter gates experienced flow-induced vibrations and failed, early in the morning of July 17, 1995. The gate operator was on the catwalk just above the gate, opening the gate. He said he felt a small steady vibration starting up, initially very light, but then quickly intensifying. Upon his initial observation, the operator pushed the stop button, thinking to close the gate. As he turned from the control panel, he saw one side of the gate moving slowly in the downstream direction, rotating about the other side of the gate, like a large hinged garage door, as detailed by Ishii (1995a).

In an earlier case, a Tainter gate, with a height of 12 m, a width of 9 m, a skinplate radius of 13 m and a mass of 37×10^3 kg at the Wachi Dam in Japan failed and was swept downstream about 140 m on July 2, 1967 (Ishii et al., 1980). The team reviewing this gate failure in Japan suggested more than 35 years

Figure 1. Side view of the 87-ton Tainter gate at the Folsom Dam in California, showing two predominant modes of vibration
From Anami et al. (2006). © 2006 ASME. Used with permission.

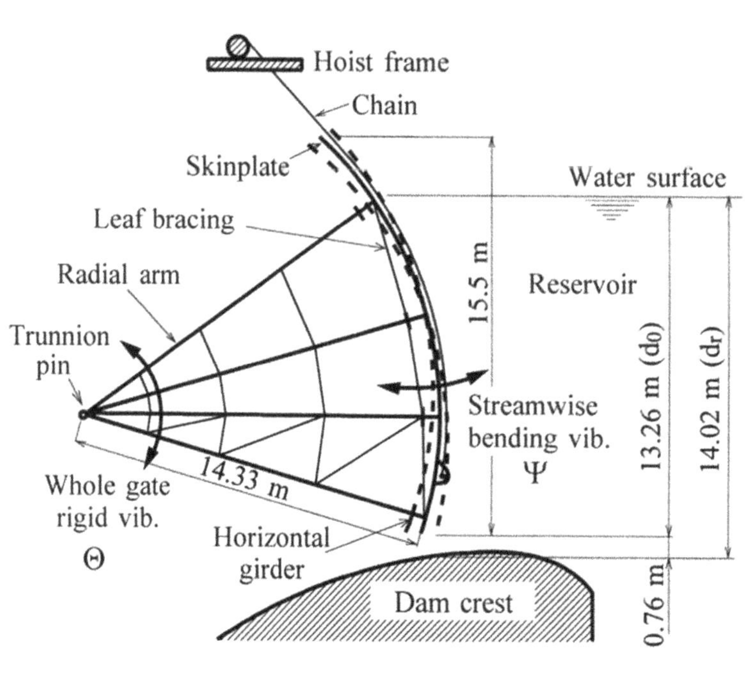

ago that flow-induced vibrations may have been a possible cause. However, at that time there was no known vibration mechanism and the structural dynamics of such large gates had not been well studied.

In examining the Wachi Dam gate failure, the eccentricity instability presented in Chapter 8 was identified. This dangerous type of flow-induced vibration of radial gates results from the eccentricity of the curved skinplate center relative to the trunnion center. A careful examination of the Wachi gate drawing revealed that the gate eccentricity was only 72 mm, and this small eccentricity was later shown to be insufficient to excite the eccentricity vibration mechanism.

Ishii (1995a) had an initial suspicion that eccentricity instability may have played a role in the Folsom Dam failure. For the onset of the eccentricity instability, however, the eccentricity of the skinplate center from the trunnion pin center would have had to have been more than 0.42 m for the Folsom gate. The radial gate at the Folsom Dam did not possess such a large eccentricity.

The Folsom Dam gate failure with an eyewitness report of vibration strongly suggested that an at-the-time unknown mechanism for a "non-eccentricity hydrodynamic instability" may have been active. The uncertainty of whether other Tainter gates with no skinplate eccentricity may be susceptible to this mechanism, made it necessary to understand how such a mechanism might function. In the interest of safety, Professor Ishii and his colleagues initiated a long-term series of experimental and theoretical studies that continued the analysis that started with eccentricity instability investigation.

As a first step in the exploration of a potential vibration mechanism for the Folsom Dam gates, in late September 1995, Ishii (1995b) conducted in-air experimental modal analysis tests on one of the remaining geometrically similar Folsom Dam gates (subsequently designated as the "original gate"), employing an impact hammer and accelerometers. From these modal analysis tests, a relatively precise assessment of the natural vibration characteristics of the failed gate was obtained. After this initial modal analysis, the remaining gates were reinforced. Subsequently, in the middle of November 1996, similar experimental modal analysis tests were conducted on one of the reinforced Folsom Dam gates in order to evaluate the dynamic effectiveness of the reinforcement (see Ishii, 1997 for details). The experimental modal analyses results can be found in Chapter 12 – Wachi Dam and Folsom Dam Failures, originally presented in detail in Anami et al. (2014).

The modal testing on the original gate identified two significant vibration modes. One mode was the whole gate rotation around the trunnion pin with a frequency of 6.88 Hz. The other mode was a streamwise bending vibration of the skinplate, associated with the deformation of the radial arm structures, with an in-air frequency of 26.9 Hz, characterized by predominantly streamwise bending of the skinplate about a horizontal nodal line corresponding to the location of the second horizontal girder, as illustrated in Figure 2.

From the modal analysis testing, and an understanding of the concepts of flow-rate variation pressure and push-and-draw pressure presented in Chapters 4 and 5, respectively, a conceptual model of the vibration mechanism can be formulated. The whole gate rotation induces a flow-rate variation pressure and a coupled inertia torque on the skinplate, as presented in Chapter 4. Both the flow-rate-variation pressure and the inertia torque excite the skinplate to rotate in a bending mode shape about a horizontal nodal line in the streamwise direction. Subsequently, the skinplate streamwise rotation results in a fluctuating push-and-draw pressure, presented in Chapters 5 and 9). The push-and draw pressure acting on the skinplate produces an additional large hydrodynamic force resulting in a significant added mass effect and an inertia torque that is fed back to excite the whole gate rotation.

Figure 2. Schematic view of the low frequency vibration mode of the Tainter gate skinplate showing a streamwise bending rotation in the vertical plane denoted as Ψ with a 1/2 wavelength bending mode shape in the spanwise direction, as well as the rigid-body rotation about the trunnion pin, denoted by Θ
From Anami et al. (2014). © 2014 IJMERR. Used with permission.

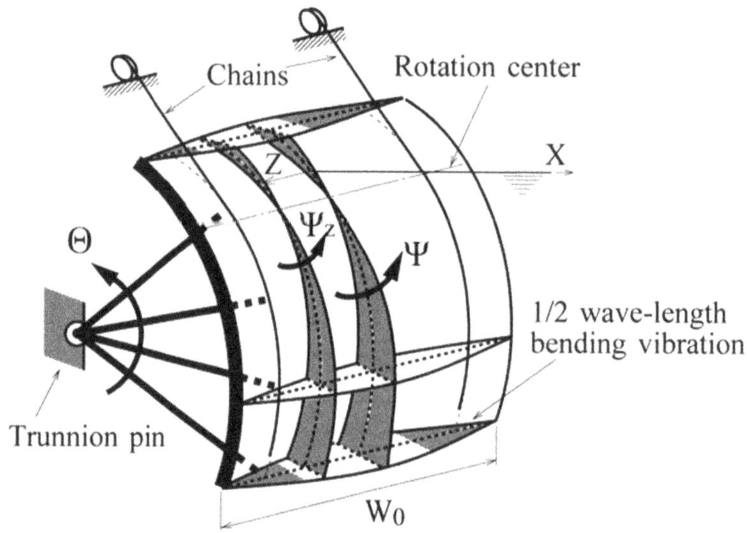

When the frequency of the bending mode approaches that of the whole-gate rigid body rotation, the displacement in one mode can provide excitation for the other. When the rigid body rotation and the skinplate bending have compatible frequencies and phasing, one mode generates the driving force for the other mode. The two vibration modes couple very effectively through the hydrodynamic forces and inertia torques, resulting in spontaneous self-excited coupled-mode vibration that grow very quickly. This mechanism may potentially excite any Tainter gate, even those with zero eccentricity between the trunnion center and the skinplate center.

Theoretical analysis of this potentially disastrous self-excited coupled-mode vibration requires the calculation of the push-and- draw pressure, as presented in Chapters 5 and 9, originally introduced by Anami & Ishii (1998a). The computation of push-and-draw pressure was subsequently undertaken for the Folsom Dam gate geometry by Anami et al. (2012a). The theoretical development of the self-excited coupled mode mechanism also requires a method to calculate the *in-water* natural vibration frequency of the skinplate rotation. Such a method was documented in Anami & Ishii (1998b).

In a later publication, Anami et al. (2012b) provide the calculation of the *in-water* natural vibration frequency for the skinplate of the Folsom Dam. Based on the measured in-air natural frequency of 26.9 Hz, Anami et al. (2012b) found the in-water natural vibration frequency of the skinplate bending mode for the Folsom Dam gate to be 6.46 Hz, under the conditions at the time of failure. The significant reduction in the bending mode frequency stems directly from large added mass effect, and produces a relatively tight frequency coherence between the skinplate bending mode and the whole gate rotational vibration around the trunnion pin at a frequency of 6.88 Hz.

In this chapter, previously developed expressions for the flow-rate-variation pressure from Chapter 4 and the push-and-draw pressure from Chapters 5 and 9 are applied to the coupled-mode vibration of a Tainter gate to derive the equations of motion assuming small amplitudes. The equations of motion are reduced to a non-dimensional form. The non-dimensional equations provide guidance on the appropriate non-dimensional parameters, such as a reduced added mass, reduced fluid damping and excitation coefficients for hydrodynamic forces, as well as the water-to-gate mass ratio and the moment-of-inertia ratio for vibrations. Ultimately, based on the derived equations of motion, a closed energy cycle is presented, which predicts the susceptibility of Tainter gates to the self-excited coupled-mode instability. The analysis is applied to the Folsom Dam Tainter gate under the conditions prevailing at the time of failure.

EQUATIONS OF MOTION

Coupled Torque of Inertia

Figure 2 provides a three-dimensional illustration of the skinplate bending vibration. In the streamwise vertical plane, the skinplate exhibits a low-frequency bending mode, essentially performing a rotational vibration around a horizontal nodal line, termed the "streamwise- rotation center line." The counter-clockwise rotation of the skinplate along the spanwise centerline, as shown in Figure 2 with the upstream reservoir to the right, is represented by Ψ. In the spanwise horizontal plane, the skinplate exhibits a half-wave-length bending mode shape with nodes at each end. Therefore, Ψ_z, the rotational angle of the skinplate at a given spanwise distance Z from the spanwise center, can be approximated by

$$\Psi_z = \Psi \sin\left(Z / W_0 + 1/2\right)\pi \tag{1}$$

where W_0 is the span of the skinplate, as shown in Figure 2.

A schematic view of the skinplate element with a small spanwise length of dZ is shown in Figure 3, where the element performs a rotation around the trunnion pin, represented by Θ, in addition to the streamwise counter-clockwise-rotation represented by Ψ_z, with a counterclockwise rotation defined as positive.

These two rotational vibrations can very effectively couple with each other through the inertia forces on the skinplate, that is, through the so-called "coupled inertia torque." The coupled inertia torque around the skinplate rotation center due to the whole gate rotational vibration around the trunnion pin is represented by $dM_{\psi\theta}$, while the coupled inertia torque around the trunnion pin due to the skinplate streamwise rotational vibration is denoted by $dM_{\theta\psi}$. Each differential coupled inertia torque can be obtained by integrating the moment due to the inertia forces on the infinitesimal skinplate element.

A small cross-section of the skinplate is indicated by the small black differential element in Figure 3. The mass of this differential element per the unit span is denoted by dm and the mass for the differential element with span dZ is $dmdZ$. Multiplying the mass by the tangential accelerations $R_\theta\ddot{\Theta}$ and $R_\Psi\ddot{\Psi}_Z$ yields the tangential inertia forces $\left(-dmdZR_\theta\ddot{\Theta}\right)$ and $\left(-dmdZR_\Psi\ddot{\Psi}_Z\right)$, which are shown as vectors in Figure 3. Here, R_θ and R_ψ represent the radii of rotation for each vibration, that is, the distance from the trunnion pin and from the skinplate center for small vibrations, and hence the centrifugal forces can be neglected as second order relative to the tangential forces.

The tangential inertia force due to the Θ vibration has an arm length of R^* from the skinplate rotation center, while the tangential inertia force due to the Ψ vibration has an arm length of $R_\psi/2$ from the

Figure 3. Cross-sectional view of skinplate undergoing counterclockwise (streamwise) bending and counterclockwise (upward) rigid-body rotational vibrations
From Anami et.al. (2014). © 2014 IJMERR. Used with permission.

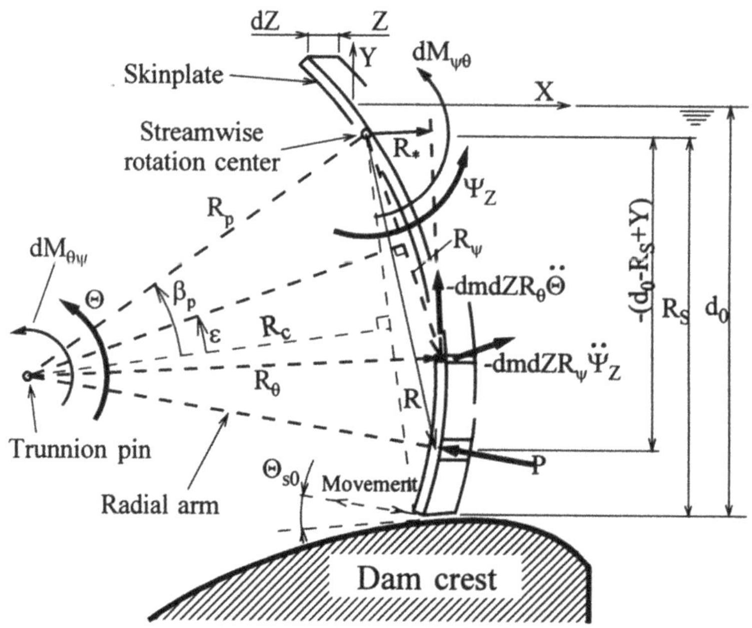

trunnion pin. Multiplying each differential tangential inertia force by the moment arm to the center of rotation for the complementary vibration and then integrating each differential tangential inertial moment over the whole skinplate with a spanwise length *dZ* yields the streamwise and up-and-down coupled inertia torques $dM_{\psi\theta}$ and $dM_{\theta\psi}$ given by the following expressions:

$$dM_{\psi\theta} = \int_m -dmdZR_\theta\ddot{\Theta}R_* \qquad \text{or} \qquad dM_{\psi\theta} = -\frac{I_{\psi\theta}}{W_0}\ddot{\Theta}dZ \tag{2}$$

$$dM_{\theta\psi} = \int_m -dmdZR_\psi\ddot{\Psi}_z\frac{R_\psi}{2} \qquad \text{or} \qquad dM_{\theta\psi} = -\frac{I_\psi}{2W_0}\ddot{\Psi}_z dZ \tag{3}$$

The subscript "*m*" on the integral represents integration over the spanwise-unit-length of the skinplate. $I_{\psi\theta}$ and I_ψ are the moments of inertia of the whole skinplate, defined respectively by

$$I_{\psi\theta} \equiv W_0\int_m R_\theta R_* dm \tag{4}$$

$$I_\psi \equiv W_0\int_m R_\psi^{\,2} dm \tag{5}$$

The moment of inertia $I_{\psi\theta}$ represents the inertia effect of the whole gate rigid-body vibration around the trunnion pin on the skinplate streamwise bending vibration. The moment of inertia I_{ψ} represents the inertia effect of the skinplate streamwise vibration on the whole gate rigid-body vibration around the trunnion pin. Together these two moments of inertia are called the "coupled moments-of-inertia" for the skinplate.

Basic Equations of Motion

Recall that we are considering gates with zero eccentricity. Therefore, the circular-arc skinplate center is assumed to coincide precisely with the trunnion pin, and the hydrodynamic pressure indicated by P in Figure 3 passes precisely through the trunnion pin center. Therefore, the hydrodynamic pressure cannot appear directly in the equation of motion of the whole gate rotational vibration around the trunnion pin. The whole gate is accelerated by the coupled inertia torque which can be obtained by integrating $dM_{\theta\psi}$ over the whole gate span, resulting in following equation of motion:

$$I_{\theta}\left(-\ddot{\Theta}-2\zeta_{a\theta}\Omega_{a\theta}\dot{\Theta}-\Omega_{a\theta}^{2}\Theta\right)+\int_{Z}dM_{\theta\psi}=0 \tag{6}$$

where I_{θ} is the moment of inertia of the whole gate around the trunnion pin, $\Omega_{a\theta}$ and $\zeta_{a\theta}$ are the in-air circular frequency and in-air damping ratio of the whole gate rotational vibration, respectively. The subscript "Z" on the integral represents integration over the gate span. Using Equations 1 and 3, the integration on Z leads to the following:

$$\int_{z}dM_{\theta\psi}=-\frac{I_{\psi}}{2W_{0}}\ddot{\Psi}\int_{-W_{0}/2}^{W_{0}/2}\sin\left(\frac{Z}{W_{0}}+\frac{1}{2}\right)\pi dZ=-\frac{I_{\psi}}{2W_{0}}\ddot{\Psi}\frac{2W_{0}}{\pi}=-\frac{I_{\psi}\ddot{\Psi}}{\pi} \tag{7}$$

With this value for the integral, Equation 6 for the rigid-body motion of the whole gate around the trunnion pin can be re-written in the following form:

$$\ddot{\Theta}+2\Omega_{a\theta}\zeta_{a\theta}\dot{\Theta}+\Omega_{a\theta}^{2}\Theta+\frac{1}{\pi}\frac{I_{\psi}}{I_{\theta}}\ddot{\Psi}=0 \tag{8}$$

In contrast to its role in the excitation of the whole gate rigid-body vibration, the hydrodynamic pressure P participates explicitly in the streamwise rotational vibration of the skinplate. The skinplate streamwise rotational vibration is excited by both the hydrodynamic force and the coupled inertia torque $dM_{\psi\theta}$, and thus is governed by the following equation of motion:

$$\left(\ddot{\Psi}_{z}+2\zeta_{a\psi}\Omega_{a\psi}\dot{\Psi}_{z}+\Omega_{a\psi}^{2}\Psi_{z}\right)dZ-\frac{W_{0}}{I_{\psi}}dM_{\psi\theta}+\frac{W_{0}}{I_{\psi}}\int_{R}PRdR\cdot dZ=0 \tag{9}$$

R represents the chord length from the skinplate rotation center to the pressure P. Integrating Equation 9 over the gate span, the equation of motion of the whole skinplate is given by

$$\int_Z \left(\ddot{\Psi}_z + 2\zeta_{a\psi}\Omega_{a\psi}\dot{\Psi}_z + \Omega_{a\psi}^2\Psi_z\right)dZ - \frac{W_0}{I_\psi}\int_Z dM_{\psi\theta} + \frac{W_0}{I_\psi}\int_Z\int_R PR\,dR\cdot dZ = 0 \tag{10}$$

As shown in Figure 2, if the skinplate vibrates in the spanwise half-wavelength bending mode with a node at each side of the skinplate, the vibration-induced hydrodynamic pressure is naturally influenced by the half-wavelength bending vibration. Therefore, the integration with respect to Z can be carried out for all terms except for the hydrodynamic pressure term, resulting in the following expression:

$$\ddot{\Psi} + 2\Omega_{a\psi}\zeta_{a\psi}\dot{\Psi} + \Omega_{a\psi}^2\Psi + \frac{\pi}{2}\frac{I_{\psi\theta}}{I_\psi}\ddot{\Theta} + \frac{\pi}{2I_\psi}\int_{-W_0/2}^{W_0/2} dZ\int_R PRdR = 0 \tag{11}$$

where Equation 2 was substituted into Equation 10. The symbols $\Omega_{a\psi}$ and $\zeta_{a\psi}$ represent the in-air circular frequency and in-air damping ratio of the skinplate streamwise natural vibration, respectively.

Hydrodynamic Pressure

When the whole gate performs rotation around the trunnion pin, the flow-rate beneath the gate varies, producing (as discussed in Chapter 4) the flow-rate variation pressure, $P_{r\theta}$. When the circular-arc skinplate performs streamwise rotational vibrations, the upstream water is pushed and drawn by the streamwise movement of the boundary, thus producing (as discussed in Chapter 5) the push-and-draw pressure, $P_{b\psi}$. If the movement of the lower edge of the skinplate is not in the tangential direction of the dam crest curve, as shown in Figure 3, an additional flow-rate variation is induced by the streamwise rotation of the skinplate, producing (see Chapter 9) the additional flow-rate-variation-pressure, $P_{r\psi}$. The subscripts "*r*" and "*b*" on these pressure terms represent, respectively, the flow-**r**ate variation under the gate and the flow-**b**oundary movement due to skinplate vibration. The second subscripts, "θ" and "ψ", represent motion generating the effect, the upward rotation of the whole gate and the streamwise rotation of the skinplate, respectively. Assuming small amplitude vibrations for both of the skinplate streamwise rotational vibration and whole gate rotational vibration around the trunnion pin, and using the superposition principle, the resultant hydrodynamic pressure fluctuation P due to both vibrations occurring simultaneously is given by the summation of these pressure components:

$$P = P_{b\psi} + P_{r\theta} + P_{r\psi} \tag{12}$$

Since the vibration amplitude in the half-wavelength mode is far smaller than the spanwise length of the skinplate, it is assumed here that the hydrodynamic pressure components are simply proportional to the streamwise and tangential vibration amplitudes of the lower-end of the skinplate. Each pressure component was analyzed, by applying the potential theory developed in Chapters 4 and 5, drawing on the work by Rayleigh (1945) for dissipative-wave radiation as applied by Ishii (1992) and Anami et al. (2012b). The potential theory yields values of pressure fluctuations that agree well with experimental results from Anami et al. (2012b), as discussed in Chapter 9. The three hydrodynamic pressure components can be reduced to the following expressions:

$$p_{b\psi} \equiv \frac{P_{b\psi} / \rho g}{R_s \Psi_0 \sin\,(Z / W_0 + 1/2)\pi} \tag{13a}$$

$$p_{r\theta} \equiv \frac{P_{r\theta} / \rho g}{\sqrt{2}\, c_f R_c \Theta_0} \tag{13b}$$

$$p_{r\psi} \equiv \frac{P_{r\psi} / \rho g}{\sqrt{2 c_f} k R_s \Psi_0 \sin\left(Z / W_0 + 1/2\right)\pi} \qquad \text{(where } k = -\sin\Theta_{s0}) \tag{13c}$$

Note that Θ_{s0} is the angle of the streamwise vibration direction relative to the tangent to the dam crest, as shown in Figure 3; k is the press-shut coefficient, which takes on a negative value for a press-open device (as the skinplate moves downward, the gate opening increases), and a positive value for a press-shut device (as the skinplate moves downward, the gate opening decreases). The symbol c_f represents the instantaneous flow-rate-variation coefficient. In Equations 13, the water head of each pressure fluctuation is divided by a corresponding representative length. The length used for the flow-rate-variation pressure (Equation 13b) was the magnitude of the change in the gate-opening height. For the push-and-draw pressure (Equation 13a) and the additional flow-rate-variation pressure (Equation 13c), the characteristic length is the streamwise vibration amplitude of the skinplate at its bottom end, since both depend on the skinplate rotation. The amplitudes of Θ and Ψ are represented by Θ_0 and Ψ_0, respectively. The radius of the skinplate is represented by R_a. The length R_c is the rotation radius of skinplate lower end about the trunnion pin, as shown in Figure 3. The length R_s is rotation radius of the lower end of the skinplate about its horizontal streamwise-rotation center.

The dimensionless pressure fluctuations defined by Equations 13 can be ultimately reduced to summations of components in phase with the respective velocities and accelerations, as given by the following expressions:

$$p_{r\theta} = p_{rs}\left(\frac{F_{a\theta}}{F}\right)\left(-\theta'\right) + p_{rp}\left(\frac{F_{a\theta}}{F}\right)^2 \theta'' \tag{14a}$$

$$p_{b\psi} = \delta_p \left\{ p_{bs}\left(\frac{F_{a\theta}}{F}\right)^2 \psi'' + p_{bp}\left(\frac{F_{a\theta}}{F}\right)\psi' \right\} \tag{14b}$$

$$p_{r\psi} = p_{rs}\left(\frac{F_{a\theta}}{F}\right)\left(-\psi'\right) + p_{rp}\left(\frac{F_{a\theta}}{F}\right)^2 \psi'' \tag{14c}$$

In Equation 14b, δ_p is the pressure correction coefficient that, as discussed in Chapter 9, accounts for the inclination of the circular-arc skinplate relative to the channel bed. The value of δ_p was obtained by model experiments, as described in Anami et al. (2012b). In the present formulation, the in-air vibration period $1/\Omega_{a\theta}$ of the whole gate vibration around the trunnion pin is adopted as the representative time scale for non-dimensionalizing the time T; further, the rotating angular amplitudes Ψ_0 and Θ_0 are adopted as amplitude scales to reduce vibration amplitudes, respectively:

$$t \equiv \Omega_{a\theta}T; \qquad \psi \equiv \Psi / \Psi_0; \quad \theta \equiv \Theta / \Theta_0 \tag{15}$$

F and $F_{a\theta}$ represent the Froude number and the basic Froude number, defined by the following expressions:

$$F \equiv \sqrt{\frac{d_0}{g}}\Omega_w \quad \text{and} \quad F_{a\theta} \equiv \sqrt{\frac{d_0}{g}}\Omega_{a\theta} \tag{16}$$

which characterize the fluctuating flow field associated with vibrations at circular frequencies of Ω_w in water and $\Omega_{a\theta}$ in air. The skinplate submergence depth is represented by d_0.

The dimensionless pressure amplitudes represented by p_{rs}, p_{rp}, p_{bs} and p_{bp} in Equation 14 were presented in Chapters 4, 5, and 9 as series summations of functions of the Froude number F, and the reduced height of the streamwise-rotation center, r_s, defined by

$$r_s = R_s / d_0 \tag{17}$$

Subscripts "s" and "p" on the pressure amplitudes in Equation 14 represent the standing-wave and progressive-wave pressure components, respectively.

Reduced Equations of Motion

Using the non-dimensional parameters introduced in Equation 15, the equation of motion of the whole gate rotational vibration around the trunnion pin, given by Equation 8, can be reduced to the following form:

$$\theta'' + 2\zeta_{a\theta}\theta' + \theta + \frac{1}{\pi}\frac{\alpha_I}{\alpha_*}\psi'' = 0 \tag{18}$$

where α_* represents the angular amplitude ratio of the whole gate rotational vibration around the trunnion pin Θ to the skinplate streamwise rotational vibration Ψ:

$$\alpha_* \equiv \Theta_0 / \Psi_0 \tag{19}$$

In addition, using the hydrodynamic pressures given by Equations 12 and 13, as well as the non-dimensional variables introduced by Equation 15, the equation of motion for the skinplate streamwise rotational vibration, Equation 11, can be written in the following form:

$$\begin{aligned} &\psi'' + 2\zeta_{a\psi}\gamma_{\psi\theta}\psi' + \gamma_{\psi\theta}^2\psi + \frac{\pi}{2}\alpha_{I\psi}\alpha_*\theta'' + \frac{\alpha_\psi}{F_{a\theta}^2}\int_{-1}^{0} p_{b\psi}^*\left(1 - \frac{1+y}{r_s}\right)dy \\ &+ \frac{\pi}{2}\sqrt{2}\,c_f\alpha_*\frac{1}{r_{sa}}\cdot\frac{\alpha_\psi}{F_{a\theta}^2}\int_{-1}^{0} p_{r\theta}\left(1 - \frac{1+y}{r_s}\right)dy + \sqrt{2}c_f k\cdot\frac{\alpha_\psi}{F_{a\theta}^2}\int_{-1}^{0} p_{r\psi}\left(1 - \frac{1+y}{r_s}\right)dy = 0 \end{aligned} \tag{20}$$

where $\gamma_{\psi\theta}$ represents the in-air natural vibration frequency ratio of the skinplate streamwise rotational vibration Ψ to the whole gate rotational vibration around the trunnion pin Θ:

$$\gamma_{\psi\theta} \equiv \Omega_{a\psi} / \Omega_{a\theta} \tag{21}$$

Further, $\alpha_{I\psi}$ represents the ratio of the coupled moment-of-inertia $I_{\psi\theta}$ to the skinplate moment-of-inertia I_ψ, and r_{sa} represents the rotation radius ratio, defined respectively as

$$\alpha_{I\psi} \equiv I_{\psi\theta} / I_\psi \quad \text{and} \quad r_{sa} \equiv R_S / R_c \tag{22}$$

In addition, α_ψ is the mass ratio of the representative water mass to the skinplate mass given by:

$$\alpha_\psi \equiv \frac{\rho d_0^2 W_0}{I_\psi / R_S^2} \tag{23}$$

By substituting the non-dimensional hydrodynamic pressures in the form given by Equations 14 into the equation of motion, Equation 20, one can arrive at the following expression:

$$\begin{aligned} &\left(1 + \delta_p\alpha_\psi\Delta m_\psi + \frac{2}{\pi}\sqrt{2}c_f k\alpha_\psi\Delta m_\theta\right)\psi'' + 2\gamma_{\psi\theta}\left(\zeta_{a\psi} + \zeta_{f\psi} - \frac{1}{\pi}\sqrt{2}c_f k\frac{\alpha_\psi\Delta c_\theta}{\gamma_{\psi\theta}F_{a\theta}}\right)\psi' + \gamma_{\psi\theta}^2\psi \\ &= \alpha_*\left\{-\left(\frac{\pi}{2}\alpha_{I\psi} + \sqrt{2}c_f\frac{\alpha_\psi\Delta m_\theta}{r_{sa}}\right)\theta'' + \sqrt{2}c_f\frac{\alpha_\psi\Delta c_\theta}{r_{sa}F_{a\theta}}\theta'\right\} \end{aligned} \tag{24}$$

where $\zeta_{f\psi}$ represents the wave-radiation damping ratio, as defined by

$$\zeta_{f\psi} \equiv \frac{\delta_p \alpha_\psi \Delta c_\psi}{2\gamma_{\psi\theta} F_{a\theta}} \tag{25}$$

The symbols Δm_ψ and Δc_ψ are the equivalent added mass and the wave-radiation damping coefficients due to the skinplate streamwise rotational vibration, respectively. They are given as

$$\Delta m_\psi(F) = \frac{1}{F^2}\int_{-1}^{0}\left(1-\frac{1+y}{r_s}\right)p_{bs}(0,y)\,dy \tag{26a}$$

$$\Delta c_\psi(F) = \frac{1}{F}\int_{-1}^{0}\left(1-\frac{1+y}{r_s}\right)p_{bp}(y)\,dy \tag{26b}$$

The symbols Δm_θ and Δc_θ are the equivalent added mass and the fluid-excitation coefficient due to the whole gate rotational vibration, respectively. They are given as

$$\Delta m_\theta = \frac{\pi}{2}\frac{1}{F^2}\int_{-1}^{0}\left(1-\frac{1+y}{r_s}\right)p_{rp}\,dy \tag{27a}$$

$$\Delta c_\theta = \frac{\pi}{2}\frac{1}{F}\int_{-1}^{0}\left(1-\frac{1+y}{r_s}\right)p_{rs}\,dy \tag{27b}$$

In Equation 24, the equation of motion for the skinplate streamwise rotational vibration, the second term on the right-hand side, the one proportional to velocity, is most significant for stability. This term results from the flow-rate variation and drives the coupled-mode self-excited vibration. The parameter $\zeta_{f\psi}$ in the second term on the left-hand side represents the wave-radiation damping effect due to dissipative waves and is dependent upon the dissipative wave-radiation damping coefficient Δc_ψ. As the Froude number increases, the dissipative wave-radiation damping coefficient Δc_ψ decreases rapidly and takes on an almost zero value for large Froude number. The parameter $\alpha_\psi \Delta m_\psi$ in the first term on the left-hand side represents the added mass effect of the fluid, which significantly reduces the in-air natural vibration frequency when the gate vibrates in water.

The calculated values for the flow-induced factors Δm_θ, Δc_θ, Δm_ψ, and Δc_ψ are shown in Figure 4. All of these flow-induced factors are functions of the Froude number. As the Froude number increases, the reduced added mass Δm_θ decreases rapidly and is nearly zero for Froude numbers F larger than about 2. As a result, the reduced added mass effect due to the whole gate rotational vibration around the trunnion pin has no effect on either the whole gate vibration or on the skinplate streamwise rotational vibration. When the Froude number F is greater than about 10, the dissipative wave-radiation damping coefficient Δc_ψ also approaches zero. In contrast, the fluid excitation coefficient Δc_θ increases rapidly from zero and

Figure 4. Equivalent added mass coefficients, Δm_ψ and Δm_θ, wave-radiation damping coefficient, Δc_ψ, and fluid-excitation coefficient, Δc_θ, for skinplate vibration in the 1/2 wavelength transverse (horizontal) bending mode
From Anami & Ishii (2001).

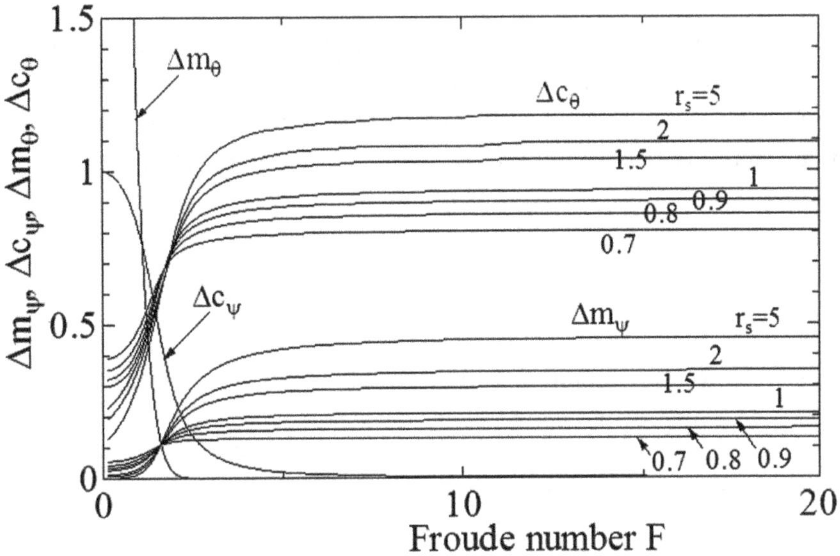

asymptotically approaches a comparatively large value, signifying a comparatively large fluid-excitation effect for sufficiently large Froude numbers. This fluid excitation drives the self-excited vibration of the coupled-mode vibration system. In addition, the reduced added mass Δm_ψ also increases rapidly from zero, and approaches an asymptotic value at large Froude numbers resulting in a significant frequency reduction for the streamwise rotational vibration relative to its in-air value.

The equation of motion, Equation 24, for the skinplate streamwise rotational vibration when arranged in standard form with the coefficient of ψ'' equal to 1.0 becomes:

$$\psi'' + 2\gamma_{\psi\theta} \frac{\zeta_{a\psi} + \zeta_{f\psi} - \frac{\sqrt{2}}{\pi} c_f k \frac{\alpha_\psi \Delta c_\theta}{F_{a\theta}\gamma_{\psi\theta}}}{1 + \delta_p \alpha_\psi \Delta m_\psi + \frac{2}{\pi}\sqrt{2}c_f k \alpha_\psi \Delta m_\theta} \psi' + \frac{\gamma^2_{\psi\theta}}{1 + \delta_p \alpha_\psi \Delta m_\psi + \frac{2}{\pi}\sqrt{2}c_f k \alpha_\psi \Delta m_\theta} \psi$$
$$= \alpha_* \frac{-\left(\frac{\pi}{2}\alpha_{I\psi} + \sqrt{2}c_f \frac{\alpha_\psi \Delta m_\theta}{r_{sa}}\right)\theta'' + \sqrt{2}c_f \frac{\alpha_\psi \Delta c_\theta}{r_{sa} F_{a\theta}} \theta'}{1 + \delta_p \alpha_\psi \Delta m_\psi + \frac{2}{\pi}\sqrt{2}c_f k \alpha_\psi \Delta m_\theta} \quad (28)$$

The coefficient of ψ gives the in-water to in-air vibration frequency ratio γ_{nw} in the following form:

$$\gamma_{nw}\left(\equiv\frac{\Omega_{nw\psi}}{\Omega_{a\theta}}\right)=\frac{\gamma_{\psi\theta}}{\sqrt{1+\delta_{p}\alpha_{\psi}\Delta m_{\psi}+\frac{2}{\pi}\sqrt{2}c_{f}k\alpha_{\psi}\Delta m_{\theta}}} \tag{29}$$

The denominator $\Omega_{a\theta}$ represents the in-air vibration frequency of the whole gate rotational vibration around the trunnion pin, which takes on essentially the same value as the in-water natural vibration frequency, since no added mass effect arises due to the whole gate vibration around the trunnion pin.

Using the in-air vibration frequency ratio $\gamma_{\psi\theta}\equiv\Omega_{a\psi}\Omega_{a\theta}$ defined by Equation 22, γ_{nw} can also be expressed as follows:

$$\gamma_{nw}=\gamma_{nw\psi}\gamma_{\psi\theta} \tag{30}$$

where $\gamma_{nw\psi}$ is the in-water to in-air vibration frequency ratio of the skinplate streamwise vibration:

$$\gamma_{nw\psi}\left(\equiv\frac{\Omega_{nw\psi}}{\Omega_{a\psi}}\right)=\frac{1}{\sqrt{1+\delta_{p}\alpha_{\psi}\Delta m_{\psi}+\frac{2}{\pi}\sqrt{2}c_{f}k\alpha_{\psi}\Delta m_{\theta}}} \tag{31}$$

With these vibration frequency ratios, the equation of motion for the skinplate, Equation 28, can be ultimately arranged as follows:

$$\begin{aligned}&\psi''+2\gamma_{nw\psi}\gamma_{nw}\left(\zeta_{a\psi}+\zeta_{f\psi}-\frac{\sqrt{2}}{\pi}c_{f}k\frac{\alpha_{\psi}\Delta c_{\theta}}{F_{a\theta}\gamma_{\psi\theta}}\right)\psi'+\gamma_{nw}^{2}\psi\\&=\alpha_{*}\gamma_{nw\psi}^{2}\left\{-\left(\frac{\pi}{2}\alpha_{I\psi}+\sqrt{2}c_{f}\frac{\alpha_{\psi}\Delta m_{\theta}}{r_{sa}}\right)\theta''+\sqrt{2}c_{f}\frac{\alpha_{\psi}\Delta c_{\theta}}{r_{sa}F_{a\theta}}\theta'\right\}\end{aligned} \tag{32}$$

Equation 32 is the final form of the equation of motion of the skinplate streamwise vibration.

APPROXIMATE SOLUTION

The coupled-mode vibration system involves two natural vibration frequencies, one for the whole gate rotational vibration around the trunnion pin, and one for the skinplate streamwise rotational vibration. The coupled-mode vibration occurs, in essence, when the two frequencies coalesce. The resulting synchronized vibration occurs with one of the natural vibration modes serving as the driving excitation for the other mode. Through the feedback of one mode driving the other, the coupled-mode self-excited vibration is built up. In the following, the driving vibration will be called the "main vibration", and the other mode that responds to the forcing of the first mode will be called the "response vibration" or the "dependent vibration". With this understanding of the physical situation, the equations of motion, Equations 18 and 32, for coupled-mode vibration can be solved approximately.

Whole Gate Mode Synchronized With Skinplate Streamwise Vibration

First, assume the following periodic solution for the main vibration, that is, for the skinplate streamwise rotational vibration:

$$\psi = \cos \gamma t \tag{33}$$

γ is the frequency ratio of the in-water streamwise rotational vibration $\Omega_{w\psi}$ to the representative in-air vibration $\Omega_{a\theta}$:

$$\gamma \equiv \Omega_{w\psi} / \Omega_{a\theta} \tag{34}$$

With this frequency, the Froude number F is given by

$$F \equiv \sqrt{\frac{d_0}{g}} \Omega_{w\psi} = F_{a\theta}\gamma \tag{35}$$

Therefore, the equation of motion, Equation 18, of the whole gate rotational vibration around the trunnion pin (response vibration) is expressed as follows:

$$\theta'' + 2\zeta_{a\theta}\theta' + \theta = \frac{1}{\pi}\frac{\alpha_I}{\alpha_*}\gamma^2 \cos \gamma t \tag{36}$$

This is the equation of motion of the so-called forced vibration (response vibration) with the cos γt term on the right-hand side. This forced vibration has a solution in the following form:

$$\theta = \cos\left(\gamma t - \varphi\right) \tag{37}$$

The problem can be reduced to one of calculating the angular-amplitude ratio α_* and phase-lag ϕ from the two expressions, Equations 36 and 37. Substituting Equation 37 into the equation of motion for the whole gate as a rigid-body, Equation 36, one can arrive at the following equation:

$$\sqrt{\left(1-\gamma^2\right)^2 + \left(2\gamma\zeta_{a\theta}\right)^2} \cos\left(\gamma t - \varphi + \varphi_\alpha\right) = \frac{1}{\pi}\frac{\alpha_I}{\alpha_*}\gamma^2 \cos \gamma t \tag{38}$$

where ϕ_α is given by

$$\varphi_\alpha = \tan^{-1}\frac{2\gamma\zeta_{a\theta}}{1-\gamma^2} \tag{39}$$

As a result, the following solutions α_* and ϕ can be derived:

$$\alpha_* = \frac{1}{\pi} \frac{\alpha_I \gamma^2}{\sqrt{\left(1-\gamma^2\right)^2 + \left(2\gamma\zeta_{a\theta}\right)^2}} \tag{40}$$

$$\varphi\left(=\varphi_\alpha\right) = \tan^{-1}\frac{2\gamma\zeta_{a\theta}}{1-\gamma^2} \tag{41}$$

Since the responding whole gate rotational vibration θ was obtained for the assumed streamwise rotational vibration ψ, the feedback process of the above solution to the main skinplate streamwise vibration must be analyzed. Upon substitution of Equation 37 into the equation of motion for the skinplate streamwise rotational vibration (the main vibration), Equation 32, the following equation of self-excited vibration can be derived:

$$\left(1+\gamma_{nw\psi}^2 \Delta m_{\theta\psi}\right)\psi'' + 2\gamma_{nw}\gamma_{nw\psi}\left(\zeta_{a\psi} + \zeta_{f\psi} - \frac{\sqrt{2}}{\pi}c_f k \frac{\alpha_\psi \Delta c_\theta}{F_{a\theta}\gamma_{\psi\theta}} - \frac{\Delta c_{\theta\psi}}{\gamma_{\psi\theta}}\right)\psi' + \gamma_{nw}^2\psi = 0 \tag{42}$$

where $\Delta m_{\theta\psi}$ and $\Delta c_{\theta\psi}$ are the reduced added mass coefficient and reduced fluid-excitation coefficient, respectively, due to the coupling of the two vibration modes:

$$\Delta m_{\theta\psi} \equiv \alpha_* \left\{\left(\frac{\pi}{2}\alpha_{I\psi} + \sqrt{2}c_f \frac{\alpha_\psi \Delta m_\theta}{r_{sa}}\right)\cos\varphi + \sqrt{2}\, c_f \frac{\alpha_\psi \Delta c_\theta}{r_{sa}F_{a\theta}}\frac{1}{\gamma}\sin\varphi\right\} \tag{43}$$

$$\Delta c_{\theta\psi} \equiv \frac{\alpha_*}{2}\left\{-\left(\frac{\pi}{2}\alpha_{I\psi} + \sqrt{2}c_f \frac{\alpha_\psi \Delta m_\theta}{r_{sa}}\right)\gamma\sin\varphi + \sqrt{2}c_f \frac{\alpha_\psi \Delta c_\theta}{r_{sa}F_{a\theta}}\cos\varphi\right\} \tag{44}$$

Subsequently, Equation 42 can be arranged as follows:

$$\psi'' + 2\gamma_{\psi\theta}\frac{\zeta_{a\psi} - \xi_{f\psi}}{1+\gamma_{nw\psi}^2\Delta m_{\theta\psi}}\psi' + \frac{\gamma_{nw}^2}{1+\gamma_{nw\psi}^2\Delta m_{\theta\psi}}\psi = 0 \tag{45}$$

where $\xi_{f\psi}$ is the fluid-excitation ratio, defined by

$$\xi_{f\psi} \equiv \frac{\Delta c_{\theta\psi}}{\gamma_{\psi\theta}} - \zeta_{f\psi} + \frac{\sqrt{2}}{\pi} c_f k \frac{\alpha_\psi \Delta c_\theta}{F_{a\theta} \gamma_{\psi\theta}} \tag{46}$$

The square root of the coefficient on ψ in Equation 45 represents the actual frequency ratio $\gamma_{w\psi\theta}$ of the in-water skinplate streamwise rotational vibration to the whole gate in-air vibration frequency:

$$\gamma_{w\psi\theta} \left(\equiv \frac{\Omega_{w\psi}}{\Omega_{a\theta}} \right) = \frac{\gamma_{nw}}{\sqrt{1 + \gamma_{nw\psi}^2 \Delta m_{\theta\psi}}} = \frac{\gamma_{\psi\theta}}{\sqrt{1 + \delta_p \alpha_\psi \Delta m_\psi + \frac{2}{\pi}\sqrt{2} c_f k \alpha_\psi \Delta m_\theta + \Delta m_{\theta\psi}}} \tag{47}$$

If the actual frequency ratio $\gamma_{w\psi\theta}$ obtained here is not in agreement with the frequency ratio γ of Equation 34 assumed first, $\gamma_{w\psi\theta}$ is replaced by γ, and the ultimate solutions of the vibration ratio $\gamma_{w\psi\theta}$ can be obtained by iterative calculations of Equations 40, 41, 43, and 47. Using the solution for the frequency ratio, the equation of motion, Equation 45, can be reduced to

$$\psi'' + 2\frac{\gamma_{w\psi\theta}^2}{\gamma_{\psi\theta}} \left(\zeta_{a\psi} - \xi_{f\psi} \right) \psi' + \gamma_{w\psi\theta}^2 \psi = 0 \tag{48}$$

From the coefficient on ψ', the actual excitation ratio ξ_ψ of the skinplate streamwise rotational vibration (main vibration) can be given by

$$\xi_\psi = \frac{\gamma_{w\psi\theta}}{\gamma_{\psi\theta}} \left(\xi_{f\psi} - \zeta_{a\psi} \right) \tag{49}$$

Skinplate Streamwise Vibration Synchronized With Whole Gate Rotational Vibration

Here, assume the following periodic solution for the whole gate rotational vibration around the trunnion pin:

$$\theta = \cos \gamma t \tag{50}$$

where γ is the frequency ratio of in-water rotational vibration around the trunnion pin $\Omega_{w\theta}$ to the representative in- air vibration $\Omega_{a\theta}$:

$$\gamma \equiv \Omega_{w\theta} / \Omega_{a\theta} \tag{51}$$

Using the in-water vibration frequency $\Omega_{w\theta}$, the Froude number F is given by

$$F \equiv \sqrt{\frac{d_0}{g}}\Omega_{w\theta} = F_{a\theta}\gamma \tag{52}$$

Using Equation 50, the right-hand side of the equation of motion, Equation 32, for the skinplate streamwise vibration can be arranged as follows:

$$\text{Right-hand-side} = \alpha_* \gamma_{nw\psi}^2 \left\{ \left(\frac{\pi}{2}\alpha_{I\psi} + \sqrt{2}c_f \frac{\alpha_\psi \Delta m_\theta}{r_{sa}} \right) \gamma^2 \cos\gamma t - \sqrt{2}c_f \frac{\alpha_\psi \Delta c_\theta}{r_{sa}F_{a\theta}} \gamma \sin\gamma t \right\} = \alpha_* f_\theta \cos\left(\gamma t + \varphi_0\right) \tag{53}$$

where f_θ and the phase-lag φ_0 are given respectively by

$$f_\theta \equiv \gamma_{nw\psi}^2 \gamma \sqrt{\left[\left(\frac{\pi}{2}\alpha_{I\psi} + \sqrt{2}c_f \frac{\alpha_\psi \Delta m_\theta}{r_{sa}} \right)\gamma\right]^2 + \left(\sqrt{2}c_f \frac{\alpha_\psi \Delta c_\theta}{r_{sa}F_{a\theta}} \right)^2} \tag{54}$$

$$\varphi_0 \equiv \tan^{-1} \frac{\sqrt{2}c_f \dfrac{\alpha_\psi \Delta c_\theta}{r_{sa}F_{a\theta}}}{\left(\dfrac{\pi}{2}\alpha_{I\psi} + \sqrt{2}c_f \dfrac{\alpha_\psi \Delta m_\theta}{r_{sa}} \right)\gamma} \tag{55}$$

Substituting Equation 53 into the right-hand side of the equation of motion, Equation 32, the following equation of motion for the forced vibration (the response vibration) can be obtained:

$$\psi'' + 2\gamma_{nw\psi}\gamma_{nw}\left(\zeta_{a\psi} + \zeta_{f\psi} - \frac{\sqrt{2}}{\pi}c_f k \frac{\alpha_\psi \Delta c_\theta}{F_{a\theta}\gamma_{\psi\theta}} \right)\psi' + \gamma_{nw}^2\psi = \alpha_* f_\theta \cos\left(\gamma t + \varphi_0\right) \tag{56}$$

The solution of this forced vibration can be assumed to be of the following form:

$$\psi = \cos\left(\gamma t + \varphi_0 - \varphi\right) \tag{57}$$

where the angular-amplitude ratio α_* and phase-lag φ are given by

$$\alpha_* = \frac{1}{f_\theta}\sqrt{\left(\gamma_{nw}^2 - \gamma^2\right)^2 + \left\{ 2\gamma_{nw}\gamma_{nw\psi}\gamma \left(\zeta_{a\psi} + \zeta_{f\psi} - \frac{\sqrt{2}}{\pi}c_f k \frac{\alpha_\psi \Delta c_\theta}{F_{a\theta}\gamma_{\psi\theta}} \right) \right\}^2} \tag{58}$$

$$\varphi = \tan^{-1} \frac{2\gamma_{nw}\gamma_{nw\psi}\gamma\left(\zeta_{a\psi} + \zeta_{f\psi} - \frac{\sqrt{2}}{\pi} c_f k \frac{\alpha_\psi \Delta c_\theta}{F_{a\theta}\gamma_{\psi\theta}}\right)}{\gamma_{nw}^2 - \gamma^2} \tag{59}$$

Since the responding skinplate streamwise vibration ψ was obtained for the main whole gate vibration θ, it is necessary to analyze the feedback process to the main whole gate vibration. Representing ψ'' in the equation of motion, Equation 18, of the main whole gate vibration in terms of θ'' and θ', respectively, the following equation of motion for the self-excited vibration system can be derived:

$$\theta'' + 2\zeta_{a\theta}\theta' + \theta + \frac{1}{\pi}\frac{\alpha_I}{\alpha_*}\left\{\cos\left(\varphi_0 - \varphi\right)\theta'' - \gamma\sin\left(\varphi_0 - \varphi\right)\theta'\right\} = 0 \tag{60}$$

which can be reduced ultimately to the following expression:

$$\left(1 + \Delta m_{\psi\theta}\right)\theta'' + 2\left(\zeta_{a\theta} - \Delta c_{\psi\theta}\right)\theta' + \theta = 0 \tag{61}$$

where $\Delta m_{\psi\theta}$ and $\Delta c_{\psi\theta}$ are the reduced added mass and the reduced dissipative wave-radiation damping coefficient, respectively, due to the coupling of the two vibration modes:

$$\Delta m_{\psi\theta} \equiv \frac{1}{\pi}\frac{\alpha_I}{\alpha_*}\cos\left(\varphi_0 - \varphi\right) \tag{62}$$

$$\Delta c_{\psi\theta} \equiv \frac{1}{2\pi}\frac{\alpha_I}{\alpha_*}\gamma\sin\left(\varphi_0 - \varphi\right) \tag{63}$$

Therefore, the equation of motion, Equation 61, can be arranged as follows:

$$\theta'' + 2\frac{\zeta_{a\theta} - \xi_{f\theta}}{1 + \Delta m_{\psi\theta}}\theta' + \frac{1}{1 + \Delta m_{\psi\theta}}\theta = 0 \tag{64}$$

where $\xi_{f\theta}$ is the fluid-excitation ratio:

$$\xi_{f\theta} = \Delta c_{\psi\theta} \tag{65}$$

The square root of the coefficient of θ in Equation 64 gives the in-water to in-air vibration frequency ratio $\gamma_{w\theta\theta}$ for the whole gate rotation around the trunnion pin:

$$\gamma_{w\theta\theta}\left(\equiv\frac{\Omega_{w\theta}}{\Omega_{a\theta}}\right)=\frac{1}{\sqrt{1+\Delta m_{\psi\theta}}} \quad (66)$$

If this frequency ratio $\gamma_{w\theta\theta}$ is not in agreement with the frequency ratio γ which was first assumed in Equation 51, $\gamma_{w\theta\theta}$ is replaced by γ, and the ultimate solution of the vibration ratio $\gamma_{w\theta\theta}$ can be obtained by iterative calculations of Equations 58, 59, 62, and 66. Using the solution of frequency ratio, the equation of motion, Equation 64 can be expressed in the following form:

$$\theta''+2\gamma_{w\theta\theta}^{2}\left(\zeta_{a\theta}-\xi_{f\theta}\right)\theta'+\gamma_{w\theta\theta}^{2}\theta=0 \quad (67)$$

From the coefficient on θ', the actual excitation ratio ξ_θ of the whole gate rotational vibration (main vibration) can be obtained as follows:

$$\xi_\theta=\gamma_{w\theta\theta}\left(\xi_{f\theta}-\zeta_{a\theta}\right) \quad (68)$$

Finally, a dynamic stability criteria can be given by conditions that force the actual excitation ratio ξ_ψ and ξ_θ to be negative. The gate is stable if the following conditions are met:

$\zeta_{a\psi} > \xi_{f\psi}$ (when synchronized with the skinplate streamwise vibration) (69)

$\zeta_{a\theta} > \xi_{f\theta}$ (when synchronized with the whole gate rotational vibration) (70)

where $\zeta_{a\psi}$ and $\zeta_{a\theta}$ are the damping ratios necessary for the skinplate streamwise vibration and for the whole gate rotation to be dynamically stable, respectively.

CHARACTERISTICS OF COUPLED-MODE SELF-EXCITED VIBRATION

The approximate solutions are calculated specifically for the case of Folsom Dam Tainter gate. The major specifications of the original Folsom Tainter gate at the time of failure are shown in Table 1. The average gate opening B was 0.762 m, which is 5.7% of the water depth d_0 of 13.26 m, and hence it can be treated as a small gate opening. The moment of inertia of the gate was calculated from the mass and dimensions of each component. The spanwise (*i.e.*, horizontal) nodal line of the skinplate, that is, the skinplate streamwise rotation center height R_s was 9.6 m. Values of the following parameters were obtained from the experimental modal analyses on a surviving original Folsom gate (see Ishii, et al., 2010 for details): the damping ratios of the whole gate vibration and the skinplate streamwise vibration, 0.012 and 0.002; the press-open angle of the lower end of the skinplate, 8.5°; and, the natural frequency of the skinplate streamwise vibration, 26.9 Hz.

The corresponding non-dimensional factors for the Folsom gate are shown in Table 2. The gate inclination coefficient is a value obtained from model tests for the gate inclined at 20° in the downstream

Table 1. Major specifications of the original Folsom Dam Tainter gate

Group	Property	Notation	Value
Gate dimensions and mass	Gate height	—	15.5 m
	Span of skinplate	Wo	12.8 m
	Radius of skinplate	R_a	14.33 m
	Gate mass	—	87 tons
Moments of inertia	Whole gate around trunnion pin	I_θ	1.31×10^7 kg·m^2
	Skinplate around its rotation center	I_ψ	1.61×10^6 kg·m^2
	Product of inertia of skinplate	$I_{\psi\theta}$	1.27×10^6 kg·m^2
Field test conditions	Depth of the upstream channel	d_r	14.02m
	Gate opening	B	0.762 m
	Depth over the gate bottom	d_0	13.26 m
	Inclination of gate bottom to water surface	Θ_s	9.5°
	In-air natural frequency of rigid-body vibration	$\Omega_{a\theta}$	6.88 Hz
	In-air natural freq. of streamwise bending vibration	$\Omega_{a\psi}$	26.9 Hz
	In-air damping ratio for rigid-body vibration	$\zeta_{a\theta}$	0.012
	In-air damping ratio for streamwise bending vibration	$\zeta_{a\psi}$	0.002
	Rotation center height	R_s	9.6 m
	Press-open angle	Θ_{s0}	8.5°
	Rotation radius of skinplate lower end	R_e	13.4 m

After Anami et al. (2014).

direction. The instantaneous flow-rate coefficient c_f is a value also obtained from the model tests. The press-shut coefficient k was calculated from Equation 13. The water-to-gate mass ratio α_ψ was 113, evidence of the light-weight design of the gate. The fluid-excitation coefficient Δc_θ and the added mass Δm_ψ are values obtained from Figure 4, where it was assumed that the Froude number takes on a value larger than about 10, which is generally true for full-scale gates.

Calculated results for the approximate solutions of equations of motion are shown in Figure 5. It was assumed that the whole gate in-air vibration frequency $\Omega_{a\theta}$ was unknown, while the skinplate streamwise in-air vibration frequency $\Omega_{a\psi}$ has been given as 26.9 Hz. In all figures, the abscissa is the frequency ratio γ_{nw} $(=\Omega_{nw\psi}/\Omega_{a\theta})$ of the in-water streamwise natural vibration of the skinplate to the in-air whole gate vibration around the trunnion pin. When this frequency ratio takes on a value of 1.0, the in-water coupled-mode vibrations of the skinplate streamwise and whole gate vibrations are in resonance.

Figure 5a shows the in-water to in-air frequency ratio $\gamma_{w\theta\theta}$ $(= \Omega_{w\theta} / \Omega_{a\theta})$ of the whole gate vibration around the trunnion pin and the frequency ratio $\gamma_{w\psi\theta}$ $(=\Omega_{w\psi}/\Omega_{a\theta})$ of the skinplate streamwise in-water vibration to the whole gate in-air vibration around the trunnion pin. The ratio $\gamma_{w\theta\theta}$ takes on a value of about 1.0, which indicates that the in -water frequency is same as the in-air frequency for the whole gate vibration around the trunnion pin, because of no added mass effect on this mode of vibration. In contrast, $\gamma_{w\psi\theta}$ takes on almost the same value as γ_{nw} on the abscissa, that is, the skinplate streamwise in-water vibration occurs at its in-water natural vibration frequency. It should be noted here that the vibrations synchronize with the skinplate streamwise vibration natural frequency in the abscissa range of $\gamma_{nw} < 1.0$ as shown by the solid line, while for $\gamma_{nw} > 1.0$ the vibrations synchronize with the whole gate natural vibration frequency. Both of the frequency ratios $\gamma_{w\theta\theta}$ and $\gamma_{w\psi\theta}$ show a tendency to decrease slightly near the resonance state at $\gamma_{nw} = 1.0$.

Table 2. Non-dimensional parameters for calculations

Non-dimensional Parameter	Notation	Value
Depth ratio ($\beta^* = d_r/d_0$)	β^*	101
Reduced rotation center height ($r_s = R_S/d_0$)	r_s	0.72
Rotation radius ratio ($r_{sa} = R_S/R_C$)	r_{sa}	0.77
Moment-of-inertia ratio ($\alpha_I = I_\psi/I_\theta$)	α_I	0.123
Moment-of-inertia ratio of skinplate ($\alpha_{I\psi} = = I_{\psi 0}/I_\psi$)	$\alpha_{I\psi}$	0.79
Water to gate mass ratio ($\alpha_\psi = \rho d_0^2 W_0/[I_\psi R_S^2]$)	α_ψ	129
Instantaneous flow-rate-coefficient	c_f	0.72
Pressure correction coefficient	δ_p	1.0
Basic Froude number ($F_{a\psi} = 2\pi\Omega_{a\psi}\sqrt{d_0/g}$)	$F_{a\psi}$	196

After Anami et al. (2014).

Figure 5b shows the fluid-excitation ratios $\xi_{f\psi}$ and $\xi_{f\theta}$, representing the dynamic instability. If the fluid-excitation ratio is a positive value, energy flows to the structural vibration, resulting in a dynamically unstable state. When the abscissa is smaller than 1.0 (γ_{nw} <1.0; $\Omega_{nw\psi} < \Omega_{a\theta}$), $\xi_{f\psi}$ takes on positive values (solid line), while $\xi_{f\theta}$ takes on negative values (dashed line). Since the press-open angle Θ_{s0} was positive, $\xi_{f\psi}$ takes on a negative value with small γ_{nw}. Because of these excitation ratio values, the self-excited vibration synchronizes with the skinplate streamwise rotational vibration. As a result, the vibration frequency synchronizes with the lower in-water inherent frequency ($\Omega_{w\psi} = \Omega_{nw\psi}$), and hence the frequency ratio $\gamma_{w\psi\theta}$ takes the same value as the in-water natural vibration frequency ratio γ_{nw} of abscissa, as shown by the solid line in Figure 5a.

In contrast, when the abscissa value is larger than 1.0 (γ_{nw} >1.0; $\Omega_{nw\psi} > \Omega_{a\theta}$), $\xi_{f\psi}$ takes negative values (dashed line), while $\xi_{f\theta}$ takes on positive values (solid line). In this case, the self-excited vibration synchronizes with the whole gate rotational vibration. As a result, the vibration frequency synchronizes with the lower in-water inherent frequency ($\Omega_{w\theta} = \Omega_{a\theta}$), and hence the frequency ratio $\gamma_{w\theta\theta}$ always takes on a value of 1.0, as shown by the solid line in Figure 5a.

Special attention should be given to the following aspects of Figure 5b:

1. When the vibration frequency ratio on the abscissa is slightly different from the resonant state (slightly higher or lower than 1.0), a large dynamic instability appears,
2. Especially, when the vibration frequency ratio on the abscissa takes on a value of about 0.96 (slightly smaller than 1.0), an intense dynamic instability appears.

The dynamic instability characteristics, shown in Figure 5b, can be explained by drawing the coupled-mode vibration trajectories. The calculated phase-lag $\Delta\phi$ and angular amplitude ratio α_* are shown in Figures 5c and 5d, respectively. The phase-lag, $\Delta\phi$, is the phase-lag of the up-and-down vibration (taken

Figure 5. Approximate solutions for self-excited coupled-mode vibration of the original Folsom Dam Tainter gate: (1) vibration frequency ratio; (b) fluid excitation ratio; (c) phase-lag; (d) vibration angle amplitude ratio; (e) press-shut trajectories; (f) press-open trajectories

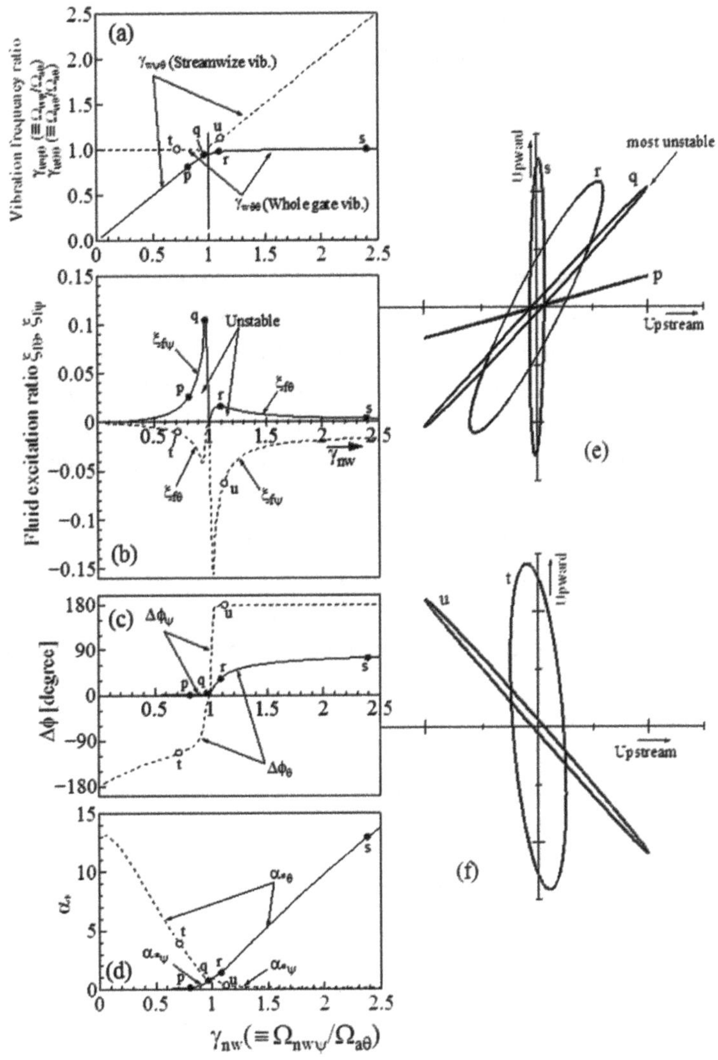

as positive upward) at the skinplate bottom end relative to the streamwise vibration (taken as positive in the upstream direction) at the same location. When the self-excited vibration synchronizes with the skinplate streamwise rotational vibration, $\Delta\phi$ is given by ϕ of Equation 41 and α_* by Equation 40, which are denoted here by $\Delta\phi_\psi$ and $\alpha_{*\psi}$, respectively. When the self-excited vibration synchronizes with the whole gate rotational vibration, $\Delta\phi$ is given by (ϕ_0 - ϕ) in Equations 55 and 59, and α_* is given by Equation 58, and are here denoted as $\Delta\phi_\theta$ and $\alpha_{*\theta}$, respectively.

Figures 5e and 5f show the coupled-mode vibration trajectories drawn using the $\Delta\phi$ and α_* data corresponding to each filled data point (denoted as "p" to "u") in Figures 5a to 5d. When the fluid-excitation ratio takes on positive values as shown by points "p" to "s", the vibration trajectories are typical of a "press-shut device", as shown in Figure 5e. To the contrary, when the fluid-excitation ratios take on negative values as shown by points "t" and "u", the vibration trajectories are typical of a "press-open device", as shown in Figure 5f.

Multiplying the equation of motion, Equation 32, for skinplate streamwise rotational vibration by $d\psi$ and then integrating, the energy equation can be derived. The integral of the θ' term (due to the flow-rate variation) indicates there is a great amount of energy, either supplied or consumed. The energy supply E_θ per cycle of vibration is given by

$$E_\theta \propto \int_{1cycle} \theta' d\psi = \int_{1cycle} \theta' \psi' dt \tag{71}$$

Therefore, when θ' and ψ' have the same signs, the energy is supplied. Conversely, when they have the opposite signs, the energy is consumed. In the case of a press-shut device, θ' and ψ' have the same signs in almost all domains, thus supplying energy from the fluid to the vibration and thus inducing the dynamic instability, as shown by the solid line in Figure 5b. To the contrary, in the case of a press-open device, θ' and ψ' have opposite signs in almost all domains, thus consuming energy from the vibration to the fluid and thus resulting in strong dynamic stability, as shown by the broken line in Figure 5b. The energy transfer will be again considered in detail in the subsequent section.

Furthermore, regarding the press-shut trajectories in Figure 5e, one may understand that as the vibration trajectory becomes flat, or as the inclination approaches 45°, the energy supply becomes larger. This can be explained as follows: Assume the streamwise displacement x of the skinplate bottom end and the up-and-down displacement y, can be represented by the following expressions:

$$x = r_0 \cos\Theta_{ps} \cdot \cos\frac{F}{F_a} t \tag{72a}$$

$$y = r_0 \sin\Theta_{ps} \cdot \cos\left(\frac{F}{F_a} t - \Delta\varphi\right) \tag{72b}$$

With these two expressions, Equation 71 yields the following result:

$$E_\theta \propto \int_{1cycle} \theta' \psi' dt \propto \int_{1cycle} y' x' dt = \frac{1}{2} r_0^2 \sin 2\Theta_{ps} \left(\frac{F}{F_a}\right) \pi \cos\Delta\varphi \tag{73}$$

From this expression, the energy supply becomes largest, when the trajectory inclination Θ_{ps} is 45° and when the trajectory thickness $\Delta\phi$ is zero.

Examining the vibration trajectories in Figure 5e on the basis of the above information, one may infer that point "q," with an inclination of nearly 45° and a comparatively flat trajectory, is the most unstable case as shown in Figure 5b. In addition, point "r", with a "thicker" trajectory (*i,e.,* not as flat as for "q") has a reduced energy supply, resulting in a moderate instability. Points "p" and "s" both have flat trajectories, but the inclination for "p" is too small, and that for point "s" is too steep, for effective cross-coupling and in both cases the resulting instability is weak.

ENERGY TRANSFER FROM VIBRATION TRAJECTORIES

The fundamental mechanism producing the vibration trajectories shown in Figures 5e and 5f and the energy transfer between the fluid and structural vibration are examined in this section. The two synchronization cases will be considered separately.

Coupled-Mode Vibration Synchronized With the Skinplate Streamwise Vibration

When the circular-arc skinplate performs streamwise rotational vibration, the whole gate is forced to vibrate around the trunnion pin by the coupled inertia torque. The coupled inertia torque appears on the right-hand side of the equation of motion, Equation 18. This coupled inertia torque on the right-hand side forces the whole gate to vibrate around the trunnion pin. Then, the component of this exciting force in the up-and-down direction, due to the coupled inertia torque, is illustrated by the dashed lines in Figure 6a, where the abscissa is the streamwise vibration displacement with positive values (the right-hand side) corresponding to the upstream direction. The exciting force attains its upward maximum, just when the skinplate moves to its most upstream position. The up-and-down vibration responds to this excitation, subject to Equation 18 with an appropriate phase-lag.

When the natural vibration frequency ratio is in the range where $\gamma_{nw} < 1.0$, the phase-lag takes on a value from 0° to 90°. Since the whole gate rotational vibration around the trunnion pin has quite a small damping ratio, however, the phase-lag takes on a small value near zero, as shown by the small arrow along the abscissa. Thus, the maximum upward vibration response appears when the skinplate has moved just slightly downstream from its most upstream position, as shown in Figure 6a. Consequently, the vibration trajectory in this case results in a very thin press-shut trajectory. The trajectory shows counter-clockwise rotation.

Now consider corresponding energy flow for the streamwise rotational vibration. The equation of motion of the streamwise rotational vibration is given by Equation 32. The first term of the right-hand side represents the inertia torque due to the coupling and the second term represents the torque due to the flow-rate variation pressure, both of which force the skinplate to vibrate in the streamwise direction. Then, the resultant exciting force in the streamwise direction due to these torques is typically illustrated by the solid line in Figure 6b, where the exciting force due to the inertia torque is shown by the dashed line ① and the exciting force due to the flow-rate variation pressure is shown by ②. The dashed line marked by ② shows the exciting force when the skinplate moves downward. This exciting force would then appear in the opposite direction when the skinplate moves upward. Figure 6c shows both the vibration trajectory and the resultant excitation force. The red lines of the vibration trajectory show when the resultant excitation force supplies energy to the streamwise vibration, while the blue lines show when the vibration energy is consumed. As is clear from Figure 6c, the supplied energy is far larger than the consumed energy, thus inducing the intense dynamic instability observed for point "q" in Figure 5b.

When the natural vibration frequency ratio $\gamma_{nw} > 1.0$, the phase-lag takes on a value from 90° to 180°. Since the damping ratio of the whole gate rotational vibration around the trunnion pin is quite small, the phase-lag takes on a large value near 180°. Then, when the skinplate reaches its most downstream posi-

Figure 6. Vibration trajectory formation and energy supply, when synchronized with the streamwise rotational vibration of the circular-arc skinplate: (a) Vibration trajectory when γ_{nw} <.1.0; (b) Exciting forces: ① Coupled inertia force, ② Flow-rate variation fluid force; (c) Energy supply by resultant force to streamwise vibration when γ_{nw} <1.0; (d) Vibration trajectory when γ_{nw} >1.0; (e) Energy supply by resultant force to streamwise vibration when γ_{nw} >1.0

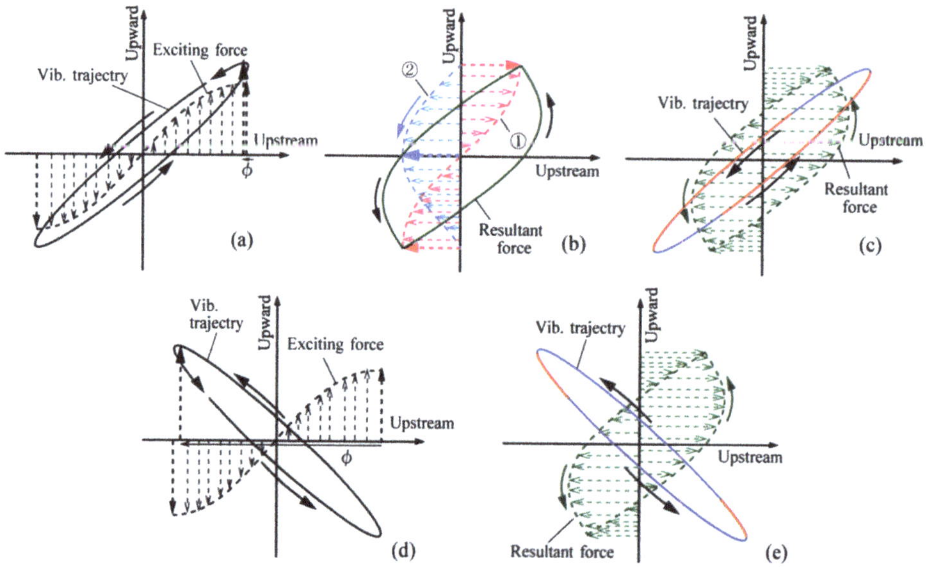

**For a more accurate representation see the electronic version.*

tion, the maximum upward vibration response appears, as shown in Figure 6d. As a result, the vibration trajectory in this case shows a thin press-open trajectory. The trajectory has a counter-clockwise rotation.

Figure 6e shows both the press-open trajectory and the resultant excitation force. The energy consumption region, shown by the absence of heavy dashed arrows outside the ellipse, is far larger than the energy supply shown by the regions with heavy dashed arrows. As a result, when the trajectory corresponds to a press-open device, the system exhibits strong dynamic stability with no chance of coupled-mode vibration, as in point "u" in Figure 5b.

Coupled-Mode Vibration Synchronized With the Whole Gate Rotational Vibration

When the whole gate performs rotational vibration around the trunnion pin, the coupled inertia torque of the skinplate and the torque due to the flow-rate variation pressure are induced, thus forcing the circular-arc skinplate to vibrate in the streamwise direction. These torques appear on the right-hand side of equation of motion, Equation 32. The streamwise resultant exciting force due to these torques is illustrated in Figure 6b and again by the dashed line in Figure 7a. The maximum exciting force in the downstream direction appears with a phase-lag between 90° and 180° from the most upward position of the skinplate. This streamwise exciting force induces the maximum vibration response in the downstream direction, however with a phase-lag which is determined by Equation 32.

When the natural vibration frequency ratio γ_{nw} <1.0, that is, when the natural vibration frequency $\Omega_{a\theta}$ of the main vibration is higher than the natural vibration frequency $\Omega_{nw\psi}$ of the response vibration, the downstream maximum vibration response appears with a phase-lag of about 180°, as shown by the arrows along the vertical axis in Figure 7a. As a result, the vibration trajectory shows a thick press-open characteristic, as illustrated by the solid line in Figure 7a. The trajectory has a clockwise rotation.

Figure 7b helps explain why the whole gate rotational vibration cannot gain energy by such vibration. The coupled inertia torque of the right-hand side of the equation of motion, Equation 18, induces the up-and-down vibration of the skinplate. The upward inertia force becomes largest when the skinplate reaches its most upstream position, as shown in Figure 6a and again by the dashed line in Figure 7b. The energy supply lines, shown in red, are far smaller than the energy consumption lines shown in blue. Thus, the system is dynamically stable, as shown by point "t" in Figure 5b.

When the natural vibration frequency ratio γ_{nw} > 1.0, that is, when the natural vibration frequency $\Omega_{a\theta}$ of the main vibration is lower than the natural vibration frequency $\Omega_{nw\psi}$ of the response vibration, the downstream maximum vibration response appears with a phase-lag of almost 0°, as shown by small

Figure 7. Vibration trajectory formation and energy supply when synchronized with the whole gate rotational vibration around the trunnion pin:(a) Vibration trajectory when γ_{nw} < 1.0; (b) Energy supply by inertial force to up-and downward vibration when γ_{nw} < 1.0; (c) Vibration trajectory when γ_{nw} >1.0; (d) Energy supply by inertial force to up-and downward vibration when γ_{nw} >1.0
From Anami et.al. (2014).

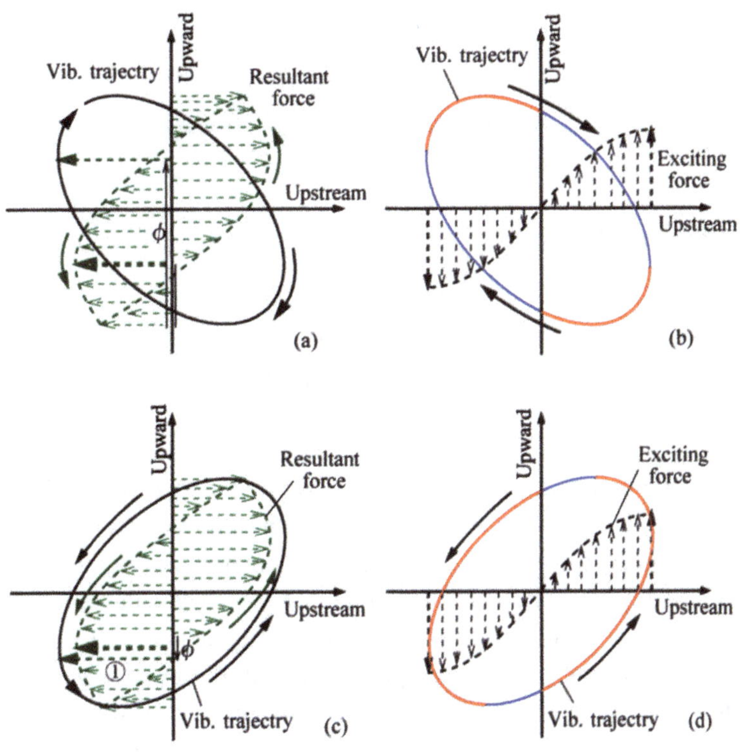

**For a more accurate representation see the electronic version.*

arrows along the vertical axis in Figure 7c. As a result, the vibration trajectory shows a thick press-open characteristic, as illustrated in Figure 7c. The trajectory shows a counter-clockwise rotation.

Figure 7d shows both of the vibration trajectory and the resultant excitation force, where the energy supply regions, again sshown by the red lined regions of the ellipse, are far larger than the energy consumption regions, indicated by the blue-lined regions. Thus, the vibration results in the dynamic instability shown by point "r" shown in Figure 5b.

INSTABILITY OF THE FOLSOM GATE FOR CONDITIONS AT FAILURE

The fluid-excitation ratios for the Folsom Dam Tainter gate under the prevailing conditions at the time of its failure were theoretically calculated, as shown in Figure 5b. The ordinate represents the damping ratio needed to prevent the dynamic instability, that is, the so-called the "stability damping ratio". Therefore, by using this calculated result and the measured vibration characteristics (vibration frequencies, damping ratios) of the Folsom Dam Tainter gate, an assessment of its dynamic stability at the time of its failure can be made.

The in-air streamwise natural vibration frequency $\Omega_{a\psi}$ of the skinplate of the original Folsom Dam Tainter gate was found to be 26.9 Hz from the experimental modal analysis by Ishii (1995a, 1995b), Anami et al. (2014). This frequency is drastically reduced by the added mass of the water, as described in Anami & Ishii (1998b) and Anami et al. (2012b). The in-water to in-air vibration frequency ratio $\Omega_{nw\psi}/\Omega_{a\psi}$ of the skinplate streamwise rotational vibration can be calculated from Equation 31. Introducing the inherent Froude number $F_{nw\psi}$ and the basic Froude number $F_{a\psi}$ for streamwise vibration of the skinplate, defined by

$$F_{nw\psi} \equiv \sqrt{\frac{d_0}{g}}\Omega_{nw\psi} \qquad \text{and} \qquad F_{a\psi} \equiv \sqrt{\frac{d_0}{g}}\Omega_{a\psi} \tag{74}$$

Equation 31 can be rewritten as follows:

$$\frac{F_{nw\psi}}{F_{a\psi}}\left(\equiv \frac{\Omega_{nw\psi}}{\Omega_{a\psi}}\right) = \frac{1}{\sqrt{1+\delta_p \alpha_\psi \Delta m_\psi (F_{nw\psi}) + \frac{2}{\pi}\sqrt{2}\, c_f k \alpha_\psi \Delta m_\theta (F_{nw\psi})}} \tag{75}$$

The reduced water added mass Δm_ψ and Δm_θ on the right side of Equation 75 are functions of the inherent Froude number $F_{nw\psi}$ and the reduced rotation center height r_s; their characteristic behavior is shown in Figure 4. Therefore, computer calculations can be performed to determine the inherent Froude number $F_{nw\psi}$ for a given value of the basic Froude number $F_{a\psi}$. The calculated results can be reduced to $F_{nw\psi}/F_{a\psi}$, which represents the in-water to in-air vibration frequency ratio $\Omega_{nw\psi}/\Omega_{a\psi}$, as shown by the solid line in Figure 8. In this figure theoretical values for both the Folsom gate and a model gate are shown as solid and dashed lines, respectively. Experimental values for the model gate are shown as well, providing evidence of the applicability of the theoretical results. More details on the laboratory-scale model testing will be presented in Chapter 11–Measurement of Coupled-Mode Instability on Model Scale

For the Folsom Dam Tainter gate, the following values were used in the calculations: the reduced rotation center height r_s was 0.72; the water-to-gate mass ratio α_ψ was 129; and the pressure correction coefficient δ_p was 1.0. With the measured in-air natural vibration frequency $\Omega_{a\psi}$ of 26.9 Hz and the gate submergence depth d_0 of 13.26 m for the Folsom Dam Tainter gate at its failure, the basic Froude number $F_{a\psi}$ takes on a large value of 196. From Figure 8, the in-water to in-air vibration frequency ratio $\Omega_{nw\psi}/\Omega_{a\psi}$ takes a constant value of 0.24 for basic Froude number $F_{a\psi}$ values greater than about 20.Therefore, the in-water to in-air vibration frequency ratio $\Omega_{nw\psi}/\Omega_{a\psi}$ is 0.24, which is far smaller than that for the model test results (see Anami et al., 2012b), shown by dashed lines in Figure 8. This is because the Folsom Tainter gate has a water-to-gate mass ratio α_ψ of 129, far larger than the value of 2.45 to 14.8 for model gate testing. The frequency ratio of 0.24 for the Folsom Dam gate means the in-water vibration frequency was slightly less than ¼ of the measured in-air natural vibration frequency $\Omega_{a\psi}$ of 26.9 Hz, or more precisely 6.46 Hz.

It is important to note here that the frequency of streamwise bending is just slightly less than the in-air natural vibration frequency $\Omega_{a\theta}$ of 6.88 Hz for the whole gate vibration around the trunnion pin. The resulting frequency ratio γ_{nw} $(= \Omega_{nw\psi}/\Omega_{a\theta})$ takes on a value of 0.94, just slightly smaller than 0.96 at which the maximum instability occurs. The measured in-air damping ratio $\zeta_{a\psi}$ of 0.002 of the skinplate streamwise vibration is plotted on the stability curve in Figure 9. For this mode of vibration, with nodes along the sides of the gate, the damping due to the side seals plays no role in the streamwise bending vibration. The difference between the fluid dynamic excitation ratio and the existing structural damping ratio is about 0.049, indicating an additional 5% damping would be needed to stabilize the gate. Increases in damping due to water have been included in the theory. As a result, one may conclude that the dynamic condition of the Folsom Dam Tainter gate under the conditions prevailing at the time of failure was in the intensely dynamically unstable region. This dynamic condition indicates the gate was readily susceptible to coupled-mode vibration, which induces a violent coupled-mode vibration synchronized with the skinplate streamwise rotational bending vibration, which would serve as the main vibration at the incipient failure.

Figure 8. In-water to in-air frequency ratio for the original Folsom Dam Tainter gate
From Anami et al. (2014). © 2014 IJMERR. Used with permission.

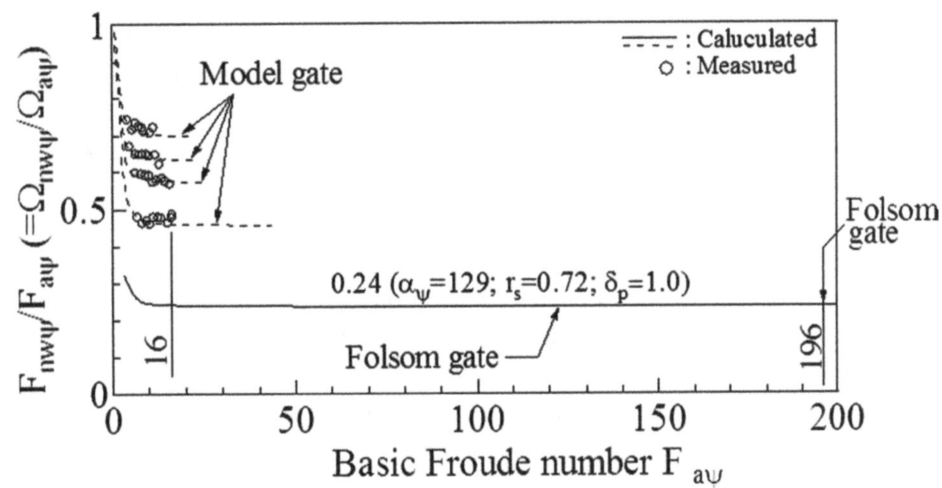

Figure 9. Dynamic instability of the failed Tainter gate at the Folsom Dam
From Anami et al. (2014). © 2014 IJMERR. Used with permission.

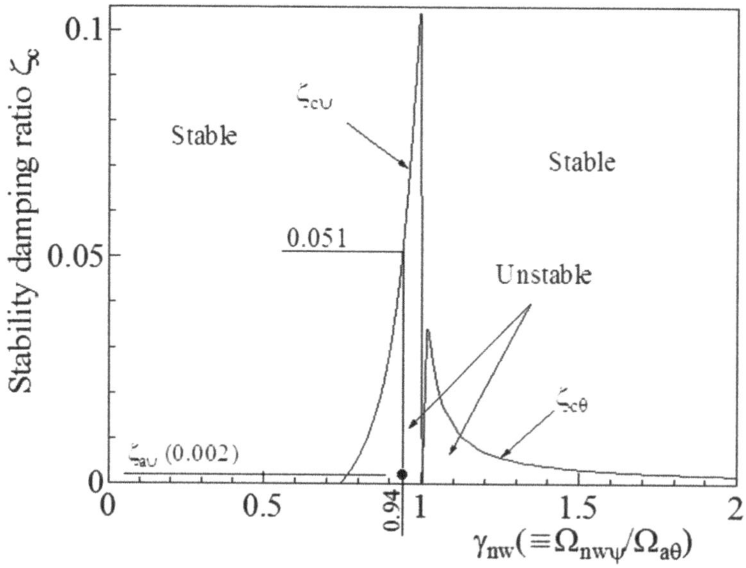

In the fundamental vibration mode, as identified by the experimental modal analysis, the skinplate performs a low frequency bending mode vibration in the streamwise and vertical plane. This lowest frequency bending mode of the skinplate is essentially a streamwise rotation of the skinplate about a point in the upper half of the skinplate. When the massive skinplate, which constitutes more than 60% of the whole gate mass, performs a rotational vibration, a large inertia torque due to its rotational vibration arises, which can then excite the whole gate to vibrate around the trunnion pin. When the whole gate vibrates around the trunnion pin, an additional inertia torque also arises, which can produce additional skinplate rotation. If the natural frequencies of these two vibration modes coincide, the streamwise rotational vibration of the skinplate and the whole gate rotational vibration around the trunnion pin can couple with each other through the inertia torques, resulting in "coupled-mode vibration."

The whole gate rotation around the trunnion pin induces a flow rate variation beneath the gate, with a corresponding flow-rate variation pressure which forces the skinplate to vibrate in the streamwise direction. Subsequently, the skinplate streamwise vibration induces a push-and- draw pressure, which significantly lowers the vibration frequency of the skinplate. Therefore, these two vibration modes can easily couple to induce violent self-excited vibrations (Anami & Ishii, 1999, 2000, 2003).

A closed loop energy cycle of the coupled-mode self-excited vibration is presented in Figure 10. If the bending rotational counter-clockwise vibration ① of the skinplate, represented by Ψ, is triggered by some random excitation, the counter-clockwise skinplate inertia torque ② appears around the trunnion pin, thus exciting the counter-clockwise whole gate vibration ③, represented by Θ. As a result, the gate opening under the gate increases, thus resulting in increased discharge water flow ④, and thus decreasing the flow-rate-variation pressure ⑤ in proportion to the whole gate vibration velocity, that feeds energy back to the skinplate bending rotational vibration ⑥, thus amplifying the bending vibration amplitude.

Figure 10. Closed energy cycle for the self-excited coupled-mode flow-induced instability of Tainter gates, even those with no skinplate center eccentricity from the trunnion pin
From Anami & Ishii (2003). © 2003 ASME. Used with permission.

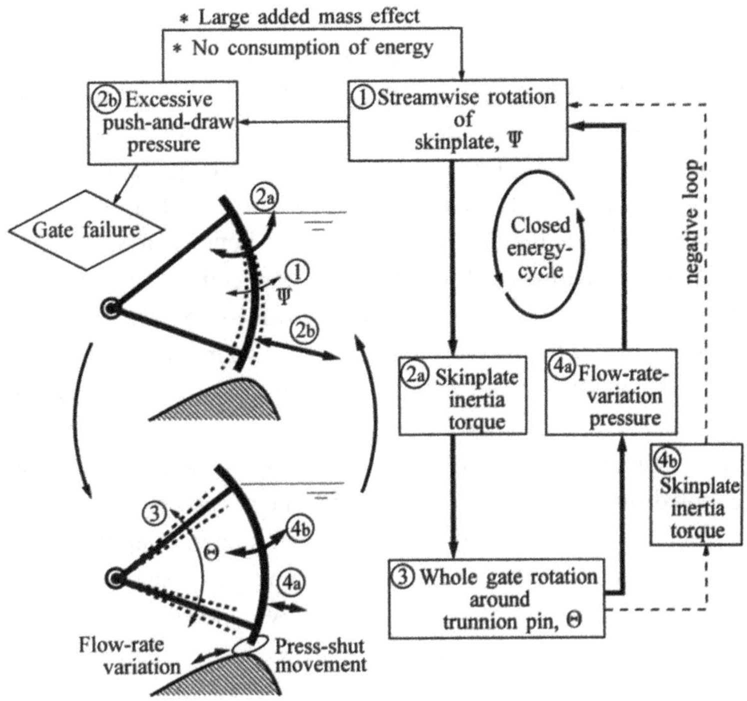

This type of self-excited coupled-mode vibration can occur naturally, even when the geometric center of the skinplate is coincident with the trunnion pin, that is, when there is no trunnion pin eccentricity.

Of special note in this closed energy-cycle is that the streamwise skinplate vibration induces an excessive push-and-draw pressure. The amplitude of this push-and-draw pressure can be more than 10 times that of the flow-rate variation pressure, and it could have caused the gate failure. The gate fails when the loading due to the excessive push-and-draw pressure exceeds the material strength.

It is also significant that the excessive push-and-draw pressure does not consume energy from the vibrating gate. The excessive push-and-draw pressure has associated with it a large added mass effect. The large added mass of water that must be accelerated with the gate as it undergoes its streamwise vibration lowers the relatively high in-air frequency of the streamwise skinplate vibration. As a result, when the in-water streamwise skinplate vibration frequency approaches the in-water whole gate rigid body rotational frequency, the system approaches resonance. Near resonance, the comparatively small amplitude flow-rate variation pressure can spontaneously amplify the amplitude of coupled-mode vibration.

CONCLUSION

Using previously derived and validated expressions for pressure loading and added mass effects, a general theory for the self-excited coupled-mode instability with no requirement on eccentricity of the skinplate center from the trunnion center was developed in this chapter. Equations of motion for the gate were derived. The theory was then applied to the failed Folsom Dam Tainter gate using experimentally determined dynamic characteristics for the gate. This experimental testing identified two predominant vibration modes. For these modes, an approximate, iterative method of solution was employed to solve the two coupled equations of motion. From the solutions, dynamic stability diagrams were derived.

The resulting solution demonstrates the existence of two regions of instability, one in which the frequency of the skinplate bending mode is just slightly less than the natural frequency of the rigid-body mode. In this region the skinplate bending acts as the main driving force and the rigid body rotation is the response mode. This is the most unstable situation. The second region of instability occurs when the frequency of the skinplate bending mode is just slightly higher than that of the rigid body rotation. In this region the rigid-body motion is the driving force and the bending mode is the response vibration. In this region, instability is still evident, but its intensity is reduced relative to that of the first region.

On the basis of this analysis, the Folsom Dam was shown to possess a strong dynamic instability for the conditions that prevailed at the instant of gate failure. There is every reason to believe that this coupled-mode self-excited instability could well have played a very strong (if not dominant) role in the failure of the Folsom gate.

More importantly, the theory provides an assessment tool to determine the susceptibility of any Tainter gate to this coupled-mode self-excited instability. The necessary inputs are the gate and sill geometry (from the drawings for the installation), the expected range of depth of the upstream reservoir, the skinplate in-air natural vibration characteristics (frequency, mode shape and damping ratio, usually from experimental modal analysis) for the lowest frequency bending mode of the skinplate, and the rigid-body rotational frequency and damping ratio.

REFERENCES

Anami, K., & Ishii, N. (1998a). In-water streamwise vibration of Folsom Dam radial gates in California. In *Proceedings of 1998 ASME/JSME Joint Pressure Vessels and Piping Conference*, (Vol. 363, pp. 87-93). New York. NY: ASME.

Anami, K., & Ishii, N. (1998b). In-air and in-water natural vibrations of Folsom dam radial gate in California. In I. M. Allison (Ed.), *Experimental Mechanics, Proceedings of 11th International Conference on Experimental Mechanics*, (pp. 29-34). Rotterdam: Balkema.

Anami, K., & Ishii, N. (2001). Analyses of FIV of Folsom Dam Tainter gates and verifications by model tests. In *Proceedings of Commemorative International Symposium for the 40th Anniversary for the Foundation of the Osaka Electro-Communication University*, (*Vol. 2*, pp. 165-181). Osaka: The Osaka Electro-Communication University.

Anami, K., Ishii, N., & Knisely, C. W. (2012a). Pressure induced by vertical planar and inclined curved weir plates undergoing streamwise rotational vibration. *Journal of Fluids and Structures*, *29*, 35–49. doi:10.1016/j.jfluidstructs.2011.11.007

Anami, K., Ishii, N., & Knisely, C. W. (2012b). Added mass and wave radiation damping for flow-induced rotational vibrations of skinplate. *Journal of Fluids and Structures*, *35*, 213–228. doi:10.1016/j.jfluidstructs.2012.07.008

Anami, K., Ishii, N., & Knisely, C. W. (2014). Retrospective consideration of a plausible vibration mechanism for the failure of the Folsom Dam Tainter gate. *International Journal of Mechanical Engineering and Robotics Research*, *3*(4), 314–345.

Ishii, N. (1992). Flow-induced vibration of long-span gates (Part I: Model development). *Journal of Fluids and Structures*, *6*(5), 539–562. doi:10.1016/0889-9746(92)90018-X

Ishii, N. (1995a). *Folsom Dam Gate Failure Evaluation and Suggestions*. Report submitted to U.S. Bureau of Reclamation.

Ishii, N. (1995b). *Folsom Dam Gate Failure Evaluation Based on Modal Analysis and Suggestion*. Report submitted to U.S. Bureau of Reclamation.

Ishii, N. (1997). *Dynamic Stability Evaluation for Folsom Dam Radial Gate and Its Inherent Non-Eccentricity Instability*. Report submitted to U.S. Bureau of Reclamation.

Rayleigh, J. W. S. (1945). *The Theory of Sound*. New York: Dover.

APPENDIX: NOMENCLATURE

B: Discharge gate opening
c_f: Instantaneous flow-rate variation coefficient
d_0: Skinplate submergence depth
$dM_{\theta\psi}$, $dM_{\psi\theta}$: Coupled moment
F: Froude number
$F_{a\theta}$, $F_{a\psi}$: Basic Froude number
$F_{nw\psi}$: Inherent Froude number
g: Acceleration due to gravity
I_θ: Moment-of-inertia of whole gate rotation around trunnion pin
I_ψ: Moment-of-inertia of skinplate streamwise rotation
$I_{\psi\theta}$: Coupled moment-of-inertia
k: Press-shut coefficient
P, $P_{b\psi}$, $P_{r\theta}$, $P_{r\psi}$: Hydraulic pressure
$p_{b\psi}$, $p_{r\theta}$, $p_{r\psi}$: Reduced hydrodynamic pressure
p_{bs}, p_{rs}: Amplitude of standing pressure wave component
p_{bp}, p_{rp}: Amplitude of progressive pressure wave component
R_a: Length of radial arm
R_c: Rotation radius of skinplate lower end
R_s, r_s: Rotation center height
r_{sa}: Ratio of rotation radius
R_ψ: Distance from skinplate rotation center to skinplate small area
T, t: Time
W_0: Spanwise length
X: Coordinate along undisturbed free surface toward upstream side
Y: Upward coordinate along skinplate
Z: Horizontal coordinate along skinplate from its center
α_*: Vibration angular amplitude ratio
α_I, $\alpha_{I\psi}$: Ratio of moment of inertia
α_ψ: Water-to-skinplate mass ratio
$\beta*$: Depth ratio
γ_{nw}: Ratio of in-water natural frequencies of coupled-mode vibration
$\gamma_{nw\psi}$: In-air to in-water natural frequency ratio of streamwise vibration
$\gamma_{w\theta\theta}$: In-air to in-water natural frequency ratio of whole gate rotational vibration around trunnion pin
$\gamma_{w\psi\theta}$: Actual vibration frequency ratio of streamwise vibration to in-air rotational vibration around trunnion pin
$\gamma_{\psi\theta}$: In-air vibration frequency ratio
$\Delta c_{\theta\psi}$: Reduced fluid excitation coefficient due to coupling
$\Delta c_{\psi\theta}$: Reduced wave-radiation damping coefficient due to coupling
ΔC_ψ, Δc_ψ: Wave-radiation damping coefficient
ΔI_ψ: Corrected moment of inertia

Δm_θ, Δm_ψ: Reduced added mass
$\Delta m_{\theta\psi}$, $\Delta m_{\psi\theta}$: Reduced added mass due to coupling
δ_p: Pressure correction coefficient
ζ_a: In-air damping ratio (without coupling)
$\zeta_{a\theta}$: In-air damping ratio of rotational vibration around trunnion pin
$\zeta_{a\psi}$: In-air damping ratio of streamwise rotational vibration
ζ_f, $\zeta_{f\psi}$: Wave-radiation damping ratio
ζ_w: In-water streamwise damping ratio (without coupling)
Θ, θ: Counter-clockwise rotation angle of whole gate rotation around trunnion pin
Θ_0, Ψ_0: Amplitude of Θ and Ψ
Θ_{ps}: Press-shut angle of vibration trajectory
Θ_{s0}: Geometric press-open angle
$\xi_{f\theta}$, $\xi_{f\psi}$: Fluid excitation ratio
ξ_θ, ξ_ψ: Actual excitation ratio
Ψ, ψ: Counter-clockwise rotation angle of skinplate streamwise vibration
Ω_a: in-air natural vibration frequency ratio (without coupling) [rad/s]
$\Omega_{a\theta}$: In-air natural vibration frequency of whole gate rotation around trunnion pin
$\Omega_{a\psi}$: In-air streamwise natural vibration frequency ratio [rad/s]
Ω_w: In-water vibration frequency ratio (without coupling) [rad/s]
$\Omega_{w\theta}$: In-water vibration frequency of rotational vibration around trunnion pin [rad/s]
$\Omega_{nw\psi}$: In-water streamwise natural vibration frequency [rad/s]
$\Omega_{w\psi}$: In-water streamwise vibration frequency [rad/s]

Subcripts

$()_a$: In-air
$()_w$: In-water
$()_\theta$: Related to the rigid body rotation around the trunnion pin
$()_\psi$: Related to the skinplate bending mode

Section 3
Validation and Applications

Chapter 11
Measurement of Coupled-Mode Instability on Model Scale:
Details of Validation

ABSTRACT

Experimental validation of the proposed mechanism of self-excited coupled-mode instability of Tainter gates, developed in Chapter 10, was obtained using laboratory models of Tainter gates. Results from both two- and three-dimensional model testing are reported. In the two-dimensional section model, the skinplate was simulated by a spring-mounted plate with the same upstream shape as the Folsom Dam gate. The simulated skinplate could undergo streamwise rotational vibration about its attachment point on the radial arm structure. The radial arm structure was also spring-mounted and could under torsional vibration about the trunnion center. The 1/13-scale three-dimensional model was carefully designed to maintain an in-water natural vibration frequency close to the same value as for the prototype Tainter gate with a similar bending mode shape to that of the Folsom gate. Both sets of model tests confirm the predictions from the theoretical development in Chapter 10.

INTRODUCTION

The coupled-mode self-excited instability mechanism, was confirmed in model tests using a small-scale Tainter gate. In this chapter a summary of both two-dimensional model testing and three-dimensional model testing is presented. Both sets of model tests confirm the predictions from the theoretical development in Chapter 10.

DOI: 10.4018/978-1-5225-3079-4.ch011

TWO-DIMENSIONAL MODEL STUDIES

Apparatus

Since construction of a full three-dimensional model which vibrates with identical vibration modes as the Folsom Dam Tainter gate would be substantially more challenging, the focus of the model scale testing was initially on a two-dimensional model of the gate. The essential feature of the coupled-mode vibration, *i.e.*, the coupling of the streamwise rotational vibration of the skinplate about a point on the upper portion of the skinplate with the whole gate rotational vibration around the trunnion pin, can be replicated in the two-dimensional model. Figure 1 shows the geometrically similar two-dimensional 1/31-scale model of Folsom Dam Tainter gate (Anami 2002, Anami & Ishii, 2003). A three-dimensional view is shown in Figure 1a, and the side view is shown in Figure 1b. Figure 2 shows photographs of the model gate.

In the model gate, the skinplate is represented as a spring-mounted rigid body which performs a rotational vibration in the streamwise direction about its attachment point on the radial arm structure. The back of the rigid skinplate rests on a horizontal coiled spring (B) that is supported by a plate mounted on the radial arm structure. The coiled spring represents the streamwise skinplate bending flexibility of the full-scale gate. The skinplate's rotation center P is placed on the radial arm structure to permit easy adjustment. The streamwise rotational vibration is denoted by Ψ. The whole gate is suspended by a coiled vertical spring (A), which represents the chain flexibility of the full-scale gate. Thus, the gate model can perform rigid body rotational vibration around the trunnion pin. The rigid body rotational vibration is denoted by Θ.

The validity of such crude modeling will be determined by comparison with results from field vibration tests. The details of the field testing are provided in Chapter 14; suffice it here to say that such crude two-dimensional treatment does indeed yield valid data concerning the vibration mechanism.

In this model, the gate can undergo rigid body rotational vibration around the trunnion pin, represented by Θ, and concurrently skinplate streamwise rotational vibration represented by Ψ. The position of both

Figure 1. Schematic view of a 1/31-scale coupled-mode vibration model of the Folsom Dam Tainter gate; (a) three-dimensional drawing of the model, and (b) a cross-sectional schematic view

From Anami (2002), Anami & Ishii (2003). © 2003 ASME. Used with permission.

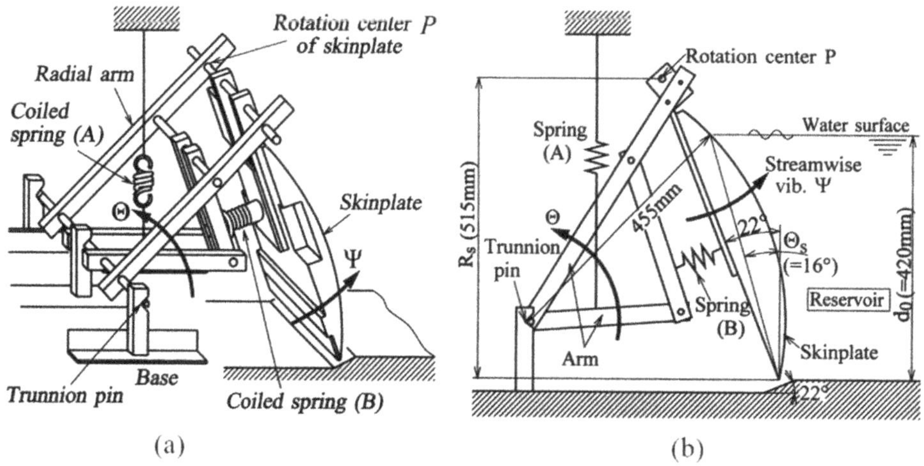

Figure 2. Photos of the 1/31-scale coupled-mode vibration model of Folsom Dam Tainter gate
From Anami (2002).

the horizontal and vertical springs can be adjusted, permitting a fine adjustment of each natural vibration frequency. The model gate was adjusted such that in static equilibrium the skinplate center coincides exactly with the trunnion center, that is, it is non-eccentric. The streamwise vibration of the skinplate naturally shifts the skinplate center from the trunnion center. However, assuming small-amplitude vibrations in this study, the eccentricity has a negligible effect in exciting the whole gate motion around the trunnion pin, when compared with that due to the inertial torque of the skinplate itself.

The major specifications for the model gate are listed in Table 1. The skinplate has a radius R_a of 455 mm and a span W_0 of 290 mm. The skinplate has a mass of 10.0 kg, and the whole gate mass is 15.1 kg. The whole gate was inserted into a 300 mm-wide channel, with a gap of about 5 mm from the channel wall on each side of the skinplate.

The trunnion shaft was replaced with a knife edge in order to reduce the friction of the trunnion pin as much as possible. In the actual Tainter gate, the whole gate is supported by the trunnion pin on each side of the gate. However, in this model gate, there is a possibility that the trunnion shaft may bend due to violent vibration of the gate, and it was supported at three spanwise points. The shaft supporting the skinplate is mounted on ball bearings to permit streamwise rotational motion. The moment of inertia of the skinplate around its streamwise rotation center, I_ψ, is 0.944 kg·m^2, and the moment of inertia of the whole gate around trunnion pin, I_θ, is 2.220 kg·m^2. These inertia moments were precisely obtained by pendulum experiments.

As shown in Figure 1b, the skinplate is inclined in the downstream direction at an angle of 22°. In addition, the channel floor just under the gate is inclined so that the press-open angle Θ_{s0} becomes zero, thus eliminating the press-open hydrodynamic pressure component. With such a simplified model gate setup, the details of the coupling of the skinplate streamwise rotational vibration and the whole gate rotational vibration around the trunnion pin can be readily obtained. The skinplate rotation center height R_S from the skinplate lower end was 515 mm. The model gate was exposed to water flow under the gate with an upstream submergence depth d_0 of 420 mm, as shown in Figure 1b. The average discharge opening was set at 5 mm, sufficiently small relative to d_0. The outflow is not submerged. The secant connecting the bottom end of the skinplate to the water surface contact point on the skinplate was inclined in the downstream direction at an angle of Θ_s=16°.

Table 1. Major specifications of 1/31-scale model of Folsom Dam Tainter gate

Major specifications	Skinplate radius, R_a	455 mm
	Spanwise length, W_0	290 mm
	Rotation center height, R_S	515 mm
	Mass of skinplate	10.0 kg
	Whole gate mass	15.1 kg
Moment of Inertia	Moment of inertia of skinplate around its streamwise rotation center, I_ψ	0.944 kgm^2
	Moment of inertia of whole gate around trunnion pin, I_θ	2.220 kgm^2
Setup conditions	Gate submergence depth, d_0	420 mm
	Average discharge opening, B	5 mm
	Skinplate inclinationangle, Θ_S	16°
	Press-open angle, Θ_{s0}	0°

After Anami et al. (2003).

Measurements

It was imperative that the in-water streamwise skinplate vibration frequency in the model test correspond closely to that of Folsom Dam Tainter gate since the instantaneous coefficient of flow-rate variation under the gate is strongly dependent upon the streamwise skinplate vibration frequency. From the field vibration tests on the Folsom Dam Tainter gate, the in-air streamwise vibration frequency of skinplate was found to be 26.9 Hz. This in-air frequency is reduced in flowing water, due to the added mass effect of water upon the skinplate. With a calculated added mass, the in-water vibration frequency was estimated to be about 6 Hz. In order to realize this in-water frequency of 6 Hz in the present model tests, the in-air streamwise vibration frequency of the model skinplate, $\Omega_{a\psi}$, was set at one of two values, 10.6 Hz or 12.3 Hz. With known values for the gate submergence depth d_0 and the skinplate in-air streamwise vibration frequency, $\Omega_{a\psi}$, the basic Froude number F_0, given by

$$F_0 \equiv 2\pi\Omega_{a\psi}\sqrt{\frac{d_0}{g}} \qquad \text{with} \qquad \Omega_{a\psi} \text{ in [Hz]} \tag{1}$$

can be calculated. The basic Froude number governs the dynamic similarity of fluctuating flow field; in these model tests the basic Froude numbers were 13.9 and 16.0. The mean gate opening height was 5 mm, which was 1.2% of the gate submergence depth d_0. In the model gate tests, the in-air whole gate vibration frequency $\Omega_{a\theta}$ was varied from 2.6 Hz to 7.3 Hz, that is, over a range that included the in-water natural streamwise vibration frequency of the skinplate at around 6 Hz.

After adjusting the gate to obtain the desired frequency combination, the gate was restrained so that it could not vibrate, and the restraint was released gently. The subsequent streamwise and vertical (up-

and-down) vibrations of the gate were measured by non-contact type gap-sensors. The measured data were recorded.

Figure 3 shows typical records of the self-excited vibration waveforms measured at the lower end of the gate at $F_0 = 16.0$. The upper waveforms show the streamwise vibration Ψ, and the lower show the vertical (up-and-down) vibration Θ. The center of the skinplate was precisely coincident with the trunnion pin. Until this test, a gate with the skinplate exactly concentric with the trunnion center *was assumed* to be stable and not undergo any self-excited vibrations. The model gate, however, clearly undergoes spontaneous vibrations, exponentially increasing in amplitude, and approaching a steady vibration with constant amplitude. These test results clearly demonstrate that Tainter gates possess a surprising inherent dynamic instability, even when the skinplate center is precisely coincident with the trunnion pin center (*i.e.*, there is no eccentricity of the skinplate center relative to the trunnion pin center).

The diagrams on the right side in Figure 3 show the vibration trajectories of the lower end of the skinplate. In all cases, the elliptic vibration trajectories exhibit the characteristic counter-clockwise behavior of a press-shut device. The gate opening is reduced when the skinplate is forced to move in the downstream direction. In Figure 3A, with an in-air frequency $\Omega_{a\theta}$ of the whole gate vibration of 4.42 Hz, the trajectory has a steep slope. The amplitude of the vertical (up-and-down) vibration is significantly

Figure 3. Self-excited vibration wave-forms and trajectories of 1/31-scale model of the Folsom Dam gate with $F_0 = 16.0$ ($d_0 = 420$ mm, $\Omega_{a\psi} = 12.3$ Hz)

From Anami & Ishii (2003). ©2003 ASME. Used with permission.

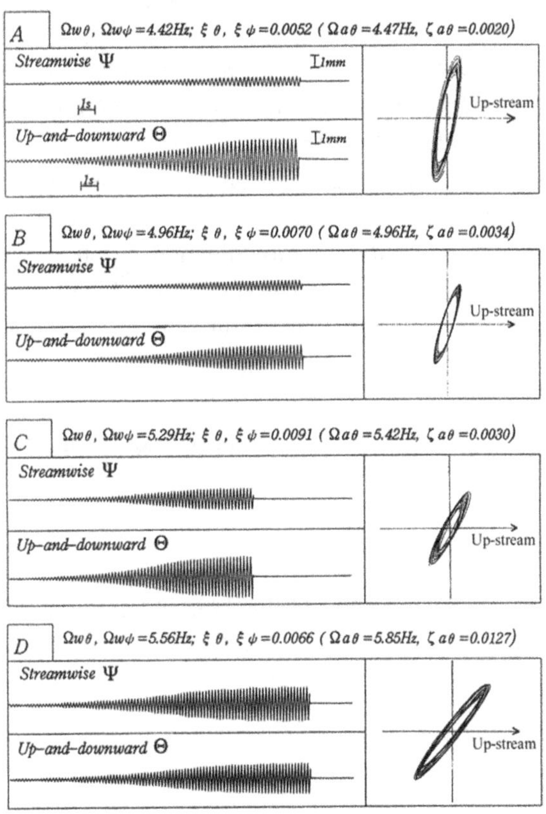

greater than that of the streamwise vibration. The excitation ratios, however, for each mode ξ_θ and ξ_ψ, measured from the exponentially increasing vibration-waveforms, were both 0.0052. The in-air damping ratio $\zeta_{a\theta}$ of 0.0020 is also shown. As $\Omega_{a\theta}$ was progressively increased to 4.96 Hz in Figure 3B and then to 5.29 Hz in Figure 3C, the streamwise vibration amplitude increased, and as a result, the inclination of the trajectory decreased. The excitation ratios increase correspondingly to 0.007 in Figure 3B and to 0.0091 in Figure 3C. Furthermore, when $\Omega_{a\theta}$ is 5.85 Hz in Figure 3D, the vertical (up-and-down) and streamwise vibrations have nearly equal amplitudes, resulting in an inclination of the trajectory of 45°. In this configuration, the gate motion can most effectively extract energy from the fluctuating flow, resulting in a relatively large excitation ratio of 0.0066 in spite of the relatively large in-air damping ratio of 0.0127.

The self-excited vibration waveforms with small amplitudes were analyzed using an FFT analyzer. The measured in-water vibration frequency data are plotted in Figure 4a, as a function of the frequency ratio γ_{nw} $(= \Omega_{nw\psi}/\Omega_{a\theta})$ of the streamwise in-water natural vibration frequency $\Omega_{nw\psi}$ to the in-air natural vibration frequency around the trunnion, $\Omega_{a\theta}$. The square and circular symbols are for basic Froude numbers of $F_0 = 13.9$ and 16.0, respectively. The frequency $\Omega_{a\theta}$ was varied from 2.6 Hz to 7.3 Hz, and $\Omega_{nw\psi}$ remained constant at 6.14 Hz for $\Omega_{a\psi} = 10.6$ Hz and 6.10 Hz for $\Omega_{a\psi} = 12.3$ Hz, respectively.

For small values of $\Omega_{a\theta}$ in Figure 4a, the frequency ratio γ_{nw} is larger than 1.0, and the two other frequency ratios plotted on the ordinate, the in-water to in-air frequency ratio $\gamma_{w\theta\theta}$ $(= \Omega_{w\theta}/\Omega_{a\theta})$, of the whole gate vibration around the trunnion, and the streamwise in-water vibration frequency to the in-air vibration frequency around the trunnion, $\gamma_{w\psi\theta}$ $(= \Omega_{w\psi}/\Omega_{a\theta})$, are both also 1.0. One may, therefore, conclude that the vibration is synchronized with the lower frequency of the whole gate natural vibration $\Omega_{a\theta}$. In contrast, for larger values of $\Omega_{a\theta}$ when $\Omega_{a\theta} > \Omega_{nw\psi}$ and γ_{nw} is slightly less than 1.0, both frequency ratios $\gamma_{w\theta\theta}$ and $\gamma_{w\psi\theta}$ take on the same value as the abscissa γ_{nw} $(= \Omega_{nw\psi}/\Omega_{a\theta})$. One may thus conclude that the vibration is synchronized with the lower frequency of skinplate streamwise natural vibration. When γ_{nw} is further reduced (at the largest values $\Omega_{a\theta}$ tested), the vibrations do not synchronize, and the frequency ratios $\gamma_{w\theta\theta}$ and $\gamma_{w\psi\theta}$ no longer have identical values.

The exponentially increasing or decreasing vibration waveforms suggest a significant characteristic of the self-excitation due to fluid motion called the "fluid-excitation ratio." At the beginning of experiment, the in-air damping ratios $\zeta_{a\theta}$ (around the trunnion pin) and $\zeta_{a\psi}$ (streamwise direction) were precisely measured. Once the in-water resultant excitation ratios ξ_θ (around the trunnion pin) and ξ_ψ (streamwise direction) were measured from any spontaneously growing vibration waveform, the fluid-excitation ratio for the whole gate vibration, $\xi_{f\theta}$, and that for the streamwise vibration, $\xi_{f\psi}$ (shown in Figure 4b), can be calculated from the following analysis.

The equation of motion for the in-air rotational vibration around the trunnion pin is given by

$$I_\theta \ddot{\Theta} + C_\theta \dot{\Theta} + K_\theta \Theta = 0 \tag{2}$$

Denoting the additional inertia moment ΔI and the fluid-excitation coefficient ΔC due to fluid motion, the equation of motion becomes:

$$\left(I_\theta + \Delta I_\theta\right)\ddot{\Theta} + \left(C_\theta - \Delta C_\theta\right)\dot{\Theta} + K_\theta \Theta = 0 \tag{3}$$

Figure 4. Test results for the model gate: (a) in-water vibration frequencies $\Omega_{w\theta}$, $\Omega_{w\psi}$, (b) fluid- excitation ratio $\xi_{c\theta}$, $\xi_{c\psi}$, (c) phase-lag $\Delta\phi$, (d) amplitude ratio α_* and (e) press-shut angle Θ_{ps}
From Anami & Ishii (2003). ©2003 ASME. Used with permission.

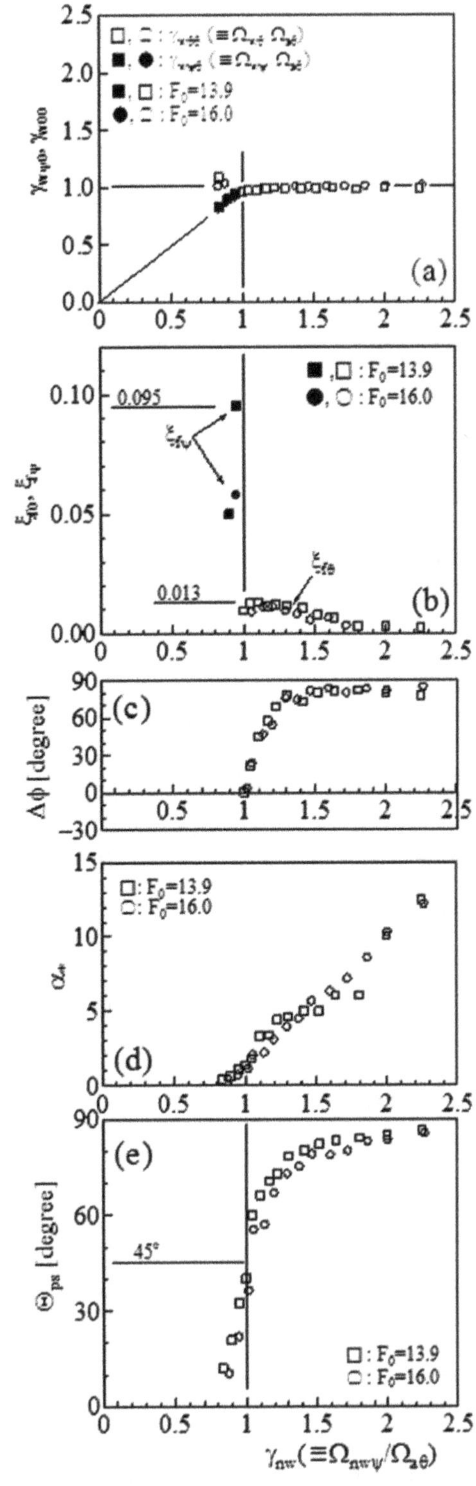

or,

$$\left(1+\frac{\Delta I_\theta}{I_\theta}\right)\ddot{\Theta}+\left(\frac{C_\theta-\Delta C_\theta}{I_\theta}\right)\dot{\Theta}+\Omega_{a\theta}^2\Theta=0 \tag{4}$$

Of course, $\Omega_{a\theta}$ is the in-air natural vibration frequency:

$$\Omega_{a\theta}\equiv\sqrt{\frac{K_\theta}{I_\theta}} \tag{5}$$

Similar to the in-air mechanical damping ratio $\zeta_{a\theta}$, defined by

$$\zeta_{a\theta}\equiv\frac{C_\theta}{2\sqrt{I_\theta K_\theta}}=\frac{C_\theta}{2I_\theta\Omega_{a\theta}} \tag{6}$$

the fluid-excitation ratio $\xi_{f\theta}$ is defined as follows:

$$\xi_{f\theta}\equiv\frac{\Delta C_\theta}{2\sqrt{I_\theta K_\theta}}=\frac{\Delta C_\theta}{2I_\theta\Omega_{a\theta}} \tag{7}$$

The equation of motion can be arranged as follows:

$$\ddot{\Theta}+2\Omega_{a\theta}\frac{\zeta_{a\theta}-\xi_{f\theta}}{1+\Delta I_\theta/I_\theta}\dot{\Theta}+\frac{\Omega_{a\theta}^2}{1+\Delta I_\theta/I_\theta}\Theta=0 \tag{8}$$

by using Equations 6 and 7. In this form, the coefficient of Θ is the square of the in-water natural vibration frequency Ω_{w0}:

$$\Omega_{w\theta}^2\equiv\frac{\Omega_{a\theta}^2}{1+\Delta I_\theta/I_\theta}. \tag{9}$$

Therefore, when the in-water to in-air vibration ratio is expressed as $\gamma_{w\theta\theta}$ ($\equiv\Omega_{w\theta}/\Omega_{a\theta}$), the equation of motion is as follows:

$$\ddot{\Theta}+2\Omega_{w\theta}\gamma_{w\theta\theta}\left(\zeta_{a\theta}-\xi_{f\theta}\right)\dot{\Theta}+\Omega_{w\theta}^2\Theta=0 \tag{10}$$

Then, in-water excitation ratio ξ_θ for the whole gate vibration is

$$\xi_{\theta} = -\gamma_{w\theta\theta}\left(\zeta_{a\theta} - \xi_{f\theta}\right) \tag{11}$$

The fluid-excitation ratio $\xi_{f\theta}$ is obtained by the following expression:

$$\xi_{f\theta}\left(\equiv \frac{\Delta C_{\theta}}{2I_{\theta}\Omega_{a\theta}}\right) = \frac{\xi_{\theta}}{\gamma_{w\theta\theta}} + \zeta_{a\theta} \tag{12}$$

A similar analysis can be made for the streamwise vibration, from which the following expression for the fluid-excitation ratio $\xi_{f\psi}$ of the streamwise rotational vibration can be derived:

$$\xi_{f\psi}\left(\equiv \frac{\Delta C_{\psi}}{2I_{\psi}\Omega_{a\psi}}\right) = \frac{\xi_{\psi}}{\gamma_{w\psi\psi}} + \zeta_{a\psi} \tag{13}$$

where ΔC_{ψ} is the fluid-excitation coefficient of the streamwise rotational vibration, and $\gamma_{w\psi\psi}$ $(= \Omega_{w\psi}/\Omega_{a\psi})$ is the in-water to in-air vibration frequency ratio for the streamwise vibration.

In the present model gate, the mean value of $\zeta_{a\theta}$ was 0.003 from in-air free vibration tests of the whole gate around the trunnion pin. The damping ratio of the skinplate streamwise vibration, $\zeta_{a\psi}$, was 0.105 for $F_0 = 13.9$ and 0.133 for $F_0 = 16.0$. In the case of model experiments, relatively large damping ratios are found for the skinplate streamwise vibration, due to leakage flows from comparatively large clearances on both sides of the skinplate. This kind of additional damping is intrinsic to a model experiment and cannot be avoided.

Consequently, in-water free vibration tests of the skinplate were made in detail to measure the damping ratio including this additional damping effect. The details of the free vibration tests were explained previously in Chapter 9 and the results are displayed in Figure 22 in Chapter 9. The corrected damping ratio obtained as a result is the in-air damping ratio $\zeta_{a\psi}$ used here.

The fluid damping ratios can be calculated from Equation 12 for synchronization with the whole gate vibration around the trunnion pin, and from Equation 13 for synchronization with the skinplate streamwise vibration. Calculated values of the fluid-excitation ratios based on experimental data are plotted in Figure 4b. For γ_{nw}, larger than 1.0, the instability synchronized with the whole gate vibration occurs and the excitation ratio data are the unfilled symbols. The maximum value of the fluid-excitation ratio $\xi_{f\theta}$ was 0.013. If γ_{nw} becomes slightly less than 1.0, the instability synchronized with the skinplate vibration and the excitation ratio data are the filled symbols. The fluid-excitation ratio $\xi_{f\psi}$ rapidly increases, showing a maximum value of 0.095 and then rapidly decreases with a further reduction in γ_{nw}.

From the measured vibration waveforms in the streamwise and vertical (up-and-down) directions recorded at the lower end of the skinplate, the phase-lag $\Delta\phi$ of the up-and-down vibration relative to streamwise vibration can be found, and is shown in Figure 4c. As γ_{nw} increases from 1.0, the phase-lag gradually increases from 0° to about 80°. This phase-lag is indicative of the fullness of the elliptical vibration trajectory (shown on the right side of Figure 3). A small phase-lag means that the vibration trajectory becomes flat

The angler amplitude ratio α_* (= Θ_0/Ψ_0) of the whole gate rotation around the trunnion pin to the skinplate streamwise vibration is shown in Figure 4d. The angler amplitude ratio α_* is about 1.0 when γ_{nw} is near 1.0. With increasing γ_{nw}, α_* increases almost linearly.

Figure 4e shows the press-shut angle Θ_{ps} defined as the inclination of the vibration trajectory. The press shut angle Θ_{ps} increases rapidly as γ_{nw} becomes large. In the range of $\gamma_{nw} < 1.0$, Θ_{ps} takes a value of 0° to 45°. When γ_{nw} exceeds 1.0, Θ_{ps} approaches 90°.

As previously predicted, the model experiments presented in this chapter confirmed that Tainter gates are indeed susceptible to the self-excited coupled-mode vibration mechanism, even when the skinplate center precisely coincides with the trunnion pin center.

Comparison With Theoretical Predictions

The theoretical analysis and its approximate solutions were presented in Chapter 10. The approximate solutions are calculated specifically for the two-dimensional coupled-mode vibration tests.

The major measurement results and the dimensionless parameters required for calculation are provided in Tables 2 and 3, respectively, where the coupled moment-of-inertia $I_{\psi\theta}$ was calculated using Equation 3.

Since the model gate is installed over a flat floor, the depth ratio β^* is 1.0. The reduced rotation center height r_s (= R_S/d_0) is 1.23, and the rotation radius ratio r_{sa} (= R_S/R_c) is 1.37. The moment-of-inertia ratios of the skinplate to the whole gate, α_I' (= I_ψ'/I_θ) and $\alpha_{I\psi}$ (= $I_{\psi\theta}/I_\psi$), take on values of 0.092 and

Table 2. Major measurement results from the two-dimensional model of Folsom Dam Tainter gate

Major specifications	Skinplate radius, R_a	455 mm	
	Spanwise length, W_0	290 mm	
	Rotation center height, R_S	515 mm	
	Rotation radius of skinplate lower end, R_c	375 mm	
	Mass of skinplate	10.0 kg	
	Whole gate mass	15.1 kg	
Moment of Inertia	Moment of inertia of skinplate around its streamwise rotation center, I_ψ	0.944 kgm^2	
	Moment of inertia of whole gate around trunnion pin, I_θ	2.220 kgm^2	
	Coupled moment of inertia, $I_{\psi\theta}$	0.289 kgm^2	
Setup conditions	Gate submergence depth, d_0	420 mm	
	Average discharge opening, B	5 mm	
	Skinplate inclinationangle, Θ_S	16°	
	Press-open angle, Θ_{s0}	0°	
	in-air streamwise vibration frequency of skinplate, $\Omega_{a\psi}$	10.6 Hz	12.3 Hz
	in-air whole gate vibration frequency, $\Omega_{a\theta}$	varied 2.7 Hz ~ 7.3 Hz	
Damping ratio	in-air damping ratio of rotational vibration around trunnion pin, $\zeta_{a\theta}$	0.003	
	in-air damping ratio of streamwise rotational vibration, $\zeta_{a\psi}$	0.105	0.133

Table 3. Non-dimensional parameters of the two-dimensional model gate for calculations

depth ratio, β^*	1.0	
reduced rotation center height, r_s	1.23	
rotation radius ratio, r_{sa} ($\equiv R_S/R_c$)	1.37	
moment-of-inertia ratios of skinplate to whole gate, α_I'($\equiv I_\psi'/I_\theta$)	0.092	
moment-of-inertia ratios of the skinplate to the whole gate, $\alpha_{I\psi}$ ($\equiv I_{\psi\theta}/I_\psi$)	0.306	
water to gate mass ratio, α_ψ ($\equiv \rho d_0^2 W_0/(I_\psi/R_s^2)$)	14.9	
instantaneous flow-rate-variation coefficient, c_f	0.7	
pressure correction coefficient, δ_p	0.83	
Basic Froude number, F_0	13.9	16.0

0.306, respectively. The water-to-gate mass ratio $\alpha_\psi \left(= \rho d_0^2 W_0 / \left(I_\psi R_S^2\right)\right)$ is comparatively small at 14.9. Since the inclination angle Θ_s of the model gate is 16°, the instantaneous flow-rate-variation coefficient c_f takes on a value of 0.7, and the pressure correction coefficient δ_p takes on a value of 0.83, from the experimental results, presented in Figures 15b and 19 in Chapter 9.

In the present two-dimensional coupled-mode vibration tests, the basic Froude numbers F_0 in the two sets of tests were 13.9 and 16.0, as shown in Table 3. Therefore, the basic Froude number $F_{a\theta}$ in the equation of motion, Equation 32 in Chapter 10, for the coupled-mode vibration is calculated by

$$F_{a\theta} = F_0 / \gamma_{\psi\theta} \tag{14}$$

The average in-air damping ratio $\zeta_{a\theta}$ of the whole gate vibration around the trunnion pin was a small value of 0.003, measured by in-air free vibration tests, and this value did not increase in-water free vibration tests. For the skinplate streamwise rotational vibration, however, the in-water damping effect became quite large, due to water leakage around the sides of the skinplate, as discussed in Chapter 9. Therefore, this large damping effect must be included in the in-air damping ratio $\zeta_{a\psi}$. Thus, $\zeta_{a\psi}$ takes on comparatively large values of 0.133 for F_0=16.0 and 0.105 for F_0=13.9, as shown in Table 3.

Given the in-air vibration ratio $\gamma_{\psi\theta}$ ($\equiv \Omega_{a\psi}/\Omega_{a\theta}$) in addition to the non-dimensional parameters shown in Table 3, the approximate solutions of the equations of motion,

$$\theta'' + 2\zeta_{a\theta}\theta' + \theta + \frac{\alpha_I'}{\alpha_*}\psi'' = 0 \tag{15}$$

and

$$\psi'' + 2\gamma_{nw\psi}\gamma_{nw}\left(\zeta_{a\psi} + \zeta_{f\psi}\right)\psi' + \gamma_{nw}^2\psi = \alpha_*\gamma_{nw\psi}^2\left\{-\left(\alpha_{I\psi} + \sqrt{2}c_f\frac{\alpha_\psi\Delta m_\theta^*}{r_{sa}}\right)\theta'' + \sqrt{2}c_f\frac{\alpha_\psi\Delta c_\theta^*}{r_{sa}F_{a\theta}}\theta'\right\} \tag{16}$$

can be calculated. The required vertical projection error Δ for the model gate, shown by the dashed line in Figure 12 in Chapter 9, was used, when calculating the equivalent added mass $\Delta m_{\psi}*$, the wave-radiation damping coefficient $\Delta c_{\psi}*$, the equivalent added mass $\Delta m_{\theta}*$, and fluid-excitation coefficient $\Delta c_{\theta}*$.

Calculated results are shown as solid lines in Figure 5, where the model gate test results previously shown in Figure 4 are also plotted. The small circles represent the data for $F_0 = 16.0$, while the squares are for $F_0 = 13.9$. The filled symbols represent vibration synchronized with the skinplate streamwise vibration, and the open symbols correspond to vibration synchronized with the whole gate rotational vibration. As for the vibration frequency ratio, the fluid-excitation ratio, the phase-lag, the vibration angler amplitude ratio and the press-shut angle, the experimental and theoretical results are generally in good agreement. A few small differences between them are also seen, but they can be discounted, or even perhaps ignored, when the complexity of the model experiments is considered resulting in some unavoidable errors in measurement.

THREE-DIMENSIONAL MODEL STUDIES

Apparatus

The fundamental structure of the model needed to be modified to permit the 1/13-scale model to maintain an in-water natural vibration frequency close to the same value as for the prototype Tainter gate. The proposed structure with low streamwise rigidity was analyzed using an FEM simulation. The technique for this FEM simulation for a complicated structure like a Tainter gate was already established by Anami (2002). The technique was applied iteratively to design a model which is a comparatively flexible structure, but can still reproduce the prototype vibration mode correctly.

In order to reproduce, on the model scale, the same vibration mode as the Folsom Dam Tainter gate, the static deformation of the skinplate due to hydrostatic load needed to be the same as the actual deformation. Various structures were examined based on this fundamental viewpoint. Consequently, it was determined that a more flexible skinplate model of the structure was optimal. The FEM model of the 1/13-scale three-dimensional equivalent model is shown in Figure 6. The 0.6 mm rectangular stainless steel plate (1199 × 990 mm) of the skinplate was supported every 160 mm in the spanwise direction with 7 circular-arc ribs (each with a thickness of 5 mm and a width of 50 mm). It was supported by 4 flat bars each with a 10 mm thick and 50 mm wide cross-section connected to the trunnion pin with a diameter of 30 mm. The structure is considerably simpler than the Folsom Dam Tainter gate (see Anami, 2002 for added detail).

The static deformation (determine by FEM) due to the hydrostatic water load at a depth of 1084 mm (1/13 the water depth at the time of the Folsom gate failure) is shown in Figure 7. The bottom center of the skinplate deforms 2.6 mm in the downstream direction, and the top center deforms in the upstream direction. The portion of the skinplate with no deformation (displacement is zero) is near the second radial arm. This static deformation distribution is in good agreement with similar FEM analysis results for the Folsom Tainter gate.

Subsequently, the natural vibration characteristics were determined via FEM analysis for the in-air skinplate streamwise bending vibration. The resultant bending vibration mode of the skinplate is shown in Figure 8a, where the bending vibration modes in the spanwise direction are shown on the right side. These mode shapes compare very well with those for the Folsom Dam Tainter gate, shown in Figure 8b.

Figure 5. Theoretical results of flow-induced coupled-mode self-excited vibration of 2-dimensional model gate, for comparison with model test results
From Anami & Ishii (2003). ©2003 ASME. Used with permission.

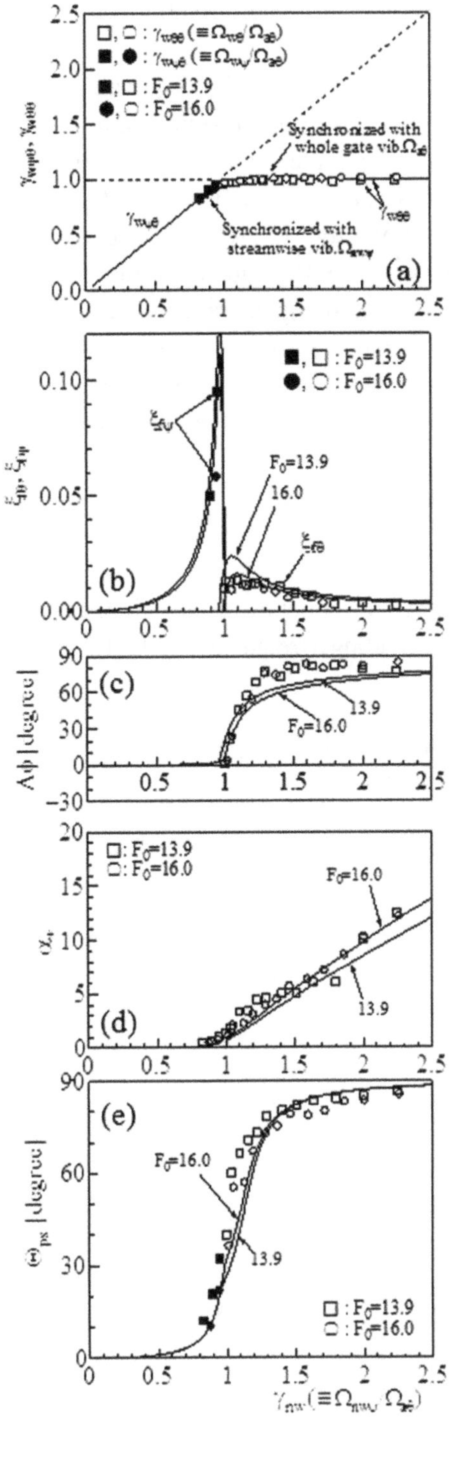

The natural frequency of the skinplate bending vibration for the three-dimensional equivalent model gate as determined by the FEM simulation was 56.0 Hz. This value is about 2.1 times higher than the measured natural vibration frequency of 26.8 Hz for the Folsom Dam Tainter gate. However, in tests with running water, the model frequency decreased to almost the same value as for the Folsom Tainter gate due to the added mass of the water.

Based on this FEM simulation, the 1/13-scale 3-D equivalent model gate was constructed. A little creativity was required for fixing the 0.7 mm-thick skinplate to the circular-arc ribs. Initially 100 temporary rivets were used to attach the skinplate to the ribs. Then the skinplate was soldered to the ribs to provide the desired rigidity. The downstream and side views of the 3-D equivalent model gate, installed in the flume, are shown in Figure 9. In order to prevent distortion of the gate structure, crossing wires are stretched between the radial arms. The 3-D equivalent model gate has a total mass of 42.5 kg. The major specifications of the 1/13-scale model gate are listed in Table 4.

In order to adjust correctly the whole gate rotational vibration frequency $\Omega_{nw\theta}$ in the experiment, leaf springs were connected to the wire cables used for raising the gate, as shown in Figure 10, where two guide bars are also attached along with the wire cables to prevent any streamwise swinging motion of the leaf spring.

Measurements

The in-air natural streamwise vibration frequency of the skinplate, $\Omega_{na\psi}$, was measured by in-air modal analysis (impulsive impact testing), yielding a frequency of 61.4 Hz, which agreed with the FEM prediction.

Figure 6. A 1/13-scale 3-D equivalent model by FEM
From Anami et al. (2006). ©2006 ASME. Used with permission.

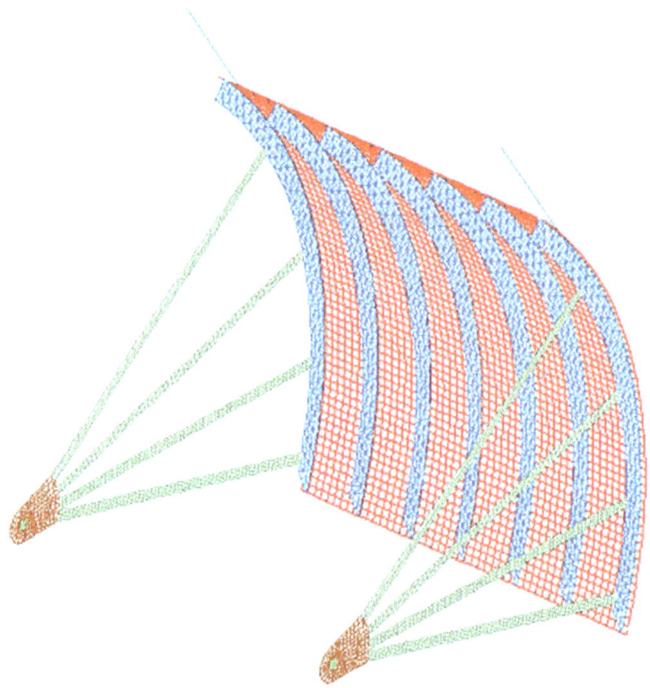

Figure 7. Static deformation by FEM analysis
From Anami et al. (2006). ©2006 ASME. Used with permission.

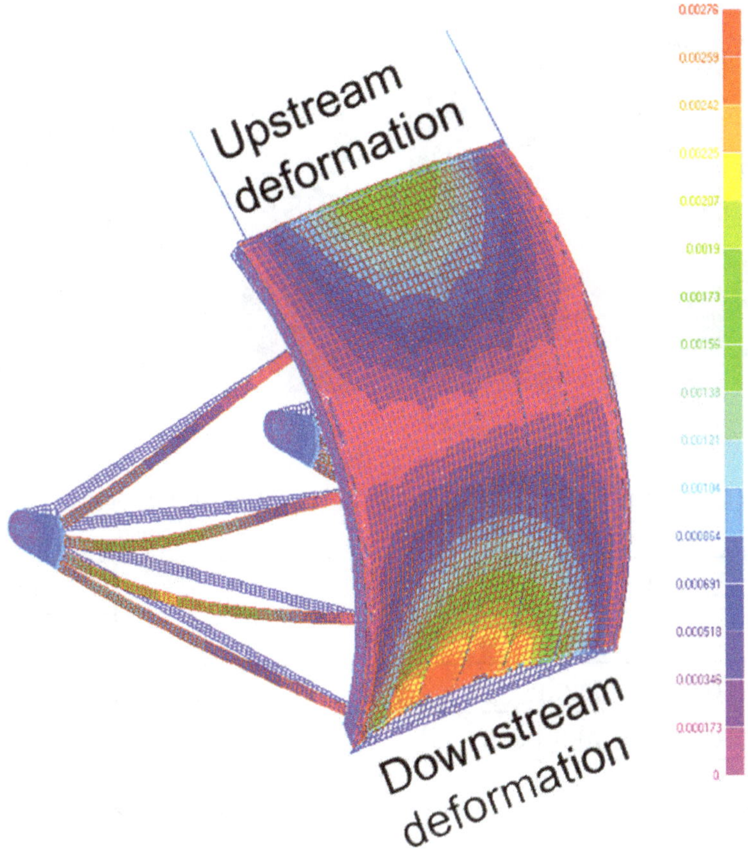

Figure 8. Skinplate streamwise bending vibration mode shape of 3-D equivalent model and Folsom gate; (a) Vibration mode shape of 3-D equivalent model by FEM; and (b) vibration mode shape of Folsom Dam Tainter gate from experimental modal analysis testing
(a) From Anami et al. (2006), ©2006 ASME. Used with permission; (b) From Anami (2002).

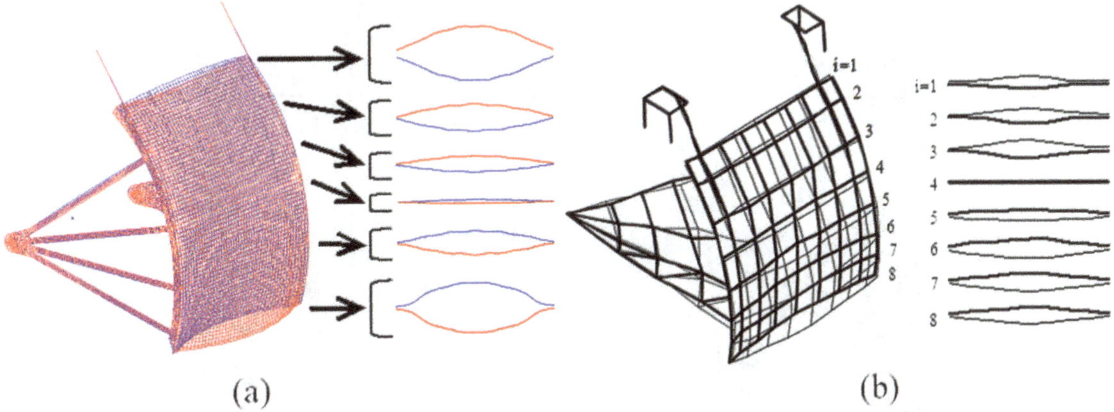

Figure 9. Photos of the three-dimensional equivalent model gate installed in the flume, (a) looking up-stream, and (b) side view of the gate
From Anami et al. (2006). ©2006 ASME. Used with permission.

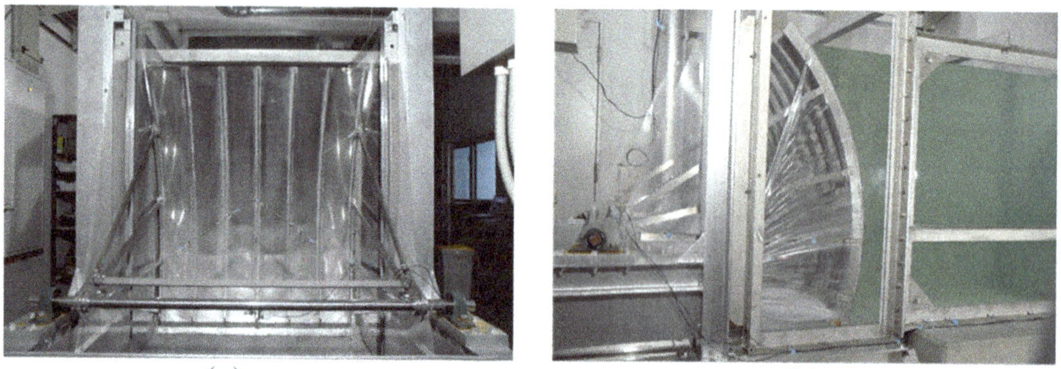

(a) (b)

Table 4. Major specifications of 1/31-scale model of Folsom Dam Tainter gate

Major specifications	Skinplate radius, R_a	1180 mm
	Spanwise length, W_0	990 mm
	Skinplate height	1190 mm
	Skinplate thickness	0.6 mm
	Whole gate mass	42.5 kg

After Anami et al. (2006).

Figure 10. Leaf spring and guide bar
From Anami et al. (2006). ©2006 ASME. Used with permission.

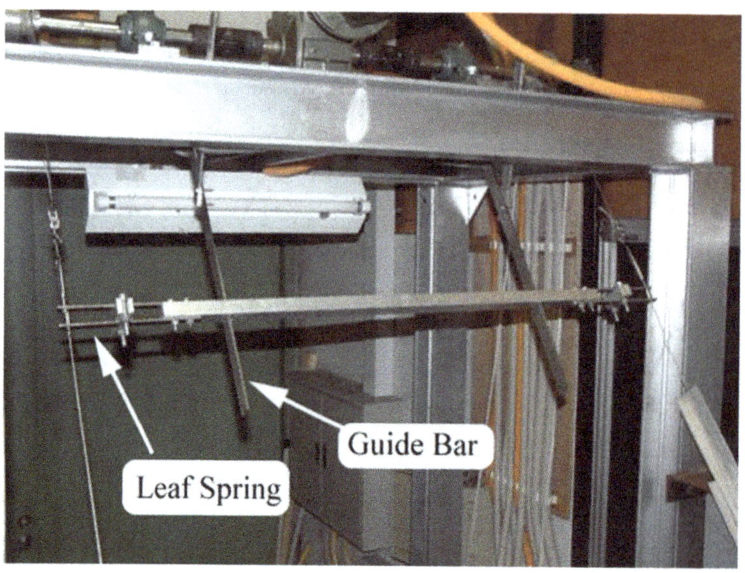

The test conditions and major mechanical constants are summarized in Table 5. The in-water natural streamwise vibration frequency $\Omega_{nw\psi}$ was lowered to 6.28 Hz with the gate submergence depth of 1.12 m, which was very close to the in-water natural vibration frequency of 6.46 Hz identified for Folsom Dam Tainter gate by our theoretical analysis. The floor just under the skinplate is inclined at 21° so that with the static deformation of the skinplate with hydraulic pressure, the gate is the neutral position between acting as a press-shut device and acting as a press-open device, that is, the so-called geometric press-shut angle was set as 0°.

Two non-contact type displacement sensors were positioned on the back-side streamwise-center bottom of the skinplate to measure the whole gate rotational vibration around the trunnion pin and the skinplate streamwise bending vibration.

From the analysis of Anami (2002) and Anami & Ishii (2003), violent coupled-mode vibration are expected to occur when the in-water natural vibration frequency of the skinplate streamwise vibration, $\Omega_{nw\psi}$, is slightly less than the natural vibration frequency of the whole gate rotational vibration around the trunnion pin, $\Omega_{nw\theta}$. Based on these analyses, the whole gate rotational vibration frequency $\Omega_{nw\theta}$ was adjusted to be 6.62 Hz, slightly larger than the skinplate streamwise bending vibration frequency of $\Omega_{nw\psi}$ = 6.28 Hz, yielding a frequency ratio of skinplate streamwise natural vibration to whole gate natural vibration, γ_{nw} ($\equiv \Omega_{nw\psi}/\Omega_{a\theta}$) of 0.99. With this condition, a very intense dynamic instability was observed, as shown in Figure 11, where the upper time history is the skinplate radial vibration and the lower time history is the skinplate tangential vibration, representing the skinplate streamwise vibration and the whole gate vibration around the trunnion pin, respectively. Violent self-excited vibrations were spontaneously induced. At an amplitude of about 13 mm the 3-D equivalent model gate struck the channel floor, and was prevented from any further increase in amplitude.

Both vibration frequencies $\Omega_{w\psi}$ and $\Omega_{w\theta}$ in Figure 11 were exactly 6.28 Hz, synchronized with the skinplate streamwise natural vibration. The excitation ratios (negative damping ratios) $\xi_{w\psi}$ and $\xi_{w\theta}$ measured from small amplitude portions of the vibration time histories were 0.015 and 0.012, respectively.

The measurement location of vibration was at the bottom spanwise center of the skinplate. The vibration trajectories are shown in the right side of the Figure 11. The upper trajectory shows comparatively small vibration, while the lower one shows comparatively large vibration. The angle of the slope of the trajectory is exactly 45° against the inclined floor just under the skinplate, thus effectively supplying energy from the fluid motion to the gate vibration, and thus inducing the observed very intense couple-mode self-excited vibration.

Similar measurements were made for various vibration frequencies $\Omega_{nw\theta}$ of whole gate rotation around the trunnion pin. The whole gate vibration frequencies $\Omega_{nw\theta}$ was varied from 4.13 Hz to 7.75 Hz. The intense coupled-mode self-excited vibration occurred in all test conditions.

Table 5. Test conditions and mechanical constants for 1/13-scale 3D equivalent model of Folsom Dam Tainter gate

		3-D model gate
depth of upstream channel	d_r	1.115 m
gate opening	B	0.005 m
depth over the gate bolttom	d_0	1.110 m
inclination angle of gate bottom to water surface	Θ_s	9.5°
natural vib. freq. of upward vib.	$\Omega_{nw\theta}$	varied
in-air natural vib. freq. of streamwise vib.	$\Omega_{na\psi}$	61.4 Hz
in-air damping ratio for upward vib	$\zeta_{a\theta}$	0.012
in-air damping ratio for streamwise vib.	$\zeta_{a\psi}$	0.007
rotation center height	R_S	0.626 m
press-open angle	Θ_{s0}	0°
rotation rasius of skinplate lower end	R_c	1.04 m

From Anami et al. (2006) ©2006 ASME. Used with permission.

Figure 11. Vibration test results with 1/13-scale 3-D equivalent model gate for vibration waveforms and trajectories at skinplate bottom center
From Anami et al. (2006) ©2006 ASME. Used with permission.

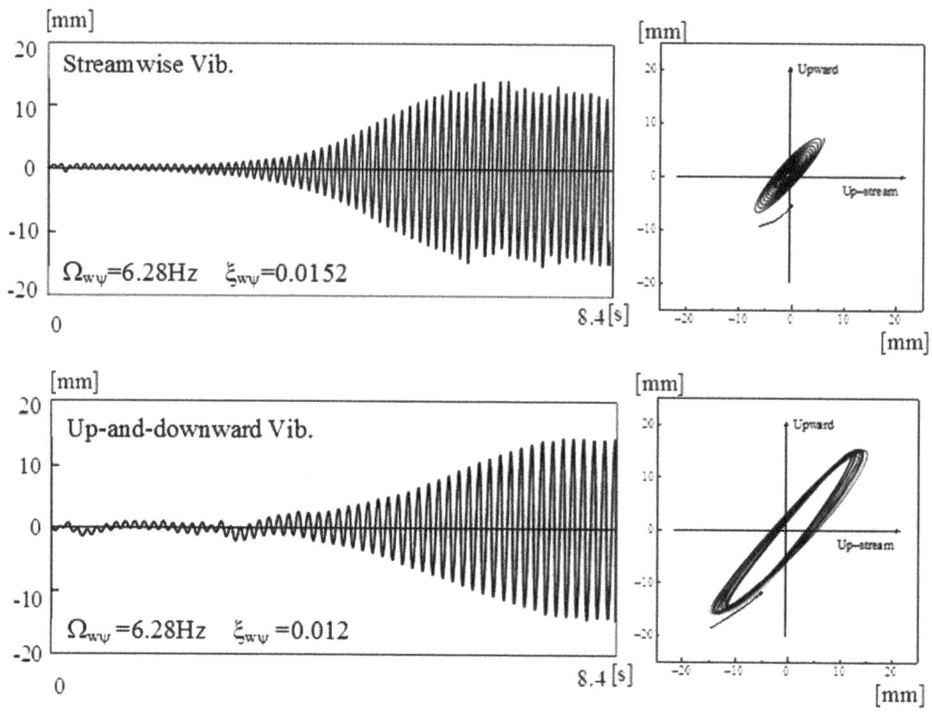

Comparison With Theoretical Predictions

The fluid excitation ratios $\xi_{f\theta}$ and $\xi_{f\psi}$ representing the level of dynamic instability due to fluid motion can be calculated from Equations 12 and 13, respectively.

The fluid excitation ratios $\xi_{f\theta}$ and $\xi_{f\psi}$ calculated from the measured vibration waveforms are shown in Figure 12. $\xi_{f\psi}$ takes on a value from 0.035 to 0.136, and $\xi_{f\theta}$ ranges from 0.013 to 0.017. When $\gamma_{nw} < 1$, or equivalently when $\Omega_{a\theta} > \Omega_{nw\psi}$, the in-air whole gate vibration frequency $\Omega_{a\theta}$ is closer to the in-water natural streamwise vibration frequency $\Omega_{nw\psi}$ of 6.28 Hz, and the magnitude of $\xi_{f\psi}$ increases quite quickly as the ratio approaches 1.0. On the other hand, for $\gamma_{nw} > 1$ or when $\Omega_{a\theta} < \Omega_{nw\psi}$, $\xi_{f\theta}$ takes on an almost almost constant value. The value remains about the same as that γ_{nw} is just slightly greater than 1.0 (which means when the in-air whole gate vibration frequency $\Omega_{a\theta}$ is approaching the in-water natural streamwise vibration frequency $\Omega_{nw\psi}$.)

The theoretically calculated results from Anami (2002) are also shown in Figure 12 as a solid line, showing close agreement with the tested data for the 3-D equivalent model of the Tainter gate. In addition, the two-dimensional model gate test results are included in Figure 12 as unfilled data points and the corresponding theoretical results by the dashed line (see Anami, 2002).

Figure 12. A stability diagram for the two- and three-dimensional model gates compared with the theoretical predictions; depending on frequency ratio a gate must have a mechanical damping ratio at least as large as the magnitude of the fluid excitation to remain stable

From Anami & Ishii (2003), Anami et al. (2006) ©2003, 2006 ASME. Used with permission.

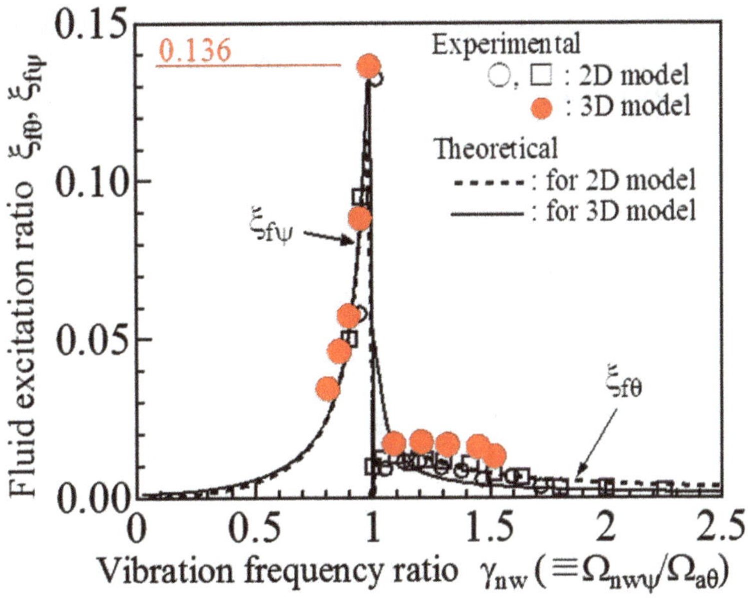

CONCLUSION

The experiments presented in this chapter confirm that Tainter gates are susceptible to the self-excited coupled-mode instability. Both the two-dimensional and three-dimensional model tests provided data on the vibration characteristics. The data from the two-dimensional model, in Figure 5, as well as the data from the three-dimensional model testing, in Figure 12, are in very good to excellent agreement with the predictions from the theory developed in Chapter 10. The two sets of experimental data also show excellent consistency. Subsequently, in Chapter 12, additional comparisons between the theoretical predictions for the behavior of the Wachi and Folsom Dam gates with observations made after these failures will be presented. Finally in Chapter 14, full-scale test data from operational Tainter gates will be compared with the theoretical predictions based on the equations in Chapter 10. The full-scale measurements will also be compared with the laboratory data presented in the present chapter.

REFERENCES

Anami, K. (2002). *Flow-Induced Coupled-Mode Self Excited Vibration of Large-Scaled Tainter Gates* (Doctoral Dissertation). Osaka Electro-Communication University, Osaka, Japan. (in Japanese)

Anami, K., & Ishii, N. (2003). Model tests for non-eccentricity dynamic instability closely related to Folsom Dam Tainter gate failure. In *ASME 2003 Pressure Vessels and Piping Conference* (pp. 213-222). New York, NY: American Society of Mechanical Engineers. doi:10.1115/PVP2003-2092

Anami, K., Ishii, N., Knisely, C. W., & Oku, T. (2006). Vibration tests with a 1/13-scaled 3-D model of the Folsom Dam Tainter gate and its prediction by theory. In *ASME 2006 Pressure Vessels and Piping/ICPVT-11 Conference* (pp. 703-710). New York, NY: American Society of Mechanical Engineers. doi:10.1115/PVP2006-ICPVT-11-93917

APPENDIX: NOMENCLATURE

c_f: Instantaneous flow-rate variation coefficient
d_0: Gate submergence depth
d_r: Depth of the reservoir
F_0: Basic Froude number
$F_{a\psi}$: Basic Froude number for whole gate rotational vibration
I_θ: Moment of inertia of the whole gate around the trunnion pin,
I_ψ: Moment-of-inertia of skinplate streamwise rotation
$I_{\psi\theta}$: Coupled moments-of-inertia
r_s, R_S: Height of skinplate rotation center [r_s is the dimensionless expression of R_s]
R_c: Rotation radius of skinplate lower end about the trunnion pin
r_{sa}: Rotation radius ratio
W_0: Spanwise length
α_*: Angular-amplitude ratio
α_I': Moment-of-inertia ratios of the skinplate to the whole gate
$\alpha_{I\psi}$: Ratio of the coupled moment-of-inertia $I_{\psi\theta}$ to the skinplate moment-of-inertia I_ψ
α_ψ: Water-to-gate mass ratio
β*: Depth ratio
γ_{nw}: In-water to in-air vibration frequency ratio
$\gamma_{w\theta\theta}$: In-water to in-air vibration frequency ratio for the whole gate rotation around the trunnion pin
$\gamma_{w\psi\theta}$: Actual frequency ratio of the in-water skinplate streamwise rotational vibration to the whole gate in-air vibration frequency
$\gamma_{w\psi\psi}$: In-water to in-air vibration frequency ratio for the streamwise vibration.
$\gamma_{\psi\theta}$: In-air natural vibration frequency ratio
δ_p: Pressure correction coefficient
Δ: Vertical projection error
Δc_θ*: Fluid-excitation coefficient
Δc_ψ*: The wave-radiation damping coefficient
Δm_θ*: The equivalent added mass
Δm_ψ*: Equivalent added mass
$\Delta\phi$: Phase-lag of the up-and-down vibration relative to streamwise vibration
ΔI: Additional inertia moment
ΔC: Fluid-excitation coefficient
ΔC_ψ: Fluid-excitation coefficient of the streamwise rotational vibration
$\zeta_{a\theta}$: In-air damping ratio of the whole gate rotational vibration,
$\zeta_{a\psi}$: In-air damping ratio for streamwise vibration of the skinplate
$\xi_{f\theta}$: Fluid-excitation ratio of whole gate rotational vibration
$\zeta_{f\psi}$: Wave-radiation damping ratio
θ, Θ: Counterclockwise rotation angle of whole rigid gate [θ is the dimensionless expression of Θ]
Θ_{ps}: Press-shut angle defined as trajectory inclination angle
Θ_s: Skinplate inclination angle

Θ_{s0}: Geometric press-open angle
ξ_{θ}: Actual excitation ratio of the whole gate rotational vibration
ξ_{ψ}: Actual excitation ratio of the skinplate streamwise rotational vibration
$\xi_{w\theta}$: Excitation ratios (negative damping ratios)
$\xi_{w\psi}$: Excitation ratios (negative damping ratios)
ρ: Density of water
ψ, Ψ: Counter-clockwise rotation angle of skinplate [ψ is the dimensionless expression of Ψ]
$\Omega_{a\theta}$: In-air whole gate rotational vibration frequency
$\Omega_{a\psi}$: In-air vibration frequency of the skinplate streamwise vibration
$\Omega_{na\psi}$: In-air natural streamwise vibration frequency
$\Omega_{nw\theta}$: Whole gate rotational vibration frequency
$\Omega_{nw\psi}$: In-water natural streamwise vibration frequency
$\Omega_{w\theta}$: In-water vibration frequency of whole gate rotational vibration
$\Omega_{w\psi}$: In-water vibration frequency of the skinplate streamwise vibration

Chapter 12
Wachi Dam and Folsom Dam:
Coupled-Mode Instabilities - Mechanism for Failures

ABSTRACT

The failures at the Wachi Dam on the Yura River in Japan on July 2, 1967 and at the Folsom Dam on the American River near Sacramento, on July 17, 1995 are analyzed in light of the development of the self-excited coupled-mode instability mechanism. In both failures vibration of the gate was either suspected or noted by eyewitnesses. An exploratory study of how the predictions from the theory of the self-excited coupled-mode instability mechanism fit with the limited known facts is undertaken. The theoretical predictions from self-excited coupled-mode instability theory are found to explain well these two failures and are consistent with the reported circumstances surrounding the failures.

INTRODUCTION

As listed in Table 1 in Chapter 1, two large Tainter gates, both with at least hints of self-excited vibration problems, failed within just a few days more than twenty-eight years between their failures on two different continents. Despite eye witness testimony in one of the two failures, no consensus has been reached concerning the exact cause of these failures consistent with the observations of the eyewitness testimony.

The failure at the Wachi Dam on the Yura River in Japan on July 2, 1967 initiated the line of investigations that are reported in this book. The investigative report concerning the Wachi Dam gate failure (Asahi Newspaper, 1967; Yano, 1968) explains that while gate No. 3 was being closed from a gate opening of about 30 cm, gate No. 4 was being opened. Spillway gate No. 3 spontaneously began vibrating and quickly vibrated to failure. The gate along with the overhead bridge was carried about 136 m downstream, leaving a giant hole at the gate bay At the time, however, there was no known mechanism for spontaneously exciting vibration with a gate opening of 30 cm. The failure report (Yano, 1968) suggested some unknown mechanism of vibration could potentially have been the cause of the failure.

DOI: 10.4018/978-1-5225-3079-4.ch012

In California some 28 years later, on July 17, 1995, during a routine opening of spillway gate No. 3 at the Folsom Dam on the American River, near Sacramento, with the gate more than 70 cm open, a sudden gate failure occurred (Japan Times, 1995; KXTV-10, 1995). The operator reported a “rumbling” and felt vibration. The forensic team investigating the failure could find no mechanism for vibration with a gate opening of 70 cm. The eccentricity instability (the focus of Chapter 8) was found to be inoperable because of the very small eccentricity between the skinplate center and the trunnion pin center. No vibration issue was included in the final forensic report.

The present chapter will examine the circumstances of these two gate failures. An exploratory study of how the predictions from the theory of the self-excited coupled-mode instability mechanism fit with the limited known facts is undertaken. The theoretical predictions from self-excited coupled-mode instability theory are found to explain well these two failures and are consistent with the reported circumstances surrounding the failures.

Although the Wachi Dam failure came first chronologically, much more data is available concerning the more recent Folsom Dam failure. We will consider the Folsom Dam failure first and then consider the Wachi Dam gate failure.

FOLSOM DAM TAINTER GATE

Observations

An upstream view of the Folsom Dam in California, constructed from 1948 to 1956, is shown in Figure 1. Eight Tainter gates are installed. The failed Tainter gate is No. 3 from the left side, which is pushed out in the downstream direction. Several views of one of the remaining intact Folsom Dam gates, identical in design with the failed gate, are shown in Figure 2. The photo in Figure 2a is a side view of the gate, showing the relatively slender radial arms, for the size of the gate, supporting the circular-arc skinplate. The failed Tainter gate is behind the remaining gate. The photo in Figure 2b shows upstream view while Figure 2c is a view looking down on the downstream side of the skinplate.

Major dimensions of the gate and its main arm members and horizontal girders are provided in Figure 3. The circular-arc skinplate has a height of 15.5 m and the span of 12.8 m, supported on each side by a trunnion pin through four radial arms. The radius of the skinplate is 14.33 m. The A-A section in Figure 3a is shown in Figure 3b. The radial arms are inclined 13.7° towards the span center, from the trunnion pin to the skinplate, in a “π-shaped” structure.

The four radial arms supporting the skinplate have H-section steel beams with a height and a width of 30 cm, and are relatively slender compared with the overall large scale of the gate. In order to strengthen the arm structure, tie beams are attached between neighboring radial arms near the mid-point, as shown in Figure 3b. Further, as shown in Figure 3a, the lower two radial arms are also reinforced by diagonal struts in addition to the lateral struts, but the upper three arms do not have these diagonal struts. The lateral and diagonal struts have T-sections with a height of 152 mm, a width of 165 mm and a thickness of 6 mm. The arm structures and the junctions between the horizontal girders and the radial arms are all bolted. The horizontal girders have I-sections with a height of 903 to 923 mm, a width of 303 to 419 mm and a thickness of 15 mm to 19 mm. The thickness of the skinplate is 15.9 mm. As shown in Figure 2 and Figure 3b, the skinplate is reinforced by 26 curved T-beams welded to the downstream surface of the skinplate.

Figure 1. Downstream-side view of Folsom Dam in California showing unrestrained discharge from the failed gate, the third gate from the left
Photograph by N. Ishii.

Figure 2. An intact remaining gate at the Folsom Dam, showing (a) a side view (with the failed gate in the rear of the photo), (b) an upstream view of the skinplate, and (c) a view looking down on the downstream side of the skinplate
Photographs by N. Ishii.

(a)

(b)

(c)

Figure 3. Dimensions of major members of 87-ton Tainter gate failed at Folsom Dam, shown in (a) a side view, and (b) cross-section A-A looking down
From Anami & Ishii (2001).

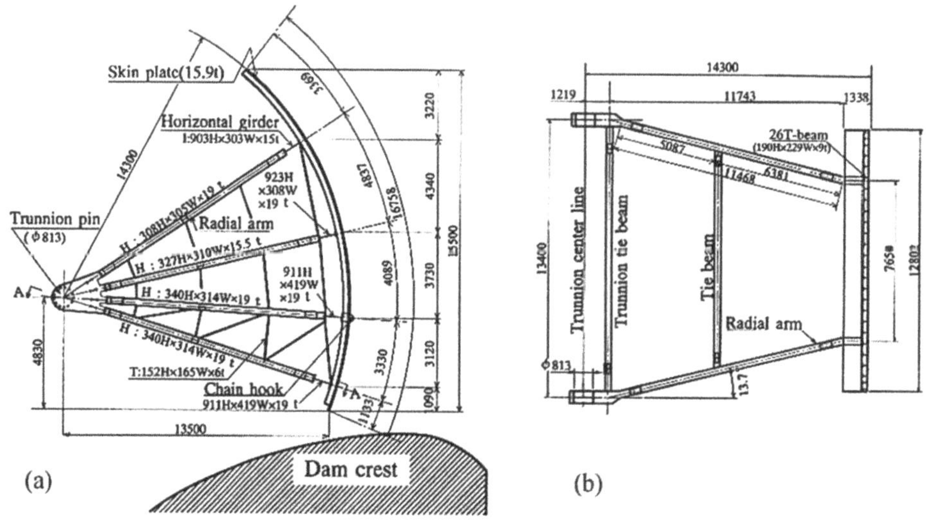

The whole mass of the Tainter gate is 87 tons. The total mass of the skinplate with T-beams, the horizontal girders and leaf bracings is 55.1 tons, about 63% of the whole gate mass.

The diameter of each of the two trunnion pins supporting the entire hydraulic load acting on the gate is about 813 mm. Such a comparatively large trunnion pin is expected to be sufficient to connect the four radial arms to the trunnion hub. The joint of trunnion hub and radial arms are fastened by bolts.

The whole Tainter gate rotates around these trunnion pins to control the discharge beneath the gate. Two enormous chains are attached to the lower portion of the skinplate on the upstream gate face on

Figure 4. The failure of the 87-ton Tainter gate at the Folsom Dam in USA on July 17, 1995
Photograph by M. Volkoff. Used with permission.

each end, and are wound by an electric motor installed on the bridge over the gate onto an elevated drum to open the gate. The full length of the chain is 16.418 m when the gate is closed.

This massive Tainter gate at the Folsom Dam experienced failure during a morning operation on July 17, 1995. A photograph, identical to the one published in "The Japan Times" newspaper on July 19, 1995, is provided in Figure 4. The Tainter gate in the picture has failed and unrestrained jets of water are spurting though the opening on each side of the gate structure. The accident occurred at about 8:00 a.m. as spillway gate No.3 was being opened to maintain flow in the river during a power plant shut down. The upstream reservoir was nearly full.

Figure 5a is a photo of the gate before the failure. Figure 5b shows the gate after the failure, hanging from the two very heavy chains hooked to the bottom spanwise ends of the skinplate. The radial arm structure on the right hand side looking downstream (left side in the photo) is severely buckled and the skinplate has opened like in the downstream direction like a large hinged garage door. Figure 5c shows the failed gate from the opposite side. The downward motion of the massive skinplate was barely restrained by the tie beams between the trunnion pins.

Immediately after the gate failure, the dam manager and gate operator clearly testified that vibrations played a role in the Folsom Dam gate failure. On July 19, 1995 the "The Japan Times" printed the following report:

A 90-ton gate at Folsom Dam buckled Monday (July 17), unleashing a thundering white-water torrent.... The problem started as operators were reopening the huge metal gate -one of eight spillways- after the flow had been temporarily shut down. "We were opening the gate, and it started to vibrate. We don't know why," said Tom Aiken, dam manager for the U.S. Bureau of Reclamation. (Japan Times, 1995)

In this newspaper report, the dam manager of USBR, Tom Aiken, was quoted as stating that vibrations played a role in the Folsom gate failure. In a televised press conference (KXTV-10, 1995) on July 17, 1995 Tom Aiken, reading from a prepared press release, stated:

Figure 5. The Tainter gate failure at the Folsom Dam (July 17, 1995), showing (a) the gate before the failure, (b) a view looking upstream from the right after the failure, and (c) a view from the left after the failure.
Photographs by N. Ishii.

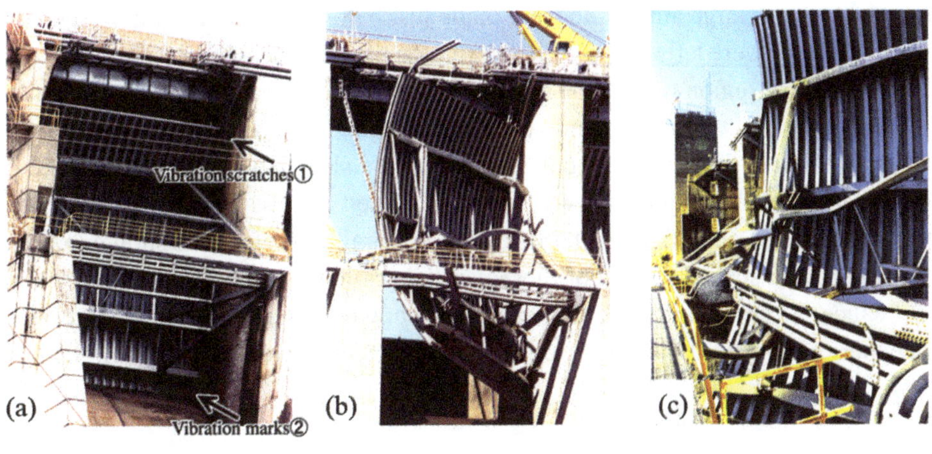

This morning at 8 o'clock we were making a routine gate change on our #3 spillway gate and our operator said he noticed something bind up a little bit and the gate vibrated and suddenly gave way and right now we have about 40,000 cubic ft/sec being released. This is below what we were at last winter during flood release so there is no immediate danger to anybody downstream. It is well contained within the levees. The cooperation that we had from the office of emergency services to notify people downstream has been outstanding. We've had excellent cooperation from the city of Folsom police department to get the road shut down so we could get our crews up there to find out what the problem was and the State parks have been excellent. They have kept people off the reservoir for boating upstream on the reservoir.

The following testimony was reported in an interview conducted by Prof. Ishii just after the gate failure which was recorded on a VCR tape. The gate operator at the time of the failure, James Taylor, was standing on the catwalk just over the gate at the time of its failure. He experienced firsthand the failure, and reported:

When spill gate No.3 reached around 2 1/2 feet I felt an unusual vibration. I immediately pushed the stop button and started to check the gate. The vibration intensified and as I turned to check the gate I could see the right side (looking down stream) of the gate separating from the structure and water coming out around the gate.

The water depth on the dam crest was 14.02 m. As the gate opening reached 0.762 m (2½ feet), the gate operator felt an unsteady vibration and stopped the gate. It is estimated that the gate failure occurred within 5 seconds after he felt the vibration.

The scratches and markings on the stainless steel side-seal plate embedded in the concrete piers of the bay provide additional evidence of the involvement of vibration in the Folsom Dam gate failure. The most pronounced markings and scratches were left behind at the top of the left side-seal plate, as shown in the text frame in Figure 5a by arrow ①. An enlarged view of this square portion is shown in Figure 6a. As evident from the severe damage on this side-seal clamp plate in the right-hand side portion, these marks and scratches were distinctly made by the line of side-seal clamp screw heads and by the clamp plate itself. Many scratches are seen in the radius direction. Most important scratches have thin half-ellipse shapes, which can be seen in lower portion of this photograph. An enlarged view of this portion is shown in Figures 6b and 6c. These elliptical trajectories show clearly that the skinplate vibrated through the coupling of the streamwise and up-and-downward motions. It should be noted that these vibration traces with large amplitude were made just before the gate failure.

Other vibration traces were left behind at the bottom on the left side-seal plate, as shown in Figure 7a which is a close up of the region marked by arrow ② in Figure 5a. Since the skinplate was being raised with considerable corrosion on the right side trunnion pin, the left side was raised more than the right side, and the skinplate had twisted in the bay. Therefore, the side-seal clamp plate on the bottom, left side of the skinplate, shown in Figure 7b, was forcibly pressed against the side-seal plate embedded in the concrete pier. Since the operation was repeated, many parallel scratches as shown in Figure 6b were left behind. The repeated striped pattern of these parallel scratches in the back and forth directions are of great significance. An enlarged view of this portion is shown in Figure 7c. These patterns suggest that the skinplate experienced streamwise vibrations at relatively small gate openings, before the gate failure.

Figure 6. Vibration marks and scratches on the upper portion of the left side-seal plate on the concrete pier: (a) heavy marks and scratches, (b) enlarged view of the heavy scratches in radial direction, and (c) enlarged view of half-elliptical marks on lower portion of (a)
Photographs by N. Ishii.

Figure 7. Marks and scratches on the lower portion of the left side-seal plate: (a) bottom portion of side-seal plate, (b) bottom portion of side-seal clamp plate, and (c) enlarged view of the scratches in the streamwise direction
Photographs by N. Ishii.

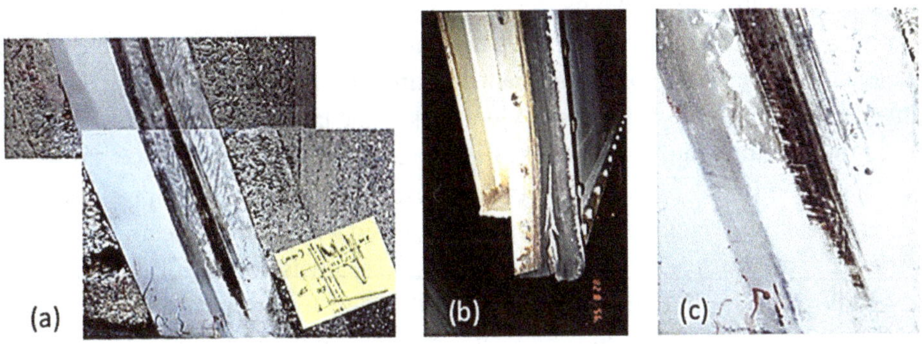

The frictional effect of corrosion on the right trunnion pin was accurately examined by Bill Nixon, a USBR engineer. Bill Nixon developed a tribometer to determine that a new trunnion would have a friction coefficient of about 0.15 while the corroded right trunnion could have had a friction coefficient as high as 0.3 at the time of gate failure. This difference in friction coefficients strongly supports the conclusion that the traces as shown in Figure 6 and 7 resulted from gate vibration.

Of course, one may suggest that the difference of friction coefficient was not fixed, and changed unstably depending on the operation. In addition, it would then be reasonable to suppose that the differences in friction coefficient might have changed rapidly, suddenly releasing the gate from the frictional constraints. Under such conditions, it would be possible for the sudden change in friction coefficient values to serve as a step-input trigger, thus initiating a violent vibration. The supposition of vibration initiation through a sudden change in the friction coefficients acting on the trunnion pin is consistent with the traces scratched into the surrounding surfaces, shown in Figures 6 and 7.

Modal Analysis of Gate

In late September 1995, an in-air experimental modal analysis was carried out on Tainter gate No. 2 (the second one from the left), as part of the failure investigation and subsequently reported in Ishii et al. (1995). Gate No. 2 is a geometrically similar to the failed Tainter gate. Subsequently, in mid-November, 1996, a similar vibration test was undertaken on a reinforced Tainter gate.

In the modal analysis testing, the gate was opened about 30 cm (with no water flow) and was struck with a 5.4 kg impact hammer (GK-086B50 by PCB Co.) which was instrumented with a force transducer (GK-206M06 by PCB Co.). The center of lower horizontal girder was struck in the radial and tangential directions, respectively, as shown by arrows in Figure 8. The impact position was chosen to induce effectively structural vibration modes associated with the bending flexibility of the skinplate and the radial arms. The radial impact was obtained by striking the flange center of the horizontal girder in the upstream direction, along the web line. The tangential impact was obtained by striking the web near the skinplate of the horizontal girder in the downward direction, perpendicular to the web. The acceleration responses of the gate to the impact force were measured with accelerometers (GK-308B piezo-type by PCB Co. for the original gate, and LS-20C servo-type by RION Co. for the reinforced gate). The acceleration responses were measured at 162 points (72 points on the skinplate, 64 points on the radial arms, 4 points on the chains, 4 points on the hoist frame, and 18 points on the cross beam between the arms). The measuring points are shown by the open circles in Figure 8.

On the skinplate, the radial (streamwise) and tangential acceleration responses were measured at each measurement location, since vibration in the spanwise direction is not so significant. On the other hand, for measurement locations on the radial arms, vibrations in the tangential and spanwise directions were measured, since the vibration in the streamwise direction is less significant.

Recorded data of the impulsive force and the acceleration responses were processed using an FFT analyzer to display real time signals and to calculate their frequency power spectra. Real time signals of the impulsive forces and the accelerations measured at the spanwise center of the skinplate are shown in upper and middle diagrams in Figure 9, where the column of figures in Figure 9a is in the tangential direction and the one in Figure 9b is in the radial direction. The upper diagrams show the impulsive forces and the middle diagrams show the acceleration responses. The impulsive forces during about 5 to 8 ms reached a maximum level of about 15 and 18 kN for tangential and radial impact, respectively. Each acceleration power spectrum was divided by the corresponding impulsive force power spectrum, to yield the corresponding transfer function. The impact tests were repeated three times at each location to permit averaging of the calculated transfer functions. The bottom diagrams in Figure 9 show samples of the averaged transfer functions.

Figure 8. Radial and tangential impact (hammering) locations and measurement locations for experimental modal analysis of the Folsom Dam Tainter gate
From Anami & Ishii (1998). © 1998 Balkema. Used with permission.

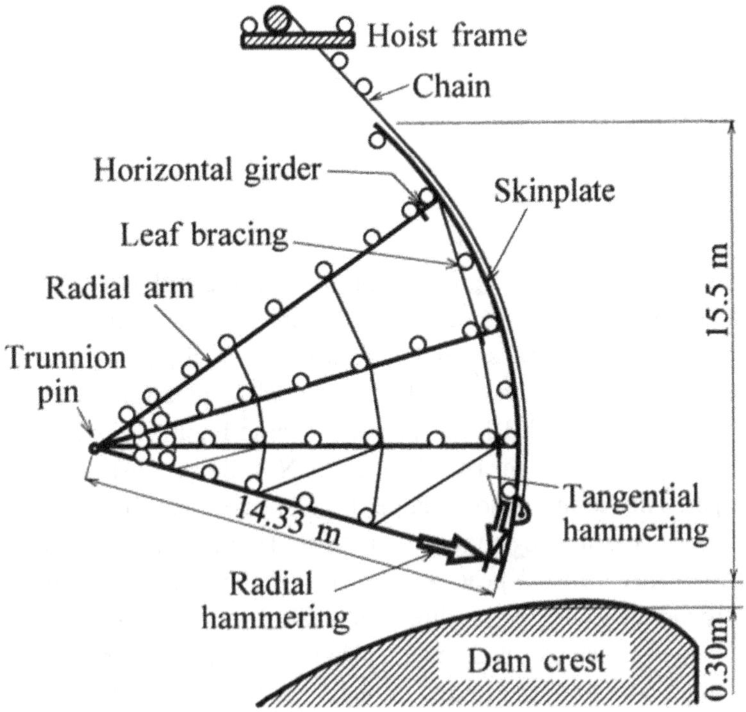

Figure 9. Representative force signals, acceleration signals and averaged transfer functions for in-air vibration of the Folsom Dam Tainter gate in (a) the tangential (up and downward) direction, and (b) the radial (streamwise) direction
From Anami & Ishii (1998). © 1998 Balkema. Used with permission.

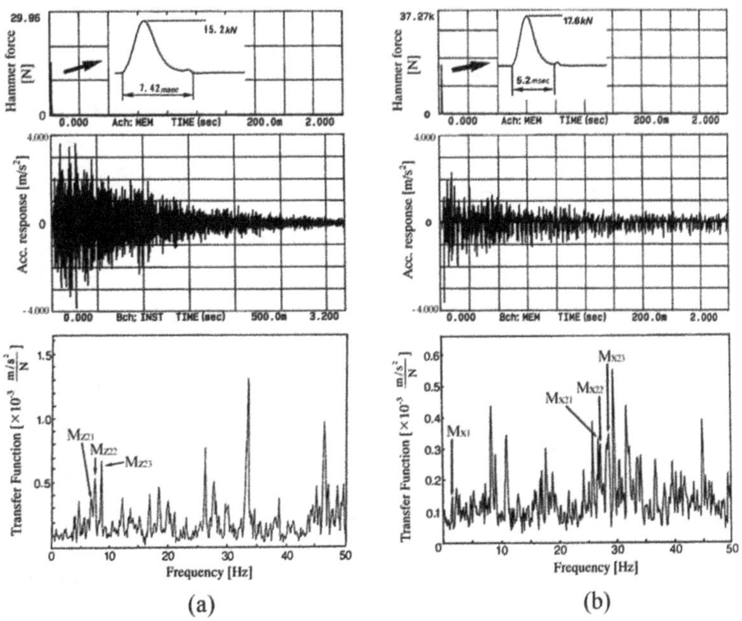

All calculated transfer functions were input into software for modal analysis (AD1461 by A&D ZONIC Co.), to calculate the inverse modal indicator functions (IMIF) as an average of all transfer functions, which is defined by

$$IMIF \equiv 1.0 - \frac{\sum_{i=1}^{n}\left[\left|Real(T_i)\right| \times \left|T_i\right|\right]}{\sum_{i=1}^{n}\left|T_i\right|^2} \tag{1}$$

where T_i represents the ith transfer function, Real(T_i) represents the real part of ith transfer function, and $|T_i|$ represents the amplitude of the transfer function. Therefore, the second term on the right-hand side represents $\cos(\phi)$ in which ϕ is the phase-lag between vibration signal and the impulsive force, taking a value of 90° at resonance. Thus, the peak level of the *IMIF* represents the strength of the natural vibration of the gate. The calculated IMIF results are shown in Figure 10, where Figure 10a is for the tangential direction and Figure 10b for the radial direction. Not all peaks in the *IMIF* necessarily provide reasonable mode shapes. The vibration mode shapes for all peaks were carefully examined, identifying meaningful mode shapes that exhibited a motion with ordered regularity. The identified mode shapes are indicated by mode names for each *IMIF* peak. A subscript z in the mode names represents the tangential (up-and-down, i.e., around the trunnion pin) direction, and subscript x indicates the radial (streamwise) direction. The mode names of natural vibration are also included in the transfer functions in Figure 9.

Major vibration modes are shown in Figures 11 and 12, where Figures 11a and 12a show side views and Figures 11b and 12b show top views of the gate. The two sets of lines show the instantaneous vibration shapes with maximum amplitude. The amplitudes are enlarged to provide easy recognition of the mode shapes.

Figure 10. Inverse modal indicator functions (IMIF) for in-air vibration of the Folsom Dam Tainter gate in (a) the tangential (up and down) direction, and (b) the radial (streamwise) direction

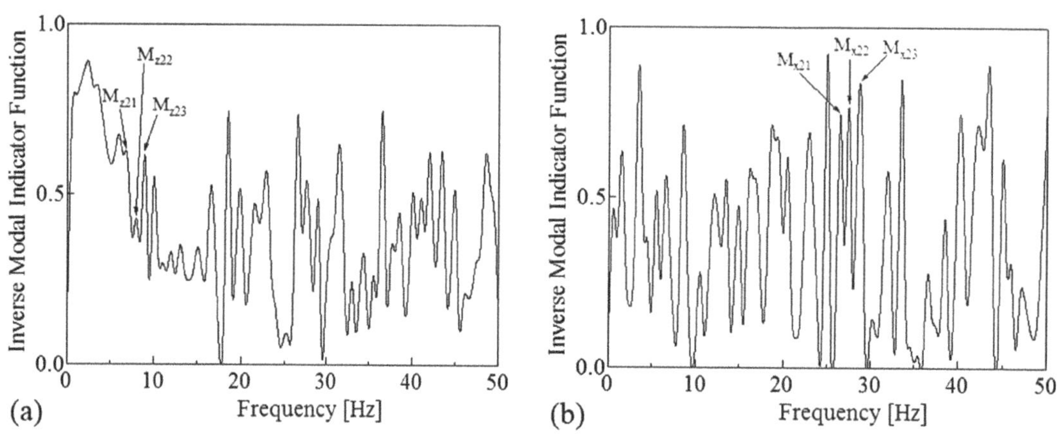

Figure 11. Fundamental mode shape of the whole gate vibration around trunnion pin of the Folsom Dam Tainter gate in Mode M_{z2} ($\Omega_{a\psi}$=6.88 Hz; $\zeta_{a\psi}$=0.012), showing (a) a side view, from Anami & Ishii (2001), and (b) a top view

From Anami & Ishii (2003). ©2003 ASME. Used with permission.

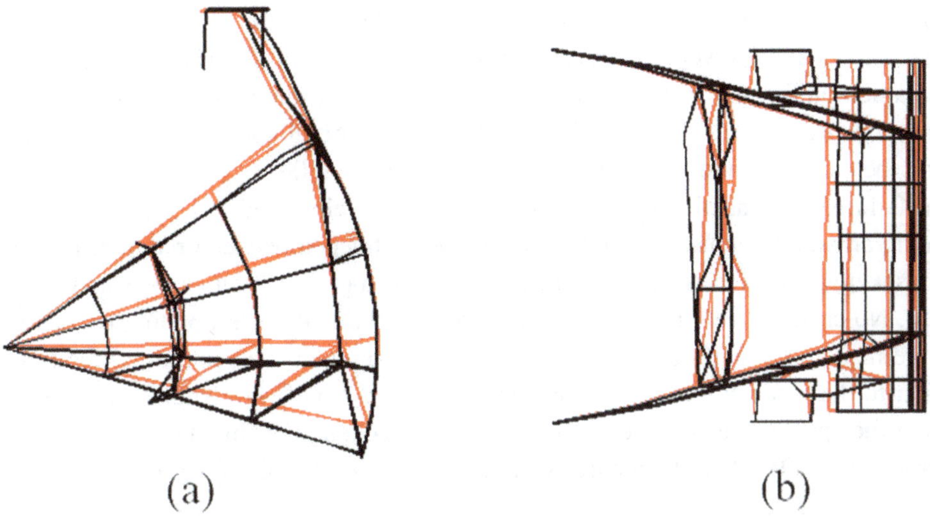

Figure 12. Fundamental mode shape of the skinplate streamwise bending vibration of the Folsom Dam Tainter gate in Mode M_{x2} ($\Omega_{a\psi}$=26.9 Hz; $\zeta_{a\psi}$=0.002), showing (a) a side view, from Anami & Ishii (2001), and (b) a top view

From Anami & Ishii (2003). ©2003 ASME. Used with permission.

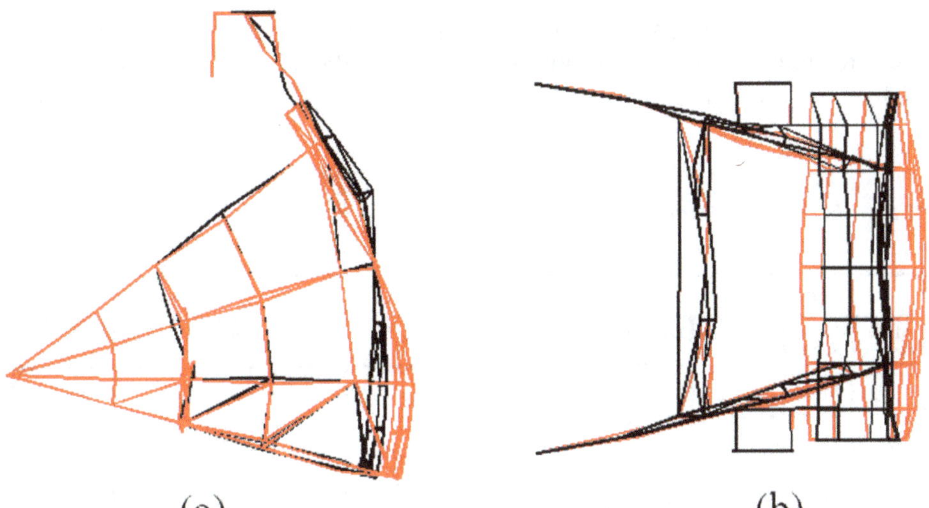

Figure 11 shows the most significant mode M_{z21}, corresponding to the rotational vibration of the whole gate around the trunnion pin. The heavy chains and the hoist frames provide elastic support. The natural vibration frequency of this mode was 6.88 Hz, and the damping ratio was 0.012.

Figure 12 shows the fundamental mode M_{x21}, which corresponds to the streamwise bending vibration of the skinplate, associated with the deformation of the radial arm structures. The skinplate amplitude corresponds to about a 3/4-wavelength bending mode, as shown in Figure 12a. In the spanwise direction there is also a non-uniformity in the vibration amplitude, approximating a half wavelength mode, with nodes at the spanwise ends, as shown in Figure 12b. The natural vibration frequency of this mode was 26.9 Hz, and the damping ratio was very small, with a value of 0.002.

After the failure, the remaining Folsom Tainter gates were reinforced. Additional plates were welded onto the radial arms, additional diagonal members were attached between radial arms, additional diagonal braces were attached between cross beams, additional girder bracings were added, and all joints were welded. No changes were made to the skinplate. Additional corresponding experimental modal analyses were made on a reinforced gate. As a result, the skinplate was found to vibrate in a bending mode, similar to the mode M_{x21}. The skinplate still exhibited a half wavelength spanwise mode shape, with nodes at the spanwise ends. The natural vibration frequency of this mode was 27.9 Hz, and the damping ratio was 0.004. Both the natural vibration frequency and damping ratio are slightly higher, relative to the original gate.

Measured natural vibration characteristics of the original and reinforced gates are summarized in Table 1, including sketches of the vibration mode-shapes. There are three apparent natural vibration modes of the whole gate in rotation around the trunnion pin, which are grouped into mode groups M_{z2} and N_{xz2} for the original and reinforced gate, respectively. Since the gate mass was increased by reinforcement, the natural vibration frequencies after reinforced were reduced, except mode N_{xz22} which showed an anomalous slight increase. In the streamwise direction, there are also three apparent bending modes for the skinplate for the original gate, but only two for the reinforced gate. These streamwise skinplate bending modes are grouped into mode groups M_{x2} and N_x, for the original and reinforced gates, respectively. The natural vibration frequencies after reinforcement increased about 1 to 2 Hz. This increase is

Table 1. Test results from experimental modal analysis of the Folsom Dam Tainter gate in-air

Original Gate				Vibration Mode Shape	Reinforced Gate			
Mode Group	Mode Name	Frequency (Hz)	Damping Ratio		Mode Group	Mode Name	Frequency (Hz)	Damping Ratio
M_{z2}	M_{z21}	6.88	0.012	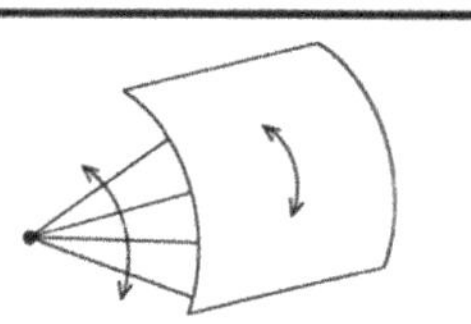	N_{xz2}	N_{xz21}	6.75	0.009
	M_{z22}	7.50	0.021			N_{xz22}	7.75	0.019
	M_{z23}	8.63	0.008			N_{xz23}	8.00	0.017
M_{x2}	M_{x21}	26.9	0.002	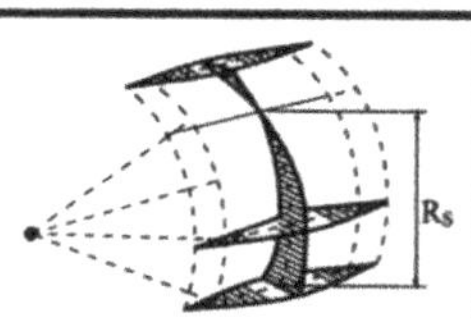	N_x	N_{x2}	27.9	0.004
	M_{x22}	27.3	0.006			N_{x3}	29.0	0.008
	M_{x23}	28.8	0.008			—	—	—

From Anami & Ishii (2001).

due to the increased bending rigidity of the skinplate resulting from the reinforcement of the radial arm structures. The damping ratios are quite small, even after reinforcement, with values of 0.008 to 0.021 for the whole gate rotational vibration around the trunnion pin, and 0.002 to 0.008 for the skinplate streamwise bending vibration.

As seen in Figures 12a, the lowest frequency bending mode of the skinplate is essentially a streamwise rotation of the skinplate about a point in the upper half of the skinplate. The height of the rotation center from the lower end of the skinplate, denoted R_s, was 9.6 m for the original gate and 12.8 m for the reinforced gate.

The spring constant due to the elastic chains and hoist frames and the moment of inertia of the whole gate around the trunnion pin determine the natural vibration frequency of 6.88 Hz, measured for the vibration mode M_{z21}.

The resultant spring constant K of the combined hoist frame/chain system (considering both sides of the skinplate) is 1.23×10^8 N/m. The moment of inertia I_θ of the whole gate around the trunnion pin can be calculated from the mass list of Folsom Dam Tainter gate, yielding 1.31×10^7 kg·m². The natural vibration frequency of the whole gate around the trunnion pin, Ω_n, can then be calculated as

$$\Omega_n = \sqrt{\frac{KR_a^2}{I_\theta}} \tag{2}$$

where the distance from the trunnion pin to the chains (radius of the skinplate), R_a, is 14.33 m. The calculated natural vibration frequency is 6.99 Hz, which is in good agreement with the measured value.

Finite Element Model

Formulation

Based on the blueprint of the Folsom Dam Tainter gate, a finite element model (frame structure) of the gate was created. Generally, in the finite element method (FEM), simplified modeling is made in many cases. Experience, however, suggests that in the case of a complicated structure such as a Tainter gate, the precision of the modeling has a significant influence on the subsequent analysis result. For this reason, the model form that was created faithfully reproduced the blueprint in precise detail.

The material properties of iron for all of the structures of Folsom Dam gate were given as follows: 208.7 GPa for Young's modulus, 78.89 GPa for the modulus of transverse elasticity, and 7,870 kg/m³ for density, and a corresponding Poisson's ratio of 0.323.

The FEM gate model was created with the "plate element", the "beam element" and the "DOF spring element". The plate element is a two-dimensional element, constituted by 3 or 4 nodes, which has the advantage that the constraint conditions are given correctly, although the number of nodes increases. Thus, in addition to the skinplate itself, the main horizontal girders of the I-beam were also defined with 3 plate elements for its flanges and web. Furthermore, the H-beams of the radial arm and cross beam were also defined as 3 plate elements. The L-angle steel struts between radial arms were defined by 2 plate elements.

The beam element can simply model a structure using 2 nodes, but it is unsuitable to give complicated constraint conditions. Therefore, in the present modeling, only the T-beams reinforcing the skinplate and the leaf-bracing L-angles were defined using the beam element, where each cross-sectional form was specified.

The massive chains which are attached to each spanwise side of the skinplate to hoist the gate, were defined as 1-DOF spring elements. The spring constant was 0.718×10^8 N/m per one chain, as calculated in the previous section.

In FEM analysis, the element division is a very significant aspect in determining the accuracy of the analysis results. It is trivial that a greater number of element divisions leads to more accurate results. The number of elements, however, will be actually constrained by the performance of computer. It is necessary to make an effective analysis with a limited number of element, where the element division was concentrated in the significant portions in the model. In the present analysis, the fine element divisions were made for the skinplate and radial arms, since the experimental modal analysis showed that they were significant factors determining the natural vibrations of the whole gate. The list of element division numbers for each component structure element is shown in Table 2. The total element number of the whole gate was 23,124, and the total nodes number was 23,967.

All structure elements were joined with bolts. Supposing high enough rigidity, the junctions were specified as rigid in the constraint condition. In this manner, a high rigidity is assigned to one independent node and two or more subordinate nodes, thus enabling the transmission of the translation force and moment produced in the independent node to the subordinate nodes.

The trunnion pin is installed in the concrete, and therefore had only one degree of freedom for movement around its rotation axis. Therefore, the flexibilities except for rotation were constrained for the trunnion pin. The chain ends have degrees of freedom in the rotational and translational directions, while the degree of freedom in the spanwise direction was constrained.

The high accuracy FEM model is shown in Figure 13, where the three-dimensional views from the downstream and upstream sides are shown in Figures 13a and 13b, respectively. The points of intersection in these views are the element nodes. The elements surrounded by the nodes on the skinplate or on the

Table 2. Number of elements for FEM analysis of the Folsom Dam Tainter gate

Name	Element	Name	Element
Skinplate	3200	Tie beam	1776
T-beam	3327	Tie beam bracing	1055
Horizontal girder	2656	Trunnion pin	40
Girder bracing	540	Trunnion Tie beam	1200
Radial arm	3720	Chains	14
Arm bracing	3552	Others	1624
Arm splicer	420	Total	23124

From Anami & Ishii (2001).

Figure 13. Precise FEM model construction of the Folsom Dam Tainter gate, showing (a) a downstream view, and (b) an upstream view
From Anami et al. (2014). © 2014 IJMERR. Used with permission.

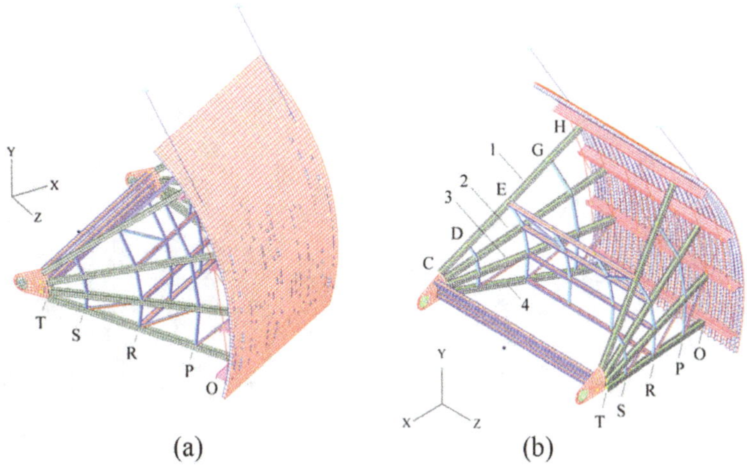

radial arms are the plate element, and the lines connecting the nodes of horizontal girders and skinplate reinforcement T-beams are beam elements. The numbers from 1 to 4 indicate the radial arms from the top. In order to identify the analysis results, junctions of the arms and the trunnion pin hub on the right hand side are labeled “C”, junctions of the arm struts and the vertical struts are labeled “D”, “E” and “G” from the trunnion pin side, and junctions of the arm struts and the horizontal girders are labeled “H”. The left hand junctions also are similarly labeled “T”, “S”, “R”, “P” and “O” as shown in Figure 13.

Natural Vibration Analysis

The natural vibration frequencies and vibration modes were analyzed using FEM analysis software (MSC/N4W, NASTRAN for Windows V3 by the MacNeal-Schwendler Co.). The analyzed frequency range was 0 to 50 Hz, the same range as the experimental modal analysis. The vibration modes obtained by calculation are shown by animation on a computer screen. The natural vibration modes were identified as those modes for which the whole gate was carrying out orderly vibration. The significant results are shown in Figures 14 and 15. Figures 14a and 15a are side views, while Figures 14b and 15b are 3-dimensional downstream views. The red and black lines show the instantaneous vibration shapes with maximum amplitude. In order to help identify deflections, the amplitudes are amplified.

Figure 14 shows the whole gate rotational vibration mode around the trunnion pin with the frequency of 7.3 Hz. This vibration mode is in good agreement with the experimental modal analysis result shown in Figure 11. Deformations are not apparent for any beams of the gate. This vibration appears to result from the elasticity of the chains that raise and lower the gate.

Figure 15 shows the skinplate streamwise vibration mode. In the streamwise vertical plane, the skinplate undergoes its lowest frequency bending mode vibration with the node near the second horizontal girder. In the horizontal plane, the skinplate undergoes a half wavelength vibration with nodes near each side of the skinplate. The natural vibration frequency is 25.4 Hz. This vibration mode is in good agreement with the measured mode shape from the experimental modal analysis shown in Figure 12.

Figure 14. Vibration mode shape of Folsom Dam Tainter gate by FEM, for whole gate vibration around trunnion pin ($\Omega_{a\psi}$=7.28 Hz), showing (a) a side view, and (b) a three-dimensional view
From Anami et al. (2014). © 2014 IJMERR. Used with permission.

(a) (b)

Figure 15. Vibration mode shape of Folsom Dam Tainter gate by FEM, for skinplate streamwise bending vibration ($\Omega_{a\psi}$=25.4 Hz); (a) side view, and (b) three-dimensional view
From Anami et al. (2014). © 2014 IJMERR. Used with permission.

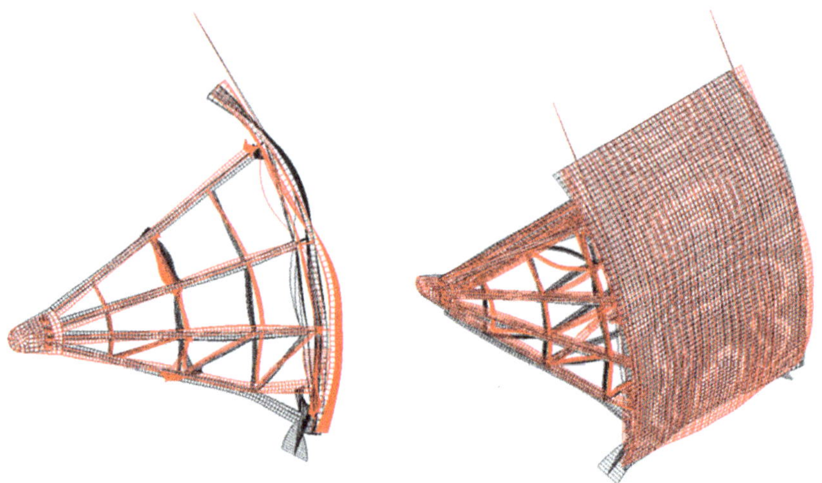

(a) (b)

The fundamental inherent vibration characteristics obtained from the FEM analysis are listed in Table 3, where the results from the experimental modal analysis are also shown for comparison. The resulting FEM natural vibration frequencies are in good agreement with the experimental modal analysis frequencies. The simulated frequencies have relative error of 5.8% for the whole gate vibration, and 5.6% for the skinplate streamwise vibration.

Table 3. Results of FEM analysis for fundamental mode vibrations of the Folsom Dam Tainter gate, and comparison with experimental modal analysis

Mode	Vibration Mode shape	Exp. Modal analysis		FEM analysis		Relative Error of Freq. [%]
		Frequency [Hz]	Damping Ratio	Frequency [Hz]	Damping Ratio	
Mz2	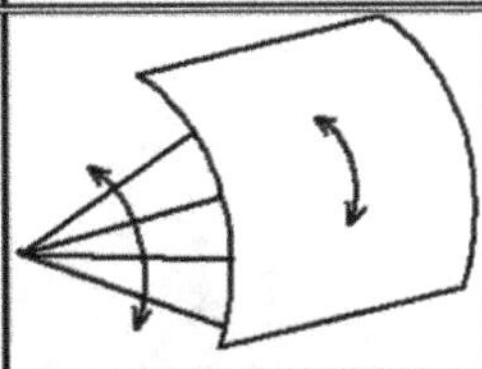	6.88	0.012	7.28	—	5.8
Mx2	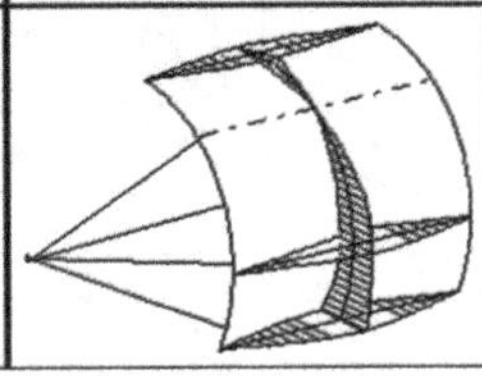	26.9	0.002	25.4	—	5.6

From Anami & Ishii (2001).

As a result, we conclude that the FEM simulation has been validated. The FEM results correlated well with the field test results, in terms of both the vibration frequencies and their corresponding mode shapes. The finite element model contains the essential physics of the large-scale Folsom Dam Tainter gate.

Static Deformation Characteristics

Structural static deformation of the Tainter gate, induced by the hydrostatic pressure, was calculated for the same FEM model that was validated by comparison with the experimental modal analysis results. It is very useful to know what level of deformation the massive skinplate experiences due to the hydrostatic pressure.

The static deformation was calculated for the gate conditions at the time of failure, that is, at a discharge opening of 0.72 m, and a gate submergence depth was 13.26 m. The whole hydraulic load is 11.2×10^6 N.

Each elemental hydrostatic load was made to act on each skinplate element submerged in water, in the direction normal to the skinplate. In addition, the gravity force was made to act downward on all elements of the whole gate.

The calculated static deformation is shown Figure 16a in a side view overlapped with the gate before deformation. Figure 16b provides a three-dimensional view from the upstream side. The upper portion of the skinplate shows no meaningful deformation; the maximum static deformation is at the bottom center of the skinplate. The deformation at each side of the skinplate is comparatively small. This is because the sides of skinplate are supported by the radial arms. The trunnion beams show large deformations, but they do not become a problem.

The most significant result of the deformation analysis is that, as shown in Figure 16a, the bottom center of the skinplate deforms19 mm in the downstream direction. The downward deformation of 4 mm is due to the structural deformation of the skinplate itself, in addition to the elastic elongation of the chains. Relatively slender and weak members were used in Folsom Dam Tainter gate especially when

Figure 16. Static structural deformation of the Folsom Dam Tainter gate with a gate submergence of 13.26 m corresponding to conditions at the time of failure, showing (a) a side view overlaid on the undeformed model, and (b) a three-dimensional view
From Anami & Ishii (2001).

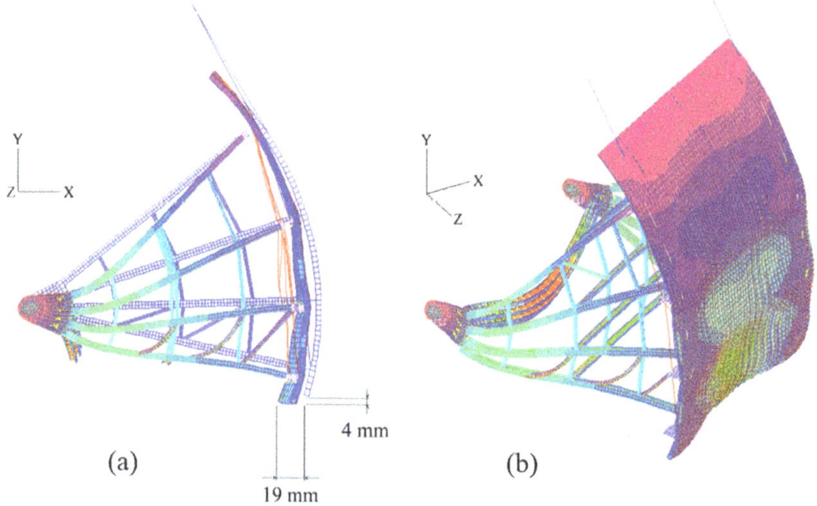

compared with its massive size, resulting in the large 19 mm streamwise deformation. The hydrodynamic load due to streamwise vibration was analyzed in detail in Chapter 9. According to the calculated result from Chapter 9, the hydrodynamic load reaches the level of the hydrostatic load with a vibration amplitude of 8.4 mm, which is less than half of 19 mm static deflection. The conclusion is that the flexibility of the massive skinplate in vibration cannot be disregarded.

Static Tensile Loads and Influence of Frictional Forces

The Coulomb friction on the trunnion pin of was quite large for the failed Folsom Dam Tainter gate. As a result, USBR concluded that the increased trunnion friction was the main factor in gate failure, that is, an excessive static load arose since the gate was raised against the trunnion friction, resulting in the incipient gate failure. In the USBR analysis of the failure, an FEM analysis was carried out for the given hydrostatic load and trunnion friction to calculate the static stress. Their results suggest that excessive tensile force developed on the diagonal member between third and fourth arms, exceeding the allowable load of the bolts. Since some doubt remains in the validity of FEM model used by USBR, it is prudent to re-examine the USBR conclusions using our previously validated FEM model.

Using the same FEM model, tensile forces acting on each member in Folsom Dam gate at the time of failure were analyzed for a gate discharge opening of 0.72 m and a gate submergence depth of 13.26 m, where the trunnion friction coefficient value was varied over a range of values, including the USBR-established maximum possible value of 0.3 at the moment of gate failure. where the trunnion friction coefficient value was varied over a range of values, including the USBR-established maximum possible value of 0.3 at the time of gate failure.

One set of calculated values, the stress computed using a trunnion friction coefficient of 0.3, is shown in Figure 17. The significance of this figure is that the maximum stress appears in the diagonal member

Figure 17. Static stress and tensile force of Folsom Dam Tainter gate with gate submergence of 13.26 m and trunnion frictional coefficient of 0.3 at failure
From Anami & Ishii (2001).

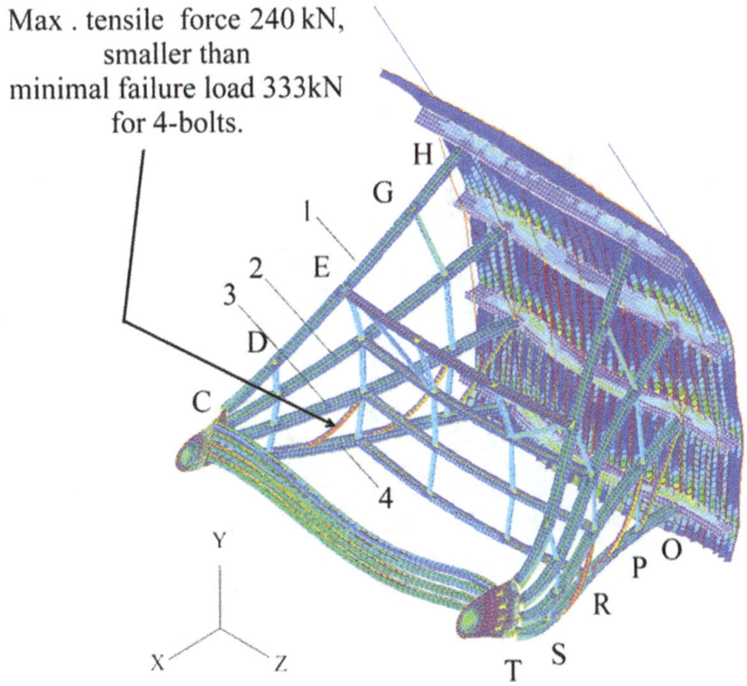

3E-4D, that is, the member between third and 4fourth arm, as indicated by the orange color. From this load, one can determine that the tensile force acting on the bolted joint of the diagonal member 3E-4D reached a maximum value of 240 kN.

The USBR Forensic Team concluded that the Folsom Dam gate failure started from the breaking of the four bolts connecting the diagonal member 3E-4D to the third and fourth radial arms. The forensic team recovered the 4-bolted members and attachment plate to make a careful tensile-loading test. They confirmed that the range of the tensile failure load was 333 kN to 363 kN. The calculated maximum tensile force of 240 kN is still sufficiently lower than the minimum failure load of 333 kN to conclude that the static load alone did not produce the observed failure.

The maximum tensile force on the diagonal members between arms was carefully calculated with increasing the trunnion friction coefficient values up to 0.7, as shown by the solid lines in Figure 18. As shown by the filled symbols, the calculated data linearly increase with increasing trunnion friction coefficient; hence, the data are connected with the solid lines. If the trunnion friction coefficient is smaller than about 0.1, the maximum tensile force appears in the center diagonal member 3G-4E, while if it exceeds about 0.1, the maximum tensile force appears in the diagonal member 3E-4D near the trunnion pin. The maximum tensile force in diagonal member 3E-4D exceeds the failure load of 333 kN, only when the trunnion friction coefficient reaches 0.42. This value is far larger than the range of 0.22 to 0.28 estimated by USBR to be possible values at failure. In addition, this value of 0.42 is still larger than the possible maximum frictional coefficient 0.3 which was obtained by the USBR using Nixon's tribometer for friction tests on the trunnion pin of the failed gate.

Figure 18. Tensile loads on strut brace with increasing trunnion friction coefficient up to 0.7
From Anami & Ishii (2001).

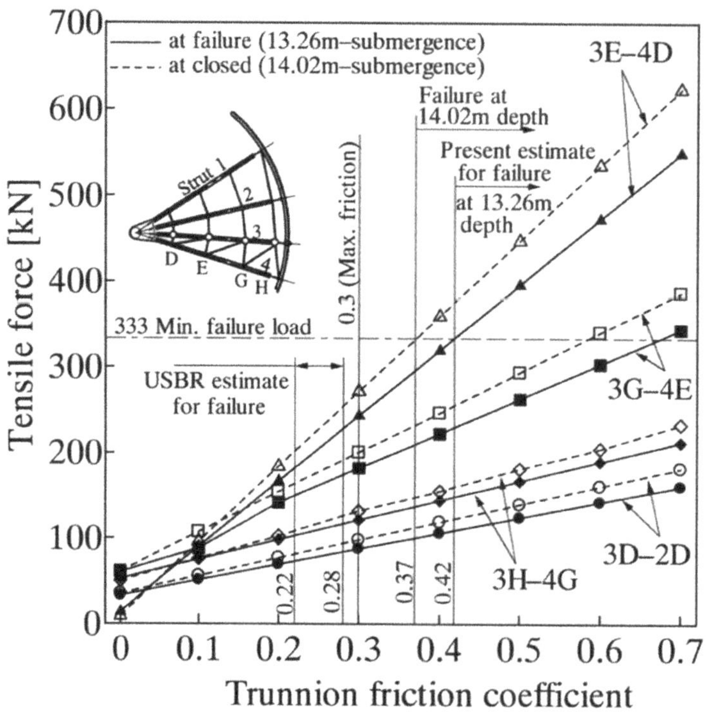

The validated FEM model results indicate that it was not possible for the Folsom Dam Tainter gate to fail due only to excessive static load resulting from the increased trunnion friction, as the Forensic Team concluded.

Supposing nonetheless that the trunnion friction was the main cause of the Folsom Dam gate failure, the accident should have occurred immediately upon beginning to open the gate, that is, when the gate submergence depth was larger than 13.26 m recorded at the time of the failure. The gate submergence depth for the closed gate would have been 14.02m, since the gate failed with a submergence depth of 13.26 m and the discharge opening was 0.76 m. Therefore, upon opening the gate, the gate submergence depth was about 6% larger than that at failure. The greater depth results in a larger hydrostatic pressure load and in a larger influence of the trunnion friction. Supposing the trunnion friction to be the cause of gate failure, one must then accept that the accident would have occurred at the initiation of the gate opening process.

Based on this reasoning, FEM calculations were made again at the condition (the gate submergence depth of 14.02 m) when the gate started to open, to re-examine the maximum tensile forces on the diagonal members. The calculated results are shown by the unfilled symbols and dashed lines in Figure 18. The tensile forces are about 12% larger than those with the gate opening of 0.76 m at failure. However, the tensile forces with in the trunnion frictional coefficient in the range of 0.22 to 0.28 are still significantly smaller than the minimum failure load of 333 kN. Moreover, not until the frictional coefficient reaches 0.37 (which is still larger than the maximum frictional coefficient of 0.3 estimated at the time of gate failure) does the maximum tensile force exceed the failure load of 333 kN.

Summarizing the above results, one may conclude that increased trunnion friction alone could not have been the only cause of the Folsom gate failure.

Stability Predictions

The in-air streamwise natural vibration frequency $\Omega_{a\psi}$ of 26.9 Hz of the skinplate of Folsom Dam Tainter gate drastically decreases due to the added mass effect of water. If the lowered in-water natural vibration frequency is close to the natural vibration frequency of whole gate vibration around the trunnion pin, the two vibrations can fall into a resonant state. In such a case, the gate will experience an intense dynamic instability. The Folsom Dam Tainter gate may have possibly fallen into such a dangerous state at the time of its failure. In order to examine this possibility, it is necessary to show clearly how much the natural vibration frequency is lowered in water. The in-water streamwise natural vibration frequency of the failed Folsom Dam Tainter gate can be calculated using the theoretical analysis shown in Chapter 10, which has been validated by the field vibration tests for installed full-scale Tainter gates (as will be discussed in Chapter 14).

As described in Chapter 10, the vertical projection error Δ can be treated as almost zero and ineligible for the Folsom Dam Tainter gate. Therefore, the in-water to in-air vibration frequency ratio $\Omega_{w\psi}/\Omega_{a\psi}$ of the skinplate streamwise rotational vibration can be calculated using

$$\frac{\Omega_{w\psi}}{\Omega_{a\psi}} = \frac{1}{\sqrt{1+\delta_p \alpha_\psi \Delta m_\psi}}, \tag{3}$$

where the reduced water added mass Δm_ψ is a function of the Froude number F and the reduced rotation center height r_s,

$$\Delta m_\psi (F) = \frac{1}{F^2} \int_{-1}^{0} \left(1 - \frac{1+y}{r_s}\right) p_{bs}(0, y)\, dy \tag{4}$$

When the Froude number F is larger than about 10, however, Δm_ψ takes on an asymptotic constant value, as shown in Chapter 10. In installed full-scale Tainter gates, the Froude number F usually takes on values from 30 to 60, far larger than 10. Therefore, the constant value of Δm_ψ depends only on the reduced rotation center height r_s. For Folsom Dam Tainter gate, the reduced rotation center height r_s is 0.72 and then Δm_ψ takes on a corresponding value of 0.13.

The water-to-gate mass ratio α_ψ, indicative of a relatively lightweight design, was 129. The pressure correction coefficient δ_p takes on a value of 1.0. With these values, the in-water to in-air vibration frequency ratio $\Omega_{w\psi}/\Omega_{a\psi}$ can be calculated from Equation 3, resulting in a value of 0.24, which is shown in Figure 19. The Folsom Dam gate has a water-to-gate mass ratio of 129. The abscissa is the basic Froude number F_0, which takes a value of 196. The vibration ratio shows a constant value over a wide range of F_0, but tends to increase in the limited range of F_0 smaller than 10. This small F_0 region is typically irrelevant for full-scale gates.

Figure 19. In-water to in-air vibration frequency ratio for Folsom Dam Tainter gate
From Ishii & Anami (2001).

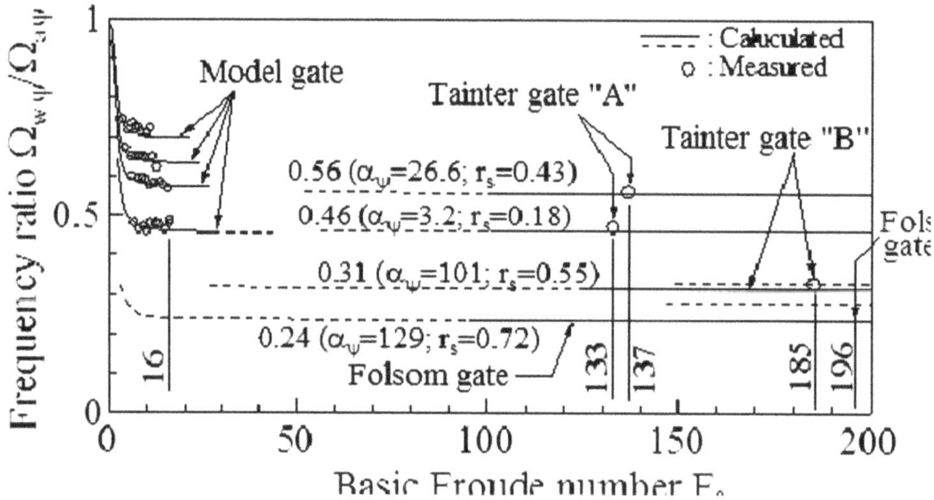

The in-water to in-air vibration frequency ratio ($\Omega_{w\psi}/\Omega_{a\psi}$) of 0.24 for Folsom Dam Tainter gate indicates that the in-air natural vibration frequency ($\Omega_{a\psi}$) of 26.9 Hz decreases to 6.46 Hz in-water. This frequency is the value of the in-water streamwise natural vibration frequency $\Omega_{nw\psi}$ of the skinplate. It should be especially noted here that this value of 6.46 Hz is quite close to the in-air natural vibration frequency ($\Omega_{a\theta}$) of 6.88 Hz of whole gate vibration around the trunnion pin.

It seems that the in-air measured rotation center position changes when the gate is submerged in water according to the distribution of hydrodynamic pressure. Due to this uncertainty in the exact value for the rotation center heights, the in-water to in-air vibration frequency ratio was calculated for a variety of rotation center heights for the Folsom Dam Tainter gate, as shown in Figure 20. It is known that the hydrodynamic pressure distribution increases toward the bottom end of the skinplate. From this pressure distribution, it is expected that the in-water rotation center value will increase upon submergence of the gate. For this reason, the rotation center height was varied from the in-air position (r_s = 0.72 for R_S = 9.6 m) up to the water surface (r_s = 1.0 for R_S = 13.26 m) in the calculations, as shown by the solid line in Figure 20.

As the rotation center position moves higher on the gate, the reduced added mass Δm_ψ increases, while the water-to-gate mass ratio α_ψ conversely decreases, resulting in a difference in the in-water to in-air vibration frequency ratio. The in-water to in-air vibration frequency ratio decreases from 0.24 down to 0.21.

As determined from the calculations in the previous section, the failed Folsom Dam Tainter gate falls into a dangerous state in which the in-water streamwise natural vibration frequency ($\Omega_{nw\psi}$) of 6.46 Hz for the skinplate is extremely close to the natural vibration frequency ($\Omega_{a\theta}$) of 6.88 Hz for the whole gate around the trunnion pin. Moreover, the frequency of $\Omega_{nw\psi}$=6.46 Hz is slightly smaller than the frequency of $\Omega_{a\theta}$=6.88 Hz, resulting in a vibration frequency ratio of $\Omega_{nw\psi}/\Omega_{a\theta}$=0.94. This value would suggest that the gate falls into the intense dynamic instability region, as was demonstrated with the theoretical results shown in Chapter 10. Furthermore, even for water-to-gate mass ratios α_ψ as large as 129 for Folsom Dam Tainter gate, the fluid damping ratio $\zeta_{f\psi}$ due to fluid motion approaches zero for basic Froude numbers F_0

Figure 20. Vibration frequency ratio, mass ratio and added mass for the Folsom Dam Tainter gate with changing rotation center height

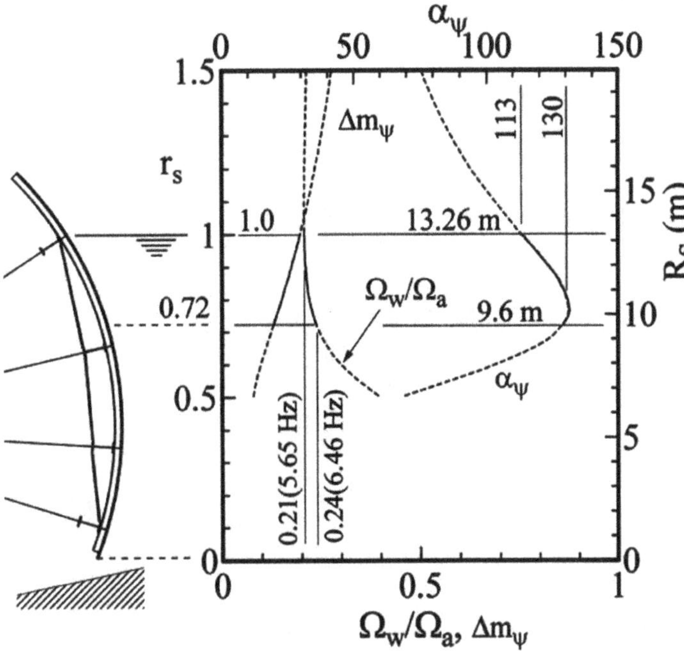

larger than 20, as described in Chapter 9. With essentially zero fluid damping to dissipate energy, supply of a small amount of energy in each cycle of oscillation will quickly drive the gate to large amplitude. If the skinplate streamwise vibration and the whole gate vibration around the trunnion pin approach each other, a resonant state could result, in which the energy due to flow-rate variation will be effectively supplied to the coupled-mode vibration system. Extremely intense self-excited vibration results.

In order to clarify further this potentially dangerous situation, numerical calculations were undertaken to obtain the approximate stability solutions for Folsom Dam Tainter gate. Figure 21 shows the dynamic stability criterion diagram calculated with the data for Folsom Dam gate. A very intense dynamic instability was found for the Folsom Dam Tainter gate under. The maximum excitation ratio level reaches 0.104.

Three natural vibration modes of M_{z21}, M_{z22}, and M_{z23} were determined for the whole gate vibration around the trunnion pin from the experimental modal analysis of Folsom Dam Tainter gate, as summarized in Table 1. The vibration frequencies $\Omega_{a\theta}$ were 6.88 Hz, 7.5 Hz, and 8.63 Hz, respectively.

The in-water streamwise natural vibration frequency of the skinplate $\Omega_{nw\psi}$ calculated in the previous section was 6.46 Hz. Therefore, the vibration frequency ratio γ_{nw} ($\equiv\Omega_{nw\psi}/\Omega_{a\theta}$) was 0.94 for Mode M_{z21}, 0.86 for Mode M_{z22}, and 0.75 for Mode M_{z23}, respectively.

Since all vibration frequency ratios are smaller than 1.0, the vibration synchronizes with the skinplate streamwise rotational vibration as the main vibration. The corresponding in-air damping ratio $\zeta_{a\psi}$ of 0.002 of this main vibration is plotted for each frequency ratio in Figure 21. The Mode M_{z23} is in the stable region narrowly, but the Mode M_{z21} and the Mode M_{z22} are in the unstable region.

The difference ($\xi_{f\psi}$-$\zeta_{a\psi}$) between the critical damping ratio $\zeta_{c\psi}$ (or $\xi_{f\psi}$) and the plotted data values $\zeta_{a\psi}$ represents the level of dynamic instability. The difference reaches its maximum value of 0.049 (= 0.051-0.002) for Mode M_{z21}. The actual excitation ratio ξ_ψ can be identified by substituting this value

Figure 21. Stability criterion curve for Folsom Dam Tainter gate at its failure (B=0.762m; d_0=13.26m)

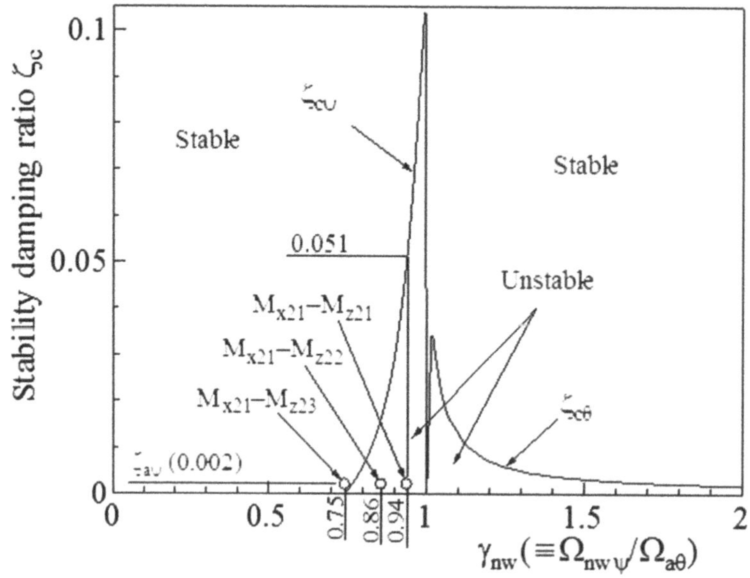

into Equation 49 in Chapter 10, where the in-water to in-air vibration frequency ratio $\gamma_{nw\psi}$ ($\equiv \Omega_{w\psi}/\Omega_{a\psi}$ or $\gamma_{w\psi\theta}/\gamma_{\psi\theta}$) of skinplate is 0.24 and $\zeta_{a\psi}$ is 0.002, resulting in ξ_{ψ} = 0.012. This value represents the actual self-excitation level at the Folsom Dam Tainter gate under prevailing conditions at the time of failure. From this analysis, the Folsom Dam Tainter gate is clearly seen to be well within the region of intense dynamic instability.

Similar calculations were made for the case just after the gate opening, with the same water depth over dam crest (d_r = 14.02 m). The discharge opening B was 0 m and the gate submergence depth d_0 was 14.02 m. The calculated results are shown by the solid line in Figure 22. The calculated results from Figure 21 are also plotted in Figure 22 as the dashed line and open circles for comparison. Immediately after the gate was opened, the in-water to in-air vibration frequency ratio $\Omega_{w\psi}/\Omega_{a\psi}$ of skinplate was 0.22, which is smaller than the previous frequency ratio of 0.24 since the effect of water added mass is larger. Therefore, the in-water streamwise natural vibration frequency is lowered to 5.92 Hz. Using this value, the vibration frequency ratio γ_{nw} becomes 0.86 for Mode M_{z21}, 0.79 for Mode M_{z22}, and 0.69 for Mode M_{z23}. These values are smaller than the corresponding values at gate failure. The in-air damping ratio $\zeta_{a\psi}$ of 0.002 is plotted as a filled-circles at these frequency ratios. Modes M_{z22} and M_{z23} are in the stable region, and only the Mode M_{z21} is in the unstable region. In addition, the point for M_{z22} is very near the critical curve. Compared with the state at failure, the gate just after opening is significantly more dynamically stable.

These results can be summarized as follows: the level of the dynamic instability is very small immediately after the gate was opened, and the strength of the dynamic instability increases as the discharge opening increases, reaching its maximum at the gate failure with the discharge opening of 0.76 m. This conclusion clearly explains why the gate failed at a larger discharge opening rather than at a smaller one, and further why the gate failed at the discharge opening of 0.76 m.

This conclusion is definitely at odds with the conclusion from the forensics team that the definitive factor in the gate failure was only excessive friction at the trunnion pin, in contradiction to eyewitness

Figure 22. Stability criterion curves for the Folsom Dam Tainter gate at the smallest gate opening (B = 0 m; d_0=14.02 m) compared with the curve under failure conditions
From Ishii & Anami (2001).

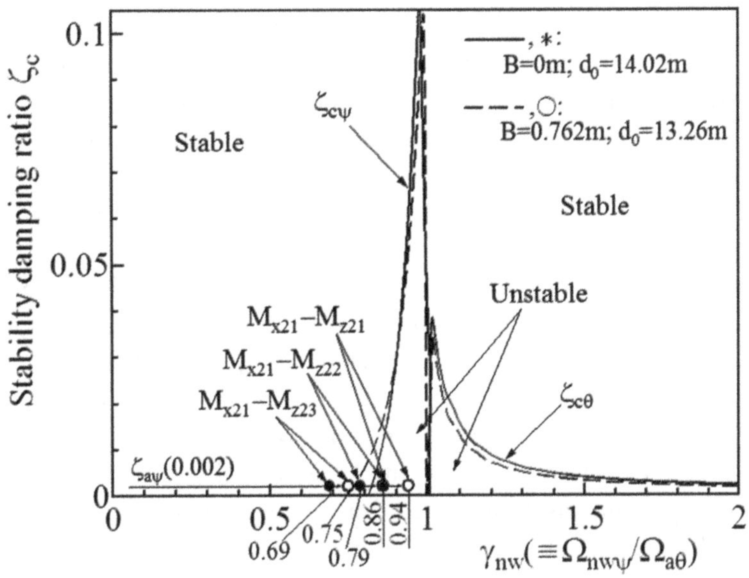

testimony. If the definitive factor of the gate failure was only excessive friction at the trunnion pin, the failure accident should have occurred immediately after the gate was opened, that is, when the hydraulic pressure was larger. However, in fact, the gate failure accident occurred at a larger discharge opening with smaller hydraulic pressure.

Estimate of Amplitude of Vibration to Cause Failure

The Folsom Dam Tainter gate was just shown to be inherently susceptible to an intense dynamic instability. In the present theoretical analysis, however, the damping effect due to large solid friction (Coulomb friction) in the Tainter gate's resistance to large hydraulic pressure has not been taken into consideration at all. In the Tainter gate with hydraulic pressure, the large hydraulic force pushes the watertight rubber seals of the skinplate on both sides against the seal plates embedded to the concrete piers. In addition, the entire hydraulic force on the skinplate is supported by the trunnion pins. Therefore, substantial Coulomb friction arises in the side seals and in the trunnion pin bearings. The damping effects due to these frictional forces have prevented the appearance of the essential dynamic instability in the routine use of the Tainter gate. This state is called "friction-maintained dynamic stability."

However, such friction-maintained stability is itself in a very dangerous, unstable state, because the essential dynamic instability may appear at any time, since the damping effect due to friction drastically decreases in inverse proportion to the vibration amplitude. If the vibration amplitude is small, the damping effect due to friction is very large, thus preventing vibrations from occurring. However, if there is a sudden displacement large enough to overcome the damping effect due to friction, the gate will begin to vibrate with an increasingly large amplitude further reducing the damping effect due to friction with each cycle of increased amplitude. In this manner an intense, and most likely catastrophic, vibration will suddenly occur.

In order to evaluate such Coulomb frictional effects experimentally, a simple experiment was conducted with the 1/31-scale Tainter gate model. With a small Coulomb friction added to each side of the skinplate, the model gate was initially forced to vibrate and its dynamic response was carefully observed. In the previous experiments for the coupled-mode self-excited vibration, the model gate was set up so that friction on the two sides of the skinplate was as small as possible. In contrast, in the present experiment a rubber sheet was attached to each side of the skinplate and in contact with the channel walls, thereby providing a small frictional resistance. The bottom end of the skin plate was fixed to the channel floor with a thin string, and the gate was raised to the point where the string suddenly broke. Thus the gate was impulsively forced to vibrate, where the initial displacement could be adjusted by changing the strength of string.

Representative records of the vibration time-histories are shown in Figure 23. Figure 23a shows the vibration waveforms without additional Coulomb friction. With no initial displacement trigger, the coupled-mode self-excited vibration occurs spontaneously. Figures 23b and 23c show the vibration waveforms with Coulomb friction on both sides of the skinplate. Figure 23b is the response with a small initial displacement trigger, where the vibration is damped with the amplitude decreasing linearly. This is the state of "friction-maintained dynamic stability" due to Coulomb friction. In contrast, Figure 23c is the response with a large initial displacement trigger, where the vibration rapidly increases exponentially in amplitude so that the influence of the Coulomb friction becomes negligible. As shown in the simple experiments presented here, friction-maintained dynamic stability can certainly be broken by a sufficiently large initial displacement trigger.

In the failed Folsom Dam Tainter gate, the right-hand trunnion pin was rusted. Since the gate was raised without noticing the rusted trunnion pin, the skinplate was twisted in the gate bay. It is presumed that when the friction in the trunnion pin was suddenly overcome under certain conditions and the gate

Figure 23. Effects of Coulomb friction on the instability of a 1/31-scaled model gate showing (a) spontaneous vibration with no added friction; (b) friction-maintained stability with added Coulomb friction and an initial soft excitation, and (c) instability despite added Coulomb friction due to initial hard excitation with reduced effective Coulomb friction due to large amplitude disturbance
From Ishii & Anami (2001).

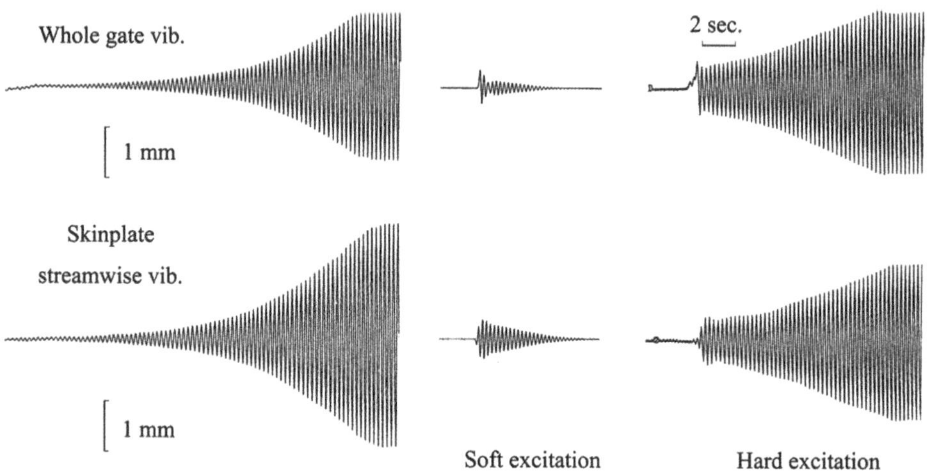

was released from its twisted configuration, the gate responded in the same manner as the model gate in Figure 23c and underwent rapidly growing vibration.

The testimony of Tom Aiken, the manager of the Folsom Dam at that time, that ".... our operator said he noticed something bind up a little bit and the gate vibrated..." (previously presented in the section on observations) gives significant credence to this presumption. There are other possible sources of hard excitation triggers in other Tainter gate installations. For example, an initial trigger might be supplied by a landslide into the upstream reservoir or an earthquake, with either source suddenly producing a large wave; by the collision of driftwood or by debris jammed in the discharge opening; or even a seiche in a long reservoir. It is incumbent upon engineers to determine what level of initial displacement trigger is needed to overcome the frictional damping effect and to plan countermeasures against such hard excitation.

The damping effect due to Coulomb friction can be represented by damping proportional to the velocity, that is, an equivalent viscous damping ratio ζ_e. With such representation of the Coulomb damping, the theoretical analysis developed in this section can be successfully used without further modification. When a structure vibrates with a vibration amplitude A and a vibration frequency Ω_n, the energy W_1 can be consumed by a damper with an equivalent viscous damping coefficient c_e. The equivalent viscous damping coefficient is related to not only the friction force F_r, but also to the vibration amplitude A and the vibration frequency Ω_n. The equivalent viscous damping ratio ζ_e can then be defined by the ratio of this equivalent viscous damping coefficient to the critical damping coefficient, as follows:

$$\zeta_e \equiv \frac{c_e}{2\sqrt{mk}} = \frac{c_e}{2m\Omega_n} = \frac{1}{2m\Omega_n} \cdot \frac{4F_r}{\pi A\Omega_n} = \frac{2}{\pi A} \cdot \frac{F_r}{k_c} \tag{5}$$

where, k_c is the spring constant of the chains. Supposing the chain is extended by A_0 due to frictional force F_r, the equivalent viscous damping ratio ζ_e results in the following expression from $F_r/k_c = A_0$:

$$\zeta_e = \frac{2}{\pi} \cdot \frac{A_0}{A} \tag{6}$$

The equivalent damping ratio ζ_e increases as the vibration amplitude A decreases, and in the inverse, ζ_e decreases as A increases.

When the hydrodynamic pressure acts on the gate and the equivalent viscous damping ratio ζ_e due to friction appears, the resultant damping ratio for the whole gate vibration around the trunnion pin naturally increases, as given by $(\zeta_{a\theta} + \zeta_e)$. Calculations for the Folsom Dam Tainter gate with $\zeta_{a\theta} = 0.012$ and $\zeta_{a\psi} = 0.002$ were made. The calculated results are shown in Figure 24. The critical value for the resultant damping ratio around the trunnion pin, $\zeta_{ca\theta}$, was 0.08, for which the stability criterion curve passes through the most unstable data point of $(M_{x21}$-$M_{z21})$. Therefore, if the following condition is satisfied:

$$\zeta_{a\theta} + \zeta_e > \zeta_{ca\theta}, \quad \therefore \quad \zeta_e > \zeta_{ca\theta} - \zeta_{a\theta} = 0.08 - 0.012 = 0.068 \tag{7}$$

the stability criterion curve passes under the data point of $(M_{x21}$-$M_{z21})$, placing the data point in the dynamically stable region.

Figure 24. Stability criterion curves at the critical damping ratio $\zeta_{ca\theta}$ of 0.08 for the Folsom Dam Tainter gate

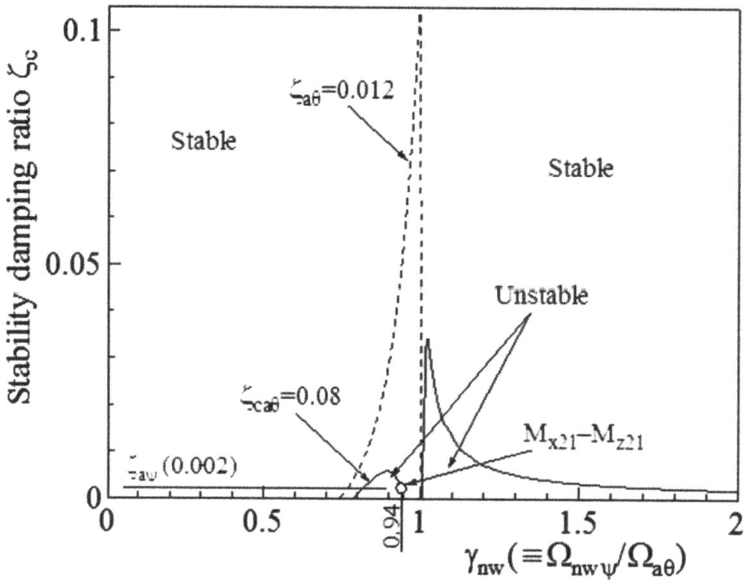

Therefore, by substituting Equation 6 into Equation 7, the vibration amplitude A for which friction-maintained dynamic stable can be obtained is as follows:

$$A < \frac{2}{\pi} \cdot \frac{A_0}{\zeta_e} = \frac{2}{\pi} \cdot \frac{A_0}{0.068}. \tag{8}$$

In order to actually calculate the values in Equation 8, it is necessary to calculate the extensional displacement A_0 (= F_r/K) of the spring elements (K is the spring constant of the chains and hoist frame, 1.23×10^8 N/m) when loaded by the frictional force F_r.

The frictional force F_r appears on the rubber seals on each side of the skinplate and on the trunnion pin. The rubber side seal for the Folsom Dam Tainter gate is P-shaped, as shown in Figure 25, where the streamwise length from the fulcrum to the contact point on the side plate is represented by a (=37.5 mm). The hydraulic pressure acts on this portion. The whole length of the rubber seal along the skinplate with submergence is represented by L (= 13.94 m).

The hydraulic force pushing the rubber seal against the seal plate on the concrete pier is given by half the calculated normal force $F_p/2$, but the resultant hydraulic force acts on both side plates embedded in the concrete piers and is given by F_p (=36.2×10^3 N). Assuming a friction coefficient μ_s at the side seal plate, the friction force F_s acting on the side plates is therefore given by

$$F_s = \mu_s F = \mu_s \rho g \frac{d_0}{2} \cdot aL \tag{9}$$

Figure 25. Side seal P-packing attached on each side of skinplate of the Folsom Dam Tainter gate

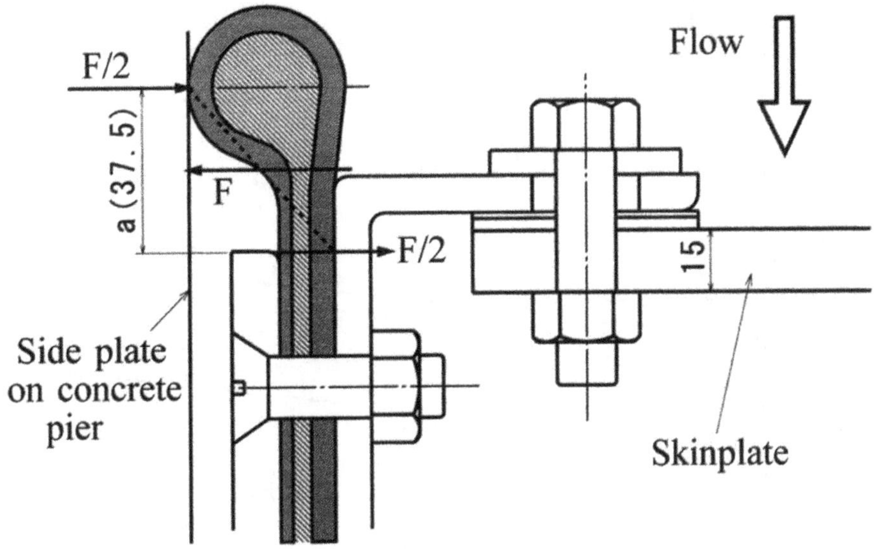

In addition, the friction force F_{tp} on the trunnion pin can be easily calculated by $F_{tp}=\mu_t P$, where μ_t is the friction coefficient at the trunnion pin and P is the total hydrostatic force. This frictional force F_{tp} can be represented by F_t acting on both sides of the skinplate:

$$F_t = \frac{D_t/2}{R_a} \cdot F_{tp} = \mu_t \frac{D_t/2}{R_a} P \tag{10}$$

where D_t is the diameter of the trunnion pin (0.813 m) and R_a is the radius of the skinplate (14.33 m). Combining these frictional effects, the total frictional force is given by

$$F_r = F_s + F_t \tag{11}$$

The above relations make it possible to calculate the extensional displacement A_0 (= F_r/K) of the chains. From the extensional displacement of the chains, Equation 8 can be evaluated, as shown in Figure 26, where the frictional coefficients μ_t at the trunnion pin are chosen to be 0.15 and 0.3, which are the values for a new trunnion in good condition and for the rusted trunnion in the failed gate, respectively. The abscissa is the frictional coefficient μ_s for the rubber seals attached to skinplate, which may change according to field conditions under routine use. If the initial displacement is smaller than the calculated result, the gate is dynamically stable, while to the contrary, if the initial displacement in the tangential direction of the skinplate is larger than the amplitude shown in the diagram, the Folsom Dam Tainter gate will certainly undergo an intense coupled-mode self-excited vibration.

Clearly, it is now essential to ascertain whether such an initial displacement trigger might have been realized at the failure of the Folsom Dam Tainter gate. Temporarily assume a friction coefficient μ_s at the rubber seals to be at a moderate value of 0.5. The initial displacement overcoming the damping ef-

Figure 26. Initial displacement trigger inducing self-excited vibration of the Folsom Dam gate, assuming μ_t = 0.15 and 0.30 at trunnion pin
From Anami et al. (2005). © 2005 ASME. Used with permission.

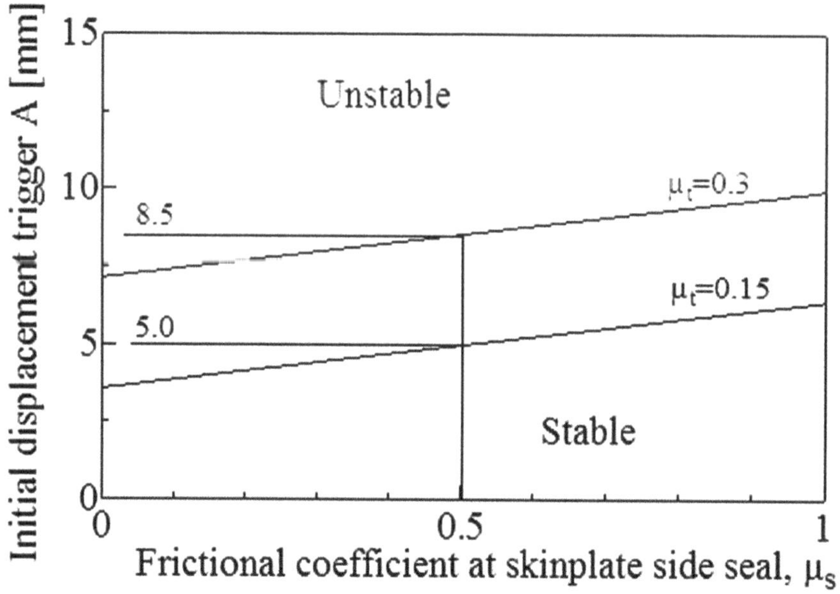

fect due to friction is only a small value of 5.0 mm for μ_t = 0.15 for a new trunnion, and an even smaller value of 8.5 mm when μ_t is 0.3 for the rusted trunnion.

It should be evident that such a small initial displacement trigger might easily have occurred, remembering that the failed Folsom Dam Tainter gate was suddenly released from its cocked position in the gate bay.

The streamwise vibration of the skinplate induces a huge push-and-draw pressure, as already determined in Chapter 10. The excessive push-and-draw hydrodynamic load is considered to have led to failure of the Folsom Dam Tainter gate.

The push-and-draw hydrodynamic load was theoretically analyzed over the streamwise vibration amplitude of the bottom end of the skinplate at its spanwise center. In addition, FEM analysis was performed to calculate the static stress, deformation and tensile load caused by the static hydraulic load on the skinplate and the Coulomb friction at the trunnion pins. By linking these two analyses, the streamwise vibration amplitude needed for the failure of the Folsom Dam Tainter gate can be determined.

The stress due to the hydrostatic load was analyzed, assuming the friction coefficient at the trunnion pin was at its presumed maximum value of 0.3 at failure. Consequently, as shown in Figures 15 to17, a maximum tensile load of 240 kN arises in the diagonal member 3E-4D between the radial arms. This tensile load was smaller than the minimum failure load of 333 kN.

A similar FEM simulation was computed to determine the resultant hydrodynamic load (the sum of hydrostatic load and the push-and-draw hydrodynamic load). From this analysis, the resultant hydrodynamic load needed to produce a tensile load exceeding the minimum failure load of 333 kN could be computed. In the present FEM calculations, however, the maximum push-and-draw hydrodynamic load was assumed to act statically on the skinplate. Thus, the tensile load on the diagonal member 3E-4D between the radial arms was calculated, as shown in Figure 27. The abscissa is the ratio of the resultant hydrodynamic load to the hydrostatic load, that is, the reduced resultant hydrodynamic load. When this reduced resultant hydrodynamic load reaches a value of 1.6, the tensile load exceeds the minimum failure load of 333kN.

The question then arises what was the necessary streamwise vibration amplitude at the bottom end of the skinplate needed to produce a resultant hydrodynamic load exceeding 1.6-times the hydrostatic load. In order to determine this value, vibration amplitudes were calculated for the Folsom Dam Tainter gate. In the failed Folsom Dam Tainter gate, the reduced rotation center height r_s was 0.72, the gate submergence depth d_0 was 13.26 m, and the in-water streamwise vibration frequency $\Omega_{w\psi}$ of the skinplate was presumed to be 6.46 Hz. The corresponding Froude number F was 47.2. With these non-dimensional values, the reduced resultant hydrodynamic load was calculated, as shown in Figure 28. The abscissa is the reduced vibration amplitude, defined by the ratio of the streamwise vibration amplitude at the bottom end of the skinplate, $R_S\Psi_0$, to the gate submergence depth d_0.

When the reduced vibration amplitude reaches a value of 0.09%, the reduced resultant hydrodynamic load on the ordinate reaches the value of 1.6. Multiplying this value of 0.09% by the gate submergence depth d_0 (=13.26 m), the actual streamwise vibration amplitude of the bottom end of the skinplate can be calculated. The value is 11.9 mm, a very small displacement especially in comparison to the large dimensions of Folsom Dam Tainter gate.

This result shows that the Folsom Dam Tainter gate could have failed due to coupled-mode vibrations with a small streamwise amplitude of only 11.9 mm at the bottom end of the spanwise-center of the skinplate. The bottom end of the spanwise-center of the skinplate had a streamwise deformation of 19 mm due to the hydrostatic load on the skinplate. A vibration amplitude of 11.9 mm, even smaller than the 19 mm static deflection, is clearly possible.

Summary of Proposed Mechanism of Failure

Synthesizing the results obtained above, a scenario resulting in the failure of the Folsom Dam Tainter gate can be summarized as follows:

- The left trunnion pin was well lubricated and was in order, and the friction coefficient was at its normal value of 0.15. However, due to insufficient maintenance of lubrication for the pin, the friction coefficient of the right trunnion pin increased to 0.3.
- Since the gate was raised without noticing the friction increase at the trunnion pin, the left side of the skinplate went up smoothly, but the right side was restrained. Consequently, the skinplate was twisted in the gate bay.
- As the gate was raised further with the skinplate cocked in the gate bay, the right trunnion pin friction was suddenly released in a number of steps. The sudden upward jerking motion as the friction was overcome initially caused only the right side of skinplate to vibrate.
- However, due to this initial vertical displacement trigger, there was a discharge fluctuation, leading to skinplate vibration and the coupled-mode vibration of the whole gate around the trunnion pin and the streamwise skinplate rotational vibration was initiated.
- The coupled-mode vibration was accompanied by a flow-rate variation under the gate, and the energy was supplied from fluid motion into the vibration system. In addition, since the natural vibration frequencies of two vibration modes were very close to each other, the coupled-mode vibration fell quickly into a dynamically unstable state.

Figure 27. Tensile force acting on the diagonal member 3E-4D depends on resultant hydrodynamic force acting on skinplate of Folsom Dam Tainter gate; failure at a reduced hydrodynamic load of 1.6
From Anami et al. (2005). © 2005 ASME. Used with permission.

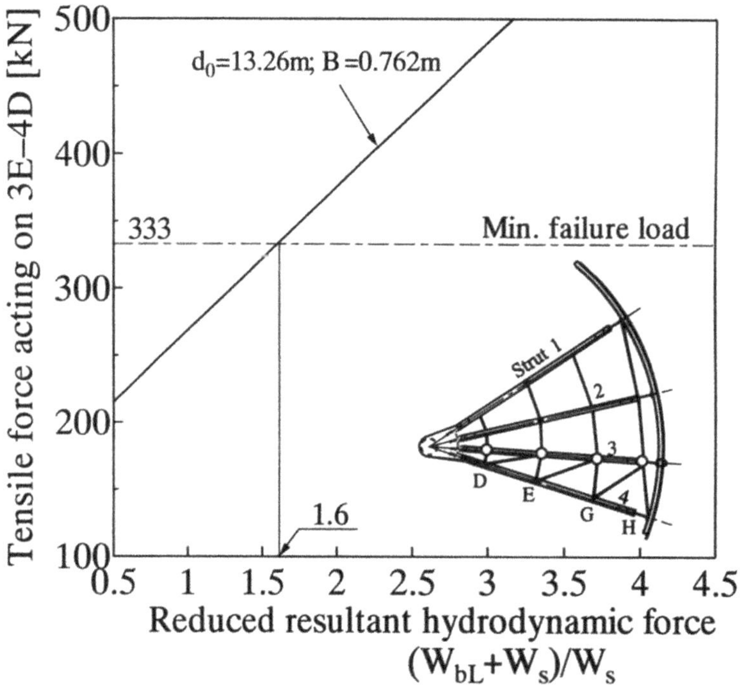

Figure 28. Relation between reduced resultant hydrodynamic force acting on skinplate and reduced amplitude of streamwise vibration of Folsom Dam Tainter gate at its failure (r_s=0.72; F=47.2)
From Ishii & Anami (2001).

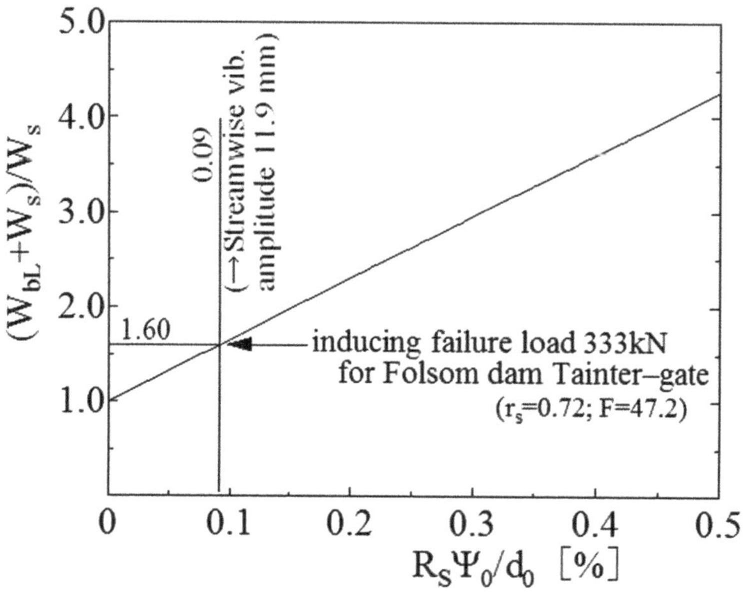

- At other times, the dynamic instability was not sufficiently strong at small discharge openings, and hence the dynamic instability was attenuated by the damping effect due to Coulomb friction both on sides of the skinplate and on the trunnion pin. As a result, the excited vibration was damped, as had been observed frequently during past gate openings.
- When the discharge opening approached 0.762 m, the in-water natural vibration frequency of the skinplate streamwise vibration became slightly lower than the natural vibration frequency of the whole gate vibration around the trunnion pin, and then, the gate was in the essential intense dynamically unstable state.
- Just at the time, the gate was suddenly released from the heavy friction of the trunnion pin, and the skinplate had a sudden upward jerking motion of more than 8.5 mm.
- Even with just a comparatively small initial displacement, the essential dynamic instability could not be restrained by the Coulomb friction.
- Therefore, the intense coupled-mode self-excited vibration occurred and the vibration amplitude increased rapidly.
- Corresponding to the rapid increase of streamwise vibration of the skinplate, an excessive push-and-draw hydrodynamic load was built up.
- When the bottom center of the skinplate reaches 11.9 mm in amplitude, the four bolts connecting the diagonal member between the No.3 and No.4 radial arms could no longer bear the sum of the excessive hydrodynamic load and the hydrostatic load. Thus, the tensile force on the diagonal member exceeded the minimal failure load of 333 kN of the four bolts, and the bolts were sheared.
- The shearing of the 4-bolts served as the initiation of failure for other bolted joints, and subsequently the complete failure of the Folsom Dam Tainter gate.

WACHI DAM TAINTER GATE

Observations

On July 2, 1967, one of the Tainter gates at Wachi gates failed, as shown in Figure 29. The failure was, reported in the Asahi Newspaper on July 3, 1967. The 37-ton Tainter gate failed and was completely swept downstream away from the gate bay no less than 136 m, creating an enormous opening in the gate bay.

Major dimensions of the gate are shown in Figure 30. The circular-arc skinplate has a height along the skinplate of 12.56 m and the span of 9 m; the skinplate is supported on each side by a trunnion pin through four radial arms. The radius of the skinplate is 13 m. The radial arms are inclined 8 degrees toward the span center, from trunnion pin to the skinplate in a "π-shaped" structure.

The gate was operated for long periods at small gate openings with an upstream gate submergence depth of 11.75 m. The gate failed while being closed from an opening of about 30 cm to zero.

The official failure investigation of the Wachi Dam Tainter gate concluded that the gate failed due to the buckling of the radial arms resulting from insufficient static strength of the gate structure (Yano, 1968). In addition, based on crack and fracture traces on the radial arms, the investigation report expressed concerns about the possible involvement of vibration-induced dynamic loading in the buckling of the arms. However, no mechanism for Tainter gate vibration was known at that time.

Figure 29. Failure of the 37-ton Tainter gate at the Wachi Dam in Japan on July 3, 1967

After the Wachi Dam Tainter gate failure, Ishii initiated an investigation of flow-induced vibration of Tainter gates. He identified an intense self-excited vibration that occurred under certain conditions (& Imaichi & Ishii, 1977; Ishii et al., 1977; Ishii & Imaichi, 1982; Ishii & Naudascher, 1992). The initial research at that time was restricted to eccentric Tainter gates in which the center of the skinplate is eccentric to the trunnion pin. Since the failed Wachi Dam gate did not have any significant eccentricity, the consensus at the time was that the self-excited vibration mechanism identified by Ishii could not have actually occurred.

With the identification of the coupled-mode self-excited vibration mechanism of Tainter gate, one naturally wonders whether the gate failure 40 years ago in Japan might have also been due to this same mechanism. A re-investigation of the Wachi gate failure was undertaken using modern sophisticated FEM technologies and our theoretical calculations developed for the gate failure at Folsom Dam to answer definitively this question

Static FEM Analyses

A precise FEM model of the Wachi Dam gate was constructed using a powerful pre- and post-processing computer tool (FEMAP Ver.5. 00), as shown in Figure 31a. The gate is represented by 21,605 plate elements and 23,150 contact points. MacNeal-Schwendler's MSC/N4W NASTRAN for Windows V3 was used in the present FEM analyses. To raise the calculation accuracy, the H method (the element subdividing method) was adopted. The balance of the load of the elements was analyzed by the displacement method. The analysis included consideration of the friction at the trunnion shaft, with the maximum load of gate submergence depth of 12 m.

The downward displacement of radial arm #3 (the lowest arm) was calculated first. Figure 31b shows the calculated deformation of the arm for values of trunnion friction coefficient ranging from 0.1 to 0.7. The largest deformation appears at vertical brace "B" and increases linearly with the trunnion friction

Figure 30. Wachi Dam Tainter gate in Kyoto, Japan showing (a) a side view (b) a top view
From Ishii et al. (2007). © 2007 JSCE. Used with permission.

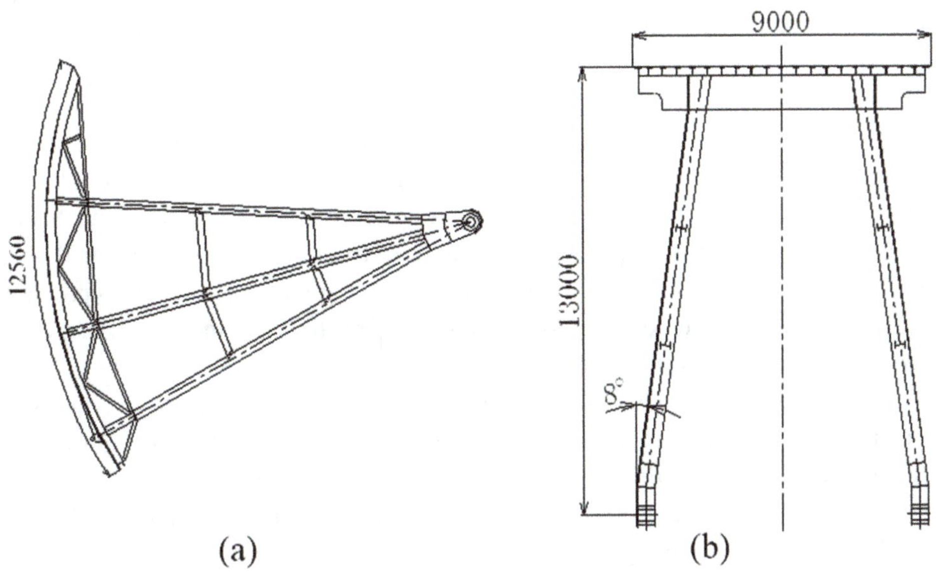

Figure 31. (a) Side view of FEM model of Wachi Dam gate showing deformation, and (b) calculated deformation of radial arm #3 to estimate trunnion friction
From Ishii et al. (2007). © 2007 JSCE. Used with permission.

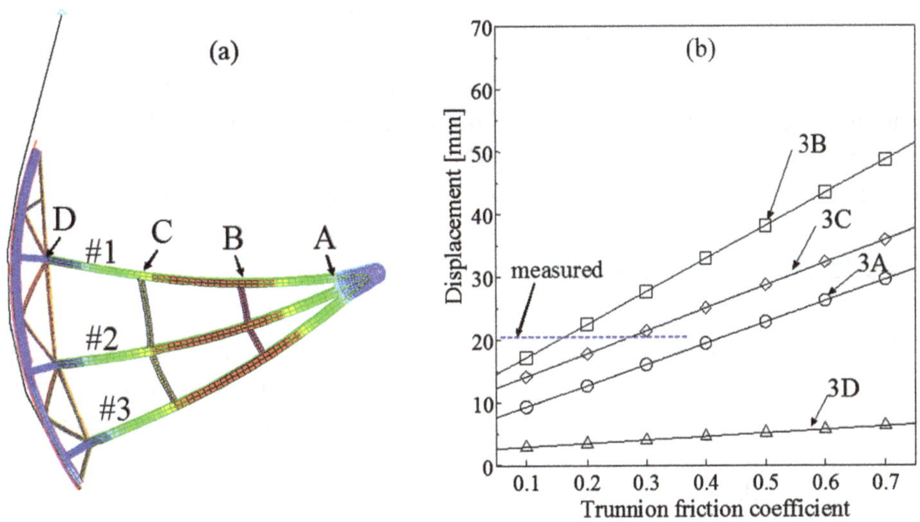

coefficient. The horizontal dashed line represents the measured 23 mm deflection of the same member in the geometrically similar and intact gate #1. The intersection of the dashed line with the calculated data line permits one to approximate the trunnion friction coefficient to be in the range from 0.1 to 0.2, at most.

The tensile load on the vertical braces can be calculated using FEM analysis. In Figure 33, the reduced tensile load relative to the failure load for the four 19 mm diameter rivets which secure the vertical braces to the radial struts is shown as a function of the hydraulic load and the trunnion friction coefficient. When the reduced tensile load exceeds 1.0, gate failure is initiated. In Figure 33a, with a fixed trunnion friction coefficient of 0.2, the reduced tensile load increases with increasing resultant hydraulic load (normalized by the hydrostatic pressure load), providing a measure of the safety factor of gate's structural strength. The tensile load on brace "2B-3B" is the largest, but is still sufficiently less the failure load of the four rivets. In Figure 32b, the reduced tensile load is shown to increase with increasing trunnion friction coefficient (for a fixed hydraulic load). For values of trunnion friction less than 0.36, the tensile load on brace "2B-3B" does not exceed the failure load of the four rivets. However, as previously shown in Figure 31b, the maximum value of trunnion friction was less than 0.2 indicating that the gate could not have failed due to increased trunnion friction.

Modern FEM analysis affords determination of both the buckling pattern of the gate structure and the safety coefficient in buckling, relative to the hydrostatic load, as shown in Figure 33. The safety coefficient for the gate with the radial struts reinforced with braces was 1.4. Even if the radial struts were not reinforced, the safety factor was 1.0, as shown in Figure 33(b). From the FEM analyses, one may conclude that the gate could not possibly have failed due solely to static loads.

FEM analysis was used to determine the natural vibration mode of the skinplate streamwise bending vibration to be 24.9 Hz, as shown in Figure 34. In this figure, the maximum amplitudes of this vibration modes are represented by two sets of lines (originally red and blue lines). The side and top views from the FEM model are shown in Figures 34a and 34b, in which the skinplate performs a streamwise bending vibration that is well-coupled with the spanwise vibration of radial arm structure, but not with the up-and-down whole gate rotational vibration. For clear identification of the bending vibration mode, eight horizontal cross-sectional views of the skinplate are shown in Figure 34c, in which the skinplate exhibits a 1-wavelength bending vibration mode with nodes at the joint of the radial arms and the skinplate. The increased amplitude toward the bottom is evident in Figure 34d where the amplitude of the spanwise center vertical cross-section of the skinplate is shown. Also in Figure 34d, a streamwise rotational vibration mode with a node at about 3/4 of the skinplate height is shown by the dashed lines.

Coupled-Mode Dynamic Instability

If the bending rotational counter-clockwise vibration of the skinplate is triggered by some random excitation, the counter-clockwise skinplate inertia torque appears around the trunnion pin, thus exciting the counter-clockwise whole gate vibration. The gate opening under the gate increases, resulting in increased discharge water flow, and thus decreasing the flow-rate-variation pressure that feeds energy back to the skinplate bending rotational vibration. This closed-energy-cycle leads the gate to a violent coupled-mode self-excited vibration.

The identified coupled-mode vibration can be analyzed with our previously established theoretical analysis. The resulting dynamic stability criterion for the Wachi Dam gate is shown in Figure 35. First, starting with the in-air natural vibration frequency of 24.9 Hz, and using the methods developed by Anami & Ishii (1999; 2000; 2003), the in-water natural vibration frequency with gate submergence depth of 12 m was calculated to be 5.68 Hz to 6.00 Hz. This in-water bending frequency corresponds to a natural vibration frequency ratio of 0.86 to 0.974 along the abscissa in Figure 35, for which there is a very high level essential dynamic instability with excitation factors in the range 0.039 to 0.125.

Figure 33. Buckling pattern and safety coefficient for Wachi Dam gate (a) with braces (safety coefficient = 1.4), and (b) without braces (safety coefficient = 1.0)
From Ishii et al. (2007). © 2007 JSCE. Used with permission.

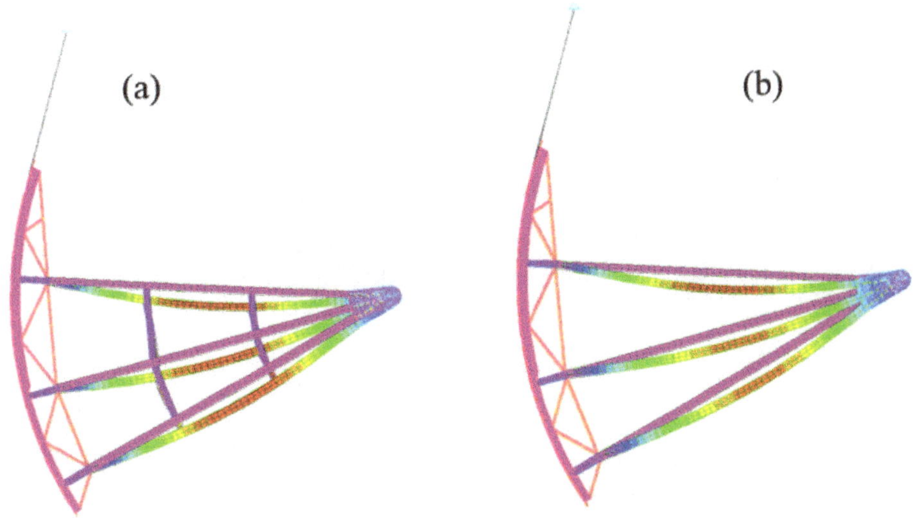

Figure 32. Tensile load on vertical braces (a) with increasing hydraulic load, and (b) with increasing trunnion friction coefficient
From Ishii et al. (2007). © 2007 JSCE. Used with permission.

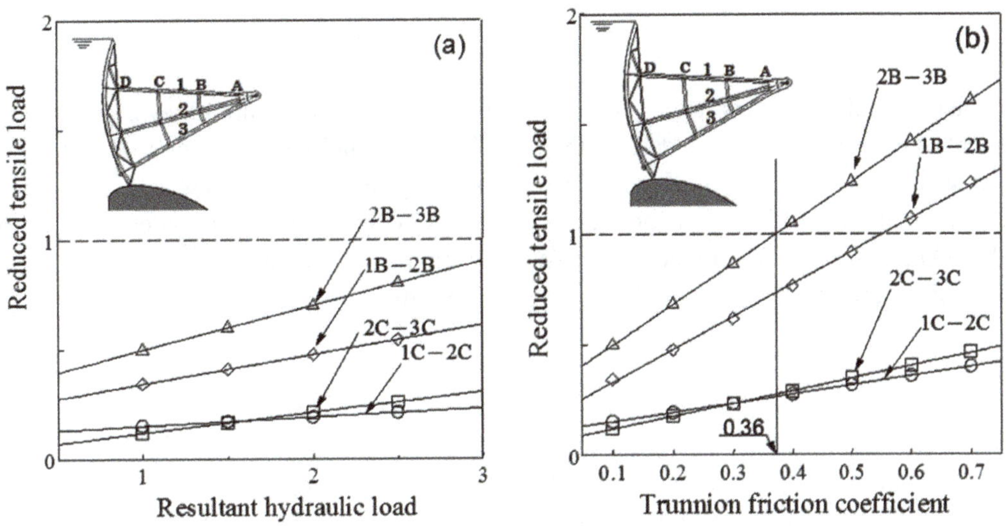

Figure 34. In-air natural vibration mode of the Wachi Dam gate's skinplate with a frequency of 24.9 Hz showing (a) a side view, (b) a three-dimensional view, (c) amplitudes at eight sections of the skinplate, and (d) amplitude at the spanwise center
From Ishii et al. (2007). © 2007 JSCE. Used with permission.

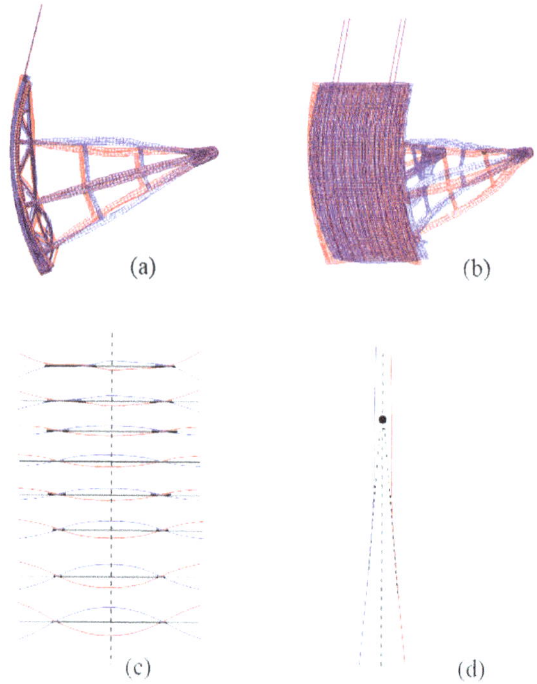

The violent skinplate streamwise vibration induces an excessive push-and-draw pressure due to the inertia effect of the water, which can destroy the gate. The resultant hydrodynamic load can be theoretically calculated for a given amplitude of skinplate motion at its bottom spanwise-center location, as shown in Figure 36. In this figure, the ordinate value of 1.41 corresponds to the safety coefficient from the FEM buckling analysis. The reduced skinplate vibration amplitude needed to reach this loading is 0.72×10^{-3} which corresponds to an amplitude of only 8.5 mm.

CONCLUSION

In this chapter the of self-excited coupled-mode instability theory was applied to the analyses of the Folsom Dam and Wachi Dam gate failures. The conclusion is that both gates could have potentially failed due to this instability mechanism. The attempt to re-examine the Wachi Dam failure was challenging due to the elapsed time and the lack of some of the physical data that was available for the Folsom Dam failure. As a point of protocol, we reported our finding on the Folsom Dam gate failure to the US Bureau of Reclamation in a series of reports in the ten years following the failure (see Ishii, 1995, Ishii et al., 1995, 1997, 2005). The USBR was aware of our findings prior to any publication.

Figure 35. Dynamic stability criterion for the Wachi Dam Tainter gate
From Ishii et al. (2007). © 2007 JSCE. Used with permission.

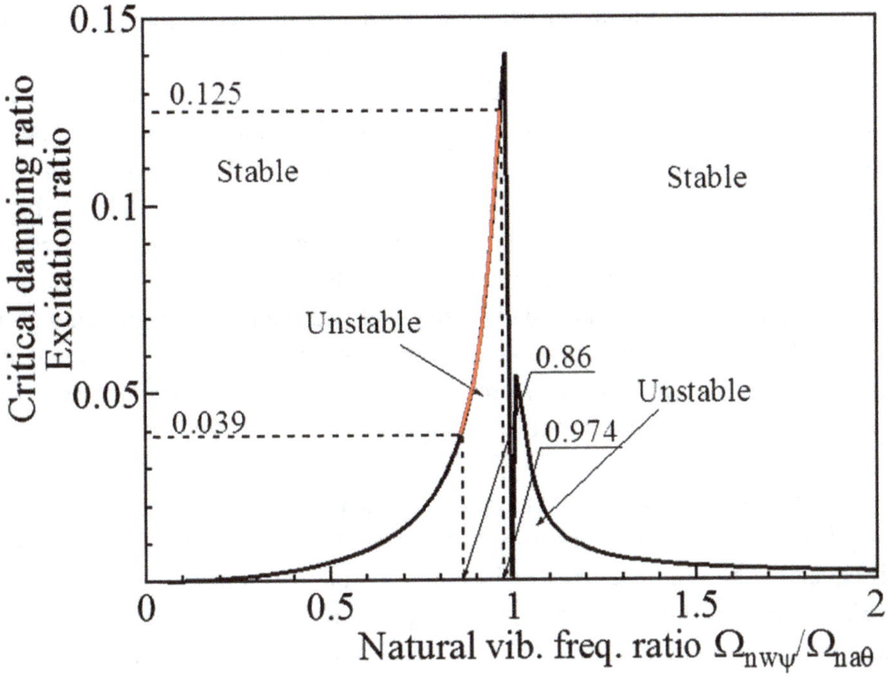

Figure 36. Resultant hydraulic load on the Wachi Dam Tainter gate, suggesting skinplate vibration for gate failure
From Ishii et al. (2007). © 2007 JSCE. Used with permission.

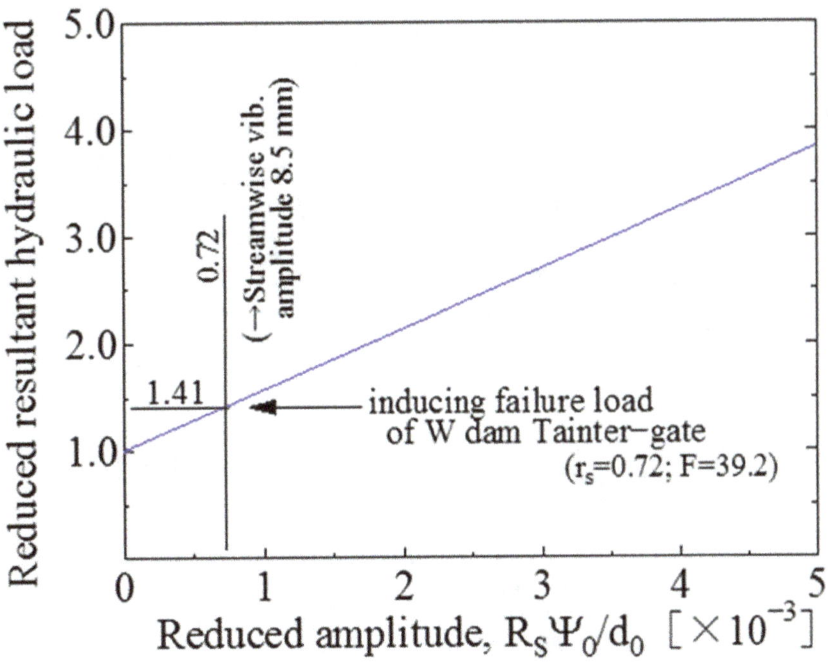

To summarize this chapter, we presented an overview of the self-excited coupled-mode instability mechanism as it relates to the Folsom Dam and Wachi Dam failure. Many of the references are to data from the Folsom Dam, because of its availability. Similar data could be obtained, in principle, from the Wachi Dam gate, but was not available at the time of writing this book.

In the fundamental vibration mode, as identified by the experimental modal analysis or from FEM analysis, the skinplate performs a low frequency bending mode vibration in the streamwise and vertical plane. This lowest frequency bending mode of the skinplate is essentially a streamwise rotation of skinplate about a point in the upper half of the skinplate. When the massive skinplate, which constitutes more than 60% of the whole gate mass (for the Folsom Dam gate), performs a rotational vibration, a large inertia torque due to its rotational vibration arises, which can then excite the whole gate to vibrate around the trunnion pin. When the whole gate vibrates around the trunnion pin, an additional inertia torque also arises, which can produce additional skinplate rotation. If the natural frequencies of these two vibration modes coincide, the streamwise rotational vibration of the skinplate and the whole gate rotational vibration around the trunnion pin can couple with each other through the inertia torques, resulting in coupled-mode vibration.

The whole gate rotation around the trunnion pin induces a flow rate variation beneath the gate, with a corresponding flow-rate variation pressure which forces the skinplate to vibrate in the streamwise direction. Subsequently, the skinplate streamwise vibration induces a push-and- draw pressure, which significantly lowers the vibration frequency of the skinplate. These two vibration modes, therefore, can readily couple, despite substantially different in-air frequencies, to produce violent self-excited vibrations (Anami & Ishii, 1999, 2000, 2003).

Figure 37. Closed energy cycle for coupled-mode flow-induced vibration, possible in Tainter gates
From Anami & Ishii (2003). ©2003 ASME. Used with permission.

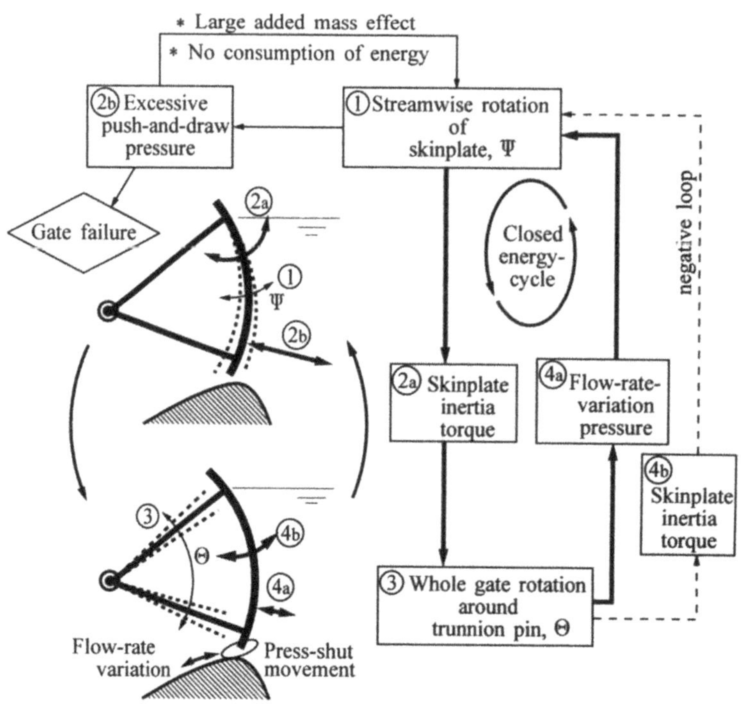

A closed loop energy cycle of the coupled-mode self-excited vibration is presented in Figure 37. If the bending rotational counter-clockwise vibration ① of the skinplate, represented by Ψ, is triggered by some random excitation, the counter-clockwise skinplate inertia torque ② appears around the trunnion pin, thus exciting the counter-clockwise whole gate vibration ③, represented by Θ. As a result, the gate opening under the gate increases, thus resulting in increased discharge water flow ④, and thus decreasing the flow-rate-variation pressure ⑤ in proportion to the whole gate vibration velocity, that feeds energy back to the skinplate bending rotational vibration ①, thus amplifying the bending vibration amplitude. This type of self-excited coupled-mode vibration can occur naturally, even when the geometric center of the skinplate is coincident with the trunnion pin, that is, when there is no trunnion pin eccentricity. Of special note in this closed energy-cycle is that the streamwise skinplate vibration induces an excessive push-and-draw pressure. The amplitude of this push-and-draw pressure can be more than 10 times that of the flow-rate variation pressure, and it could have caused the gate failures at the Folsom and Wachi Dams. The gate fails when the loading due to the excessive push-and-draw pressure exceeds the material strength.

It is also significant that the excessive push-and-draw pressure does not consume energy from the vibrating gate. The excessive push-and-draw pressure has associated with it a large added mass effect. The large added mass of water that must be accelerated with the gate as it undergoes its streamwise vibration lowers the relatively high in-air frequency of the streamwise skinplate vibration. As a result, when the in-water streamwise skinplate vibration frequency approaches the in-water whole gate rigid body rotational frequency, the system approaches resonance. Near resonance, the comparatively small amplitude flow-rate variation pressure can spontaneously amplify the amplitude of coupled-mode vibration.

REFERENCES

Anami, K., & Ishii, N. (1998). In-air and in-water natural vibrations of Folsom dam radial gate in California. In I. M. Allison (Ed.), *Experimental Mechanics, Proceedings of 11th International Conference on Experimental Mechanics*, (pp. 29-34). Rotterdam: Balkema.

Anami, K., & Ishii, N. (1999). *Flow-induced coupled-mode vibration of Folsom Dam radial gates in California. In Proceedings of 1999 ASME Pressure Vessel & Piping Conference* (Vol. 369, pp. 343–350). New York, NY: American Society of Mechanical Engineers.

Anami, K., & Ishii, N. (2000). Flow-induced dynamic instability closely related to Folsom Dam Tainter gate failure in California. In S. Ziada & T. Staubli (Eds.), *Flow Induced Vibration* (pp. 205–212). Rotterdam: Balkema.

Anami, K., & Ishii, N. (2001). Field and model investigations of FIV of Folsom Dam Tainter gate. In *Proceedings of Commemorative International Symposium for the 40th Anniversary for the Foundation of the Osaka Electro-Communication University*, Vol. 2, (pp.165-181). Osaka, Japan: Osaka Electro-Communication University.

Anami, K., & Ishii, N. (2003). Model tests for non-eccentricity dynamic instability closely related to Folsom Dam Tainter gate failure. *Symposium on Flow-Induced Vibration 2003, Proceedings of ASME Pressure Vessels and Piping Conference*, Vol.465, (pp. 213-221). New York, NY: American Society of Mechanical Engineers.

Anami, K., Ishii, N., & Knisely, C. W. (2014). Retrospective consideration of a plausible vibration mechanism for the failure of the Folsom Dam Tainter gate. *International Journal of Mechanical Engineering and Robotics Research*, *3*(4), 314–345.

Asahi Newspaper. (1967, July 3). Hydraulic gate blows away at newly constructed Wachi dam of Kansai Electric Power Company. *Asahi Newspaper*.

Imaichi, K.& Ishii, N. (1977). Instability of an idealized Tainter gate system without damping caused by surface waves on the backwater of dam. *Transactions of. Japan Society. of Mechanical Engineers*, *20*(146), 963-970. (in Japanese)

Ishii, N. (1995). Folsom Dam Gate Failure Evaluation and Suggestions. Report submitted to Bureau of Reclamation, U.S. Department of the Interior.

Ishii, N., & Anami, K. (2001). Field and model investigations of FIV of Folsom Dam Tainter gate. In *Proceedings of Commemorative International Symposium for the 40th Anniversary for the Foundation of the Osaka Electro-Communication University*, Vol. 2, (pp. 143-164). Osaka, Japan: Osaka Electro-Communication University.

Ishii, N., Anami, K., & Knisely, C. W. (1995). Folsom Dam Gate Failure Evaluation Based on Modal Analysis and Suggestion. Report submitted to Bureau of Reclamation, U.S. Department of the Interior.

Ishii, N., Anami, K., & Knisely, C. W. (1997). Dynamic Stability Evaluation for Folsom Dam Radial Gate and Its Inherent Non-Eccentricity Instability. Report submitted to Bureau of Reclamation, U.S. Department of the Interior.

Ishii, N., Anami, K., & Knisely, C. W. (2005). *Lessons from the Folsom Dam* Tainter gate *Failure and Future Work for Subsequent Generations*. Presented at USBR Office, the Folsom Dam, Folsom.

Ishii, N., Anami, K., Makihata, T., Oku, T., & Ohara, T. (2007). Tainter gate failure in Japan. *Journal of Structural Engineering (JSCE), 53A*, 589-596. (in Japanese)

Ishii, N., & Imaichi, K. (1982). Instability of elastically suspended Tainter gate system caused by surface waves on the reservoir of a dam (Approximate Solution). *Transactions of the Japan Society of Mechanical Engineers, 48*(428), 640-647. (in Japanese)

Ishii, N., Imaichi, K., & Hirose, A. (1977). Instability of elastically suspended Tainter gate system caused by surface waves on the reservoir of a dam. *Journal of Fluids Engineering*, *99*(4), 699–708. doi:10.1115/1.3448888

Ishii, N., & Naudascher, E. (1992). A design criterion for dynamic stability of Tainter gates. *Journal of Fluids and Structures*, *6*(1), 67–84. doi:10.1016/0889-9746(92)90056-9

Japan Times. (1995, July 19). Water pours through gate at dam in Calif. *Japan Times*.

KXTV-10. (1995, July 17). *The Folsom Dam Gate Failure* [Television broadcast]. KXTV-10 News.

Yano, K. (1968). *Wachi Dam Tainter Gate Failure*. Annual report of Disaster Prevention Research Institute, Kyoto University, No.11, B, 203-219. (in Japanese)

APPENDIX: NOMENCLATURE

a: Streamwise length from the fulcrum to the contact point on the side plate
A: Vibration amplitude (initial disturbance shock)
A_0: Extensional displacement of the spring elements
B: Discharge opening
c_e: Equivalent viscous damping coefficient
d_0: Gate submergence depth
d_r: Depth of the reservoir
D_t: Diameter of the trunnion pin
F: Froude number
F_0: Basic Froude number
F_p: Resultant hydraulic force acts on both side plates embedded in the concrete piers
F_r: Friction force
F_s: Friction force acting on the side plates
F_t: Friction force acting on both sides of the skinplate
F_{tp}: Friction force on the trunnion pin
g: Acceleration of gravity
I_θ: Moment of inertia of the whole gate around the trunnion pin,
K: Resultant spring constant combined hoist frame/chain system
k_c: Spring constant of the chains
L: Whole length of the rubber seal along the skinplate with submergence
p_{bs}: Reduced amplitude of standing pressure-wave component of push and draw pressure
P: Total hydrostatic force
R_a: Radius of circular arc skinplate
r_s, R_S: Height of skinplate rotation center [r_s is the dimensionless expression of R_s]
T_i: ith Transfer function
α_ψ: Water-to-gate mass ratio
γ_{nw}: In-water to in-air vibration frequency ratio
$\gamma_{w\psi\theta}$: Actual frequency ratio of the in-water skinplate streamwise rotational vibration to the whole gate in-air vibration frequency
$\gamma_{\psi\theta}$: In-air natural vibration frequency ratio
δ_p: Pressure correction coefficient
$\zeta_{a\theta}$: In-air damping ratio of the whole gate rotational vibration,
$\zeta_{a\psi}$: In-air damping ratio for streamwise vibration of the skinplate
$\zeta_{ca\theta}$: Critical value for the resultant damping ratio around the trunnion pin
$\zeta_{c\psi}$: Critical damping ratio
ζ_e: Equivalent viscous damping ratio
$\zeta_{f\psi}$: Wave-radiation damping ratio
μ_s: Friction coefficient at the side seal plate,
μ_t: Friction coefficient at the trunnion pin
$\xi_{f\psi}$: Fluid excitation ratio

ξ_{ψ}: Actual excitation ratio of the skinplate streamwise rotational vibration
ρ: Density of water
ϕ: Phase-lag of vibration to impulsive force
$\Omega_{a\theta}$: In-air whole gate rotational vibration frequency
$\Omega_{a\psi}$: In-air vibration frequency of the skinplate streamwise vibration
Ω_{n}: Vibration frequency
$\Omega_{w\psi}$: In-water vibration frequency of the skinplate streamwise vibration
$\Omega_{nw\psi}$: In-water natural streamwise vibration frequency

Chapter 13
Analogies Between Coupled-Mode Gate Vibration and Coupled-Mode Flutter:
The Need for Dynamic Design

ABSTRACT

In this chapter the similarities between the Tacoma Narrows Bridge failure in 1940 and the Folsom Dam gate failure in 1995 are examined. In both cases, static design guidelines were followed in the design of the structure under the assumption that large, massive structures would not be susceptible to dynamic excitation. Fundamentals of two-dimensional coupled mode flutter are presented. The frequency mode coalescence that occurs in two-dimensional flutter is noted. It is seen to have some resemblance to the mode-coupling in the coupled-mode instability of Tainter gate. The need for development of dynamic design guidelines for Tainter gates is argued to be parallel to the need for dynamic design guidelines for suspension bridges in the wake of the Tacoma Narrows failure.

INTRODUCTION

In the previous chapter, we described how fluid-structure interaction caused gates at the Folsom Dam and Wachi Dam to fail. Forty-five years before the Folsom gate failure, the first Tacoma Narrows Bridge collapsed due to the effects of moderate winds, well below the design wind speed. What do these disasters have in common? We suggest that, in both cases, the structures were designed for static stability without taking into account the dynamic forces exerted by the fluid on the structure. Table 1 shows at least 10 major suspension bridge failures in the 19th century due to the wind. It is unlikely that the bridge builders who were professionally active in the mid-1930s and 1940s had any first-hand knowledge of these earlier failures. In fact, the deflection theory of bridge design that was popular at the time probably lulled the younger generation of engineers into a false sense of confidence that increasingly elegant bridges could be constructed with less material and longer spans. However, paraphrasing George Santayana, those who do not learn from the mistakes of history are bound to repeat them.

DOI: 10.4018/978-1-5225-3079-4.ch013

When the Tacoma Narrows Bridge was opened in 1940 with a span of nearly 860 m, it was the third longest suspension bridge in the world at that time. Carrying only two lanes of traffic, the bridge was very slender. The bridge deck was supported by 8-ft high, solid girders, giving it an H-shaped cross section.

On 7 November 1940, four months after the opening of the bridge, with a moderately strong wind of about 42 mph (18.8 m/s) blowing, the bridge underwent large amplitude heaving (vertical) vibration. After more than an hour of heaving vibrations, the vibration mode abruptly changed from a heaving mode to an anti-symmetric, full-span torsional mode with a torsional node at the mid-point of the center span. Initially, the torsional mode frequency was 14 cycles/min (0.233 Hz), decreasing later to about 12 cycles/min (0.20 Hz). In this torsional mode, the bridge failed in a little more than one hour. The conclusion of the forensic report by Ammann et al. (1941) was that a center cable tie became loose in the heaving mode, allowing the cables on opposite sides of the center span to become asymmetric. The strong, steady wind blowing across the bridge increased the torsional motion that lead to complete destruction of the bridge.

The Tacoma Narrows Bridge had already shown a propensity for vertical (heaving) wind-induced vibrations while under construction. These vibrations continued, even with relatively low speed winds, over the four months the bridge was open to traffic. The vibrations were not predictable; sometimes they were present and other times not, under apparently similar conditions. Methods of attenuation were explored and a number were implemented, including adding diagonal ties at the center span, with mixed success. The most common mode of vibration was a half-wavelength heaving mode shape between the towers, with a maximum mid-span amplitude of about 2 ft (0.61 m). With increasing wind speed, the vibration mode often, but not always, changed to one with a shorter spatial wavelength with an increase in frequency and a corresponding increase in the number of nodes, up to nine nodes. A maximum peak-to- peak amplitude of about 5 ft (1.52 m) was observed in a mode with two nodes between the towers at a frequency of vibration of about 12 cycles/minute.

Table 1. Failures of suspension bridges in the 1800s, followed by a gap of more than 50 years between the failure of Keefer's Niagara Bridge and the Tacoma Narrows Bridge

Bridge	Designer	Span (ft)	Span (m)	Failure Date
Dryburgh Abbey (Scotland)	John and William Smith	260	79.25	1818
Union (England)	Samuel Brown.	449	136.85	1821
Nassau (Germany)	Lossen and Wolf	245	74.68	1834
Brighton Chair Pier (England)	Samuel Brown.	255	77.72	1836
Montrose (Scotland)	Samuel Brown	432	131.67	1838
Menai Strait (Wales)	Thomas Telford	580	176.78	1839
Roche-Beruard (France)	P. LeBlanc	641	195.38	1852
Wheeling (USA)	Charles Ellet	1010	307.85	1854
Lewiston-Queenston (USA)	Edward Serrell	1041	317.30	1864
Niagara-Clifton (USA)	Samuel Keeler	1260	384.05	1889
Tacoma Narrows (USA)	Leon Moisseiff	2800	853.44	1940

Adapted from Ge &Tanaka (2013).

There is substantial disagreement concerning the precise mechanism of the failure of the Tacoma Narrows Bridge (Scott, 2001; Gazzola, 2015). In a general interest article entitled "The Strangest, Most Spectacular Bridge Collapse (And How We Got It Wrong)," Pasternak (2015) recounts the varied, and (in hindsight) frequently incorrect or incomplete explanations of the failure. Billah & Scanlan (1991) argue convincingly for the need to update introductory physics books to provide a more complete explanation of the mechanism of failure. Gazzola (2015) tells us that every explanation of the Tacoma Narrows Bridge failure has detractors, and there is no commonly "accepted" explanation. After more than 75 years of research and analysis, the engineering community has still not arrived at universal agreement on the precise details of the mechanism of failure of the Tacoma Narrows Bridge (see Arioli & Gazzola, 2015; Plaut, 2008; McKenna & Ó Tuama, 2001; McKenna, 1999; and others).

In the intervening years, our understanding of unsteady bluff body aerodynamics has advanced. Concepts from unsteady aircraft aerodynamics, including coupled-mode flutter, have been adapted to describe the time-dependent, separating flow around bluff bodies and the resulting pressure fluctuations and unsteady forces acting on those bodies. Scott (2001) provides an overview of these developments, while Simiu & Scanlan (1996), Ge & Tanaka (2013), Holmes (2015), and Dowell et al. (2015) provide a more mathematical treatment of these analytical analyses. Just as the collapse of the Tacoma Narrows Bridge challenged the prevailing design principles of the mid-20th century, the failure of the gate at the Folsom Dam suggests the need for a dynamic stability analysis of hydraulic structures in the design stage.

In this chapter, we will first present an analysis of the consensus mechanism that caused the Tacoma Narrows Bridge to fail. We will then compare and contrast this mechanism with that of coupled-mode airfoil flutter. Finally, we will show how the mechanism of coupled-mode flutter is both similar to and different from the self-excited coupled-mode gate instability that we identified for the Folsom Dam gate. Ultimately, we suggest that our new understanding of the mechanisms of self-excited vibrations of these structures must be reflected in extended design criteria for dynamic stability in addition to those for static stability.

ANALYSIS OF THE TACOMA NARROWS BRIDGE FAILURE

Professor F. B. Farquharson, who was at the time investigating the vibration of the bridge, arranged to have local photographers B. Elliot and H. Monroe film the wildly oscillating bridge. Walter Miles of the Pacific Bridge Company and Farquharson also contributed photographic documentation. Figure 1 shows several still photographs extracted from the films.

Details of the Failure

A Board of Engineers was convened by John M. Carmody, Administrator of the Federal Works Agency, to investigate the Tacoma Narrows Bridge failure. The Board of Engineers' report (Ammann et al., 1941), commonly called the "Carmody Report," includes first-hand observations of the events on 7 November 1940 by several individuals. Early in the morning, the bridge was undergoing the accustomed vertical vibrations of about 2 ft (0.61 m) with eight or nine nodes at frequencies of 36 to 38 cycles/minute with a wind speed of about 42 mph (18.8 m/s). *Lateral* displacement at the main span center was estimated to be about 2 ft (0.61 m).

Figure 1. Photographs extracted from films showing (a) torsional vibration of the main span, (b) an on-bridge axial view of the torsional vibration, and (c) collapse of the bridge
Photograph from (a) Ammann et al. (1941), (b) and (c) Elliot (1940).

About 9:30 a.m. the center tie on the north cable began to move longitudinally; by 10:00 a.m. one of the diagonal braces broke and the center cable tie (band) began to slip over the cable, moving back and forth with a double amplitude of about 42 in (1.07 m). After the bridge collapsed, about 500 wires were found to be broken by the movements of the band; no one knows if these wires failed during heaving vibration or during the torsional vibrations. The moving cable tie meant that the diagonal brace was ineffective. Did the large amplitude heaving vibration produce a large enough effective angle of attack to initiate the torsional mode? No one knows; the mechanism of transition from heaving to torsional mode or to a torsional mode coupled with heaving has not been elucidated. In any event, the torsional vibration mode was no longer constrained, and was extracting energy from the wind. The vibration frequency abruptly changed from the approximately 37 cycles/min in the vertical mode to about 14 cycles/min in the torsional mode. The frequency of the torsional mode subsequently decreased to 12 cycles/min (consistent with the frequency of the previously observed heaving mode with the greatest amplitude). The anti-symmetric torsional mode had a node at the center of the main span. The lateral displacement of the main span was said to increase, but was not measured. The torsional mode continued with an amplitude up to about 45° with a main cable movement of about 28 feet (8.53 m) for about 30 minutes before a section of the concrete roadway slab fell away from the center of the bridge.

About 30 minutes later, a large 600 ft (183 m) section of the span near the west quarter portion of the main span dropped into the water, followed shortly by most of the main span. Without the weight of the center span, the side spans sagged and the two towers deflected some 25 ft (7.62 m) towards their respective shorelines.

Analysis of the Tacoma Narrows Bridge Failure

The sequence of events leading up to the failure is quite well documented through eyewitness testimony and film recording of the collapse. The controversy began with the explanations proposed for the mechanism by which a relatively steady wind, well below the design speed, could serve as the source of the fluctuating pressure loading needed to deliver sufficient energy to drive the torsional mode of the bridge to failure.

One explanation considered in the original Carmody Report (Ammann et al., 1941) was that the fluctuating loading resulted from von Kármán vortex shedding, which occurs in the wake of a bluff body when placed in a fluid stream. The shedding of these concentrated patches of vorticity into the wake results in a fluctuating cross-stream force. The frequency of vortex shedding varies linearly with the wind velocity (see Chapter 3).

However, the authors of the Carmody Report were aware of the wind tunnel tests of scaled models of suspension bridges carried out by Prof. Farquharson, and, in part based on those results, rejected the idea that vortex-induced vibrations could have caused the failure:

It is very improbable that resonance with alternating vortices plays an important role in the oscillation of suspension bridges. First, it was found that there is no sharp correlation between wind velocity and oscillation frequency such as is required in case of resonance with vortices whose frequency depends on the wind velocity. Secondly, there is no evidence for the formation of alternating vortices at a cross section similar to that used in the Tacoma Bridge, at least as long as the structure is not oscillating. It seems that it is more correct to say that the vortex formation and frequency is determined by the oscillation of the structure than that the oscillatory motion is induced by the vortex formation. (pp. 130-131)

With the same conviction that vortex shedding could *not* have been the cause of the bridge failure, the authors suggest that self-excited torsional instability was to blame:

There appears no doubt as to the self-induced character of the torsional oscillations of the Tacoma Narrows Bridge. (p. 125)

The fact that, after the beginning of a torsional motion, the angular amplitude reached the value of ±45° in a very short time, indicates that the wind velocity was above the critical limit and previously the development of torsional oscillations was suppressed by the internal resistance of the structure. (p. 126)

...convincing evidence has been found by oscillatory tests that beyond a certain wind velocity negative aerodynamic damping is to be expected in almost any suspended bridge structure when it oscillates torsionally... Its appearance and destructive effect in the Tacoma Narrows Bridge resulted from its narrow width and lack of structural damping. (p. 130)

Thus, as noted by Billah & Scanlan (1991), there is no apparent mystery surrounding the Tacoma Narrows Bridge failure. The failure was attributed to self-excited torsional instability by the Board of Engineers, and was subsequently confirmed by Scanlan & Tomko (1971). Resonant von Kármán vortex shedding was not the cause of the failure. Gazzola (2015) responds that the analysis of Scanlan & Tomko (1971) was for linearized equations of motion and thus cannot be used to form conclusions about large amplitude bridge motions. There is room to question whether the vibration was truly a single degree-of-freedom torsional instability, or if it was a coupled-mode heaving-torsional vibration. Gazzola (2015) notes that careful observation of the film of the collapse shows that the bridge is still undergoing heaving motion while large amplitude torsional vibration were occurring.

Despite the unambiguous statement of the "Carmody Report" (Ammann, 1941), many introductory physics books, as detailed in Billah & Scanlan (1991), continue to state that the Tacoma Bridge failure was due to resonance of the structure with the frequency of the shedding of von Kármán vortices, providing a false explanation for generations of students. Unfortunately, for many students, the version of the Tacoma Bridge failure as presented in introductory physics textbooks is the only explanation of the Tacoma Narrows failure they will ever encounter.

Vertical Vibration Modes of the Tacoma Narrows Bridge

Because the bridge had already been oscillating vertically during construction, studies to understand and counteract the problem were already underway when the bridge collapsed. Farquharson (1941) presented wind tunnel test data on a 1:50 scale model of the Tacoma Narrows Bridge. Figure 2 is based on these data and shows the amplitude of the vertical modes of vibration as a function of the wind velocity. Note that each vertical mode looks like a rounded peak, whereby the amplitude first increases with wind speed and then decreases with further increases in wind speed. This amplitude-limited behavior of the vertical modes is due to the sign change of the fluid damping with increasing amplitude, as shown in Figure 3. Similar sign changes in fluid damping or alternatively, fluid excitation, are also found in the characteristics of both long-span gates (Chapter 7) and Tainter gates (Chapters 10, 11 and 12).

Ammann et al. (1941) reported that the data shown in Figure 3 for the lab model with an equivalent 13-ft girder are consistent with the damping of the full-scale Tacoma Narrows Bridge with an 8-ft girder ("wind tunnel tests at the University of Washington [indicate] a negative damping of the order of magnitude of that calculated for the case (c) would be obtained for the Tacoma Narrows Bridge as built." p. 113). These results are significant because they demonstrate that vertical modes of vibration will be amplitude-limited and thus cannot cause the bridge to fail.

Torsional Vibration Modes of the Tacoma Narrows Bridge

Eyewitness testimony and photographic evidence clearly indicate that the bridge collapsed after the onset of torsional vibrations with or without heaving-mode (vertical mode) coupling. In other words, the aerodynamic moment with respect to the roadway axis caused the bridge to fail.

In a fashion similar to the fluid damping and fluid excitation discussed throughout this book in reference to coupled-mode dynamic instability of hydraulic gates, Ammann et al. (1941) observe that for bridge sections,

The most important aerodynamic effect induced by the torsional oscillation of the roadway is the moment with respect to the axis which may act as positive or negative damping. Since no information was available on this question, the aerodynamic experiments described in Appendix VIII were conducted at the California Institute of Technology with a 1:80 scale model of a section of the suspended structure of the Tacoma Narrows Bridge, in order to determine the direction and magnitude of the moment induced by the oscillatory motion of the structure...It was found in all cases that the aerodynamic damping is

Figure 2. Dependence of modal vibration amplitudes on wind speed for a 1/50 scale model in wind tunnel; Vertical modes: 0-NV denotes vertical bending vibration with no node; 1-NV vertical bending vibration with one node; 2-NV vertical bending vibration with two nodes; 3-NV vertical bending vibration with three nodes; 4-NV vertical bending vibration with four nodes. Torsional mode: 1-NT denotes torsional mode with one transverse nodal line at center span.
From Farquharson (1941), as cited in Kawada (2010).

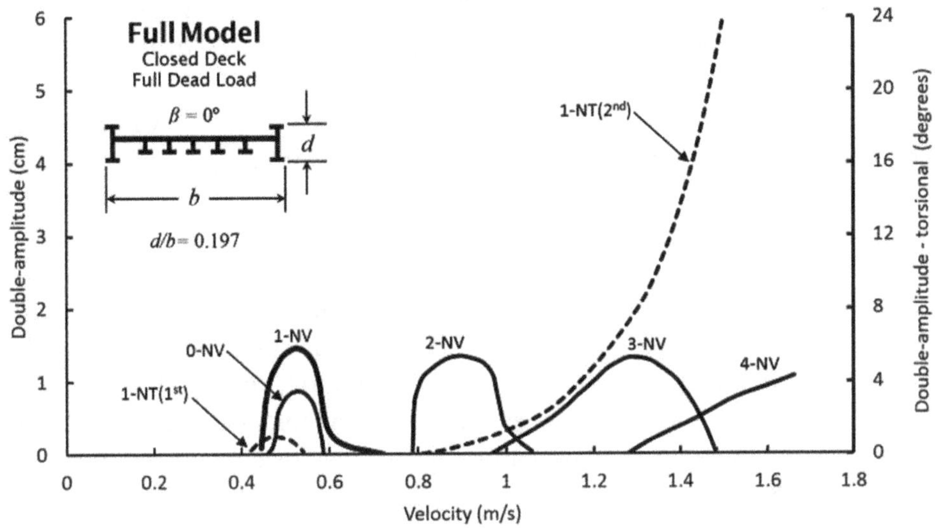

Figure 3. Amplitude-limited vertical mode behavior results from the sign change of fluid damping with increasing vibration amplitude. Circles represent a wind velocity of 6.8 mph; triangles represent a velocity of 41 mph. As the vibration amplitude increases, net negative damping (i.e., excitation) changes to net positive damping. Note for small damping, the damping ratio $\zeta = \delta/2\pi$ where δ is the log decrement.
From Ammann et al. (1941)

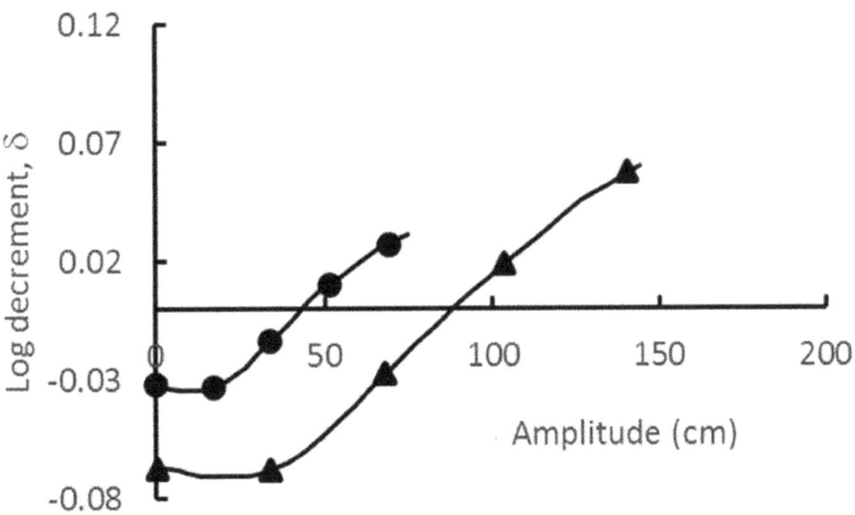

positive below a certain velocity and becomes negative beyond this limit. From the experiments the rate of energy dissipation and input was calculated.

The torsional characteristics of the scaled model of the Tacoma Narrow Bridge in a wind tunnel are shown in Figure 2. Notice that at low wind velocities, the amplitude of the torsional mode with one transverse nodal line at center span (1-NT(1st)) is lower than that of the vertical modes of vibration at any of the tested velocities. However, the torsional vibration 1-NT(2nd) that begins at a higher wind velocity shows exponentially increasing amplitude. According to Kawada (2010), this mode caused the bridge to fail.

Figure 4a shows the log decrement as a function of non-dimensional velocity measured by von Kármán & Dunn using wind tunnel section models (see Appendix VIII in Ammann et al., 1941). Figure 4b shows a dimensional version of the curve-fit in Figure 4a converted to correspond to the full-scale Tacoma Narrows Bridge. The damping at wind speeds above about 25 mph is clearly negative, indicating energy input from the wind to the structural vibration.

With the technology available in 1941, the Carmody report explained the failure as a single degree-of-freedom instability. Successful adaptation of aircraft flutter mechanisms to bridge decks waited until the late 1960s and early 70s. However, even when Scanlan & Sabzevari (1967, 1969) and Scanlan & Tomko (1971) applied aircraft flutter theory to small amplitude bluff body flutter, they came to the same conclusion: the Tacoma Narrows Bridge failed due to a single degree-of-freedom instability. Recent advances in nonlinear analysis have reopened the discussion of single-mode failure versus coupled-mode failure (see Gazzola, 2015; Arioli & Gazzola, 2015; Plaut, 2008; McKenna & Ó Tuama, 2001; McKenna, 1999; and references cited in these works). According to Gazzola (2015), using more complicated, nonlinear

Figure 4. Non-dimensional torsional damping from (a) wind tunnel testing; and (b) converted to dimensional values corresponding to the Tacoma Narrows Bridge
From Ammann et al., 1941.
Note the negative scale on the y-axis.

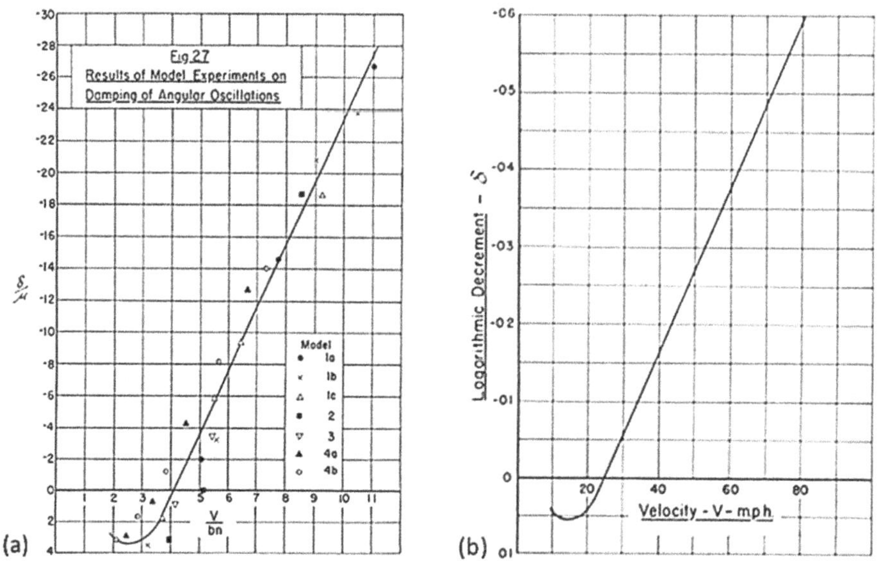

analysis may provide more accurate predictions of mechanisms of failure than the simpler, linearized analysis. With the further development of nonlinear computational methods coupled with enhanced computing power, future analyses may not only be able to predict the mechanisms involved in the onset of vibrations, but also those involved in structural failure.

In the next section, we will introduce the theory of coupled-mode airfoil flutter, and then explain how this theory can be adapted to understand bluff body flutter.

FLUTTER THEORY FOR AIRFOILS

From the early days of flight, aerodynamicists have been aware of the importance of the interaction between airflow and elastic deformation of the aircraft, as documented in the historical review article by Garrick & Reed (1981). These authors cite the work of several aerodynamic researchers from 1916 onwards; however, the development by Theodorsen (1935) has been most widely accepted in the U.S.

We will focus on classical coupled plunge-torsion flutter of aircraft wings or airfoils, although many types of flutter have been analyzed. We refer interested readers to modern aeroelasticity texts such as Dowell et al. (2015), Hodges & Pierce (2014), Wright & Cooper (2007); as well as the classic texts by Fung (1968), Bisplinghoff & Ashley (1996), Bisplinghoff et al. (2013) or Scanlan & Rosenbaum (1951).

We will model the airfoil section, shown in Figure 5, as a two-dimensional system with a unit depth into the page. The airfoil chord is $c = 2b$; the positive downward directed displacement is h; the angular displacement is α, defined as positive when producing a nose-up orientation, as shown. The flow is assumed incompressible ($Ma = 0$) with known density ρ_∞. The lift is assumed to act upward at the quarter chord location where $x_{AC} = b/2 = c/4$. The dimensional distance from the leading edge to the center of gravity (CG) is $(1 + e)b$, where $-1 \leq e \leq 1$, while that from the leading edge to the elastic axis is $(1 + a)$ b, where $-1 \leq a \leq 1$; when a or e are zero, the elastic axis or the center of gravity, respectively, is located at mid-chord. The dimensional distance from the center of gravity to the elastic axis is $X_\theta = (e - a)b$, non-dimensionalized as $x_\theta = X_\theta/b = e\text{-}a$. This parameter, x_θ, called the static unbalance is positive if the center of gravity is between the elastic axis and the trailing edge and is negative if it is between the leading edge and the elastic axis.

The equations of motion for the airfoil in Figure 5, subjected to fluid flow, can be found in a variety of sources, including Scanlan & Rosenbaum (1951), Bisplinghoff & Ashley (1996), Bisplinghoff et al. (2012), Fung (2008), and Hodges & Pierce (2011), as well as others. In this chapter, the two-dimensional coupled flutter equations will be written as:

$$m\ddot{h} + \left[m x_\theta b \ddot{\theta} \right] + c_h \dot{h} + k_h h = -L \tag{1}$$

$$I\ddot{\theta} + \left[m x_\theta b \ddot{h} \right] + c_\theta \dot{\theta} + k_\theta \theta = M_{c/4} + b\left(\frac{1}{2} + a \right) L = M \tag{2}$$

Figure 5. Two-dimensional thin airfoil section of chord 2b supported on vertical and torsional springs, where AC is the aerodynamic center, CG is the center of mass, and P is the location of the elastic axis
After Hodges & Pearce, 2011.

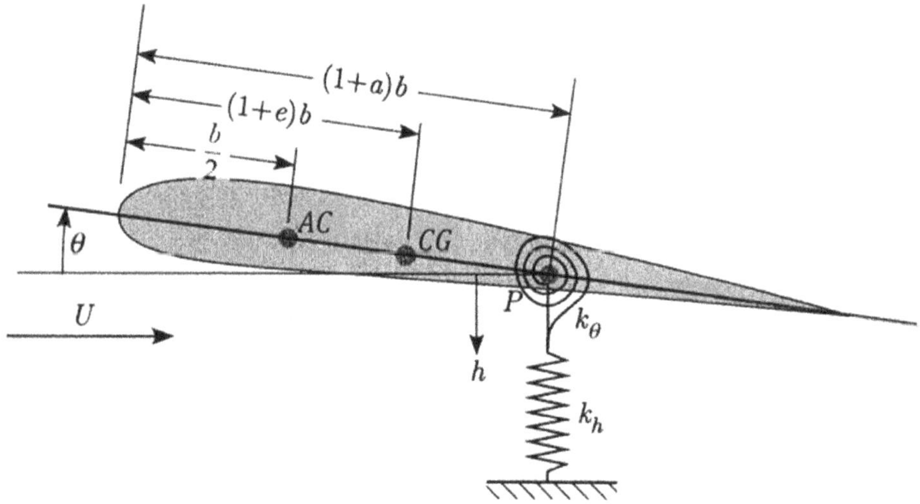

The second term, enclosed in square brackets, in each of the two equations of motion, Equations 1 and 2, will disappear if the center of gravity coincides with the elastic axis since x_θ would then be zero. We will subsequently see that this coincidence of the center of gravity with the elastic axis is common for bridges, but unusual for airfoils. In simple coupled-flutter analysis, mechanical damping is usually neglected. We will assume $c_h = c_\theta = 0$.

The individual uncoupled, no flow, undamped natural frequencies can be written as

$$\omega_h = \sqrt{\frac{k_h}{m}} \text{ and } \omega_\theta = \sqrt{\frac{k_\theta}{I}} \tag{3}$$

and will be used to re-write the equations of motion.

If we assume simple harmonic motion in which

$$h = \bar{h}\exp(i\omega t) \text{ and } \theta = \bar{\theta}\exp(i\omega t) \tag{4}$$

where the overbar indicates a complex amplitude, for example, $\bar{h} = h_o \exp\left(i\varphi_h\right)$, where h_o is the amplitude and φ_h is the phase, then we can write the lift and moment as

$$L = \bar{L}\exp(i\omega t) \text{ and } M = \bar{M}\exp(i\omega t) \tag{5}$$

Heaving amplitude h_o is measured at the chord-wise location of the elastic axis.

Neglecting damping terms and inserting the four expressions in Equations 4 and 5 into Equations 1 and 2 yields

$$-\omega^2 m\bar{h} - \omega^2 mbx_\theta\bar{\theta} + m\omega_h^2\bar{h} = -\bar{L} \tag{6}$$

$$-\omega^2 mbx_\theta\bar{h} - \omega^2 I\bar{\theta} + I\omega_\theta^2\bar{\theta} = \bar{M} \tag{7}$$

In flutter analysis, the flutter frequency, ω, is commonly non-dimensionalized using the semi-chord, b, and the freestream velocity, U. The non-dimensional frequency, k, given by

$$k = \frac{b\omega}{U} \tag{8}$$

varies between zero and one in most aeronautical applications.

In a low-speed flow where compressibility is insignificant ($Ma \approx 0$), we will write lift and moment as complex linear functions of k, $\bar{h}\,/\,b$ and $\bar{\theta}$ as follows:

$$\bar{L} = -\pi\rho_\infty b^3\omega^2\left\{L_h(k)\frac{\bar{h}}{b} + L_\theta(k)\bar{\theta}\right\} \tag{9}$$

$$\bar{M} = \pi\rho_\infty b^4\omega^2\left\{M_h\left(k\right)\frac{\bar{h}}{b} + M_\theta\left(k\right)\bar{\theta}\right\} \tag{10}$$

The terms in curly brackets account for amplitude and phase changes in lift and moment due to the frequency and the behavior of the complex vibration amplitudes in plunge and torsion.

These lift and moment expressions, Equations 9 and 10, can be substituted into Equations 6 and 7, respectively. After this substitution, the two equations of motion can be normalized; we will divide Equation 6 by $-\pi\rho_\infty b^3\omega^2$ and Equation 7 by $-\pi\rho_\infty b^4\omega^2$ to yield

$$\left\{\frac{m}{\pi\rho_\infty b^2}\left[1-\left(\frac{\omega_h}{\omega}\right)^2\right] + L_h\left(k\right)\right\}\frac{\bar{h}}{b} + \left[\frac{mx_\theta}{\pi\rho_\infty b^2} + L_\theta\left(k\right)\right]\bar{\theta} = 0 \tag{11}$$

$$\left\{\frac{mx_\theta}{\pi\rho_\infty b^2} + M_h\left(k\right)\right\}\frac{\bar{h}}{b} + \left\{\frac{I}{\pi\rho_\infty b^4}\left[1-\left(\frac{\omega_\theta}{\omega}\right)^2\right] + M_\theta\left(k\right)\right\}\bar{\theta} = 0 \tag{12}$$

where the aerodynamic lift and moment expressions have been moved to the left side to produce homogeneous equations.

For a given application, the geometry and materials provide definitions for m, I, ω_h, ω_θ, x_θ, b and the air flow conditions define ρ_∞. Because Equations 11 and 12 are linear and homogeneous in $\bar{h} / b$ and $\bar{\theta}$, the nontrivial solutions to these equations of motion can be found by setting the determinant of their coefficients equal to zero.

To create coefficients that are explicit in a single common term, the factor $\sigma = \omega_h / \omega_\theta$ is introduced in Equation 14 below. In addition, the following two coefficients, the mass ratio μ_m and the mass radius of gyration r_I, are substituted to keep the notation compact:

$$\mu_m = \frac{m}{\pi \rho_\infty b^2} \text{ and } r_I = \sqrt{\frac{I}{mb^2}} \tag{13}$$

$$\begin{vmatrix} \mu_m \left[1 - \sigma^2 \left(\frac{\omega_\theta}{\omega} \right)^2 \right] + L_h(k) & \mu_m x_\theta + L_\theta(k) \\ \mu_m x_\theta + M_h(k) & \mu_m r_I^2 \left[1 - \left(\frac{\omega_\theta}{\omega} \right)^2 \right] + M_\theta(k) \end{vmatrix} = 0 \tag{14}$$

The determinant yields a second-order polynomial in $\lambda = \left(\frac{\omega_\theta}{\omega} \right)^2$. The two unknowns $\frac{\omega_\theta}{\omega}$ and k can be determined from the single second-order polynomial in λ since it will contain real and imaginary components due to the complex aerodynamic coefficients, L_h, L_θ, M_h, and M_θ. Solution methods for solving the flutter determinant have been developed and can be found in aeroelasticity texts such as Hodges & Pearce (2011), Wright & Cooper (2014), and many others.

The roots of the characteristic equation from the determinant are complex. The real part represents the square of the aerodynamic frequency ratio, while the imaginary part represents the damping. The onset reduced frequency for flutter is found when the imaginary part of one of the roots equals zero. The square root of the corresponding real part provides the frequency ratio, $\frac{\omega_\theta}{\omega}$.

To solve the characteristic equation, a model of the aerodynamic lift and moment must be developed or selected from existing theories to determine the aerodynamic coefficients, L_h, L_θ, M_h, and M_θ.

There are a number of ways to model the aerodynamic force and moment. One option is to assume steady-flow aerodynamic lift in phase with the torsional motion, $L = 2\pi \rho_\infty U^2 \theta$, and zero quarter chord moment (see Blevins, 1994 and Hodges & Pearce, 2011 for details). A more realistic option is to use an unsteady aerodynamic formulation that permits a phase shift between the two airfoil degrees-of-freedom, as well as between the airfoil torsional motion and the resulting lift and moment. For our purposes we will use Theodorsen's (1934) unsteady aerodynamics theory because it has been used to adapt airfoil flutter to an analysis of bridge flutter.

Theodorsen (1934) used linear potential flow theory to analyze small amplitude harmonic oscillations of a thin airfoil modeled as a flat plate. Details of Theodorsen's derivation can be found in his original paper as well as in the text by Bisplinghoff et al. (2013). The airfoil of semi-chord b is assumed to oscillate in plunge and torsion at a frequency ω, with no flow separation from the airfoil surface.

Lift and quarter-chord moment are given in terms of the complex Theodorsen function given as

$$C\left(k\right)=\frac{H_1^{(2)}\left(k\right)}{H_1^{(2)}\left(k\right)+iH_0^{(2)}\left(k\right)}=\frac{K_1\left(ik\right)}{K_0\left(ik\right)+K_1\left(k\right)}=F\left(k\right)+iG\left(k\right) \tag{15}$$

where $H_n^{(2)}$ are nth-order Hankel functions of the second kind and K_n are n^{th}-order Bessel functions of the second kind. The Theodorsen function attenuates the amplitude of the quasi-static lift and introduces a phase shift relative to the body motion, as shown in Figure 6. Although these Hankel and Bessel functions may seem quite complicated, computational engineering software such as Mathematica and MATLAB makes computations with these functions as simple as evaluation of trigonometric functions.

The expressions for lift and quarter-chord moment are

$$L=\pi\rho_\infty b^2\left[\ddot{z}+U\dot{\theta}-ba\ddot{\theta}\right]+2\pi\rho_\infty UbC\left(k\right)\left[\dot{z}+U\theta+b\left(\frac{1}{2}-a\right)\dot{\theta}\right] \tag{16}$$

Figure 6. The complex Theodorsen function (a) serves to attenuate the quasi-static lift and introduce a phase shift, (b) consists of real and imaginary components C(k) = F(k) + iG(k), and (c) can be approximated algebraically
(a) and (b) from Bisplinghoff et al. (2013) and (c) from Scanlan (2002).

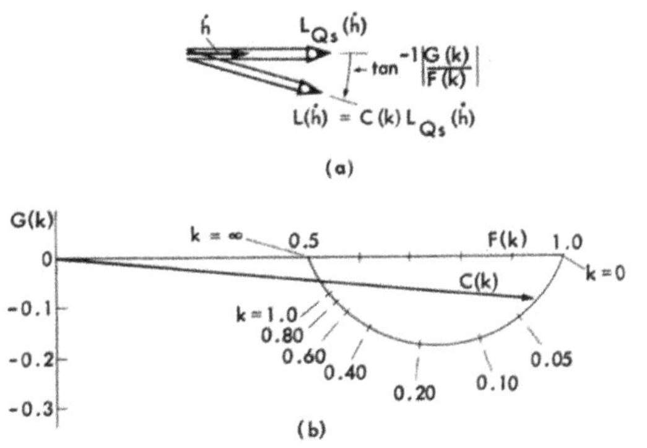

(c) If we let a = 0.165, b = 0.0455, c = 0.335, and d = 0.300, then the real and imaginary parts of the Theodorsen function can be approximated as

$$F(k)=1-a-c+\frac{ab^2}{b^2+k^2}+\frac{cd^2}{d^2+k^2}$$

$$G(k)=-k\left[\frac{ab}{b^2+k^2}+\frac{cd}{d^2+k^2}\right]$$

$$M = \pi\rho_\infty b^2\left[ba\ddot{z} + Ub\left(\frac{1}{2} - a\right)\dot{\theta} - b^2\left(\frac{1}{8} + a^2\right)\ddot{\theta}\right] + 2\pi\rho_\infty Ub^2\left(\frac{1}{2} + a\right)C\left(k\right)\left[\dot{z} + U\theta + b\left(\frac{1}{2} - a\right)\dot{\theta}\right] \quad (17)$$

Substituting the expressions for oscillatory motion in plunge and torsion, Equation 4, and the Theodorsen function, Equation 15, into Equations 16 and 17 leads to

$$L = \rho_\infty bU^2\left[\left(L_{Th} + ikL_{T\dot{h}}\right)\frac{h_o}{b} + \left(L_{T\theta} + ikL_{T\dot{\theta}}\right)\theta_o\right]e^{i\omega t} \quad (18)$$

$$M = \rho_\infty b^2U^2\left[\left(M_{Th} + ikM_{T\dot{h}}\right)\frac{h_o}{b} + \left(M_{T\theta} + ikM_{T\dot{\theta}}\right)\theta_o\right]e^{i\omega t} \quad (19)$$

where the non-dimensional *oscillatory aerodynamic derivatives* of the Theodorsen function are given by

$$L_{Th} = 2\pi\left(\frac{-k^2}{2} - Gk\right) \quad (20)$$

$$L_{T\dot{h}} = 2\pi F \quad (21)$$

$$L_{T\theta} = 2\pi\left[\frac{k^2 a}{2} + F - Gk\left(\frac{1}{2} - a\right)\right] \quad (22)$$

$$L_{T\dot{\theta}} = 2\pi\left[\frac{1}{2} + F\left(\frac{1}{2} - a\right) + \frac{G}{k}\right] \quad (23)$$

$$M_{Th} = 2\pi\left[-\frac{k^2 a}{2} - k\left(\frac{1}{2} - a\right)G\right] \quad (24)$$

$$M_{T\dot{h}} = 2\pi\left(\frac{1}{2} + a\right)F \quad (25)$$

$$M_{T\theta} = 2\pi\left[\frac{k^2}{2}\left(\frac{1}{8}-a^2\right)+F\left(\frac{1}{2}+a\right)-Gk\left(\frac{1}{2}-a\right)\left(\frac{1}{2}+a\right)\right] \tag{26}$$

$$M_{T\dot{\theta}} = 2\pi\left[-\frac{k}{2}\left(\frac{1}{2}-a\right)+kF\left(\frac{1}{2}+a\right)\left(\frac{1}{2}-a\right)-\frac{G}{k}\left(\frac{1}{2}+a\right)\right] \tag{27}$$

The parameter a in Equations 20 to 27 is the non-dimensional distance from mid-chord to the elastic axis, with positive values aft of the mid-chord position, as shown in Figure 5. Inserting the non-dimensional flutter derivatives from Theodorsen into the lift and moment equations, Equations 18 and 19, it is possible to manipulate these expressions into the complex amplitude form given in Equations 9 and 10 and follow through the analysis from Equations 11 to 14 to the solution. The details of the solution are not trivial, but are rigorously explained in the cited aeroelastic texts.

Characteristics of Coupled-Mode Airfoil Solutions

The relevance of coupled mode airfoil flutter to self-excited coupled-mode gate vibration can most easily be understood by looking at plots of frequency ratio and damping as a function of air speed. Two such plots are shown in Figure 7. In Figure 7a, the torsional frequency is about five times the plunge frequency (labeled as the bending frequency) with no flow. With increasing reduced velocity, the torsional frequency decreases while the plunge frequency remains almost constant until just before the onset of flutter. The onset of flutter is when one of the two damping ratios goes to zero. The two frequencies approach each other as flutter begins, but do not merge at the flutter onset velocity. This frequency merging behavior is often identified as "coalescence flutter." Note the damping ratio for the torsional mode goes to zero, but the damping ratio for the plunge (bending mode) increases substantially as the reduced frequency increases to the onset of flutter. This damping ratio behavior might lead us to expect increasing torsional amplitudes with constant or decreasing heaving amplitudes.

In Figure 7b the torsional frequency is approximately 10% higher than the plunge (bending) frequency with no flow. In this case, the plunge (bending) frequency initially decreases slightly with increasing reduced velocity before beginning to increase towards the torsional frequency that decreases monotonically with increasing reduced velocity. Note that despite the slight difference in the behavior of the plunge mode, the coalescence of frequencies observed in Figure 7a is also present in Figure 7b. A more substantial difference is in the damping ratio curves in Figure 7b. The behavior of the two damping ratios is opposite that found Figure 7a: the bending mode damping ratio goes to zero and the torsional mode damping ratio increases. Thus, we might expect large amplitude plunge vibration with steady or decreasing amplitude torsional vibration.

With respect to amplitude expectations, we must realize, however, that the flutter analysis reviewed in this chapter is for small amplitude vibrations, using linearized equations of motion, so the comments about amplitude must be regarded as speculative.

An example of the motion of an airfoil in plunge-torsion coupled-mode flutter, also referred to as classical flutter, is given in Figure 8. The flutter of a wing on an aircraft and of the whole aircraft are complicated by the need for three-dimensional analysis and many possible modes of vibration.

Figure 7. Effect of frequency ratio σ on flutter solution with all other parameters held constant for (a) frequency ratio σ = 0.227, and (b) frequency ratio σ = 0.909. Other parameters are x_θ = 0.024, r_l = 0.62, μ_m = 38 and a = -0.36
From Bisplinghoff & Ashley, 1996.

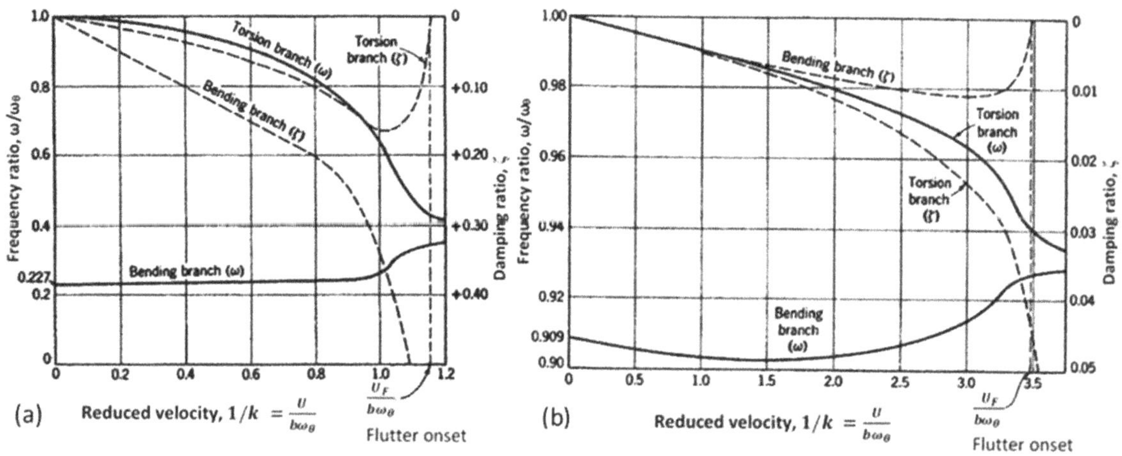

A fundamental assumption in classical flutter analysis is that the vibration amplitudes are small. For the torsional modes this limitation means the airfoil does not reach its stall angle where the flow separates from the leading edge with a sudden decrease in lift. When the angle of attack is sufficiently close to the stall angle, a small angular oscillation will be accompanied by a relatively large change in the lift as the airfoil stalls. When stall occurs a type of flutter known as stall flutter, a single degree-of-freedom torsional instability, can occur. Dowell et al. (2015) note that torsional single degree-of-freedom bridge deck instability is akin to stall flutter of airfoils.

A characteristic of stall flutter is that it can appear as a spontaneous vibration with infinitesimally small soft-excitation, or it may appear as an instability requiring a larger value of the controlling system parameter. Figure 9 portrays these two types of stall flutter. Note that once a system requiring hard excitation begins vibrating, the vibrations will continue with decreasing values of the parameter below the onset value for hard excitation, generating a characteristic hysteresis loop. Similar behavior has been noted in the excitation of gate vibrations, as well.

Figure 8. Example of coupled mode flutter trajectories in which the torsional mode leads the plunge mode by 90°
From Wardlaw, 1994. © Springer 1994. Used with permission.

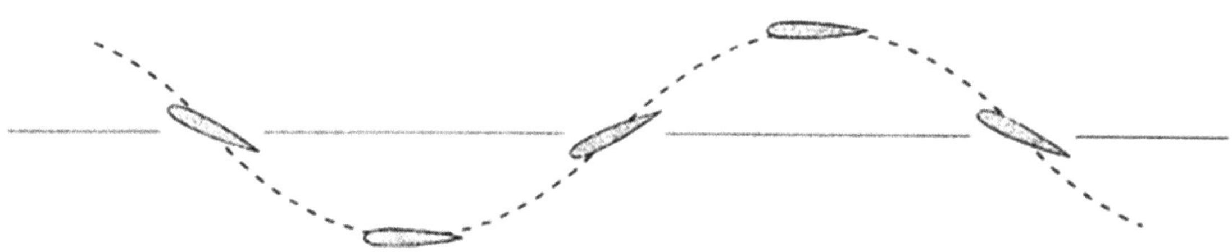

Figure 9. Stall flutter can be initiated by both soft excitation and hard excitation, with loops. The parameter α_{ss} is one known to participate in the instability
From Dowell et al., 2015. © 2015 Springer. Used with permission.

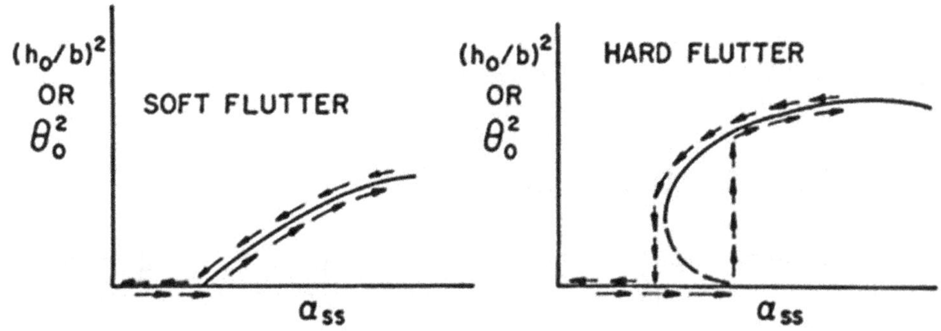

BLUFF BODY FLUTTER ANALYSIS

Early attempts to apply the Theodorsen theory for unsteady aerodynamics to bridge sections were undertaken by Bleich (1948), Bleich et al. (1951) and Selberg (1957, 1965). Bleich assumed the road deck could be modeled as a flat plate and then developed correction methods to account for the leading edge effects. Selberg (1957) based his model on the work of Bleich and developed from it a relatively simple empirical correlation for the flutter onset velocity:

$$\frac{U_F}{B\left(\frac{\omega_\theta}{2\pi}\right)} = 3.7\left[\left(1-\frac{\omega_h}{\omega_\theta}\right)^2 \frac{mr}{\rho B^3}\right]^{1/2} \tag{28}$$

where U_F is the flat plate flutter onset velocity, B is the bridge width, ω_θ if the torsional frequency, ω_h is the heaving (plunge)frequency, r is the bridge radius of gyration, and m is the bridge deck mass per unit length.

The bluff leading edge of H-section bridge decks clearly violated the assumption of no flow separation from the surface in Theodorsen's analysis. In addition to this obvious difference, Wardlaw (1994) enumerates additional differences between bridge flutter and airfoil flutter:

- The center of gravity for bridges is typically on the center line due to symmetry, while for the airfoils the center of gravity is typically ahead of the mid-chord. Inertial coupling that is inevitably a factor in airfoil flutter is seldom a consideration in bridge flutter.
- The aerodynamic moment due to the lift force acting near the quarter-chord point (in both systems) is a greater factor for bridges due to the greater distance between the quarter-chord and the elastic axis.

- Bridges are not positively damped for pure torsional motion and can undergo single degree-of-freedom torsional instability that is not usually possible for airfoils due to their positive damping in torsion.

Wardlaw (1994) provides a concise summary of the concepts for bridge flutter and torsional instability developed by Scanlan, his students, and collaborators. Scanlan (2002; 2015) adds his own perspective to the development of methods for bridge flutter analysis.

Formulation of Bridge Flutter Analysis

Informed by linear, first-order sectional lift and moment expressions for airfoils, as Scanlan (2002) recalls, he initially formulated analogous bluff body sectional lift and moment expressions as linear, first- order equations, given as follows:

$$L = \frac{1}{2}\rho U^2 B\left(a_1\dot{h} + a_2\dot{\theta} + a_3\theta\right) \tag{29}$$

$$M = \frac{1}{2}\rho U^2 B^2\left(b_1\dot{h} + b_2\dot{\theta} + ab_3\theta\right) \tag{30}$$

where ρ = air density; U = cross-wind velocity; B = bridge width in the wind direction; a_i and b_i are appropriate constants; and h and α are vertical and torsional coordinates, respectively. In the formulation of Equations 29 and 30, any influence of the h value on lift and moment and any streamwise deflection of the bridge and its influence on lift and moment were neglected.

Scanlan & Tomko (1971) reworked these equations into non-dimensional form. Later a fourth term accounting for the effect of bridge deflection h on the lift and moment was included:

$$L = \frac{1}{2}\rho U^2 B\left(KH_1^*(K)\frac{\dot{h}}{U} + KH_2^*(K)\frac{B\dot{\theta}}{U} + K^2H_3^*(K)\theta + K^2H_4^*(K)\frac{h}{B}\right) \tag{31}$$

$$M = \frac{1}{2}\rho U^2 B^2\left(KA_1^*(K)\frac{\dot{h}}{U} + KA_2^*(K)\frac{B\dot{\theta}}{U} + K^2A_3^*(/UK)\theta + K^2A_4^*(K)\frac{h}{B}\right) \tag{32}$$

The non-dimensional frequency $K = \omega B / U = 2\pi NB/U$ (for N in Hz) is similar to Theodorsens's k, differing by a factor of two, that is, $K = 2k$. The eight coefficients H_i^* and A_i^*, known as flutter derivatives (also called Scanlan coefficients), must be evaluated experimentally for all but the case of a flat plate bridge deck.

Later these equations were extended to include fluctuating drag and deflection in the wind direction and the effects of this streamwise motion on the lift, drag and aerodynamic moment. These equations include 18 flutter derivatives. See Singh et al. (1994) or Ge & Tanaka (2013) for further details on the three-dimensional extension of the bluff body flutter formulation.

Comparing Equations 32 and 33 with the corresponding Theodorsen function formulations, Equations 18 and 19 along with the accompanying Equations 20 to 27, Simiu & Scanlan (1996) provide equivalent expressions for the Scanlan flutter derivatives for a flat-plate bridge deck (with the center of gravity at the chord-wise location of the elastic axis) in terms of the real and imaginary parts of the Theodorsen function: $K^2H_1^* = -2\pi F\left(K\right)$

$$K^2H_2^* = \frac{-\pi K}{2}\left[1 + \frac{4G\left(K\right)}{K} + F\left(K\right)\right]$$

$$K^2H_3^* = -\pi\left[2F\left(K\right) - \frac{KG\left(K\right)}{2}\right]$$

$$K^2H_4^* = \frac{\pi}{2}K^2\left[1 + \frac{4G\left(K\right)}{K}\right]$$

$$K^2A_1^* = -\frac{\pi}{2}KF\left(K\right) \tag{33a}$$

$$K^2A_2^* = \frac{-\pi}{2}\left[\frac{K}{4} - G\left(K\right) - \frac{KF\left(K\right)}{4}\right] \tag{33b}$$

$$K^2A_3^* = \frac{\pi}{2}\left[\frac{K^2}{32} + F\left(K\right) - \frac{KG\left(K\right)}{4}\right] \tag{33c}$$

$$K^2A_4^* = \frac{-\pi}{2}\left[KG\left(K\right)\right] \tag{33d}$$

Scanlan & Tomko (1971) measured values of H_i and A_i for an NACA 0012 airfoil in a wind tunnel and showed acceptable agreement between their measured data and Theodorsen's theoretical equations. They also measured the flutter derivatives for nine bridge cross-sections, some with slight modifications, and provided graphs of curves fitted to their empirical data along with the curve fit for their measured airfoil data. Figure 10 contains data for the airfoil, the Tacoma Narrows model (bridge deck 1), and three other truss-stiffened bridge decks.

Figure 10. Scanlan & Tomko (1971) curve fits to measured flutter derivatives for bridge decks compared with corresponding airfoil values. The parameter on the abscissa is reduced velocity, given as $2\pi/K = U / [B(\omega/2\pi)]$. A_i are for torsional mode and H_i heaving mode
After Wardlaw, (1994). © Springer 1994. Used with permission.

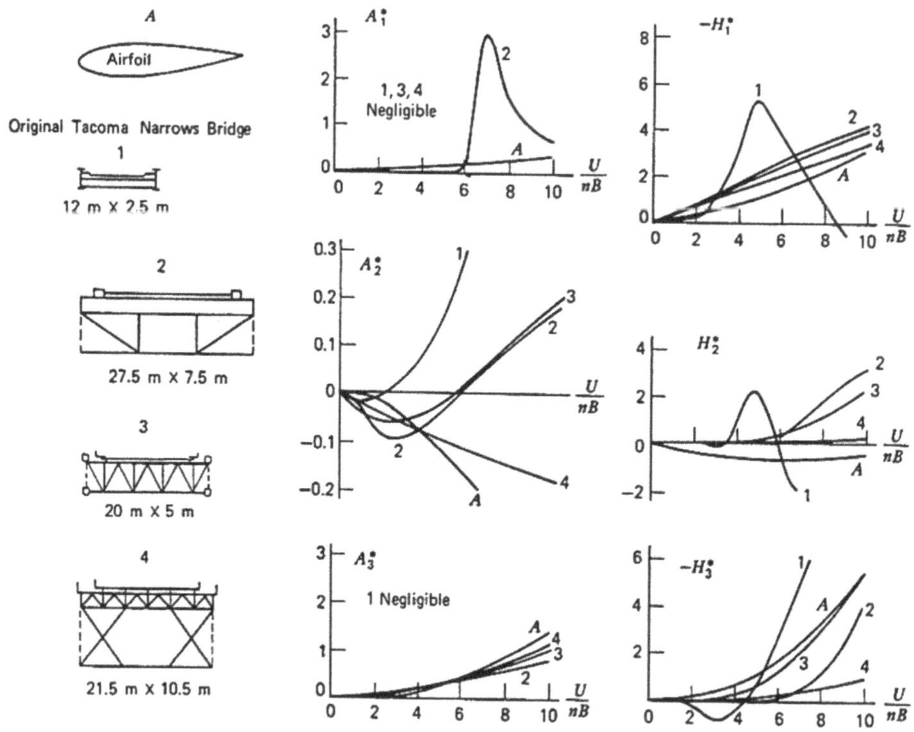

In Figure 10 the parameter on the abscissa is a reduced velocity given as $2\pi/K = U / [B(\omega/2\pi)]$ using the notation in this chapter. Scanlan & Tomko used V (same as our U) for the freestream wind velocity and n for the frequency in Hz (same as our $\omega/2\pi$).

The behavior of the A_2^* and H_2^* flutter coefficients is of particular interest with regard to damping. First of all, there is a significant difference in the behavior of the coefficients for bridge decks and the airfoil. Secondly, the A_2^* data indicate that for bridge decks 1 (Tacoma Narrows model), 2, and 3, the coefficients change from negative to positive values with increasing reduced velocity, corresponding to changes from positive to negative damping. These three bridge sections are susceptible to single degree-of-freedom torsional instability since A_2^* is the coefficient on $\dot{\theta}$ in Equation 32 that, when taken to the other side, contributes negative damping for positive values of A_2^*. The most unstable of the sections tested was the Tacoma Narrows model. Note the similarity between the Tacoma Narrows A_2^* curve and the measured log decrement curve given in Figure 4. On the other hand, beyond a reduced velocity of 6, the H_2^* coefficients for the Tacoma Narrows deck become negative. This result is consistent with the observation that the vertical or heaving mode of the bridge did not cause the structure to fail.

Two of the plots in Figure 10 show the negative of the flutter coefficients, namely, $-H_1^*$ and $-H_3^*$. In these plots, positive values of H_1^* indicate negative damping and possible instability in the heaving mode with no torsional coupling. The airfoil and all bridge decks, except the Tacoma Narrows model, show negative values of H_1^* indicating positive aerodynamic damping. The Tacoma Narrows model shows a change from negative to positive values only for a much higher reduced velocity than that for a similar change in the uncoupled torsional damping characteristic.

Values for A_4^* and H_4^* were not reported. Freestream turbulence, buffeting and streamwise motion, and possible Reynolds number effects must also be considered in an analysis of flutter, but will not be included in this chapter. We refer interested readers to Simiu & Scanlan (1994), Holmes (2015), or Ge and Tanaka (2013).

Simiu & Scanlan (1994) also present a method for solving the equations of motion to determine the flutter onset velocity using measured flutter derivatives from wind tunnel testing. Scanlan (2002) comments that despite the limit on small amplitude vibration, this bluff body flutter formulation has been used with success. He writes, "In practice, however, if the relevant structural displacements remain small, as with low-amplitude vibrations, most studies involving flutter derivatives yield reasonably accurate results for depicting overall system dynamics."

Juxtapose this explanation of small amplitude of about 3° with the failure of the Tacoma Narrows with torsional amplitudes of about 45°. We cannot use the Scanlan & Tomko analysis for large amplitude vibration. Clearly, nonlinear effects played a role, but methods of nonlinear analysis are still being developed, and ultimately may help clarify whether the failure of the Tacoma Narrows Bridge was a single degree-of-freedom torsional instability, or whether the heaving mode vibrations served to initiate the torsional mode and continued to contribute to the energy flow into the bridge motion, albeit at substantially smaller amplitudes than the torsional amplitude.

One Degree-of Freedom Torsional Instability

For a one degree-of-freedom torsional instability of a bridge deck with no heaving, using Equation 32 for the moment, and assuming the flutter derivatives are known, Equation 2 can be written as.

$$\ddot{\theta} + 2\zeta_\theta \omega_\theta \dot{\theta} + \omega_\theta^2 \theta = \frac{\rho U^2 B^2}{I}\left(KA_2^*\left(K\right)\frac{B\dot{\theta}}{U} + K^2 A_3^*\left(K\right)\theta\right) \tag{34}$$

Noting that $U^2K^2 = \omega^2 B^2$ in the coefficient term in front of $A_3^*\left(K\right)$, and taking the terms on the right side to the left we can write

$$\ddot{\theta} + 2\omega_\theta\left(\zeta_\theta - \frac{1}{2}\frac{\rho U B^3 K A_2^*\left(K\right)}{I\omega_\theta}\right)\dot{\theta} + \left(\omega_\theta^2 - \frac{\rho\omega^2 B^4}{I}A_3^*\left(K\right)\right)\theta = 0 \tag{35}$$

From Equation 35 we can conclude that when the net system damping, that is, the coefficient on the first derivative term, becomes negative, instability will occur. Sufficiently large positive values of $A_2^*\left(K\right)$ will produce negative damping. The needed minimum magnitude of $A_2^*\left(K\right)$ will be

$$A_2^*\left(K\right) \geq \frac{2\zeta_\theta I\omega_\theta}{\rho UB^3K} \tag{36}$$

The onset velocity must be found iteratively. By selecting a value of K, the values of A_2^* and A_3^* can be found from the assumed tabulated data. The value of ω can be found by setting the coefficient in front of θ equal to ω^2. With known K and ω values, the value of the velocity is determined from $U = B\omega/K$. The iteration on K continues until the condition in Equation 36 is met.

COMPARISON OF SELF-EXCITED COUPLED-MODE GATE INSTABILITY AND TWO-DIMENSIONAL BRIDGE FLUTTER

In Chapter 10, we described a mechanism for the self-excited coupled-mode dynamic instability of Tainter gates, which is operative even in the absence of any gate eccentricity. The self-excitation mechanism for this dynamic instability was presented in Figure 10 in Chapter 10, but is shown again here as Figure 11 for the reader's convenience.

The counter-clockwise bending vibration ① of the skinplate, represented by ψ, triggered by some random excitation, generates a counter-clockwise skinplate inertia torque ② around the trunnion pin. The unbalanced moment of inertia produces counter-clockwise rigid-body gate rotation ③, represented by θ. As a result, the gate opening under the gate increases, resulting in increased discharge ④ with a corresponding decrease in the flow-rate-variation pressure proportional to rigid body angular velocity,

Figure 11. Mechanism of self-excited coupled-mode Tainter gate instability
From Anami & Ishii (2003), © 2003 ASME. Used with permission.

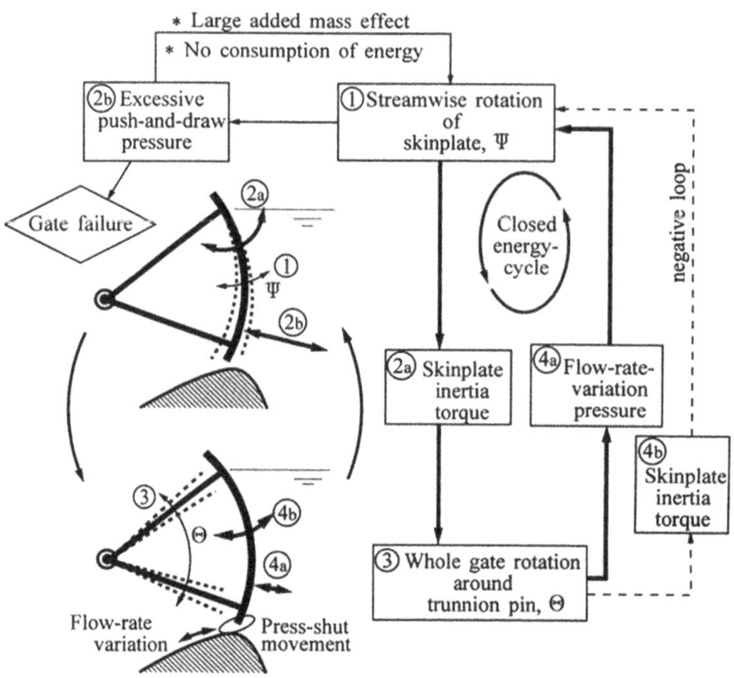

$\dot{\theta}$, of the gate. The decreased flow-rate-variation pressure feeds energy back to the bending vibration of the skinplate ①.

This closed-energy cycle leading the Tainter gate to a dangerous spontaneous self-excited vibration involves two degrees-of-freedom: the skinplate bending rotational vibration and the rigid body gate vibration around the trunnion pin. Neither mode, if existing without the other, would lead to instability. The coupled-mode dynamic instability of Tainter gates is a "two degrees-of-freedom dynamic instability," which in many ways is fundamentally similar to the "two degrees-of-freedom coupled-mode flutter" to which H-section bridge decks are particularly susceptible. As discussed in this chapter, the mechanism that lead to the destruction of the Tacoma Narrows Bridge in 1940 may be an example of coupled-mode flutter, while others consider it a single degree-of-freedom torsional instability.

The closed-energy cycle for the self-excitation mechanism of the hydraulic Tainter gate structure is compared with the two degrees-of-freedom bridge instability in air in Figure 12, where the bridge is represented by an H-section. The clockwise torsional vibration ① of the bridge is represented by ϕ, corresponding to the skinplate bending rotational vibration Ψ of the Tainter gate. If the torsional vibration ϕ is triggered by a disturbance, the angle of attack of the uniform air flow increases, and produces lift ② on the bridge girder, which corresponds to the skinplate inertia torque. The lift drives the heaving vibrations ③, represented by η, which corresponds to the whole gate vibration velocity $\dot{\theta}$. As a result, the flow over the bridge is significantly accelerated, as shown by ④, corresponding to the increased water flow under the gate. This locally accelerated flow is further enhanced by the torsional vibration of the bridge deck. (In the hydraulic gate, the discharge water flow is further increased, if the gate is a geometrical press-shut device.) As a result of the bridge torsional motion, the intense vortex generated behind the upstream bridge flange produces a strong force due to the accelerating vortex flow ⑤, corresponding to the flow-rate-variation pressure for the hydraulic gate. The vortex flow feeds energy back into the bridge torsional vibrations ①. Thus, all the significant components in two degrees-of-freedom flutter of bridge decks correspond to those in the self-excited coupled-mode dynamic instability of a hydraulic Tainter gate.

Eyewitness accounts and film records document that large amplitude vibrations were needed to destroy the Tacoma Narrows Bridge deck. Torsional mode vibration started about 10 a.m. on 7 November 1940. Complete failure of the center span occurred by 11:10. The torsional vibration was initially at about 14 cycles/min (0.233 Hz), but quickly transitioned to 12 cycles/min (0.20 Hz), with a maximum amplitude of about 35° to 45°. Assuming that the torsional vibration frequency was at 12 cycles/minute and that the bridge collapsed after approximately 60 to 70 minutes, we can calculate that about 780 (±60) cycles were needed for bridge failure.

On the other hand, no crew was on hand to film the Folsom Dam gate failure. Eyewitness reports indicate that the gate failed about 15 seconds after the noticeable onset of vibrations at about 6.5 Hz (Ishii et al., 2014). Based on the available data, it was calculated that about 100 cycles with an estimated maximum centerline skinplate deflection of about 0.347 in (11.9 mm) were needed for gate failure (Anami et al., 2005). Thus, compared to the bridge, the gate required smaller amplitude vibrations over a smaller number of cycles for failure.

In general, fluid dynamic forces scale with the density of the fluid, the square of the velocity, and the surface area over which the pressure acts. The ratio of standard water density to standard air density is 832. The estimated water velocity of the Folsom Dam was about 38 ft/sec (11.58 m/s) (with an assumed discharge coefficient of 0.72), while the wind velocity at the Tacoma Narrows is often cited as 42 mph (or 61.6 ft/s) at the time of failure, although it may have been a bit higher (Ammann et al., 1940; Plaut, 2008). Thus, the ratio of the dynamic pressure at the Folsom Dam to that of the Tacoma Narrows was about 318. To estimate the ratio of steady fluid forces, characteristic areas must be selected. For the Folsom Dam gate, the area will be chosen as the depth of the water above the gate bottom (43.6 ft or 13.29 m) times the span of the gate (42.0 ft or 12.80 m). For the Tacoma Narrows Bridge, the representative area will be taken as the main span (2800 ft or 853.44 m) times the girder depth (8 ft or 2.44 m), although the wind also acted on the cables and other bridge surfaces as well. Thus, the steady fluid forces on the Folsom Dam gate were, very crudely estimated, 30 times greater than the steady fluid loading on the windward face of the bridge. This estimate cannot be regarded as more than an order of magnitude approximation. These steady forces do not drive vibration, but the ratio is indicative of the order of the maximum fluctuating force on the gate relative to the maximum fluctuating force on the bridge assuming a similar order of magnitude of the fluctuating force to mean force ratios on the two structures. Although very crude, this ratio of steady forces helps make sense of the order of magnitude difference in the number of cycles of vibration from onset to failure for these two structures. The hydrodynamic pressure caused by the water-added mass, the so-called "push-and-draw pressure," induced by the skinplate streamwise bending vibration, is very large; large enough to destroy the much heavier gate structure with just small amplitude displacements.

In the mechanisms for both coupled-mode flutter and coupled-mode gate instability, the frequencies of the two involved modes approach a common value at the onset of the vibration, though in neither mechanism do they become exactly equal. The water system is more strongly dependent upon added-mass effects, but in both systems negative damping serves to excite the structure.

In addition, the coupled-mode dynamic gate instability has been shown to possess both soft and hard excitation characteristics (see Anami et al., 2005). The characteristics for gates are much like those for stall flutter of an airfoil (see Figure 9). The mechanism usually requires a triggering displacement that

Figure 12. Similarities between the two degrees-of-freedom self-excitation mechanisms proposed for the failure of the Folsom Dam Tainter gate and the Tacoma Narrows Bridge
From Ishii, et al. (2010). © 2010 Techno Press. Used with permission.

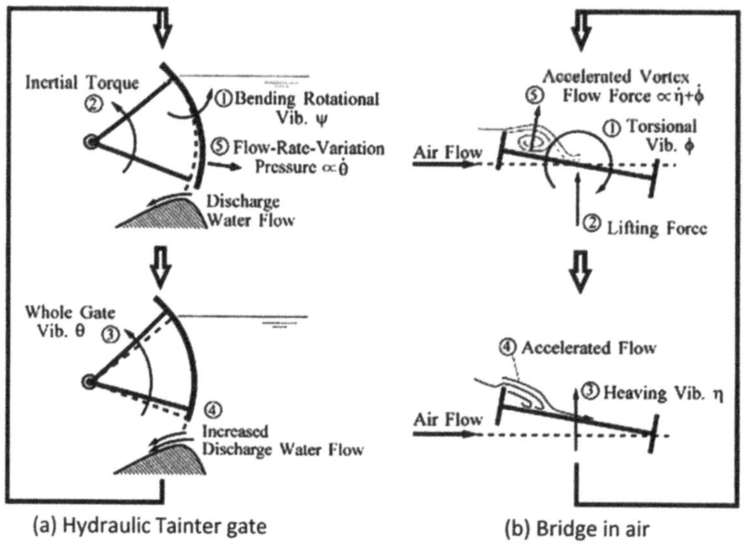

is sufficiently large to overcome the mechanical damping. The source of this triggering displacement may be as apparently benign as corrosion on the trunnion pin or as overtly dangerous as an earthquake. In any event, the resultant vibrations can be avoided through proper design/retrofitting of the skinplate to give it sufficient rigidity so that the frequency of the in-water bending mode of the skinplate cannot become coincident with the rigid body rotational frequency of the gate.

CONCLUSION

More than 70 years after the Tacoma Narrows Bridge collapse, there is still no universal agreement on the mechanism of failure. In their original report, Ammann et al. (1941) attributed the failure to self-excited torsional instability and ruled out resonant von Kármán vortex shedding. This conclusion was supported by Farquharson's wind tunnel studies (1941) and models of single degree-of-freedom instability (Scanlan & Tomko, 1971). Gazzola (2015), however, suggests that Scanlan & Tomko's analysis was for linearized equations of motion and thus cannot be used to form conclusions about large amplitude bridge vibration. Furthermore, Gazzola notes that careful study of the film of the collapse shows that the bridge is still undergoing heaving motion while large amplitude torsional vibration were occurring. Thus there is uncertainty whether the vibration was truly a single degree-of-freedom torsional instability, or whether it was a coupled-mode heaving-torsional vibration.

Billah & Scanlan (1991) clearly state that the mechanisms of bluff-body flutter are not the same as airfoil flutter. For example, the single-mode plunge (heaving) flutter and single-mode torsional flutter seen in bluff bodies do not occur for most wings and airfoils (see Simiu and Scanlan, 1996). Other than stall flutter, aircraft wing flutter occurs as a classical coupled-mode vibration with input from a vertical (plunge) mode and a torsional mode, each mode serving to excite the other mode. Bridge decks can undergo torsional flutter without coupling to a vertical mode. Bridge decks, however, can also undergo coupled-mode flutter with coupling between vertical and torsional modes.

Engineers in the field of aerodynamics have been aware of the importance of the interaction between airflow and elastic deformation of the aircraft for more than a century. However, it took the Tacoma Narrows Bridge disaster for engineers to begin applying aerodynamic principles to bluff bodies such as bridges. Some sources suggest that the two degrees-of-freedom flutter of H-section bridge decks was the cause of the Tacoma Narrows Bridge failure. For this reason, we spend more than half of this chapter showing how coupled-mode airfoil flutter is relevant to the coupling between the heaving and torsional modes of the bridge.

Furthermore, we suggest that the self-excitation mechanism for the dynamic instability of a massive hydraulic Tainter gate is analogous to the two degrees-of-freedom flutter of H-section bridge decks. In other words, the mechanism that led to the failure of the Tacoma Narrows Bridge in 1940 may also account for the failure of the Folsom Dam Tainter gate in 1995. Both failures involved structures that were exposed to naturally-occurring fluid flow, and both structures were constructed according to static design criteria. A significant difference in the two coupled-mode systems is the fluid density. The water density for the Folsom Dam gate was about 1000 times larger than that of air; the fluid forces were correspondingly larger and failure occurred much more quickly than for the coupled-mode Tacoma Narrows bridge vibration in air. There was no time for photographic documentation, making the analysis of the Folsom Dam gate failure even more complicated than that for the Tacoma Narrows bridge failure.

The dynamic instability of hydraulic gates needs to be recognized and precautions taken to prevent future gate failures by this mechanism. Just as an understanding of the mechanism of self-excited vibrations of bridge decks spawned the requirement for including aerodynamic stability analyses in bridge designs, we expect that hydrodynamic stability analysis will become a requirement in the design of hydraulic gates. The mechanisms of coupled-mode gate instability need to be incorporated into extended design criteria. It is in the interest of society for gate designers to address not just static, but dynamic, gate stability before another failure occurs.

REFERENCES

Ammann, O.H., von Kármán, T., & Woodruff, G.B. (1941). *The failure of the Tacoma Narrows Bridge: A report to the Honorable John M. Carmody, Administrator, Federal Works Agency, Washington, D.C.* Report by the Board of Engineers.

Anami, K., & Ishii, N. (2003). Model tests for non-eccentricity dynamic instability closely related to Folsom Dam Tainter gate failure. In *ASME 2003 Pressure Vessels and Piping Conference* (pp. 213-222). New York, NY: American Society of Mechanical Engineers. doi:10.1115/PVP2003-2092

Anami, K., Ishii, N., Knisely, C. W., & Oku, T. (2005). Hydrodynamic pressure load on Folsom Dam Tainter gate at onset of failure due to flow-induced vibrations. In *ASME 2005 Pressure Vessels and Piping Conference* (pp. 557-564). New York, NY: American Society of Mechanical Engineers. doi:10.1115/PVP2005-71444

Arioli, G., & Gazzola, F. (2015). A new mathematical explanation of what triggered the catastrophic torsional mode of the Tacoma Narrows Bridge. *Applied Mathematical Modelling*, *39*(2), 901–912. doi:10.1016/j.apm.2014.06.022

Billah, K. Y., & Scanlan, R. H. (1991). Resonance, Tacoma Narrows Bridge failure, and undergraduate physics textbooks. *American Journal of Physics*, *59*(2), 118–124. doi:10.1119/1.16590

Bisplinghoff, R. L., & Ashley, H. (1996). *Principles of Aeroelasticity*. Mineola, NY: Dover Publications.

Bisplinghoff, R. L., Ashley, H., & Halfman, R. L. (2013). *Aeroelasticity*. Mineola, NY: Dover Publications.

Bleich, F. (1948). Dynamic instability of truss-stiffened suspension bridges under wind action. *Proceedings of the American Society of Civil Engineers*, *74*(8), 1269–1314.

Bleich, F., McCullough, C. B., Rosecrons, R., & Vincent, G. S. (1950). *The Mathematical Theory of Vibration in Suspension Bridge*. Washington, DC: United States Government Printing Office.

Blevins, R. D. (1994). *Flow-Induced Vibrations* (2nd ed.). Malabar, FL: Krieger.

Dowell, E. H. (Ed.). (2015). *A Modern Course in Aeroelasticity* (5th ed.). Cham, Switzerland: Springer. doi:10.1007/978-3-319-09453-3

Elliot, B. (1940). *Color Film of Bridge Failure*. The Camera Shop Tacoma.

Farquharson, F. B. (1941). Dynamic tests on bridge models. *Engineering News Record*, *127*(1), 37.

Fung, Y. C. (1998). *An Introduction to the Theory of Aeroelasticity*. Mineola, NY: Dover Publications.

Garrick, I. E., & Reed, W. H. III. (1981). Historical development of aircraft flutter. *Journal of Aircraft, 18*(11), 897–912. doi:10.2514/3.57579

Gazzola, F. (2015). *Mathematical Models of Suspension Bridges -- Nonlinear Structural Instability*. Berlin, Germany: Springer. doi:10.1007/978-3-319-15434-3

Ge, Y., & Tanaka, H. (2013). Long-span bridge aerodynamics. In Y. Tamura & A. Kareem (Eds.), *Advanced Structural Wind Engineering*. Berlin, Germany: Springer. doi:10.1007/978-4-431-54337-4_4

Hodges, D. H., & Pierce, G. A. (2014). Introduction to Structural Dynamics and Aeroelasticity (2nd ed.). Cambridge, UK: Cambridge University Press.

Holmes, J. D. (2015). *Wind Loading of Structures* (3rd ed.). Boca Raton, FL: CRC Press.

Ishii, N., Anami, K., & Knisely, C. W. (2014). Retrospective consideration of a plausible vibration mechanism for the failure of the Folsom Dam Tainter gate. *International Journal of Mechanical Engineering and Robotics Research, 3*(4), 314–345.

Ishii, N., Anami, K., Knisely, C. W., Tsuji, T., & Oku, T. (2010). Analogy between the coupled-mode vibration failure of the Folsom Dam Tainter gate and the Tacoma Narrows Bridge failure. In C. K. Choi, & Y. B. Yang (Eds.), *Proceedings of the First International Conference on Advances in Multiscale Mechanics (AIMMl10)* (pp. 629-641). Jeju, South Korea: Techno Press.

Kawada, T. (2010). *History of the Modern Suspension Bridge*. Reston, VA: American Society of Civil Engineers. doi:10.1061/9780784410189

McKenna, P. J. (1999). Large torsional oscillations in suspension bridges revisited: Fixing an old approximation. *The American Mathematical Monthly, 106*(1), 1–18. doi:10.2307/2589581

McKenna, P. J., & Tuama, Ó. (2001). Large torsional oscillations in suspension bridges visited again: Vertical forcing creates torsional response. *The American Mathematical Monthly, 108*(8), 738–745. doi:10.2307/2695617

Pasternack, A. (2015). The strangest, most spectacular bridge collapse (and how we got it wrong). *Vice Magazine*. Accessed 23 May 2016 at motherboard.vice.com/read/the-myth-of-galloping-gertie

Plaut, R. H. (2008). Snap loads and torsional oscillations of the original Tacoma Narrows Bridge. *Journal of Sound and Vibration, 309*(3-5), 613–636. doi:10.1016/j.jsv.2007.07.057

Scanlan, R. H. (2002). Observations on low-speed aeroelasticity. *Journal of Engineering Mechanics, 128*(12), 1254–1258. doi:10.1061/(ASCE)0733-9399(2002)128:12(1254)

Scanlan, R. H. (2015). Aeroelastic problems of civil engineering structures. In E. H. Dowell (Ed.), *A Modern Course in Aeroelasticity* (5th ed.; pp. 285–347). Cham, Switzerland: Springer.

Scanlan, R. H., & Rosenbaum, R. (1951). Aircraft Vibration and Flutter. New York, NY: Macmillan.

Scanlan, R. H., & Sabzevari, A. (1967). Suspension bridge flutter revisited. *ASCE Structural Engineering Conference*. New York, NY: American Society of Civil Engineers.

Scanlan, R. H., & Sabzevari, A. (1969). Experimental aerodynamic coefficients in the analytical study of suspension bridge flutter. *Journal of Mechanical Engineering Science*, *11*(3), 234–242. doi:10.1243/JMES_JOUR_1969_011_031_02

Scanlan, R. H., & Tomko, J. J. (1971). Airfoil and bridge deck flutter derivatives. *Journal of the Engineering Mechanics Division*, *97*(6), 1717–1737.

Simiu, E., & Scanlan, R. H. (1996). *Wind Effects on Structures* (3rd ed.). New York, NY: Wiley.

Singh, L., Jones, N. P., Scanlan, R. H., & Lorendeaux, O. (1994). Simultaneous identification of 3-DOF aeroelastic parameters. In *Wind Engineering, Retrospect and Prospect* (Vol. 2, pp. 972–981). New York, NY: Wiley.

Theodorsen, T. (1935). *General theory of aerodynamic instability and the mechanism of flutter. NACA Technical Report, No. 496*. Langley, VA: National Advisory Committee for Aeronautics.

Thorby, D. (2009). *Structural Dynamics and Vibration in Practice*. Oxford, UK: Butterworth-Heinemann.

Wardlaw, R. L. (1994). Flutter and torsional instability. In H. Sockel (Ed.), *Wind-Excited Vibrations of Structures* (pp. 293–319). Vienna, Austria: Springer. doi:10.1007/978-3-7091-2708-7_6

Wright, J. R., & Cooper, J. E. (2015). *Introduction to Aircraft Aeroelasticity and Loads* (2nd ed.). New York, NY: Wiley.

Chapter 14
Full-Scale Gate Measurements:
Extended Validation of Theory

ABSTRACT

In this chapter, dynamic field measurements of three full-scale operational Tainter gates are documented. Measurements of the in-air gate structural dynamics were made when feasible or were simulated using FEM analysis when physical measurement was not possible. In-water structural dynamics were documented using experimental modal analysis with gate excitation supplied by the sudden tension failure of a carefully machined steel rod. The field measurements permitted assessments of the stability of each gate and these assessments are compared with theoretical predictions. Two gates were found to be stable and one was determined to have a potential instability. Theoretical predictions are shown to agree with the experimental results from field measurements.

INTRODUCTION

Field testing on three full-scale operational Tainter gates in Japan are reported in this chapter. On two of the gates, in-air modal analysis was conducted to establish the in-air dynamic characteristics of the gate. On the third gate, in-air testing was not possible, so a finite element model of the gate was used to determine in-air mode shapes and frequencies. In-water testing was carried out to document the vibration characteristics of the gates. External excitation in the form of the sudden failure of a precisely machined steel rod/bar was used for the gates that did not show self-excited instability. The dynamic conditions of the three gates were determined for the in-air and in-water data. When plotted on the dynamic stability criterion curve for each respective gate, the stability prediction from the theoretical curve agreed with the observed conditions. Two gates were stable and one was unstable.

DOI: 10.4018/978-1-5225-3079-4.ch014

TAINTER GATE "A"

Field vibration tests were undertaken on a 29-ton Tainter gate, hereafter called Tainter gate "A" (see Anami et al., 2004a). A photograph of Tainter gate "A," taken from the downstream side, is shown in Figure 1. The configuration and major dimensions are shown in Figure 2. In Figure 2a, the skinplate height is specified as 8.3 m and the skinplate radius is 8.5 m. The span of the skinplate is 8.75 m, as shown in Figure 2b. The skinplate is supported by two truss-structure radial arms. The gate is portal-shaped with the radial arms parallel to the streamwise direction. As shown in Figure 1, many reinforcing bracings are attached on the back side of the skinplate, and many additional bracings are attached between the horizontal girders. The gate elements are riveted together. The entire skinplate structure including the bracing and the horizontal girders has a mass of 21,091 kg and the radial arms have a mass of 7,941 kg. The skinplate mass is about 73% of the whole gate mass of 29,032 kg (about 29 tons).

In order to determine the in-air natural vibration characteristics, stop logs were installed on the upstream side of the gate to isolate the gate from the water, and experimental modal analysis was carried out.

The gate was raised so it was no longer in contact with the sill. The gate was struck with the instrumented impact hammer in the central portion of the lower horizontal girder in the tangential direction (vertical direction) and the radial direction (streamwise direction) of the skinplate, as shown by the arrows in Figure 2. The acceleration responses of the gate to the impact force were measured at 48 points on the gate (6 points on each arm and 36 points on the skinplate), as shown by the filled circles in Figures 3a and 3b. Two significant directions of vibration were measured at each measurement location. The gate acceleration responses to the impact force were measured with servo-type accelerometers (LS-20C by RION Co.).

Figure 1. View of downstream side of Tainter gate "A"
Photograph by K. Anami.

Figure 2. Two views of the 29-ton Tainter gate "A" in Japan showing (a) side view, and (b) the A-A cross-sectional top view
From Anami et al. (2004a) ©2004 ASME. Used with permission.

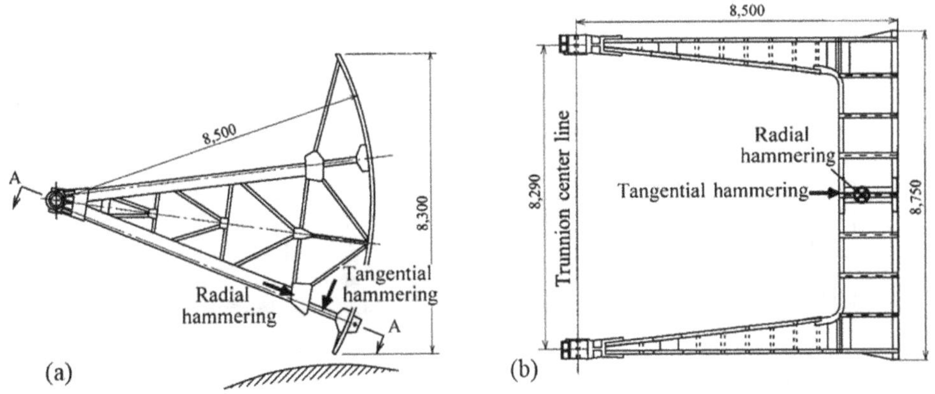

Figure 3. Measurement locations for in-air modal analysis tests on Tainter gate "A" (a) upstream view of skinplate, and (b) side view

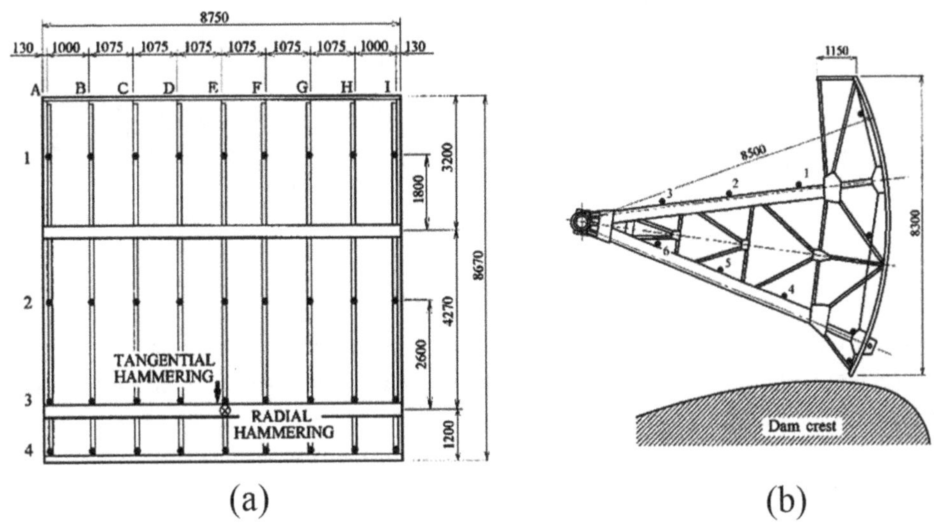

In the upstream view of the skinplate, in Figure 3a, the measurement locations are denoted by a letter and a number according to the following scheme: spanwise locations A, B, C,…I starting from the left, and 1, 2, 3, 4 in descending order from the top; R and L indicate locations on the arms looking downstream. The notations L1 to L6, and R1 to R6 are used to denote the measurement locations on each arm.

The recorded impulsive force and acceleration response data were processed on an FFT analyzer to display their real time waveforms and to calculate the corresponding frequency power spectra. An example of the real time impulsive force and acceleration waveforms is shown in Figure 4, where Figure 4a shows the real time wave forms for tangential impact and Figure 4b shows the traces for radial impact. The upper diagrams are the impulsive forces, while the middle diagrams show the acceleration responses at measurement location 1E. Each acceleration power spectrum was divided by the corresponding impulsive

Figure 4. Modal analysis test data for impact forces (top), acceleration responses (middle), and transfer functions (bottom) for Tainter gate "A" with (a) tangential impact, and (b) radial impact
From Ishii & Anami (2001).

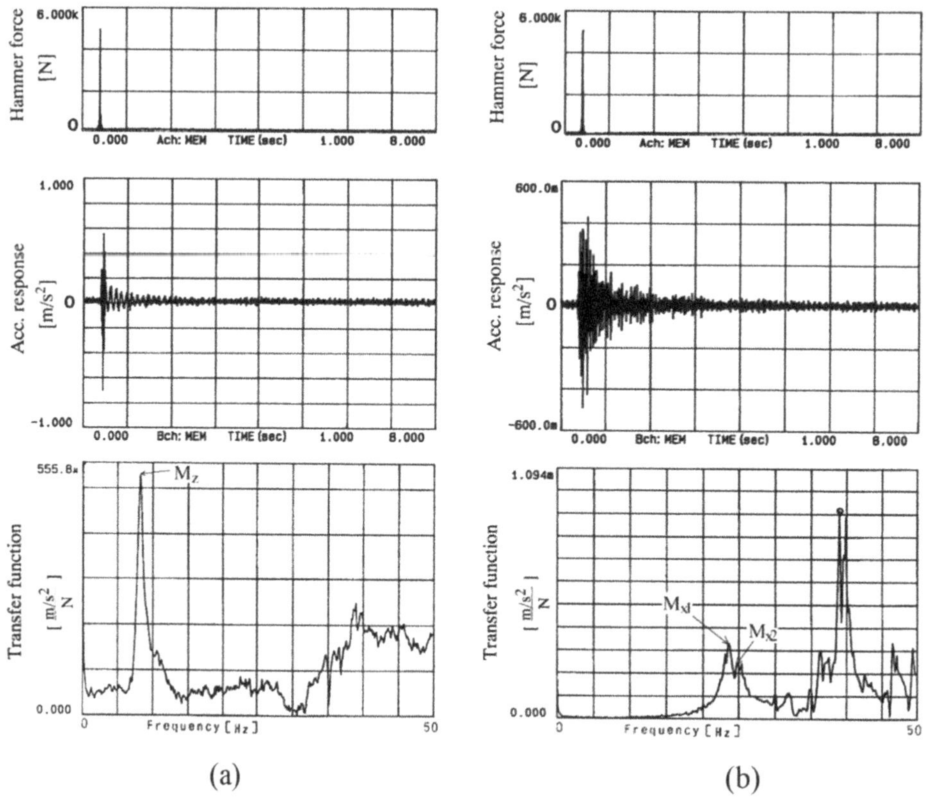

force power spectrum, to compute each transfer function, as shown in the lower diagrams of Figure 4. The impact tests were repeated three times at each point to measure three acceleration responses, allowing an average transfer functions to be calculated.

Figure 5a shows the inverse mode indicator functions (IMIF) resulting from tangential impact, and Figure 5b is the IMIF resulting from radial impact. Detailed examination of each vibration mode shape was made for all peaks in the IMIF to identify meaningful mode shapes exhibiting motion with ordinary regularity. The identified peaks are each associated with a mode name label. The mode name M_z represents the tangential vibration (*i.e.*, around the trunnion pin), and M_x the streamwise (radial) vibration. The transfer function peaks corresponding to identified mode shapes are marked on the transfer functions at the bottom of Figure 4.

The side view of the significant vibration mode M_z around the trunnion pin is shown in Figure 6, where the upper radial arm and the skinplate have not been connected, since there were not enough measured data points. This figure clearly shows that the whole gate undergoes a rotational vibration around the trunnion pin. The natural vibration frequency was 8.13 Hz and the damping ratio was 0.009.

The significant vibration modes of the skinplate streamwise vibration are M_{x1} and M_{x2}, The side views of these vibration modes are shown in Figure 7. The skinplate shows the streamwise lower-mode bending vibration with a node at about 1/3 of the gate height from the lower end of the skinplate. These vibration

Figure 5. Inverse modal indicator functions for Tainter gate "A" with (a) tangential impact, and (b) radial impact

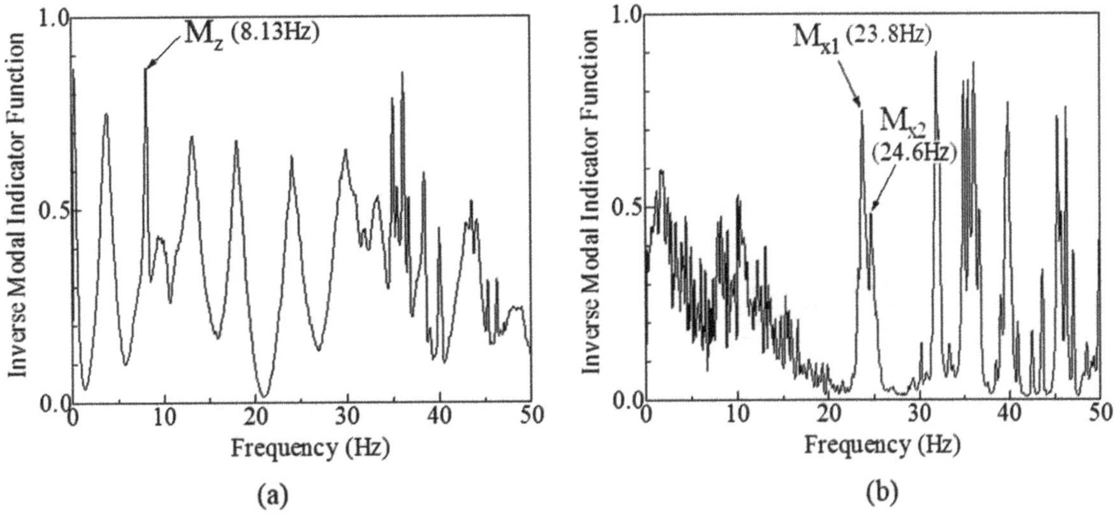

modes can be interpreted as streamwise rotational vibration modes. The rotation center height from the lower end of the skinplate is 1.98 m for mode M_{x1}, and 2.03 m for mode M_{x2}. The skinplate vibrates in the spanwise half wavelength bending mode with nodes at its spanwise ends.

In order to measure the in-water spontaneous vibrations or alternatively the in-water vibrations impulsively excited by the impact hammer, the gate was exposed to flowing water with an upstream water depth of 7.72 m above the dam crest.

The discharge opening height B was changed in a stepwise manner from 15 mm to 150 mm, each with its corresponding upstream gate submergence depth d_0 of (7.72-B) m. The vibrations were measured using six servo-type accelerometers, which were among the ones used in in-air tests. The vibration responses were measured at three locations (1E, 2E, and 4E) on the central part of the skinplate, and at two locations (4A and 4C) on the bottom end of the skinplate. The vibration responses in the vertical direction, the streamwise direction, and the spanwise direction were measured at measurement location 4E at the bottom center of the skinplate. Only streamwise vibrations were measured at the four other locations.

The vibration responses were measured with and without impact hammering. The gate was struck with the impact hammer instrumented with a force transducer. All accelerometers, impact hammer, and instrumentation were previously used in the in-air modal analysis.

Though the gate was dynamically stable at all gate openings, uncoupled vibrations were observed resulting from impulsive hammering. The frequency of the measured vibration waveforms were determined using the FFT analyzer. Representative transfer functions at point 4E for tangential impact and radial impact, respectively, for a discharge opening of 2.5 cm are shown in Figure 8. Although the level is low, some sharp peaks are obtained. These peaks represent the in-water vibration characteristics of the gate.

Figure 8a for the tangential vibration of the skinplate exhibits a sharp peak at 8.38 Hz, close to the in-air natural vibration frequency $\Omega_{a\theta}$ of 8.13 Hz, which represents the whole gate rotational vibration around the trunnion pin. The in-water vibration frequency and the in-air natural vibration frequency take on almost the same value, since the rotational vibration around the trunnion pin is not significantly influenced by the water added mass. Figure 8b for the streamwise vibration of the skinplate exhibits

Figure 6. Vibration mode shape for Mode M_z (Ω_a =8.13Hz; ζ_a =0.009)
From Anami et al. (2004a). © 2004 ASME. Used with permission.

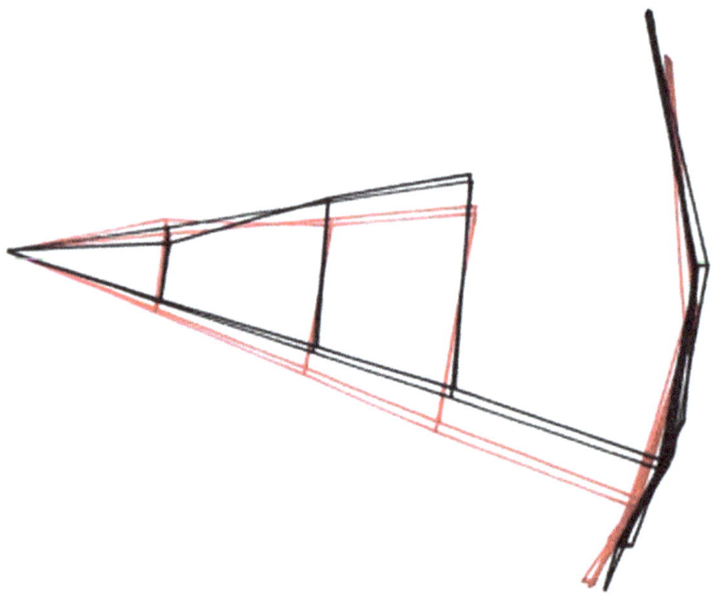

Figure 7. Mode shape of vibration group M_x for Tainter gate "A" showing (a) Mode M_{x1} (Ω_a =23.8Hz; ζ_a = 0.026), and (b) Mode M_{x2} (Ω_a = 24.6Hz; ζ_a = 0.011)
From Anami et al. (2004a) ©2004 ASME. Used with permission.

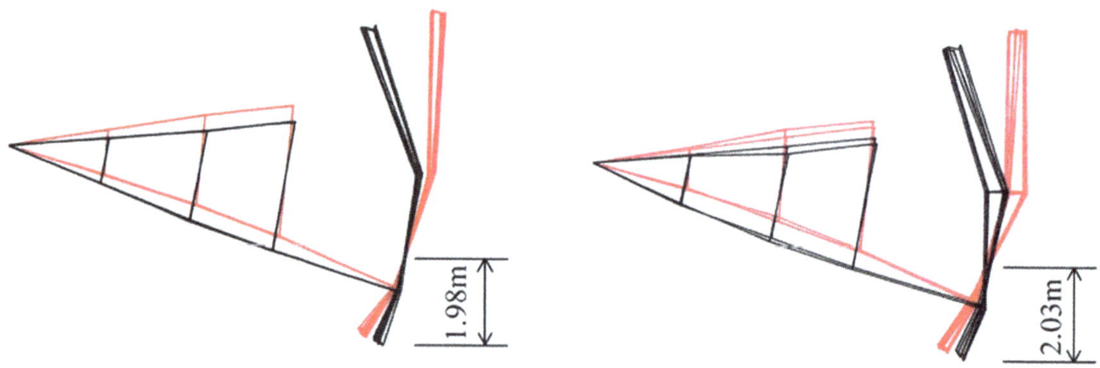

sharp peaks at 11.1 Hz and 13.8 Hz, which represent the skinplate streamwise vibration. In order to identify the mode of this vibration component, the vibration amplitudes along the skinplate spanwise center were estimated using measured data at points 1E, 2E, and 4E. The natural vibration frequency was selected with a bandpass filter and the vibration mode was identified from the amplitude and phase. Figure 9 shows the identified vibration mode shapes, where Figure 9a is for 11.1 Hz and Figure 9b is for 13.8 Hz. The negative value of vibration displacement on the abscissa means that the vibration has a 180° phase-lag relative to the vibration at the skinplate top center (1E). The vertical axis shows the

Figure 8. Transfer functions for in-water tests of Tainter gate "A" (a) in the tangential direction, and (b) in the radial direction
From Ishii & Anami (2001).

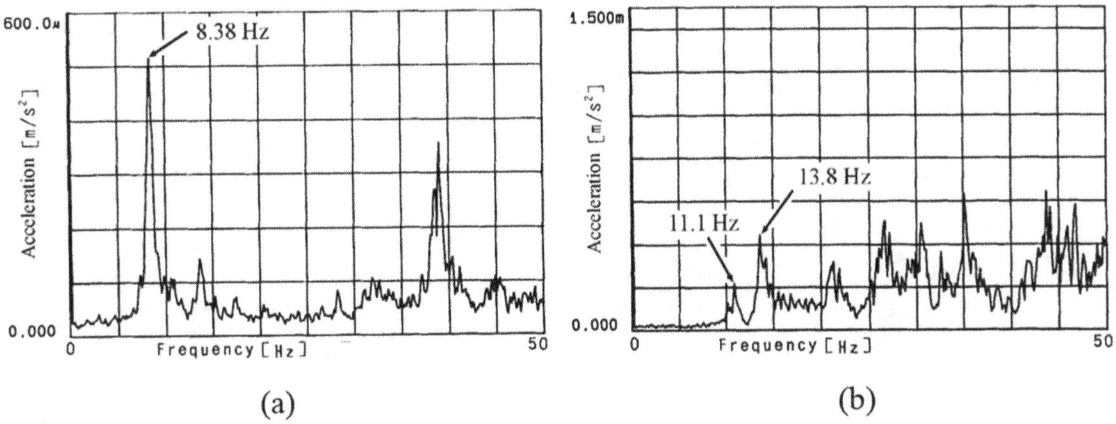

Figure 9. Streamwise in-water vibration mode of the skinplate for three gate openings of Tainter gate "A" (a) for 11.1 Hz, (b) for 13.8 Hz
From Anami et al. (2004a).© 2004 ASME. Used with permission.

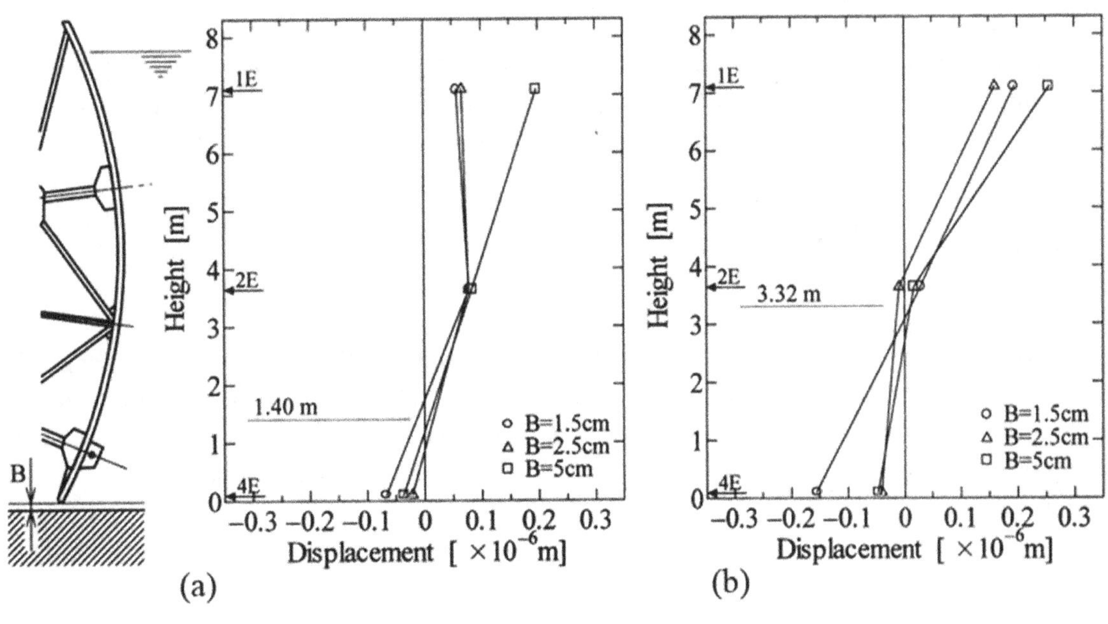

position of the measurement locations. The mode shape is analyzed for comparatively small discharge openings B of 1.5 cm, 2.5 cm and 5 cm.

Although the number of data points was limited and the vibrations were very small in amplitude, the associated mode of vibration is clearly identified, as shown in Figure 9. The natural vibrations at both 11.1 Hz and 13.8 Hz are streamwise rotational vibration modes, with the rotation center height at 1.40 m for the 11.1 Hz and 3.32 m for the 13.8 Hz vibration, respectively. These vibration modes show good

Table 1. Field vibration test results on 29-ton Tainter gate "A"

Mode Name	Viberation Mode Shape	In-air			In-water of 7.72m			Basic Froude numbe	Froude numbe	Freq. ratio
		Freq. Ω_a (Hz)	Dampin Ratio ζ_a	Height of rotation center R_s (m)	Freq. Ω_w (Hz)	Height of rotation center R_s (m)	Reduced hight of rotation center, r_s	F_0	F	Ω_w/Ω_a
M_z	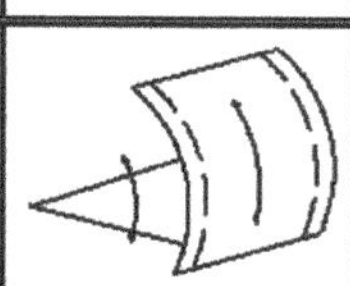	8.13	0.009	—	8.38	—	—	45	47	1.03
M_{x1}		23.8	0.026	1.98	11.1	1.40	0.18	133	62	0.47
M_{x2}		24.6	0.011	2.03	13.8	3.32	0.43	137	77	0.56

From Ishii & Anami (2001)

agreement with the in-air natural vibration modes M_{x1} and M_{x2} previously shown in Figure 7, obtained by experimental modal analysis, although the rotation center height shows a slight difference. The in-water natural vibrations (11.1 Hz and 13.8 Hz) correspond to the in-air natural vibration modes of M_{x1} (23.8 Hz) and M_{x2} (24.6 Hz).

The in-air and in-water experimental modal analysis results are summarized in Table 1. The basic Froude number F_0 and the Froude number F were calculated with the in-air natural vibration frequency Ω_a and the in-water vibration frequency Ω_w, respectively, using the following equations:

$$F_0 \equiv \sqrt{\frac{d_0}{g}}\Omega_a \text{ and } F \equiv \sqrt{\frac{d_0}{g}}\Omega_w \quad (1)$$

Since the skinplate submergence depth d_0 was 7.72 m, the value of the Froude number F ranges from 47 to 77. The basic Froude number F_0 is 45 for the whole gate rotational vibration around the trunnion pin and 133 to 137 for the skinplate streamwise vibration, respectively. Accordingly, the in-water to in-air frequency ratios Ω_w/Ω_a (= F/F_0) were calculated. The reduced rotation center height r_s was also calculated as follows:

$$r_s \equiv \frac{R_S}{d_0} \quad (2)$$

The reduced rotation center height was 0.18 for Mode M_{x1} and 0.43 for Mode M_{x2}.The major specifications of Tainter gate "A" are shown in Table 2. The field test conditions and the major mechanical constants are shown in Table 3. In-water vibration tests were conducted for a water depth over dam crest d_r of 7.72 m and the discharge opening B varied from 0.015 m to 0.15 m. The corresponding gate submergence depth d_0 in the vibration tests is from 7.71 m to 7.57 m. The theoretical calculations for

Table 2. Major specifications of Tainter gate "A"

			M_{x1}	M_{x2}
Gate dimensions and mass	gate height	—	8.3 m	
	spanwise length of skinplate	W_0	8.75 m	
	radius of skinplate	R_a	8.5 m	
	gate mass	—	29 ton	
Moment of inertia	whole gate around trunnion pin	I_θ	1.53×10^6 kgm^2	
	skinplate around its rotation center	I_ψ	3.21×10^5 kgm^2	2.16×10^5 kgm^2
	product of inertia of skinplate	$I_{\psi\theta}$	0.84×10^5 kgm^2	0.18×10^5 kgm^2

this gate were made for B=0.025 m and d_0 = 7.695 m, with measurements repeated three times. The skinplate inclination angle Θ_s was -3°, indicating the skinplate was inclined toward the upstream side.

Regarding the streamwise vibration, two modes, Modes M_{x1} and $M_{x2,}$ were considered. The corresponding natural vibration frequency $\Omega_{a\psi}$, the damping ratio $\zeta_{a\psi}$ and the in-water rotation center height R_S for each vibration mode were used. The moment-of-inertia I_ψ around the skinplate rotation center and the coupled moment-of-inertia $I_{\psi\theta}$ were calculated using Equations 5 and 4 from Chapter 10, respectively. The rotation radius R_c of the bottom end of skinplate was measured from the blueprints.

The geometrical press-open angle Θ_{s0} of the bottom end of the skinplate was measured from the experimental modal analysis results. The dam crest beneath the bottom end is almost flat, while the bottom end of the skinplate exhibits press-open characteristics. The natural vibration frequency $\Omega_{a\theta}$, the damping ratio $\zeta_{a\theta}$ and the moment of inertia I_θ of the whole gate vibration around the trunnion pin are, of course, common to both modes.

The dimensionless variables calculated from the specifications in Tables 2 and 3 are shown in Table 4, where all definitions are also presented. The depth ratio β* of the upstream reservoir to the gate submergence is 1.0 for both vibration modes, M_{x1} and M_{x2}. The reduced rotation center height r_s is 0.18 and 0.43 for the two modes. The water-to-gate mass ratio α_ψ is 3.2 for Mode M_{x1} and 26.6 for Mode M_{x2}.

Since the skinplate inclination angle Θ_s was -3°, c_f takes on a value of 0.76 and δ_p takes on a value of about 1.1 at the inclination angle of zero. The basic Froude number ranges from 133 to 137, since the in-air natural vibration frequency $\Omega_{a\psi}$ from the experimental modal analysis was 23.8 Hz and 24.6 Hz. It is about an order of magnitude larger than the basic Froude number of 16 for model experiments.

The dynamic stability criterion curve calculated with these non-dimensional factors is shown in Figure 10. The curve in the range of γ_{nw} < 1.0 represents the critical damping ratio $\zeta_{c\psi}$ for the skinplate streamwise vibration, and the curve in γ_{nw} > 1.0 represents the critical damping ratio $\zeta_{c\theta}$ for the whole gate vibration around the trunnion pin. The solid line shows the calculated result for Mode M_{x2}, and the dashed line for the Mode $M_{x1.}$ The unstable region is especially small for the Mode M_{x1}, since the rotation center is comparatively low.

The frequency ratio γ_{nw} (= $\Omega_{nw\psi}$ /$\Omega_{a\theta}$) of the measured skinplate streamwise in-water vibration (13.75 Hz, 11.13 Hz) to the whole gate in-air vibration around the trunnion pin (7.89 Hz) was 1.70 for Mode M_{x1} and 1.37 for Mode $M_{x2.}$ In both modes, γ_{nw} is larger than 1.0. In this range of frequency ratios, the vibration synchronizes with the whole gate vibration around the trunnion pin, which is the main vibra-

Table 3. Field test conditions of Tainter gate "A" and the major mechanical constants

			M_{x1}	M_{x2}
Field test conditions	depth of upstream channel	d_r	7.72 m	
	gate opening	B	0.025m (0.015~0.15 m)	
	depth over the gate bolttom	d_0	7.695m (7.71~7.57 m)	
	inclination angle of gate bottom to water surface	Θ_s	-3°	
	natural vib. freq. of upward vib. in air	$\Omega_{a\theta}$	8.13 Hz	
	natural vib. freq. of streamwise vib. in air	$\Omega_{a\psi}$	23.8 Hz	24.6 Hz
	damping ratio of upward vib. in air	$\zeta_{a\theta}$	0.009	
	damping ratio of streamwise vib. in air	$\zeta_{a\psi}$	0.026	0.011
	rotation center height	R_S	1.40 m	3.32 m
	press-open angle	Θ_{s0}	23°	21°
	rotation rasius of skinplate lower end	R_c	8.88 m	8.72 m

Table 4. Non-dimensional factors for calculations

		M_{x1}	M_{x2}
depth ratio ($\beta^* \equiv d_r/d_0$)	β^*	1.00	
reduced rotation center height ($r_s \equiv R_S/d_0$)	r_s	0.18	0.43
rotation radius ratio ($r_{sa} \equiv R_S/R_c$)	r_{sa}	0.16	0.38
press-shut coefficient ($k \equiv -\sin \Theta_{s0}$)	k	-0.39	-0.35
moment-of-inertia ratio ($\alpha_I \equiv I_\psi/I_\theta$)	α_I	0.21	0.14
moment-of-inertia ratio of skinplate ($\alpha_{I\psi} \equiv I_{\psi\theta}/I_\psi$)	$\alpha_{I\psi}$	0.26	0.083
water to gate mass ratio ($\alpha_\psi \equiv \rho d_0^2 W_0/(I_\psi/R_S^2)$)	α_ψ	3.2	26.6
instantaneous flow-rate-coefficient	c_f	0.76	
pressure correction coefficient	δ_p	1.1	
basic Froude number $\left(F_0 \equiv 2\pi\Omega_{a\psi} \times \sqrt{d_0/g}\right)$	F_0	133	137

tion. The in-air damping ratio $\zeta_{a\theta}$ of the main vibration takes a value of 0.009, which is plotted over each frequency ratio on the abscissa γ_{nw}. The plotted data for Modes M_{x1} and M_{x2} are both in the stable region.

Consequently, the dynamic stability of Tainter gate "A" has been theoretically confirmed and is in agreement with the experimental evidence.

Figure 10. Stability criterion curve for Tainter gate "A"

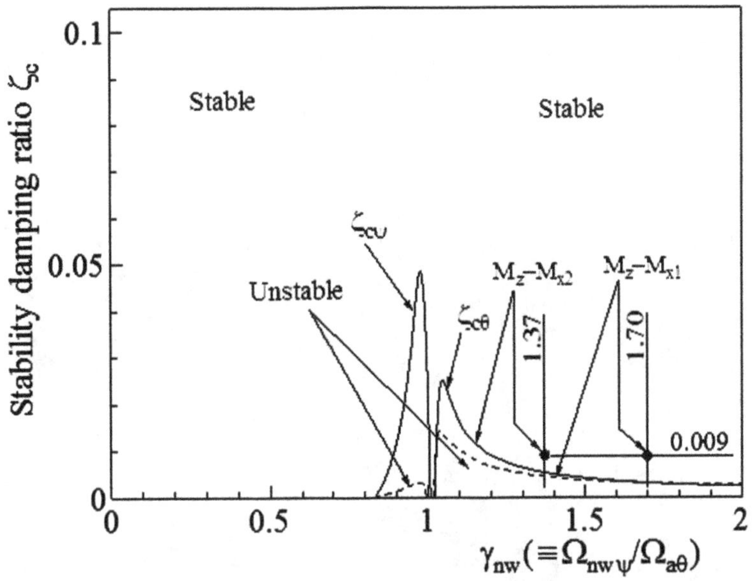

TAINTER GATE "B"

Field vibration tests were undertaken on a 34-ton Tainter gate, hereafter called Tainter gate "B" (see Anami et al., 2004b). In-air and in-water experimental modal analyses were made first, and in-air and in-water steel bar breaking tests were subsequently undertaken. In order to measure the natural vibration characteristics of such a huge Tainter gate, a new steel bar breaking test was developed as a new type of impulsive excitation. Furthermore, in the in-water tests, a pressure transducer was directly attached to the skinplate permitting the measurement of the hydrodynamic pressure.

A photograph of Tainter gate "B," taken from the downstream side, is shown in Figure 11. The configuration and major dimensions are shown in Figure 12. The skinplate has a height of 9.8 m, a radius of 11.0 m, and a span of 9.0 m. The whole gate mass reaches 33,874 kg (about 34 tons). The skinplate is supported by two truss structure radial arms on each side. This gate is π-shaped, with 5° inclined radial arms. Gates with π-shape structures are quite common, since the bending moment at the junction of arm and skinplate is reduced. The gate elements are riveted together.

In order to determine the in-air natural vibration characteristics, experimental modal analysis was undertaken. The gate was raised so it was no longer in contact with the sill. The gate was struck with the instrumented impact hammer in the central portion of the lower horizontal girder in the tangential direction (vertical direction) and the radial direction (streamwise direction) of the skinplate, as shown by the arrows in Figure 12. The vibration acceleration responses of the gate to the impulsive force were measured. The acceleration responses, made using servo-type accelerometers, were measured at 80 points (56 points on the skinplate and 24 points on the radial arms).

When the gate was struck in the tangential direction, the natural vibration around the trunnion pin could not be determined; the side rollers on one side of the skinplate were tightly pressed against the pier and the skinplate was hard to move in the tangential direction. However, precise data were recorded for the streamwise direction. As a result, the Mode M_x with the natural vibration frequency of 30.8 Hz and

Figure 11. View of the downstream side of Tainter gate "B"
Photograph by K. Anami.

Figure 12. Two views of the 34-ton Tainter gate "B" in Japan, showing (a) a side view, and (b) the A-A cross-sectional top view
From Anami et al. (2004b).© 2004 Centre Poly-Média. Used with permission.

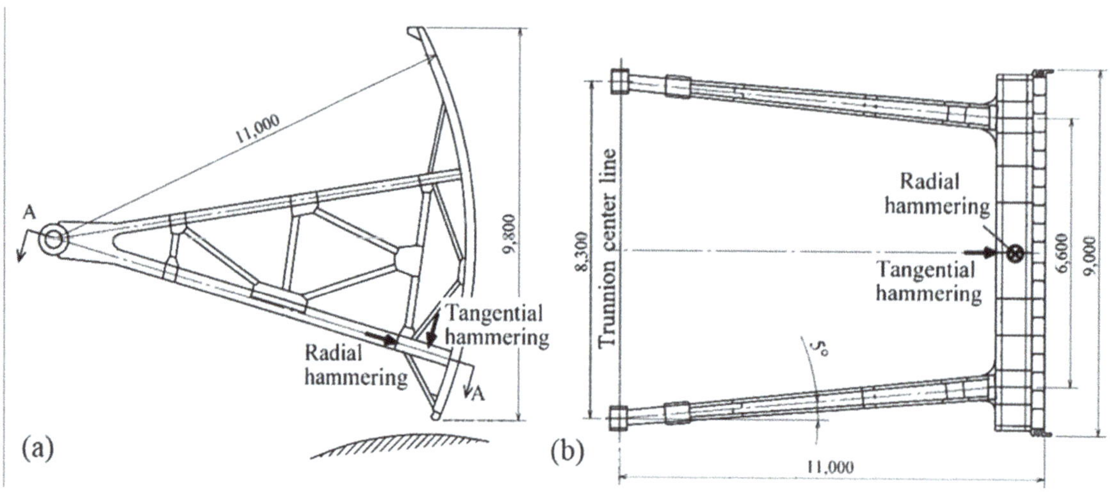

a damping ratio of 0.018 was obtained. The side view of this vibration mode is shown in Figure 13. The radial arms bend in the vertical direction and the skinplate performs a streamwise rotational vibration. The rotation center height from the bottom end of the skinplate was 6.9 m.The skinplate vibrates in the spanwise half-wavelength bending mode, with the nodes at each spanwise end.

As mentioned in the introduction to Tainter gate "B," a second set of in-air vibration tests were undertaken. At that time, a new type of in-air excitation, the breaking of a steel bar, was used in its first application. As shown in Figure 14, the steel bar (S45C; carbon steels for machine structural use) was fixed between the skinplate bottom member at spanwise center of the skinplate and the dam crest sill. The gate was raised in a slow, stepwise manner until the bar suddenly failed, triggering vibration due to the sudden release of the tension force. The acceleration responses of the gate were measured. This test was the first time such excitation due to breaking a steel bar was used for a full-scale operational Tainter gate. The diameter of steel bar was between 3 and 8 mm, The tensile breaking force for the steel bar and the extension of the lifting cable were about 3.9 kN and 0.27 mm for the 3 mm-diameter steel bar, and about 28 kN and 1.89 mm for the 8 mm-diameter bar. Figure 15 shows the setup of the steel bar. The enlarged view of steel bar is shown in Figure 16, where Figure 16a is before breaking and Figure 16b is after breaking.

In order to measure the vibration responses, accelerometers were installed at the center of the skinplate for the tangential and streamwise directions. In addition, the tangential acceleration was measured at the right side of the skinplate. Figure 17 shows the acceleration responses at the skinplate center position (denoted by "4D"), triggered by the sudden failure of the steel-bar with a diameter of 5 mm. The additional character "V" represents the vertical (tangential) direction and "R" the radial (streamwise) direction. Figure 17a shows the real time waveforms of the acceleration responses and Figure 17b the corresponding frequency power spectrum. The upper diagrams show the tangential vibration and lower diagrams the streamwise vibration. The tangential vibration corresponds very closely to the ideal free damping vibration waveform. The triggered vibration in the vertical direction, immediately after break-

Figure 13. Vibration mode shape of Mode M_x
From Anami et al. (2004b). © 2004 Centre Poly-Média. Used with permission.

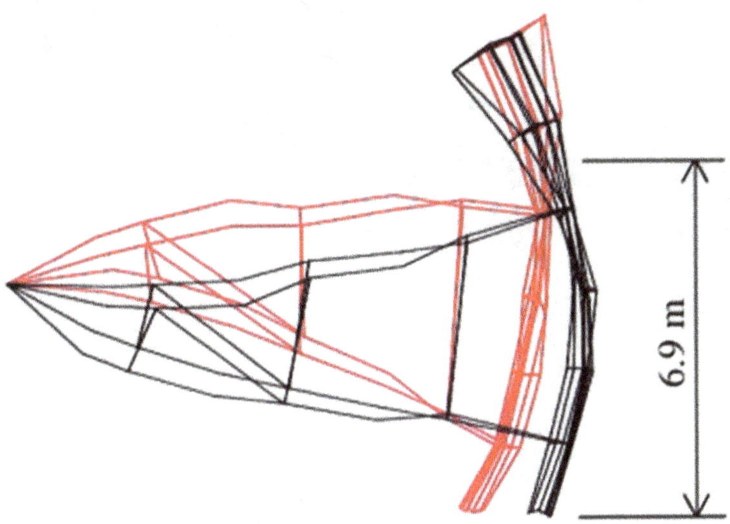

Figure 14. Schematic set up for the steel rod used for triggering vibration of Tainter gate "B"

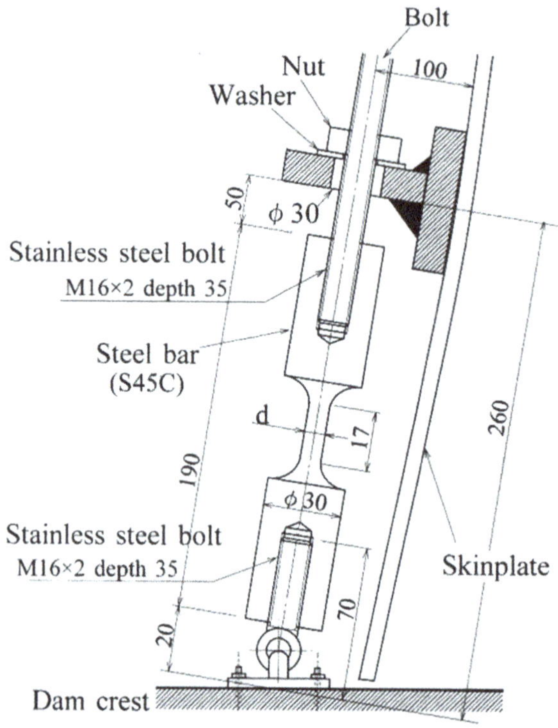

Figure 15. Steel rod connected between the lower portion of skinplate and dam crest sill
Photograph by N. Ishii

Figure 16. Enlarged view of steel rod: (a) before breaking, and (b) after being broken
Photographs by N. Ishii.

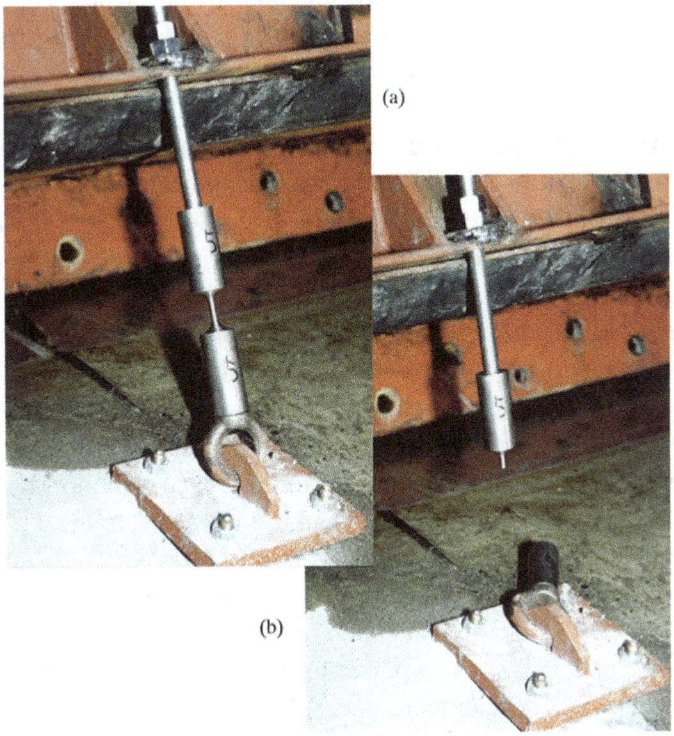

Figure 17. In-air vibration responses and frequency spectra triggered by breaking a steel bar with a diameter d of 5 mm: (a) in-air vibration waveforms, and (b) corresponding frequency spectra
From Anami et al. (2004b). © 2004 Centre Poly-Média. Used with permission.

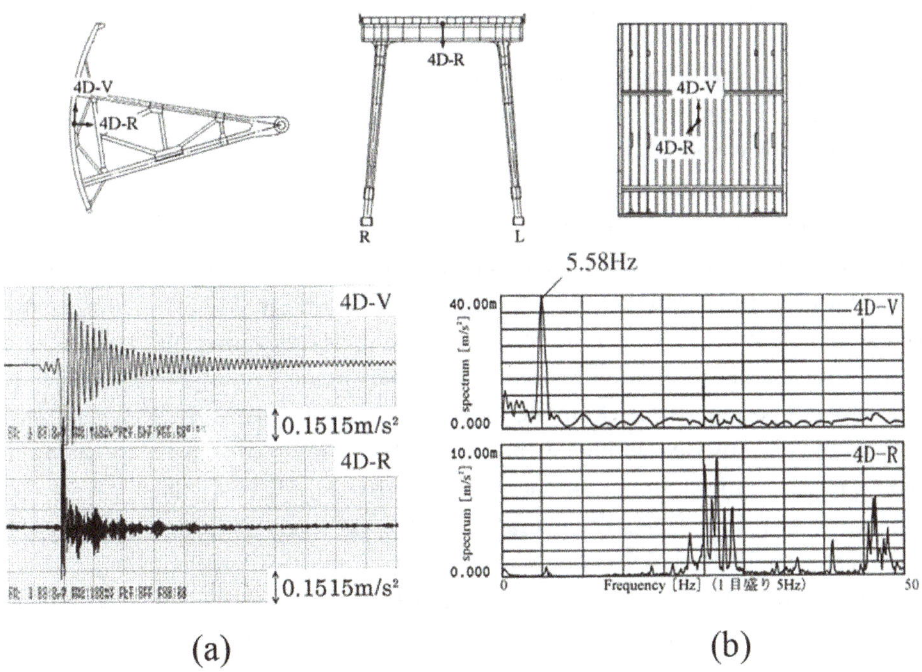

ing the steel bar, was about 0.35 mm in amplitude and decreased exponentially, while the vibration in the radial direction exhibited rather random behavior at higher frequencies. The exponentially damped natural vibration had a predominant frequency of 7.89 Hz, with the damping ratio of 0.011. In streamwise direction, many 25 Hz to 30 Hz vibrations are overlapping and beating.

In order to measure the in-water spontaneous vibrations, the gate was exposed to flowing water with an upstream water depth of 9.24 m above the dam crest. No spontaneous vibration was measured. Subsequently, in-water vibrations were triggered using the steel rod breaking excitation and were successfully recorded. In addition to the vibration measurements, the hydrodynamic pressure also was measured by a transducer, directly attached to the bottom portion of the skinplate at its spanwise-center, as a first attempt at such measurement.

The vibration acceleration responses were measured at along the skinplate center, and on the radial arm. The discharge opening at the instant of the steel rod breaking was measured at both spanwise ends of the skinplate. The steel bar and the underflow jet are shown in the Figure 18. The gate was initially closed (Figure 18a). The gate is then opened slightly and a thin water jet appears (Figure 18b). The water jet increases with increasing the discharge opening (Figure 18c) and a bit more until just before the steel bar breaks (Figure 18d). Figure 18e shows the moment just after the bar breaks. The M16 nut binding the steel bar was accelerated upward to near the top of this figure.

Representative real-time waveforms of vibration and their corresponding frequency spectra are shown in Figure 19. The diameter of steel rod was 6 mm. Figure 19a shows the real-time waveforms of vibration and Figure 19b the corresponding frequency power spectra. The measurement locations are shown by the arrows in the upper sketches. The additional letter "V" indicates vibration in the vertical (tangential) direction and "R" the radial (streamwise) direction. Each frequency spectrum contains distinct dominant peaks at either or both frequencies: 7.89 Hz and 9.99 Hz.

In the spectrum of 1D-R, the peak at 7.89 Hz is very small, and 9.99 Hz is dominant, Therefore, it can be judged that 7.89 Hz represents the whole gate rotational vibration around the trunnion pin (vertical vibration of skinplate), and 9.99 Hz represents the skinplate streamwise vibration,. The waveform of 1D-R decreases almost exponentially; the waveform of 7D-R and 7D-V are clearly beating. This behavior is the result of the tangential and streamwise vibrations with close frequencies affecting each other resulting in the coupling of the two closely spaced vibration frequencies.

The damping ratios of vibration were analyzed and averaged from all the experimental data. The damping ratio for these in-water damped vibrations was $\zeta_{w\theta} = 0.018$ for the vertical vibration and $\zeta_{w\psi} = 0.011$ for the streamwise vibration.

The skinplate streamwise vibration mode can be identified by analyzing the vibrations measured at 7 positions on the skinplate at its spanwise center. With the measured data, the streamwise vibration mode was identified. The vibration amplitude of each point was analyzed and the phase-lag relative to Point 1D was also analyzed. Figure 20 shows the resulting streamwise vibration mode shape. Although the vibration amplitude is small (50 μm at maximum), the lower-mode-bending vibration has been clearly identified. The top and bottom of the skinplate vibrate with a 180° phase-lag, All data show a similar vibration mode for the streamwise rotational bending vibration with its rotation center height in the range of 4.93 m to 5.66 m. Since the skinplate submergence depth d_0 was 9.24 m, the reduced rotation center height is in the range of 0.53 to 0.61. The average of rotation center height is 5.11 m, and the average reduced rotation center height is 0.55.

All in-air and in-water vibration test results are summarized in Table 5. The mode name M_z denotes the whole gate rotational vibration around the trunnion pin, and the mode name M_x denotes the skinplate

Figure 18. Breaking of a steel bar and underflow water jet: (a) before opening; (b) first appearance of the water jet; (c) more water flowing; (d) just before breaking; (e) just after breaking
Photographs by N. Ishii.

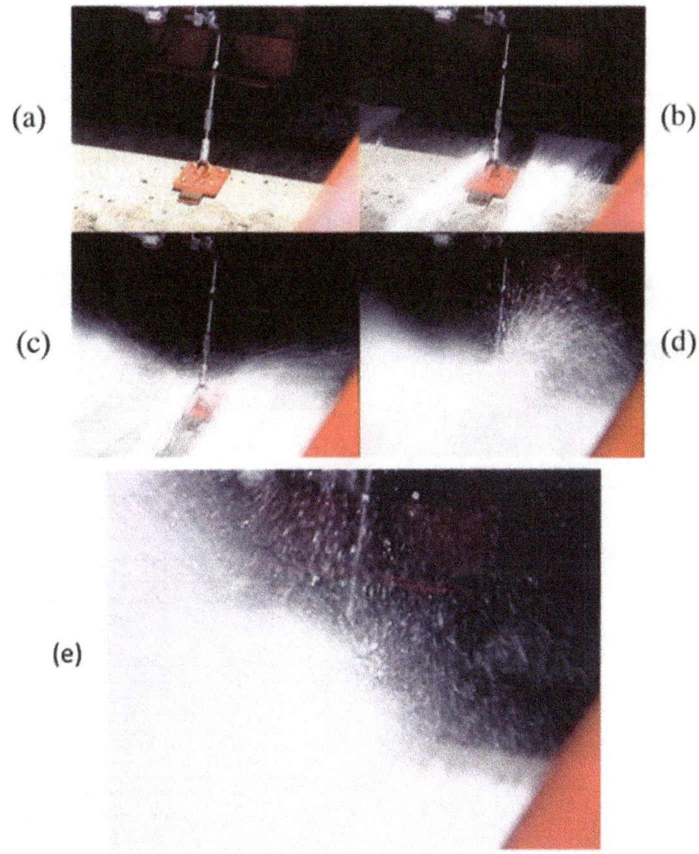

streamwise rotational vibration. The basic Froude number F_0, Froude number F, and in-water to in-air vibration frequency ratio are also listed together.

In addition to the measurement of vibration induced by the steel rod breaking for Gate "B," the hydrodynamic pressure due to the gate vibration could also be measured simultaneously. A pressure tap with a diameter of 5.3 mm was bored in the skinplate to attach a pressure transducer. The pressure tap is located 700 mm from the bottom end of the skinplate and 225 mm left of the skinplate center. The attached pressure transducer (PMS-5M by Toyoda Machine Works Co.) has a dynamic range up to 50 m of water head.

Figure 21 shows the lower portion of the skinplate. The pressure transducer was attached to the upper region included in Figure 21a. Figure 21b shows an enlarged view of the pressure transducer attached to the skinplate, and Figure 21c is a sketch of the pressure transducer setup. A steel plate with a diameter of 50 mm and a thickness of 30 mm is welded onto the skinplate to reinforce the area around the small pressure tap. The pressure transducer is attached to the center of the reinforcing plate. There is a space with a diameter of 5.3 mm and a length of 33 mm in front of the pressure transducer. This test is the first known instance of drilling a small hole in an installed Tainter gate in order to measure the hydrodynamic pressure.

Figure 19. Representative in-water accelerations with excitation from breaking steel bar (d = 6 mm), showing (a) real time wave forms, and (b) corresponding frequency spectra
From Ishii & Anami (2001).

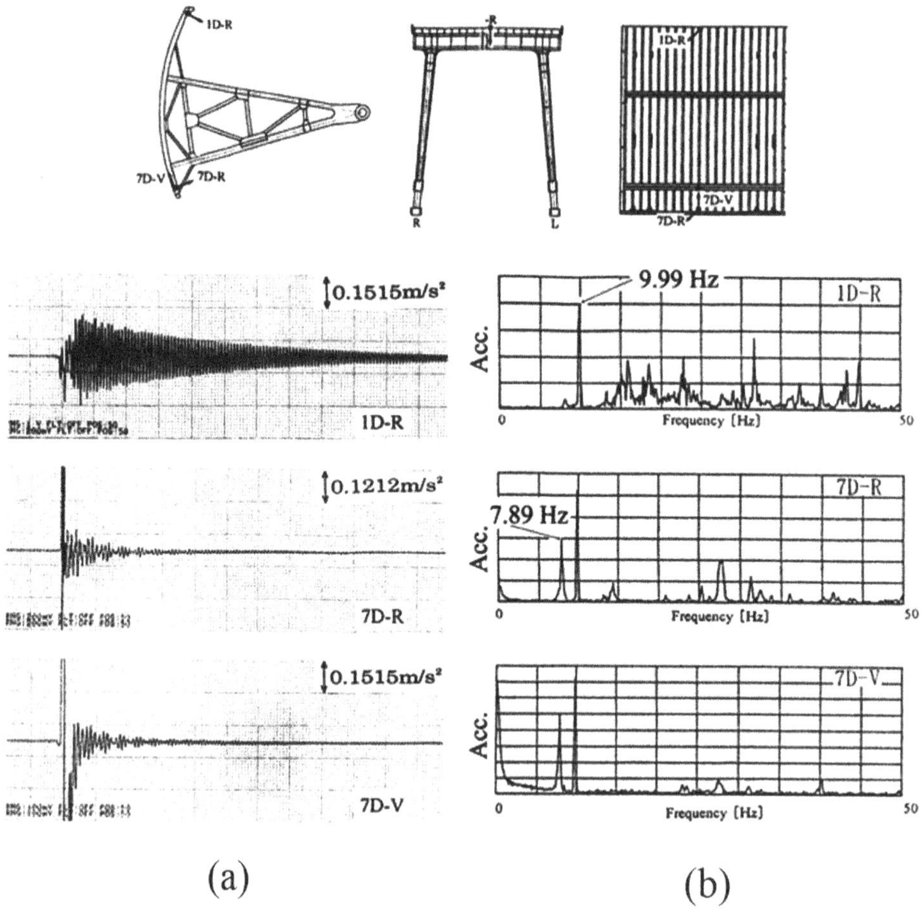

The measured hydrodynamic pressure is shown in Figure 22. The acceleration waveforms at Point 7D, shown in Figure 19, is also shown simultaneously. As shown in the pressure spectrum, the two sharp peaks have frequencies that are in good agreement with the natural vibration frequency of 7.89 Hz for the vertical direction and 9.99 Hz for the streamwise direction, respectively. The hydrodynamic pressure fluctuations are clearly associated with the skinplate vibrations.

The measured hydrodynamic pressure and the theoretically calculated pressure are shown in Figure 23. For the streamwise vibration, the Froude number F was 61 and the average of reduced rotation center height r_s was 0.55, as shown in Table 5. With these values, the push-and-draw pressure amplitude p_{b0} was calculated, as shown by the solid line in Figure 23a. Since the range of r_s was from 0.53 to 0.61, calculated results are also shown by dashed lines for the maximum and minimum r_s values. For the tangential vibration, the Froude number was 48. With this value, the flow-rate-variation pressure p_{r0} was calculated, as shown in Figure 23b.

Figure 20. In-water streamwise vibration mode shape of Mode M_x (Ω_w = 9.99Hz)
From Anami et al. (2004b). © 2004 Centre Poly-Média. Used with permission.

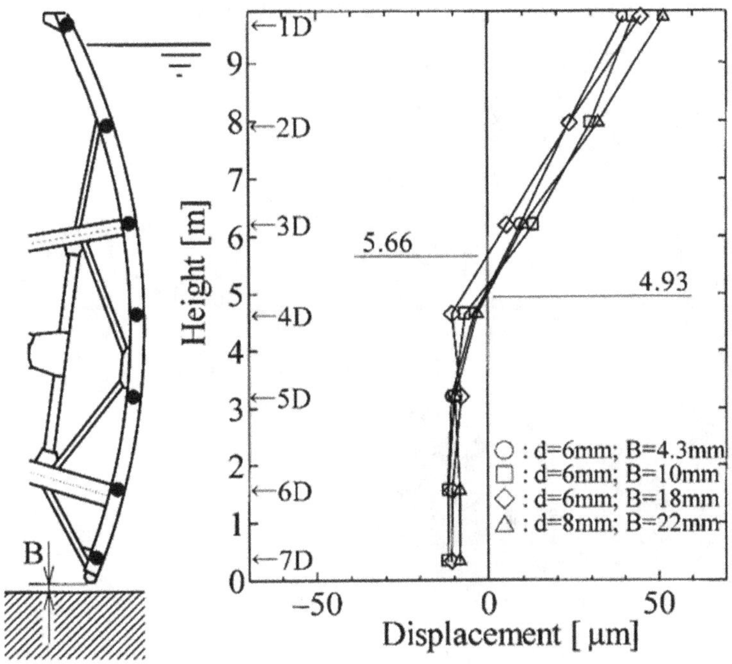

Table 5. In-air and in-water field vibration test results of Tainter gate "B"

Mode Name	Vibration Mode Shape	In-air			In-water of 9.24				Basic Froud numbe	Froud numbe	Freq. ratio
		Freq. Ω_a (Hz)	Dampin Ratio ζ_a	Height of rotation center R_s (m)	Freq. Ω_w (Hz)	Dampin Ratio ζ_w	Height of rotation center R_s (m)	Reduced hight of rotation center, r_s	F_0	F	Ω_w/Ω_a
M_z	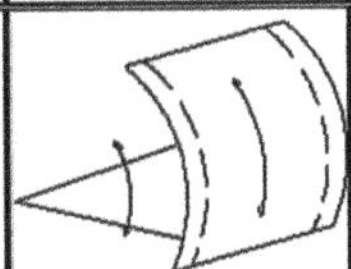	7.89	0.011	—	7.91	0.018	—	—	48	48	1.0
M_x	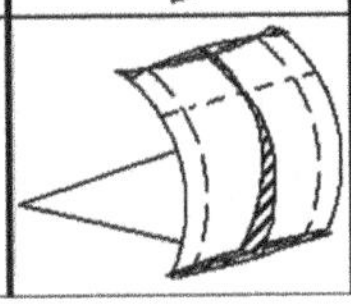	30.38	0.018	6.9	9.99	0.011	4.93~5.66	0.55 (0.53~0.61)	185	61	0.33

From Ishii & Anami (2001).

To plot the experimental data in Figure 23, the push-and-draw and flow-rate variation pressures were extracted from the measured resultant pressure P_0, The leading vibration for 9.99 Hz is the streamwise rotational vibration. Even if the flow-rate variation pressure is completely ignored, it has little effect on the data, since the flow-rate variation pressure is about 1/10 of the push-and-draw pressure. The measured data at 9.99 Hz were reduced to non-dimensional data, as plotted in Figure 23a, using a pressure correction coefficient δ_p of 1.1. The data in Figure 23a are plotted at y ($\equiv Y/d_0$)=-0.924, since the measurement

Figure 21. Lower portion of skinplate of Tainter gate "B" showing (a) the skinplate with the pressure transducer, steel bar and accelerometers, (b) enlarged view of the pressure transducer attached to the skinplate, and (c) the setup for the pressure transducer on the skinplate

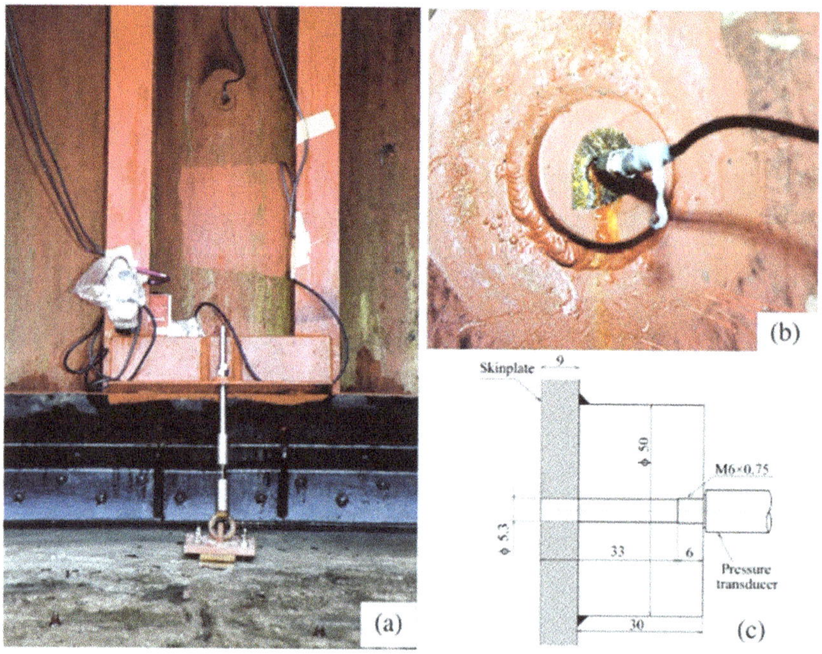

location was Y =-8.54 m and the gate submergence depth d_0 was 9.24 m. The plotted measured data are in good agreement with the theoretical results. In the theoretical analyses, it was assumed that the streamwise lower-mode bending vibration of the skinplate could be treated as a rotational rigid vibration. Since the theoretical predictions show good agreement with the field test results, one may therefore infer that such a simplified treatment of the bending vibration is valid in practical applications, as long as the vibration amplitude remains small.

The leading vibration for 7.35 Hz is the tangential vibration. It was quite difficult to extract the small flow-rate-variation pressure from the measured data of resultant pressure. In this study, however, the theoretical push-and-draw pressure has been verified. Therefore, the flow rate variation pressure was calculated by subtracting the calculated push-and-draw pressure from the resultant pressure for 7.35 Hz. The non-dimensional flow-rate variation pressure is plotted in Figure 23b, where the instantaneous flow-rate variation coefficient c_f takes on a value of 0.77. Despite some scatter in the measured data, the data show substantial agreement with the theoretical results, as seen in this figure.

The present experimental results are extremely valuable since they represent the first occasion in which the theoretical analysis of the hydrodynamic pressure acting on the gate was validated on an installed, full-scale Tainter gate.

The calculated dynamic stability criterion curves are shown in Figure 24. There was only one significant streamwise natural vibration mode M_x, and the dynamic stability criterion curves were calculated for both the maximum and minimum rotation center heights, The solid line shows the calculated result for the maximum rotation center height of 5.66 m, and the dashed line for the minimum rotation center height of 4.93 m. No large difference is seen between these two cases, but the unstable region becomes

Figure 22. Examples of measured hydrodynamic pressure showing (a) real time wave forms, and (b) frequency spectra
From Ishii & Anami (2001).

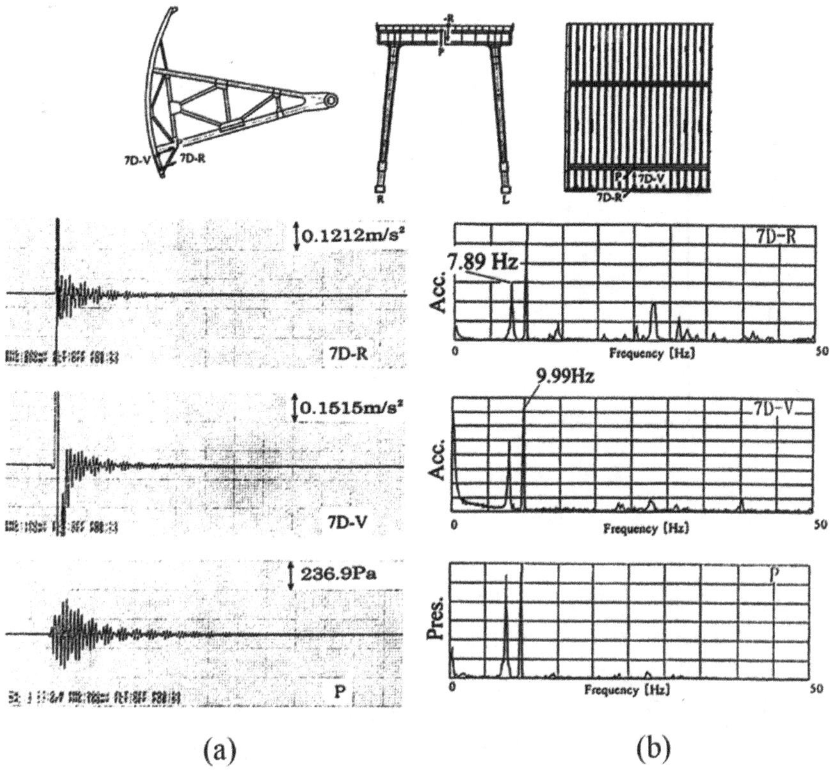

Figure 23. Hydrodynamic pressure acting on the lower center portion of the skinplate measured on the full-scale Tainter gate "B" showing (a) push-and-draw pressure amplitude, and (b) flow-rate-variation pressure amplitude
From Ishii & Anami (2001).

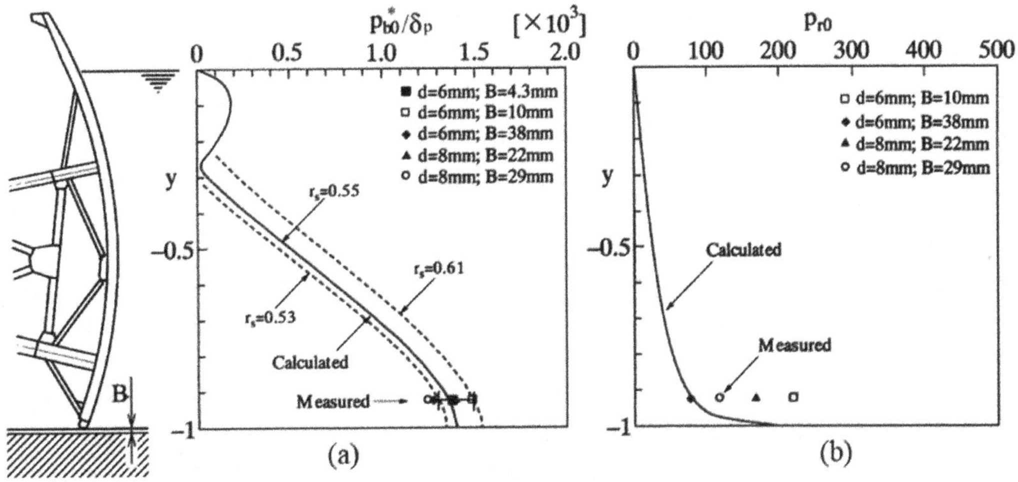

Figure 24. Stability criterion curve for Tainter gate "B"

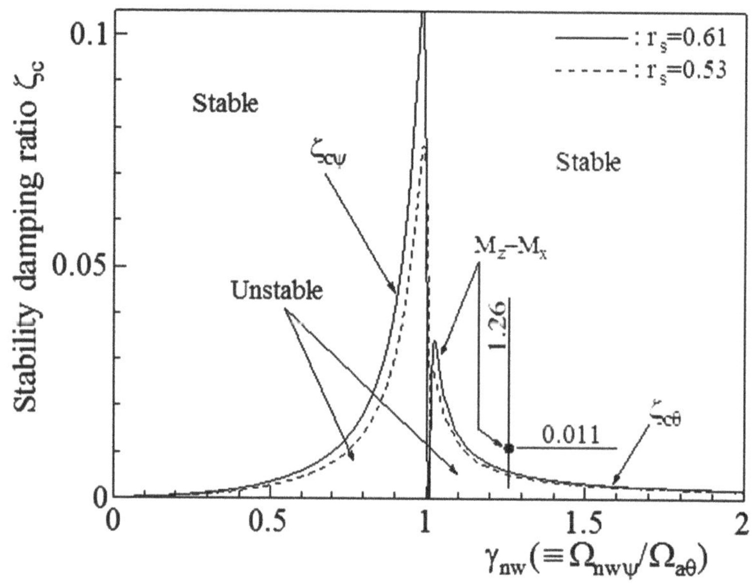

a bit larger for the higher rotation center position. Compared with Tainter gate "A," the mass ratio α_I for gate "B", representing the degree of easy coupling of the two vibration modes, is about the half as large, but the unstable region is still larger, since the rotation center height is larger.

The frequency ratio γ_{nw} ($= \Omega_{nw\psi}/\Omega_{a\theta}$) of the measured in-water streamwise vibration (9.99 Hz) to the in-air vibration around the trunnion pin (7.89 Hz) was 1.26. The in-air damping ratio $\zeta_{a\theta}$ of 0.011 for the whole gate rotational vibration is plotted versus γ_{nw}. The plotted data lies in the stable region.

The dynamic stability of Tainter gate "B" has been theoretically confirmed, in agreement with the experimental evidence.

TAINTER GATE "C"

The third full-scale Tainter gate under investigated will be denoted as Tainter gate "C" (see Anami *et. al.*, 2007a, 2007b), shown in Figure 25. This Tainter gate is a π-shaped gate with the radial arms inclined about 10° toward the center of the skinplate. This type of gate is quite similar to the Folsom Dam Tainter gate, which had an inclination angle of 13.7°. The Folsom Dam Tainter gate has tie beams between the right and left radial arms, while Tainter gate "C" has no tie beam. The skinplate structure (skinplate, horizontal girders and bracings) of Tainter gate "C" contains about 63% of the whole gate mass. This value is also in agreement with 63% for the Folsom Dam Tainter gate.

In-water vibration tests were conducted for a small discharge opening with gate submergence depths of 9.54 m to 9.00 m. There was no measurable change in the dam water depth during individual sets of measurements.

In this experiment, the vibrational accelerations, the hydrodynamic pressure and the controlled discharge openings were measured. For measurement of vibration accelerations, 27 accelerometers were attached on the gate, on the bridge and on the concrete pier near the trunnion. The hydrodynamic pressure

Figure 25. View of the downstream side of Tainter gate "C"
Photograph by N. Ishii.

induced by gate vibration was measured with a pressure transducer (PMS-5M-1M by Toyoda Machine Works Co.). The pressure transducer was attached to a magnet stand (as shown in Figure 26), which was pulled into position underwater using 2 ropes on either side of the concrete pier over the skinplate. The magnet stand was fixed on the skinplate at predetermined depths.

When the gate submergence depth was 9.54 m and 9.27 m, intense self-excited vibrations were induced at small discharge openings. In both cases, the ideal self-excited vibration could be measured. Surface waves appeared on the water surface, rapidly spreading into the whole upstream reservoir. With small vibration amplitudes, immediately after vibration started, plane waves arose along the skinplate/water interface and propagated uniformly upstream. The still water surface, with no discharge is shown in Figure 26a. As the vibration amplitude became larger, wave fronts perpendicular to the skinplate, the so-called "streamwise waves" appeared, spreading along the skinplate from its spanwise center toward each side, as shown in Figure 26b. The vibration frequency was comparatively high (6.79 Hz), It is believed the relatively high frequency is responsible for the generation of these streamwise waves. Near the central portion of the skinplate, plane waves and streamwise waves overlapped.

The analysis of the measured vibration data on the skinplate indicated that the gate did not undergo spanwise vibration, but did undergo streamwise (radial) and vertical (tangential) vibrations which increased exponentially in amplitude, approaching steady amplitude vibration.

At some measurement locations on the skinplate, vertical and streamwise vibrations were measured. These vertical and streamwise vibration waveforms were compounded to obtain the vibration trajectories. Representative vibration waveforms and the trajectory measured at the bottom center of the skinplate are shown in Figure 28, where low-pass filter processing was performed to remove the high frequency components of vibration. Figure 28a clearly shows that the vibrations increase exponentially, while Figure 28b shows that the vibration trajectory of the bottom end of the skinplate exhibits the characteristic of a press-shut device, a significant aspect of this vibration. When the skinplate vibrates with this press-shut

Figure 26. Waves generated on the upstream water surface, Tainter gate "C" (a) with no discharge opening, (b) with violent, self-excited vibration of 50-ton small gate opening
From Anami et al. (2007b). © 2007 ASME. Used with permission.

(a)

(b)

Figure 27. Pressure transducer fixed on the magnet block
Photograph by K. Anami.

trajectory, the vibration-induced hydrodynamic pressure feeds back the energy to the streamwise vibration of the skinplate, thus inducing the violent self-excited vibration. The press-shut angle was 43°, very close to the optimal 45° angle for which energy is most efficiently fed back. Completely similar results were obtained at the other measurement locations.

The press-shut device vibration trajectory shown in Figure 28b is naturally a result of the phasing between the vertical and streamwise vibrations. This press-shut behavior indicates that for Tainter gate

Figure 28. Self-excited vibration waveforms and press-shut trajectory at skinplate center-bottom point, with a gate submergence d_0 of 9.27 m and a gate opening B of 5 mm to 7 mm, showing (a) the growing transient vibrations, and (b) the press-shut trajectory
From Anami et al. (2007b). © 2007 ASME. Used with permission.

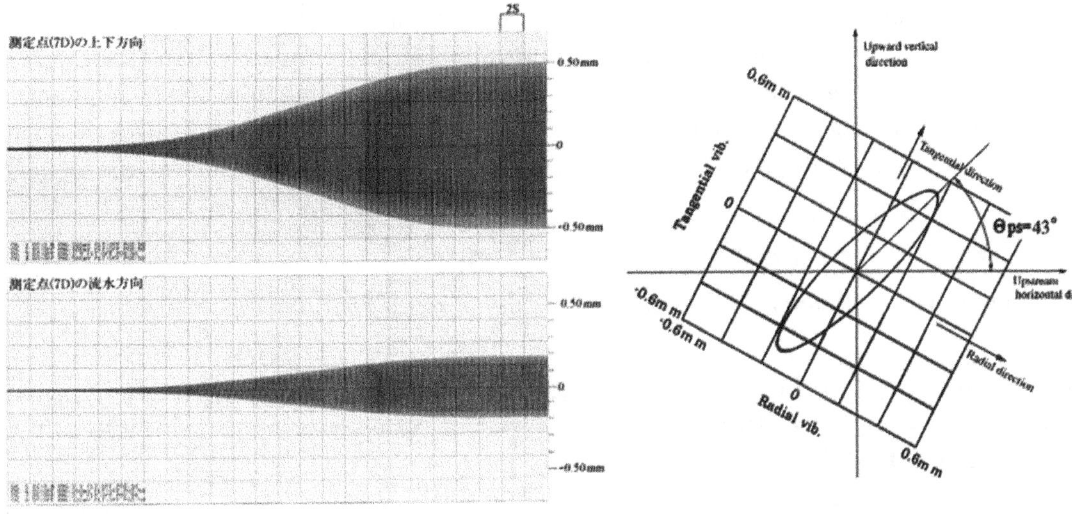

(a) (b)

"C," the whole gate vibration around the trunnion pin and the skinplate streamwise vibration are coupled with each other, accompanied by the flow-rate-variation under the gate, to generate the self-excited coupled-mode vibration system.

Significant data such as in-water natural vibration frequency, excitation ratio, vibration amplitude and vibration mode can be obtained by analyzing the self-excited vibration waveforms shown in Figure 28. The steady vibration was analyzed using an FFT analyzer to obtain the frequency spectrum, where a sharp peak was observed at 6.79 Hz for a gate submergence depth d_0 of 9.27 m and 6.71 Hz for $d_0 = 9.54$ m. The resulting values are listed in Table 6. The excitation ratios were analyzed from the exponentially growing waveforms with small amplitudes. The excitation ratio was consequently found to be 0.0034 for $d_0 = 9.27$ m and 0.0052 for $d_0 = 9.54$ m.

For the gate submergence depth of 9.00 m, intense self-excited vibrations were not observed, but small beating vibrations were measured. Analyzing the vibration accelerations measured at the skinplate bottom center, the two natural vibrations, that is, the whole gate vibration around the trunnion pin and skinplate streamwise vibration could be identified. The measured frequency of the whole gate vibration around the trunnion pin was 7.18 Hz, as shown in Table 6. The skinplate streamwise vibration exhibited two sharp peaks at 7.18 Hz and 11.3 Hz. The vibration at 7.18 Hz is due to the coupling with the whole gate vibration around the trunnion pin, while the vibration at 11.3 Hz is due to the skinplate streamwise natural vibration. Since the gate submergence depth d_0 was decreased to 9.00 m, the water added mass effect decreased, and thus skinplate streamwise vibration frequency was lowered only to 11.3 Hz, a significantly smaller frequency reduction than was measured for $d_0 = 9.27$ m (6.79 Hz) and for $d_0 = 9.54$ m (6.71 Hz). As a result, this skinplate streamwise natural vibration at a frequency of 11.3 Hz was higher than the whole gate natural vibration frequency of 7.18 Hz, and thus no self-excited vibration could occur.

For the gate submergence depth of 9.0 m, furthermore, modal response impact testing was conducted for the skinplate in the streamwise direction with a discharge opening of 4 mm. The damping ratio was calculated from the measured transfer function to be comparatively small value of 0.011, as shown in Table 6.

The in-water streamwise vibration amplitudes measured on the skinplate are shown in Figure 29, where several instantaneous vibration-shapes with different amplitudes from the small amplitude to steady state are presented. The skinplate clearly undergoes uniform parallel vibration for small amplitudes at the beginning of the self-excited vibration. As the vibration amplitude grows, the displacement in the lower half of skinplate increases more rapidly and becomes larger than the amplitude on the top half.

Table 6. List of in-water vibration frequencies and excitation ratios for Tainter gate "C"

<table>
<tr><td>gate submergence depth</td><td>d_0</td><td>9.54 m</td><td>9.27 m</td><td>9.00 m</td></tr>
<tr><td>vib. freq. around trunnion pin</td><td>$\Omega_{w\theta}$</td><td rowspan="2">6.706 Hz</td><td rowspan="2">6.791 Hz</td><td>7.18 Hz</td></tr>
<tr><td>streamwise vib. freq.</td><td>Ω_{wx}</td><td>11.31 Hz</td></tr>
<tr><td>excitation ratio</td><td>ξ_x</td><td>0.0052</td><td>0.0034</td><td>—</td></tr>
<tr><td>damping ratio</td><td>ζ_x</td><td>—</td><td>—</td><td>0.011</td></tr>
</table>

From Anami et al. (2007b). © 2007 ASME. Used with permission.

Figure 29. Streamwise parallel vibration mode of skinplate of Tainter gate "C", at d_0=9.27 m
From Anami et al. (2007b). © 2007 ASME. Used with permission.

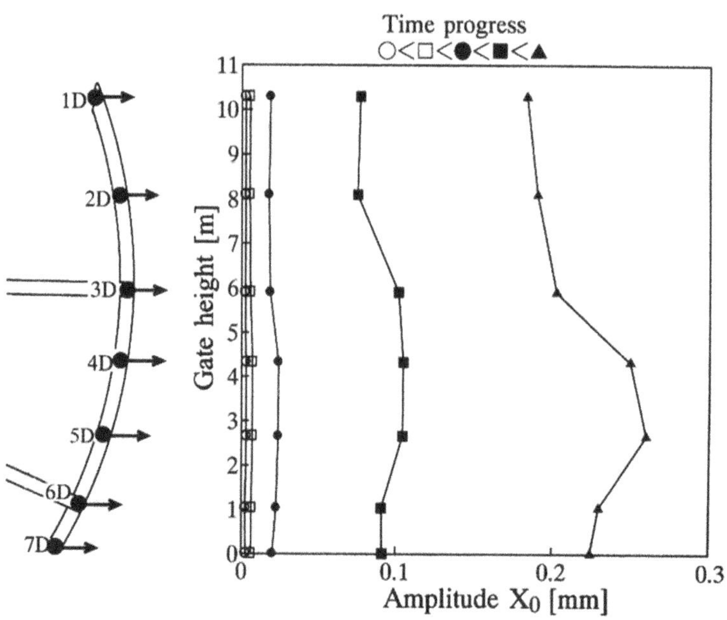

In-air experimental modal analysis could not be conducted for Tainter gate "C," so the in-air natural vibration characteristics were determined using FEM analysis. The computer and analysis software are the same as used in the FEM analysis for Folsom Dam Tainter gate. The FEM model without truss structures is shown in Figure 30. The total number of elements is 23,893.

An in-air natural vibration analysis was performed with the same FEM model, where the frequency range of analysis was from 0 Hz to 50 Hz. The natural vibration modes of the whole gate exhibiting an orderly movement were identified. Among the analyzed modes, the mode in which the skinplate center carries out a parallel vibration is shown in Figure 31, where the two sets of lines (originally red and blue lines) show the instantaneous vibration shapes with maximum amplitude. Figure 31b shows a cross-sectional view at the skinplate center. The vibration frequency was 29.38 Hz,

As seen from this figure, the skinplate performs a streamwise parallel vibration in the spanwise half-wavelength bending mode with a node at each side of the skinplate, where the upper center portion of skinplate has a large amplitude and the whole skinplate vibrates in-phase. Interlocked with this skinplate bending vibration, both radial arms perform spanwise bending vibration, as shown in Figure 31c. Therefore, the whole skinplate is considered to have a ready susceptibility to such bending vibration, because of the π-shaped structure of radial arms.

In this field test, the hydrodynamic pressure was also measured. The flow-rate variation pressure is very small and only about 1/10 of the push-and-draw pressure. So, the measured hydrodynamic pressure is dominated by the push-and-draw pressure. As determined in Figure 29, the skinplate performs almost parallel vibration in the streamwise direction, The theoretical analysis of the push-and-draw pressure in such a case has been already established in Chapter 5. The hydrodynamic pressure head $P_b/\rho g$ of the push-and-draw pressure is divided by the amplitude X_0 of the parallel vibration of the skinplate, to define the reduced push-and-draw pressure.

Figure 30. FEM model of Tainter gate "C"
From Anami et al. (2007b). © 2007 ASME. Used with permission.

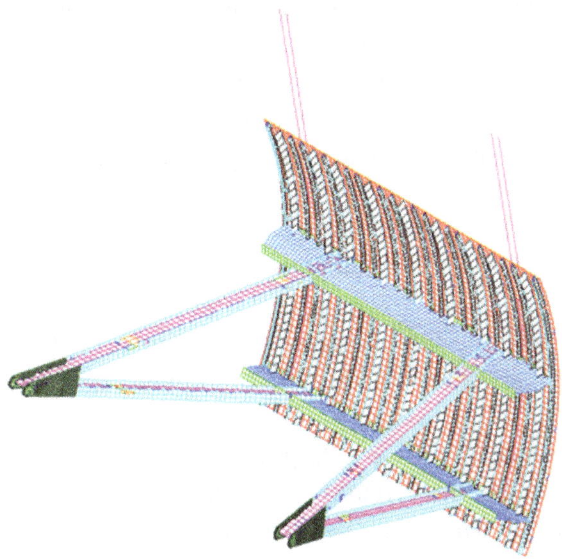

Figure 31. In-air vibration mode shape of Tainter gate "C" at 29.38 Hz showing (a) a three-dimensional view; (b) a skinplate center cut view; and (c) a top view
From Anami et al. (2007b). © 2007 ASME. Used with permission.

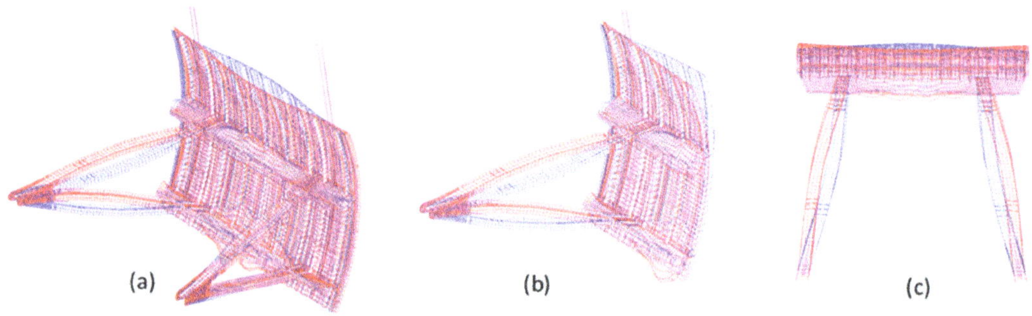

The present field test results can be used to verify the theoretical analysis of push-and-draw pressure, Comparison of the theoretical results with the experiment data is shown in Figure 32, where the vertical axis is the reduced upward coordinate y and the abscissa is the reduced amplitude of the resultant hydrodynamic pressure, p_{b0}. The parameter is the Froude number F, which takes on the following values:

$$d_0 = 9.54 \text{ m}, \Omega_w = 6.706 \text{ Hz} \rightarrow F_{9.54} = 41.6$$

$$d_0 = 9.27 \text{ m}, \Omega_w = 6.791 \text{ Hz} \rightarrow F_{9.27} = 41.5$$

The pressure amplitude p_{b0} theoretically calculated for these Froude numbers are shown by the solid line in Figure 32, where the theoretical results for the two Froude are indistinguishable.

The reduced hydrodynamic pressure amplitude data are plotted at their corresponding reduced depth by the small filled circles in Figure 32. The theoretical result shown by solid line is in close agreement with the field test data. This validation of the theoretical analysis by the massive, installed, full-scale Tainter gate is of great value.

From Figure 32, it is possible to calculate the maximum pressure amplitude considered to have been generated at the bottom center of the skinplate. The maximum vibration amplitude was 0.55 mm. The theoretical reduced amplitude was 1280 at the bottom end of the skinplate, giving a maximum pressure amplitude of 704 mm water head. Thus, it is known that a fairly large push-and-draw pressure was present.

Tainter gate "C" has the two following fundamental vibration modes:

1. Streamwise parallel vibration of the skinplate,
2. Whole gate rotational rigid vibration around the trunnion pin.

It is postulated that the coupling between these two vibration modes through hydrodynamic and inertial forces produced the press-shut trajectory shown in Figure 28, resulting in the "coupled-mode self-excited vibration."

A special feature of Tainter gate "C" is that the gravitational center of whole skinplate is lower than the position of the trunnion pin. Therefore, if the whole skinplate performs streamwise parallel vibration, the inertial force induces a significant torque around the trunnion pin. This inertia torque, in turn, drives

Figure 32. Hydrodynamic pressure acting on the skinplate, calculated for Froude number F of 41.5 and 41.6, and plotted data measured for Tainter gate "C"
From Anami et al. (2007b). © 2007 ASME. Used with permission.

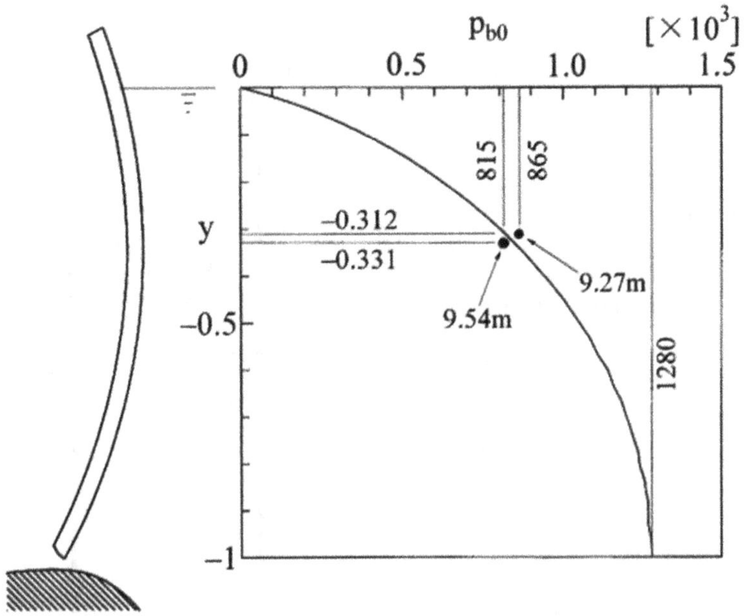

the whole gate vibration around the trunnion pin. These processes are expressed with "1" and "2a" in the closed energy cycle shown in Figure 33

When the whole gate undergoes vibration around the trunnion pin due to the inertial torque, the discharge opening will naturally change, inducing a flow-rate variation pressure. If the bottom end of the skinplate has the vibration trajectory of a press-shut device, the flow-rate variation pressure supplies energy to the skinplate parallel vibration. The closed energy cycle, shown in Figure 33, results in the effective coupling of the skinplate parallel vibration and the whole gate vibration around the trunnion pin, producing coupled-mode self-excited vibration. The mechanism of this coupled-mode self-excited vibration is identical with that of the Folsom Dam Tainter gate. The only difference is that the skinplate vibrates in a rotational vibration mode for the Folsom Dam Tainter gate whereas the skinplate vibration is in a parallel streamwise mode for Tainter gate "C."

Figure 34 shows the calculated dynamic stability criterion diagram for Tainter gate "C." The vertical axis is the critical damping ratio ζ_c required for the stability. The abscissa is the in-water natural vibration frequency ratio γ_{nw}. As listed in Table 6, the measured value of Ω_{nwx} was 6.706 Hz for $d_0 = 9.54$ m, and 6.791 Hz for $d_0 = 9.27$ m. The value of $\Omega_{a\theta}$ in the denominator of the abscissa in Figure 34 is the natural vibration frequency of the whole gate vibration around the trunnion pin, which takes a on value of 7.18 Hz, as listed in the same table. Therefore, the in-water natural vibration frequency ratio γ_{nw} takes a value of 0.93 and 0.95, values just less than one. The damping ratio of Tainter gate "C", measured from the in-water impact tests, was determined to be 0.011 for the skinplate streamwise vibration. This value is considered to be the representative value of in-air damping ratio, which is plotted at the vibration frequency ratios of 0.93 and 0.95, as shown in Figure 34. The plotted data are precisely in the most intensely unstable region.

Figure 33. Closed energy cycle of coupled-mode self-excited vibration with skinplate parallel vibration
From Anami& Ishii (2007b). © 2007 ASME. Used with permission.

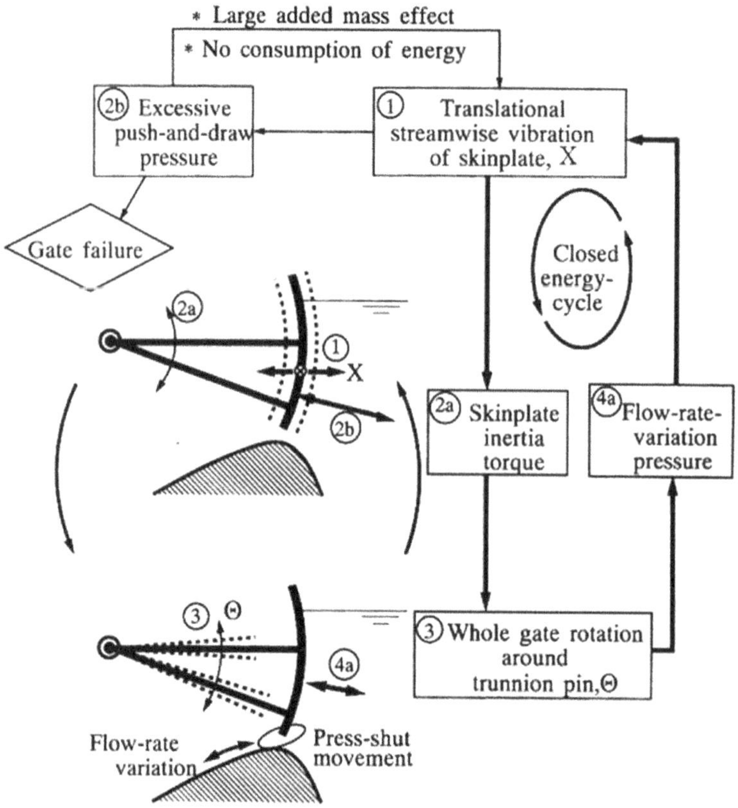

Clearly, the intense self-excited vibration of Tainter gate "C" was entirely explained by the present theoretical analysis and the resulting dynamic stability analysis. The conclusion then is that the violent self-excited vibration of Tainter gate "C" resulted from the coupling of the skinplate streamwise parallel vibration with the whole gate vibration around the trunnion pin through the hydrodynamic and inertia forces.

As shown in Figure 34, Tainter gate "C" fell into the region of intense dynamic instability. However, this gate did not fail immediately, and has not yet failed. Detailed observations of the intense self-excited vibration Tainter gate "C" have identified the following circumstances:

1. Soft rubber seals were attached to each side of the skinplate and were pressed against the side seal plate embedded in the concrete pier by the large hydraulic pressure. The rubber seals themselves experienced repeated elastic expansion and compression in the up-and-down direction.
2. However, the rubber seals were not sliding over the side seal plate, but rather were tightly fixed to it.

Such a state can be analyzed as follows:

Figure 34. Stability criterion curve and plotted streamwise damping ratio data, of Tainter gate "C" with gate submergence d_0 of 9.27 m and 9.54 m
From Anami et al. (2007b). © 2007 ASME. Used with permission.

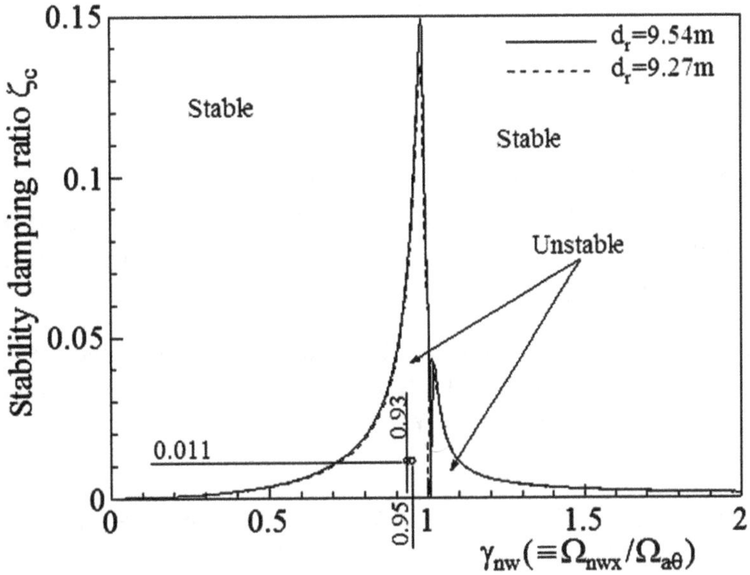

1. The rubber seals were soft, thus acting as a spring and allowing the spontaneous vibration.
2. As the vibration increases up to some level, the elasticity of rubber seal becomes significantly reduced, and thus the Coulomb friction between the rubber seal and the side seal plate began to break the amplification of vibration, thus maintaining a steady vibration with a moderate amplitude.

Thus, the observed vibration was a "friction-maintained steady vibration" or a case of "friction-maintained stability," as previously discussed in Chapter 10 in relation to the Folsom Dam gate. As noted there, friction-maintained stability is dangerous. It allows for complacency; small amplitude vibrations becomes routine and are accepted. If some sufficiently large trigger displacement had overcome the damping effect due to the Coulomb friction, during the present field vibration tests, or does so at some future time, Tainter gate "C" would have been or would be in danger of complete failure.

CONCLUSION

Testing for dynamic stability on three large Tainter gates was undertaken to determine their susceptibility to the self-excited coupled-mode instability mechanism and to validate or refute the theory developed to predict the stability of Tainter gates. The agreement between theory and measurements was very good. The theory correctly predicted the stability or instability of all three structures.

The agreement in detail concerning pressure measurements of the skinplate face were acceptable, but showed some unavoidable variance due to the limited amount of data that could collected. The measurement on full-scale gates is time-consuming and expensive both in person-hours as well as in the sophisticated equipment and instruments needed to conduct such testing.

In order to prevent further accidents similar to the Folsom Dam gate failure and to promote long-term safe operation of Tainter gates, the dynamic stability of operational Tainter gates should assessed and precautions should be taken for those suspected of instability. Any gate with confirmed susceptibility to the self-excited coupled-mode instability needs to be retrofitted to maintain the safety of the public.

REFERENCES

Anami, K., & Ishii, N. (2003). Model tests for non-eccentricity dynamic instability closely related to Folsom Dam Tainter gate failure. *Symposium on Flow-Induced Vibration 2003, Proceedings of ASME Pressure Vessels and Piping Conference*, Vol.465, (pp. 213-221). New York, NY: American Society of Mechanical Engineers. doi:10.1115/PVP2003-2092

Anami, K., Ishii, N., & Knisely, C. W. (2004a) Field vibration tests and dynamic stability analysis of a full-scaled Tainter gate, *Proceedings of 2004 ASME Pressure Vessels and Piping Conference (San Diego, California, USA)*, Vol.488, (pp. 81-88).). New York, NY: American Society of Mechanical Engineers. doi:10.1115/PVP2004-3008

Anami, K., Ishii, N., & Knisely, C. W. (2004b) Validation of theoretical analysis of Tainter gate instability using full-scale field tests, *Proceedings of 2004 Flow-Induced Vibration Conference (Paris)*, Vol. 1, (pp. 409-414). Paris, France: Centre Poly-Média.

Anami, K., Ishii, N., Knisely, C. W., & Oku, T. (2007a). Full-scale verification of coupled-mode excitation of a 50-ton Tainter gate. *Modernization and Optimization of Existing Dams and Reservoirs, Proceedings of 27th Annual USSD Conference*, (pp. 647-658). Philadelphia, PA: United States Society on Dams.

Anami, K., Ishii, N., Knisely, C. W., & Oku, T. (2007b) Field measurement of dynamic instability of a 50-ton Tainter gate. Proceedings of 2007, A.SME Pressure Vessels and Piping Division Conference. New York, NY: American Society of Mechanical Engineers.

Ishii, N., & Anami, K. (2001). Application of experiences and of recent findings with Tainter gates in Japan on the Folsom Dam gate stability. *Proceedings of Commemorative International Symposium for the 40th Anniversary for the Foundation of the OECU*, (vol. 2, pp. 193-206). Osaka, Japan: Osaka Electro-Communication University. doi:10.1115/PVP2007-26543

APPENDIX: NOMENCLATURE

B: Discharge opening
c_f: Instantaneous flow-rate variation coefficient
d: Diameter of steel rod
d_0: Gate submergence depth
d_r: Depth of the reservoir
F: Froude number
F_0: Basic Froude number
g: Acceleration of gravity
I_θ: Moment of inertia of the whole gate around the trunnion pin,
I_ψ: Moment-of-inertia of skinplate streamwise rotation
$I_{\psi\theta}$: Coupled moments-of-inertia
P_0: Measured resultant pressure
p_{b0}: Reduced push-and-draw pressure amplitude
R_c: Radius of the bottom end of the weir plate
r_s, R_S: Height of skinplate rotation center [r_s is the dimensionless expression of R_s]
W_0: Spanwise length
y, Y: Vertical coordinate [y is the dimensionless expression of Y (m)]
$\alpha_{I\psi}$: Ratio of the coupled moment-of-inertia $I_{\psi\theta}$ to the skinplate moment-of-inertia I_ψ
α_ψ: Water-to-gate mass ratio
β^*: Depth ratio
γ_{nw}: In-water to in-air vibration frequency ratio
$\gamma_{nw\psi}$: In-water to in-air vibration frequency ratio of the skinplate streamwise vibration
δ_p: Pressure correction coefficient
ζ_a, $\zeta_{a\psi}$: In-air damping ratio for streamwise vibration of the skinplate
$\zeta_{a\theta}$: In-air damping ratio of the whole gate rotational vibration,
$\zeta_{c\theta}$: Critical damping ratio of the whole gate rotational vibration
$\zeta_{c\psi}$: Critical damping ratio of the skinplate streamwise vibration
$\zeta_{w\theta}$: In-water damping ratio of the whole gate rotational vibration,
$\zeta_{w\psi}$: In-water damping ratio of the skinplate streamwise vibration
ζ_x: In-water damping ratio
ξ_x: Excitation ratio
Θ_s: Gate inclination angle
Θ_{s0}: Geometric press-open angle
ρ: Density of water
Ω_a, $\Omega_{a\psi}$: In-air vibration frequency of the skinplate streamwise vibration
$\Omega_{a\theta}$: In-air whole gate rotational vibration frequency
Ω_w: In-water vibration frequency of the skinplate streamwise vibration
Ω_{wx}: In-water vibration frequency of the skinplate streamwise parallel vibration

Chapter 15
Reflections and Conclusions

ABSTRACT

In this chapter, reflection on the discovery of gate instability mechanisms is provided. Stemming from Ishii's encounter as an undergraduate with the Wachi gate failure and his subsequent development of the theory for eccentricity instability, a framework for future analyses of other gate instabilities was established. The study of two degrees-of-freedom instabilities of long-span gates created a paradigm for mode coupling in hydraulic gate vibrations. The Folsom failure occurred with eyewitness testimony claiming to have heard and felt vibration. The path to understanding the mechanism that could produce vibration of the Folsom gate was the realization that the skinplate can be easily excited to undergo streamwise vibration. To counteract such vibration, dynamic design criteria for Tainter gates are needed. A draft formulation of dynamic guidelines for Tainter gate design is developed. We hope for feedback from those who use the guidelines to provide for the continuing improvement of the guidelines.

PREAMBLE

Engineers, in the fulfillment of their professional duties, shall hold paramount the safety, health, and welfare of the public. (NSPE Code of Ethics for Engineers, Fundamental Canon Number 1)

INTRODUCTION

In many respects this book has been shaped by the experiences of three people, but it also bears the influences of many people who taught us basic engineering principles, many colleagues with whom we interacted, and the students whom we taught. Along the way, the authors came to understand, in a very personal way, the first fundamental canon of the engineers' code of ethics cited in the preamble to this final chapter.

We, the authors, might find great consolation in discovering that our work contains errors that make it superfluous because the instabilities about which we write would never occur. What we fear most, however, is a massive gate failure somewhere in the world at an intolerable cost of human lives, despite

DOI: 10.4018/978-1-5225-3079-4.ch015

our best efforts to make our work as accurate and as error-free as possible. We would be extremely frustrated to see our work dismissed by decision makers for various reasons, only to have an avoidable catastrophe occur. Politics and personal preferences should not dominate reasoned attempts to improve the human condition.

This book is an international collaborative effort that has involved meetings at various venues around the world and millions of bytes of data transmitted and received initially via postal carriers and now via internet carriers. The framework of this final chapter is to put the previous fourteen chapters in a chronological context to provide a big picture view of how these topics were developed.

BEGINNINGS

Ishii became interested in gate vibrations in 1967, when he was in his junior year at the university. As he recalls in his reflections at the end of this chapter, he first learned the details of the Wachi Dam gate failure in a lecture in his hydraulics class presented by Professor Emeritus T. Uematsu. A few years later, in 1971, when he had enrolled as a doctoral candidate, Ishii's major professor, Professor K. Imaichi, suggested that Ishii undertake the study of the Wachi Dam gate failure as his thesis topic. His dissertation work on the failure at the Wachi Dam was completed in 1975. His research led to the identification of the eccentricity instability mechanism for Tainter gates (see Ishii, 1975; Imaichi & Ishii, 1977; Ishii et al. 1977a, 1977b, 1980, 1983; Ishii & Imaichi, 1982). In 1983 he was awarded a von Humboldt Fellowship to work with Professor E. Naudascher at the Institute of Hydromechanics in Karlsruhe, Germany. Several papers resulted from Ishii's stay in Karlsruhe (Ishii & Naudascher, 1984; Ishii et al., 1987, Ishii & Naudascher, 1992).

Knisely was employed at the Institute of Hydromechanics in Karlsruhe and was asked to help Ishii get settled. Although these two authors met and developed a friendship while they were both working with Professor E. Naudascher in Karlsruhe, their research collaboration did not begin until a few years later, after Ishii had returned to Japan and Knisely accepted a position working with Professor H. Nakagawa at Kyoto University in 1987.

Once Knisely settled in Japan, Ishii and Knisely began a collaboration on the flow-induced vibration of long-span gates with an initial focus on the effects of the shear layer beneath wide gates with submerged discharge (see Ishii & Knisely, 1988, 1989, 1990). After the work on shear layers, Ishii and Knisely moved on to consider streamwise vibration of long-span gates and then long-span gates with two degrees- of-freedom, and then long-span gates with simultaneous over- and underflow (see Ishii, 1990a, 1990b, 1992; Ishii & Knisely, 1992; Ishii et al., 1994, 1995a, 1995b; Nakata et al., 1996).

REFLECTIONS ON THE PROCESS

There were several milestones in our quest to understand gate instability and, ultimately, to develop methods of predicting the flow-induced vibration behavior of hydraulic gates. The *first was the development of the method of applying Rayleigh's dissipative wave theory by Ishii* in his thesis work and subsequent publications noted above. We have chosen to write an entire chapter on the fundamental nature of this solution, because it has guided almost all subsequent development. Despite the substantial simplification of the flow field needed to make the boundary conditions tractable and the subsequent complicated mathematical solution, this potential flow analysis provided a method of predicting fluctuating pressure in the reservoir due to discharge fluctuations. The *second milestone in understanding gate instability is* that, for Froude numbers great than about 2.0 (see Figure 6 in Chapter 7), *the flow-rate- variation pressure acts to provide fluid damping or fluid excitation*. The predictive capability of this formulation was validated in laboratory tests, in which the physics of pressure fluctuations due to discharge fluctuation through small openings was correctly described. This analysis became the cornerstone for the methods that followed.

The observation that full-scale long-span gates vibrated in the streamwise direction and our desire to understand the origin of these vibrations required a method to predict the loading on a gate resulting from push-and-draw pressure fluctuations. After successfully predicting flow-rate-variation pressure fluctuations by applying the potential flow model to a simplified geometry, the same geometry was used to analyze streamwise gate displacement. The solution of the mathematical formulation made it possible to predict the fluctuating pressure field in the reservoir due to the streamwise motion of the gate. Provided the Froude number exceeds a value of about 5.0 (see Figure 7 in Chapter 7), the push-and-draw pressure neither provides energy to nor extracts energy from the vibrating body that generated the pressure fluctuation. Therefore, *the third essential element in understanding gate instability is that push-and-draw pressure acts as added mass to reduce the vibration frequency of the structure*.

Shell-type long-span gates with two degrees-of-freedom pose a different problem, that of coupled vibration. First we considered only shell-type gates subjected to underflow, with an inclined portion to the upstream weir plate. We needed a method of predicting the vertical gate response to the dynamic loading on the inclined surface resulting from streamwise gate motion. We assumed that the horizontal motion with a large push-and-draw pressure serves as the driver for the vertical motion. As shown in Figure 1b (a copy of Figure 1 from Chapter 2), the phase between the pressure load and the vertical motion depends on their frequency ratio. The forcing load and the system response vary from being in-phase for low frequency ratios to 180° out-of-phase for high frequency ratios. The phase relationships among the horizontal motion, the push-and-draw pressure, and the vertical response to the pressure loading dictate whether or not the vibration will be sustained. When the horizontal and vertical motions effectively couple, the trajectory of the gate motion is observed to be an inclined ellipse, traversed in

Figure 1. (a) Amplitude and (b) phase response for a second-order system undergoing forced vibrations

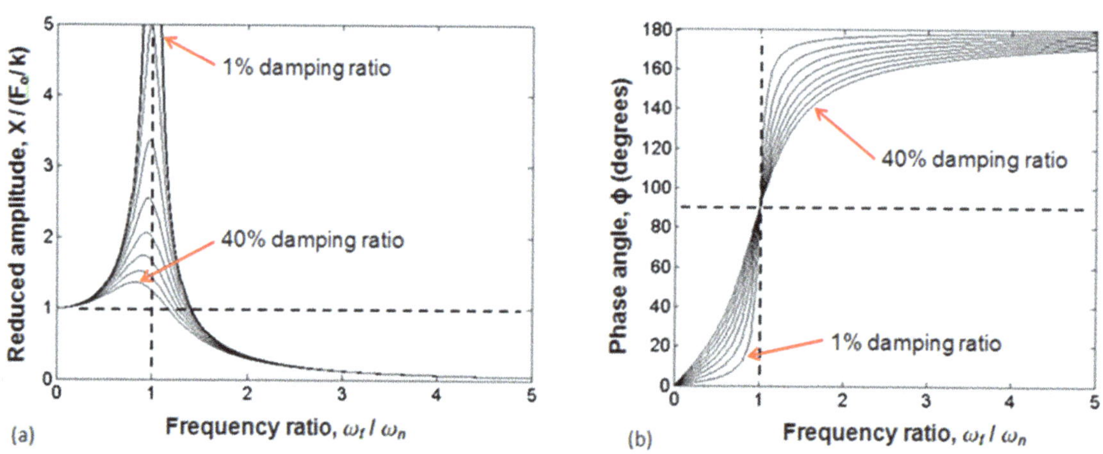

the clockwise direction (assuming the positive x-direction is upstream), as shown in the trajectory on the right side of Figure 2

This inclined ellipse traversed in the clockwise direction is indicative of press-shut behavior. The inclined ellipse indicates the gate moves downward and downstream, then upward and upstream. A gate with such a trajectory can be likened to a plate gate, inclined in the downstream direction, subjected to a sudden increase in upstream pressure loading. The gate tends to close, or press shut. Gates that behave as press-shut devices are susceptible to coupled-mode vibrations. Thus, *the fourth essential element in understanding gate instability is that the frequency ratio and phasing between the two directions of motion play an important role in two degrees-of-freedom systems.*

For shell-type long-span gates with overflow, both an inclined upstream weir plate and the gate crest with overflow can experience vertical force components. To predict the gate response in such a situation, these two vertically loaded areas had to be analyzed separately. For this reason, we examined gates with overflow and no inclined weir plate as well as gates with inclined weir plates and no overflow. The *fifth essential element in understanding gate instability is thus to break a complicated problem into several simpler problems* if you can do so without changing the fundamental physics of the problem.

IN-WATER FREQUENCY AND DYNAMIC PRESSURES

The problem of vibration of gates with two-degrees of freedom in underflow was analyzed by applying the one degree-of-freedom analysis to the main driving vibration in the x-direction and treating the y-direction vibration as a response vibration to the fluctuating pressure load, as represented in the flow chart in Figure 3. The in-water x-direction vibration frequency was determined by methods from Chapters 5 and 7 and by knowing the in-air vibration characteristics.

To analyze the two degree-of-freedom vibration of gates with overflow, this method had to be adapted to take into account the effect of the overflow on the added mass calculation. Since the gate height is smaller than the upstream submergence, d_0, the length scale was corrected. As a crude approximation, half of the gate height was assumed to be effective in producing added mass. With this length scale correction, the calculated results agreed well with the measured values for the laboratory-scale studies. We

Figure 2. An example of coupled-mode vibration for a two degrees-of-freedom shell-type gate with a 30° inclined upstream weir plate in underflow showing characteristic press-shut behavior
From Nakata et al. (1996). © 1996 JSME. Used with permission.

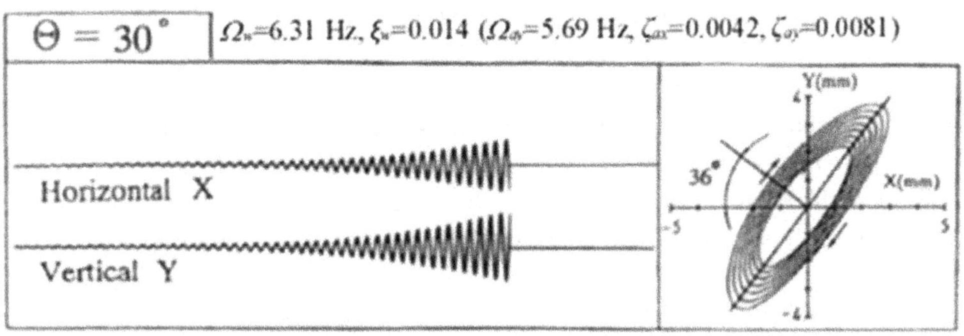

Figure 3. Flow chart of coupled mode vibration of two-degrees-of-freedom vibration of a shell-type long-span gate in underflow
From Ishii & Knisely (1992). ©1992 Elsevier. Used with permission

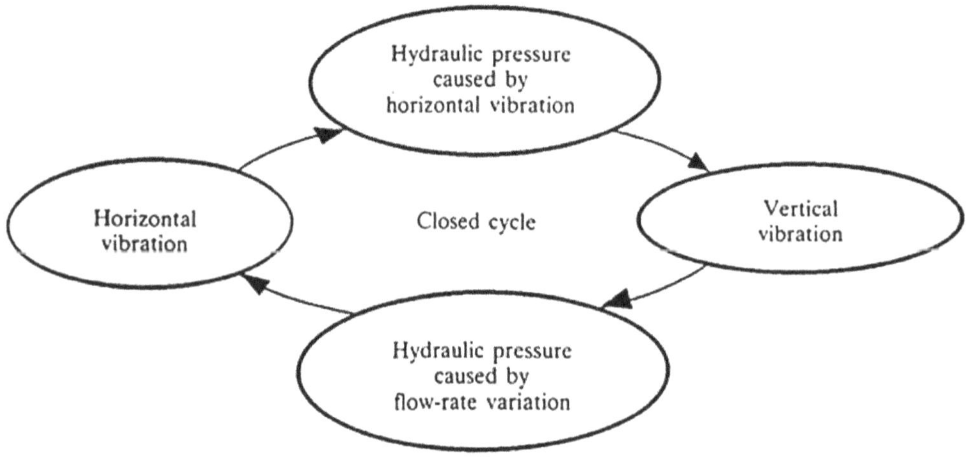

expect our approximation to be applicable to most smoothly curved overflow crests, although further validation is needed.

The crest pressure and the loading on the gate could be estimated by using the corrected length scale to calculate added mass and fluctuating pressure magnitude. To our knowledge, the estimation of dynamic loading on a gate with overflow presented in Chapter 7 is still the only simple analytical (non-CFD) method for doing so.

SUMMER 1995: FAILURE DRIVES SUCCESS

During the summer of 1995, Knisely was in Japan to continue his collaboration with Ishii. On July 17, the Folsom Dam gate failed. Neither Ishii nor Knisely realized how significant this failure would be on their research direction for the next twenty years. At this time, the second author of this book, Keiko Anami, was a third-year undergraduate student. She became fascinated with Ishii's description of the Folsom Dam gate failure in one of her undergraduate classes, and subsequently took on the topic of coupled-mode vibration of Tainter gates for her PhD dissertation after she completed her undergraduate studies and entered graduate school.

Ishii was a member of the forensics team for the Folsom Dam failure. He carefully calculated the gate eccentricity and found that eccentricity instability was not an option. He spent many hours walking the roadway above the dam crest trying to understand the contradiction between the reported vibration at the time of failure and the lack of any known mechanism to explain the vibration. Ultimately, the forensics team issued a report, ascribing the failure to static loading on the gate due to excessive trunnion friction. Ishii was still bothered by the contradiction between the eyewitness testimony and the forensics team's conclusion, but, unable to offer a mechanism for the observed vibration, he had to concede to the majority opinion.

The answer to Ishii's question about the mechanism of vibration came very slowly with substantial investments of time and money. Chapters 9 to 12 summarize the studies that were undertaken to understand the Folsom Dam failure. Although we recognized some similarities in the vibration mechanisms between the two degrees-of-freedom shell-type long-span gates and Tainter gates, there were also fundamental differences: one was the role played by the unknown two-dimensional mode shape of the very large skinplate, and the second was the parametric coupling between the skinplate deformation and the moment of inertia about the trunnion pin. The Folsom Dam gate had about 63% of its mass concentrated in the skinplate. Small deflections of the skinplate made substantial changes in the moment of inertia. The parametric coupling made the vibration analysis of the Folsom Dam much more challenging.

Over the course of a five-year period, Anami developed the theory of self-excited coupled-mode vibration for her dissertation (see Anami 2002). The theory she developed is presented in Chapter 10. The flow chart of the mechanism of self-excited coupled-mode dynamic instability of Tainter gates is replicated in Figure 4 as a symbol of the culmination of many years of hard work.

The desire to truly understand the failure of the Folsom Dam gate drove us to find a resolution to the inconsistencies between the eyewitness testimony and the report of the forensics team. The self-excited coupled-mode instability mechanism provided not only a way to reconcile the Folsom Dam discrepancies, but also offered a plausible answer to the lingering question about the Wachi Dam gate failure.

An understanding of the self-excited coupled-mode instability opens the dialogue on possible countermeasures. The traditional countermeasure of stiffening the gate by adding mass is problematic on many fronts. The additional mass changes the requirements for hoist motors that open and close the gate, and often the additional gate mass means the moment of inertia also increases, resulting in a frequency ratio that remains almost unchanged. Possible countermeasures that would change the stiffness without adding mass include using different materials for the skinplate or changing the elasticity of the chains (see Anami et al., 2013). In addition, understanding the mechanism may also help gate owners understand the range of operational parameters (see Anami et al., 2015).

The optimal solution to potential Tainter gate instability issues would be to solve the problem of dynamic stability in the design stage. The next section of this final chapter incorporates the findings presented in this book into a coherent dynamic design process that we hope will help alleviate Tainter gate stability issues in the design stage.

DEVELOPMENT OF DESIGN STAGE DYNAMIC STABILITY ANALYSIS

Need for Dynamic Design Analysis

Despite the advantage of Tainter gates having small hoisting loads, that permit operating these gates with relatively low capacity hoist motors, they have one substantial shortcoming. As discussed in the previous chapters in this book, all Tainter gates are potentially susceptible the coupled-mode dynamic instability mechanism. Since the mechanism depends primarily on the ratio of two in-water vibration frequencies –one frequency is that of the streamwise rotational vibration of the skinplate in a bending mode, and the second is that of the whole gate torsional vibration around the trunnion pin–this susceptibility can be eliminated by judicious adjustment of the natural vibration frequency ratio in the design stage (Anami et al., 2017).

Figure 4. Closed energy cycle for the self-excited coupled-mode flow-induced instability of Tainter gates From Anami & Ishii (2001).

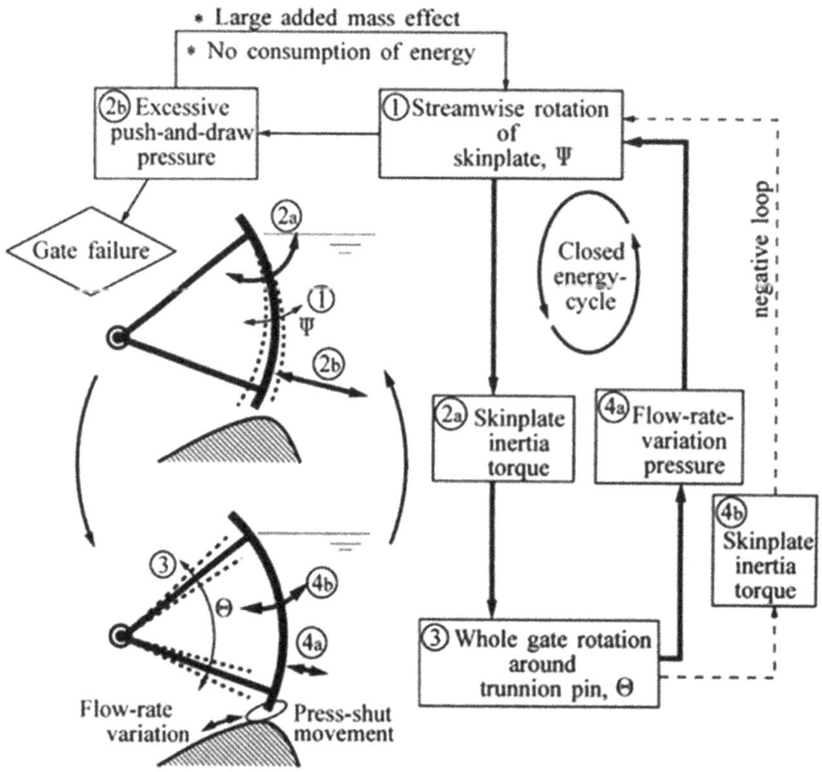

Currently Tainter gate designers are required to adhere to static design criteria, focusing on the static structural strength of the gate. No dynamic design criteria are required; such dynamic criteria would focus on the inherent vibration characteristics of the gate. Careful consideration of gate dynamics in the design stage would result in final gate designs that are both statically and dynamically stable.

To ensure both the dynamic and static stability of newly constructed Tainter gates, methods of dynamic analysis and dynamic design criteria are needed to complement existing static design criteria to assure the dynamic stability of any new Tainter gate.

A detailed draft procedure for the dynamic design of Tainter gates is proposed. The intent is for this procedure to be applied to preliminary designs by gate engineers to assure the dynamic stability of their gates. The steps in this procedure are:

1. Undertake a finite element method (FEM) analysis of the preliminary design to determine the in-air structural vibrations characteristics of the gate and, thereby, identify the lowest frequency bending mode of the skinplate. Such a mode exhibits a rotational or parallel vibratory displacement distribution along the vertical spanwise centerline.
2. Approximate the gate's structural damping characteristics using our previously determined in-water damping characteristics for nine types of full-scale Tainter gates, presented below.

3. Compute the theoretical stability diagram using the specifications for the preliminary gate design. The stability diagram shows the in-water damping as a function of the in-water vibration frequency ratio.
4. Adjust the gate specifications following the dynamic design guide (provided below) to move the stability characteristic away from any identified unstable range of in-water frequency ratios.

The mechanism of the coupled-mode instability is presented schematically in Figure 4. The details of the mechanism were presented near the end of Chapter 10, as well as several times in subsequent chapters. An understanding of this mechanism will help designers understand the dynamic design process, and for this reason the previous summary of the mechanism is repeated here.

When the in-water frequencies of streamwise bending and rigid body rotation about the trunnion pin are approximately equal the mechanism is activated. The whole gate rotation around the trunnion pin induces a flow rate variation beneath the gate, with a corresponding flow-rate variation pressure which forces the skinplate to vibrate in the streamwise direction. Subsequently, the skinplate streamwise vibration induces a push-and- draw pressure, which significantly lowers the in-air vibration frequency of the skinplate. Therefore, these two vibration modes can easily couple to induce violent self-excited vibrations (Anami & Ishii, 1999, 2000, 2003).

Figure 4 presents the closed loop energy cycle of the coupled-mode self-excited vibration. If the bending rotational counter-clockwise vibration ① of the skinplate, represented by Ψ, is triggered by some random excitation, the counter-clockwise skinplate inertia torque ② appears around the trunnion pin, thus exciting the counter-clockwise whole gate vibration ③, represented by Θ. As a result, the gate opening under the gate increases, thus resulting in increased discharge water flow ④, and thus decreasing the flow-rate-variation pressure ⑤ in proportion to the whole gate vibration velocity, that feeds energy back to the skinplate bending rotational vibration ⑥, thus amplifying the bending vibration amplitude. The changing inertia torque due to skinplate deflection serves to drive the whole gate vibration and the cycle closes.

Step 1: FEM Analysis to Identify the In-Air Natural Vibrations Characteristics

In order to investigate the dynamic stability of the gate, it is very important to determine the natural vibration characteristics of the gate. For installed gates, the actual vibration characteristics can be found from experimental modal analysis. However, conducting experimental modal analysis requires great expenditures of both time and money. For a preliminary paper design, any physical testing such as experimental modal analysis is clearly impossible. The alternative is computer-based FEM analysis. Dynamic FEM analysis, however, is generally considered not especially well-suited for large complex structures such as dam gates, since it includes uncertainties involved in the element selection in model creation and the determination of the rigidity of each member. Due to these uncertainties, an FEM analysis may result in poor comparisons with experimental modal analysis results. To overcome these difficulties, the authors have conducted dynamic FEM analyses for the natural vibrations of installed Tainter gates, and subsequently compared the FEM results with those from experimental modal analysis. By careful adherence to certain guidelines in FEM model creation, the FEM results were confirmed by the experimental data, demonstrating the potential applicability of FEM for the dynamic analysis of large, complicated structures such as Tainter gates (see Ishii et al. 2009a; 2009b; Anami et al. 2014) The following section outlines the creation of an FEM model.

Formulation of Finite Element Model

Based on the blueprint of the Tainter gate, a detailed finite element model of the gate must be created. Experience suggests that for complicated structures such as Tainter gates, the precision of the modeling has a significant influence on the subsequent results. For this reason, the form of the model must faithfully replicate the precise details of the blueprint. The FEM gate model is created using plate elements, beam elements, and spring elements.

In FEM analysis, the element division is a very significant aspect in achieving accurate analysis results. An accurate analysis can be obtained with a limited number of elements, but the element division should be concentrated in the significant portions in the model where loads and moments are transferred. In our FEM analysis of the Folsom Tainter gate, fine element divisions were especially important for the connections between the skinplate and radial arms, since the experimental modal analysis showed that they were significant factors in determining the natural vibrations of the whole gate.

All structure elements in actual gates are joined with bolts, rivets or welds. To simulate sufficient rigidity, the junctions in the FEM model should be specified as rigid in the constraint characteristic. In this manner, a high rigidity is assigned to one independent node and two or more subordinate nodes, thus enabling the transmission of the translation force and moment produced in the independent node to the subordinate nodes.

The trunnion pin, installed in the concrete, is modelled with one degree of freedom for movement around its rotation axis. Therefore, all other trunnion pin flexibilities should be constrained. The chain or wire cable ends are assigned degrees of freedom in the rotational and translational directions, while the spanwise direction can be constrained.

Representative examples of high accuracy FEM models of installed Tainter gates are shown in Figure 5, where the element divisions are shown. The points of intersection in these views are the element nodes. The elements surrounded by the nodes on the skinplate and on the radial arms are plate elements,

Figure 5. Precise FEM model construction of representative full-scale Tainter gates: (a) 87-ton Folsom Tainter gate, from Anami et al (2014). © 2014 IJMERR. Used with permission; and, (b) 24-ton Tainter gate "T"

From Ishii et al. (2009a). © 2009 Aqua~Media International Ltd. Used with permission.

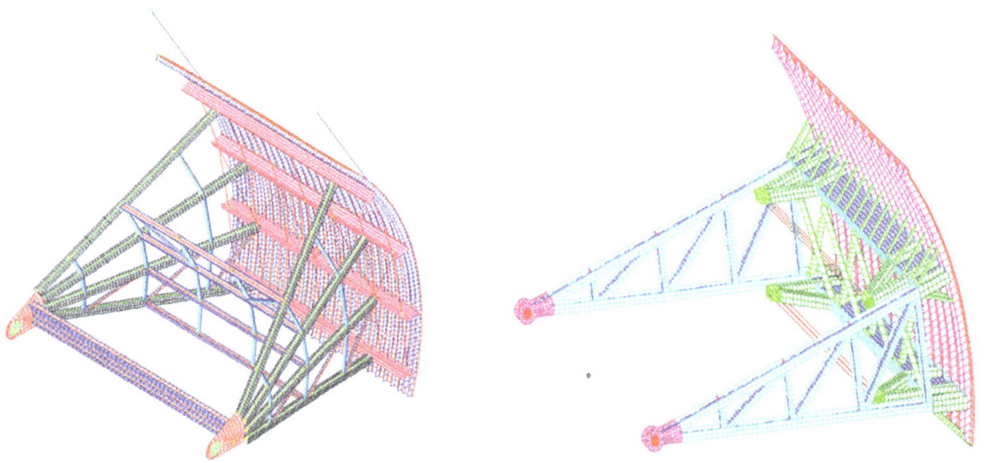

while the lines connecting the nodes of the horizontal girders and skinplate reinforcement T-beams are modeled using beam elements.

Natural Vibration Analysis

The natural vibration frequencies and vibration modes were analyzed using FEM analysis software (MSC, NASTRAN for Windows). The analyzed frequency range was 0 to 50 Hz, because it is known from experience that the natural vibration frequency of these large gate structures is around 30 Hz. The calculated vibration modes can be shown by animation on a computer screen, to determine the natural vibration modes as those modes for which the whole gate was carrying out orderly vibration.

The results obtained from the FEM analysis for the Folsom Dam gate are shown in Figure 6, where the two sets of lines (originally red and black lines) show the instantaneous vibration shapes with maximum amplitude. In order to help identify deflections, the amplitudes are amplified.

Figure 6(a) shows the whole gate rotational vibration mode around the trunnion pin with the frequency of 7.3 Hz. Deformations are not apparent for any of the gate beams. This vibration is apparently caused by the elastic support of the chains which raise and lower the gate. Figure 6(b) shows the skinplate streamwise bending vibration mode. In the streamwise vertical plane, the skinplate is performing its lowest frequency bending mode vibration with the node near the second horizontal girder. In the horizontal plane, the skinplate is performing a half wavelength vibration with nodes near each side of the skinplate. The natural vibration frequency is 25.4 Hz.

These resulting FEM natural vibration modes and frequencies are in good agreement with the experimental modal analysis results from the field vibration tests undertaken for the Folsom Dam Tainter gate with the same design as the failed gate in 1995, given in Figures 11 and 12 in Chapter 12.

A similar FEM analysis undertaken for Tainter gate "T", using the high precision FEM model shown in Figure 5(b). The inherent vibration characteristics are listed in Table 1, where the in-water field vibra-

Figure 6. Vibration mode shapes of Folsom Dam Tainter gate, calculated by FEM analyses: (a) Whole gate rotational vibration, $\Omega_{a\theta}$ = 7.28 Hz, (b) skinplate bending vibration, $\Omega_{a\psi}$ = 25.4 Hz
From Anami et al. (2014). © 2014 IJMERR. Used with permission.

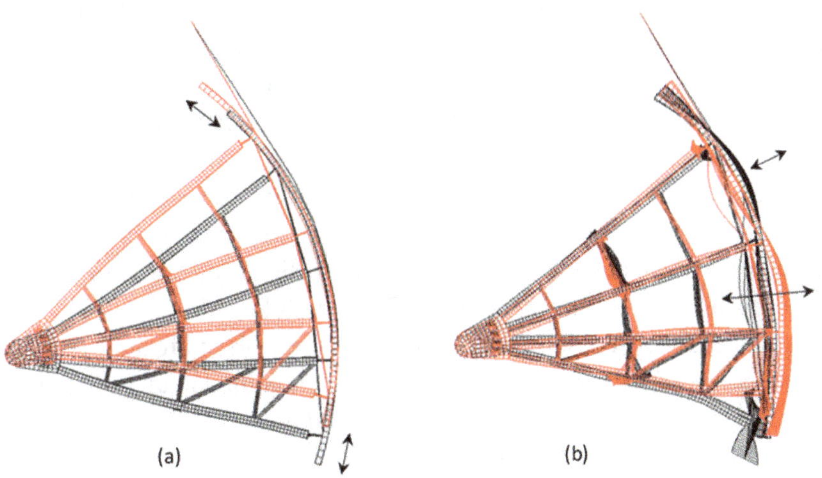

Table 1. Results of FEM analysis for fundamental vibration modes of installed Tainter gate "T", and comparison with those from field vibration tests

Mode	FEM analysis		Field vibration tests		
	Vibration Mode shape	freq. (Hz)	Vibration Mode shape	frequency (Hz) measured (in-water)	frequency (Hz) calculated (in-air)
Mz	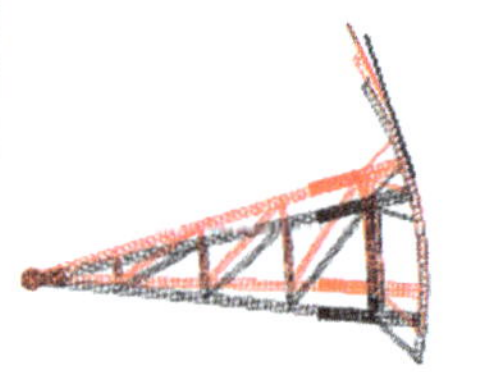	$\Omega_{a\theta}$ 6.5	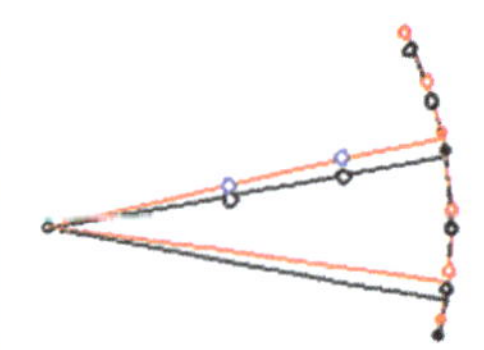	$\Omega_{w\theta}$ 6.5	$\Omega_{a\theta}$ 6.5 relative error 0 %
Mx1	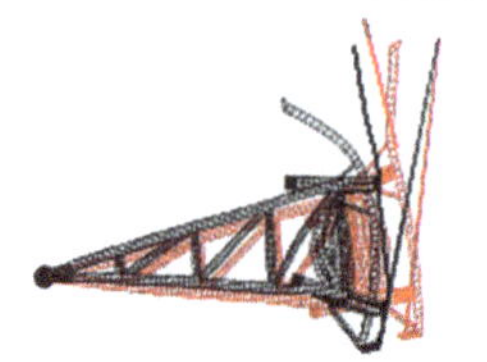	$\Omega_{a\psi}$ 20.8	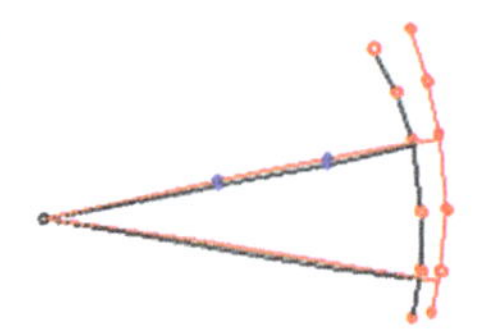	$\Omega_{w\psi}$ 5.5	$\Omega_{a\psi}$ 20.9 relative error 0.48 %
Mx2		$\Omega_{a\psi}$ 13.7	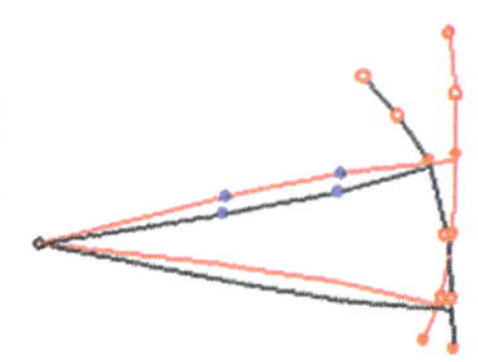	$\Omega_{w\psi}$ 6.5	$\Omega_{a\psi}$ 14.9 relative error 8.0 %

From Ishii et al. (2009a).

tion test results are also shown for comparison. Here, attention is directed to the skinplate streamwise vibration modes which are particularly important in the discussion of the dynamic stability of the gate. The in-air and in-water vibration mode shapes are quite similar, suggesting the weak effect of upstream submergence on vibration modes (Anami et al., 2014). The in-air and in-water frequencies of the skinplate streamwise vibration modes are naturally different, depending on the added mass effect of water; those for the whole gate rotational vibration are almost the same in-air and in-water, providing evidence of almost no added mass effect of the water on the rigid-body gate rotational mode.

The good agreement between computed and measured frequencies validate the FEM simulations, showing they are well-suited for determining the vibration modes. Finite element models constructed following the above guidelines contain the essential physics of the large-scaled installed Tainter gates. With the appropriate selection of elements in the FEM model, the natural vibration characteristics can be determined using FEM simulations for preliminary gate designs and for other gates with limited access for which experimental modal analysis would not be possible.

It should be noted here that the damping ratios for both natural vibration modes cannot be determined from FEM analysis, but are readily identified from experimental modal analysis (Anami et al., 2014). Suggested reasonable damping values may be assumed based on multiple full-scale gate testing. These values are given in the next section of this chapter.

Step 2: Determine Dynamic Characteristics Using Major Parameters

The natural vibration characteristics obtained by FEM analysis or experimental modal analysis permit the calculation of the basic parameters necessary to determine the gate's dynamic stability.

Basic Froude Number and Rotation Center Height

The first parameter is called the "basic Froude number" F_0. It is defined using the natural vibration frequency $\Omega_{a\psi}$ of the skinplate streamwise vibration and the gate submergence depth d_0 as

$$F_0 = 2\pi\sqrt{d_0 / g}\Omega_{a\psi} \text{ (with } \Omega_{a\psi} \text{ in [Hz])} \quad (1)$$

In the determination of dynamic stability, this basic Froude number is an important parameter characterizing the dynamic similarity of the flow field.

The skinplate streamwise vibration modes, shown in Figure 6(b) and Table 1, are typically the lowest-order bending vibration mode in the streamwise direction. In these modes the skinplate performs a rotational vibration around a horizontal nodal line, termed the streamwise-rotation center line, where the skinplate rotation center height R_S from the bottom of the skinplate is significant in dynamic stability determination. This is because not only the moment of inertia of the gate, but also the hydrodynamic pressure acting on the skinplate depends on the rotational center height. The hydrodynamic pressure induced by the skinplate streamwise rotation has been theoretically analyzed and fully verified by model and field experiments (Anami et al. 2012a; 2014) As a result, the second parameter needed is the reduced rotation center height r_s,

$$r_s = R_S / d_0 \quad (2)$$

which is defined as the ratio of the skinplate rotation center height R_S to the gate submergence depth d_0. The rotation center height R_S can be determined from FEM analysis or by experimental modal analysis.

Water-to-Gate Mass Ratio

The hydrodynamic pressure acting on the skinplate produces an added mass effect. Due to the added mass effect of the water, the in-air natural vibration frequency of the skinplate is significantly decreased (Anami et al. 2012b). The third parameter, the water-to-gate mass ratio, α_ψ, is needed to determine the decrease in the in-water gate vibration frequency relative to its in-air value. The water-to-gate mass ratio is defined as

$$\alpha_\psi = \frac{\rho d_0^2 W_0}{I_\psi / R_S^2} \quad (3)$$

The numerator is the representative mass of water in a rectangular parallelepiped given by the product of the square of one side d_0 and the spanwise length W_0. The denominator is the equivalent mass determined by the moment of inertia of the skinplate around its rotation center I_ψ divided by the square of rotation center height R_S. Thus, larger values of this water-to-gate mass ratio correspond to more lightweight designs in which the water mass is larger than the skinplate mass and I_ψ can be calculated assuming that the skinplate, the reinforcing bracings of the skinplate, and the horizontal girders perform rotational motion, according to our previous experience.

Structural Damping Ratio

A reasonably accurate estimate of the in-air damping ratio for all identified natural vibration modes is needed to determine the dynamic stability of the gate. However, damping ratios cannot be determined using FEM analysis. Field vibration tests in which the damping ratio of the gate can be definitively measured are, therefore, necessary.

To collect empirical data for the structural damping ratio, we have conducted field vibration tests on nine different Tainter gates, some of which were presented in Chapter 14. Table 2 shows the field vibration test results (Anami et al., 2007, Ishii et al., 2009a; 2009b) for the frequencies and damping ratios of the inherent vibrations obtained in the in-air and in-water field vibration tests, along with the test conditions. Blank spaces indicate that the corresponding value could not be measured. Simple mean values are listed at the bottom of the table. For the streamwise vibration mode, damping ratios showed no significant difference between in-air (0.014) and in-water (0.015) average values, ranging from a maximum of 0.040 to a minimum of 0.003. Note, however, for Gate B where both in-air and in-water measurements were made, there is a 38% reduction in the damping ratio in water relative to the in-air value. In contrast, the damping ratios for the whole gate rotational vibration around the trunnion pin were substantially different in air and in water. The in-water values ranged from a maximum of 0.064 to a minimum of 0.011, with an average damping ratio of 0.024. The average value in air, 0.011, is less than half the in-water value, and ranged between 0.026 and 0.002. The increased in-water damping ratio for the whole gate rotation is definitely due to increased Coulomb friction on the trunnion pin and on the side rubber seals due to hydraulic pressure loads.

Therefore, a reasonable estimate of in-water damping ratio for the whole gate rotational vibration mode around the trunnion pin would be 0.024 and for the streamwise vibration mode a value of 0.015, as shown bolded in the last row of Table 2. More conservative estimates of damping ratios would be the minimum observed values of 0.011 for the whole gate rotation mode and 0.003 for the bending mode assessments. It should be noted here that these small damping ratios are due to the internal damping effects for small amplitude vibrations.

Theoretical Analysis for Dynamic Stability

The coupled-mode vibration system involves two natural vibration frequencies, one for the whole gate rotational vibration around the trunnion pin and the other for the skinplate streamwise bending vibration. Basically the coupled-mode vibration occurs when the two frequencies synchronize with one of the natural vibration modes serving as the driving excitation for the other mode. Through the feedback of one mode driving the other, the coupled-mode self-excited vibration is built up. With this understanding the physical state, the equations of motion of the coupled-mode vibration can be solved approximately (Anami et al., 2014).

Table 2. Field vibration test results for natural frequencies and damping ratios, from experimental modal analyses or steel rod breaking excitation tests

Full-scale gate	Gate mass (ton)	In-air				In-water					
		Frequency (Hz)		Damping ratio		Water depth, d_0 (m)	reduced rotation center height, r_s	Frequency (Hz)		Damping ratio	
		$\Omega_{a\theta}$	$\Omega_{a\psi,x}$	$\zeta_{a\theta}$	$\zeta_{a\psi,x}$			$\Omega_{w\theta}$	$\Omega_{nw\psi,x}$	$\zeta_{w\theta}$	$\zeta_{w\psi,x}$
Folsom dam Tainter gate 15.5m heigh	87	6.88	26.9	0.012	0.002	(13.26)	(0.72)				
Tainter gate "A" 8.3 m heigh	29	8.13	23.8 24.6	0.009	0.026 0.011	7.72	0.18 0.43	8.38	11.1 13.8	-	- -
Tainter gate "B" 9.8 m heigh	34	7.89	30.38	0.011	0.018	9.24	0.55	7.34	9.99	0.018	0.011
Tainter gate "C" 10.3 m heigh	47					9.27	∞	7.18	6.79	0.011	0.011
Tainter gate "T" 6.7 m heigh	24					6.38	∞ 0.32	6.5	5.5 6.5	0.011	0.030 0.012
Tainter gate "P" 12.5 m heigh	77					12.02	∞ 0.27	8.3	4.6 6.3	0.018	0.040 0.021
Tainter-gate "D" 10.9 m heigh	83.7					5.85	0.92 0.16	4.75	16.1 20.7	0.036	0.006 0.006
Tainter-gate "E" 14.7 m heigh	73					11.19	1.07 0.41	6.63	8.63 13.63	0.012	0.014 0.003
Tainter-gate "F" 9.3 m heigh	52					3.8	0.92	5.94	16.7	0.064	0.009
Representative value				0.011	0.014					0.024	0.015

In the approximate solution method, the periodic vibration of main vibration is assumed and the solution of the response vibration is obtained. Thereafter, the feedback process for reinforcing the main vibration is analyzed. This analysis is carried out when the skinplate streamwise vibration and whole gate rotational vibration becomes main vibration, respectively.

Dynamic Stability Criterion Diagram

A Tainter gate is dynamically stable if the gate has a damping ratio larger than the magnitude of the fluid excitation ratio $\xi_{f\theta}$ and $\xi_{f\psi}$. Thus, depending on the driving vibration, $\xi_{f\theta}$ or $\xi_{f\psi}$ represent the critical damping ratio necessary for the dynamic stability. The dynamic stability criteria can be given by the following expressions:

$$\zeta_{a\psi} > \xi_{f\psi} \text{ (when synchronized with the skinplate streamwise vibration)} \tag{4}$$

$$\zeta_{a\theta} > \xi_{f\theta} \text{ (when synchronized with the whole gate rotational vibration)}$$

where, the fluid excitation ratios, $\xi_{f\psi}$ and $\xi_{f\theta}$, that serve as indicators for dynamic stability, are defined by

$$\xi_{f\psi} = \frac{\Delta c_{\theta\psi}}{\gamma_{\psi\theta}} - \frac{\delta_p \alpha_\psi \Delta c_\psi}{2F_{a\theta}\gamma_{\psi\theta}} + \frac{\sqrt{2}}{\pi} c_f k \frac{\alpha_\psi \Delta c_\theta}{F_{a\theta}' \gamma_{\psi\theta}} \tag{5a}$$

or for F> 10 when $\Delta m_\theta \rightarrow 0$ and $\Delta c_\psi \rightarrow 0$,

$$\xi_{f\psi} = \frac{\Delta c_{\theta\psi}}{\gamma_{\psi\theta}} + \frac{\sqrt{2}}{\pi} c_f k \frac{\alpha_\psi \Delta c_\theta}{F_{a\theta} \gamma_{\psi\theta}} \tag{5b}$$

$$\xi_{f\theta} = \Delta c_{\psi\theta} \tag{6}$$

The symbols $\Delta c_{\theta\psi}$ and $\Delta c_{\psi\theta}$ are the fluid-excitation coefficient and dissipative wave-radiation damping coefficient due to the coupling of the two vibration modes, respectively, given for $F > 10$ by

$$\Delta c_{\theta\psi} = \frac{1}{2\pi} \frac{\alpha_I \gamma^2}{\sqrt{\left(1-\gamma^2\right)^2 + \left(2\gamma\zeta_{a\theta}\right)^2}} - \left(\frac{\pi}{2}\alpha_{I\psi}\right)\gamma \sin\varphi + \sqrt{2}c_f \frac{\alpha_\psi \Delta c_\theta}{r_{sa} F_{a\theta}} \cos\varphi \tag{7}$$

$$\Delta c_{\psi\theta} \equiv \frac{1}{2\pi}\frac{\alpha_I}{\alpha_*}\gamma \sin\left(\varphi_0 - \varphi\right) \tag{8}$$

where, α_* and ϕ are the angular-amplitude ratio and phase-lag. The ratios of the moments of inertia of the gate are given by α_I and $\alpha_{I\psi}$, and γ is the in-water to in-air vibration frequency ratio. Computation of the dynamic stability curve follows directly from equations in Chapter 10, specifically Equations 40, 41, 43, 44, and 46 for $\xi_{f\psi}$ and Equations 54, 55, 58, 59, 62, 63 and 65 for $\xi_{f.\theta}$ The symbols are all defined in Chapter 10, as well as in Anami et al. (2014)

In Equation 5, the symbol Δc_θ is the fluid-excitation coefficient, Δc_ψ is the small wave-radiation damping coefficient, and Δm_θ is the small equivalent water added mass without coupling. The major characteristics of the parameters Δc_θ, Δc_ψ and Δm_θ are shown as functions of the Froude number F in Figure 7, along with the equivalent added mass of Δm_ψ, which significantly reduces the vibration frequency in water.

Figure 7. Equivalent added mass Δm_ψ and Δm_θ, wave-radiation damping coefficient Δc_ψ and fluid-excitation coefficient Δc_θ for practical Tainter gates (previously presented as Figure 4 in Chapter 10)

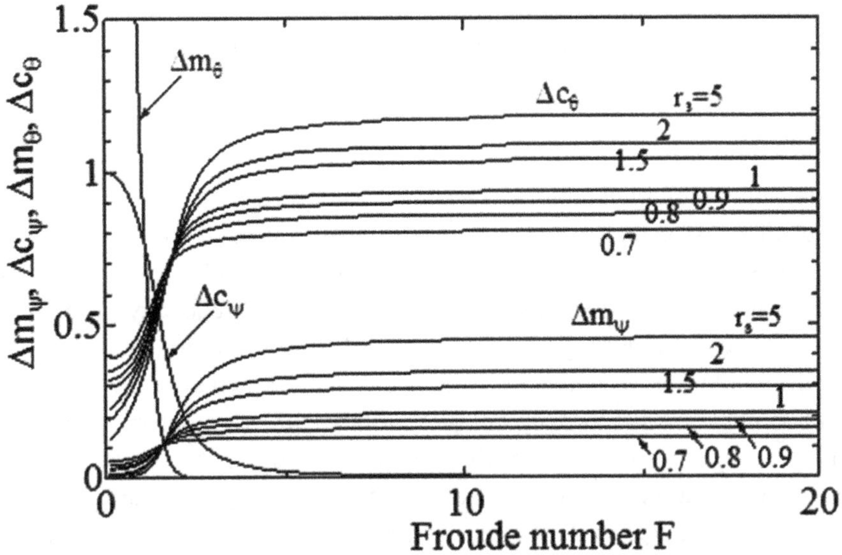

These are obtained by integrating the progressive and standing pressure amplitudes acting on the gate, over the gate submergence depth, respectively (Anami et al., 2012b). The Froude number F is defined by

$$F = 2\pi\sqrt{d_0 / g}\,\Omega_{nw\psi} \tag{9}$$

which governs the dynamic similarity of the flow field.

As the Froude number F increases, the equivalent added mass Δm_θ decreases rapidly and approaches zero. This means that the reduced added mass effect due to the whole gate rotational vibration around the trunnion pin does not affect the skinplate streamwise rotational vibration. The dissipative wave-radiation damping coefficient Δc_ψ also approaches zero with increasing the Froude number. In contrast, the fluid excitation coefficient Δc_θ increases rapidly from zero and approaches a comparatively large asymptotic value, signifying a comparatively large fluid-excitation effect for sufficiently large Froude numbers. This is the significant factor which drives the self-excited vibration of the coupled-mode vibration system. In addition, the equivalent added mass Δm_ψ also increases rapidly from zero, and approaches an asymptotic value, at large Froude numbers resulting in a significant frequency reduction for the streamwise rotational vibration relative to its in-air value. Note the asymptotic values for these two functions depend upon the reduced rotation height r_s.

To compute the dynamic stability curve for a full-scale gate, first select a submergence depth d_0 from the range of anticipated depths to be encountered. Then the following values need to be evaluated:

- F_o from Equation1 (using $\Omega_{a\psi}$ from FEM analysis)
- r_s from Equation 2 (using R_s from FEM analysis)
- F assumed to be greater than 10
- $\Delta m_\theta \rightarrow 0$ and $\Delta c_\psi \rightarrow 0$ for F >10k = -sinΘ_{s0} (see Chapter 10, Equation 13c)

- c_f estimate for gate (typically $0.7 \leq c_f \leq 0.8$)
- δ_p read from Figure 19 in Chapter 9 ($0.7 \leq \delta_p \leq 1.2$)
- Δm_ψ and Δc_θ read from Figure 7 as functions of r_s Once these values have been established and all gate properties have been evaluated, follow carefully equations in Chapter 10 to compute $\xi_{f\psi}$ and $\xi_{f\theta}$. Plotting these two functions versus the frequency ratio $\gamma_{nw} = \Omega_{nw\psi}/\Omega_{a\theta}$ produces the stability criterion diagram.

Examples of dynamic stability diagrams for installed Tainter gates "C" and "E" are shown in Figures 8 and 9, respectively. The reduced rotation height, as shown in Table 2, of $r_s \rightarrow \infty$ for gate "C", represents parallel skinplate vibration, while the value of $r_s = 1.07$ for gate "E" indicates skinplate rotational vibration about a point just above the free surface. The abscisa is the frequency ratio γ_{nw} of the in-water streamwise natural vibration frequency Ω_{nwx} of the skinplate to the in-air rotational natural vibration frequency $\Omega_{a\theta}$ of the whole gate. The gate may become unstable when synchronized with the skinplate streamwise vibration for $\gamma_{nw} < 1.0$, or may become unstable when synchronized with the whole gate rotational vibration for $\gamma_{nw} > 1.0$. Therefore the dynamic stability criterion is given by the skinplate rotational fluid excitation ratio $\xi_{f\psi}$ (given by Equation 5) for $\gamma_{nw} < 1.0$, while by the whole gate rotational fluid excitation ratio $\xi_{f\theta}$ (given by Equation 6) for $\gamma_{nw} > 1.0$. The solid lines in Figures 8 and 9 show the calculated stability damping ratio, required for dynamic stability. The region beneath the solid line is the unstable region. The very intense dynamic instability that appears in the region where

Figure 8. Dynamic stability diagram for Tainter gate "C" showing the gate characteristic point within the unstable region (previously shown as Figure 34 in Chapter 14)

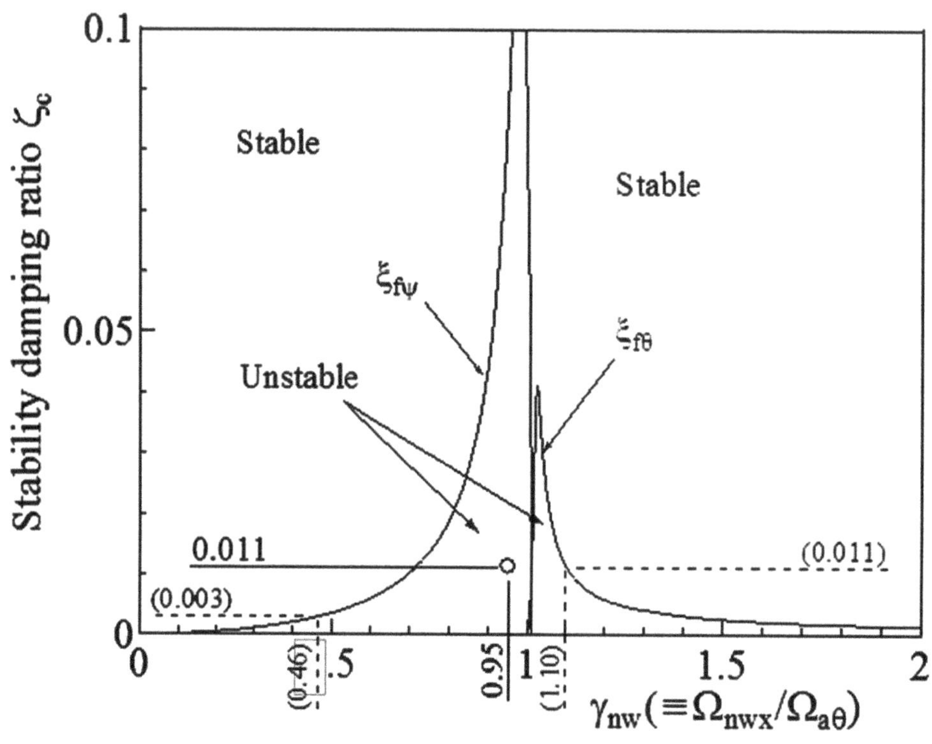

Figure 9. Dynamic stability diagram for Mode M_{X1} of Tainter gate "E" showing the gate characteristic point above the unstable region

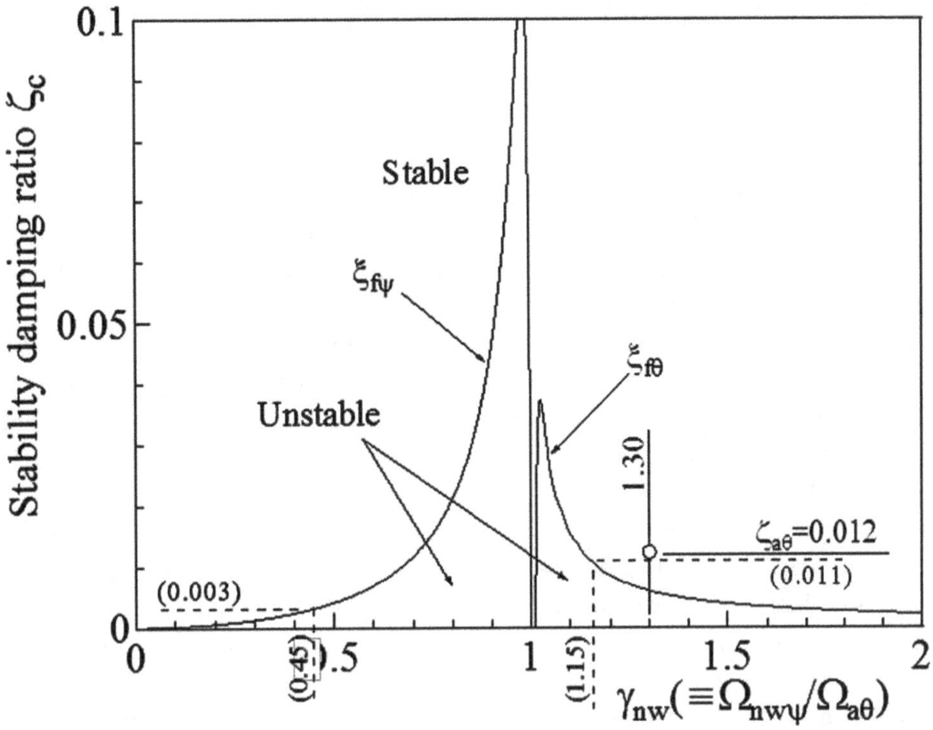

the frequency ratio γ_{nw} is just slightly smaller than 1.0, should be carefully noted. This region should definitely be avoided.

Determination of Gate Dynamic Stability

In order to examine whether the gate is dynamically stable or unstable, the "gate characteristic point" must be plotted on the stability diagrams shown in Figures 8 and 9. The gate characteristic point is determined by the vibration frequency ratio γ_{nw} (= $\Omega_{nw\psi}/\Omega_{a\theta}$) and the damping ratio.

For a new gate design, FEM analysis can be undertaken to determine the in-air vibration frequencies $\Omega_{a\theta}$ and $\Omega_{a\psi}$. From the in-air streamwise skinplate frequency, the skinplate in-water vibration frequency $\Omega_{nw\psi}$ can be calculated for any arbitrary gate submergence depth, using the following expression:

$$\frac{\Omega_{nw\psi}}{\Omega_{a\psi}}\left(=\frac{F}{F_0}\right)=\frac{1}{\sqrt{1+\delta_p\alpha_\psi\Delta m_\psi(F)}} \tag{10}$$

where δ_p is a pressure correction factor which takes a value of 0.83 from model experiments, α_ψ is the mass ratio defined in Equation 3(as a function of the gate submergence depth), and Δm_ψ is read from Figure 7. As a result, the vibration frequency ratio γ_{nw} can be calculated for any gate submergence depth.

When field vibration tests are undertaken for a Tainter gate at a fixed gate submergence depth, the measured vibration frequency ratio γ_{nw} may be used in Equation 10 to calculate the in-air skinplate vibration frequency $\Omega_{a\psi}$. For other gate submergence depths, skinplate in-water vibration frequency, $\Omega_{nw\psi}$, can then be calculated using the value of the right hand side of Equation 10 along with the value of the in-air skinplate vibration frequency $\Omega_{a\psi}$ determined from the measurement undertaken at a single submergence depth.

In contrast to the variation of the streamwise bending frequency with depth, the vibration frequency of whole gate rotation around the trunnion pin remains the same, independent of the upstream gate submergence depth. This is because the the circular-arc skinplate of Tainter gate and the hydrodynamic pressure are both centered on the trunnion pin, resulting in the water added mass having no effect on the whole gate rotational vibration.

As a result, the vibration frequency ratio γ_{nw} is determined by the ratio of the in-water skinplate streamwise vibration frequency $\Omega_{nw\psi}$, obtained from Equation 10, to the whole gate vibration frequency around the trunnion pin, $\Omega_{a\theta}$, obtained from either FEM analysis or field vibration tests. The value γ_{nw} for under the test conditions for Tainter gate "C" was found to be 0.95 (see Figure 8) and for Tainter gate "E" was found to be 1.30 (see Figure 9).

The options for determining the value of the damping ratio for a given gate are rather limited: it must either be measured or it must be assumed. When the main vibration is the skinplate streamwise vibration for $\gamma_{nw} < 1$ the streamwise damping ratio is used. When the whole gate vibration is the main vibration for $\gamma_{nw} > 1$, the corresponding rotational damping ratio is used. The appropriate damping ratio for Tainter gate "C" (Figure 8) with $\gamma_{nw} = 0.95 < 1.0$ is the value of $\zeta_{w\psi} = 0.011$ for the skinplate streamwise vibration as the main vibration. On the other hand, for Tainter gate "E" (Figure 9) with $\gamma_{nw} = 1.30 > 1.0$, the appropriate value is $\zeta_{w\theta} = 0.012$ for the whole gate rotational vibration as the main vibration. If the damping ratio cannot be measured, the reasonable representative value, given in last line at the bottom of Table 2, may be adopted for each damping ratio. More conservative values would be the minimum listed value.

The respective gate characteristic points for Tainter gates "C" and "E" were then plotted in Figures 8 and 9. For Tainter gate "C" in Figure 8, the gate characteristic point lies within the very intense dynamically unstable region, while for Tainter gate "E" in Figure 9, the gate characteristic point lies above the dynamic stability criteria curve, thus confirming the dynamic stability of the gate.

Countermeasure Design Concepts for Dynamic Stability of Tainter Gates

If the gate characteristic point is in the unstable region (as shown, for example, in Figure 8 for Tainter gate "C"), a countermeasure must be developed to achieve dynamic stability for the gate. The procedure for acheiving dynamic stability for an existing gate or for the preliminary design of a new gate under design is as follows:

Step A: Determine the in-air natural vibration characteristics of the gate from FEM analysis or field tests.
Step B: Compute and plot the dynamic stability diagram
Step C: Plot the gate characteristic point on the dynamic stability diagram using (a) representative damping values, (b) more conservative minimum values, or (c) measured values of the damping ratio. If the gate condition falls within the instability region, structural modification are needed to move the gate condition point to satisfy either of the following two conditions:

$$\gamma_{nw} (= \Omega_{nw\psi} / \Omega_{a\theta}) < 0.45 \text{ or } \gamma_{nw} (= \Omega_{nw\psi} / \Omega_{a\theta}) > 1.15 \quad (11)$$

If the criteria in Equation 11 are not satisfied, the structural design of the gate or of the lifting wire/chain design should be modified so the criteria in Equation 11 are satisfied. If modifications are made to the gate structure, return to STEP 1, and use FEM analysis to determine the new in-air natural vibration characteristics.

If changes are needed to satisfy the criteria in Equation 11, adjustment of either the in-water streamwise vibration frequency of the skinplate $\Omega_{nw\psi,x}$ or the whole gate rotational vibration frequency around the trunnion pin $\Omega_{a\theta}$ are needed. For large-sized Tainter gates, it may be very difficult to set the whole gate rotational vibration frequency at sufficiently high values. Therefore, in order to satisfy the characteristic by making $\gamma_{nw} < 0.45$, the in-water streamwise natural vibration frequency of the skinplate would need to be lowered, that is, the skinplate must be designed as a soft structure. Such a soft structure, however, is unable meet the original requirements of supporting the hydrostatic pressure. Therefore, it is almost impossible to modify the design to satisfy $\gamma_{nw} < 0.45$.

Therefore, the remaining possibility for achieving dynamic stability is to modify the gate to satisfy $\gamma_{nw} > 1.15$. The dynamic structural design requirement should then be that the skinplate streamwise in-water natural vibration frequency $\Omega_{nw\psi}$ must be greater than 1.15 times the whole gate in-air vibration frequency $\Omega_{a\theta}$. In the skinplate design, sufficient increases in the rigidity of the reinforcing vertical beams, of the horizontal girders and those of the truss structure would be needed. In addition, since deformation of the radial arms is closely related to the skinplate streamwise vibration frequency, an increase in the rigidity of radial arm structure would also be needed.

In the coupled mode vibration mechanism, the whole gate rotational vibration and the skinplate streamwise vibration become coupled. In the resulting coupled-mode vibration system, the whole gate rotational vibration produces a flow-rate-variation beneath the gate that supplies energy for the skinplate vibration which, in turn, drives the whole gate rotational vibration, leading self-excited vibration.

Therefore, a method to control the whole gate vibration around the trunnion pin may prove far more effective. Replacing the gate hoisting mechanism with a softer one with a reduced spring constant might suifficiently lower the whole gate frequency, $\Omega_{a\theta}$, and thereby raise the frequency ratio γ_{nw} to a value that exceeds 1.15.

Continued Development of the Design Guide

In order to ensure the dynamic stability of a gate, the essential aspects of a procedure to determine the dynamic stability of a Tainter gate was provided. The procedure can be used both for new preliminary gate designs as well as for problematic installed gates.

In order to design a Tainter gate with complete dynamic stability under any operational characteristics, the gate should be so designed that the skinplate streamwise in-water natural vibration frequency always remains larger than at least 1.15 times the whole gate rotational vibration frequency around the trunnion pin. For this design criterion to be satisfied, it is most certainly necessary to increase the structural rigidity of the skinplate and radial arms or replace the gate hoisting mechanism with a softer one.

Given the worldwide application of Tainter gates, it would seem logical to strive for both static and dynamic stability considerations in the design stage. This draft process is, we hope, a step in the direction of developing dynamic design criteria for Tainter gates. The authors would greatly appreciate feedback from designers who may try this process, have problems with implementaqtion, or have other thoughts on development of a set of standard dynamic design criteria fro Tainter gates.

REFLECTIONS BY PROFESSOR EMERITUS ISHII

I recall many moments in the past, both before and just after I started my life's work on flow-induced vibration of hydraulic gates. I was a 3rd year student at Osaka University when the 3rd gate on Wachi Dam failed in 1967. Professor Emeritus Tokio Uematsu presented the details of this serious gate failure to us in our class in hydraulics. When I was in my first year as a Ph.D. student in 1971, my adviser, Professor Kensaku Imaichi, suggested that I investigate the Wachi Dam gate failure. I note with sadness the passing of Professor Imaichi earlier this year at 86 years of age.

Before I started my research efforts in flow-induced vibration, I was reading the German textbook *Grenzschicht Theorie* by H. Schlichting, in my quest for a topic for my dissertation. I was completely exhausted by reading about the breadth of the research work reported by Schlichting, and so many experimental projects to consider. Around the time I was overwhelmed by the difficulty of finding my dissertation topic among the topics in Schlichting's book, Professor Imaichi's suggestion to pursue the subject of the mechanism of flow-induced vibrations in the Wachi Dam gate failure was just like a message from God to help me. When I started to study flow-induced vibration problems in many books and journal papers, I was again exhausted by the multitude of so many experimental works.

One day I found, in a flash of inspiration, that a complete mathematical analysis might be possible for the type of flow-induced vibration involved in the Wachi Dam gate failure. Since the subject seemed to me to be quite independent of the non-linear hydrodynamic effects such as those caused by shedding vortices, I started to study many mathematical techniques from many books by Lord Rayleigh, H. Lamb, J. J. Stoker and others. I studied topics such as Fourier integrals and Fourier transforms, Laplace transforms, the Cauchy-Riemann complex integral theorem and so on. In addition, I studied intensely Rayleigh's wave theory. As a result, I was able to identify theoretically and experimentally the essential flow-induced dynamic instability that eccentric Tainter gates possess, with great supports by Mr. Shinji Takemoto, Mr. Masahiro Yamasaki, Mr. Atsushi Tanaka and Mr. Masagazu Morinaga. I completed my dissertation and was awarded my Ph.D. degree from Osaka University in 1975.

I was very pleased with the recognition of my efforts when my research work was reported in the Asahi Newspaper on March 3, 1977. My initial joy at the newspaper report, however, changed to deep despair when is I experienced extraordinarily intense pushback and condemnation from several gate companies, which insisted that they did not construct eccentric Tainter gates. I was forced to stop my research on the dynamic instability of Tainter gates. At that time I could find no way forward with my research, except to change the focus from Tainter gates to long-span gates. With serious disappointment, I sought opportunities to go abroad and was fortunate to receive an Alexander von Humboldt Foundation fellowship for one year in Germany (West Germany, at the time) starting in November 1983. My host was Professor Eduard Naudascher in the Institute of Hydrodynamics at the University of Karlsruhe. It was in Karlsruhe where my future best friend in my life, Professor Charles W. Knisely, was waiting for me.

I suffered from stomach cancer in 1993 and was fortunate for early detection. Through surgical removal of the cancer, I was able to survive. I was greatly motivated to work on flow-induced vibrations after surviving this scare.

In the summer of 1995, the Japan Times reported the failure of the Number 3 gate at the Folsom Dam in California. I made quick contact with the USBR a few days after the gate failure and was asked to be a member of the Forensics Team. I restarted my investigation of the flow-induced vibrations of Tainter gates, thinking of what might have been the cause of the failure. I presented the details of the Folsom Dam gate failure to my third year undergraduates in their class on hydraulics at Osaka Electro-Communication University. To my great pleasure, one of my students in that class, Keiko Anami, expressed an intense interest in investigating the cause of the failure with me. She was able to easily master the difficulties of my previous theoretical work on the instability of eccentric Tainter gates. She undertook the study of the Folsom Dam gate failure as her thesis topic under my direction when she enrolled as a graduate student. She successfully applied theoretical techniques to identify the essential flow-induced coupled-mode dynamic instability that any non-eccentric Tainter gate possesses. She was awarded her Ph.D. degree in 2002 from Osaka Electro-Communication University.

I am very pleased to note that two International Symposia related to Tainter gate vibrations were held at Osaka Electro-Communication University. The first Symposium was the Commemorative International Symposium for the 40th Anniversary of the Foundation of the Osaka Electro-Communication University, which was held November 4 to 5, 2001. The topic of the symposium was *Flow Induced Vibration of Engineering Structures: Lessons from Recent Tainter Gate Failures.* Invited speakers from overseas were Professor Eduard Naudascher of University of Karlsruhe, Professor David Weaver of McMaster University, Professor Charles W. Knisely of Bucknell University and Senior Researcher Robert Todd from the United States Bureau of Reclamation (Denver). Domestic invited speakers were Professor Fumio Hara of Tokyo Science University, Professor Masaru Matsumoto of Kyoto University and Professor Shigehiko Kaneko of Tokyo University. A commemorative closing address was presented by Professor Naudascher, included as an appendix to this final chapter. Figure 10 is a photograph of Professor Naudascher at our 2001 Symposium.

The second International Symposium with the theme *Impending Plans for Assuring the Dynamic Stability of U.S. Tainter Gates* was held on March 27, 2006. Invited speakers were Dr. Robert Todd of the USBR, Supervising Engineer Mark Schultz from the Department of Water Resources, State of California, Manager Dr. Stephen Slinkard of United States Army of Corps of Engineers, and Professor Charles W. Knisely of Bucknell University. The domestic invited speakers were Professor Emeritus Hiroji Nakagawa of Kyoto University, Professor Shigehiko Kaneko of Tokyo University, Professor Tetsuya Sumi of Kyoto University, Manager Koki Kataoka of Kansai Electric Power Company, Professor Toru Iijima of Muroran Institute of Technology and Professor Yoshinobu Tsujimoto of Osaka University. A commemorative photograph from this second International Symposium is shown in Figure 11.

Our research work over the past 45 years are presented in this book. The most significant feature of this book is the successful development of an entirely theoretical method to predict the flow-induced vibration problems of hydraulic structures. A second very important development is the bold simplified modeling of the flow fields and vibration systems, which made possible the theoretical development. Small errors between the theory and measurement could be accounted for by small systemic empirical coefficients that modified the predictions for better agreement with model and field experiments.

The delay in the completion of this manuscript resulted from my extremely slow work pace as I faced my impending retirement. I was encouraged by my colleague, Professor Knisely, who was always very gentle and very patient with me. I trust him deeply from the bottom of my heart. The publication of this book would have never been realized without a great effort by Professor Knisely and the continuing kind support by his wife Karin Knisely, to both of whom I would like to express my sincere gratitude.

Figure 10. A visit to the gate vibrations laboratory during the 2001 Symposium Flow Induced Vibration of Engineering Structures: Lessons from Recent Tainter Gate Failures, November 5, 2001.
From left to right, R. Todd, D. Weaver, E. Naudascher and N. Ishii.

Figure 11. Commemorative photograph from the International Symposium Impending Plans for Assuring the Dynamic Stability of U.S. Tainter Gates at Osaka Electro-Communication University, March 27, 2006

My sincere apology for the delay in the preparation is extended to the IGI Global publishing company people, represented by Jan Travers, Keith Greenberg and others.

We sincerely hope that this book is effective in preventing further human-made disasters such as Wachi and Folsom Dam gate failures for successive generations and, in particular, for Prof. Ishii's grandchild Masaki Ishii born on March 10, 2017.

CLOSING

This book documents a combined investment of over 100 person-years of effort. For each of the authors, this effort has been both a challenge and a pleasure. It has been immensely rewarding to have this opportunity to reflect on developments over a career and to try to formulate a logical explanation of what transpired.

APPENDIX: CLOSING ADDRESS FOR 2001 INTERNATIONAL SYMPOSIUM

Flow Induced Vibration of Engineering Structures: Lessons From Recent Tainter Gate Failures

Given by Eduard Naudascher, Professor Emeritus, University of Karlsruhe, Germany, November 5, 2001.

"It is an honor and a pleasure to present the closing address at this outstanding Symposium. I'm sure you all will agree that our expectations in following the invitation from Osaka were more than fulfilled. I have learned a lot, and it was so good to meet colleagues and friends and to have had the opportunity for exchanges with them.

Permit me to focus on the central topic of our Special Symposium II, "Lessons from Recent Tainter Gate Failures" and to start with a personal remark. I have often been marveling about Professor Ishii's untiring pursuit of that subject since the later 70ies and particularly during the past 6 years. Today I cannot but express my deep respect for the outcome of this effort and for its significance. To be sure, the safety of these gates ranges among the most important concerns of hydraulic engineers in view of the great number of installations that utilize Tainter gates worldwide. So I was not surprised when I learned that the U.S. Bureau of Reclamation called on Professor Ishii in July 1995 asking him for consultation on the failure of one of the 50-foot-high Tainter gates of Folsom dam. Fortunately, the incidence did not cause losses of human lives as in cases of similar catastrophes; but the loss of the greatest part of precious water stored behind Folsom dam for the supply on Northern California was serious enough.

I have known Professor Ishii since his participation at the 2nd International Vibration Symposium in Karlsruhe 1979 and since his time as a visiting Humboldt Research Fellow at our University in 1983/84. For *more than 25 years* he has now been studying flow-induced gate vibrations, and he has done that disregarding his health, until in 1992 he was diagnosed with an early stage of cancer. In one of his reports on the Folsom-dam gate failure he remarked that luckily he lived on to help uncover the causes of this failure. I hope he'll permit me to quote from one of these reports: "The 3 months since the gate failure", he writes, "have passed in an incredible flurry of activity; but for [me] these 3 months have been the best of [my] life." (Looking back on the 3 months preceding this conference, I'm afraid, he could undoubtedly make a similar statement today. – Take good care of yourself, dear Noriaki, in the months to come!)

But be assured, it was worth your while! For the discovery of a new and most dangerous instability of Tainter gates, you and Keiko Anami are to be highly congratulated! Congratulations not only for the detective kind of way in which you applied forensic observations, field tests, computer simulations etc. to prove that hydrostatic loading and friction *alone* could not have destroyed the Folsom-dam gate, as was initially suspected; but compliments also to you and Keiko-san for your impressive theory on a previously unknown dynamic instability of Tainter gates, complete with convincing verifications by model tests and the prove that an instability of that type must have been involved in the Folsom-dam catastrophe.

As if all this hadn't been enough, you insisted that there should be additional proof by the evidence from field tests. So you, with the help of Keiko-san, planned and prepared for such tests, not minding that they could only be carried out a few days prior to this symposium! All of us who know of your tireless zeal and determination, and who also know about the frustrations possible with field tests, we do share your joy and pride. You certainly deserve the victor's laurel that the successful field tests have bestowed on you!

After all, the importance of your discovery of a new gate instability rests in the fact that its danger – like in case of time bombs – may go undetected for 30, 40, – even 50 years as the Folsom dam failure has demonstrated. Since the newly found instability can also lead to *hard* in contrast to soft excitation, gate vibrations may be excited only if a certain threshold amplitude is initially exceeded. At the Folsom-dam gate 3, this vibration-triggering event occurred most probably after an intermediate blockage of the gate during the lifting process (caused by excessive friction) was suddenly overcome. Therefore, measures that exclude excessive friction – as claimed by the Bureau of Reclamation in its Forensic Report of Oct. 17, 1996 – may well have been necessary to reduce extra static loading on the radial arm structure. But they will also reduce structural damping and thus increase the danger for self-excited gate vibration! To be sure, the triggering amplitude for such vibrations to come about could be generated in other ways, e.g. by waves induced in cases of landslides into the reservoir, by boats or logs bumping into a gate, or by obstacles intermediately jamming the gate lifting or closure.

I can only hope that managers and engineers responsible for the operation and maintenance of Tainter gate installations will immediately take notice and action. In particular I hope the Bureau of Reclamation will update its Forensic Report, as well as subsequent publications and memos based on it. But of course, these new findings call also for a revision of many other publications that serve as design guides. A case in point is the 1994 IAHR Hydraulic Design Manual on Flow Induced Vibrations by Don Rockwell and myself; it's high time for its first revised edition!

Let me come to the main purpose of this address, the closure of our meeting. I'm sure I speak in the name of all participants if I extend our sincere thanks to all of you at this University who helped to organize this great Symposium and make our stay here so pleasant. We have learned a great deal from the various presentations and very much enjoyed the wonderful Japanese hospitality. It was a great meeting! Thank you, Professor Ishii, Professor Sasaki, and all around you who made this possible and who contributed to its success!

Please convey our gratitude to Dr. Fukuda and the University Board of Directors for sponsoring this Symposium and for the generous support to the contributors from abroad; and, last not least, our thanks also to your many friendly assistants, most of them graduate students of this university. It was an excellent idea to broaden the scope of the Symposium topic to include FIV of gates other than Tainter gates, and we are indebted to all invited contributions for bringing us up to date on the latest advances in their research fields.

We congratulate this University to its 40th Anniversary and wish for continued success, and good fortune and prosperity for the future!"

REFERENCES

Anami, K. (2002). *Flow-Induced Coupled-Mode Self Excited Vibration of Large-Scaled Tainter-Gates* (Ph.D. Dissertation). Osaka Electro-Communication University, Osaka, Japan. (in Japanese.)

Anami, K., & Ishii, N. (2000). Flow-induced dynamic instability closely related to Folsom Dam Tainter-gate failure in California. In S. Ziada & T. Staubli (Eds.), *Flow Induced Vibration* (pp. 205–212). Rotterdam: Balkema.

Anami, K., & Ishii, N. (2001). Analyses of FIV of Folsom Dam Tainter gates and verifications by model tests. In *Proceedings of Commemorative International Symposium for the 40th Anniversary for the Foundation of the Osaka Electro-Communication University*, (*Vol. 2*, pp.165-181). Osaka: The Osaka Electro-Communication University.

Anami, K., Ishii, N., Knisely, C., Oku, T., & Tsuji, T. (2013). Dynamic stabilization of Folsom Dam Tainter-gates by replacing hoist chains with cables. *11th International Conference on Vibration Problems*, Lisbon, Portugal.

Anami, K., Ishii, N., & Knisely, C. W. (2012a). Pressure Induced by Vertical Planar and Inclined Curved Weir Plates Undergoing Streamwise Rotational Vibration. *Journal of Fluids and Structures*, *29*, 25–49. doi:10.1016/j.jfluidstructs.2011.11.007

Anami, K., Ishii, N., & Knisely, C. W. (2012b). Added mass and wave radiation damping for flow-induced rotational vibrations of skinplates of hydraulic gates. *Journal of Fluids and Structures*, *35*, 213–228. doi:10.1016/j.jfluidstructs.2012.07.008

Anami, K., Ishii, N., & Knisely, C. W. (2014). Theory of coupled-mode self-excited vibration of Tainter gates. *International Journal of Mechanical Engineering and Robotics Research*, *3*(4), 678–707. http://www.ijmerr.com/

Anami, K., Ishii, N., & Knisely, C. W. (2015). Dynamic stability identification of a full-scale Tainter-gate and extension of gate-opening range for operation. In *ASME 2015 Pressure Vessels and Piping Conference*. New York, NY: American Society of Mechanical Engineers. doi:10.1115/PVP2015-45418

Anami, K., Ishii, N., Knisely, C. W., & Oku, T. (2007). Field measurement of dynamic instability of a 50-ton Tainter-gate. In *ASME 2007 Pressure Vessels and Piping Conference* (pp. 449-456). New York, NY: American Society of Mechanical Engineers.

Anami, K., Ishii, N., Knisely, C. W., & Oku, T. (2017). Design guidelines for dynamic stability of Tainter gates. In *ASME 2017 Pressure Vessels and Piping Conference*. New York, NY: American Society of Mechanical Engineers.

Imaichi, K., & Ishii, N. (1977). Instability of an Idealized Tainter-Gate System without Damping Caused by Surface Waves on the Back Water of Dam. *Bulletin of the JSME*, *20*(146), 963–970. doi:10.1299/jsme1958.20.963

Ishii, N. (1975). *Instability of Elastically Suspended Tainter-Gate System Caused by Surface Waves on the Reservoir of Dam* (Ph.D. Dissertation). Osaka University, Toyonaka, Osaka, Japan.

Ishii, N. (1990a). Flow-induced vibration of long-span gate (4th report, vibration frequency ratio and fluid damping ratio). *Trans. Japan Soc. Mech. Eng. (Ser. C)*, *56*(531), 2887-2892.

Ishii, N. (1990b). Flow-Induced Vibration of Long-Span Gates: Verification of Added Mass and Fluid Damping. *JSME International Journal. Series 2, Fluids Engineering, Heat Transfer, Power, Combustion, Thermophysical Properties*, *33*(4), 642–648.

Ishii, N. (1992). Flow-induced vibration of long-span gates part I: Model development. *Journal of Fluids and Structures*, *6*(5), 539–562. doi:10.1016/0889-9746(92)90018-X

Ishii, N., Anami, K., Knisely, C. W., & Oku, T. (2009a). Field tests concerning the dynamic stability of Tainter gates at the Thermalito Diversion Dam. Proceedings of WaterPower XVI.

Ishii, N., Anami, K., Knisely, C. W., Oku, T., & Aslam, M. (2009b). Field vibration tests concerning the dynamic stability of Tainter gates at the Poe Power Dam. *Proceedings of Hydro 2009 Conference* (CD-ROM) (No. 8-10).

Ishii, N., Imaichi, J. K., & Hirose, A. (1980). Dynamic instability of Tainter-gates. In *Practical Experiences with Flow-induced Vibrations: Symposium*. Springer. doi:10.1007/978-3-642-81528-7_49

Ishii, N., & Imaichi, K. (1982). Instability of elastically suspended Tainter-gate system caused by surface waves on the reservoir of a dam (Approximate Solution) [in Japanese]. *Transactions of the Japan Society of Mechanical Engineers, Series B*, *48*(428), 640–647. doi:10.1299/kikaib.48.640

Ishii, N., Imaichi, K., & Hirose, A. (1977a). Instability of elastically suspended Tainter-gate system caused by surface waves on the reservoir of a dam. *Journal of Fluids Engineering*, *99*(4), 699–708. doi:10.1115/1.3448888

Ishii, N., Imaichi, K., & Takemoto, S. (1977b). Water Waves and Structural Loads Caused by a Periodic Change of Discharge from a Sluice-Gate. *Bulletin of the JSME*, *20*(146), 998–1007. doi:10.1299/jsme1958.20.998

Ishii, N., Imaichi, K., & Yamasaki, M. (1983). Flow-induced vibration of single arm gates. In *Proceedings 20th IAHR Congress* (pp. 317-324). Delft: IAHR.

Ishii, N., & Knisely, C. W. (1988). Shear layer behavior under a streamwise vibrating gate and its significance to gate vibration. *Proc. 2nd Intern. Symp. on Fluid Control, Measurement, Mechanics and Flow Visualization*, 45-49.

Ishii, N., & Knisely, C. W. (1989). *Shear layer dynamics and unsteady discharge beneath a long span gate undergoing streamwise vibrations. In Proceedings Hydrocomp* (pp. 500–509). Rotterdam: Elsevier.

Ishii, N., & Knisely, C.W. (1990). Shear layer behavior under a streamwise vibrating gate and its significance to gate vibration. *JSME International Journal. Ser. 3, Vibration, control engineering, engineering for industry*, *33*(2), 131-138.

Ishii, N., & Knisely, C. W. (1992). Flow-induced vibration of shell-type long-span gates. *Journal of Fluids and Structures*, *6*(6), 681–703. doi:10.1016/0889-9746(92)90003-L

Ishii, N., Knisely, C. W., & Nakata, A. (1994). Coupled-mode vibration of gates with simultaneous over- and underflow. *Journal of Fluids and Structures*, *8*(5), 455–469. doi:10.1006/jfls.1994.1022

Ishii, N., Knisely, C. W., & Nakata, A. (1995a). Field study of a long-span shell-type gate undergoing flow-induced vibrations. *Journal of Fluids and Structures*, *9*(1), 19–41. doi:10.1006/jfls.1995.1002

Ishii, N., Nakata, A., Irie, K., Moriya, Y., & Ogawa, T. (1995b). *Effects of inclined weir plate angle of shell-type long-span gates upon flow-induced vibrations (No. CONF-950740--)*. New York, NY: American Society of Mechanical Engineers.

Ishii, N., & Naudascher, E. (1984). Field tests on natural vibration modes of a Tainter gate. In Channels and Channel Control Structures, (pp. 209-222). Springer. doi:10.1007/978-3-662-11300-4_17

Ishii, N., & Naudascher, E. (1992). A design criterion for dynamic stability of Tainter gates. *Journal of Fluids and Structures*, *6*(1), 67–84. doi:10.1016/0889-9746(92)90056-9

Ishii, N., Naudascher, E., & Thang, N. (1987). A flow-induced vibration of long-span gates,In *Proceedings of the International Conference on Flow-Induced Vibrations*, (pp. 293-303). BHRA.

Nakata, A., Ishii, N., Knisely, C. W., Irie, K., & Moriya, Y. (1996). Effects of weir plate inclination angle on flow-induced vibrations of long-span, shell-type gates. *JSME International Journal. Series C, Dynamics, Control, Robotics, Design and Manufacturing*, *39*(2), 203–211. doi:10.1299/jsmec1993.39.203

About the Authors

Noriaki Ishii is Professor Emeritus, Osaka Electro-Communication University. He graduated from Kannonji-Daiichi High School in 1966 and entered the Department of Engineering Science of Osaka University as a member of their sixth graduating class. At Osaka University, he was well educated in using various methods to solve engineering problems, especially in the use of mathematical techniques. He studied the dynamic performance of centrifugal pumps in his B.S. thesis project and was awarded his B.S. degree in 1969. He then focused on the dynamics and performance optimization of reciprocating refrigerant compressors in collaboration with the Panasonic Company for his M.S. degree in 1971. After entering the doctoral Engineering Science course, he continued to study rotary rolling-piston refrigerant compressors and later scroll compressors. Ishii was not satisfied with studying small-sized engineering compressors, and so began his interest in investigating the dynamics of large hydraulic Tainter gates in 1972. His dissertation was on the failure analysis of Wachi-dam Tainter gate in Japan for which he received his Ph.D. from Osaka University in 1975. Immediately after his graduation, he joined the faculty at OECU. His main areas of research are in refrigerant compressor dynamics and optimization, flow-induced vibrations, and hydraulic gate instabilities. He has authored or co-authored over 97 journal articles and 146 conference papers. He holds Japan, European and United State patents entitled *Hybrid Vehicle Driving System, Hybrid Vehicle, and Driving Method.*

Keiko Anami is Associate Professor of Mechanical Engineering, Osaka Electro-Communication University. Anami studied mechanical engineering at Osaka Electro-Communication University, receiving a B.S. degree in 1997 and an M.S. degree in 1999. She then received her Ph.D. in Mechanical Engineering from Osaka Electro-Communication University in 2002. She held positions at Kyoto University (Kyoto) and Ashikaga Institute of Technology (Tochigi prefecture) before returning to Osaka Electro-Communication University. Her main areas of research are flow-induced vibrations, fluid dynamics, compressor optimization, wind energy conversion, and hydraulic gate instabilities. She has authored or co-authored about 30 journal articles and over 50 international conference papers. She is a recipient of *The Commendation for Science and Technology by the Minister of Education, Culture, Sports, Science and Technology* in 2013.

Charles W. Knisely is a Professor of Mechanical Engineering, Bucknell University. Knisely studied mechanical engineering at Bucknell University, receiving a B.S. degree in 1975 and an M.S. degree in 1978. He then received his Ph.D. in Mechanical Engineering from Lehigh University in Bethlehem, PA in 1980. He subsequently worked in industrial research in Switzerland, and later at a government research laboratory near Bethesda, MD. He held positions at the University of Karlsruhe (Germany) and at Kyoto University in Japan before returning to Bucknell University. During sabbatical leaves and over several summers he held visiting academic positions at the Osaka Electro-Communication University (Japan), the University of Siegen (Germany), and the University of Canterbury (New Zealand). In 2015, he served as a Distinguished Fulbright Chair at the Federal Universities in Uberlândia (UFU) and Goiás (UFG) in Brazil. His main areas of research are flow-induced vibrations, fluid dynamics, compressor optimization, wind energy conversion, and hydraulic gate instabilities. He has authored or co-authored over 100 journal articles and conference papers. He holds seven patents. He and his wife, Karin, authored a book, *Engineering Communication*, published by Cengage Learning.

Index

A

acceleration response 17, 277, 593, 595
added mass 17, 30, 31, 74-77, 122, 133, 143, 195, 201-205, 213, 218-232, 271, 276, 285-287, 298-309, 316, 317, 335-340, 345-352, 363-385, 394-398, 403-407, 423-434, 441, 444, 451, 456-467, 474-483, 490-497, 502, 510, 512, 519, 540-544, 560, 597, 617, 628-630, 636-644, 651, 652
aeration 19, 117, 132, 231-238, 254-270, 288, 360
approximate solution 51, 381, 403, 404, 463, 476, 562, 639, 652
avoidable catastrophe 626, 627

B

Bernoulli equation 19, 94, 103-107, 112, 121
bi-stable 1, 23-26, 122, 132, 133
bluff body 122, 131, 135, 565-573, 581-585
boundary layer 94, 113, 121-127, 138
Bridge failure 565-570, 587-591

C

circular-arc skinplate 387-390, 407, 408, 436-451, 460, 461, 469-472, 487, 488, 522, 553, 644
closed loop energy cycle 463, 492, 560, 633
coalescence 316, 565, 579
Colebrook equation 94, 103-106
comparison 90, 192, 193, 255, 270, 290, 312, 313, 345-348, 372, 373, 394, 401, 447, 499, 500, 508, 511, 517, 535-537, 544, 550, 586, 619, 636
continuous systems 44, 45
countermeasure 28, 38, 133, 231, 236, 256, 258, 263-268, 289, 626, 631, 644
countermeasure design concept 626
coupled equations of motion 463, 494
coupled system 28, 50, 54, 231, 270, 275, 319
coupled-mode 1, 9, 15-18, 28-33, 50, 61, 64, 77, 90, 297, 316-324, 355-361, 370, 371, 376-381, 459-467, 474-477, 482-501, 508-522, 543-560, 565-573, 579, 586-591, 616, 621-624, 629-633, 638-653
coupled-mode instability 9, 15, 31, 61, 316-324, 380, 459-467, 491, 494, 499, 518-522, 559, 565, 623, 631, 633

D

dam failure 1-13, 34-42, 387, 465, 522, 559, 630, 631, 650
damping ratio 45-49, 56, 61, 73, 79, 131, 151, 227, 278, 279, 301-312, 322, 336-343, 357, 362, 364, 372, 378, 380, 386-405, 429-435, 440-442, 451, 452, 457-462, 469-473, 487-497, 504-509, 517, 519, 530, 531, 543-547, 563, 571, 579, 596-617, 622-626, 638-645, 652
damping–Coulomb 44
damping–hysteretic 44
damping–viscous 44
Darcy-Weisbach equation 94, 105, 106
design criteria 1, 139, 228, 565, 567, 589, 590, 626, 632, 646
design guide 626, 633, 645
dimensional analysis 94, 98
discharge coefficients 94, 113-116
discharge fluctuation 1, 31, 61, 141, 153-157, 195, 204, 319, 359, 379, 552, 628
dispersive wave 140
dynamic design 375, 377, 565, 626, 631-633, 646
dynamic instability 15, 17, 31-37, 89, 189, 192, 294, 308, 314, 315, 320, 342, 378-381, 387, 389, 395, 400-405, 484-494, 503, 515-521, 540-545, 552-561, 570, 586-590, 622-624, 631, 643-652
dynamic pressure 245, 252, 253, 258-261, 588, 626

dynamic stability 15, 37, 87-91, 151, 297, 308, 320, 336, 356, 379, 382, 390, 399-404, 463, 482-495, 521, 543-546, 557-567, 593, 602, 613, 622-624, 631-653
dynamic stability criterion curve 593, 602
dynamic stability diagram 463, 642-645

E

eccentric gate 387
eccentricity 1, 13, 30, 31, 89, 200, 353, 387-405, 463-469, 493, 494, 501, 503, 522, 553, 560, 586, 626-630
eccentricity instability 1, 13, 30, 31, 89, 200, 387-389, 403, 463, 465, 522, 626-630
energy 9, 11, 17-22, 27, 31, 42-45, 62, 69-78, 96, 97, 102-105, 116, 126, 133-143, 152, 154, 189-195, 205, 217-220, 225-231, 246, 247, 259, 268, 298, 306, 321, 333, 334, 341, 344, 352-359, 366, 395, 401, 424-433, 463, 467, 484-493, 504, 515, 543, 546, 552-560, 565-572, 585, 587, 616, 621, 628-633, 645
energy flow 44, 69, 71, 355, 487, 585
energy transfer 22, 27, 225, 227, 246, 333, 334, 341, 344, 353, 366, 486, 487
essential dynamic instability 521, 545, 552, 557
ethics 626
excitation 1, 11-27, 32, 44, 71-77, 89, 90, 122-131, 136-143, 151-154, 189, 194, 195, 201-205, 225-228, 241, 263-269, 279, 280, 294-329, 334-349, 357, 363-406, 423, 466-469, 474-499, 504-507, 515-520, 543-546, 556-565, 570, 571, 580, 581, 586, 588, 593, 603-609, 616, 618, 624-628, 633, 639-643, 650
excitation mechanism 32, 140, 269, 324
experimental modal analysis 31, 44, 45, 76, 77, 368, 441, 465, 490-494, 513, 528-537, 543, 559, 593, 594, 599-603, 617, 633-637

F

failure 1-22, 29-43, 63, 77, 88, 127, 268, 387, 400, 441, 460-467, 482, 490-495, 510, 518-531, 536-573, 585-593, 605, 623-631, 646-651
failure scenario 521
FEM 14, 76, 510-515, 521, 533-540, 550-559, 593, 617-619, 632-645
FEM analysis 510-513, 534-537, 550-559, 593, 617, 633-645
FFT 75, 499, 504, 528, 595, 597, 616
field vibration test 593, 600, 610, 635-639
flap gate 28-33, 40, 42, 67, 118, 134, 231-238, 243, 269-292, 316, 336, 358, 459, 460
flow instability 87, 122, 132, 133, 269, 328
flow rate 22, 102, 140, 141, 151, 199-205, 236, 261, 266, 272, 273, 297, 298, 351, 353, 376, 379, 387, 389, 409, 442, 456, 492, 560, 613, 633
flow rate variation 140, 141, 151, 202, 203, 351, 353, 442, 456, 492, 560, 613, 633
flow-rate variation pressure 143, 152-154, 181-203, 228, 297-304, 375, 391, 405-408, 418, 427, 428, 445, 450, 456, 461-465, 470, 487-493, 560, 612, 613, 619, 621, 633
fluctuating load 387
fluctuating pressure 24-29, 84, 126, 140, 193, 194, 199-201, 221, 222, 241-252, 263-272, 319, 333, 334, 342, 352-367, 376-379, 384, 418, 569, 628-630
fluid dynamics 15-23, 89, 94-99, 104, 113
fluid excitation ratio 335-339, 344, 385, 485, 497, 499, 563, 640, 643
fluid properties 95
flutter 50, 90, 565, 567, 572-592
Folsom Dam 6-19, 31-37, 42, 43, 287, 424, 441-444, 449, 456-467, 482-485, 490-503, 508-554, 559-567, 587-591, 614-624, 630-635, 647-651
Fourier transform 75, 76, 168, 208, 210, 322, 411-415
free discharge 114-119, 140, 201, 294, 317, 323, 329, 336, 343, 355, 356, 378, 379
free shear layers 23, 94, 123, 125
free surface 39, 75-82, 107, 108, 128, 131, 140, 141, 150-160, 173-186, 194-208, 233, 246, 268, 290, 298, 351, 373, 376, 385, 395, 397, 405-413, 425, 445, 461, 496, 643
friction-maintained dynamic stability 90, 521, 545, 546
Froude number 114, 154, 164, 174, 176, 181-202, 213-227, 244, 261, 283-285, 300, 305-314, 337-343, 351-358, 363, 372-378, 384, 391, 396-405, 419, 424-435, 440-446, 451-461, 472-483, 490, 491, 496-502, 509, 519, 541, 550, 563, 600, 602, 611, 612, 620, 625, 628, 637, 641
full-scale gate 194, 276, 287, 319, 368-379, 453, 500, 593, 637, 642
full-scale measurements 294, 518
full-scale test 372, 407, 518

G

gate excitation 24, 25, 44, 89, 140, 143, 153, 204, 317, 593
gate failure 1, 6-16, 29-36, 42, 43, 387, 400, 464, 465, 493-495, 510, 518-527, 537-545, 553-565, 587-590, 623-631, 646-649

gate vibration 1, 7-18, 27, 32-40, 61, 64, 76, 84, 91, 135, 140-143, 151-154, 174, 180-195, 204, 205, 212-231, 269-275, 281, 297-301, 306, 307, 318, 329-337, 345, 351-358, 363-366, 375-387, 397, 401, 453, 472-476, 481, 483, 491-493, 502-509, 515, 517, 527-547, 552-560, 565, 579, 587, 601, 602, 611-622, 633, 637, 644, 645, 650, 652

H

Haaland equation 103-106
hard excitation 15-19, 71, 89, 546, 580, 581, 588
heaving 50, 566-574, 579-589
historical record 1
hydrodynamic pressure 15, 142, 195, 213, 221, 227, 231, 276, 281-287, 297-301, 362, 375-378, 390, 392, 405-408, 417-433, 438-446, 453-460, 469-471, 496, 501, 542, 547, 588, 590, 603, 605, 611-620, 637, 644

I

in-air frequency 283, 286, 309, 338, 339, 349, 352, 362, 363, 379, 407, 433-436, 465, 483, 491, 493, 502-504, 560, 600
in-air modal analysis 372, 515, 593-597
incompressible 94-97, 102-105, 112, 113, 120, 137, 142-149, 154, 206, 573
induced hydrodynamic pressure 231, 282
inflatable weir 235, 236
instability 1, 9-38, 61, 63, 71, 77-89, 122-136, 189, 192, 200, 201, 228, 240-242, 252, 263-269, 288-294, 308, 314-329, 342-345, 355-359, 378-381, 387-389, 395, 400-405, 459-467, 484-507, 515-522, 540-546, 552-572, 580-593, 622-633, 643-652
in-water frequency 213, 220, 225, 285, 337-339, 346-352, 358-363, 373, 385, 502, 629, 633
in-water to in-air vibration frequency ratio 307-312, 372-378, 396, 430, 435, 441, 457, 458, 475, 476, 482, 490, 491, 507, 519, 541-544, 563, 611, 625, 641
in-water vibration frequency 203, 214, 227, 231, 285, 287, 309-314, 337, 343, 368-378, 386, 406, 451, 480, 491, 497-504, 520, 564, 597, 600, 625, 633, 643, 644

K

kinetic energy 22, 78, 96, 103, 105

L

laboratory measurements 294, 349
long-span gate 1, 29, 30, 141, 142, 151-155, 189, 191, 202-205, 227, 294-299, 310, 339, 356-363, 368, 374-381, 630, 652
loss coefficients 19, 94, 103, 104

M

Manning 106
mathematical description 31, 140
mechanical vibration 92
mechanism 6-33, 38, 64, 77, 84, 90, 127-132, 140, 196, 205, 225, 231-241, 254, 268-271, 279, 310, 319-328, 336, 355-361, 376-380, 387, 400, 463-466, 487, 495-500, 508, 521, 522, 552, 553, 559, 561, 567-569, 586-592, 621-633, 645, 646
modal analysis 31, 36, 44, 45, 76, 77, 368, 372, 441, 463, 465, 490-495, 513, 515, 528-537, 543, 559, 561, 593-603, 617, 633-637
mode 9, 19, 29, 31, 44, 50-69, 76-90, 128, 205, 231, 242, 244, 251-254, 264-281, 286, 296-301, 316-324, 329, 336, 346, 358, 361, 368-372, 387, 389, 458, 463-470, 475-477, 483, 491-499, 504, 510-513, 530-537, 543, 544, 555-559, 565-572, 579-621, 626-645
mode shape 54, 58, 64-68, 76, 77, 90, 278, 297, 324, 329, 463-467, 494, 499, 513, 531-536, 566, 596, 598, 605-610, 619, 631
model experiment 450, 507
Moody diagram 103, 106

N

nappe 1, 19, 28, 29, 38-42, 81, 89, 90, 109, 117, 132, 138, 139, 231-279, 287-293, 360, 361, 368, 369
nappe oscillation 38-42, 132, 138, 139, 232-245, 251-258, 263-276, 287-291
natural vibration 37, 45, 212, 213, 220, 277, 281, 301, 352, 382-393, 400-404, 418, 429, 450-454, 459-476, 483-519, 530-535, 540-544, 552-557, 563, 594-605, 611-622, 631-646, 653
Newtonian 94, 95, 112

O

observations 11, 36, 154, 179, 186, 189, 239-246, 274, 287, 334, 341, 518-522, 546, 553, 567, 591, 622, 649
one-dimensional flow 112, 123

open channel 105-107, 134, 288
Orr-Sommerfeld 123, 126
overflow 28, 29, 89, 90, 108, 113-119, 131-138, 231-243, 252-271, 276-282, 287, 288, 294, 358-363, 368-384, 629, 630

P

parabolic nappe 236, 237, 252, 255, 266, 267
parametric coupling 44, 49, 60, 631
potential energy 22, 78, 96, 103
potential flow 19, 82, 94, 120-122, 134, 141, 195, 200, 222, 287, 297, 316, 379, 408, 459, 577, 628
potential flow theory 19, 141, 222, 287, 297, 316, 459, 577
power spectra 75, 528, 595, 608
press-open 1, 20-23, 28, 88, 189, 295, 318, 341-343, 353, 355, 362-368, 389, 471, 483-490, 497, 501, 515, 520, 601, 625
press-open device 20, 21, 28, 88, 189, 342, 353, 355, 362, 389, 471, 486, 488, 515
press-shut 1, 17-23, 28-30, 88, 89, 189-195, 294, 297, 316, 318, 341-344, 353, 355, 362-370, 379, 387, 389, 471, 483-487, 496, 497, 503-510, 515, 519, 587, 616-621, 629
press-shut device 20, 28, 30, 189-195, 297, 316, 342, 353, 355, 362, 365, 379, 387, 389, 471, 486, 503, 515, 587, 616, 621
pressure correction factor 440, 442, 457, 644
pressure drop 94, 261
pressure loading 18, 27, 88, 201, 225, 266, 294, 297, 319, 342, 359, 376, 407, 442, 459, 494, 569, 628, 629
progressive-wave pressure component 218, 220, 226, 300-303, 341, 354, 385, 394, 405, 422-426, 441
push-and-draw pressure 205, 206, 212-227, 297-306, 351, 352, 375, 384, 408, 418, 424-428, 437-456, 461-471, 493, 550, 557, 560, 588, 612, 613, 619, 620, 625, 628

R

Rayleigh wave theory 140-143, 151-154
reattachment 15, 23, 25, 122, 125
reservoir pressure 20, 140-143, 152, 181
resonator 19, 23, 28, 44, 77-90, 132, 236
rigid-body vibration 387, 389, 469
rotational bending vibration 492, 610

S

seiche 82, 546
seismic 5, 16-18, 35, 36
self-excited coupled-mode dynamic instability 15, 33, 586, 587, 631
self-excited coupled-mode instability 9, 320, 380, 459-467, 494, 499, 518-522, 559, 623, 631
self-excited vibration 33, 64, 226, 227, 276-280, 286, 288, 309, 311, 325, 335, 343, 347, 361, 363, 370, 388, 460, 474-485, 492, 503, 504, 511, 515, 521, 543-560, 587, 614-622, 633, 639-645, 651
separation 15, 23-28, 42, 67, 122-130, 240-250, 263-268, 316, 324, 325, 336, 577, 581
shear layer 23-28, 33, 37, 122-135, 240, 241, 316-319, 330-336, 377-381, 627, 652
shear layer instability 24, 28, 122-128, 240, 319
simultaneous over- and underflow 10, 28-30, 90, 118, 119, 232-234, 294, 316, 358-369, 376, 381, 627
skinplate 13-18, 30, 31, 45, 61-64, 76, 90, 115, 116, 140-142, 151, 153, 191, 197, 204, 205, 221, 276-286, 320-330, 387-392, 401-408, 419, 436-564, 586-589, 594-626, 631-646
soft excitation 546, 581, 650
spectral analysis 76, 276
stability criterion 308, 336, 339, 356, 357, 463, 543-547, 557, 558, 593, 602, 603, 613, 614, 622, 640-643
stability diagram 463, 517, 633, 642-645
Stability Predictions 540
standing waves 81-86, 131, 174, 179-184, 201, 228, 236, 383, 424
standing-wave pressure component 185, 217-220, 300-303, 385, 393, 405, 422-427, 440, 441
static deformation 510-515, 536, 537
static stability 4, 15, 87-89, 98, 390, 565, 567, 632
static tensile loads 537
streamwise bending vibration of the skinplate 465, 531
streamwise gate motion 205, 217, 297, 362, 628
streamwise rotation 407, 459, 465, 470, 482, 492, 496, 501, 519, 532, 559, 625, 637
streamwise vibration 27, 137-141, 192, 204, 224-227, 287, 296, 299, 339, 362, 374, 383, 390, 408, 432, 442, 447, 449, 455-459, 469, 471, 476-510, 515-520, 535, 537, 543, 550-552, 557-564, 596-617, 622-627, 633-645
submerged discharge 9, 27-33, 114-119, 131, 294, 297, 316-319, 329, 336, 379, 380, 627
surface waves 19, 20, 37, 77, 83, 128, 140, 143, 181-186, 200-205, 217, 295, 297, 380, 381, 392, 404, 561, 562, 614, 651, 652
system stability 87

T

Tacoma Narrows 18, 565-572, 583-591
Tainter gate 1, 10-14, 30-45, 61-64, 114, 116, 140-142, 151-153, 189-195, 201, 205, 276, 295, 313, 319, 360, 380-382, 387-392, 399-409, 440-467, 482-485, 490-502, 508-565, 586-626, 631-653
Theodorsen 573-583, 592
theoretical analysis 40, 143, 153, 195, 205, 220, 221, 314, 320, 372, 379, 390, 397, 436-449, 456, 458, 466, 508, 515, 540, 545, 546, 557, 613, 619-624, 638
theoretical development 154, 200, 205, 294, 297, 466, 499, 647
theoretical results 153, 154, 195, 220-228, 298, 305, 312, 313, 361, 372, 377, 380, 403, 444, 451-456, 491, 510, 511, 517, 543, 612, 613, 619, 620
three-dimensional equivalent model 510-514
torsion 63, 274, 575-582
torsional vibration 54, 274, 499, 568-572, 579, 587, 589, 631
transfer function 76, 77, 372, 528, 530, 563, 595, 596, 617
transition 23, 94, 122-126, 353, 568
trunnion pin 14-17, 30, 31, 90, 116, 140, 141, 152, 189, 191, 297, 324, 387-393, 401-409, 463-510, 515-563, 586-589, 596-603, 609-625, 631-646
turbulence 20, 75, 94, 122-138, 297, 316, 585
two-dimensional model 298, 378, 379, 428, 436, 437, 499, 500, 508, 509, 517, 518

U

underflow 10, 28-30, 38, 90, 113-119, 131, 139, 232-235, 276, 281, 294, 316, 317, 336, 355-370, 376-382, 608, 627-630, 653

V

validation of theory 593
vertical flat-and-rigid weir plate 408
vertical vibration 27, 30, 38, 319, 337-345, 351-370, 376, 386, 408, 456, 459, 570, 609
vertical water curtain 240, 246, 266, 267
vibration 1, 7-78, 84-93, 126-143, 151-155, 174-205, 212-235, 243, 265-301, 306-410, 415-580, 585-653
vibration characteristics 29, 33, 42, 227, 228, 270, 277, 288, 348, 359, 368, 378, 403, 453, 460, 465, 490, 494, 512, 518, 531, 535, 593-597, 603, 617, 629-637, 645
vibration mode 205, 272, 281, 296, 297, 323, 324, 358, 368, 387, 389, 466, 492, 510-513, 530-536, 555-559, 566, 568, 596-621, 635-638
vibration theory 89, 204, 271, 339, 397
viscous effects 94, 107, 140, 141, 316, 408
vortex shedding excitation 294

W

Wachi Dam 11, 30, 43, 387, 400, 464, 465, 521, 522, 553-565, 627, 631, 646
water mass 204, 222, 225, 285, 286, 292, 306, 393, 430-432, 473, 638
wave-radiation damping 218, 220, 225, 432-435, 440, 441, 451, 457-459, 473-475, 481, 496, 497, 510, 519, 563, 640, 641
weir 28, 29, 68, 84, 108-111, 117, 119, 131, 153, 189-195, 200-205, 232-269, 294-298, 336, 342-348, 355-370, 375-382, 407-412, 417-449, 457-461, 495, 625-629, 651, 653
weir plate 29, 153, 189-195, 200-205, 294-298, 336, 342-348, 355-365, 370, 375-382, 407-412, 417-449, 457-461, 625-629, 653
whole gate rotation around the trunnion pin 465, 482, 492, 508, 515, 519, 560, 633, 644

www.ingramcontent.com/pod-product-compliance
Lightning Source LLC
LaVergne TN
LVHW060251011026
849161LV00011B/2689
* 9 7 8 1 5 2 2 5 3 0 7 9 4 *